Jorge Angeles

Fundamentals of Robotic Mechanical Systems

Theory, Methods, and Algorithms

With 132 Figures

Jorge Angeles
Department of Mechanical Engineering
and Centre for Intelligent Machines
McGill University
Montréal, Quebec
CANADA

Series Editor
Frederick F. Ling
Ernest F. Gloyna Regents Chair in Engineering
Department of Mechanical Engineering
The University of Texas at Austin
Austin, TX 78712-1063 USA
and
William Howard Hart Professor Emeritus
Department of Mechanical Engineering,
Aeronautical Engineering and Mechanics
Rensselaer Polytechnic Institute
Troy, NY 12180-3590 USA

Library of Congress Cataloging-in-Publication Data
Angeles, Jorge, 1943–
Fundamentals of robotic mechanical systems : theory, methods, and algorithms / Jorge Angeles.
p. cm. — (Mechanical engineering series)
ISBN 0-387-94540-7 (alk. paper)
1. Robotics. I. Title. II. Series: Mechanical engineering series (Berlin, Germany)
TJ211.A43 1997
629.8′92—dc20 96-32218
CIP

Printed on acid-free paper.

Production managed by Steven Pisano; manufacturing supervised by Joe Quatela.
Photocomposed copy prepared from the author's LaTeX files.
Printed and bound by Maple-Vail Book Manufacturing Group, York, PA.
Printed in the United States of America.

9 8 7 6 5 4 3 2 1

ISBN 0-387-94540-7 Springer-Verlag New York Berlin Heidelberg SPIN 10502266

To Anne-Marie, who has given me
not only her love, but also her precious time,
without which this book would not have been possible.

Series Preface

Mechanical engineering, an engineering discipline borne of the needs of the industrial revolution, is once again asked to do its substantial share in the call for industrial renewal. The general call is urgent as we face profound issues of productivity and competitiveness that require engineering solutions, among others. The Mechanical Engineering Series features graduate texts and research monographs intended to address the need for information in contemporary areas of mechanical engineering.

The series is conceived as a comprehensive one that covers a broad range of concentrations important to mechanical engineering graduate education and research. We are fortunate to have a distinguished roster of consulting editors on the advisory board, each an expert in one the areas of concentration. The names of the consulting editors are listed on the next page of this volume. The areas of concentration are: applied mechanics; biomechanics; computational mechanics; dynamic systems and control; energetics; mechanics of materials; processing; thermal science; and tribology.

Austin, Texas — Frederick F. Ling

Mechanical Engineering Series

Frederick F. Ling
Series Editor

Preface

No todos los pensamientos son algorítmicos.

—Mario Bunge[1]

The beginnings of modern robotics can be traced back to the late sixties with the advent of the microprocessor, which made possible the computer control of a multiaxial manipulator. Since those days, robotics has evolved from a technology developed around this class of manipulators for the replaying of a preprogrammed task to a multidiscipline encompassing many branches of science and engineering. Research areas such as computer vision, artificial intelligence, and speech recognition play key roles in the development and implementation of robotics; these are, in turn, multidisciplines supported by computer science, electronics, and control, at their very foundations. Thus we see that robotics covers a rather broad spectrum of knowledge, the scope of this book being only a narrow band of this spectrum, as outlined below.

Contemporary robotics aims at the design, control, and implementation of systems capable of performing a task defined at a high level, in a language resembling those used by humans to communicate among themselves. Moreover, robotic systems can take on forms of all kinds, ranging from the most intangible, such as interpreting images collected by a space sound, to the most concrete, such as cutting tissue in a surgical operation. We can,

[1] *Not all thinking processes are algorithmic*—translation of the author—personal communication during the *Symposium on the Brain-Mind Problem. A Tribute to Professor Mario Bunge on His 75th Birthday*, Montreal, September 30, 1994.

therefore, notice that motion is not essential to a robotic system, for this system is meant to replace humans in many of their activities, moving being but one of them. However, since robots evolved from early programmable manipulators, one tends to identify robots with motion and manipulation. Certainly, robots may rely on a mechanical system to perform their intended tasks. When this is the case, we can speak of *robotic mechanical systems*, which are the subject of this book. These tasks, in turn, can be of a most varied nature, mainly involving motions such as manipulation, but they can also involve locomotion. Moreover, manipulation can be as simple as displacing objects from a belt conveyor to a magazine. On the other hand, manipulation can also be as complex as displacing these objects while observing constraints on both motion and force, e.g., when cutting live tissue of vital organs. We can, thus, distinguish between plain manipulation and dextrous manipulation. Furthermore, manipulation can involve locomotion as well.

The task of a robotic mechanical system is, hence, intimately related to motion control, which warrants a detailed study of mechanical systems as elements of a robotic system. The aim of this book can, therefore, be stated as *establishing the foundations on which the design, control, and implementation of robotic mechanical systems are based.*

The book evolved from sets of lecture notes developed at McGill University over the last twelve years, while I was teaching a two-semester sequence of courses on robotic mechanical systems. For this reason, the book comprises two parts, an introductory and an intermediate part on robotic mechanical systems. Advanced topics, such as redundant manipulators, manipulators with flexible links and joints, and force control, are omitted. The feedback control of robotic mechanical systems is also omitted, although the book refers the reader, when appropriate, to the specialized literature. An aim of the book is to serve as a textbook in a one-year robotics course; another aim is to serve as a reference to the practicing engineer.

The book assumes some familiarity with the mathematics taught in any engineering or science curriculum in the first two years of college. Familiarity with elementary mechanics is helpful, but not essential, for the elements of this science needed to understand the mechanics of robotic systems are covered in the first three chapters, thereby making the book self-contained. These three chapters, moreover, are meant to introduce the reader to the notation and the basics of mathematics and rigid-body mechanics needed in the study of the systems at hand. The material covered in the same chapters can thus serve as reading material for a course on the mathematics of robotics, intended for sophomore students of science and engineering, prior to a more formal course on robotics.

The first chapter is intended to give the reader an overview of the subject matter and to highlight the major issues in the realm of robotic mechanical systems. Chapter 2 is devoted to notation, nomenclature, and the basics of linear transformations to understand best the essence of rigid-body kinematics, an area that is covered in great detail throughout the book. A unique feature of this chapter is the discussion of the hand-eye calibration problem: Many a paper has been written in an attempt to solve this fundamental problem, always leading to a cumbersome solution that invokes nonlinear-equation solving, a task that invariably calls for an iterative procedure; moreover, within each iteration, a singular-value decomposition, itself iterative as well, is required. In Chapter 2, a novel approach is introduced, that resorts to invariant properties of rotations and leads to a *direct solution*, involving straightforward matrix and vector multiplications. Chapter 3 reviews, in turn, the basic theorems of rigid-body kinetostatics and dynamics. The viewpoint here represents a major departure from most existing books on robotic manipulators: proper orthogonal matrices can be regarded as coordinate transformations indeed, but they can also be regarded as representations, once a coordinate frame has been selected, of rigid-body rotations. I adopt the latter viewpoint, and hence, fundamental concepts are explained in terms of their invariant properties, i.e., properties that are independent of the coordinate frame adopted. Hence, matrices are used first and foremost to represent the physical motions undergone by rigid bodies and systems thereof, and are to be interpreted as such when studying the basics of rigid-body mechanics in this chapter. Chapter 4 is the first chapter entirely devoted to robotic mechanical systems, properly speaking. This chapter covers extensively the kinematics of robotic manipulators of the serial type. However, as far as displacement analysis is concerned, the chapter limits itself to the simplest robotic manipulators, namely, those with a *decoupled architecture*, i.e., those that can be decomposed into a *regional architecture* for the positioning of one point of their end-effector (EE), and a *local architecture* for the orientation of their EE. In this chapter, the notation of Denavit and Hartenberg is introduced and applied consistently throughout the book. Jacobian matrices, workspaces, singularities, and kinetostatic performance indices are concepts studied in this chapter. A novel algorithm is included for the determination of the workspace boundary of positioning manipulators. Furthermore, Chapter 5 is devoted to the topic of trajectory planning, while limiting its scope to problems suitable to a first course on robotics; this chapter thus focuses on pick-and-place operations. Chapter 6, moreover, introduces the dynamics of robotic manipulators of the serial type, while discussing extensively the recursive Newton-Euler algorithm and laying the foundations of multibody dynamics, with an introduction to the Euler-Lagrange formulation. The latter is used to derive the general algebraic structure of the mathematical models of the systems under study, thus completing the introductory part of the book.

The intermediate part comprises four chapters. Chapter 7 is devoted to the increasingly important problem of determining the angular velocity and the angular acceleration of a rigid body, when the velocity and acceleration of a set of its points are known. Moreover, given the intermediate level of the chapter, only the theoretical aspects of the problem are studied, and hence, perfect measurements of point position, velocity, and acceleration are assumed, thereby laying the foundations for the study of the same problems in the presence of noisy measurements. This problem is finding applications in the control of parallel manipulators, which is the reason why it is included here. If time constraints so dictate, this chapter can be omitted, for it is not needed in the balance of the book.

The formulation of the inverse kinematics of the most general robotic manipulator of the serial type, leading to a monovariate polynomial of the 16th degree, not discussed in previous books on robotics, is included in Chapter 8. Likewise, the direct kinematics of the platform manipulator popularly known as the *Stewart platform*, a.k.a. the *Stewart-Gough platform*, leading to a 16th-degree monovariate polynomial, is also given due attention in this chapter. Moreover, an alternative approach to the monovariate-polynomial solution of the two foregoing problems, that is aimed at solving them *semigraphically*, is introduced in this chapter. With this approach, the underlying multivariate algebraic system of equations is reduced to a system of two nonlinear bivariate equations that are trigonometric rather than polynomial. Each of these two equations, then, leads to a contour in the plane of the two variables, the desired solutions being found as the coordinates of the intersections of the two contours.

Discussed in Chapter 9 is the problem of trajectory planning as pertaining to continuous paths, which calls for some concepts of differential geometry, namely, the Frenet-Serret equations relating the tangent, normal, and binormal vectors of a smooth curve to their rates of change with respect to the arc length. The chapter relies on cubic parametric splines for the synthesis of the generated trajectories in joint space, starting from their descriptions in Cartesian space. Finally, Chapter 10 completes the discussion initiated in Chapter 6, with an outline of the dynamics of parallel manipulators and rolling robots. Here, a multibody dynamics approach is introduced, as in the foregoing chapter, that eases the formulation of the underlying mathematical models.

Two appendices are included: Appendix A summarizes a series of facts from the kinematics of rotations, that are available elsewhere, with the purpose of rendering the book self-contained; Appendix B is devoted to the numerical solution of over- and underdetermined linear algebraic systems, its purpose being to guide the reader to the existing robust techniques for the computation of least-square and minimum-norm solutions. The book concludes with a set of problems, along with a list of references, for all ten chapters.

On Notation

The important issue of notation is given due attention. In figuring out the notation, I have adopted what I call the C^3 *norm*. Under this norm, the notation should be

1. *C*omprehensive,
2. *C*oncise, and
3. *C*onsistent.

Within this norm, I have used boldface fonts to indicate vectors and matrices, with uppercases reserved for matrices and lowercases for vectors. In compliance with the invariant approach adopted at the outset, I do not regard vectors solely as arrays, but as geometric or mechanical objects. Regarding such objects as arrays is necessary only when it is required to perform operations with them for a specific purpose. An essential feature of vectors in a discussion is their dimension, which is indicated with a single number, as opposed to the convention whereby vectors are regarded as matrix arrays of numbers; in this convention, the dimension has to be indicated with two numbers, one for the number of columns, and one for the number of rows; in the case of vectors, the latter is always one, and hence, need not be mentioned. Additionally, calligraphic literals are reserved for sets of points or of other objects. Since variables are defined every time that they are introduced, and the same variable is used in the book to denote different concepts in different contexts, a list of symbols is not included.

How to Use the Book

The book can be used as a reference or as a text for the teaching of the mechanics of robots to an audience that ranges from junior undergraduates to doctoral students. In an introductory course, the instructor may have to make choices regarding what material to skip, given that the duration of a regular semester does not allow to cover all that is included in the first six chapters. Topics that can be skipped, if time so dictates, are the discussions, in Chapter 4, of workspaces and performance indices, and the section on simulation in Chapter 6. Under strict time constraints, the whole Chapter 5 can be skipped, but then, the instructor will have to refrain from assigning problems or projects that include calculating the inverse dynamics of a robot performing pick-and-place operations. None of these has been included in Section 6 of the Exercises.

If sections of Chapters 4 and 5 have been omitted in a first course, it is highly advisable to include them in a second course, prior to discussing the chapters included in the intermediate part of the book.

Acknowledgements

For the technical support received during the writing of this book, I am indebted to many people: First and foremost, Eric Martin and Ferhan Bulca, Ph.D. candidates under my cosupervision, are deeply thanked for their invaluable help and unlimited patience in the editing of the manuscript and the professional work displayed in the production of the drawings. With regard to this task, Dr. Max A. González-Palacios, currently Assistant Professor of Mechanical Engineering at Universidad Iberoamericana at León, Mexico, is due special recognition for the high standards he set while working on his Ph.D. at McGill University. My colleagues Ken J. Waldron, Clément Gosselin, and Jean-Pierre Merlet contributed with constructive criticism. Dr. Andrés Kecskeméthy proofread major parts of the manuscript during his sabbatical leave at McGill University. In doing this, Dr. Kecskeméthy corrected a few derivations that were flawed. Discussions on geometry and analysis held with Dr. Manfred Husty, of Leoben University, in Austria, also a sabbaticand at McGill University, were extremely fruitful in clearing up many issues in Chapters 2 and 3. An early version of the manuscript was deeply scrutinized by Meyer Nahon, now Associate Professor at the University of Victoria, when he was completing his Ph.D. at McGill University. Discussions with Farzam Ranjbaran, a Ph.D. candidate at McGill University, on kinetostatic performance indices, helped clarify many concepts around this issue. Dr. Kourosh Etemadi-Zanganeh contributed with ideas for a more effective discussion of the parametric representation of paths in Chapter 8 and with some of the examples in Chapters 4 and 8 during his work at McGill University as a Ph.D. student. The material supplied by Clément Gosselin on trajectory planning helped me start the writing of Chapter 5. All individuals and institutions who contributed with graphical material are given due credit in the book. Here, they are all deeply acknowledged.

A turning point in writing this manuscript was the academic year 1991–1992, during which I could achieve substantial progress while on sabbatical leave at the Technical University of Munich under an Alexander von Humboldt Research Award. Deep gratitude is expressed here to both the AvH Foundation and Prof. Friedrich Pfeiffer, Director of the Institute B for Mechanics and my host in Munich. Likewise, Prof. Manfred Broy, of the Computer Science Institute at the Techincal University of Munich, is herewith acknowledged for having given me access to his Unix network when the need arose. The intellectual environment at the Technical University of Munich was a source of encouragement and motivation to pursue the writing of the book.

Moreover, financial support from NSERC[2] and Quebec's FCAR,[3] in the form of research and strategic grants, are duly acknowledged. IRIS,[4] a network of Canadian centers of excellence, supported this work indirectly through project grants in the areas of robot design and robot control. An invaluable tool in developing material for the book proved to be RVS, the McGill Robot Visualization System, developed in the framework of an NSERC Strategic Grant on robot design, and the two IRIS project grants mentioned above. RVS was developed by John Darcovich, a Software Engineer at McGill University for about four years, and now at CAE Electronics Ltd., of Saint-Laurent, Quebec. While RVS is user-friendly and available upon request, no technical support is offered. For further details on RVS, the reader is invited to look at the home page of the McGill University Centre for Intelligent Machines:

`http://www.cim.mcgill.ca/~rvs`

Furthermore, Lenore Reismann, a professional technical editor based in Redwood City, California, proofread parts of the manuscript and edited its language with great care. Lenore's professional help is herewith highly acknowledged. Dr. Rüdiger Gebauer, mathematics editor at Springer-Verlag New York, is gratefully acknowledged for his encouragement in pursuing this project. Springer-Verlag's Dr. Thomas von Foerster is likewise acknowledged for the care with which he undertook the production of the book, while his colleague Steve Pisano, for his invaluable help in the copyediting of the final draft. Steve and his staff not only took care of the fine points of the typesetting, but also picked up a few technical flaws in that draft. Last, but not least, may I acknowledge the excellent facilities and research environment provided by the Centre for Intelligent Machines, the Department of Mechanical Engineering of McGill University, and McGill University as a whole, which were instrumental in completing this rather lengthy project.

[2]Natural Sciences and Engineering Research Council, of Canada

[3]*Fonds pour la formation de chercheurs et l'aide à la recherche*

[4]Institute for Robotics and Intelligent Systems

Contents

Series Preface vii

Preface ix

1 An Overview of Robotic Mechanical Systems 1
1.1 Introduction 1
1.2 The General Structure of Robotic Mechanical Systems . . 3
1.3 Serial Manipulators 6
1.4 Parallel Manipulators 8
1.5 Robotic Hands 11
1.6 Walking Machines 14
1.7 Rolling Robots 16

2 Mathematical Background 19
2.1 Preamble 19
2.2 Linear Transformations 20
2.3 Rigid-Body Rotations 25
2.4 Composition of Reflections and Rotations 47
2.5 Coordinate Transformations and Homogeneous Coordinates 48
2.6 Similarity Transformations 58
2.7 Invariance Concepts 63

3 Fundamentals of Rigid-Body Mechanics 71
3.1 Introduction 71
3.2 General Rigid-Body Motion and Its Associated Screw . . 72
3.3 Rotation of a Rigid Body About a Fixed Point 83
3.4 General Instantaneous Motion of a Rigid Body 84
3.5 Acceleration Analysis of Rigid-Body Motions 91
3.6 Rigid-Body Motion Referred to Moving Coordinate Axes . 93
3.7 Static Analysis of Rigid Bodies 95
3.8 Dynamics of Rigid Bodies 100

4 Kinetostatics of Simple Robotic Manipulators **105**
4.1 Introduction 105
4.2 The Denavit-Hartenberg Notation 106
4.3 The Kinematics of Six-Revolute Manipulators 114
4.4 The IKP of Decoupled Manipulators 118
4.5 Velocity Analysis of Serial Manipulators 139
4.6 Acceleration Analysis of Serial Manipulators 158
4.7 Static Analysis of Serial Manipulators 162
4.8 Planar Manipulators 164
4.9 Kinetostatic Performance Indices 174

5 Trajectory Planning: Pick-and-Place Operations **191**
5.1 Introduction 191
5.2 Background on PPO 192
5.3 Polynomial Interpolation 194
5.4 Cycloidal Motion 201
5.5 Trajectories with Via Poses 203
5.6 Synthesis of PPO Using Cubic Splines 205

6 Dynamics of Serial Robotic Manipulators **213**
6.1 Introduction 213
6.2 Inverse vs. Forward Dynamics 213
6.3 Fundamentals of Multibody System Dynamics 215
6.4 Recursive Inverse Dynamics 225
6.5 The Natural Orthogonal Complement in Robot Dynamics 236
6.6 Manipulator Forward Dynamics 246
6.7 Incorporation of Gravity into the Dynamics Equations . . 270
6.8 The Modeling of Dissipative Forces 271

7 Special Topics in Rigid-Body Kinematics **275**
7.1 Introduction 275
7.2 Computation of Angular Velocity from Point-Velocity Data 276
7.3 Computation of Angular Acceleration from Point-Acceleration Data 282

8 Kinematics of Complex Robotic Mechanical Systems **289**
8.1 Introduction 289
8.2 The IKP of General Six-Revolute Manipulators 290
8.3 Kinematics of Parallel Manipulators 315
8.4 Multifingered Hands 336
8.5 Walking Machines 341
8.6 Rolling Robots 345

9 Trajectory Planning: Continuous-Path Operations **355**
9.1 Introduction 355
9.2 Curve Geometry 356
9.3 Parametric Path Representation 362
9.4 Parametric Splines in Trajectory Planning 375
9.5 Continuous-Path Tracking 380

10 Dynamics of Complex Robotic Mechanical Systems **393**
10.1 Introduction 393
10.2 Classification of Robotic Mechanical Systems with Regard to Dynamics 394
10.3 The Structure of the Dynamics Models of Holonomic Systems 395
10.4 Dynamics of Parallel Manipulators 398
10.5 Dynamics of Rolling Robots 409

A Kinematics of Rotations: A Summary **429**

B The Numerical Solution of Linear Algebraic Systems **437**
B.1 The Overdetermined Case 438
B.2 The Underdetermined Case 443

Exercises **447**
1 An Overview of Robotic Mechanical Systems 447
2 Mathematical Background 449
3 Fundamentals of Rigid-Body Mechanics 457
4 Kinetostatics of Simple Robotic Manipulators 463
5 Trajectory Planning: Pick-and-Place Operations 471
6 Dynamics of Serial Robotic Manipulators 475
7 Special Topics on Rigid-Body Kinematics 480
8 Kinematics of Complex Robotic Mechanical Systems . . . 483
9 Trajectory Planning: Continuous-Path Operations 485
10 Dynamics of Complex Robotic Mechanical Systems 489

References **493**

Index **505**

1

An Overview of Robotic Mechanical Systems

1.1 Introduction

In defining the scope of our subject, we have to establish the genealogy of robotic mechanical systems. These are, obviously, a subclass of the much broader class of *mechanical systems*. Mechanical systems, in turn, constitute a subset of the more general concept of *dynamic systems*. Therefore, in the final analysis, we must have an idea of what a *system*, in general, is.

The *Concise Oxford Dictionary* defines system as a "complex whole, set of connected things or parts, organized body of material or immaterial things", whereas the *Random House College Dictionary* defines the same as "an assemblage or combination of things or parts forming a complex or unitary whole." *Le Petit Robert*, in turn, defines system as "Ensemble possédant une structure, constituant un tout organique", which can be loosely translated as "A structured assemblage constituting an organic whole." In the foregoing definitions, we note that the underlying idea is that of a set of elements interacting as a whole.

On the other hand, a *dynamic system* is a subset of the set of systems. For our purposes, we can dispense with a rigorous definition of this concept. Suffice it to say that a dynamic system is a system in which one can distinguish three elements, namely, a *state*, an *input*, and an *output*, in addition to a rule of transition from one current state to a future one. Moreover, the state is a *functional* of the input and a function of a *previous* state. In this concept, then, the idea of order is important, and can be taken into account by properly associating each state value with time. The state at every instant is a functional, as opposed to a function, of the input, which is characteristic of dynamic systems. This means that the state of a dynamic system at a certain instant is determined not only by the value of the input at that instant, but also by the past history of that input. By virtue of this property, dynamic systems are said to have *memory*.

On the contrary, systems whose state at a given instant is only a *function* of the input at the current time are static and are said to have no memory. Additionally, since the state of a dynamic system is a result of all the past history of the input, the future values of this having no influence on the state, dynamic systems are said to be *nonanticipative* or *causal*. By the same token, systems whose state is the result of future values of the input are said to be *anticipative* or *noncausal*. In fact, we will not need to worry about the latter, and hence, all systems we will study can be assumed to be causal.

Obviously, a mechanical system is a system composed of mechanical elements. If this system complies with the definition of dynamic system, then we end up with a *dynamic mechanical system*. For brevity, we will refer to such systems as *mechanical systems*, the dynamic property being taken for granted throughout the book. Mechanical systems of this type are those that occur whenever the inertia of their elements is accounted for. Static mechanical systems are those in which inertia is neglected. Moreover, the elements constituting a mechanical system are rigid and deformable solids, compressible and incompressible fluids, and inviscid and viscous fluids.

From the foregoing discussion, then, it is apparent that mechanical systems can be constituted either by lumped-parameter or by distributed-parameter elements. The former reduce to particles; rigid bodies; massless, conservative springs; and massless, nonconservative dashpots. The latter appear whenever bodies are modeled as continuous media. In this book, we will focus on lumped-parameter mechanical systems.

Furthermore, a mechanical system can be either natural or man-made, the latter being the subject of our study. Man-made mechanical systems can be either controlled or uncontrolled. Most engineering systems are controlled mechanical systems, and hence, we will focus on these. Moreover, a controlled mechanical system may be *robotic* or nonrobotic. The latter are systems supplied with primitive controllers, mostly analog, such as thermostats, servovalves, etc. Robotic mechanical systems, in turn, can be *programmable*, such as most current industrial robots, or *intelligent*, as discussed below. Programmable mechanical systems obey motion commands either stored in a memory device or generated on-line. In either case, they need primitive sensors, such as joint encoders, accelerometers, and dynamometers.

Intelligent robots or, more broadly speaking, *intelligent machines*, are yet to be demonstrated, but have become the focus of intensive research. If intelligent machines are ever feasible, they will depend highly on a sophisticated sensory system and the associated hardware and software for the processing of the information supplied by the sensors. The processed information would then be supplied to the actuators in charge of producing the desired motion of the robot. Contrary to programmable robots, whose operation is limited to *structured environments*, intelligent machines should be capable of reacting to unpredictable changes in an *unstructured environment*. Thus, intelligent machines should be supplied with decision-making capabilities aimed at mimicking the natural decision-making process of living organisms. This is the reason why such systems are termed intelligent in the first place. Thus, intelligent machines are expected to *perceive* their environment and draw conclusions based on this perception. What is supposed to make these systems *intelligent* is their capability of perceiving, which involves a certain element of subjectivity. By far, the most complex of perception tasks, both in humans and machines, is visual (Levine, 1985; Horn, 1986).

In summary, then, an intelligent machine is expected to (*i*) *perceive* the environment; (*ii*) *reason* about the perceived information; (*iii*) *make decisions* based on this perception; and (*iv*) *act* according to a plan specified at a very high level. What the latter means is that the motions undergone by the machine are decided upon based on instructions similar to those given to a human being, like *bring me a glass of water without spilling the water.*

Whether intelligent machines with all the above features will be one day possible or not is still a subject of discussion, sometimes at a philosophical level. Penrose (1994) wrote a detailed discussion refuting the claim that intelligent machines are possible.

A genealogy of mechanical systems, including robotic ones, is given in Fig. 1.1. In that figure, we have drawn a dashed line between mechanical systems and other systems, both man-made and natural, in order to emphasize the interaction of mechanical systems with electrical, thermal, and other systems, including the human system, which is present in telemanipulators, to be discussed below.

1.2 The General Structure of Robotic Mechanical Systems

From Section 1.1, then, a robotic mechanical system is composed of a few subsystems, namely, (*i*) a mechanical subsystem composed in turn of both rigid and deformable bodies, although the systems we will study here are composed only of the former; (*ii*) a sensing subsystem; (*iii*) an actuation subsystem; (*iv*) a controller; and (*v*) an information-processing subsystem.

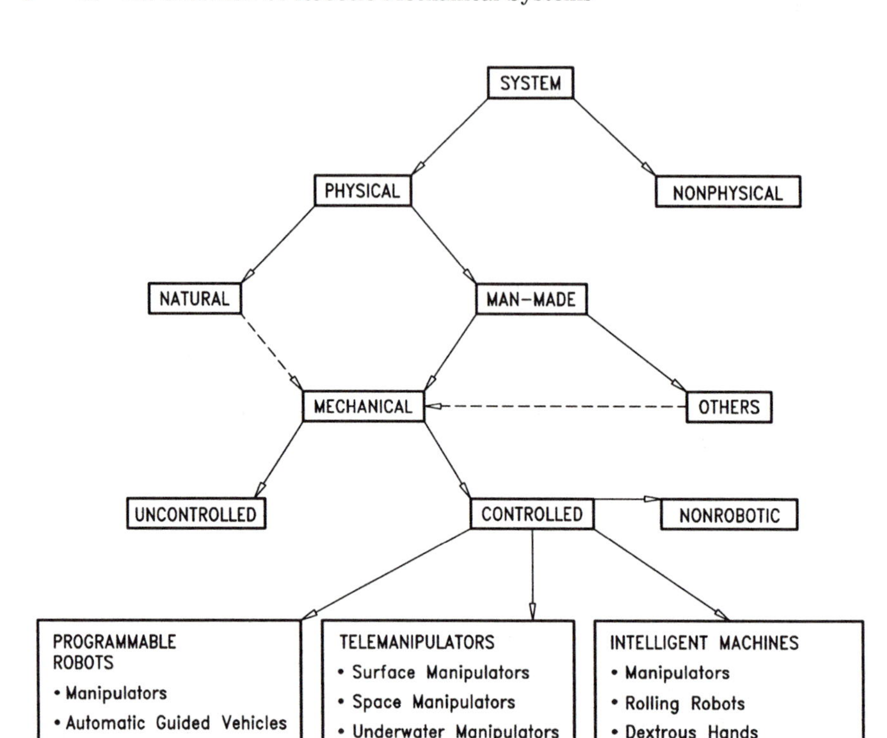

FIGURE 1.1. A genealogy of robotic mechanical systems.

Additionally, these subsystems communicate among themselves via *interfaces*, whose function consists basically of decoding the transmitted information from one medium to another. Figure 1.2 shows a block-diagram representation of a typical robotic mechanical system. Its input is a *prescribed task*, which is defined either on the spot or off-line. The former case is essential for a machine to be called intelligent, while the latter is present in programmable machines. Thus, tasks would be described to intelligent machines by a software system based on techniques of artificial intelligence (AI). This system would replace the human being in the decision-making process. Programmable robots require human intervention either for the coding of preprogrammed tasks at a very low level or for *telemanipulation*. A very low level of programming means that the motions of the machine are specified as a sequence of either joint motions or Cartesian coordinates associated with landmark points of that specific body performing the task at hand. The output of a robotic mechanical system is the *actual task*, which is monitored by the sensors. The sensors, in turn, transmit task information in the form of feedback signals, to be compared with the prescribed task. The errors between the prescribed and the actual task are then fed back

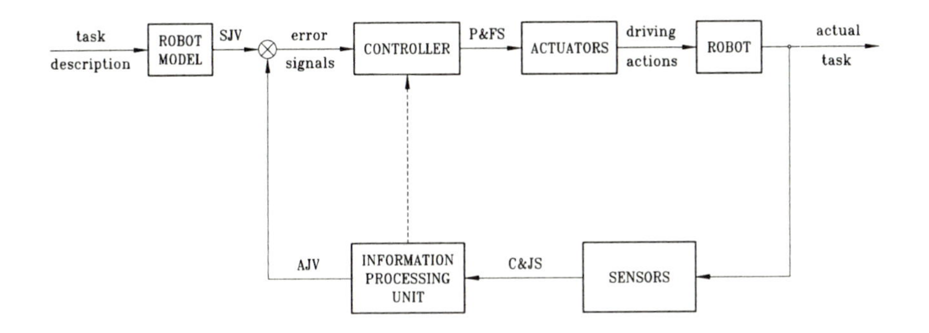

FIGURE 1.2. Block diagram of a general robotic mechanical system.

into the controller, which then synthesizes the necessary corrective signals. These are, in turn, fed back into the actuators, which then drive the mechanical system through the required task, thereby closing the loop. The problem of robot control has received extensive attention in the literature, and will not be pursued here. The interested reader is referred to the excellent works on the subject, e.g., those of Samson, Le Borgne, and Espiau (1991) and, at a more introductory level, of Spong and Vidyasagar (1989). Of special relevance to robot control is the subject of nonlinear control at large, a pioneer here being Isidori (1989).

Robotic mechanical systems with a human being in their control loop are called *telemanipulators.* Thus, a telemanipulator is a robotic mechanical system in which the task is controlled by a human, possibly aided by sophisticated sensors and display units. The human operator is then a central element in the block diagram loop of Fig. 1.2. Based on the information displayed, the operator makes decisions about corrections in order to accomplish the prescribed task. Shown in Fig. 1.3 is a telemanipulator to be used in space applications, namely, the *Space Station Remote Manipulator System*, along with the *Special Purpose Dextrous Manipulator* (SPDM), both mounted on the *Mobile Servicing System* (MSS). Moreover, a detailed view of the SPDM is shown in Fig. 1.4. In the manipulators of these two figures, the human operator is an astronaut who commands and monitors the motions of the robot from inside the EVA (extravehicular activity) workstation. The number of controlled axes of each of these manipulators being larger than six, both are termed *redundant.* The challenge here is that the mapping from task coordinates to joint motions is not unique, and hence, among the infinitely many joint trajectories that the operator has at his or

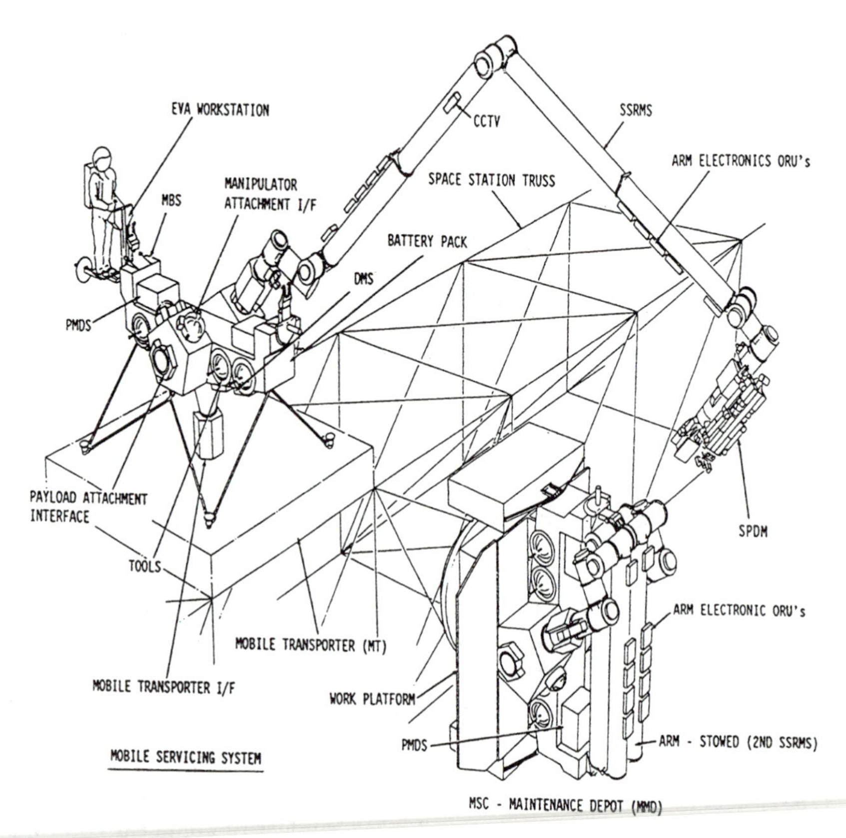

FIGURE 1.3. Space Station Remote Manipulator System and Special-Purpose Dextrous Manipulator (courtesy of the Canadian Space Agency.)

her disposal for a given task, an on-board processor must evaluate the best one according to a performance criterion.

While the manipulators of Figs. 1.3 and 1.4 are still at the development stage, examples of robotic mechanical systems in operation are the well-known six-axis industrial manipulators, six-degree-of-freedom flight simulators, walking machines, mechanical hands, and rolling robots. We outline the various features of these systems below.

1.3 Serial Manipulators

Among all robotic mechanical systems mentioned above, robotic manipulators deserve special attention, for various reasons. One is their relevance in industry. Another is that they constitute the simplest of all robotic mechanical systems, and hence, appear as constituents of other, more complex

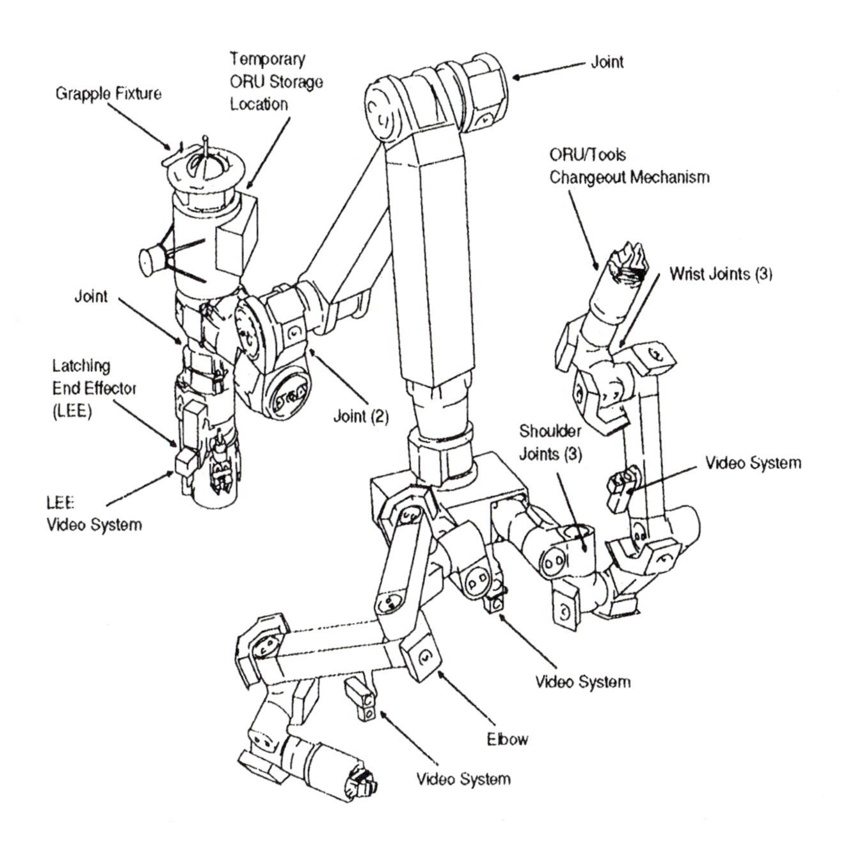

FIGURE 1.4. The Special-Purpose Dextrous Manipulator (SPDM) System (courtesy of the Canadian Space Agency.)

robotic mechanical systems, as will become apparent in later chapters. A manipulator, in general, is a mechanical system aimed at *manipulating* objects. Manipulating, in turn, means to move something with one's hands, as it derives from the Latin *manus*, meaning *hand.* The basic idea behind the foregoing concept is that hands are among the organs that the human brain can control mechanically with the highest accuracy, as the work of an artist like Picasso, of an accomplished guitar player, or of a surgeon can attest.

Hence, a manipulator is any device that helps man perform a manipulating task. Although manipulators have existed ever since man created the first tool, only very recently, namely, by the end of World War II, have manipulators developed to the extent that they are now capable of actually mimicking motions of the human arm. In fact, during WWII, the need for manipulating probe tubes containing radioactive substances arose. This led to the first six-degree-of-freedom (DOF) manipulators.

Shortly thereafter, the need for manufacturing workpieces with high accuracy arose in the aircraft industry, which led to the first numerically-controlled (NC) machine tools. The synthesis of the six-DOF manipulator and the NC machine tool produced what became the *robotic manipulator*. Thus, the essential difference between the early manipulator and the evolved *robotic* manipulator is the term *robotic*, which has only recently, as of the late sixties, come into the picture. A robotic manipulator is to be distinguished from the early manipulator by its capability of lending itself to *computer control*. Whereas the early manipulator needed the presence of a manned *master manipulator*, the robotic manipulator can be programmed once and for all to repeat the same task for ever and ever. Programmable manipulators have existed for about thirty years, namely, since the advent of microprocessors, which allowed a human master to *teach* the manipulator by actually driving the manipulator itself, or a replica thereof, through a desired task, while recording all motions undergone by the master. Thus, the manipulator would later on repeat the identical task by mere playback. However, the capabilities of industrial robots are fully exploited only if the manipulator is programmed with software, rather than actually driving it through its task trajectory, which many a time, e.g., in car-body spot-welding, requires separating the robot from the production line for more than a week. One of the objectives of this book is to develop tools for the programming of robotic manipulators.

However, the capabilities offered by robotic mechanical systems go well beyond the mere playback of preprogrammed tasks. Current research aims at providing robotic systems with software and hardware that will allow them to make decisions on the spot and learn while performing a task. The implementation of such systems calls for task-planning techniques that fall beyond the scope of this book and, hence, will not be treated here. For a glimpse of such techniques, the reader is referred to the work of Latombe (1991) and the references therein.

1.4 Parallel Manipulators

Robotic manipulators first appeared as mechanical systems constituted by a structure consisting of very robust links coupled by either rotational or translating joints, the former being called *revolutes*, the latter *prismatic joints*. Moreover, these structures are a *concatenation* of links, thereby forming an *open kinematic chain*, with each link coupled to a predecessor and a successor, except for the two end links, which are coupled only to either a predecessor or to a successor, but not to both. Because of the *serial* nature of the coupling of links in this type of manipulator, even though they are supplied with structurally robust links, their load-carrying

FIGURE 1.5. A six-degree-of-freedom flight simulator (courtesy of CAE Electronics Ltd.)

capacity and their stiffness is too low when compared with the same properties in other multiaxis machines, such as NC machine tools. Obviously, a low stiffness implies a low positioning accuracy. In order to remedy these drawbacks, *parallel manipulators* have been proposed to withstand higher payloads with lighter links. In a parallel manipulator, we distinguish one *base platform*, one *moving platform*, and various *legs*. Each leg is, in turn, a kinematic chain of the serial type, whose end links are the two platforms. Contrary to serial manipulators, all of whose joints are actuated, parallel manipulators contain unactuated joints, which brings about a substantial difference between the two types. The presence of unactuated joints makes the analysis of parallel manipulators, in general, more complex than that of their serial counterparts.

A paradigm of parallel manipulators is the flight simulator, consisting of six legs actuated by hydraulic pistons, as displayed in Fig. 1.5. Recently, an explosion of novel designs of parallel manipulators has occurred aimed at fast assembly operations, namely, the Delta robot (Clavel, 1988), developed at the Lausanne Federal Polytechnic Institute, shown in Fig. 1.6; the Hexa robot (Pierrot, Fournier, and Dauchez, 1991), developed at the University of Montpellier; and the Star robot (Hervé and Sparacino, 1992), developed at the *Ecole Centrale* of Paris. One more example of parallel manipulator is the *Trussarm*, developed at the University of Toronto Institute of Aerospace Studies (UTIAS), shown in Fig. 1.7a (Hughes, Sincarsin and Carroll, 1991). Merlet (1990), of the *Institut National de Recherche en Informatique et en Automatique*, of Sophia-Antipolis, France, developed a six-axis parallel robot, called in French a *main gauche*, or left hand, shown in Fig. 1.7b, to be used as an aid to another robot, possibly of the serial type, to enhance its dexterity. Hayward, of McGill University, designed and constructed a parallel manipulator to be used as a shoulder module for orientation tasks (Hayward, 1994); the module is meant for three-degree-of-freedom motions but is provided with four hydraulic actuators, which gives it redundant actuation—Fig. 1.7c.

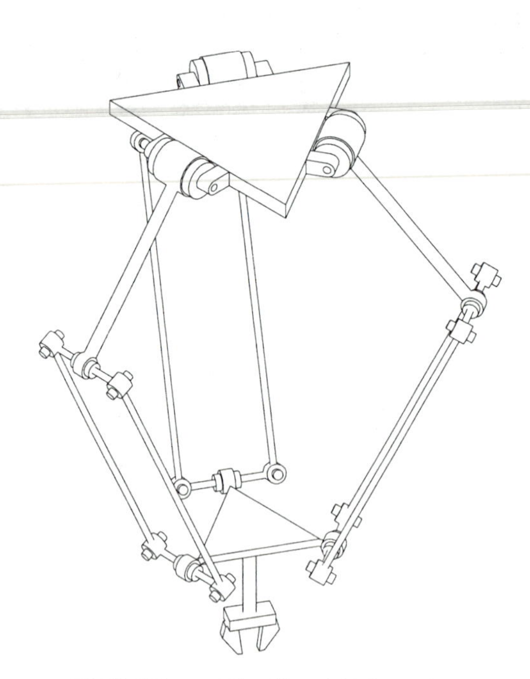

FIGURE 1.6. The Clavel Delta robot.

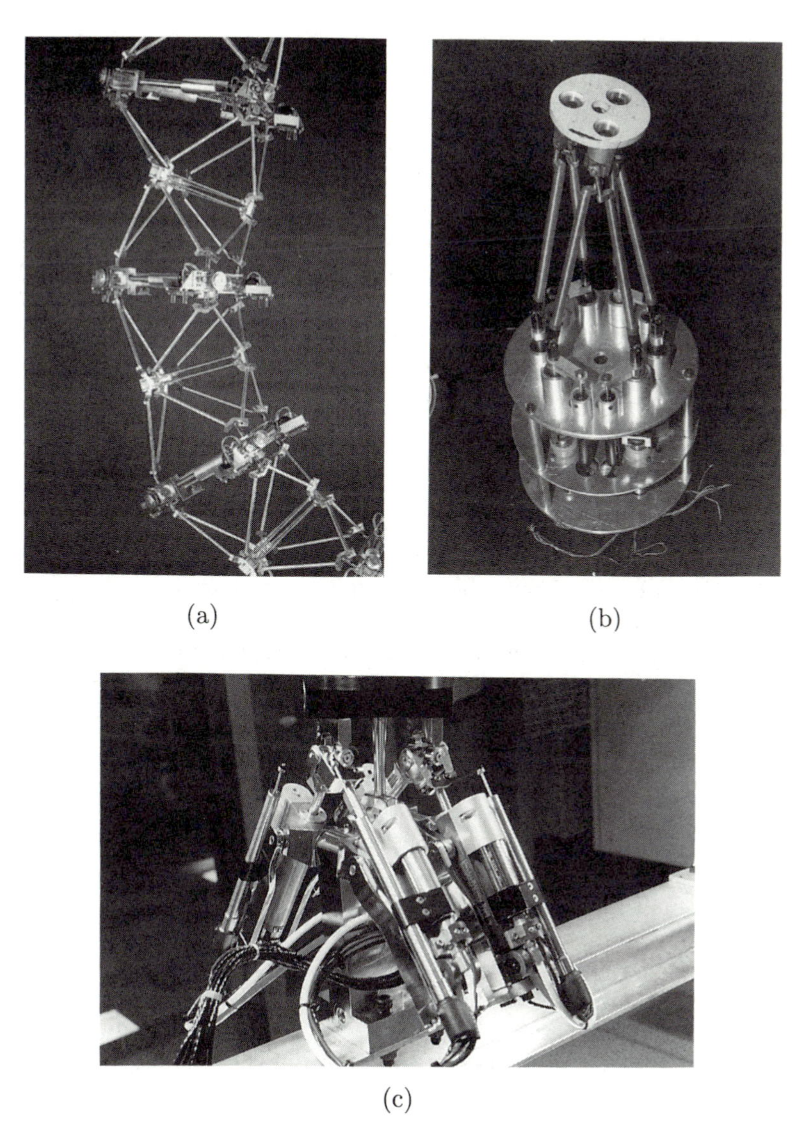

(a) (b) (c)

FIGURE 1.7. A sample of parallel manipulators: (a) The UTIAS Trussarm (courtesy of Prof. P. C. Hughes); (b) the Merlet left hand (courtesy of Dr. J.-P. Merlet); and (c) the Hayward shoulder module (courtesy of Prof. V. Hayward.)

1.5 Robotic Hands

As stated above, the hand can be regarded as the most complex mechanical subsystem of the human manipulation system. Other mechanical subsystems constituting this system are the arm and the forearm. Moreover, the shoulder, coupling the arm with the torso, can be regarded as a *spherical joint*, i.e., the concatenation of three revolute joints with intersecting axes. Furthermore, the arm and the forearm are coupled via the elbow, with the forearm and the hand finally being coupled by the wrist. Frequently, the wrist is modeled as a spherical joint as well, while the elbow is modeled as a simple revolute joint. Robotic mechanical systems mimicking the motions of the arm and the forearm constitute the manipulators discussed in the previous section. Here we outline more sophisticated manipulation systems that aim at producing the motions of the human hand, i.e., robotic mechanical hands. These robotic systems are meant to perform *manipulation* tasks, a distinction being made between *simple manipulation* and *dextrous manipulation*. What the former means is the simplest form, in which the fingers play a minor role, namely, by serving as simple static structures that keep an object rigidly attached with respect to the palm of the hand—when the palm is regarded as a rigid body. As opposed to simple manipulation, dextrous manipulation involves a controlled motion of the grasped object with respect to the palm. Simple manipulation can be achieved with the aid of a manipulator and a gripper, and need not be further discussed here. The discussion here is, then, about dextrous manipulation.

In dextrous manipulation, the grasped object is required to move with respect to the palm of the grasping hand. This kind of manipulation appears in performing tasks that require high levels of accuracy, like handwriting or cutting tissue with a scalpel. Usually, grasping hands are multifingered, although some grasping devices exist that are constituted by a simple, open, highly redundant *kinematic chain* (Pettinato and Stephanou, 1989). The kinematics of grasping is discussed in Chapter 4. The basic kinematic structure of a multifingered hand consists of a palm, which plays the role of the base of a simple manipulator, and a set of fingers. Thus, kinematically speaking, a multifingered hand has a *tree topology*, i.e., it entails a common rigid body, the palm, and a set of jointed bodies emanating from the palm. Upon grasping an object with all the fingers, the chain becomes closed with multiple loops. Moreover, the architecture of the fingers is that of a simple manipulator. It consists of a number—two to four—of revolute-coupled links playing the role of phalanges. However, unlike manipulators of the serial type, whose joints are all independently actuated, those of a mechanical finger are not and, in many instances, are driven by one single master actuator, the remaining joints acting as slaves. Many versions of multifingered hands exist: Stanford/JPL; Utah/MIT; TU Munich; Karlsruhe; Bologna; Leuven; Milan; Belgrade; and U. of Toronto, among others. Of these, the Utah/MIT Hand (Jacobsen et al., 1984; 1986) is commercially available. It consists of four fingers, one of which is opposed to the other three and hence, plays the role of the human thumb. Each finger consists, in turn, of four phalanges coupled by revolute joints; each of these is driven by two tendons that can deliver force only when in tension, each being actuated independently. The TU Munich Hand, shown in Fig. 1.8, is designed with four identical fingers laid out symmetrically on a hand palm. This hand is

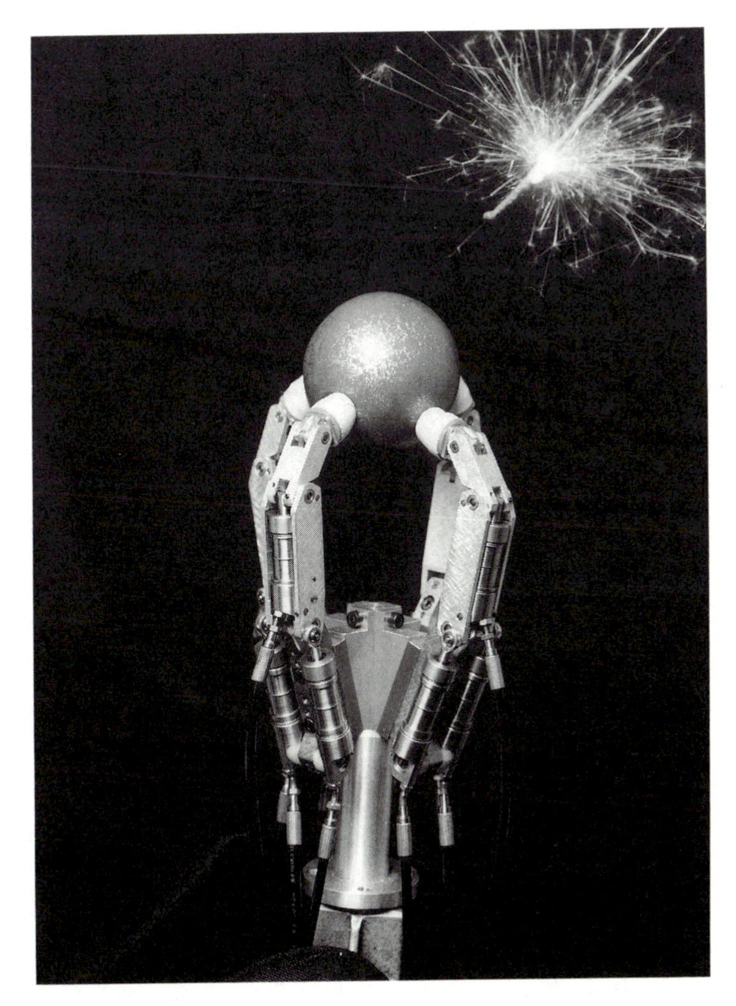

FIGURE 1.8. The four-fingered hydraulically actuated TU Munich Hand (courtesy of Prof. F. Pfeiffer.)

hydraulically actuated, and provided with a very high payload-to-weight ratio. Indeed, each finger weighs only 1.470 N, but can exert a force of up to 30 N.

We outline below some problems and research trends in the area of dextrous hands. A key issue here is the programming of the motions of the fingers, which is a much more complicated task than the programming of a six-axis manipulator. In this regard, Liu, Iberall, and Bekey (1989) introduced a task-analysis approach meant to program robotic hand motions at a higher level. They use a heuristic, knowledge-based approach. From an analysis of the various modes of grasping, they conclude that the requirements for grasping tasks are (*i*) stability, (*ii*) manipulability, (*iii*) torquability, and (*iv*) radial rotatability. Stability is defined as a measure of the tendency of an object to return to its original position after disturbances. Manipulability, as understood in this context, is the ability to impart motion to the object while keeping the fingers in contact with the object. Torquability, or tangential rotatability, is the ability to rotate the long axis of an object—here the authors must assume that the manipulated objects are convex and can be approximated by three-axis ellipsoids, thereby distinguishing between a longest and a shortest axis—with a minimum force, for a prescribed amount of torque. Finally, radial rotatability is the ability to rotate the grasped object about its long axis with minimum torque about the axis.

Furthermore, Allen, Michelman, and Roberts (1989) introduced an integrated system of both hardware and software for dextrous manipulation. The system consists of a Sun-3 workstation controlling a Puma 500 arm with VAL-II. The Utah/MIT hand is mounted on the end-effector of the arm. The system integrates force and position sensors with control commands for both the arm and the hand. To demonstrate the effectiveness of their system, the authors implemented a task consisting of removing a light bulb from its socket. Finally, Rus (1992) reports a paradigm allowing the high-level, task-oriented manipulation control of planar hands. Whereas technological aspects of dextrous manipulation are highly advanced, theoretical aspects are still under research in this area. An extensive literature survey, with 405 references on the subject of manipulation, is given by Reynaerts (1995).

1.6 Walking Machines

We focus here on multilegged walking devices, i.e., machines with more than two legs. In walking machines, stability is the main issue. One distinguishes between two types of stability, static and dynamic. Static stability refers to the ability of sustaining a configuration from reaction forces only, unlike dynamic stability, which refers to that ability from both reaction and inertia forces. Intuitively, it is apparent that static stability requires more contact points and, hence, more legs, than dynamic stability. Hopping devices (Raibert, 1986) and bipeds (Vukobratovic and Stepanenko, 1972) are examples of walking machines whose motions are dependent upon dynamic stability. For static balance, a walking machine requires a kinematic structure capable of providing the ground reaction forces needed to balance the weight of the machine. A biped is not capable of static equilibrium because during the swing phase of one leg, the body is supported by a single contact point, which is incapable of producing the necessary balancing forces to keep it in equilibrium. For motion on a horizontal surface, a minimum of three legs is required to produce static stability. Indeed, with three legs, one of these can undergo swing while the remaining two legs are in contact with the ground, and hence, two contact points are present to provide the necessary balancing forces from the ground reactions. By the same token, the minimum number of legs required to sustain static stability in general is four, although a very common architecture of walking machines is the hexapod, examples of which are the Ohio State University (OSU) Hexapod (Klein, Olson, and Pugh, 1983) and the OSU Adaptive Suspension Vehicle (ASV) (Song and Waldron, 1989), shown in Fig. 1.10. A six-legged walking machine with a design that mimics the locomotion system of the *Carausius morosus* (Graham, 1972), also known as the *walking stick*, has been developed at the Technical University of Munich (Pfeiffer, Eltze, and Weidemann, 1995). A prototype of this machine, known as the *TUM Hexapod*, is included in Fig. 1.9. The legs of the TUM Hexapod are operated under neural-network control, which gives them a reflex-like response when encountering obstacles. Upon sensing an obstacle, the leg bounces back and tries again to move forward, but raising the foot to a higher level.

Other machines that are worth mentioning are the Sutherland, Sprout and Associates Hexapod (Sutherland and Ullner, 1984), the Titan series of quadrupeds (Hirose, Masui, and Kikuchi, 1985) and the Odetics series of axially symmetric hexapods (Russell, 1983).

A survey of walking machines, of a rather historical interest by now, is given in (Todd, 1985), while a more recent comprehensive account of walking machines is available in a special issue of *The International Journal of Robotics Research* (Volume 9, No. 2).

Walking machines appear as the sole means of providing locomotion in highly unstructured environments. In fact, the unique adaptive suspension provided by these machines allows them to navigate on uneven terrain.

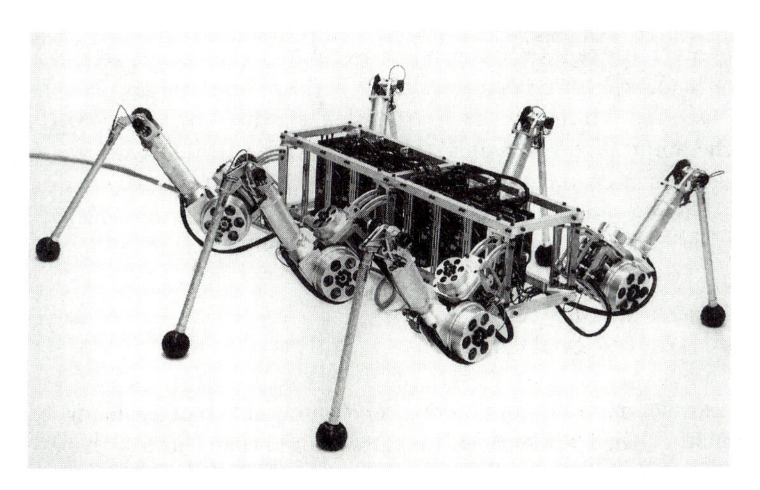

FIGURE 1.9. A prototype of the TU Munich Hexapod (Courtesy of Prof. F. Pfeiffer. Reproduced with permission of TSI Enterprises, Inc.)

FIGURE 1.10. The OSU ASV. An example of a six-legged walking machine (courtesy of Prof. K. Waldron. Reproduced with permission of The MIT Press.)

However, walking machines cannot navigate on every type of uneven terrain, for they are of limited dimensions. Hence, if terrain irregularities such as a crevasse wider than the maximum horizontal leg reach or a cliff of depth greater than the maximum vertical leg reach are present, then the machine is prevented from making any progress. This limitation, however, can be overcome by providing the machine with the capability of attaching its feet to the terrain in the same way as a mountain climber goes up a cliff. Moreover, machine functionality is limited not only by the topography of the terrain, but also by its constitution. Whereas hard rock poses no serious problem to a walking machine, muddy terrain can hamper its operation to the point that it may jam the machine. Still, under such adverse conditions, walking machines offer a better maneuverability than other vehicles. Some walking machines have been developed and are operational, but their operation is often limited to slow motions. It can be said, however, that like research on multifingered hands, the pace of theoretical research on walking machines has been much slower than that of their technological developments. The above-mentioned OSU ASV and the TU Munich Hexapod are among the most technologically developed walking machines.

1.7 Rolling Robots

While parallel manipulators indeed solve many inherent problems of serial manipulators, their workspaces are more limited than those of the latter. As a matter of fact, even serial manipulators have limited workspaces due to the finite lengths of their links. Manipulators with limited workspaces can be enhanced by mounting them on rolling robots. These are systems evolved from earlier systems called *automatic guided vehicles*, or AGVs for short. AGVs in their most primitive versions are four-wheeled electrically powered vehicles that perform moving tasks with a certain degree of autonomy. However, these vehicles are usually limited to motions along predefined tracks that are either railways or magnetic strips glued to the ground.

The most common rolling robots use conventional wheels, i.e., wheels consisting basically of a pneumatic tire mounted on a hub that rotates about an axle fixed to the platform of the robot. Thus, the operation of these machines does not differ much from that of conventional terrestrial vehicles. An essential difference between rolling robots and other robotic mechanical systems is the kinematic constraints between wheel and ground in the former. These constraints are of a type known as *nonholonomic*, as discussed in detail in Chapter 6. Nonholonomic constraints are kinematic relations between point velocities and angular velocities that cannot be integrated in the form of algebraic relations between translational and rotational displacement variables. The outcome of this lack of integrability leads to a lack of a one-to-one relationship between Cartesian variables and joint variables. In fact, while angular displacements read by joint encoders of serial manipulators determine uniquely the position and orientation of their end-effector, the angular displacement of the wheels of rolling machines do not determine the position and orientation of the vehicle body. As a matter of fact, the control of rolling robots bears common features with that of the redundancy resolution of manipulators of the serial type at the joint-rate level. In these manipulators, the number of actuated joints is greater than the dimension of the task space. As a consequence, the task velocity does not determine the joint rates. Not surprisingly, the two types of problems are being currently solved using the same tools, namely, differential geometry and Lie algebra (De Luca and Oriolo, 1995).

As a means to supply rolling robots with 3-dof capabilities, omnidirectional wheels (ODW) have been proposed. An example of ODWs are those that bear the name of *Mekanum* wheels, consisting of a hub with rollers on its periphery that roll freely about their axes, the latter being oriented at a constant angle with respect to the hub axle. In Fig. 1.11, a Mekanum wheel is shown, along with a rolling robot supplied with this type of wheels. Rolling robots with ODWs are, thus, 3-dof vehicles, and hence, can translate freely in two horizontal directions and rotate independently about a vertical axis. However, like their 2-dof counterparts, 3-dof rolling robots are also nonholonomic devices, and thus, pose the same problems for their control as the former.

Recent developments in the technology of rolling robots have been reported that incorporate alternative types of ODWs. For example, Killough and Pin (1992) developed a rolling robot with what they call *orthogonal ball wheels*, consisting basically of spherical wheels that can rotate about two mutually orthogonal axes. West and Asada (1995), in turn, designed a rolling robot with *ball wheels*, i.e., balls that act as omnidirectional wheels;

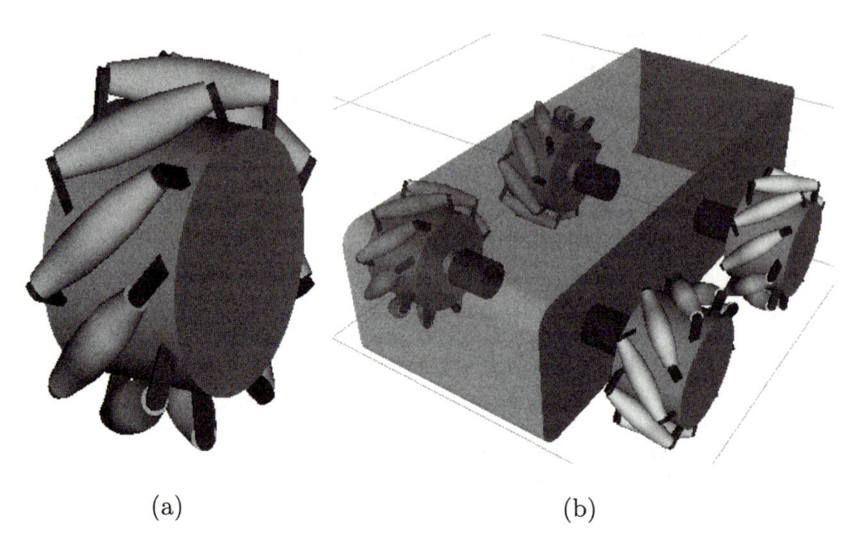

FIGURE 1.11. (a) A Mekanum wheel; (b) A rolling robot supplied with Mekanum wheels.

each ball being mounted on a set of rollers, one of which is actuated; hence, three such wheels are necessary to fully control the vehicle. The unactuated rollers serve two purposes, i.e., to provide stability to the wheels and the vehicle and to measure the rotation of the ball, thereby detecting slip. Furthermore, Borenstein (1993) proposed a mobile robot with four degrees of freedom; these were achieved with two chassis coupled by an extensible link, each chassis being driven by two actuated conventional wheels.

2
Mathematical Background

2.1 Preamble

First and foremost, the study of motions undergone by robotic mechanical systems or, for that matter, by mechanical systems at large, requires a suitable motion representation. Now, the motion of mechanical systems involves the motion of the particular links comprising those systems, which in this book are supposed to be rigid. The assumption of rigidity, although limited in scope, still covers a wide spectrum of applications, while providing insight into the motion of more complicated systems, such as those involving deformable bodies.

The most general kind of rigid-body motion consists of both translation and rotation. While the study of the former is covered in elementary mechanics courses and is reduced to the mechanics of particles, the latter is more challenging. Indeed, point translation can be studied simply with the aid of 3-dimensional vector calculus, while rigid-body rotations require the introduction of *tensors*, i.e., entities mapping vector spaces into vector spaces.

Emphasis is placed on *invariant* concepts, i.e., items that do not change upon a change of coordinate frame. Examples of invariant concepts are geometric quantities such as distances and angles between lines. Although we may resort to a coordinate frame and vector algebra to compute distances and angles and represent vectors in that frame, the final result will be independent of how we choose that frame. The same applies to quantities whose evaluation calls for the introduction of tensors. Here, we must distinguish between the physical quantity represented by a vector or a tensor and the representation of that quantity in a coordinate frame using a 1-dimensional array of components in the case of vectors, or a 2-dimensional array in the case of tensors. It is unfortunate that the same word is used in English to denote a vector and its array representation in a given coordinate frame. Regarding tensors, the associated arrays are called *matrices*. By abuse of terminology, we will refer to both tensors and their arrays as matrices, although keeping in mind the essential conceptual differences involved.

2.2 Linear Transformations

The physical 3-dimensional space is a particular case of a *vector space*. A vector space is a set of objects, called *vectors*, that follow certain algebraic rules. Throughout the book, vectors will be denoted by boldface lowercase characters, whereas tensors and their matrix representations are denoted by boldface uppercase characters. Let $\mathbf{v}$, $\mathbf{v}_1$, $\mathbf{v}_2$, $\mathbf{v}_3$, and $\mathbf{w}$ be elements of a given vector space $\mathcal{V}$, which is *defined over the real field*, and let α and β be two elements of this field, i.e., α and β are two real numbers. Below we summarize the aforementioned rules:

(*i*) The sum of $\mathbf{v}_1$ and $\mathbf{v}_2$, denoted by $\mathbf{v}_1 + \mathbf{v}_2$, is itself an element of $\mathcal{V}$ and is *commutative*, i.e., $\mathbf{v}_1 + \mathbf{v}_2 = \mathbf{v}_2 + \mathbf{v}_1$;

(*ii*) $\mathcal{V}$ contains an element $\mathbf{0}$, called the *zero* vector of $\mathcal{V}$, which, when added to any other element $\mathbf{v}$ of $\mathcal{V}$, leaves it unchanged, i.e., $\mathbf{v}+\mathbf{0} = \mathbf{v}$;

(*iii*) The sum defined in (*i*) is *associative*, i.e., $\mathbf{v}_1+(\mathbf{v}_2+\mathbf{v}_3) = (\mathbf{v}_1+\mathbf{v}_2)+\mathbf{v}_3$;

(*iv*) For every element $\mathbf{v}$ of $\mathcal{V}$, there exists a corresponding element, $\mathbf{w}$, also of $\mathcal{V}$, which, when added to $\mathbf{v}$, produces the zero vector, i.e., $\mathbf{v} + \mathbf{w} = \mathbf{0}$. Moreover, $\mathbf{w}$ is represented as $-\mathbf{v}$;

(*v*) The product $\alpha\mathbf{v}$, or $\mathbf{v}\alpha$, is also an element of $\mathcal{V}$, for every $\mathbf{v}$ of $\mathcal{V}$ and every real α. This product is associative, i.e., $\alpha(\beta\mathbf{v}) = (\alpha\beta)\mathbf{v}$;

(*vi*) If α is the real unity, then $\alpha\mathbf{v}$ is identically $\mathbf{v}$;

(*vii*) The product defined in (*v*) is *distributive* in the sense that (*a*) $(\alpha + \beta)\mathbf{v} = \alpha\mathbf{v} + \beta\mathbf{v}$ and (*b*) $\alpha(\mathbf{v}_1 + \mathbf{v}_2) = \alpha\mathbf{v}_1 + \alpha\mathbf{v}_2$.

Although vector spaces can be defined over other fields, we will deal with vector spaces over the real field unless explicit reference to another field is made. Moreover, vector spaces can be either finite- or infinite-dimensional, but we will not need the latter. In geometry and elementary mechanics, the dimension of the vector spaces needed is usually three, but when studying multibody systems, an arbitrary finite dimension will be required. The concept of *dimension* of a vector space is discussed in more detail later.

A *linear transformation*, represented as an *operator* $\mathbf{L}$, of a vector space $\mathcal{U}$ into a vector space $\mathcal{V}$, is a rule that assigns to every vector $\mathbf{u}$ of $\mathcal{U}$ at least one vector $\mathbf{v}$ of $\mathcal{V}$, represented as $\mathbf{v} = \mathbf{L}\mathbf{u}$, with $\mathbf{L}$ endowed with two properties:

(*i*) *homogeneity*: $\mathbf{L}(\alpha\mathbf{u}) = \alpha\mathbf{v}$; and

(*ii*) *additivity*: $\mathbf{L}(\mathbf{u}_1 + \mathbf{u}_2) = \mathbf{v}_1 + \mathbf{v}_2$.

In the foregoing discussion, $\mathcal{U}$ and $\mathcal{V}$ need not be identical, of course. A linear transformation of a vector space $\mathcal{V}$ into itself is called an *homeomorphism*. It assigns to every vector $\mathbf{v}$ of $\mathcal{V}$ another vector $\mathbf{w}$ of $\mathcal{V}$. Note that, in the foregoing definitions, no mention has been made of components, and hence, vectors and their transformations should not be confused with their *array representations*.

Particular types of linear transformations of the 3-dimensional Euclidean space that will be encountered frequently in this context are *projections*, *reflections*, and *rotations*. One further type of transformation, which is not linear, but nevertheless appears frequently in kinematics, is the one known as *affine transformation*. The foregoing transformations are defined below. It is necessary, however, to introduce additional concepts pertaining to general linear transformations before expanding into these definitions.

The *range* of a linear transformation $\mathbf{L}$ of $\mathcal{U}$ into $\mathcal{V}$ is the set of vectors $\mathbf{v}$ into which some vector $\mathbf{u}$ of $\mathcal{U}$ is mapped, i.e., the range of $\mathbf{L}$ is defined as the set of $\mathbf{v} = \mathbf{L}\mathbf{u}$, for every vector $\mathbf{u}$ of $\mathcal{U}$. The *kernel* of $\mathbf{L}$ is the set of vectors $\mathbf{u}_N$ of $\mathcal{U}$ that are mapped by $\mathbf{L}$ into the zero vector $\mathbf{0} \in \mathcal{V}$. It can be readily proven (see Exercises 2.1–2.3) that the kernel and the range of a linear transformation are both vector subspaces of $\mathcal{U}$ and $\mathcal{V}$, respectively, i.e., they are themselves vector spaces, but of a dimension smaller than or equal to that of their associated vector spaces. Moreover, the kernel of a linear transformation is often called the *nullspace* of the said transformation.

Henceforth, the 3-dimensional Euclidean space is denoted by $\mathcal{E}^3$. Having chosen an origin O for this space, its geometry can be studied in the context of general vector spaces. Hence, points of $\mathcal{E}^3$ will be identified with vectors of the associated 3-dimensional vector space. Moreover, lines and planes passing through the origin are subspaces of dimensions 1 and 2, respectively, of $\mathcal{E}^3$. Clearly, lines and planes not passing through the origin of $\mathcal{E}^3$ are not subspaces but can be handled with the algebra of vector spaces, as will be shown here.

An *orthogonal projection* $\mathbf{P}$ of $\mathcal{E}^3$ onto itself is a linear transformation of the said space onto a plane Π passing through the origin and having a unit normal $\mathbf{n}$, with the properties:

$$\mathbf{P}^2 = \mathbf{P}, \quad \mathbf{P}\mathbf{n} = \mathbf{0} \tag{2.1a}$$

Any matrix with the first property above is termed *idempotent*. For $n \times n$ matrices, it is sometimes necessary to indicate the lowest integer l for which an analogous relation follows, i.e., for which $\mathbf{P}^l = \mathbf{P}$. In this case, the matrix is said to be idempotent of degree l.

Clearly, the projection of a position vector $\mathbf{p}$, denoted by $\mathbf{p}'$, onto a plane Π of unit normal $\mathbf{n}$, is $\mathbf{p}$ itself minus the component of $\mathbf{p}$ along $\mathbf{n}$, i.e.,

$$\mathbf{p}' = \mathbf{p} - \mathbf{n}(\mathbf{n}^T\mathbf{p}) \tag{2.1b}$$

where the superscript T denotes either vector or matrix transposition and $\mathbf{n}^T\mathbf{p}$ is equivalent to the usual *dot product* $\mathbf{n} \cdot \mathbf{p}$.

Now, the *identity* matrix $\mathbf{1}$ is defined as the isomorphism of a vector space into itself leaving every vector $\mathbf{v}$ of $\mathcal{V}$ unchanged, i.e.,

$$\mathbf{1}\mathbf{v} = \mathbf{v} \tag{2.2}$$

Thus, $\mathbf{p}'$, as given by eq.(2.1b), can be rewritten as

$$\mathbf{p}' = \mathbf{1}\mathbf{p} - \mathbf{n}\mathbf{n}^T\mathbf{p} \equiv (\mathbf{1} - \mathbf{n}\mathbf{n}^T)\mathbf{p} \tag{2.3}$$

and hence, the *orthogonal projection* $\mathbf{P}$ onto Π can be represented as

$$\mathbf{P} = \mathbf{1} - \mathbf{n}\mathbf{n}^T \tag{2.4}$$

where the product $\mathbf{n}\mathbf{n}^T$ amounts to a 3×3 matrix.

Now we turn to reflections. Here we have to take into account that reflections occur frequently accompanied by rotations, as yet to be studied. Since reflections are simpler to represent, we first discuss these, rotations being discussed in full detail in Section 2.3. What we shall discuss in this section is *pure reflections*, i.e., those occurring without any concomitant rotation. Thus, all reflections studied in this section are pure reflections, but for the sake of brevity, they will be referred to simply as *reflections*.

A *reflection* $\mathbf{R}$ of $\mathcal{E}^3$ onto a plane Π passing through the origin and having a unit normal $\mathbf{n}$ is a linear transformation of the said space into itself such that a position vector $\mathbf{p}$ is mapped by $\mathbf{R}$ into a vector $\mathbf{p}'$ given by

$$\mathbf{p}' = \mathbf{p} - 2\mathbf{n}\mathbf{n}^T\mathbf{p} \equiv (\mathbf{1} - 2\mathbf{n}\mathbf{n}^T)\mathbf{p}$$

Thus, the reflection $\mathbf{R}$ can be expressed as

$$\mathbf{R} = \mathbf{1} - 2\mathbf{n}\mathbf{n}^T \tag{2.5}$$

From eq.(2.5) it is then apparent that a pure reflection is represented by a linear transformation that is symmetric and whose square equals the identity matrix, i.e., $\mathbf{R}^2 = \mathbf{1}$. Indeed, symmetry is apparent from the equation above; the second property is readily proven below:

$$\begin{aligned}\mathbf{R}^2 &= (\mathbf{1} - 2\mathbf{n}\mathbf{n}^T)(\mathbf{1} - 2\mathbf{n}\mathbf{n}^T)\\ &= \mathbf{1} - 2\mathbf{n}\mathbf{n}^T - 2\mathbf{n}\mathbf{n}^T + 4(\mathbf{n}\mathbf{n}^T)(\mathbf{n}\mathbf{n}^T) = \mathbf{1} - 4\mathbf{n}\mathbf{n}^T + 4\mathbf{n}(\mathbf{n}^T\mathbf{n})\mathbf{n}^T\end{aligned}$$

which apparently reduces to $\mathbf{1}$ because $\mathbf{n}$ is a unit vector. Note that from the second property above, we find that pure reflections observe a further interesting property, namely,

$$\mathbf{R}^{-1} = \mathbf{R}$$

i.e., every pure reflection equals its inverse. This result can be understood intuitively by noticing that, upon doubly reflecting an image using two mirrors, the original image is recovered.

Further, we take to deriving the *orthogonal decomposition* of a given vector $\mathbf{v}$ into two components, one along and one normal to a unit vector $\mathbf{e}$. The component of $\mathbf{v}$ along $\mathbf{e}$, termed here the *axial component*, $\mathbf{v}_\parallel$—read *v-par*—is simply given as

$$\mathbf{v}_\parallel \equiv \mathbf{e}\mathbf{e}^T\mathbf{v} \tag{2.6a}$$

while the corresponding *normal component*, $\mathbf{v}_\perp$—read *v-perp*—is simply the difference $\mathbf{v} - \mathbf{v}_\parallel$, i.e.,

$$\mathbf{v}_\perp \equiv \mathbf{v} - \mathbf{v}_\parallel \equiv (\mathbf{1} - \mathbf{e}\mathbf{e}^T)\mathbf{v} \tag{2.6b}$$

the matrix in parentheses in the foregoing equation being rather frequent in kinematics. This matrix will appear when studying rotations. Rotations are more complicated transformations that deserve special attention, the entire Section 2.3 being devoted to these transformations.

Further concepts are now recalled: The *basis* of a vector space $\mathcal{V}$ is a set of *linearly independent* vectors of $\mathcal{V}$, $\{\mathbf{v}_i\}_1^n$, in terms of which any vector $\mathbf{v}$ of $\mathcal{V}$ can be expressed as

$$\mathbf{v} = \alpha_1\mathbf{v}_1 + \alpha_2\mathbf{v}_2 + \cdots + \alpha_n\mathbf{v}_n, \tag{2.7}$$

where the elements of the set $\{\alpha_i\}_1^n$ are all elements of the field over which $\mathcal{V}$ is defined, i.e., they are real numbers in the case at hand. The number n of elements in the set $\mathcal{B} = \{\mathbf{v}_i\}_1^n$ is called *the dimension* of $\mathcal{V}$. Note that *any* set of n linearly independent vectors of $\mathcal{V}$ can play the role of a basis of this space, but once this basis is defined, the set of real coefficients $\{\alpha_i\}_1^n$ for expressing a given vector $\mathbf{v}$ is *unique*.

Let $\mathcal{U}$ and $\mathcal{V}$ be two vector spaces of dimensions m and n, respectively, and $\mathbf{L}$ a linear transformation of $\mathcal{U}$ into $\mathcal{V}$, and define bases $\mathcal{B}_U$ and $\mathcal{B}_V$ for $\mathcal{U}$ and $\mathcal{V}$ as

$$\mathcal{B}_U = \{\mathbf{u}_j\}_1^m, \quad \mathcal{B}_V = \{\mathbf{v}_i\}_1^n \tag{2.8}$$

Since each $\mathbf{L}\mathbf{u}_j$ is an element of $\mathcal{V}$, it can be represented uniquely in terms of the vectors of $\mathcal{B}_V$, namely, as

$$\mathbf{L}\mathbf{u}_j = l_{1j}\mathbf{v}_1 + l_{2j}\mathbf{v}_2 + \cdots + l_{nj}\mathbf{v}_n, \quad j = 1, \ldots, m \tag{2.9}$$

Consequently, in order to represent the *images* of the m vectors of $\mathcal{B}_U$, namely, the set $\{\mathbf{L}\mathbf{u}_j\}_1^m$, $n \times m$ real numbers l_{ij}, for $i = 1, \ldots, n$ and $j = 1, \ldots, m$, are necessary. These real numbers are now arranged in the $n \times m$ array $[\,\mathbf{L}\,]_{\mathcal{B}_U}^{\mathcal{B}_V}$ defined below:

$$[\,\mathbf{L}\,]_{\mathcal{B}_U}^{\mathcal{B}_V} \equiv \begin{bmatrix} l_{11} & l_{12} & \cdots & l_{1m} \\ l_{21} & l_{22} & \cdots & l_{2m} \\ \vdots & \vdots & \ddots & \vdots \\ l_{n1} & l_{n2} & \cdots & l_{nm} \end{bmatrix} \tag{2.10}$$

The foregoing array is thus called the *matrix representation of* $\mathbf{L}$ *with respect to* $\mathcal{B}_U$ *and* $\mathcal{B}_V$. We thus have an important definition, namely,

Definition 2.2.1 *The* j*th column of the matrix representation of* $\mathbf{L}$ *with respect to the bases* $\mathcal{B}_U$ *and* $\mathcal{B}_V$ *is composed of the* n *real coefficients* l_{ij} *of the representation of the image of the* j*th vector of* $\mathcal{B}_U$ *in terms of* $\mathcal{B}_V$.

The notation introduced in eq.(2.10) is rather cumbersome, for it involves one subscript and one superscript. Moreover, each of these is subscripted. In practice, the bases involved are self-evident, which makes an explicit mention of these unnecessary. In particular, when the mapping $\mathbf{L}$ is an isomorphism, i.e., a one-to-one mapping of $\mathcal{U}$ onto itself, then a single basis suffices to represent $\mathbf{L}$ in matrix form. In this case, its bracket will bear only a subscript, and no superscript, namely, $[\,\mathbf{L}\,]_{\mathcal{B}}$. Moreover, we will use, henceforth, the concept of basis and coordinate frame interchangeably, since one implies the other.

Two different bases are unavoidable when the two spaces under study are physically distinct, which is the case in velocity analyses of manipulators. As we will see in Chapter 4, in these analyses we distinguish between the velocity of the manipulator in Cartesian space and that in the joint-rate space. While the Cartesian-space velocity—or Cartesian velocity, for brevity—consists, in general, of a 6-dimensional vector containing the 3-dimensional angular velocity of the end-effector and the translational velocity of one of its points, the latter is an n-dimensional vector. Moreover, if the manipulator is coupled by revolute joints only, the units of the joint-rate vector are all s^{-1}, whereas the Cartesian velocity contains some components with units of s^{-1} and others with units of ms^{-1}.

Further definitions are now recalled. Given an homeorphism $\mathbf{L}$ of an n-dimensional vector space $\mathcal{V}$, a nonzero vector $\mathbf{e}$ that is mapped by $\mathbf{L}$ into a multiple of itself, $\lambda\mathbf{e}$, is called an *eigenvector* of $\mathbf{L}$, the scalar λ being called an *eigenvalue* of $\mathbf{L}$. The eigenvalues of $\mathbf{L}$ are determined by the equation

$$\det(\lambda\mathbf{1} - \mathbf{L}) = 0 \tag{2.11}$$

Note that the matrix $\lambda\mathbf{1} - \mathbf{L}$ is *linear* in λ, and since the determinant of an $n \times n$ matrix is a homogeneous nth-order function of its entries, the left-hand side of eq.(2.11) is an nth-degree polynomial in λ. The foregoing polynomial is termed *the characteristic polynomial of* $\mathbf{L}$. Hence, every $n \times n$ matrix $\mathbf{L}$ has n complex eigenvalues, even if $\mathbf{L}$ is defined over the real field. If it is, then its complex eigenvalues appear in conjugate pairs. Clearly, the eigenvalues of $\mathbf{L}$ are the roots of its characteristic polynomial, while eq.(2.11) is called the *characteristic equation* of $\mathbf{L}$.

Example 2.2.1 *What is the representation of the reflection* $\mathbf{R}$ *of* $\mathcal{E}^3$ *into itself, with respect to the x-y plane, in terms of unit vectors parallel to the* X, Y, Z *axes that form a coordinate frame* $\mathcal{F}$*?*

Solution: Note that in this case, $\mathcal{U} = \mathcal{V} = \mathcal{E}^3$ and, hence, it is not necessary to use two different bases for $\mathcal{U}$ and $\mathcal{V}$. Now, let $\mathbf{i}$, $\mathbf{j}$, $\mathbf{k}$, be unit vectors parallel to the X, Y, and Z axes of a frame $\mathcal{F}$. Clearly,

$$\mathbf{Ri} = \mathbf{i}$$
$$\mathbf{Rj} = \mathbf{j}$$
$$\mathbf{Rk} = -\mathbf{k}$$

Thus, the representations of the images of $\mathbf{i}$, $\mathbf{j}$ and $\mathbf{k}$ under $\mathbf{R}$, in $\mathcal{F}$, are

$$[\mathbf{Ri}]_{\mathcal{F}} = \begin{bmatrix} 1 \\ 0 \\ 0 \end{bmatrix}, \quad [\mathbf{Rj}]_{\mathcal{F}} = \begin{bmatrix} 0 \\ 1 \\ 0 \end{bmatrix}, \quad [\mathbf{Rk}]_{\mathcal{F}} = \begin{bmatrix} 0 \\ 0 \\ -1 \end{bmatrix}$$

where subscripted brackets are used to indicate the representation frame. Hence, the matrix representation of $\mathbf{R}$ in $\mathcal{F}$, denoted by $[\mathbf{R}]_{\mathcal{F}}$, is

$$[\mathbf{R}]_{\mathcal{F}} = \begin{bmatrix} 1 & 0 & 0 \\ 0 & 1 & 0 \\ 0 & 0 & -1 \end{bmatrix}$$

2.3 Rigid-Body Rotations

A *linear isomorphism*, i.e., a one-to-one linear transformation mapping a space $\mathcal{V}$ onto itself, is called an *isometry* if it preserves distances between any two points of $\mathcal{V}$. If $\mathbf{u}$ and $\mathbf{v}$ are regarded as the position vectors of two such points, then the distance d between these two points is defined as

$$d \equiv \sqrt{(\mathbf{u} - \mathbf{v})^T(\mathbf{u} - \mathbf{v})} \tag{2.12}$$

The volume V of the tetrahedron defined by the origin and three points of the 3-dimensional Euclidean space of position vectors $\mathbf{u}$, $\mathbf{v}$, and $\mathbf{w}$ is obtained as one sixth of the absolute value of the *double mixed product* of these three vectors,

$$V \equiv \frac{1}{6}|\mathbf{u} \times \mathbf{v} \cdot \mathbf{w}| = \frac{1}{6}|\det[\mathbf{u} \quad \mathbf{v} \quad \mathbf{w}]| \tag{2.13}$$

i.e., if a 3×3 array $[\mathbf{A}]$ is defined in terms of the components of $\mathbf{u}$, $\mathbf{v}$, and $\mathbf{w}$, in a given basis, then the first column of $[\mathbf{A}]$ is given by the three components of $\mathbf{u}$, the second and third columns being defined analogously.

Now, let $\mathbf{Q}$ be an isometry mapping the triad $\{\mathbf{u},\ \mathbf{v},\ \mathbf{w}\}$ into $\{\mathbf{u}',\ \mathbf{v}',\ \mathbf{w}'\}$. Moreover, the distance from the origin to the points of position vectors $\mathbf{u}$, $\mathbf{v}$, and $\mathbf{w}$ is given simply as $\|\mathbf{u}\|$, $\|\mathbf{v}\|$, and $\|\mathbf{w}\|$, which are defined as

$$\|\mathbf{u}\| \equiv \sqrt{\mathbf{u}^T\mathbf{u}}, \quad \|\mathbf{v}\| \equiv \sqrt{\mathbf{v}^T\mathbf{v}}, \quad \|\mathbf{w}\| \equiv \sqrt{\mathbf{w}^T\mathbf{w}} \tag{2.14}$$

Clearly,

$$\|\mathbf{u}'\| = \|\mathbf{u}\|, \quad \|\mathbf{v}'\| = \|\mathbf{v}\|, \quad \|\mathbf{w}'\| = \|\mathbf{w}\| \tag{2.15a}$$

and

$$\det[\,\mathbf{u}' \quad \mathbf{v}' \quad \mathbf{w}'\,] = \pm\det[\,\mathbf{u} \quad \mathbf{v} \quad \mathbf{w}\,] \tag{2.15b}$$

If, in the foregoing relations, the sign of the determinant is preserved, the isometry represents a *rotation*; otherwise, it represents a reflection. Now, let $\mathbf{p}$ be the position vector of any point of $\mathcal{E}^3$, its image under a rotation $\mathbf{Q}$ being $\mathbf{p}'$. Hence, distance preservation requires that

$$\mathbf{p}^T\mathbf{p} = \mathbf{p}'^T\mathbf{p}' \tag{2.16}$$

where

$$\mathbf{p}' = \mathbf{Q}\mathbf{p} \tag{2.17}$$

condition (2.16) thus leading to

$$\mathbf{Q}^T\mathbf{Q} = \mathbf{1} \tag{2.18}$$

where $\mathbf{1}$ was defined in Section 2.2 as the *identity* 3×3 *matrix*, and hence, eq.(2.18) states that $\mathbf{Q}$ is an *orthogonal matrix*. Moreover, let $\mathbf{T}$ and $\mathbf{T}'$ denote the two matrices defined below:

$$\mathbf{T} = [\,\mathbf{u} \quad \mathbf{v} \quad \mathbf{w}\,], \quad \mathbf{T}' = [\,\mathbf{u}' \quad \mathbf{v}' \quad \mathbf{w}'\,] \tag{2.19}$$

from which it is clear that

$$\mathbf{T}' = \mathbf{Q}\mathbf{T} \tag{2.20}$$

Now, for a rigid-body rotation, eq.(2.15b) should hold with the positive sign, and hence,

$$\det(\mathbf{T}) = \det(\mathbf{T}') \tag{2.21a}$$

and, by virtue of eq.(2.20), we conclude that

$$\det(\mathbf{Q}) = +1 \tag{2.21b}$$

Therefore, $\mathbf{Q}$ is a *proper orthogonal matrix*, i.e., it is a proper isometry. Now we have

Theorem 2.3.1 *The eigenvalues of a proper orthogonal matrix* $\mathbf{Q}$ *lie on the unit circle centered at the origin of the complex plane.*

Proof: Let λ be one of the eigenvalues of $\mathbf{Q}$ and $\mathbf{e}$ the corresponding eigenvector, so that

$$\mathbf{Q}\mathbf{e} = \lambda\mathbf{e} \tag{2.22}$$

In general, $\mathbf{Q}$ is not expected to be symmetric, and hence, λ is not necessarily real. Thus, λ is considered complex, in general. In this light, when transposing both sides of the foregoing equation, we will need to take the complex conjugates as well. Henceforth, the complex conjugate of a vector or a matrix will be indicated with an asterisk as a superscript. As well, the conjugate of a complex variable will be indicated with a bar over the said variable. Thus, the transpose conjugate of the latter equation takes on the form

$$\mathbf{e}^*\mathbf{Q}^* = \overline{\lambda}\mathbf{e}^* \tag{2.23}$$

Multiplying the corresponding sides of the two previous equations yields

$$\mathbf{e}^*\mathbf{Q}^*\mathbf{Q}\mathbf{e} = \overline{\lambda}\lambda\mathbf{e}^*\mathbf{e} \tag{2.24}$$

However, $\mathbf{Q}$ has been assumed real, and hence, $\mathbf{Q}^*$ reduces to $\mathbf{Q}^T$, the foregoing equation thus reducing to

$$\mathbf{e}^*\mathbf{Q}^T\mathbf{Q}\mathbf{e} = \overline{\lambda}\lambda\mathbf{e}^*\mathbf{e} \tag{2.25}$$

But $\mathbf{Q}$ is orthogonal by assumption, and hence, it obeys eq.(2.18), which means that eq.(2.25) reduces to

$$\mathbf{e}^*\mathbf{e} = |\lambda|^2\mathbf{e}^*\mathbf{e} \tag{2.26}$$

where $|\cdot|$ denotes the *modulus* of the complex variable within it. Thus, the foregoing equation leads to

$$|\lambda|^2 = 1 \tag{2.27}$$

thereby completing the intended proof. As a direct consequence of Theorem 2.3.1, we have

Corollary 2.3.1 *A proper orthogonal* 3×3 *matrix has at least one eigenvalue that is* $+1$.

Now, let $\mathbf{e}$ be the eigenvector of $\mathbf{Q}$ associated with the eigenvalue $+1$. Thus,

$$\mathbf{Q}\mathbf{e} = \mathbf{e} \tag{2.28}$$

What eq.(2.28) states is summarized as a theorem below:

Theorem 2.3.2 (Euler, 1776) *A rigid-body motion about a point* O *leaves fixed a set of points lying on a line* $\mathcal{L}$ *that passes through* O *and is parallel to the eigenvector* $\mathbf{e}$ *of* $\mathbf{Q}$ *associated with the eigenvalue* $+1$.

A further result, that finds many applications in robotics and, in general, in system theory, is given below:

Theorem 2.3.3 (Cayley-Hamilton) *Let $P(\lambda)$ be the characteristic polynomial of an $n \times n$ matrix $\mathbf{A}$, i.e.,*

$$P(\lambda) = \det(\lambda \mathbf{1} - \mathbf{A}) = \lambda^n + a_{n-1}\lambda^{n-1} + \cdots + a_1\lambda + a_0 \tag{2.29}$$

Then $\mathbf{A}$ satisfies its characteristic equation, i.e.,

$$\mathbf{A}^n + a_{n-1}\mathbf{A}^{n-1} + \cdots + a_1\mathbf{A} + a_0\mathbf{1} = \mathbf{O} \tag{2.30}$$

where $\mathbf{O}$ is the $n \times n$ zero matrix.

Proof: See Halmos (1974).

What the Cayley-Hamilton Theorem states is that any power $p \geq n$ of the $n \times n$ matrix $\mathbf{A}$ can be expressed as a linear combination of the first n powers of $\mathbf{A}$—the 0th power of $\mathbf{A}$ is, of course the $n \times n$ identity matrix $\mathbf{1}$. An important consequence of this result is that any *analytic* matrix function of $\mathbf{A}$ can be expressed not as an infinite series, but as a sum, namely, a linear combination of the first n powers of $\mathbf{A}$: $\mathbf{1}$, $\mathbf{A}$, ..., $\mathbf{A}^{n-1}$. An *analytic* function $f(x)$ of a real variable x is, in turn, a function with a series expansion. Moreover, an analytic matrix function of a matrix argument $\mathbf{A}$ is defined likewise, an example of which is the exponential function. From the previous discussion, then, the exponential of $\mathbf{A}$ can be written as a linear combination of the first n powers of $\mathbf{A}$. It will be shown later that any proper orthogonal matrix $\mathbf{Q}$ can be represented as the exponential of a skew-symmetric matrix derived from the unit vector $\mathbf{e}$ of $\mathbf{Q}$, of eigenvalue $+1$, and the associated angle of rotation, as yet to be defined.

2.3.1 The Cross-Product Matrix

Prior to introducing the matrix representation of a rotation, we will need a few definitions. We will start by defining the partial derivative of a vector with respect to another vector. This is a matrix, as described below: In general, let $\mathbf{u}$ and $\mathbf{v}$ be vectors of spaces $\mathcal{U}$ and $\mathcal{V}$, of dimensions m and n, respectively. Furthermore, let t be a real variable and f be real-valued function of t, $\mathbf{u} = \mathbf{u}(t)$ and $\mathbf{v} = \mathbf{v}(\mathbf{u}(t))$ being m- and n-dimensional vector functions of t as well, with $f = f(\mathbf{u}, \mathbf{v})$. The derivative of $\mathbf{u}$ with respect to t, denoted by $\dot{\mathbf{u}}(t)$, is an m-dimensional vector whose ith component is the derivative of the ith component of $\mathbf{u}$ in a given basis, u_i, with respect to t. A similar definition follows for $\dot{\mathbf{v}}(t)$. The partial derivative of f with respect to $\mathbf{u}$ is an m-dimensional vector whose ith component is the partial derivative of f with respect to u_i, with a corresponding definition for the partial derivative of f with respect to $\mathbf{v}$. The foregoing derivatives, as all other vectors, will be assumed, henceforth, to be *column* arrays. Thus,

$$\frac{\partial f}{\partial \mathbf{u}} \equiv \begin{bmatrix} \partial f/\partial u_1 \\ \partial f/\partial u_2 \\ \vdots \\ \partial f/\partial u_m \end{bmatrix}, \quad \frac{\partial f}{\partial \mathbf{v}} \equiv \begin{bmatrix} \partial f/\partial v_1 \\ \partial f/\partial v_2 \\ \vdots \\ \partial f/\partial v_n \end{bmatrix} \tag{2.31}$$

Furthermore, the partial derivative of $\mathbf{v}$ with respect to $\mathbf{u}$ is an $n \times m$ array whose (i, j) entry is defined as $\partial v_i/\partial u_j$, i.e.,

$$\frac{\partial \mathbf{v}}{\partial \mathbf{u}} \equiv \begin{bmatrix} \partial v_1/\partial u_1 & \partial v_1/\partial u_2 & \cdots & \partial v_1/\partial u_m \\ \partial v_2/\partial u_1 & \partial v_2/\partial u_2 & \cdots & \partial v_2/\partial u_m \\ \vdots & \vdots & \ddots & \vdots \\ \partial v_n/\partial u_1 & \partial v_n/\partial u_2 & \cdots & \partial v_n/\partial u_m \end{bmatrix} \tag{2.32}$$

Hence, the total derivative of f with respect to $\mathbf{u}$ can be written as

$$\frac{df}{d\mathbf{u}} = \frac{\partial f}{\partial \mathbf{u}} + (\frac{\partial \mathbf{v}}{\partial \mathbf{u}})^T \frac{\partial f}{\partial \mathbf{v}} \tag{2.33}$$

If, moreover, f is an explicit function of t, i.e., if $f = f(\mathbf{u}, \mathbf{v}, t)$ and $\mathbf{v} = \mathbf{v}(\mathbf{u}, t)$, then, one can write the total derivative of f with respect to t as

$$\frac{df}{dt} = \frac{\partial f}{\partial t} + \left(\frac{\partial f}{\partial \mathbf{u}}\right)^T \frac{d\mathbf{u}}{dt} + \left(\frac{\partial f}{\partial \mathbf{v}}\right)^T \frac{\partial \mathbf{v}}{\partial t} + \left(\frac{\partial f}{\partial \mathbf{v}}\right)^T \frac{\partial \mathbf{v}}{\partial \mathbf{u}} \frac{d\mathbf{u}}{dt} \tag{2.34}$$

The total derivative of $\mathbf{v}$ with respect to t can be written, likewise, as

$$\frac{d\mathbf{v}}{dt} = \frac{\partial \mathbf{v}}{\partial t} + \frac{\partial \mathbf{v}}{\partial \mathbf{u}} \frac{d\mathbf{u}}{dt} \tag{2.35}$$

Example 2.3.1 *Let the components of* $\mathbf{v}$ *and* $\mathbf{x}$ *in a certain reference frame* $\mathcal{F}$ *be given as*

$$[\mathbf{v}]_{\mathcal{F}} = \begin{bmatrix} v_1 \\ v_2 \\ v_3 \end{bmatrix}, \quad [\mathbf{x}]_{\mathcal{F}} = \begin{bmatrix} x_1 \\ x_2 \\ x_3 \end{bmatrix} \tag{2.36a}$$

Then

$$[\mathbf{v} \times \mathbf{x}]_{\mathcal{F}} = \begin{bmatrix} v_2 x_3 - v_3 x_2 \\ v_3 x_1 - v_1 x_3 \\ v_1 x_2 - v_2 x_1 \end{bmatrix} \tag{2.36b}$$

Hence,

$$\left[\frac{\partial(\mathbf{v} \times \mathbf{x})}{\partial \mathbf{x}}\right]_{\mathcal{F}} = \begin{bmatrix} 0 & -v_3 & v_2 \\ v_3 & 0 & -v_1 \\ -v_2 & v_1 & 0 \end{bmatrix} \tag{2.36c}$$

Henceforth, the partial derivative of the cross product of any 3-dimensional vectors $\mathbf{v}$ and $\mathbf{x}$ will be denoted by the 3×3 matrix $\mathbf{V}$. For obvious reasons, $\mathbf{V}$ is termed the *cross-product matrix* of vector $\mathbf{v}$. Sometimes the cross-product matrix of a vector $\mathbf{v}$ is represented as $\widetilde{\mathbf{v}}$, but we do not follow this notation for the sake of consistency, since we decided at the outset to represent matrices with boldface uppercase letters. Thus, the foregoing cross product admits the alternative representations

$$\mathbf{v} \times \mathbf{x} = \mathbf{V}\mathbf{x} \tag{2.37}$$

Now, the following is apparent:

Theorem 2.3.4 *The cross-product matrix* $\mathbf{A}$ *of any 3-dimensional vector* $\mathbf{a}$ *is skew-symmetric, i.e.,*

$$\mathbf{A}^T = -\mathbf{A}$$

and, as a consequence,

$$\mathbf{a} \times (\mathbf{a} \times \mathbf{b}) = \mathbf{A}^2\mathbf{b} \tag{2.38}$$

where $\mathbf{A}^2$ *can be readily proven to be*

$$\mathbf{A}^2 = -\|\mathbf{a}\|^2\mathbf{1} + \mathbf{a}\mathbf{a}^T \tag{2.39}$$

with $\| \cdot \|$ *denoting the* Euclidean norm *of the vector inside it.*

Note that given *any* 3-dimensional vector $\mathbf{a}$, its cross-product matrix $\mathbf{A}$ is *uniquely* defined. Moreover, this matrix is skew-symmetric. The converse also holds, i.e., given any 3×3 skew-symmetric matrix $\mathbf{A}$, its associated *vector* is uniquely defined as well. This result is made apparent from Example 2.3.1 and will be discussed further when we define the *axial vector* of an arbitrary 3×3 matrix below.

2.3.2 The Rotation Matrix

In deriving the matrix representation of a rotation, we should recall Theorem 2.3.2, which suggests that an explicit representation of $\mathbf{Q}$ in terms of its eigenvector $\mathbf{e}$ is possible. Moreover, this representation must contain information on the amount of the rotation under study, which is nothing but the *angle of rotation.* Furthermore, line $\mathcal{L}$, mentioned in *Euler's Theorem,* is termed the *axis of rotation* of the motion of interest. In order to derive the aforementioned representation, consider the rotation depicted in Fig. 2.1 of angle ϕ about line $\mathcal{L}$.

From Fig. 2.1(a), clearly, one can write

$$\mathbf{p}' = \overrightarrow{OQ} + \overrightarrow{QP'} \tag{2.40}$$

where $\overrightarrow{OQ}$ is the axial component of $\mathbf{p}$ along vector $\mathbf{e}$, which is derived as in eq.(2.6a), namely,

$$\overrightarrow{OQ} = \mathbf{e}\mathbf{e}^T\mathbf{p} \tag{2.41}$$

Furthermore, from Fig. 2.1b,

$$\overrightarrow{QP'} = (\cos\phi)\overrightarrow{QP} + (\sin\phi)\overrightarrow{QP''} \tag{2.42}$$

with $\overrightarrow{QP}$ being nothing but the *normal component* of $\mathbf{p}$ with respect to $\mathbf{e}$, as introduced in eq.(2.6b), i.e.,

$$\overrightarrow{QP} = (\mathbf{1} - \mathbf{e}\mathbf{e}^T)\mathbf{p} \tag{2.43}$$

and $\overrightarrow{QP''}$ given as

$$\overrightarrow{QP''} = \mathbf{e} \times \mathbf{p} \equiv \mathbf{E}\mathbf{p} \tag{2.44}$$

Substitution of eqs.(2.44) and (2.43) into eq.(2.42) leads to

$$\overrightarrow{QP'} = \cos\phi(\mathbf{1} - \mathbf{e}\mathbf{e}^T)\mathbf{p} + \sin\phi\mathbf{E}\mathbf{p} \tag{2.45}$$

If now eqs.(2.41) and (2.45) are substituted into eq.(2.40), one obtains

$$\mathbf{p}' = \mathbf{e}\mathbf{e}^T\mathbf{p} + \cos\phi(\mathbf{1} - \mathbf{e}\mathbf{e}^T)\mathbf{p} + \sin\phi\mathbf{E}\mathbf{p} \tag{2.46}$$

Thus, eq.(2.40) reduces to

$$\mathbf{p}' = [\mathbf{e}\mathbf{e}^T + \cos\phi(\mathbf{1} - \mathbf{e}\mathbf{e}^T) + \sin\phi\mathbf{E}]\mathbf{p} \tag{2.47}$$

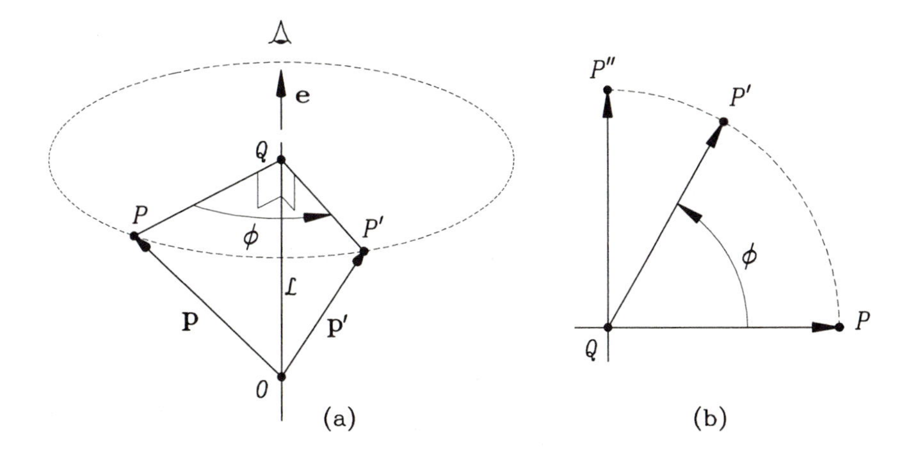

FIGURE 2.1. Rotation of a rigid body about a line.

From eq.(2.47) it is apparent that $\mathbf{p}'$ is a linear transformation of $\mathbf{p}$, the said transformation being given by the matrix inside the brackets, which is the rotation matrix $\mathbf{Q}$ sought, i.e.,

$$\mathbf{Q} = \mathbf{e}\mathbf{e}^T + \cos\phi(\mathbf{1} - \mathbf{e}\mathbf{e}^T) + \sin\phi\mathbf{E} \tag{2.48}$$

A special case arises when $\phi = \pi$, in which case

$$\mathbf{Q} = -\mathbf{1} + 2\mathbf{e}\mathbf{e}^T, \quad \text{for } \phi = \pi \tag{2.49}$$

whence it is apparent that $\mathbf{Q}$ is symmetric if $\phi = \pi$. Of course, $\mathbf{Q}$ becomes symmetric also when $\phi = 0$, but this is a rather obvious case, leading to $\mathbf{Q} = \mathbf{1}$. Except for these two cases, the rotation matrix is not symmetric. However, under no circumstance does the rotation matrix become skew-symmetric, for a 3×3 skew-symmetric matrix is by necessity singular, which contradicts the property of proper orthogonal matrices of eq.(2.21b).

Now one more representation of $\mathbf{Q}$ in terms of $\mathbf{e}$ and ϕ is given. For a fixed axis of rotation, i.e., for a fixed value of $\mathbf{e}$, the rotation matrix $\mathbf{Q}$ is a function of the angle of rotation ϕ, only. Thus, the series expansion of $\mathbf{Q}$ in terms of ϕ is

$$\mathbf{Q}(\phi) = \mathbf{Q}(0) + \mathbf{Q}'(0)\phi + \frac{1}{2!}\mathbf{Q}''(0)\phi^2 + \ldots + \frac{1}{k!}\mathbf{Q}^{(k)}(0)\phi^k + \ldots \tag{2.50}$$

where the superscript (k) stands for the kth derivative of $\mathbf{Q}$ with respect to ϕ. Now, from the definition of $\mathbf{E}$, one can readily prove the relations below:

$$\mathbf{E}^{(2k+1)} = (-1)^k\mathbf{E}, \quad \mathbf{E}^{2k} = (-1)^k(\mathbf{1} - \mathbf{e}\mathbf{e}^T) \tag{2.51}$$

Furthermore, using eqs.(2.48) and (2.51), one can readily show that

$$\mathbf{Q}^{(k)}(0) = \mathbf{E}^k \tag{2.52}$$

with $\mathbf{E}$ defined already as the cross-product matrix of $\mathbf{e}$. Moreover, from eqs.(2.50) and (2.52), $\mathbf{Q}(\phi)$ can be expressed as

$$\mathbf{Q}(\phi) = \mathbf{1} + \mathbf{E}\phi + \frac{1}{2!}\mathbf{E}^2\phi^2 + \ldots + \frac{1}{k!}\mathbf{E}^k\phi^k + \ldots$$

whose right-hand side is nothing but the exponential of $\mathbf{E}\phi$, i.e.,

$$\mathbf{Q}(\phi) = e^{\mathbf{E}\phi} \tag{2.53}$$

Equation (2.53) is the exponential representation of the rotation matrix in terms of its *natural invariants*, $\mathbf{e}$ and ϕ. The foregoing parameters are termed *invariants* because they are clearly independent of the coordinate axes chosen to represent the rotation under study. The adjective *natural* is necessary to distinguish them from other invariants that will be introduced presently. This adjective seems suitable because the said invariants stem naturally from Euler's Theorem.

Now, in view of eqs.(2.51), the above series can be written as

$$\mathbf{Q}(\phi) = \mathbf{1} + \left[-\frac{1}{2!}\phi^2 + \frac{1}{4!}\phi^4 - \cdots + \frac{1}{(2k)!}(-1)^k\phi^{2k} + \cdots\right](\mathbf{1} - \mathbf{e}\mathbf{e}^T)$$
$$+ \left[\phi - \frac{1}{3!}\phi^3 + \cdots + \frac{1}{(2k+1)!}(-1)^k\phi^{2k+1} + \cdots\right]\mathbf{E}$$

The series inside the first brackets is apparently $\cos\phi - 1$, while that in the second is $\sin\phi$. We have, therefore, an alternative representation of $\mathbf{Q}$, namely,

$$\mathbf{Q} = \mathbf{1} + \sin\phi\mathbf{E} + (1 - \cos\phi)\mathbf{E}^2 \tag{2.54}$$

which is an expected result in view of the Cayley-Hamilton Theorem.

The Canonical Forms of the Rotation Matrix

The rotation matrix takes on an especially simple form if the axis of rotation coincides with one of the coordinate axes. For example, if the X axis is parallel to the axis of rotation, i.e., parallel to vector $\mathbf{e}$, in a frame that we will label $\mathcal{X}$, then, we will have

$$[\mathbf{e}]_{\mathcal{X}} = \begin{bmatrix} 1 \\ 0 \\ 0 \end{bmatrix}, \quad [\mathbf{E}]_{\mathcal{X}} = \begin{bmatrix} 0 & 0 & 0 \\ 0 & 0 & -1 \\ 0 & 1 & 0 \end{bmatrix}, \quad [\mathbf{E}^2]_{\mathcal{X}} = \begin{bmatrix} 0 & 0 & 0 \\ 0 & -1 & 0 \\ 0 & 0 & -1 \end{bmatrix}$$

In the $\mathcal{X}$-frame, then,

$$[\mathbf{Q}]_{\mathcal{X}} = \begin{bmatrix} 1 & 0 & 0 \\ 0 & \cos\phi & -\sin\phi \\ 0 & \sin\phi & \cos\phi \end{bmatrix} \tag{2.55a}$$

Likewise, if we define the coordinate frames $\mathcal{Y}$ and $\mathcal{Z}$ so that their Y and Z axes, respectively, coincide with the axis of rotation, then

$$[\mathbf{Q}]_{\mathcal{Y}} = \begin{bmatrix} \cos\phi & 0 & \sin\phi \\ 0 & 1 & 0 \\ -\sin\phi & 0 & \cos\phi \end{bmatrix} \tag{2.55b}$$

and

$$[\mathbf{Q}]_{\mathcal{Z}} = \begin{bmatrix} \cos\phi & -\sin\phi & 0 \\ \sin\phi & \cos\phi & 0 \\ 0 & 0 & 1 \end{bmatrix} \tag{2.55c}$$

The representations of eqs.(2.55a–c) can be called the X-, Y-, and Z-*canonical forms* of the rotation matrix. In many instances, a rotation matrix cannot be directly derived from information on the original and the final orientations of a rigid body, but the overall motion can be readily decomposed into a sequence of simple rotations taking the above canonical forms—see, e.g., Exercise 18.

2.3.3 The Linear Invariants of a 3×3 Matrix

Now we introduce two *linear invariants* of 3×3 matrices. Given *any* 3×3 matrix $\mathbf{A}$, its *Cartesian decomposition*, the counterpart of the Cartesian representation of complex numbers, consists of the sum of its symmetric part, $\mathbf{A}_S$, and its skew-symmetric part, $\mathbf{A}_{SS}$, defined as

$$\mathbf{A}_S \equiv \frac{1}{2}(\mathbf{A} + \mathbf{A}^T), \quad \mathbf{A}_{SS} \equiv \frac{1}{2}(\mathbf{A} - \mathbf{A}^T) \tag{2.56}$$

The *axial vector* or for brevity, the *vector* of $\mathbf{A}$, is the vector $\mathbf{a}$ with the property

$$\mathbf{a} \times \mathbf{v} \equiv \mathbf{A}_{SS}\mathbf{v} \tag{2.57}$$

for any 3-dimensional vector $\mathbf{v}$. The *trace* of $\mathbf{A}$ is the sum of the eigenvalues of $\mathbf{A}_S$, which are real. Since no coordinate frame is involved in the above definitions, these are invariant. When calculating these invariants, of course, a particular coordinate frame must be used. Let us assume that the entries of matrix $\mathbf{A}$ in a certain coordinate frame are given by the array of real numbers a_{ij}, for $i, j = 1, 2, 3$. Moreover, let $\mathbf{a}$ have components a_i, for $i = 1, 2, 3$, in the same frame. The above-defined invariants are thus calculated as

$$\operatorname{vect}(\mathbf{A}) \equiv \mathbf{a} \equiv \frac{1}{2}\begin{bmatrix} a_{32} - a_{23} \\ a_{13} - a_{31} \\ a_{21} - a_{12} \end{bmatrix}, \quad \operatorname{tr}(\mathbf{A}) \equiv a_{11} + a_{22} + a_{33} \tag{2.58}$$

From the foregoing definitions, the following is now apparent:

Theorem 2.3.5 *The vector of a 3×3 matrix vanishes if and only if it is symmetric, whereas the trace of an $n \times n$ matrix vanishes if the matrix is skew symmetric.*

Other useful relations are given below. For any 3-dimensional vectors $\mathbf{a}$ and $\mathbf{b}$,

$$\operatorname{vect}(\mathbf{a}\mathbf{b}^T) = -\frac{1}{2}\mathbf{a} \times \mathbf{b} \tag{2.59}$$

and

$$\operatorname{tr}(\mathbf{a}\mathbf{b}^T) = \mathbf{a}^T\mathbf{b} \tag{2.60}$$

The second relation is quite straightforward, but the first one is less so; a proof of the first relation is given below: Let $\mathbf{w}$ denote $\operatorname{vect}(\mathbf{a}\mathbf{b}^T)$. From Definition (2.57), for any 3-dimensional vector $\mathbf{v}$,

$$\mathbf{w} \times \mathbf{v} = \mathbf{W}\mathbf{v} \tag{2.61}$$

where $\mathbf{W}$ is the skew-symmetric component of $\mathbf{a}\mathbf{b}^T$, namely,

$$\mathbf{W} \equiv \frac{1}{2}(\mathbf{a}\mathbf{b}^T - \mathbf{b}\mathbf{a}^T) \tag{2.62}$$

and hence,

$$\mathbf{W}\mathbf{v} = \mathbf{w} \times \mathbf{v} = \frac{1}{2}[(\mathbf{b}^T\mathbf{v})\mathbf{a} - (\mathbf{a}^T\mathbf{v})\mathbf{b}] \tag{2.63}$$

Now, let us compare the last expression with the double cross product $(\mathbf{b} \times \mathbf{a}) \times \mathbf{v}$, namely,

$$(\mathbf{b} \times \mathbf{a}) \times \mathbf{v} = (\mathbf{b}^T\mathbf{v})\mathbf{a} - (\mathbf{a}^T\mathbf{v})\mathbf{b} \tag{2.64}$$

from which it becomes apparent that

$$\mathbf{w} = \frac{1}{2}\mathbf{b} \times \mathbf{a} \tag{2.65}$$

and the aforementioned relation readily follows.

Note that Theorem 2.3.5 states a *necessary and sufficient* condition for the vanishing of the vector of a 3×3 matrix, but only a sufficient condition for the vanishing of the trace of an $n \times n$ matrix. What this implies is that the trace of an $n \times n$ matrix can vanish without the matrix being necessarily skew symmetric, but the trace of a skew-symmetric matrix necessarily vanishes. Also note that whereas the vector of a matrix is defined *only* for 3×3 matrices, the trace can be defined more generally for $n \times n$ matrices.

2.3.4 The Linear Invariants of a Rotation

From the invariant representations of the rotation matrix, eqs.(2.48) and (2.54), it is clear that the first two terms of $\mathbf{Q}$, $\mathbf{e}\mathbf{e}^T$ and $\cos\phi(\mathbf{1} - \mathbf{e}\mathbf{e}^T)$, are symmetric, whereas the third one, $\sin\phi\mathbf{E}$, is skew-symmetric. Hence,

$$\operatorname{vect}(\mathbf{Q}) = \operatorname{vect}(\sin\phi\,\mathbf{E}) = \sin\phi\,\mathbf{e} \tag{2.66}$$

whereas

$$\operatorname{tr}(\mathbf{Q}) = \operatorname{tr}[\mathbf{e}\mathbf{e}^T + \cos\phi(\mathbf{1} - \mathbf{e}\mathbf{e}^T)] \equiv \mathbf{e}^T\mathbf{e} + \cos\phi(3 - \mathbf{e}^T\mathbf{e}) = 1 + 2\cos\phi \tag{2.67}$$

from which one can readily solve for $\cos\phi$, namely,

$$\cos\phi = \frac{\operatorname{tr}(\mathbf{Q}) - 1}{2} \tag{2.68}$$

Henceforth, the vector of $\mathbf{Q}$ will be denoted by $\mathbf{q}$ and its components in a given coordinate frame by q_1, q_2, and q_3. Moreover, rather than using $\operatorname{tr}(\mathbf{Q})$ as the other linear invariant, $q_0 \equiv \cos\phi$ will be introduced to refer to the *linear invariants of the rotation matrix*. Hence, the rotation matrix is fully defined by *four scalar parameters*, namely $\{q_i\}_0^3$, which will be conveniently stored in the 4-dimensional array $\boldsymbol{\lambda}$, defined as

$$\boldsymbol{\lambda} \equiv [q_1,\ q_2,\ q_3,\ q_0]^T \tag{2.69}$$

Note, however, that the four components of $\boldsymbol{\lambda}$ are not independent, for they obey the relation

$$\|\mathbf{q}\|^2 + q_0^2 \equiv \sin^2\phi + \cos^2\phi = 1 \tag{2.70}$$

Thus, eq.(2.70) can be written in a more compact form as

$$\|\boldsymbol{\lambda}\|^2 \equiv q_1^2 + q_2^2 + q_3^2 + q_0^2 = 1 \tag{2.71}$$

What eq.(2.70) states has a straightforward geometric interpretation: As a body rotates about a fixed point, its motion can be described in a 4-dimensional space by the motion of a point of position vector $\boldsymbol{\lambda}$ that moves on the surface of the unit sphere centered at the origin of the said space. Alternatively, one can conclude that, as a rigid body rotates about a fixed point, its motion can be described in a 3-dimensional space by the motion of position vector $\mathbf{q}$, which moves within the unit solid sphere centered at the origin of the said space. Given the dependence of the four components of vector $\boldsymbol{\lambda}$, one might be tempted to solve for, say, q_0 from eq.(2.70) in terms of the remaining components, namely, as

$$q_0 = \pm\sqrt{1 - (q_1^2 + q_2^2 + q_3^2)} \tag{2.72}$$

This, however, is not a good idea because the sign ambiguity of eq.(2.72) leaves angle ϕ undefined, for q_0 is nothing but $\cos\phi$. Moreover, the three components of vector $\mathbf{q}$ alone, i.e., $\sin\phi\,\mathbf{e}$, do not suffice to define the rotation represented by $\mathbf{Q}$. Indeed, from the definition of $\mathbf{q}$, one has

$$\sin\phi = \pm\|\mathbf{q}\|, \quad \mathbf{e} = \mathbf{q}/\sin\phi \tag{2.73}$$

from which it is clear that $\mathbf{q}$ alone does not suffice to define the rotation under study, since it leaves angle ϕ undefined. Indeed, the vector of the rotation matrix provides no information about $\cos\phi$. Yet another representation of the rotation matrix is displayed below, in terms of its linear invariants, that is readily derived from representations (2.48) and (2.54), namely,

$$\mathbf{Q} = \frac{\mathbf{q}\mathbf{q}^T}{\|\mathbf{q}\|^2} + q_0(\mathbf{1} - \frac{\mathbf{q}\mathbf{q}^T}{\|\mathbf{q}\|^2}) + \overline{\mathbf{Q}} \tag{2.74a}$$

in which $\overline{\mathbf{Q}}$ is the cross-product matrix of vector $\mathbf{q}$, i.e.,

$$\overline{\mathbf{Q}} \equiv \frac{\partial(\mathbf{q}\times\mathbf{x})}{\partial\mathbf{x}}$$

for any vector $\mathbf{x}$.

Note that by virtue of eq.(2.70), the representation of $\mathbf{Q}$ given in eq.(2.74a) can be expressed alternatively as

$$\mathbf{Q} = q_0\mathbf{1} + \overline{\mathbf{Q}} + \frac{\mathbf{q}\mathbf{q}^T}{1+q_0} \tag{2.74b}$$

From either eq.(2.74a) or eq.(2.74b) it is apparent that linear invariants are not suitable to represent a rotation when the associated angle is either π or close to it. Note that a rotation through an angle ϕ about an axis given by vector $\mathbf{e}$ is identical to a rotation through an angle $-\phi$ about an axis given by vector $-\mathbf{e}$. Hence, changing the sign of $\mathbf{e}$ does not change the rotation matrix, provided that the sign of ϕ is also changed. Henceforth, we will choose the sign of the components of $\mathbf{e}$ so that $\sin\phi \geq 0$, which is equivalent to assuming that $0 \leq \phi \leq \pi$. Thus, $\sin\phi$ is calculated as $\|\mathbf{q}\|$, while $\cos\phi$ as indicated in eq.(2.68). Obviously, $\mathbf{e}$ is simply $\mathbf{q}$ normalized, i.e., $\mathbf{q}$ divided by its Euclidean norm.

2.3.5 *Examples*

The examples below are meant to stress the foregoing ideas on rotation invariants.

Example 2.3.2 *If* $[\mathbf{e}]_{\mathcal{F}} = [\sqrt{3}/3,\ -\sqrt{3}/3,\ \sqrt{3}/3]^T$ *in a given coordinate frame* $\mathcal{F}$ *and* $\phi = 120°$, *what is* $\mathbf{Q}$ *in* $\mathcal{F}$*?*

Solution: From the data,

$$\cos\phi = -\frac{1}{2}, \quad \sin\phi = \frac{\sqrt{3}}{2}$$

Moreover, in the $\mathcal{F}$ frame,

$$[\mathbf{e}\mathbf{e}^T]_{\mathcal{F}} = \frac{1}{3}\begin{bmatrix} 1 \\ -1 \\ 1 \end{bmatrix}\begin{bmatrix} 1 & -1 & 1 \end{bmatrix} = \frac{1}{3}\begin{bmatrix} 1 & -1 & 1 \\ -1 & 1 & -1 \\ 1 & -1 & 1 \end{bmatrix}$$

and hence,

$$[\mathbf{1} - \mathbf{e}\mathbf{e}^T]_{\mathcal{F}} = \frac{1}{3}\begin{bmatrix} 2 & 1 & -1 \\ 1 & 2 & 1 \\ -1 & 1 & 2 \end{bmatrix}, \quad [\mathbf{E}]_{\mathcal{F}} \equiv \frac{\sqrt{3}}{3}\begin{bmatrix} 0 & -1 & -1 \\ 1 & 0 & -1 \\ 1 & 1 & 0 \end{bmatrix}$$

Thus, from eq.(2.48),

$$[\mathbf{Q}]_{\mathcal{F}} = \frac{1}{3}\begin{bmatrix} 1 & -1 & 1 \\ -1 & 1 & -1 \\ 1 & -1 & 1 \end{bmatrix} - \frac{1}{6}\begin{bmatrix} 2 & 1 & -1 \\ 1 & 2 & 1 \\ -1 & 1 & 2 \end{bmatrix} + \frac{3}{6}\begin{bmatrix} 0 & -1 & -1 \\ 1 & 0 & -1 \\ 1 & 1 & 0 \end{bmatrix}$$

i.e.,

$$[\mathbf{Q}]_{\mathcal{F}} = \begin{bmatrix} 0 & -1 & 0 \\ 0 & 0 & -1 \\ 1 & 0 & 0 \end{bmatrix}$$

Example 2.3.3 *The matrix representation of a linear transformation* $\mathbf{Q}$ *in a certain reference frame* $\mathcal{F}$ *is given below. Find out whether the said transformation is a rigid-body rotation. If it is, find its natural invariants.*

$$[\,\mathbf{Q}\,]_{\mathcal{F}} = \begin{bmatrix} 0 & 1 & 0 \\ 0 & 0 & 1 \\ 1 & 0 & 0 \end{bmatrix}$$

Solution: First the given array is tested for orthogonality:

$$[\,\mathbf{Q}\,]_{\mathcal{F}}[\,\mathbf{Q}^T\,]_{\mathcal{F}} = \begin{bmatrix} 0 & 1 & 0 \\ 0 & 0 & 1 \\ 1 & 0 & 0 \end{bmatrix} \begin{bmatrix} 0 & 0 & 1 \\ 1 & 0 & 0 \\ 0 & 1 & 0 \end{bmatrix} = \begin{bmatrix} 1 & 0 & 0 \\ 0 & 1 & 0 \\ 0 & 0 & 1 \end{bmatrix}$$

thereby showing that the said array is indeed orthogonal. Thus, the linear transformation could represent a reflection or a rotation. In order to decide which one this represents, the determinant of the foregoing array is computed:

$$\det(\,\mathbf{Q}\,) = +1$$

which makes apparent that $\mathbf{Q}$ indeed represents a rigid-body rotation. Now, its natural invariants are computed. The unit vector $\mathbf{e}$ can be computed as the eigenvector of $\mathbf{Q}$ associated with the eigenvalue $+1$. This requires, however, finding a nontrivial solution of a homogeneous linear system of three equations in three unknowns. This is not difficult to do, but it is cumbersome and is not necessary. In order to find $\mathbf{e}$ and ϕ, it is recalled that $\text{vect}(\mathbf{Q}) = \sin\phi\,\mathbf{e}$, which is readily computed with differences only, as indicated in eq.(2.58), namely,

$$[\,\mathbf{q}\,]_{\mathcal{F}} \equiv \sin\phi\,[\mathbf{e}]_{\mathcal{F}} = -\frac{1}{2}\begin{bmatrix} 1 \\ 1 \\ 1 \end{bmatrix}$$

Under the assumption that $\sin\phi \geq 0$, then,

$$\sin\phi \equiv \|\mathbf{q}\| = \frac{\sqrt{3}}{2}$$

and hence,

$$[\mathbf{e}]_{\mathcal{F}} = \frac{[\mathbf{q}]_{\mathcal{F}}}{\|\mathbf{q}\|} = -\frac{\sqrt{3}}{3}\begin{bmatrix} 1 \\ 1 \\ 1 \end{bmatrix}$$

and

$$\phi = 60^\circ \quad \text{or} \quad 120^\circ$$

The foregoing ambiguity is resolved by the trace of $\mathbf{Q}$, which yields

$$1 + 2\cos\phi \equiv \text{tr}(\mathbf{Q}) = 0, \quad \cos\phi = -\frac{1}{2}$$

The negative sign of $\cos\phi$ indicates that ϕ lies in the second quadrant—it cannot lie in the third quadrant because of our assumption about the sign of $\sin\phi$—, and hence,

$$\phi = 120^\circ$$

Example 2.3.4 *A coordinate frame X_1, Y_1, Z_1 is rotated into a configuration X_2, Y_2, Z_2 in such a way that*

$$X_2 = -Y_1, \quad Y_2 = Z_1, \quad Z_2 = -X_1$$

Find the matrix representation of the rotation in X_1, Y_1, Z_1 coordinates. From this representation, compute the direction of the axis and the angle of rotation.

Solution: Let $\mathbf{i}_1$, $\mathbf{j}_1$, $\mathbf{k}_1$ be unit vectors parallel to X_1, Y_1, Z_1, respectively, $\mathbf{i}_2$, $\mathbf{j}_2$, $\mathbf{k}_2$ being defined correspondingly. One has

$$\mathbf{i}_2 = -\mathbf{j}_1, \quad \mathbf{j}_2 = \mathbf{k}_1, \quad \mathbf{k}_2 = -\mathbf{i}_1$$

and hence, from Definition 2.2.1, the matrix representation $[\,\mathbf{Q}\,]_1$ of the rotation under study in the X_1, Y_1, Z_1 coordinate frame is readily derived:

$$[\,\mathbf{Q}\,]_1 = \begin{bmatrix} 0 & 0 & -1 \\ -1 & 0 & 0 \\ 0 & 1 & 0 \end{bmatrix}$$

from which the linear invariants follow, namely,

$$[\mathbf{q}]_1 \equiv [\,\text{vect}(\mathbf{Q})\,]_1 = \sin\phi\,[\,\mathbf{e}\,]_1 = \frac{1}{2}\begin{bmatrix} 1 \\ -1 \\ -1 \end{bmatrix}, \quad \cos\phi = \frac{1}{2}[\,\text{tr}(\mathbf{Q}) - 1\,] = -\frac{1}{2}$$

Under our assumption that $\sin\phi \geq 0$, we obtain

$$\sin\phi = \|\mathbf{q}\| = \frac{\sqrt{3}}{2}, \quad [\,\mathbf{e}\,]_1 = \frac{[\mathbf{q}]_1}{\sin\phi} = \frac{\sqrt{3}}{3}\begin{bmatrix} 1 \\ -1 \\ -1 \end{bmatrix}$$

From the foregoing values for $\sin\phi$ and $\cos\phi$, angle ϕ is computed uniquely as

$$\phi = 120^\circ$$

Example 2.3.5 *Show that the matrix $\mathbf{P}$ given in eq.(2.4) satisfies properties (2.1a).*

Solution: First, we prove idempotency, i.e.,

$$\begin{aligned} \mathbf{P}^2 &= (\mathbf{1} - \mathbf{n}\mathbf{n}^T)(\mathbf{1} - \mathbf{n}\mathbf{n}^T) \\ &= \mathbf{1} - 2\mathbf{n}\mathbf{n}^T + \mathbf{n}\mathbf{n}^T\mathbf{n}\mathbf{n}^T = \mathbf{1} - \mathbf{n}\mathbf{n}^T = \mathbf{P} \end{aligned}$$

thereby showing that $\mathbf{P}$ is, indeed, idempotent. Now we prove that $\mathbf{n}$ is an eigenvector of $\mathbf{P}$ with eigenvalue 0, and hence, $\mathbf{n}$ spans the nullspace of $\mathbf{P}$. In fact,

$$\mathbf{Pn} = (\mathbf{1} - \mathbf{nn}^T)\mathbf{n} = \mathbf{n} - \mathbf{nn}^T\mathbf{n} = \mathbf{n} - \mathbf{n} = \mathbf{0}$$

thereby completing the proof.

Example 2.3.6 *The representations of three linear transformations in a given coordinate frame $\mathcal{F}$ are given below:*

$$[\mathbf{A}]_{\mathcal{F}} = \frac{1}{3}\begin{bmatrix} 2 & 1 & 2 \\ -2 & 2 & 1 \\ -1 & -2 & 2 \end{bmatrix}$$

$$[\mathbf{B}]_{\mathcal{F}} = \frac{1}{3}\begin{bmatrix} 2 & 1 & 1 \\ 1 & 2 & -1 \\ 1 & -1 & 2 \end{bmatrix}$$

$$[\mathbf{C}]_{\mathcal{F}} = \frac{1}{3}\begin{bmatrix} 1 & 2 & 2 \\ 2 & 1 & -2 \\ 2 & -2 & 1 \end{bmatrix}$$

One of the foregoing matrices is an orthogonal projection, one is a reflection, and one is a rotation. Identify each of these and give its invariants.

Solution: From representations (2.48) and (2.54), it is clear that a rotation matrix is symmetric if and only if $\sin\phi = 0$. This means that a rotation matrix cannot be symmetric unless its angle of rotation is either 0 or π, i.e., unless its trace is either 3 or -1. Since $[\mathbf{B}]_{\mathcal{F}}$ and $[\mathbf{C}]_{\mathcal{F}}$ are symmetric, they cannot be rotations, unless their traces take on the foregoing values. Their traces are thus evaluated below:

$$\text{tr}(\mathbf{B}) = 2, \quad \text{tr}(\mathbf{C}) = 1$$

which thus rules out the foregoing matrices as suitable candidates for rotations. Thus, $\mathbf{A}$ is the only candidate left for proper orthogonality, its suitability being tested below:

$$[\mathbf{AA}^T]_{\mathcal{F}} = \frac{1}{9}\begin{bmatrix} 9 & 0 & 0 \\ 0 & 9 & 0 \\ 0 & 0 & 9 \end{bmatrix}, \quad \det(\mathbf{A}) = +1$$

and hence, $\mathbf{A}$ indeed represents a rotation. Its natural invariants are next computed:

$$\sin\phi\,[\mathbf{e}]_{\mathcal{F}} = [\text{vect}(\mathbf{A})]_{\mathcal{F}} = \frac{1}{2}\begin{bmatrix} -1 \\ 1 \\ -1 \end{bmatrix}, \quad \cos\phi = \frac{1}{2}[\text{tr}(\mathbf{A}) - 1] = \frac{1}{2}(2-1) = \frac{1}{2}$$

We assume, as usual, that $\sin\phi \geq 0$. Then,

$$\sin\phi = \|\text{vect}(\mathbf{A})\| = \frac{\sqrt{3}}{2}, \quad \text{i.e.,} \ \phi = 60^\circ$$

Moreover,

$$[\mathbf{e}]_{\mathcal{F}} = \frac{[\text{vect}(\mathbf{A})]_{\mathcal{F}}}{\|\text{vect}(\mathbf{A})\|} = \frac{\sqrt{3}}{3}\begin{bmatrix} -1 \\ 1 \\ -1 \end{bmatrix}$$

Now, one matrix of $\mathbf{B}$ and $\mathbf{C}$ is an orthogonal projection and the other is a reflection. To be a reflection, a matrix has to be orthogonal. Hence, each matrix is tested for orthogonality:

$$[\mathbf{B}\mathbf{B}^T]_{\mathcal{F}} = \frac{1}{9}\begin{bmatrix} 6 & 3 & 3 \\ 3 & 6 & -3 \\ 3 & -3 & 6 \end{bmatrix} = [\mathbf{B}^2]_{\mathcal{F}} = [\mathbf{B}]_{\mathcal{F}}, \ [\mathbf{C}\mathbf{C}^T]_{\mathcal{F}} = \frac{1}{9}\begin{bmatrix} 9 & 0 & 0 \\ 0 & 9 & 0 \\ 0 & 0 & 9 \end{bmatrix}$$

thereby showing that $\mathbf{C}$ is orthogonal and $\mathbf{B}$ is not. Furthermore, $\det(\mathbf{C}) = -1$, which confirms that $\mathbf{C}$ is a reflection. Now, if $\mathbf{B}$ is a projection, it is bound to be singular and idempotent. From the orthogonality test it is clear that it is idempotent. Moreover, one can readily verify that $\det(\mathbf{B}) = 0$, and hence $\mathbf{B}$ is singular. The unit vector $[\mathbf{n}]_{\mathcal{F}} = [n_1, n_2, n_3]^T$ spanning its nullspace is determined from the general form of projections, eq.(2.1a), whence it is apparent that

$$\mathbf{n}\mathbf{n}^T = \mathbf{1} - \mathbf{B}$$

Therefore, if a solution $\mathbf{n}$ has been found, then $-\mathbf{n}$ is also a solution, i.e., *the problem admits two solutions*, one being the negative of the other. These two solutions are found below, by first rewriting the above system of equations in component form:

$$\begin{bmatrix} n_1^2 & n_1 n_2 & n_1 n_3 \\ n_1 n_2 & n_2^2 & n_2 n_3 \\ n_1 n_3 & n_2 n_3 & n_3^2 \end{bmatrix} = \frac{1}{3}\begin{bmatrix} 1 & -1 & -1 \\ -1 & 1 & 1 \\ -1 & 1 & 1 \end{bmatrix}.$$

Now, from the diagonal entries of the above matrices, it is apparent that the three components of $\mathbf{n}$ have identical absolute values, i.e., $\sqrt{3}/3$. Moreover, from the off-diagonal entries of the same matrices, the second and third components of $\mathbf{n}$ bear equal signs, but we cannot tell whether positive or negative, because of the quadratic nature of the problem at hand. The two solutions are thus obtained as

$$\mathbf{n} = \pm\frac{\sqrt{3}}{3}\begin{bmatrix} 1 \\ -1 \\ -1 \end{bmatrix}$$

which is the only invariant of $\mathbf{B}$.

We now look at $\mathbf{C}$, which is a reflection, and hence, bears the form

$$\mathbf{C} = \mathbf{1} - 2\mathbf{n}\mathbf{n}^T$$

In order to determine $\mathbf{n}$, note that

$$\mathbf{n}\mathbf{n}^T = \frac{1}{2}(\mathbf{1} - \mathbf{C})$$

or in component form,

$$\begin{bmatrix} n_1^2 & n_1 n_2 & n_1 n_3 \\ n_1 n_2 & n_2^2 & n_2 n_3 \\ n_1 n_3 & n_2 n_3 & n_3^2 \end{bmatrix} = \frac{1}{3} \begin{bmatrix} 1 & -1 & -1 \\ -1 & 1 & 1 \\ -1 & 1 & 1 \end{bmatrix}$$

which is identical to the matrix equation derived in the case of matrix $\mathbf{B}$. Hence, the solution is the same, i.e.,

$$\mathbf{n} = \pm\frac{\sqrt{3}}{3} \begin{bmatrix} 1 \\ -1 \\ -1 \end{bmatrix}$$

thereby finding the invariant sought.

Example 2.3.7 *The vector and the trace of a rotation matrix $\mathbf{Q}$, in a certain reference frame $\mathcal{F}$, are given as*

$$[\operatorname{vect}(\mathbf{Q})]_{\mathcal{F}} = \frac{1}{2} \begin{bmatrix} -1 \\ 1 \\ -1 \end{bmatrix}, \quad \operatorname{tr}(\mathbf{Q}) = 2$$

Find the matrix representation of $\mathbf{Q}$ in the given coordinate frame and in a frame having its Z axis parallel to $\operatorname{vect}(\mathbf{Q})$.

Solution: We shall resort to eq.(2.74a) to determine the rotation matrix $\mathbf{Q}$. The quantities involved in the aforementioned representation of $\mathbf{Q}$ are readily computed, as shown below:

$$[\mathbf{q}\mathbf{q}^T]_{\mathcal{F}} = \frac{1}{4} \begin{bmatrix} 1 & -1 & 1 \\ -1 & 1 & -1 \\ 1 & -1 & 1 \end{bmatrix}, \quad \|\mathbf{q}\|^2 = \frac{3}{4}, \quad [\overline{\mathbf{Q}}]_{\mathcal{F}} = \frac{1}{2} \begin{bmatrix} 0 & 1 & 1 \\ -1 & 0 & 1 \\ -1 & -1 & 0 \end{bmatrix}$$

from which $\mathbf{Q}$ follows:

$$[\mathbf{Q}]_{\mathcal{F}} = \frac{1}{3} \begin{bmatrix} 2 & 1 & 2 \\ -2 & 2 & 1 \\ -1 & -2 & 2 \end{bmatrix}$$

in the given coordinate frame. Now, let $\mathcal{Z}$ denote a coordinate frame whose Z axis is parallel to $\mathbf{q}$. Hence,

$$[\mathbf{q}]_{\mathcal{Z}} = \frac{\sqrt{3}}{2}\begin{bmatrix} 0 \\ 0 \\ 1 \end{bmatrix}, \quad [\mathbf{q}\mathbf{q}^T]_{\mathcal{Z}} = \frac{3}{4}\begin{bmatrix} 0 & 0 & 0 \\ 0 & 0 & 0 \\ 0 & 0 & 1 \end{bmatrix}, \quad [\overline{\mathbf{Q}}]_{\mathcal{Z}} = \frac{\sqrt{3}}{2}\begin{bmatrix} 0 & -1 & 0 \\ 1 & 0 & 0 \\ 0 & 0 & 0 \end{bmatrix}$$

which readily leads to

$$[\mathbf{Q}]_{\mathcal{Z}} = \begin{bmatrix} 1/2 & -\sqrt{3}/2 & 0 \\ \sqrt{3}/2 & 1/2 & 0 \\ 0 & 0 & 1 \end{bmatrix}$$

and is in the Z-canonical form.

Example 2.3.8 *A procedure for trajectory planning produced a matrix representing a rotation for a certain* pick-and-place operation, *as shown below:*

$$[\mathbf{Q}] = \begin{bmatrix} 0.433 & -0.500 & z \\ x & 0.866 & -0.433 \\ 0.866 & y & 0.500 \end{bmatrix}$$

where x, y, and z are entries that are unrecognizable due to failures in the printing hardware. Knowing that $\mathbf{Q}$ is in fact a rotation matrix, find the missing entries.

Solution: Since $\mathbf{Q}$ is a rotation matrix, the product $\mathbf{P} \equiv \mathbf{Q}^T\mathbf{Q}$ should equal the 3×3 identity matrix, and $\det(\mathbf{Q})$ should be $+1$. The foregoing product is computed first:

$$[\mathbf{P}]_{\mathcal{F}} = \begin{bmatrix} 0.437 + z^2 & 0.433(x - z - 1) & 0.5(-y + z) + 0.375 \\ * & 0.937 + x^2 & 0.866(x + y) - 0.216 \\ * & * & 1 + y^2 \end{bmatrix}$$

where the entries below the diagonal have not been printed because the matrix is symmetric. Upon equating the diagonal entries of the foregoing array to unity, we obtain

$$x = \pm 0.250, \quad y = 0, \quad z = \pm 0.750$$

while the vanishing of the off-diagonal entries leads to

$$x = 0.250, \quad y = 0, \quad z = -0.750$$

which can be readily verified to produce $\det(\mathbf{Q}) = +1$.

2.3.6 The Euler-Rodrigues Parameters

The invariants defined so far, namely, the natural and the linear invariants of a rotation matrix, are not the only ones that are used in kinematics. Additionally, one has the *Euler parameters*, or *Euler-Rodrigues parameters*, as Cheng and Gupta (1989) propose that they should be called, represented here as $\mathbf{r}$ and r_0. The Euler-Rodrigues parameters are defined as

$$\mathbf{r} \equiv \sin(\frac{\phi}{2})\,\mathbf{e}, \quad r_0 = \cos(\frac{\phi}{2}) \tag{2.75}$$

One can readily show that $\mathbf{Q}$ takes on a quite simple form in terms of the Euler-Rodrigues parameters, namely,

$$\mathbf{Q} = (r_0{}^2 - \mathbf{r}\cdot\mathbf{r})\mathbf{1} + 2\mathbf{r}\mathbf{r}^T + 2r_0\mathbf{R} \tag{2.76}$$

in which $\mathbf{R}$ is the cross-product matrix of $\mathbf{r}$, i.e.,

$$\mathbf{R} \equiv \frac{\partial(\mathbf{r}\times\mathbf{x})}{\partial\mathbf{x}}$$

for arbitrary $\mathbf{x}$.

Note that the Euler-Rodrigues parameters appear quadratically in the rotation matrix. Hence, these parameters cannot be computed with simple sums and differences. A closer inspection of eq.(2.74b) reveals that the linear invariants appear *almost linearly* in the rotation matrix. This means that the rotation matrix, as given by eq.(2.74b), is composed of two types of terms, namely, linear and rational. Moreover, the rational term is composed of a quadratic expression in the numerator and a linear expression in the denominator, the ratio thus being linear, which explains why the linear invariants can be obtained by sums and differences from the rotation matrix.

The relationship between the linear invariants and the Euler-Rodrigues parameters can be readily derived, namely,

$$r_0 = \pm\sqrt{\frac{1+q_0}{2}}, \quad \mathbf{r} = \frac{\mathbf{q}}{2r_0}, \quad \phi \neq \pi \tag{2.77}$$

Furthermore, note that, if $\phi = \pi$, then $r_0 = 0$, and formulae (2.77) fail to produce $\mathbf{r}$. However, from eq.(2.75),

$$\text{For } \phi = \pi: \quad \mathbf{r} = \mathbf{e}, \quad r_0 = 0 \tag{2.78}$$

We now derive invariant relations between the rotation matrix and the Euler-Rodrigues parameters. To do this, we resort to the concept of *matrix square root.* As a matter of fact, the square root of a square matrix is nothing but a particular case of an *analytic function* of a square matrix, discussed in connection with Theorem 2.3.3 and the exponential representation of the rotation matrix. Indeed, the square root of a square matrix is an analytic function of that matrix, and hence, admits a series expansion in powers of the matrix. Moreover, by virtue of the Cayley-Hamilton Theorem (Theorem 2.3.3) the said square root should be, for a 3×3 matrix, a linear combination of the identity matrix $\mathbf{1}$, the matrix itself, and its square, the coefficients being found using the eigenvalues of the matrix.

Furthermore, from the geometric meaning of a rotation through the angle ϕ about an axis parallel to the unit vector $\mathbf{e}$, it is apparent that the square of the matrix representing the foregoing rotation is itself a rotation about the same axis, but through the angle 2ϕ. By the same token, the square root of the same matrix is again a rotation matrix about the same axis, but through an angle $\phi/2$. Now, while the square of a matrix is unique, its square root is not. This fact is apparent for diagonalizable matrices, whose diagonal entries are their eigenvalues. Each eigenvalue, whether positive or negative, admits two square roots, and hence, a diagonalizable $n \times n$ matrix admits as many square roots as there are combinations of the two possible roots of individual eigenvalues, disregarding rearrangements of the latter. Such a number is 2^n, and hence, a 3×3 matrix admits eight square roots. For example, the eight square roots of the identity 3×3 matrix are displayed below:

$$\begin{bmatrix} 1 & 0 & 0 \\ 0 & 1 & 0 \\ 0 & 0 & 1 \end{bmatrix}, \quad \begin{bmatrix} 1 & 0 & 0 \\ 0 & 1 & 0 \\ 0 & 0 & -1 \end{bmatrix}, \quad \begin{bmatrix} 1 & 0 & 0 \\ 0 & -1 & 0 \\ 0 & 0 & 1 \end{bmatrix}, \quad \begin{bmatrix} -1 & 0 & 0 \\ 0 & 1 & 0 \\ 0 & 0 & 1 \end{bmatrix},$$

$$\begin{bmatrix} 1 & 0 & 0 \\ 0 & -1 & 0 \\ 0 & 0 & -1 \end{bmatrix}, \quad \begin{bmatrix} -1 & 0 & 0 \\ 0 & 1 & 0 \\ 0 & 0 & -1 \end{bmatrix}, \quad \begin{bmatrix} -1 & 0 & 0 \\ 0 & -1 & 0 \\ 0 & 0 & 1 \end{bmatrix}, \quad \begin{bmatrix} -1 & 0 & 0 \\ 0 & -1 & 0 \\ 0 & 0 & -1 \end{bmatrix}$$

In fact, the foregoing result can be extended to orthogonal matrices as well and, for that matter, to any square matrix with n linearly independent eigenvectors. That is, an $n \times n$ orthogonal matrix admits 2^n square roots. However, not all eight square roots of a 3×3 orthogonal matrix are orthogonal. In fact, not all eight square roots of a 3×3 proper orthogonal matrix are proper orthogonal either. Of these square roots, nevertheless, there is one that is proper orthogonal, the one representing a rotation of $\phi/2$. We will denote this particular square root of $\mathbf{Q}$ by $\sqrt{\mathbf{Q}}$. The Euler-Rodrigues parameters of $\mathbf{Q}$ can thus be expressed as the linear invariants of $\sqrt{\mathbf{Q}}$, namely,

$$\mathbf{r} = \operatorname{vect}(\sqrt{\mathbf{Q}}), \quad r_0 = \frac{\operatorname{tr}(\sqrt{\mathbf{Q}}) - 1}{2} \tag{2.79}$$

It is important to recognize the basic differences between the linear invariants and the Euler-Rodrigues parameters. Whereas the former can be readily derived from the matrix representation of the rotation involved by simple additions and subtractions, the latter require square roots and entail sign ambiguities. However, the former fail to produce information on the axis of rotation whenever the angle of rotation is π, whereas the latter produce that information *for any value of the angle of rotation.*

The Euler-Rodrigues parameters are nothing but the *quaternions* invented by Sir William Rowan Hamilton (1844) in an extraordinary moment of creativity on Monday, 16 October 1843, as "Hamilton, accompanied by Lady Hamilton, was walking along the Royal Canal in Dublin towards the Royal Irish Academy, where Hamilton was to preside a meeting." (Altmann, 1989).

Moreover, the Euler-Rodrigues parameters should not be confused with the *Euler angles*, which are not invariant and hence, admit multiple definitions. The foregoing means that no single set of Euler angles exists for a given rotation matrix, the said angles depending on how the rotation is decomposed into three simpler rotations. For this reason, Euler angles will not be stressed here. The reader is referred to Exercise 18 for a short discussion of Euler angles; Synge (1960) includes a classical treatment of the same.

Example 2.3.9 *Find the Euler-Rodrigues parameters of the proper orthogonal matrix* $\mathbf{Q}$ *given as*

$$\mathbf{Q} = \frac{1}{3}\begin{bmatrix} -1 & 2 & 2 \\ 2 & -1 & 2 \\ 2 & 2 & -1 \end{bmatrix}$$

Solution: Since the given matrix is symmetric, its angle of rotation is π and its vector linear invariant vanishes, which prevents us from finding the direction of the axis of rotation from the linear invariants; moreover, expressions (2.77) do not apply. However, we can use eq.(2.49) to find the unit vector $\mathbf{e}$ parallel to the axis of rotation, i.e.,

$$\mathbf{e}\mathbf{e}^T = \frac{1}{2}(\mathbf{1} + \mathbf{Q})$$

or in component form,

$$\begin{bmatrix} e_1^2 & e_1 e_2 & e_1 e_3 \\ e_1 e_2 & e_2^2 & e_2 e_3 \\ e_1 e_3 & e_2 e_3 & e_3^2 \end{bmatrix} = \frac{1}{3}\begin{bmatrix} 1 & 1 & 1 \\ 1 & 1 & 1 \\ 1 & 1 & 1 \end{bmatrix}$$

A simple inspection of the components of the two sides of the above equation reveals that all three components of $\mathbf{e}$ are identical and moreover, of the same sign, but we cannot tell which sign this is. Therefore,

$$\mathbf{e} = \pm\frac{\sqrt{3}}{3}\begin{bmatrix} 1 \\ 1 \\ 1 \end{bmatrix}$$

Moreover, from the symmetry of $\mathbf{Q}$, we know that $\phi = \pi$, and hence,

$$\mathbf{r} = \mathbf{e}\sin\left(\frac{\phi}{2}\right) = \pm\frac{\sqrt{3}}{3}\begin{bmatrix}1\\1\\1\end{bmatrix}, \quad r_0 = \cos\left(\frac{\phi}{2}\right) = 0$$

2.4 Composition of Reflections and Rotations

As pointed out in Section 2.2, reflections occur often accompanied by rotations. The effect of this combination is that the rotation destroys the two properties of pure reflections—symmetry and idempotency. Indeed, let $\mathbf{R}$ be a pure reflection, taking on the form appearing in eq.(2.5), and $\mathbf{Q}$ an arbitrary rotation, taking on the form of eq.(2.48). The product of these two transformations, $\mathbf{QR}$, denoted by $\mathbf{T}$, is apparently neither symmetric nor idempotent, as the reader can readily verify. Likewise, the product of these two transformations in the reverse order is neither symmetric nor idempotent.

As a consequence of the foregoing discussion, an improper orthogonal transformation that is not symmetric can always be decomposed into the product of a rotation and a pure reflection, the latter being symmetric and idempotent. Moreover, this decomposition can take on the form of any of the two possible orderings of the rotation and the reflection. Note, however, that once the order has been selected, the decomposition is not unique. Indeed, if we want to decompose $\mathbf{T}$ in the above paragraph in to the product $\mathbf{QR}$, then we can freely choose the unit normal $\mathbf{n}$ of the plane of reflection and write

$$\mathbf{R} \equiv \mathbf{1} - 2\mathbf{n}\mathbf{n}^T$$

vector $\mathbf{n}$ then being found from

$$\mathbf{n}\mathbf{n}^T = \frac{1}{2}(\mathbf{1} - \mathbf{R})$$

Hence, the factor $\mathbf{Q}$ of that decomposition is obtained as

$$\mathbf{Q} = \mathbf{T}\mathbf{R}^{-1} \equiv \mathbf{T}\mathbf{R} = \mathbf{T} - 2(\mathbf{T}\mathbf{n})\mathbf{n}^T$$

where use has been made of the idempotency of $\mathbf{R}$.

Example 2.4.1 *Join the palms of your two hands in the position adopted by swimmers when preparing for plunging, while holding a sheet of paper between them. The sheet defines a plane in each hand that we will call the* hand plane, *its unit normal, pointing outside of the hand, being called the* hand normal *and represented as vectors $\mathbf{n}_R$ and $\mathbf{n}_L$ for the right and left hand, respectively. Moreover, let $\mathbf{o}_R$ and $\mathbf{o}_L$ denote unit vectors pointing in the direction of the finger axes of each of the two hands. Thus, in the* *swimmer position described above,* $\mathbf{n}_L = -\mathbf{n}_R$ *and* $\mathbf{o}_L = \mathbf{o}_R$. *Now, without moving your right hand, let the left hand attain a position whereby the left-hand normal lies at right angles with the right-hand normal, the palm pointing downwards and the finger axes of the two hands remaining parallel. Find the representation of the transformation carrying the right hand to the final configuration of the left hand, in terms of the unit vectors* $\mathbf{n}_R$ *and* $\mathbf{o}_R$.

Solution: Let us regard the desired transformation $\mathbf{T}$ as the product of a rotation $\mathbf{Q}$ by a pure reflection $\mathbf{R}$, in the form $\mathbf{T} = \mathbf{QR}$. Thus, the transformation occurs so that the reflection takes place first, then the rotation. The reflection is simply that mapping the right hand into the left hand, and hence, the reflection plane is simply the hand plane, i.e.,

$$\mathbf{R} = \mathbf{1} - 2\mathbf{n}_R\mathbf{n}_R^T$$

Moreover, the left hand rotates from the swimmer position about an axis parallel to the finger axes through an angle of 90° clockwise from your viewpoint, i.e., in the positive direction of vector $\mathbf{o}_R$. Hence, the form of the rotation involved can be readily derived from eq.(2.48) and the above information, namely,

$$\mathbf{Q} = \mathbf{o}_R\mathbf{o}_R^T + \mathbf{O}_R$$

where $\mathbf{O}_R$ is the cross-product matrix of $\mathbf{o}_R$. Hence, upon performing the product $\mathbf{QR}$, we have

$$\mathbf{T} = \mathbf{o}_R\mathbf{o}_R^T + 2\mathbf{O}_R - 2(\mathbf{o}_R \times \mathbf{n}_R)\mathbf{n}_R^T$$

which is the transformation sought.

2.5 Coordinate Transformations and Homogeneous Coordinates

Crucial to robotics is the unambiguous description of the geometrical relations among the various bodies in the environment surrounding a robot. These relations are established by means of *coordinate frames*, or *frames*, for brevity, attached to each rigid body in the scene, including the robot links. The origins of these frames, moreover, are set at landmark points and orientations defined by key geometric entities like lines and planes. For example, in Chapter 4 we attach two frames to every moving link of a serial robot, with origin at a point on each of the axis of the two joints coupling this link with its two neighbors. Moreover, the Z-axis of each frame is defined, according to the Denavit-Hartenberg notation, introduced in that chapter, along each joint axis, while the X-axis of the frame closer to the base—termed the fore frame—is defined along the common perpendicular to the two joint axes. The origin of the same frame is thus defined as the intersection of the fore axis with the common perpendicular to the two axes. This section is devoted to the study of the coordinate transformations of vectors when these are represented in various frames.

2.5.1 Coordinate Transformations Between Frames with a Common Origin

We will refer to two coordinate frames in this section, namely, $\mathcal{A} = \{X, Y, Z\}$ and $\mathcal{B} = \{\mathcal{X}, \mathcal{Y}, \mathcal{Z}\}$. Moreover, let $\mathbf{Q}$ be the rotation carrying $\mathcal{A}$ into $\mathcal{B}$, i.e.,

$$\mathbf{Q}: \quad \mathcal{A} \quad \rightarrow \quad \mathcal{B} \tag{2.80}$$

The purpose of this subsection is to establish the relation between the representations of the position vector of a point P in $\mathcal{A}$ and $\mathcal{B}$, denoted by $[\mathbf{p}]_{\mathcal{A}}$ and $[\mathbf{p}]_{\mathcal{B}}$, respectively. Let

$$[\mathbf{p}]_{\mathcal{A}} = \begin{bmatrix} x \\ y \\ z \end{bmatrix} \tag{2.81}$$

We want to find $[\mathbf{p}]_{\mathcal{B}}$ in terms of $[\mathbf{p}]_{\mathcal{A}}$ and $\mathbf{Q}$, when the latter is represented in either frame. The coordinate transformation can best be understood if we regard point P as attached to frame $\mathcal{A}$, as if it were a point of a box with sides of lengths x, y, and z, as indicated in Fig. 2.2a. Now, frame $\mathcal{A}$ undergoes a rotation $\mathbf{Q}$ about its origin that carries it into a new attitude, that of frame $\mathcal{B}$, as illustrated in Fig. 2.2b. Point P in its rotated position is labeled Π, of position vector $\boldsymbol{\pi}$, i.e.,

$$\boldsymbol{\pi} = \mathbf{Q}\mathbf{p} \tag{2.82}$$

It is apparent that the relative position of point P with respect to its box does not change under the foregoing rotation, and hence,

$$[\boldsymbol{\pi}]_{\mathcal{B}} = \begin{bmatrix} x \\ y \\ z \end{bmatrix} \tag{2.83}$$

Moreover, let

$$[\boldsymbol{\pi}]_{\mathcal{A}} = \begin{bmatrix} \xi \\ \eta \\ \zeta \end{bmatrix} \tag{2.84}$$

The relation between the two representations of the position vector of any point of the 3-dimensional Euclidean space is given by

Theorem 2.5.1 *The representations of the position vector $\boldsymbol{\pi}$ of any point in two frames $\mathcal{A}$ and $\mathcal{B}$, denoted by $[\boldsymbol{\pi}]_{\mathcal{A}}$ and $[\boldsymbol{\pi}]_{\mathcal{B}}$, respectively, are related by*

$$[\boldsymbol{\pi}]_{\mathcal{A}} = [\mathbf{Q}]_{\mathcal{A}}[\boldsymbol{\pi}]_{\mathcal{B}} \tag{2.85}$$

Proof: Let us write eq.(2.82) in $\mathcal{A}$:

$$[\boldsymbol{\pi}]_{\mathcal{A}} = [\mathbf{Q}]_{\mathcal{A}}[\mathbf{p}]_{\mathcal{A}} \tag{2.86}$$

Now, from Fig. 2.2b and eqs.(2.81) and (2.83) it is apparent that

$$[\boldsymbol{\pi}]_{\mathcal{B}} = [\mathbf{p}]_{\mathcal{A}} \tag{2.87}$$

Upon substituting eq.(2.87) into eq.(2.86), we obtain

$$[\boldsymbol{\pi}]_{\mathcal{A}} = [\mathbf{Q}]_{\mathcal{A}}[\boldsymbol{\pi}]_{\mathcal{B}} \tag{2.88}$$

q.e.d. Moreover, we have

Theorem 2.5.2 *The representations of $\mathbf{Q}$ carrying $\mathcal{A}$ into $\mathcal{B}$ in these two frames are identical, i.e.,*

$$[\mathbf{Q}]_{\mathcal{A}} = [\mathbf{Q}]_{\mathcal{B}} \tag{2.89}$$

Proof: Upon substitution of eq.(2.82) into eq.(2.85), we obtain

$$[\mathbf{Q}\mathbf{p}]_{\mathcal{A}} = [\mathbf{Q}]_{\mathcal{A}}[\mathbf{Q}\mathbf{p}]_{\mathcal{B}}$$

or

$$[\mathbf{Q}]_{\mathcal{A}}[\mathbf{p}]_{\mathcal{A}} = [\mathbf{Q}]_{\mathcal{A}}[\mathbf{Q}\mathbf{p}]_{\mathcal{B}}$$

Now, since $\mathbf{Q}$ is orthogonal, it is nonsingular, and hence, $[\mathbf{Q}]_{\mathcal{A}}$ can be deleted from the foregoing equation, thus leading to

$$[\mathbf{p}]_{\mathcal{A}} = [\mathbf{Q}]_{\mathcal{B}}[\mathbf{p}]_{\mathcal{B}} \tag{2.90}$$

However, by virtue of Theorem 2.5.1, the two representations of $\mathbf{p}$ observe the relation

$$[\mathbf{p}]_{\mathcal{A}} = [\mathbf{Q}]_{\mathcal{A}}[\mathbf{p}]_{\mathcal{B}} \tag{2.91}$$

the theorem being proved upon equating the right-hand sides of eqs.(2.90) and (2.91).

Note that the foregoing theorem states a relation valid only for the conditions stated therein. The reader should not conclude from this result that rotation matrices have the same representations in every frame. This point is stressed in Example 2.5.1. Furthermore, we have

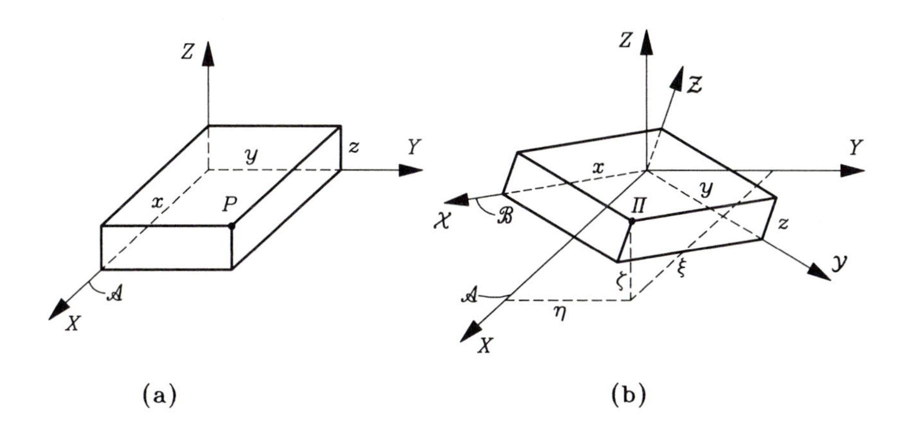

FIGURE 2.2. Coordinate transformation: (a) coordinates of point P in the $\mathcal{A}$-frame; and (b) relative orientation of frame $\mathcal{B}$ with respect to $\mathcal{A}$.

Theorem 2.5.3 *The inverse relation of Theorem 2.5.1 is given by*

$$[\,\boldsymbol{\pi}\,]_{\mathcal{B}} = [\,\mathbf{Q}^T\,]_{\mathcal{B}}[\,\boldsymbol{\pi}\,]_{\mathcal{A}} \tag{2.92}$$

Proof: This is straightforward in light of the two foregoing theorems, and is left to the reader as an exercise.

Example 2.5.1 *Coordinate frames $\mathcal{A}$ and $\mathcal{B}$ are shown in Fig. 2.3. Find the representations of $\mathbf{Q}$ rotating $\mathcal{A}$ into $\mathcal{B}$ in these two frames and show that they are identical. Moreover, if $[\,\mathbf{p}\,]_{\mathcal{A}} = [\,1,\ 1,\ 1\,]^T$, find $[\,\mathbf{p}\,]_{\mathcal{B}}$.*

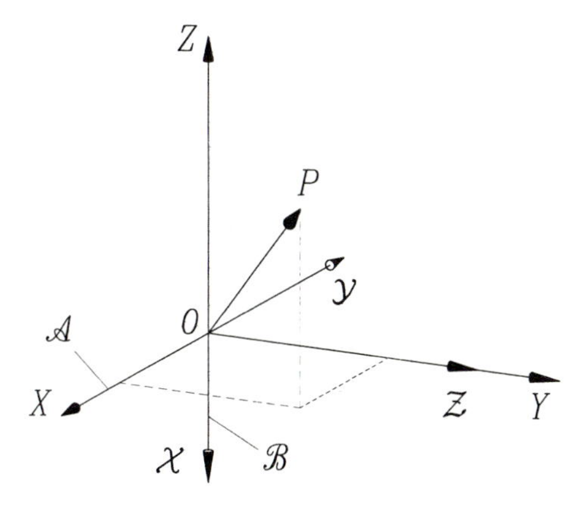

FIGURE 2.3. Coordinate frames $\mathcal{A}$ and $\mathcal{B}$ with a common origin.

Solution: Let $\mathbf{i}$, $\mathbf{j}$, and $\mathbf{k}$ be unit vectors in the directions of the X-, Y-, and Z-axes, respectively; unit vectors $\boldsymbol{\iota}$, $\boldsymbol{\gamma}$, and $\boldsymbol{\kappa}$ are defined likewise as parallel to the $\mathcal{X}$-, $\mathcal{Y}$-, and $\mathcal{Z}$-axes of Fig. 2.3. Therefore,

$$\mathbf{Q}\,\mathbf{i} \equiv \boldsymbol{\iota} = -\mathbf{k}, \quad \mathbf{Q}\,\mathbf{j} \equiv \boldsymbol{\gamma} = -\mathbf{i}, \quad \mathbf{Q}\,\mathbf{k} \equiv \boldsymbol{\kappa} = \mathbf{j}$$

Therefore, using Definition 2.2.1, the matrix representation of $\mathbf{Q}$ carrying $\mathcal{A}$ into $\mathcal{B}$, in $\mathcal{A}$, is given by

$$[\,\mathbf{Q}\,]_{\mathcal{A}} = \begin{bmatrix} 0 & -1 & 0 \\ 0 & 0 & 1 \\ -1 & 0 & 0 \end{bmatrix}$$

Now, in order to find $[\,\mathbf{Q}\,]_{\mathcal{B}}$, we apply $\mathbf{Q}$ to the three unit vectors of $\mathcal{B}$, $\boldsymbol{\iota}$, $\boldsymbol{\gamma}$, and $\boldsymbol{\kappa}$. Thus, for $\boldsymbol{\iota}$, we have

$$\mathbf{Q}\boldsymbol{\iota} = \begin{bmatrix} 0 & -1 & 0 \\ 0 & 0 & 1 \\ -1 & 0 & 0 \end{bmatrix} \begin{bmatrix} 0 \\ 0 \\ -1 \end{bmatrix} = \begin{bmatrix} 0 \\ -1 \\ 0 \end{bmatrix} = -\mathbf{j} = -\boldsymbol{\kappa}$$

Likewise,

$$\mathbf{Q}\boldsymbol{\gamma} = -\boldsymbol{\iota}, \quad \mathbf{Q}\boldsymbol{\kappa} = \boldsymbol{\gamma}$$

Again, from Definition 2.2.1, we have

$$[\,\mathbf{Q}\,]_{\mathcal{B}} = \begin{bmatrix} 0 & -1 & 0 \\ 0 & 0 & 1 \\ -1 & 0 & 0 \end{bmatrix} = [\,\mathbf{Q}\,]_{\mathcal{A}}$$

thereby confirming Theorem 2.5.2. Note that the representation of this matrix in any other coordinate frame would be different. For example, if we represent this matrix in a frame whose X-axis is directed along the axis of rotation of $\mathbf{Q}$, then we end up with the X-canonical representation of $\mathbf{Q}$, namely,

$$[\,\mathbf{Q}\,]_{\mathcal{X}} = \begin{bmatrix} 1 & 0 & 0 \\ 0 & \cos\phi & -\sin\phi \\ 0 & \sin\phi & \cos\phi \end{bmatrix}$$

with the angle of rotation ϕ being readily computed as $\phi = 150°$, which thus yields

$$[\,\mathbf{Q}\,]_{\mathcal{X}} = \begin{bmatrix} 1 & 0 & 0 \\ 0 & -1/2 & -\sqrt{3}/2 \\ 0 & \sqrt{3}/2 & -1/2 \end{bmatrix}$$

which apparently has different entries from those of $[\,\mathbf{Q}\,]_{\mathcal{A}}$ and $[\,\mathbf{Q}\,]_{\mathcal{B}}$ found above.

Now, from eq.(2.92),

$$[\,\mathbf{p}\,]_{\mathcal{B}} = \begin{bmatrix} 0 & 0 & -1 \\ -1 & 0 & 0 \\ 0 & 1 & 0 \end{bmatrix} \begin{bmatrix} 1 \\ 1 \\ 1 \end{bmatrix} = \begin{bmatrix} -1 \\ -1 \\ 1 \end{bmatrix}$$

a result that can be readily verified by inspection.

2.5.2 Coordinate Transformation with Origin Shift

Now, if the coordinate origins do not coincide, let $\mathbf{b}$ be the position vector of $\mathcal{O}$, the origin of $\mathcal{B}$, from O, the origin of $\mathcal{A}$, as shown in Fig. 2.4. The corresponding coordinate transformation from $\mathcal{A}$ to $\mathcal{B}$, the counterpart of Theorem 2.5.1, is given below.

Theorem 2.5.4 *The representations of the position vector* $\mathbf{p}$ *of a point* P *of the Euclidean 3-dimensional space in two frames* $\mathcal{A}$ *and* $\mathcal{B}$ *are related by*

$$[\,\mathbf{p}\,]_{\mathcal{A}} = [\,\mathbf{b}\,]_{\mathcal{A}} + [\,\mathbf{Q}\,]_{\mathcal{A}}[\,\boldsymbol{\pi}\,]_{\mathcal{B}} \tag{2.93a}$$

$$[\,\boldsymbol{\pi}\,]_{\mathcal{B}} = [\,\mathbf{Q}^T\,]_{\mathcal{B}}([\,-\mathbf{b}\,]_{\mathcal{A}} + [\,\mathbf{p}\,]_{\mathcal{A}}) \tag{2.93b}$$

with $\mathbf{b}$ *defined as the vector directed from the origin of* $\mathcal{A}$ *to that of* $\mathcal{B}$, *and* $\boldsymbol{\pi}$ *the vector directed from the origin of* $\mathcal{B}$ *to* P, *as depicted in Fig. 2.4.*

Proof: We have, from Fig. 2.4,

$$\mathbf{p} = \mathbf{b} + \boldsymbol{\pi} \tag{2.94}$$

If we express the above equation in the $\mathcal{A}$-frame, we obtain

$$[\,\mathbf{p}\,]_{\mathcal{A}} = [\,\mathbf{b}\,]_{\mathcal{A}} + [\,\boldsymbol{\pi}\,]_{\mathcal{A}}$$

where $\boldsymbol{\pi}$ is assumed to be readily available in $\mathcal{B}$, and so the foregoing equation must be expressed as

$$[\,\mathbf{p}\,]_{\mathcal{A}} = [\,\mathbf{b}\,]_{\mathcal{A}} + [\,\mathbf{Q}\,]_{\mathcal{A}}[\,\boldsymbol{\pi}\,]_{\mathcal{B}}$$

which thus proves eq.(2.93a). To prove eq.(2.93b), we simply solve eq.(2.94) for $\boldsymbol{\pi}$ and apply eq.(2.92) to the equation thus resulting, which readily leads to the desired relation.

Example 2.5.2 *If* $[\,\mathbf{b}\,]_{\mathcal{A}} = [\,-1, -1, \,-1\,]^T$ *and* $\mathcal{A}$ *and* $\mathcal{B}$ *have the relative orientations given in Example 2.5.1, find the position vector, in* $\mathcal{B}$, *of a point* P *of position vector* $[\,\mathbf{p}\,]_{\mathcal{A}}$ *given as in the same example.*

Solution: What we obviously need is $[\,\boldsymbol{\pi}\,]_{\mathcal{B}}$, which is given in eq.(2.93b). We thus compute first the sum inside the parentheses of that equation, i.e.,

$$[\,-\mathbf{b}\,]_{\mathcal{A}} + [\,\mathbf{p}\,]_{\mathcal{A}} = \begin{bmatrix} 2 \\ 2 \\ 2 \end{bmatrix}$$

We need further $[\,\mathbf{Q}^T\,]_{\mathcal{B}}$, which can be readily derived from $[\,\mathbf{Q}\,]_{\mathcal{B}}$. We do not have as yet this matrix, but we have $[\,\mathbf{Q}^T\,]_{\mathcal{A}}$, which is identical to $[\,\mathbf{Q}\,]_{\mathcal{B}}$ by virtue of Theorem 2.5.2. Therefore,

$$[\,\boldsymbol{\pi}\,]_{\mathcal{B}} = \begin{bmatrix} 0 & 0 & -1 \\ -1 & 0 & 0 \\ 0 & 1 & 0 \end{bmatrix} \begin{bmatrix} 2 \\ 2 \\ 2 \end{bmatrix} = \begin{bmatrix} -2 \\ -2 \\ 2 \end{bmatrix}$$

a result that the reader is invited to verify by inspection.

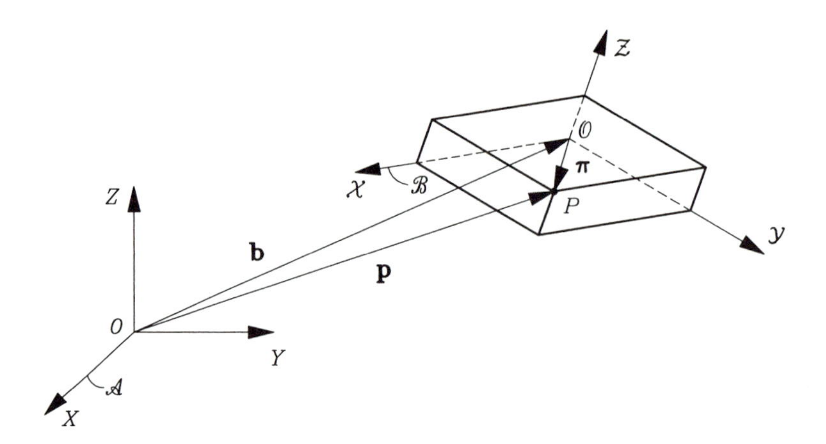

FIGURE 2.4. Coordinate frames with different origins.

2.5.3 *Homogeneous Coordinates*

The general coordinate transformation, involving a shift of the origin, is not linear, in general, as can be readily realized by virtue of the *nonhomogeneous* term involved, i.e., the first term of the right-hand side of eq.(2.93a), which is independent of $\mathbf{p}$. Such a transformation, nevertheless, can be represented in homogeneous form if *homogeneous coordinates* are introduced. These are defined below: Let $[\,\mathbf{p}\,]_{\mathcal{M}}$ be the coordinate array of a *finite* point P in reference frame $\mathcal{M}$. What we mean by a finite point is one whose coordinates are all finite. We are thus assuming that the point P at hand is not *at infinity*, points at infinity being dealt with later. The homogeneous coordinates of P are those in the 4-dimensional array $\{\mathbf{p}\}_{\mathcal{M}}$, defined as

$$\{\mathbf{p}\}_{\mathcal{M}} \equiv \begin{bmatrix} [\,\mathbf{p}\,]_{\mathcal{M}} \\ 1 \end{bmatrix} \tag{2.95}$$

The *affine transformation* of eq.(2.93a) can now be rewritten in homogeneous-coordinate form as

$$\{\mathbf{p}\}_{\mathcal{A}} = \{\mathbf{T}\}_{\mathcal{A}}\{\boldsymbol{\pi}\}_{\mathcal{B}} \tag{2.96}$$

where $\{\mathbf{T}\}_{\mathcal{A}}$ is defined as a 4×4 array, i.e.,

$$\{\mathbf{T}\}_{\mathcal{A}} \equiv \begin{bmatrix} [\,\mathbf{Q}\,]_{\mathcal{A}} & [\,\mathbf{b}\,]_{\mathcal{A}} \\ [\,\mathbf{0}^T\,]_{\mathcal{A}} & 1 \end{bmatrix} \tag{2.97}$$

The inverse transformation of that defined in eq.(2.97) is derived from eq.(2.93a), i.e.,

$$\{\,\mathbf{T}^{-1}\,\}_{\mathcal{B}} = \begin{bmatrix} [\,\mathbf{Q}^T\,]_{\mathcal{B}} & [\,-\mathbf{b}\,]_{\mathcal{B}} \\ [\,\mathbf{0}^T\,]_{\mathcal{B}} & 1 \end{bmatrix} \tag{2.98}$$

Furthermore, homogeneous transformations can be concatenated. Indeed, let $\mathcal{F}_k$, for $k = i-1,\, i,\, i+1$, denote three coordinate frames, with origins at O_k. Moreover, let $\mathbf{Q}_{i-1}$ be the rotation carrying $\mathcal{F}_{i-1}$ into an orientation coinciding with that of $\mathcal{F}_i$. If a similar definition for $\mathbf{Q}_i$ is adopted, then $\mathbf{Q}_i$ denotes the rotation carrying $\mathcal{F}_i$ into an orientation coinciding with that of $\mathcal{F}_{i+1}$. First, the case in which all three origins coincide is considered. Clearly,

$$[\,\mathbf{p}\,]_i = [\,\mathbf{Q}_{i-1}^T\,]_{i-1}[\,\mathbf{p}\,]_{i-1} \tag{2.99}$$

$$[\,\mathbf{p}\,]_{i+1} = [\,\mathbf{Q}_i^T\,]_i[\,\mathbf{p}\,]_i = [\,\mathbf{Q}_i^T\,]_i[\,\mathbf{Q}_{i-1}^T\,]_{i-1}[\,\mathbf{p}\,]_{i-1} \tag{2.100}$$

the inverse relation of that appearing in eq.(2.100) being

$$[\,\mathbf{p}\,]_{i-1} = [\,\mathbf{Q}_{i-1}\,]_{i-1}[\,\mathbf{Q}_i\,]_i[\,\mathbf{p}\,]_{i+1} \tag{2.101}$$

If now the origins do not coincide, let $\mathbf{a}_{i-1}$ and $\mathbf{a}_i$ denote the vectors $\overrightarrow{O_{i-1}O_i}$ and $\overrightarrow{O_iO_{i+1}}$, respectively. The homogeneous-coordinate transformations $\{\mathbf{T}_{i-1}\}_{i-1}$ and $\{\mathbf{T}_i\}_i$ thus arising are obviously

$$\{\mathbf{T}_{i-1}\}_{i-1} = \begin{bmatrix} [\,\mathbf{Q}_{i-1}\,]_{i-1} & [\,\mathbf{a}_{i-1}\,]_{i-1} \\ [\,\mathbf{0}^T\,]_{i-1} & 1 \end{bmatrix}, \quad \{\mathbf{T}_i\}_i = \begin{bmatrix} [\,\mathbf{Q}_i\,]_i & [\,\mathbf{a}_i\,]_i \\ [\,\mathbf{0}^T\,]_i & 1 \end{bmatrix} \tag{2.102}$$

whereas their inverse transformations are

$$\{\mathbf{T}_{i-1}^{-1}\}_i = \begin{bmatrix} [\,\mathbf{Q}_{i-1}^T\,]_i & [\,\mathbf{Q}_{i-1}^T\,]_i[\,-\mathbf{a}_{i-1}\,]_{i-1} \\ [\,\mathbf{0}^T\,]_i & 1 \end{bmatrix} \tag{2.103}$$

$$\{\mathbf{T}_i^{-1}\}_{i+1} = \begin{bmatrix} [\,\mathbf{Q}_i^T\,]_{i+1} & [\mathbf{Q}_i^T\,]_{i+1}[\,-\mathbf{a}_i\,]_i \\ [\,\mathbf{0}^T\,]_{i+1} & 1 \end{bmatrix} \tag{2.104}$$

Hence, the coordinate transformations involved are

$$\{\mathbf{p}\}_{i-1} = \{\mathbf{T}_{i-1}\}_{i-1}\{\mathbf{p}\}_i \tag{2.105}$$

$$\{\mathbf{p}\}_{i-1} = \{\mathbf{T}_{i-1}\}_{i-1}\{\mathbf{T}_i\}_i\{\mathbf{p}\}_{i+1} \tag{2.106}$$

the corresponding inverse transformations being

$$\{\,\mathbf{p}\,\}_i = \{\mathbf{T}_{i-1}^{-1}\}_i\{\,\mathbf{p}\,\}_{i-1} \tag{2.107}$$

$$\{\,\mathbf{p}\,\}_{i+1} = \{\mathbf{T}_i^{-1}\}_{i+1}\{\,\mathbf{p}\,\}_i = \{\mathbf{T}_i^{-1}\}_{i+1}\{\mathbf{T}_{i-1}^{-1}\}_i\{\,\mathbf{p}\,\}_{i-1} \tag{2.108}$$

Now, if P lies at infinity, we can express its homogeneous coordinates in a simpler form. To this end, we rewrite expression (2.95) in the form

$$\{\mathbf{p}\}_{\mathcal{M}} \equiv \|\mathbf{p}\| \begin{bmatrix} [\,\mathbf{e}\,]_{\mathcal{M}} \\ 1/\|\mathbf{p}\| \end{bmatrix}$$

and hence,

$$\lim_{\|\mathbf{p}\|\to\infty} \{\mathbf{p}\}_{\mathcal{M}} = \left(\lim_{\|\mathbf{p}\|\to\infty} \|\mathbf{p}\|\right)\left(\lim_{\|\mathbf{p}\to\infty\|} \begin{bmatrix} [\,\mathbf{e}\,]_{\mathcal{M}} \\ 1/\|\mathbf{p}\| \end{bmatrix}\right)$$

or

$$\lim_{\|\mathbf{p}\|\to\infty} \{\mathbf{p}\}_{\mathcal{M}} = (\lim_{\|\mathbf{p}\|\to\infty} \|\mathbf{p}\|) \begin{bmatrix} [\mathbf{e}]_{\mathcal{M}} \\ 0 \end{bmatrix}$$

We now define the *homogeneous coordinates of a point P lying at infinity* as the 4-dimensional array appearing in the foregoing expression, i.e.,

$$\{\mathbf{p}_\infty\}_{\mathcal{M}} \equiv \begin{bmatrix} [\mathbf{e}]_{\mathcal{M}} \\ 0 \end{bmatrix} \tag{2.109}$$

which means that a point at infinity, in homogeneous coordinates, has only a direction, given by the unit vector $\mathbf{e}$, but an undefined location. When working with objects within the atmosphere of the earth, for example, stars can be regarded as lying at infinity, and hence, their location is completely specified simply by their longitude and latitude, which suffice to define the direction cosines of a unit vector in spherical coordinates.

On the other hand, a rotation matrix can be regarded as composed of three columns, each representing a unit vector, e.g.,

$$\mathbf{Q} = [\mathbf{e}_1 \quad \mathbf{e}_2 \quad \mathbf{e}_3]$$

where the triad $\{\mathbf{e}_k\}_1^3$ is orthonormal. We can thus represent $\{\mathbf{T}\}_{\mathcal{A}}$ of eq.(2.97) in the form

$$\{\mathbf{T}\}_{\mathcal{A}} = \begin{bmatrix} \mathbf{e}_1 & \mathbf{e}_2 & \mathbf{e}_3 & \mathbf{b} \\ 0 & 0 & 0 & 1 \end{bmatrix} \tag{2.110}$$

thereby concluding that the columns of the 4×4 matrix $\mathbf{T}$ represent the homogeneous coordinates of a set of corresponding points, the first three of which are at infinity.

Example 2.5.3 *An ellipsoid is centered at a point $O_{\mathcal{B}}$ of position vector $\mathbf{b}$, its three axes $\mathcal{X}$, $\mathcal{Y}$, and $\mathcal{Z}$ defining a coordinate frame $\mathcal{B}$. Moreover, its semiaxes have lengths $a = 1$, $b = 2$, and $c = 3$, the coordinates of $O_{\mathcal{B}}$ in a coordinate frame $\mathcal{A}$ being $[\mathbf{b}]_{\mathcal{A}} = [1,\ 2,\ 3]^T$. Additionally, the direction cosines of $\mathcal{X}$ are $(0.933,\ 0.067,\ -0.354)$, whereas $\mathcal{Y}$ is perpendicular to $\mathbf{b}$ and to the unit vector $\mathbf{u}$ that is parallel to the $\mathcal{X}$ axis. Find the equation of the ellipsoid in $\mathcal{A}$. (This example has relevance in collision-avoidance algorithms, some of which approximate manipulator links as ellipsoids, thereby easing tremendously the computational requirements.)*

Solution: Let $\mathbf{u}$, $\mathbf{v}$, and $\mathbf{w}$ be unit vectors parallel to the $\mathcal{X}$, $\mathcal{Y}$, and $\mathcal{Z}$ axes, respectively. Then,

$$[\mathbf{u}]_{\mathcal{A}} = \begin{bmatrix} 0.933 \\ 0.067 \\ -0.354 \end{bmatrix}, \quad \mathbf{v} = \frac{\mathbf{u} \times \mathbf{b}}{\|\mathbf{u} \times \mathbf{b}\|}, \quad \mathbf{w} = \mathbf{u} \times \mathbf{v}$$

and hence,

$$[\mathbf{v}]_{\mathcal{A}} = \begin{bmatrix} 0.243 \\ -0.843 \\ 0.481 \end{bmatrix}, \quad [\mathbf{w}]_{\mathcal{A}} = \begin{bmatrix} -0.266 \\ -0.535 \\ -0.803 \end{bmatrix}$$

from which the rotation matrix $\mathbf{Q}$, rotating the axes of $\mathcal{A}$ into orientations coinciding with those of $\mathcal{B}$, can be readily represented in $\mathcal{A}$, or in $\mathcal{B}$ for that matter, as

$$[\mathbf{Q}]_{\mathcal{A}} = [\mathbf{u},\, \mathbf{v},\, \mathbf{w}]_{\mathcal{A}} = \begin{bmatrix} 0.933 & 0.243 & -0.266 \\ 0.067 & -0.843 & -0.535 \\ -0.354 & 0.481 & -0.803 \end{bmatrix}$$

On the other hand, if the coordinates of a point P in $\mathcal{A}$ and $\mathcal{B}$ are $[\mathbf{p}]_{\mathcal{A}} = [p_1,\, p_2,\, p_3]^T$ and $[\boldsymbol{\pi}]_{\mathcal{B}} = [\pi_1,\, \pi_2,\, \pi_3]^T$, respectively, then the equation of the ellipsoid in $\mathcal{B}$ is clearly

$$\mathcal{B}\colon \quad \frac{\pi_1^2}{1^2} + \frac{\pi_2^2}{2^2} + \frac{\pi_3^2}{3^2} = 1$$

Now, what is needed in order to derive the equation of the ellipsoid in $\mathcal{A}$ is simply a relation between the coordinates of P in $\mathcal{B}$ and those in $\mathcal{A}$. These coordinates are related by eq.(2.93b), which requires $[\mathbf{Q}^T]_{\mathcal{B}}$, while we have $[\mathbf{Q}]_{\mathcal{A}}$. Nevertheless, by virtue of Theorem 2.5.2

$$[\mathbf{Q}^T]_{\mathcal{B}} = [\mathbf{Q}^T]_{\mathcal{A}} = \begin{bmatrix} 0.933 & 0.067 & -0.354 \\ 0.243 & -0.843 & 0.481 \\ -0.266 & -0.535 & -0.803 \end{bmatrix}$$

Hence,

$$[\boldsymbol{\pi}]_{\mathcal{B}} = \begin{bmatrix} 0.933 & 0.067 & -0.354 \\ 0.243 & -0.843 & 0.481 \\ -0.266 & -0.535 & -0.803 \end{bmatrix} \left(\begin{bmatrix} -1 \\ -2 \\ -3 \end{bmatrix} + \begin{bmatrix} p_1 \\ p_2 \\ p_3 \end{bmatrix} \right)$$

Therefore,

$$\begin{aligned} \pi_1 &= 0.933p_1 + 0.067p_2 - 0.354p_3 - 0.005 \\ \pi_2 &= 0.243p_1 - 0.843p_2 + 0.481p_3 \\ \pi_3 &= -0.266p_1 - 0.535p_2 - 0.803p_3 + 3.745 \end{aligned}$$

Substitution of the foregoing relations into the ellipsoid equation in $\mathcal{B}$ leads to

$$\begin{aligned} \mathcal{A}\colon \quad & 32.1521{p_1}^2 + 7.70235{p_2}^2 + 9.17286{p_3}^2 - 8.30524p_1 - 16.0527p_2 \\ & -23.9304p_3 + 9.32655p_1p_2 + 9.02784p_2p_3 - 19.9676p_1p_3 + 20.101 = 0 \end{aligned}$$

which is the equation sought, and which was obtained using computer algebra.

2.6 Similarity Transformations

Transformations of the position vector of points under a change of coordinate frame involving both a translation of the origin and a rotation of the coordinate axes was the main subject of Section 2.5. In this section, we study the transformations of components of vectors other than the position vector, while extending the concept to the transformation of matrix entries. How these transformations take place is the subject of this section.

What is involved in the present discussion is a *change of basis* of the associated vector spaces, and hence, this is not limited to 3-D vector spaces. That is, n-dimensional vector spaces will be studied in this section. Moreover, only isomorphisms, i.e., transformations $\mathbf{L}$ of the n-dimensional vector space $\mathcal{V}$ onto itself will be considered. Let $\mathcal{A} = \{\mathbf{a}_i\}_1^n$ and $\mathcal{B} = \{\mathbf{b}_i\}_1^n$ be two *different* bases of the same space $\mathcal{V}$. Hence, any vector $\mathbf{v}$ of $\mathcal{V}$ can be expressed in either of two ways, namely,

$$\mathbf{v} = \alpha_1\mathbf{a}_1 + \alpha_2\mathbf{a}_2 + \cdots + \alpha_n\mathbf{a}_n \tag{2.111}$$

$$\mathbf{v} = \beta_1\mathbf{b}_1 + \beta_2\mathbf{b}_2 + \cdots + \beta_n\mathbf{b}_n \tag{2.112}$$

from which two representations of $\mathbf{v}$ are readily derived, namely,

$$[\,\mathbf{v}\,]_{\mathcal{A}} = \begin{bmatrix} \alpha_1 \\ \alpha_2 \\ \vdots \\ \alpha_n \end{bmatrix}, \quad [\,\mathbf{v}\,]_{\mathcal{B}} = \begin{bmatrix} \beta_1 \\ \beta_2 \\ \vdots \\ \beta_n \end{bmatrix} \tag{2.113}$$

Furthermore, let the two foregoing bases be related by

$$\mathbf{b}_j = a_{1j}\mathbf{a}_1 + a_{2j}\mathbf{a}_2 + \cdots + a_{nj}\mathbf{a}_n, \quad j = 1, \ldots, n \tag{2.114}$$

Now, in order to find the relationship between the two representations of eq.(2.113), eq.(2.114) is substituted into eq.(2.112), which yields

$$\begin{aligned} \mathbf{v} = {} & \beta_1(a_{11}\mathbf{a}_1 + a_{21}\mathbf{a}_2 + \cdots + a_{n1}\mathbf{a}_n) \\ & + \beta_2(a_{12}\mathbf{a}_1 + a_{22}\mathbf{a}_2 + \cdots + a_{n2}\mathbf{a}_n) \\ & \vdots \\ & + \beta_n(a_{1n}\mathbf{a}_1 + a_{2n}\mathbf{a}_2 + \cdots + a_{nn}\mathbf{a}_n) \end{aligned} \tag{2.115}$$

This can be rearranged to yield

$$\begin{aligned} \mathbf{v} = {} & (a_{11}\beta_1 + a_{12}\beta_2 + \cdots + a_{1n}\beta_n)\mathbf{a}_1 \\ & + (a_{21}\beta_1 + a_{22}\beta_2 + \cdots + a_{2n}\beta_n)\mathbf{a}_2 \\ & \vdots \\ & + (a_{n1}\beta_1 + a_{n2}\beta_2 + \cdots + a_{nn}\beta_n)\mathbf{a}_n \end{aligned} \tag{2.116}$$

Comparing eq.(2.116) with eq.(2.111), one readily derives

$$[\,\mathbf{v}\,]_{\mathcal{A}} = [\,\mathbf{A}\,]_{\mathcal{A}}[\,\mathbf{v}\,]_{\mathcal{B}} \tag{2.117}$$

where

$$[\,\mathbf{A}\,]_{\mathcal{A}} \equiv \begin{bmatrix} a_{11} & a_{12} & \cdots & a_{1n} \\ a_{21} & a_{22} & \cdots & a_{2n} \\ \vdots & \vdots & \ddots & \vdots \\ a_{n1} & a_{n2} & \cdots & a_{nn} \end{bmatrix} \tag{2.118}$$

which are the relations sought. Clearly, the inverse relationship of eq.(2.117) is

$$[\,\mathbf{v}\,]_{\mathcal{B}} = [\,\mathbf{A}^{-1}\,]_{\mathcal{A}}[\,\mathbf{v}\,]_{\mathcal{A}} \tag{2.119}$$

Next, let $\mathbf{L}$ have the representation in $\mathcal{A}$ given below:

$$[\,\mathbf{L}\,]_{\mathcal{A}} = \begin{bmatrix} l_{11} & l_{12} & \cdots & l_{1n} \\ l_{21} & l_{22} & \cdots & l_{2n} \\ \vdots & \vdots & \ddots & \vdots \\ l_{n1} & l_{n2} & \cdots & l_{nn} \end{bmatrix} \tag{2.120}$$

Now we aim at finding the relationship between $[\,\mathbf{L}\,]_{\mathcal{A}}$ and $[\,\mathbf{L}\,]_{\mathcal{B}}$. To this end, let $\mathbf{w}$ be the image of $\mathbf{v}$ under $\mathbf{L}$, i.e.,

$$\mathbf{L}\mathbf{v} = \mathbf{w} \tag{2.121}$$

which can be expressed in terms of either $\mathcal{A}$ or $\mathcal{B}$ as

$$[\,\mathbf{L}\,]_{\mathcal{A}}[\,\mathbf{v}\,]_{\mathcal{A}} = [\,\mathbf{w}\,]_{\mathcal{A}} \tag{2.122}$$

$$[\,\mathbf{L}\,]_{\mathcal{B}}[\,\mathbf{v}\,]_{\mathcal{B}} = [\,\mathbf{w}\,]_{\mathcal{B}} \tag{2.123}$$

Now we assume that the image vector $\mathbf{w}$ of the transformation of eq.(2.121) is *identical* to that of vector $\mathbf{v}$ in the range of $\mathbf{L}$, which is not always the case. Our assumption is, then, that similar to eq.(2.117),

$$[\,\mathbf{w}\,]_{\mathcal{A}} = [\,\mathbf{A}\,]_{\mathcal{A}}[\,\mathbf{w}\,]_{\mathcal{B}} \tag{2.124}$$

Now, substitution of eq.(2.124) into eq.(2.122) yields

$$[\,\mathbf{A}\,]_{\mathcal{A}}[\,\mathbf{w}\,]_{\mathcal{B}} = [\,\mathbf{L}\,]_{\mathcal{A}}[\,\mathbf{A}\,]_{\mathcal{A}}[\,\mathbf{v}\,]_{\mathcal{B}} \tag{2.125}$$

which can be readily rearranged in the form

$$[\,\mathbf{w}\,]_{\mathcal{B}} = [\,\mathbf{A}^{-1}\,]_{\mathcal{A}}[\,\mathbf{L}\,]_{\mathcal{A}}[\,\mathbf{A}\,]_{\mathcal{A}}[\,\mathbf{v}\,]_{\mathcal{B}} \tag{2.126}$$

Comparing eq.(2.123) with eq.(2.126) readily leads to

$$[\,\mathbf{L}\,]_{\mathcal{B}} = [\,\mathbf{A}^{-1}\,]_{\mathcal{A}}[\,\mathbf{L}\,]_{\mathcal{A}}[\,\mathbf{A}\,]_{\mathcal{A}} \tag{2.127}$$

which upon rearrangement, becomes

$$[\mathbf{L}]_{\mathcal{A}} = [\mathbf{A}]_{\mathcal{A}}[\mathbf{L}]_{\mathcal{B}}[\mathbf{A}^{-1}]_{\mathcal{A}} \tag{2.128}$$

Relations (2.117), (2.119), (2.127), and (2.128) constitute what are called *similarity transformations*. These are important because they preserve *invariant* quantities such as the eigenvalues and eigenvectors of matrices, the magnitudes of vectors, the angles between vectors, and so on. Indeed, one has

Theorem 2.6.1 *The characteristic polynomial of a given $n \times n$ matrix remains unchanged under a similarity transformation. Moreover, the eigenvalues of two matrix representations of the same $n \times n$ linear transformation are identical, and if $[\mathbf{e}]_{\mathcal{B}}$ is an eigenvector of $[\mathbf{L}]_{\mathcal{B}}$, then under the similarity transformation (2.128), the corresponding eigenvector of $[\mathbf{L}]_{\mathcal{A}}$ is $[\mathbf{e}]_{\mathcal{A}} = [\mathbf{A}]_{\mathcal{A}}[\mathbf{e}]_{\mathcal{B}}$.*

Proof: From eq.(2.11), the characteristic polynomial of $[\mathbf{L}]_{\mathcal{B}}$ is

$$P(\lambda) = \det(\lambda[\mathbf{1}]_{\mathcal{B}} - [\mathbf{L}]_{\mathcal{B}}) \tag{2.129}$$

which can be rewritten as

$$\begin{aligned} P(\lambda) &\equiv \det(\lambda[\mathbf{A}^{-1}]_{\mathcal{A}}[\mathbf{1}]_{\mathcal{A}}[\mathbf{A}]_{\mathcal{A}} - [\mathbf{A}^{-1}]_{\mathcal{A}}[\mathbf{L}]_{\mathcal{A}}[\mathbf{A}]_{\mathcal{A}}) \\ &= \det([\mathbf{A}^{-1}]_{\mathcal{A}}(\lambda[\mathbf{1}]_{\mathcal{A}} - [\mathbf{L}]_{\mathcal{A}})[\mathbf{A}]_{\mathcal{A}}) \\ &= \det([\mathbf{A}^{-1}]_{\mathcal{A}})\det(\lambda[\mathbf{1}]_{\mathcal{A}} - [\mathbf{L}]_{\mathcal{A}})\det([\mathbf{A}]_{\mathcal{A}}) \end{aligned}$$

But

$$\det([\mathbf{A}^{-1}]_{\mathcal{A}})\det([\mathbf{A}]_{\mathcal{A}}) = 1$$

and hence, the characteristic polynomial of $[\mathbf{L}]_{\mathcal{A}}$ is identical to that of $[\mathbf{L}]_{\mathcal{B}}$. Since both representations have the same characteristic polynomial, they have the same eigenvalues. Now, if $[\mathbf{e}]_{\mathcal{B}}$ is an eigenvector of $[\mathbf{L}]_{\mathcal{B}}$ associated with the eigenvalue λ, then

$$[\mathbf{L}]_{\mathcal{B}}[\mathbf{e}]_{\mathcal{B}} = \lambda[\mathbf{e}]_{\mathcal{B}}$$

Next, eq.(2.127) is substituted into the foregoing equation, which thus leads to

$$[\mathbf{A}^{-1}]_{\mathcal{A}}[\mathbf{L}]_{\mathcal{A}}[\mathbf{A}]_{\mathcal{A}}[\mathbf{e}]_{\mathcal{B}} = \lambda[\mathbf{e}]_{\mathcal{B}}$$

Upon rearrangement, this equation becomes

$$[\mathbf{L}]_{\mathcal{A}}[\mathbf{A}]_{\mathcal{A}}[\mathbf{e}]_{\mathcal{B}} = \lambda[\mathbf{A}]_{\mathcal{A}}[\mathbf{e}]_{\mathcal{B}} \tag{2.130}$$

whence it is apparent that $[\mathbf{A}]_{\mathcal{A}}[\mathbf{e}]_{\mathcal{B}}$ is an eigenvector of $[\mathbf{L}]_{\mathcal{A}}$ associated with the eigenvalue λ, q.e.d.

Theorem 2.6.2 *If $[\mathbf{L}]_{\mathcal{A}}$ and $[\mathbf{L}]_{\mathcal{B}}$ are related by the similarity transformation (2.127), then*

$$[\mathbf{L}^k]_{\mathcal{B}} = [\mathbf{A}^{-1}]_{\mathcal{A}}[\mathbf{L}^k]_{\mathcal{A}}[\mathbf{A}]_{\mathcal{A}} \tag{2.131}$$

for any integer k.

Proof: This is done by induction. For $k = 2$, one has

$$\begin{aligned}[\mathbf{L}^2]_{\mathcal{B}} &\equiv [\mathbf{A}^{-1}]_{\mathcal{A}}[\mathbf{L}]_{\mathcal{A}}[\mathbf{A}]_{\mathcal{A}}[\mathbf{A}^{-1}]_{\mathcal{A}}[\mathbf{L}]_{\mathcal{A}}[\mathbf{A}]_{\mathcal{A}} \\ &= [\mathbf{A}^{-1}]_{\mathcal{A}}[\mathbf{L}^2]_{\mathcal{A}}[\mathbf{A}]_{\mathcal{A}}\end{aligned}$$

Now, assume that the proposed relation holds for $k = n$. Then,

$$\begin{aligned}[\mathbf{L}^{n+1}]_{\mathcal{B}} &\equiv [\mathbf{A}^{-1}]_{\mathcal{A}}[\mathbf{L}^n]_{\mathcal{A}}[\mathbf{A}]_{\mathcal{A}}[\mathbf{A}^{-1}]_{\mathcal{A}}[\mathbf{L}]_{\mathcal{A}}[\mathbf{A}]_{\mathcal{A}} \\ &= [\mathbf{A}^{-1}]_{\mathcal{A}}[\mathbf{L}^{n+1}]_{\mathcal{A}}[\mathbf{A}]_{\mathcal{A}}\end{aligned}$$

i.e., the relation also holds for $k = n + 1$, thereby completing the proof.

Theorem 2.6.3 *The trace of an $n \times n$ matrix does not change under a similarity transformation.*

Proof: A preliminary relation will be needed: Let $[\mathbf{A}]$, $[\mathbf{B}]$ and $[\mathbf{C}]$ be three different $n \times n$ matrix arrays, in a given reference frame, that need not be indicated with any subscript. Moreover, let a_{ij}, b_{ij}, and c_{ij} be the components of the said arrays, with indices ranging from 1 to n. Hence, using standard index notation,

$$\mathrm{tr}([\mathbf{A}][\mathbf{B}][\mathbf{C}]) \equiv a_{ij}b_{jk}c_{ki} = b_{jk}c_{ki}a_{ij} \equiv \mathrm{tr}([\mathbf{B}][\mathbf{C}][\mathbf{A}]) \tag{2.132}$$

Taking the trace of both sides of eq.(2.127) and applying the foregoing result produces

$$\mathrm{tr}([\mathbf{L}]_{\mathcal{B}}) = \mathrm{tr}([\mathbf{A}^{-1}]_{\mathcal{A}}[\mathbf{L}]_{\mathcal{A}}[\mathbf{A}]_{\mathcal{A}}) = \mathrm{tr}([\mathbf{A}]_{\mathcal{A}}[\mathbf{A}^{-1}]_{\mathcal{A}}[\mathbf{L}]_{\mathcal{A}}) = \mathrm{tr}([\mathbf{L}]_{\mathcal{A}}) \tag{2.133}$$

thereby proving that the trace remains unchanged under a similarity transformation.

Example 2.6.1 *We consider the equilateral triangle sketched in Fig. 2.5, of side length equal to 2, with vertices P_1, P_2, and P_3, and coordinate frames $\mathcal{A}$ and $\mathcal{B}$ of axes X, Y and X', Y', respectively, both with origin at the centroid of the triangle. Let $\mathbf{P}$ be a 2×2 matrix defined by*

$$\mathbf{P} = [\,\mathbf{p}_1 \quad \mathbf{p}_2\,]$$

with $\mathbf{p}_i$ denoting the position vector of P_i in a given coordinate frame. Show that matrix $\mathbf{P}$ does not obey a similarity transformation upon a change of frame, and compute its trace in frames $\mathcal{A}$ and $\mathcal{B}$ to make it apparent that this matrix does not comply with the conditions of Theorem 2.6.3.

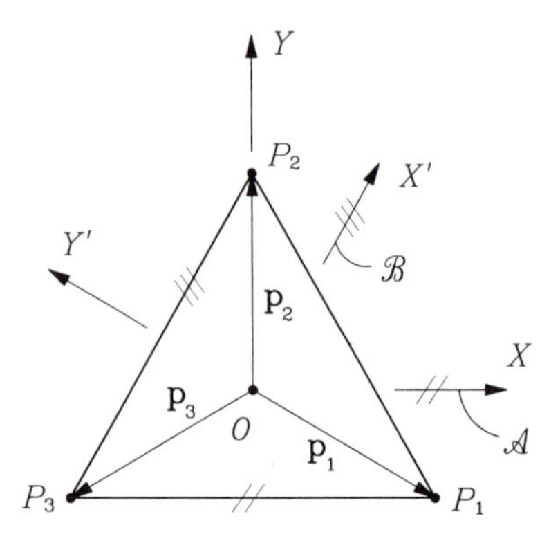

FIGURE 2.5. Two coordinate frames used to represent the position vectors of the corners of an equilateral triangle.

Solution: From the figure it is apparent that

$$[\mathbf{P}]_{\mathcal{A}} = \begin{bmatrix} 1 & 0 \\ -\sqrt{3}/3 & 2\sqrt{3}/3 \end{bmatrix}, \quad [\mathbf{P}]_{\mathcal{B}} = \begin{bmatrix} 0 & 1 \\ -2\sqrt{3}/3 & \sqrt{3}/3 \end{bmatrix}$$

Apparently,

$$\mathrm{tr}([\mathbf{P}]_{\mathcal{A}}) = 1 + \frac{2\sqrt{3}}{3} \neq \mathrm{tr}([\mathbf{P}]_{\mathcal{B}}) = \frac{\sqrt{3}}{3}$$

The reason why the trace of this matrix did not remain unchanged under a coordinate transformation is that the matrix does not obey a similarity transformation under a change of coordinates. Indeed, vectors $\mathbf{p}_i$ change as

$$[\mathbf{p}_i]_{\mathcal{A}} = [\mathbf{Q}]_{\mathcal{A}}[\mathbf{p}_i]_{\mathcal{B}}$$

under a change of coordinates from $\mathcal{B}$ to $\mathcal{A}$, with $\mathbf{Q}$ denoting the rotation carrying $\mathcal{A}$ into $\mathcal{B}$. Hence,

$$[\mathbf{P}]_{\mathcal{A}} = [\mathbf{Q}]_{\mathcal{A}}[\mathbf{P}]_{\mathcal{B}}$$

which is different from the similarity transformation of eq.(2.128). However, if we now define

$$\mathbf{R} \equiv \mathbf{P}\mathbf{P}^T$$

then

$$[\mathbf{R}]_{\mathcal{A}} = \begin{bmatrix} 1 & -\sqrt{3}/3 \\ -\sqrt{3}/3 & 5/3 \end{bmatrix}, \quad [\mathbf{R}]_{\mathcal{B}} = \begin{bmatrix} 1 & \sqrt{3}/3 \\ \sqrt{3}/3 & 5/3 \end{bmatrix}$$

and hence,

$$\mathrm{tr}([\mathbf{R}]_{\mathcal{B}}) = \frac{8}{3}$$

thereby showing that $\mathbf{R}$ does comply with the conditions of Theorem 2.6.3. Indeed, under a change of frame, matrix $\mathbf{R}$ changes as

$$[\mathbf{R}]_{\mathcal{A}} = [\mathbf{P}\mathbf{P}^T]_{\mathcal{A}} = [\mathbf{Q}]_{\mathcal{A}}[\mathbf{P}]_{\mathcal{B}}([\mathbf{Q}]_{\mathcal{A}}[\mathbf{P}]_{\mathcal{B}})^T = [\mathbf{Q}]_{\mathcal{A}}[\mathbf{P}\mathbf{P}]_{\mathcal{B}}^T[\mathbf{Q}^T]_{\mathcal{A}}$$

which is indeed a similarity transformation.

2.7 Invariance Concepts

From Example 2.6.1 it is apparent that certain properties, like the trace, of certain square matrices do not change under a coordinate transformation. For this reason, a matrix like $\mathbf{R}$ of that example is said to be *frame-invariant*, or simply *invariant*, whereas matrix $\mathbf{P}$ of the same example is not. In this section, we formally define the concept of *invariance* and highlight its applications and its role in robotics. Let a scalar, a vector, and a matrix function of the position vector $\mathbf{p}$ be denoted by $f(\mathbf{p})$, $\mathbf{f}(\mathbf{p})$ and $\mathbf{F}(\mathbf{p})$, respectively. The representations of $\mathbf{f}(\mathbf{p})$ in two different coordinate frames, labelled $\mathcal{A}$ and $\mathcal{B}$, will be indicated as $[\mathbf{f}(\mathbf{p})]_{\mathcal{A}}$ and $[\mathbf{f}(\mathbf{p})]_{\mathcal{B}}$, respectively, with a similar notation for the representations of $\mathbf{F}(\mathbf{p})$. Moreover, let the two frames differ both in the location of their origins and in their orientations. Additionally, let the *proper orthogonal* matrix $[\mathbf{Q}]_{\mathcal{A}}$ denote the rotation of coordinate frame $\mathcal{A}$ into $\mathcal{B}$. Then, the scalar function $f(\mathbf{p})$ is said to be frame invariant, or invariant for brevity, if

$$f([\mathbf{p}]_{\mathcal{B}}) = f([\mathbf{p}]_{\mathcal{A}}) \tag{2.134}$$

Moreover, the vector quantity $\mathbf{f}$ is said to be invariant if

$$[\mathbf{f}]_{\mathcal{A}} = [\mathbf{Q}]_{\mathcal{A}}[\mathbf{f}]_{\mathcal{B}} \tag{2.135}$$

and finally, the matrix quantity $\mathbf{F}$ is said to be invariant if

$$[\mathbf{F}]_{\mathcal{A}} = [\mathbf{Q}]_{\mathcal{A}}[\mathbf{F}]_{\mathcal{B}}[\mathbf{Q}^T]_{\mathcal{A}} \tag{2.136}$$

Thus, the difference in origin location becomes irrelevant in this context, and hence, will no longer be considered. From the foregoing discussion, it is clear that the same vector quantity has different components in different coordinate frames; moreover, the same matrix quantity has different entries in different coordinate frames. However, certain scalar quantities associated with vectors, e.g., the inner product, and matrices, e.g., the matrix *moments*, to be defined presently, remain unchanged under a change of frame. Additionally, such vector operations as the cross product of two vectors are invariant. In fact, the scalar product of two vectors $\mathbf{a}$ and $\mathbf{b}$ remains unchanged under a change of frame, i.e.,

$$[\mathbf{a}]_{\mathcal{A}}^T[\mathbf{b}]_{\mathcal{A}} = [\mathbf{a}]_{\mathcal{B}}^T[\mathbf{b}]_{\mathcal{B}} \tag{2.137}$$

Additionally,

$$[\mathbf{a} \times \mathbf{b}]_{\mathcal{A}} = [\mathbf{Q}]_{\mathcal{A}} [\mathbf{a} \times \mathbf{b}]_{\mathcal{B}} \tag{2.138}$$

The kth *moment* of an $n \times n$ matrix $\mathbf{T}$, denoted by $\mathcal{I}_k$, is defined as (Leigh, 1968)

$$\mathcal{I}_k \equiv \text{tr}(\mathbf{T}^k), \quad k = 0, 1, \ldots \tag{2.139}$$

where $\mathcal{I}_0 = \text{tr}(\mathbf{1}) = n$. Now we have

Theorem 2.7.1 *If the trace of an $n \times n$ matrix $\mathbf{T}$ is invariant, then so are its moments.*

Proof: This is straightforward. Indeed, from Theorem 2.6.2, we have

$$[\mathbf{T}^k]_{\mathcal{B}} = [\mathbf{A}^{-1}]_{\mathcal{A}} [\mathbf{T}^k]_{\mathcal{A}} [\mathbf{A}]_{\mathcal{A}} \tag{2.140}$$

Now, if the trace of $\mathbf{T}$ is invariant, the invariance of the moments follows from that of the trace, q.e.d.

Furthermore,

Theorem 2.7.2 *An $n \times n$ matrix has only n linearly independent moments.*

Proof: Let the characteristic polynomial of $\mathbf{T}$ be

$$P(\lambda) = a_0 + a_1\lambda + \cdots + a_{n-1}\lambda^{n-1} + \lambda^n = 0 \tag{2.141}$$

Upon application of the Cayley-Hamilton Theorem, eq.(2.141) leads to

$$a_0\mathbf{1} + a_1\mathbf{T} + \cdots + a_{n-1}\mathbf{T}^{n-1} + \mathbf{T}^n = \mathbf{0} \tag{2.142}$$

where $\mathbf{1}$ denotes the $n \times n$ identity matrix.

Now, if we take the trace of both sides of eq.(2.142), and Definition (2.139) is recalled, one has

$$a_0\mathcal{I}_0 + a_1\mathcal{I}_1 + \cdots + a_{n-1}\mathcal{I}_{n-1} + \mathcal{I}_n = 0$$

from which it is apparent that $\mathcal{I}_n$ can be expressed as a linear combination of the first n moments of $\mathbf{T}$. By simple induction, one can likewise prove that the mth moment is dependent upon the first n moments if $m \geq n$, thereby completing the proof.

The vector invariants of an $n \times n$ matrix are its eigenvectors, which have physical significance in the case of symmetric matrices. The eigenvalues of these matrices are all real, its eigenvectors being also real and mutually orthogonal. Skew-symmetric matrices, in general, need not have either real eigenvalues or real eigenvectors. However, if we limit ourselves to 3×3 skew-symmetric matrices, exactly one of their eigenvalues, and its associated eigenvector, are both real. The eigenvalue of interest is 0, and the associated vector is the axial vector of the matrix under study.

Although in general, the n linearly independent moments of an $n \times n$ arbitrary matrix do not suffice to characterize the transformation represented by this matrix, they do suffice in the case of symmetric matrices. That is, if two symmetric $n \times n$ matrices have their first n moments identical, then they represent the same transformation, although in different coordinate frames.

In order to show that the moments do not fully characterize $n \times n$ matrices if these are not symmetric, we produce below a counterexample. Consider the matrix

$$\mathbf{A} = \begin{bmatrix} 1 & 1 \\ 0 & 1 \end{bmatrix}$$

Its two first moments are $\mathcal{I}_0 = 2$, $\mathcal{I}_1 = \operatorname{tr}(\mathbf{A}) = 2$, which happen to be the first two moments of the 2×2 identity matrix as well. However, while the identity matrix leaves all 2-dimensional vectors unchanged, the matrix $\mathbf{A}$ does not.

Now, if two symmetric matrices, say $\mathbf{A}$ and $\mathbf{B}$, represent the same transformation, they are related by a similarity transformation, i.e., a nonsingular matrix $\mathbf{T}$ exists such that

$$\mathbf{B} = \mathbf{T}^{-1}\mathbf{A}\mathbf{T}$$

Given $\mathbf{A}$ and $\mathbf{T}$, then, finding $\mathbf{B}$ is trivial, a similar statement holding if $\mathbf{B}$ and $\mathbf{T}$ are given; however, if $\mathbf{A}$ and $\mathbf{B}$ are given, finding $\mathbf{T}$ is more difficult. The latter problem occurs frequently in robotics in the context of *calibration*, to be discussed in Subsection 2.7.1.

Example 2.7.1 *Two symmetric matrices are displayed below. Find out whether they are related by a similarity transformation.*

$$\mathbf{A} = \begin{bmatrix} 1 & 0 & 1 \\ 0 & 1 & 0 \\ 1 & 0 & 2 \end{bmatrix}, \quad \mathbf{B} = \begin{bmatrix} 1 & 0 & 0 \\ 0 & 2 & -1 \\ 0 & -1 & 1 \end{bmatrix}$$

Solution: The traces of the two matrices are apparently identical, namely, 4. What we are left with is to determine if their second moments are also identical. To accomplish this, we need the square of the two matrices, from which it is straightforward to compute their traces. Thus, from

$$\mathbf{A}^2 = \begin{bmatrix} 2 & 0 & 3 \\ 0 & 1 & 0 \\ 3 & 0 & 5 \end{bmatrix}, \quad \mathbf{B}^2 = \begin{bmatrix} 1 & 0 & 0 \\ 0 & 5 & -3 \\ 0 & -3 & 2 \end{bmatrix}$$

we readily obtain

$$\operatorname{tr}(\mathbf{A}^2) = \operatorname{tr}(\mathbf{B}^2) = 8$$

and hence, the two matrices are related by a similarity transformation. Moreover, their third- and higher-order moments are identical as well. As a matter of verification, we compute their third moments below. From

$$\mathbf{A}^3 = \begin{bmatrix} 5 & 0 & 8 \\ 0 & 1 & 0 \\ 8 & 0 & 13 \end{bmatrix}, \quad \mathbf{B}^3 = \begin{bmatrix} 1 & 0 & 0 \\ 0 & 13 & -8 \\ 0 & -8 & 5 \end{bmatrix}$$

we obtain

$$\operatorname{tr}(\mathbf{A}^3) = \operatorname{tr}(\mathbf{B}^3) = 19$$

Example 2.7.2 *Same as Example 2.7.1, for the two matrices displayed below:*

$$\mathbf{A} = \begin{bmatrix} 1 & 0 & 2 \\ 0 & 1 & 0 \\ 2 & 0 & 0 \end{bmatrix}, \quad \mathbf{B} = \begin{bmatrix} 1 & 1 & 1 \\ 1 & 1 & 0 \\ 1 & 0 & 0 \end{bmatrix}$$

Solution: As in the previous example, the traces of these matrices are identical, i.e., 2. However, $\operatorname{tr}(\mathbf{A}^2) = 10$, while $\operatorname{tr}(\mathbf{B}^2) = 6$. We thus conclude that the two matrices cannot be related by a similarity transformation.

2.7.1 Applications to Redundant Sensing

A sensor, such as a camera or a range finder, is often mounted on a robotic end-effector to determine the *pose*—i.e., the position and orientation, as defined in Subsection 3.2.3—of an object. If redundant sensors are introduced, and we attach frames $\mathcal{A}$ and $\mathcal{B}$ to each of these, then each sensor can be used to determine the orientation of the end-effector with respect to a reference configuration. This is a simple task, for all that is needed is to measure the rotation $\mathbf{R}$ that each of the foregoing frames underwent from the reference configuration, in which these frames are denoted by $\mathcal{A}_0$ and $\mathcal{B}_0$, respectively. Let us assume that these measurements produce the orthogonal matrices $\mathbf{A}$ and $\mathbf{B}$, representing $\mathbf{R}$ in $\mathcal{A}$ and $\mathcal{B}$, respectively. With this information we would like to determine the relative orientation $\mathbf{Q}$ of frame $\mathcal{B}$ with respect to frame $\mathcal{A}$, a problem that is called here *instrument calibration.*

We thus have $\mathbf{A} \equiv [\mathbf{R}]_{\mathcal{A}}$ and $\mathbf{B} \equiv [\mathbf{R}]_{\mathcal{B}}$, and hence, the algebraic problem at hand consists in determining $[\mathbf{Q}]_{\mathcal{A}}$ or equivalently, $[\mathbf{Q}]_{\mathcal{B}}$. The former can be obtained from the similarity transformation of eq.(2.136), which leads to

$$\mathbf{A} = [\mathbf{Q}]_{\mathcal{A}} \mathbf{B} [\mathbf{Q}^T]_{\mathcal{A}}$$

or

$$\mathbf{A}[\mathbf{Q}]_{\mathcal{A}} = [\mathbf{Q}]_{\mathcal{A}} \mathbf{B}$$

This problem could be solved if we had three invariant vectors associated with each of the two matrices $\mathbf{A}$ and $\mathbf{B}$. Then, each corresponding pair of vectors of these triads would be related by eq.(2.135), thereby obtaining three such vector equations that should be sufficient to compute the nine components of the matrix $\mathbf{Q}$ rotating frame $\mathcal{A}$ into $\mathcal{B}$. However, since $\mathbf{A}$ and $\mathbf{B}$ are orthogonal matrices, they admit only one real invariant vector, namely, their axial vector, and we are short of two vector equations. We thus need two more invariant vectors, represented in both $\mathcal{A}$ and $\mathcal{B}$, to determine $\mathbf{Q}$. The obvious way of obtaining one additional vector in each frame is to take not one, but two measurements of the orientation of $\mathcal{A}_0$ and $\mathcal{B}_0$ with respect to $\mathcal{A}$ and $\mathcal{B}$, respectively. Let the matrices representing these orientations be given, in each of the two coordinate frames, by $\mathbf{A}_i$ and $\mathbf{B}_i$, for $i = 1, 2$. Moreover, let $\mathbf{a}_i$ and $\mathbf{b}_i$, for $i = 1, 2$, be the axial vectors of matrices $\mathbf{A}_i$ and $\mathbf{B}_i$, respectively.

Now we have two possibilities: (i) neither of $\mathbf{a}_1$ and $\mathbf{a}_2$ and, consequently, neither of $\mathbf{b}_1$ and $\mathbf{b}_2$, is zero; and (ii) at least one of $\mathbf{a}_1$ and $\mathbf{a}_2$, and consequently, the corresponding vector of the $\{\mathbf{b}_1, \mathbf{b}_2\}$ pair, vanishes. In the first case, nothing prevents us from computing a third vector of each set, namely,

$$\mathbf{a}_3 = \mathbf{a}_1 \times \mathbf{a}_2, \quad \mathbf{b}_3 = \mathbf{b}_1 \times \mathbf{b}_2 \tag{2.143}$$

In the second case, however, we have two more possibilities, i.e., the angle of rotation of that orthogonal matrix, $\mathbf{A}_1$ or $\mathbf{A}_2$, whose axial vector vanishes is either 0 or π. If the foregoing angle vanishes, then $\mathcal{A}$ underwent a pure translation from $\mathcal{A}_0$, the same holding, of course, for $\mathcal{B}$ and $\mathcal{B}_0$. This means

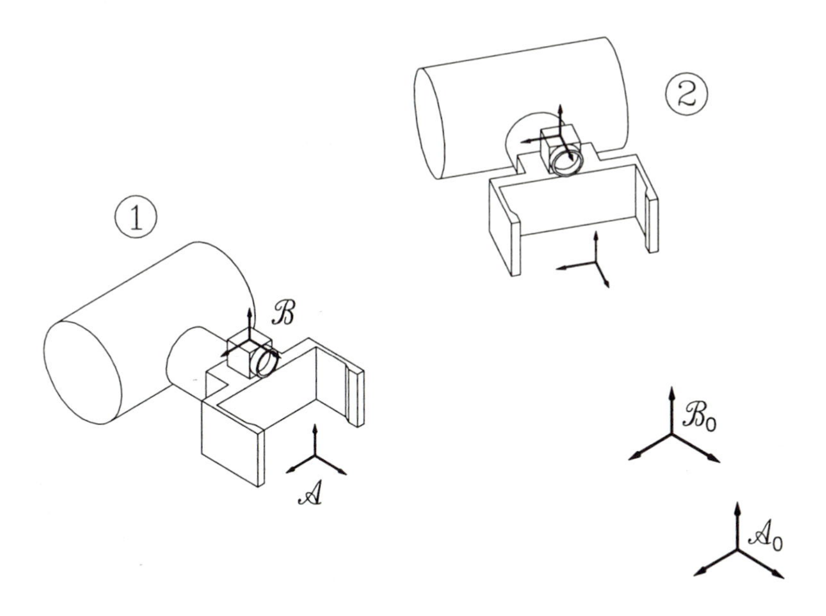

FIGURE 2.6. Measuring the orientation of a camera-fixed coordinate frame with respect to a frame fixed to a robotic end-effector.

that the corresponding measurement becomes useless for our purposes, and a new measurement is needed, involving a rotation. If, on the other hand, the same angle is π, then the associated rotation is symmetric and the unit vector $\mathbf{e}$ parallel to its axis can be determined from eq.(2.49) in both $\mathcal{A}$ and $\mathcal{B}$. This unit vector, then, would play the role of the vanishing axial vector, and we would thus end up, in any event, with two pairs of nonzero vectors, $\{\mathbf{a}_i\}_1^2$ and $\{\mathbf{b}_i\}_1^2$. As a consequence, we can always find two triads of nonzero vectors, $\{\mathbf{a}_i\}_1^3$ and $\{\mathbf{b}_i\}_1^3$, that are related by

$$\mathbf{a}_i = [\mathbf{Q}]_{\mathcal{A}}\,\mathbf{b}_i, \quad \text{for } i = 1, 2, 3 \tag{2.144}$$

The problem at hand now reduces to computing $[\mathbf{Q}]_{\mathcal{A}}$ from eq.(2.144). In order to perform this computation, we write the three foregoing equations in matrix form, namely,

$$\mathbf{E} = [\mathbf{Q}]_{\mathcal{A}}\,\mathbf{F} \tag{2.145}$$

with $\mathbf{E}$ and $\mathbf{F}$ defined as

$$\mathbf{E} \equiv [\mathbf{a}_1 \quad \mathbf{a}_2 \quad \mathbf{a}_3], \quad \mathbf{F} \equiv [\mathbf{b}_1 \quad \mathbf{b}_2 \quad \mathbf{b}_3] \tag{2.146}$$

Now, by virtue of the form in which the two vector triads were defined, none of the two above matrices is singular, and hence, we have

$$[\mathbf{Q}]_{\mathcal{A}} = \mathbf{E}\mathbf{F}^{-1} \tag{2.147}$$

Moreover, note that the inverse of $\mathbf{F}$ can be expressed in terms of its columns explicitly, without introducing components, if the concept of *reciprocal bases* is recalled (Brand, 1955). Thus,

$$\mathbf{F}^{-1} = \frac{1}{\Delta}\begin{bmatrix} (\mathbf{b}_2 \times \mathbf{b}_3)^T \\ (\mathbf{b}_3 \times \mathbf{b}_1)^T \\ (\mathbf{b}_1 \times \mathbf{b}_2)^T \end{bmatrix}, \quad \Delta \equiv \mathbf{b}_1 \times \mathbf{b}_2 \cdot \mathbf{b}_3 \tag{2.148}$$

Therefore,

$$[\mathbf{Q}]_{\mathcal{A}} = \frac{1}{\Delta}[\mathbf{a}_1(\mathbf{b}_2 \times \mathbf{b}_3)^T + \mathbf{a}_2(\mathbf{b}_3 \times \mathbf{b}_1)^T + \mathbf{a}_3(\mathbf{b}_1 \times \mathbf{b}_2)^T] \tag{2.149}$$

thereby completing the computation of $[\mathbf{Q}]_{\mathcal{A}}$ *directly and with simple vector operations.*

Example 2.7.3 (Hand-Eye Calibration) *Determine the relative orientation of a frame $\mathcal{B}$ attached to a camera mounted on a robot end-effector, with respect to a frame $\mathcal{A}$ fixed to the latter, as shown in Fig. 2.6. It is assumed that two measurements of the orientation of the two frames with respect to frames $\mathcal{A}_0$ and $\mathcal{B}_0$ in the reference configuration of the end-effector are available. These measurements produce the orientation matrices $\mathbf{A}_i$ of the frame fixed to the camera and $\mathbf{B}_i$ of the frame fixed to the end-effector, for $i = 1, 2$. The numerical data of this example are given below:*

$$\mathbf{A}_1 = \begin{bmatrix} -0.92592593 & -0.37037037 & -0.07407407 \\ 0.28148148 & -0.80740741 & 0.51851852 \\ -0.25185185 & 0.45925926 & 0.85185185 \end{bmatrix}$$

$$\mathbf{A}_2 = \begin{bmatrix} -0.83134406 & 0.02335236 & -0.55526725 \\ -0.52153607 & 0.31240270 & 0.79398028 \\ 0.19200830 & 0.94969269 & -0.24753503 \end{bmatrix}$$

$$\mathbf{B}_1 = \begin{bmatrix} -0.90268482 & 0.10343126 & -0.41768659 \\ 0.38511568 & 0.62720266 & -0.67698060 \\ 0.19195318 & -0.77195777 & -0.60599932 \end{bmatrix}$$

$$\mathbf{B}_2 = \begin{bmatrix} -0.73851280 & -0.54317226 & 0.39945305 \\ -0.45524951 & 0.83872293 & 0.29881721 \\ -0.49733966 & 0.03882952 & -0.86668653 \end{bmatrix}$$

Solution: Shiu and Ahmad (1987) formulated this problem in the form of a matrix linear homogeneous equation, while Chou and Kamel (1988) solved the same problem using quaternions and very cumbersome numerical methods that involve singular-value computations; the latter require an iterative procedure within a Newton-Raphson method, itself iterative, for nonlinear-equation solving. Other attempts to solve the same problem have been reported in the literature, but these also resorted to extremely complicated numerical procedures for nonlinear-equation solving (Chou and Kamel, 1991). More recently, Horaud and Dornaika (1995) proposed a more concise method based on quaternions, a.k.a. Euler-Rodrigues parameters, that nevertheless is computationally costlier than the method we use here. The approach outlined in this subsection is essentially the same as that proposed earlier (Angeles, 1989), although here we have adopted a simpler procedure than that of the foregoing reference.

First, the vector of matrix $\mathbf{A}_i$, represented by $\mathbf{a}_i$, and the vector of matrix $\mathbf{B}_i$, represented by $\mathbf{b}_i$, for $i = 1, 2$, are computed from simple differences of the off-diagonal entries of the foregoing matrices, followed by a division by 2 of all the entries thus resulting, which yields

$$\mathbf{a}_1 = \begin{bmatrix} -0.02962963 \\ 0.08888889 \\ 0.32592593 \end{bmatrix}, \quad \mathbf{a}_2 = \begin{bmatrix} 0.07784121 \\ -0.37363778 \\ -0.27244422 \end{bmatrix}$$

$$\mathbf{b}_1 = \begin{bmatrix} -0.04748859 \\ -0.30481989 \\ 0.14084221 \end{bmatrix}, \quad \mathbf{b}_2 = \begin{bmatrix} -0.12999385 \\ 0.44869636 \\ 0.04396138 \end{bmatrix}$$

In the calculations below, 16 digits were used, but only eight are displayed. Furthermore, with the foregoing vectors, we compute $\mathbf{a}_3$ and $\mathbf{b}_3$

from cross products, thus obtaining

$$\mathbf{a}_3 = \begin{bmatrix} 0.09756097 \\ 0.01730293 \\ 0.00415020 \end{bmatrix}$$

$$\mathbf{b}_3 = \begin{bmatrix} -0.07655343 \\ -0.01622096 \\ -0.06091842 \end{bmatrix}$$

Furthermore, Δ is obtained as

$$\Delta = 0.00983460$$

while the individual *rank-one matrices* inside the brackets of eq.(2.149) are calculated as

$$\mathbf{a}_1(\mathbf{b}_2 \times \mathbf{b}_3)^T = \begin{bmatrix} 0.00078822 & 0.00033435 & -0.00107955 \\ -0.00236467 & -0.00100306 & 0.00323866 \\ -0.00867044 & -0.00367788 & 0.01187508 \end{bmatrix}$$

$$\mathbf{a}_2(\mathbf{b}_3 \times \mathbf{b}_1)^T = \begin{bmatrix} -0.00162359 & 0.00106467 & 0.00175680 \\ 0.00779175 & -0.00510945 & -0.00843102 \\ 0.00568148 & -0.00372564 & -0.00614762 \end{bmatrix}$$

$$\mathbf{a}_3(\mathbf{b}_1 \times \mathbf{b}_2)^T = \begin{bmatrix} -0.00746863 & -0.00158253 & -0.00594326 \\ -0.00132460 & -0.00028067 & -0.00105407 \\ -0.00031771 & -0.00006732 & -0.00025282 \end{bmatrix}$$

whence $\mathbf{Q}$ in the $\mathcal{A}$ frame is readily obtained as

$$[\mathbf{Q}]_{\mathcal{A}} = \begin{bmatrix} -0.84436553 & -0.01865909 & -0.53545750 \\ 0.41714750 & -0.65007032 & -0.63514856 \\ -0.33622873 & -0.75964911 & 0.55667078 \end{bmatrix}$$

thereby completing the desired computation.

3

Fundamentals of Rigid-Body Mechanics

3.1 Introduction

The purpose of this chapter is to lay down the foundations of the kinetostatics and dynamics of rigid bodies, as needed in the study of multibody mechanical systems. With this background, we study the kinetostatics and dynamics of robotic manipulators of the serial type in Chapters 4 and 6, respectively, while devoting Chapter 5 to the study of trajectory planning. The latter requires, additionally, the background of Chapter 4. A special feature of this chapter is the study of the relations between the angular velocity of a rigid body and the time-rates of change of the various sets of rotation invariants introduced in Chapter 2. Similar relations between the angular acceleration and the second time-derivatives of the rotation invariants are also recalled, the corresponding derivations being outlined in Appendix A.

Furthermore, an introduction to the very useful analysis tool known as *screw theory* (Roth, 1984) is included. In this context, the concepts of twist and wrench are introduced, that prove in subsequent chapters to be extremely useful in deriving the kinematic and static, i.e., the *kinetostatic*, relations among the various bodies of multibody mechanical systems.

3.2 General Rigid-Body Motion and Its Associated Screw

In this section, we analyze the general motion of a rigid body. Thus, let A and P be two points of the same rigid body $\mathcal{B}$, the former being a particular reference point, whereas the latter is an arbitrary point of $\mathcal{B}$. Moreover, the position vector of point A in the original configuration is $\mathbf{a}$, and the position vector of the same point in the displaced configuration, denoted by A', is $\mathbf{a}'$. Similarly, the position vector of point P in the original configuration is $\mathbf{p}$, while in the displaced configuration, denoted by P', its position vector is $\mathbf{p}'$. Furthermore, $\mathbf{p}'$ is to be determined, while $\mathbf{a}$, $\mathbf{a}'$, and $\mathbf{p}$ are given, along with the rotation matrix $\mathbf{Q}$. Vector $\mathbf{p} - \mathbf{a}$ can be considered to undergo a rotation $\mathbf{Q}$ about point A throughout the motion taking the body from the original to the final configuration. Since vector $\mathbf{p} - \mathbf{a}$ is mapped into $\mathbf{p}' - \mathbf{a}'$ under the above rotation, one can write

$$\mathbf{p}' - \mathbf{a}' = \mathbf{Q}(\mathbf{p} - \mathbf{a}) \tag{3.1}$$

and hence,

$$\mathbf{p}' = \mathbf{a}' + \mathbf{Q}(\mathbf{p} - \mathbf{a}) \tag{3.2}$$

which is the relationship sought. Moreover, let $\mathbf{d}_A$ and $\mathbf{d}_P$ denote the displacements of A and P, respectively, i.e.,

$$\mathbf{d}_A \equiv \mathbf{a}' - \mathbf{a}, \quad \mathbf{d}_P \equiv \mathbf{p}' - \mathbf{p} \tag{3.3}$$

From eqs.(3.2) and (3.3) one can readily obtain an expression for $\mathbf{d}_P$, namely,

$$\begin{aligned}\mathbf{d}_P &= \mathbf{a}' - \mathbf{p} + \mathbf{Q}(\mathbf{p} - \mathbf{a}) \\ &= \mathbf{a}' - \mathbf{a} - \mathbf{p} + \mathbf{Q}(\mathbf{p} - \mathbf{a}) + \mathbf{a} \qquad (3.4) \\ &= \mathbf{d}_A + (\mathbf{Q} - \mathbf{1})(\mathbf{p} - \mathbf{a}) \qquad (3.5)\end{aligned}$$

What eq.(3.5) states is that the displacement of an arbitrary point P of a rigid body whose position vector in an original configuration is $\mathbf{p}$ is determined by the displacement of one certain point A and the concomitant rotation $\mathbf{Q}$. Clearly, once the displacement of P is known, its position vector $\mathbf{p}'$ can be readily determined. An interesting result in connection with the foregoing discussion is summarized below:

Theorem 3.2.1 *The component of the displacements of all the points of a rigid body undergoing a general motion along the axis of the underlying rotation is a constant.*

Proof: Multiply both sides of eq.(3.5) by $\mathbf{e}^T$, the unit vector parallel to the axis of the rotation represented by $\mathbf{Q}$, thereby obtaining

$$\mathbf{e}^T\mathbf{d}_P = \mathbf{e}^T\mathbf{d}_A + \mathbf{e}^T(\mathbf{Q} - \mathbf{1})(\mathbf{p} - \mathbf{a})$$

Now, the second term of the right-hand side of the above equation vanishes because $\mathbf{Q}\mathbf{e} = \mathbf{e}$, and hence, $\mathbf{Q}^T\mathbf{e} = \mathbf{e}$, by hypothesis, the said equation thus leading to

$$\mathbf{e}^T\mathbf{d}_P = \mathbf{e}^T\mathbf{d}_A \equiv d_0 \tag{3.6}$$

thereby showing that the displacements of all points of the body have the same projection d_0 onto the axis of rotation, q.e.d.

As a consequence of the foregoing result, we have the classical *Mozzi-Chasles Theorem* (Mozzi, 1763; Chasles, 1830; Ceccarelli, 1995), namely,

Theorem 3.2.2 (Mozzi, 1763; Chasles, 1830) *Given a rigid body undergoing a general motion, a set of its points located on a line $\mathcal{L}$ undergo identical displacements of minimum magnitude. Moreover, line $\mathcal{L}$ and the minimum-magnitude displacement are parallel to the axis of the rotation involved.*

Proof: The proof is straightforward in light of Theorem 3.2.1, which allows us to express the displacement of an arbitrary point P as the sum of two orthogonal components, namely, one parallel to the axis of rotation, independent of P and denoted by $\mathbf{d}_{\parallel}$, and one perpendicular to this axis, denoted by $\mathbf{d}_{\perp}$, i.e.,

$$\mathbf{d}_P = \mathbf{d}_{\parallel} + \mathbf{d}_{\perp} \tag{3.7a}$$

where

$$\mathbf{d}_{\parallel} = \mathbf{e}\mathbf{e}^T\mathbf{d}_P = d_0\mathbf{e}, \quad \mathbf{d}_{\perp} = (\mathbf{1} - \mathbf{e}\mathbf{e}^T)\mathbf{d}_P \tag{3.7b}$$

and clearly, d_0 is a constant that is defined as in eq.(3.6), while $\mathbf{d}_{\parallel}$ and $\mathbf{d}_{\perp}$ are mutually orthogonal. Indeed,

$$\mathbf{d}_{\parallel} \cdot \mathbf{d}_{\perp} = d_0\mathbf{e}^T(\mathbf{1} - \mathbf{e}\mathbf{e}^T)\mathbf{d}_P = d_0(\mathbf{e}^T - \mathbf{e}^T)\mathbf{d}_P = 0$$

Now, by virtue of the orthogonality of the two components of $\mathbf{d}_P$, it is apparent that

$$\|\mathbf{d}_P\|^2 = \|\mathbf{d}_{\parallel}\|^2 + \|\mathbf{d}_{\perp}\|^2 = d_0^2 + \|\mathbf{d}_{\perp}\|^2$$

for the displacement $\mathbf{d}_P$ of any point of the body. Now, in order to minimize $\|\mathbf{d}_P\|$ we have to make $\|\mathbf{d}_{\perp}\|$, and hence, $\mathbf{d}_{\perp}$ itself, equal to zero, i.e., we must have $\mathbf{d}_P$ parallel to $\mathbf{e}$:

$$\mathbf{d}_P = \alpha\mathbf{e}$$

for a certain scalar α. That is, the displacements of minimum magnitude of the body under study are parallel to the axis of $\mathbf{Q}$, thereby proving the first part of the Mozzi-Chasles Theorem. The second part is also readily proven by noticing that if P^* is a point of minimum magnitude of position vector $\mathbf{p}^*$, its component perpendicular to the axis of rotation must vanish, and hence,

$$\begin{aligned}\mathbf{d}_{\perp}^* &\equiv (\mathbf{1} - \mathbf{e}\mathbf{e}^T)\mathbf{d}_{P^*} \\ &= (\mathbf{1} - \mathbf{e}\mathbf{e}^T)\left[\mathbf{d}_A + (\mathbf{1} - \mathbf{e}\mathbf{e}^T)(\mathbf{Q} - \mathbf{1})(\mathbf{p}^* - \mathbf{a})\right] = \mathbf{0}\end{aligned}$$

Upon expansion of the above expression for $\mathbf{d}_\perp^*$, the foregoing equation leads to

$$(\mathbf{1} - \mathbf{e}\mathbf{e}^T)\mathbf{d}_A + (\mathbf{Q} - \mathbf{1})(\mathbf{p}^* - \mathbf{a}) = \mathbf{0}$$

Now it is apparent that if we define a line $\mathcal{L}$ passing through P^* and parallel to $\mathbf{e}$, then the position vector $\mathbf{p}^* + \lambda\mathbf{e}$ of any of its points P satisfies the foregoing equation. As a consequence, all points of minimum magnitude lie in a line parallel to the axis of rotation of $\mathbf{Q}$, q.e.d.

An important implication of the foregoing theorem is that a rigid body can attain an arbitrary configuration from a given original one, following a screw-like motion of axis $\mathcal{L}$ and pitch p, the latter being defined presently. Thus, it seems appropriate to call $\mathcal{L}$ the *screw axis* of the rigid-body motion.

Note that d_0, as defined in eq.(3.6), is an invariant of the motion at hand. Thus, associated with a rigid-body motion, one can then define a *screw* of axis $\mathcal{L}$ and pitch p. Of course, the pitch is defined as

$$p \equiv \frac{d_0}{\phi} = \frac{\mathbf{d}_P^T\mathbf{e}}{\phi} \quad \text{or } p \equiv \frac{2\pi d_0}{\phi} \tag{3.8}$$

which has units of m/rad or correspondingly, of m/turn. Moreover, the angle ϕ of the rotation involved can be regarded as one more feature of this motion. This angle is, in fact, the *amplitude* associated with the said motion. We will come across screws in discussing velocities and forces acting on rigid bodies, along with their pitches and amplitudes. Thus, it is convenient to introduce this concept at this stage.

3.2.1 The Screw of a Rigid-Body Motion

The screw axis $\mathcal{L}$ is totally specified by a given point P_0 of $\mathcal{L}$ that can be defined, for example, as that lying closest to the origin, and a unit vector $\mathbf{e}$ defining its direction. Expressions for the position vector of P_0, $\mathbf{p}_0$, in terms of $\mathbf{a}$, $\mathbf{a}'$ and $\mathbf{Q}$, are derived below:

If P_0 is defined as above, i.e., as the point of $\mathcal{L}$ lying closest to the origin, then, obviously, $\mathbf{p}_0$ is perpendicular to $\mathbf{e}$, i.e.,

$$\mathbf{e}^T\mathbf{p}_0 = 0 \tag{3.9}$$

Moreover, the displacement $\mathbf{d}_0$ of P_0 is parallel to the vector of $\mathbf{Q}$, and hence, is identical to $\mathbf{d}_\parallel$ defined in eq.(3.7b), i.e., it satisfies

$$(\mathbf{Q} - \mathbf{1})\mathbf{d}_0 = \mathbf{0}$$

where $\mathbf{d}_0$ is given as in eq.(3.5), namely, as

$$\mathbf{d}_0 = \mathbf{d}_A + (\mathbf{Q} - \mathbf{1})(\mathbf{p}_0 - \mathbf{a}) \tag{3.10a}$$

Now, since $\mathbf{d}_0$ is identical to $\mathbf{d}_\parallel$, we have, from eq.(3.7b),

$$\mathbf{d}_A + (\mathbf{Q} - \mathbf{1})(\mathbf{p}_0 - \mathbf{a}) = \mathbf{d}_\parallel \equiv \mathbf{e}\mathbf{e}^T\mathbf{d}_0$$

But from Theorem 3.2.1,

$$\mathbf{e}^T\mathbf{d}_0 = \mathbf{e}^T\mathbf{d}_A$$

and so

$$\mathbf{d}_A + (\mathbf{Q} - \mathbf{1})(\mathbf{p}_0 - \mathbf{a}) = \mathbf{e}\mathbf{e}^T\mathbf{d}_A$$

or after rearranging terms,

$$(\mathbf{Q} - \mathbf{1})\mathbf{p}_0 = (\mathbf{Q} - \mathbf{1})\mathbf{a} - (\mathbf{1} - \mathbf{e}\mathbf{e}^T)\mathbf{d}_A \tag{3.10b}$$

Furthermore, in order to find an expression for $\mathbf{p}_0$, eq.(3.9) is adjoined to eq.(3.10b), thereby obtaining

$$\mathbf{A}\mathbf{p}_0 = \mathbf{b} \tag{3.11}$$

where $\mathbf{A}$ is a 4×3 matrix and $\mathbf{b}$ is a 4-dimensional vector, being given by

$$\mathbf{A} \equiv \begin{bmatrix} \mathbf{Q} - \mathbf{1} \\ \mathbf{e}^T \end{bmatrix}, \quad \mathbf{b} \equiv \begin{bmatrix} (\mathbf{Q} - \mathbf{1})\mathbf{a} - (\mathbf{1} - \mathbf{e}\mathbf{e}^T)\mathbf{d}_A \\ 0 \end{bmatrix} \tag{3.12}$$

Equation (3.11) cannot be solved for $\mathbf{p}_0$ directly, because $\mathbf{A}$ is not a square matrix. In fact, that equation represents an *overdetermined* system of four equations and three unknowns. Thus, in general, that system does not admit a solution. However, the four equations are compatible, and hence in this particular case, a solution of that equation, which turns out to be *unique*, can be determined. In fact, if both sides of eq.(3.11) are multiplied from the left by $\mathbf{A}^T$, we have

$$\mathbf{A}^T\mathbf{A}\mathbf{p}_0 = \mathbf{A}^T\mathbf{b} \tag{3.13}$$

Moreover, if the product $\mathbf{A}^T\mathbf{A}$, which is a 3×3 matrix, is invertible, then $\mathbf{p}_0$ can be computed from eq.(3.13). In fact, the said product is not only invertible, but also admits an inverse that is rather simple to derive, as shown below. Now the rotation matrix $\mathbf{Q}$ is recalled in terms of its *natural invariants*, namely, the unit vector $\mathbf{e}$ parallel to its axis of rotation and the angle of rotation ϕ about this axis, as given in eq.(2.48), reproduced below for quick reference:

$$\mathbf{Q} = \mathbf{e}\mathbf{e}^T + \cos\phi(\mathbf{1} - \mathbf{e}\mathbf{e}^T) + \sin\phi\mathbf{E}$$

where $\mathbf{1}$ represents the 3×3 identity matrix and $\mathbf{E}$ the *cross-product matrix* of $\mathbf{e}$, as introduced in eq.(2.37). Further, eq.(2.48) is substituted into eq.(3.12), which yields

$$\mathbf{A}^T\mathbf{A} = 2(1 - \cos\phi)\mathbf{1} - (1 - 2\cos\phi)\mathbf{e}\mathbf{e}^T \tag{3.14}$$

It is now apparent that the foregoing product is a linear combination of $\mathbf{1}$ and $\mathbf{e}\mathbf{e}^T$. This suggests that its inverse is very likely to be a linear combination of these two matrices as well. If this is in fact true, then one can write

$$(\mathbf{A}^T\mathbf{A})^{-1} = \alpha\mathbf{1} + \beta\mathbf{e}\mathbf{e}^T \tag{3.15}$$

coefficients α and β being determined from the condition that the product of $\mathbf{A}^T\mathbf{A}$ by its inverse should be $\mathbf{1}$, which yields

$$\alpha = \frac{1}{2(1-\cos\phi)}, \quad \beta = \frac{1-2\cos\phi}{2(1-\cos\phi)} \tag{3.16}$$

and hence,

$$(\mathbf{A}^T\mathbf{A})^{-1} = \frac{1}{2(1-\cos\phi)}\mathbf{1} + \frac{1-2\cos\phi}{2(1-\cos\phi)}\mathbf{e}\mathbf{e}^T \tag{3.17}$$

On the other hand,

$$\mathbf{A}^T\mathbf{b} = (\mathbf{Q}-\mathbf{1})^T[(\mathbf{Q}-\mathbf{1})\mathbf{a} - \mathbf{d}_A] \tag{3.18}$$

Upon solving eq.(3.13) for $\mathbf{p}_0$ and substituting relations (3.17) and (3.18) into the expression thus resulting, one finally obtains

$$\mathbf{p}_0 = \frac{(\mathbf{Q}-\mathbf{1})^T(\mathbf{Q}\mathbf{a}-\mathbf{a}')}{2(1-\cos\phi)}, \quad \text{for } \phi \neq 0 \tag{3.19}$$

We have thus defined a line $\mathcal{L}$ of the rigid body under study that is completely defined by its point P_0 of position vector $\mathbf{p}_0$ and a unit vector $\mathbf{e}$ determining its direction. Moreover, we have already defined the pitch of the associated motion, eq.(3.8). The line thus defined, along with the pitch, determines the screw of the motion under study.

3.2.2 The Plücker Coordinates of a Line

Alternatively, the screw axis, and any line for that matter, can be defined more conveniently by its *Plücker coordinates.* In motivating this concept, we recall the equation of a line $\mathcal{L}$ passing through two points P_1 and P_2 of position vectors $\mathbf{p}_1$ and $\mathbf{p}_2$, as shown in Fig. 3.1.

If point P lies in $\mathcal{L}$, then, it must be collinear with P_1 and P_2, a property that is expressed as

$$(\mathbf{p}_2 - \mathbf{p}_1) \times (\mathbf{p} - \mathbf{p}_1) = \mathbf{0}$$

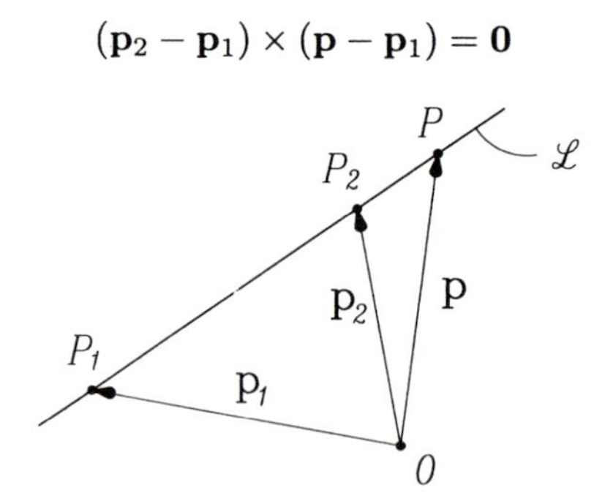

FIGURE 3.1. A line $\mathcal{L}$ passing through two points.

or upon expansion,

$$(\mathbf{p}_2 - \mathbf{p}_1) \times \mathbf{p} + \mathbf{p}_1 \times (\mathbf{p}_2 - \mathbf{p}_1) = \mathbf{0} \tag{3.20}$$

If we now introduce the cross-product matrices $\mathbf{P}_1$ and $\mathbf{P}_2$ of vectors $\mathbf{p}_1$ and $\mathbf{p}_2$ in the above equation, we have an alternative expression for the equation of the line, namely,

$$(\mathbf{P}_2 - \mathbf{P}_1)\mathbf{p} + \mathbf{p}_1 \times (\mathbf{p}_2 - \mathbf{p}_1) = \mathbf{0}$$

The above equation can be regarded as a linear equation in the homogeneous coordinates of point P, namely,

$$[\mathbf{P}_2 - \mathbf{P}_1 \quad \mathbf{p}_1 \times (\mathbf{p}_2 - \mathbf{p}_1)] \begin{bmatrix} \mathbf{p} \\ 1 \end{bmatrix} = \mathbf{0} \tag{3.21}$$

It is now apparent that the line is defined completely by two vectors, the difference $\mathbf{p}_2 - \mathbf{p}_1$, or its cross-product matrix for that matter, and the cross product $\mathbf{p}_1 \times (\mathbf{p}_2 - \mathbf{p}_1)$. We will thus define a 6-dimensional array $\gamma_{\mathcal{L}}$ containing these two vectors, namely,

$$\gamma_{\mathcal{L}} \equiv \begin{bmatrix} \mathbf{p}_2 - \mathbf{p}_1 \\ \mathbf{p}_1 \times (\mathbf{p}_2 - \mathbf{p}_1) \end{bmatrix} \tag{3.22}$$

whose six scalar entries are the Plücker coordinates of $\mathcal{L}$. Moreover, if we let

$$\mathbf{e} \equiv \frac{\mathbf{p}_2 - \mathbf{p}_1}{\|\mathbf{p}_2 - \mathbf{p}_1\|}, \qquad \mathbf{n} \equiv \mathbf{p}_1 \times \mathbf{e} \tag{3.23}$$

then we can write

$$\gamma_{\mathcal{L}} = \|\mathbf{p}_2 - \mathbf{p}_1\| \begin{bmatrix} \mathbf{e} \\ \mathbf{n} \end{bmatrix}$$

The six scalar entries of the above array are the *normalized Plücker coordinates* of $\mathcal{L}$. Vector $\mathbf{e}$ determines the direction of $\mathcal{L}$, while $\mathbf{n}$ determines its location; $\mathbf{n}$ can be interpreted as the moment of a unit force parallel to $\mathbf{e}$ and of line of action $\mathcal{L}$. Hence, $\mathbf{n}$ is called the *moment* of $\mathcal{L}$. Henceforth, only the normalized Plücker coordinates of lines will be used. For brevity, we will refer to these simply as the Plücker coordinates of the line under study. The Plücker coordinates thus defined will be thus stored in a *Plücker array* $\boldsymbol{\kappa}_{\mathcal{L}}$ in the form

$$\boldsymbol{\kappa} = \begin{bmatrix} \mathbf{e} \\ \mathbf{n} \end{bmatrix} \tag{3.24}$$

where for conciseness, we have dropped the subscript $\mathcal{L}$, while assuming that the line under discussion is self-evident.

Note, however, that the six components of the Plücker array, i.e., the *Plücker coordinates* of line $\mathcal{L}$, are not independent, for they obey

$$\mathbf{e} \cdot \mathbf{e} = 1, \quad \mathbf{n} \cdot \mathbf{e} = 0 \tag{3.25}$$

and hence, any line $\mathcal{L}$ has only four independent Plücker coordinates. In the foregoing paragraphs, we have talked about the Plücker array of a line, and not about the Plücker vector; the reason for this distinction is given below. The set of Plücker arrays is a clear example of an array of real numbers not constituting a vector space. What disables Plücker arrays from being vectors are the two constraints that their components must satisfy, namely, (*i*) the sum of the squares of the first three components of a Plücker array is unity, and (*ii*) the unit vector of a line is normal to the moment of the line. Nevertheless, we can perform with Plücker arrays certain operations that pertain to vectors, as long as we keep in mind the essential differences. For example, we can multiply Plücker arrays by matrices of the suitable dimension, with entries having appropriate units, as we will show presently.

It must be pointed out that a Plücker array is dependent upon the location of the point with respect to which the moment of the line is measured. Indeed, let $\boldsymbol{\kappa}_A$ and $\boldsymbol{\kappa}_B$ denote the Plücker arrays of the same line $\mathcal{L}$ when its moment is measured at points A and B, respectively. Moreover, this line passes through a point P of position vector $\mathbf{p}$ for a particular origin O. Now, let the moment of $\mathcal{L}$ with respect to A and B be denoted by $\mathbf{n}_A$ and $\mathbf{n}_B$, respectively, i.e.,

$$\mathbf{n}_A \equiv (\mathbf{p}-\mathbf{a}) \times \mathbf{e}, \quad \mathbf{n}_B \equiv (\mathbf{p}-\mathbf{b}) \times \mathbf{e} \tag{3.26}$$

and hence,

$$\boldsymbol{\kappa}_A \equiv \begin{bmatrix} \mathbf{e} \\ \mathbf{n}_A \end{bmatrix}, \quad \boldsymbol{\kappa}_B \equiv \begin{bmatrix} \mathbf{e} \\ \mathbf{n}_B \end{bmatrix} \tag{3.27}$$

Obviously,

$$\mathbf{n}_B - \mathbf{n}_A = (\mathbf{a}-\mathbf{b}) \times \mathbf{e} \tag{3.28}$$

i.e.,

$$\boldsymbol{\kappa}_B = \begin{bmatrix} \mathbf{e} \\ \mathbf{n}_A + (\mathbf{a}-\mathbf{b}) \times \mathbf{e} \end{bmatrix} \tag{3.29}$$

which can be rewritten as

$$\boldsymbol{\kappa}_B = \mathbf{U}\boldsymbol{\kappa}_A \tag{3.30}$$

with the 6×6 matrix $\mathbf{U}$ defined as

$$\mathbf{U} \equiv \begin{bmatrix} \mathbf{1} & \mathbf{O} \\ \mathbf{A}-\mathbf{B} & \mathbf{1} \end{bmatrix} \tag{3.31}$$

while $\mathbf{A}$ and $\mathbf{B}$ are, respectively, the cross-product matrices of vectors $\mathbf{a}$ and $\mathbf{b}$, and $\mathbf{O}$ denotes the 3×3 zero matrix. Given the lower-triangular structure of matrix $\mathbf{U}$, its determinant is simply the product of its diagonal entries, which are all unity. Hence,

$$\det(\mathbf{U}) = 1 \tag{3.32}$$

$\mathbf{U}$ thus belonging to the *unimodular group* of 6×6 matrices. These matrices are rather simple to invert. In fact, as one can readily prove,

$$\mathbf{U}^{-1} = \begin{bmatrix} \mathbf{1} & \mathbf{O} \\ \mathbf{B} - \mathbf{A} & \mathbf{1} \end{bmatrix} \tag{3.33}$$

Relation (3.30) can then be called the *Plücker-coordinate transfer formula.*

Note that upon multiplication of both sides of eq.(3.28) by $(\mathbf{a} - \mathbf{b})$,

$$(\mathbf{a} - \mathbf{b})^T \mathbf{n}_B = (\mathbf{a} - \mathbf{b})^T \mathbf{n}_A \tag{3.34}$$

and hence, the moments of the same line $\mathcal{L}$ with respect to two points are not independent, for they have the same component along the line joining the two points.

A special case of a line, of interest in kinematics, is a *line at infinity.* This is a line with undefined orientation, but with a defined direction of its moment; this moment is, moreover, *independent* of the point with respect to which it is measured. Very informally, the Plücker coordinates of a line at infinity can be derived from the general expression, eq.(3.24), if we rewrite it in the form

$$\boldsymbol{\kappa} = \|\mathbf{n}\| \begin{bmatrix} \mathbf{e}/\|\mathbf{n}\| \\ \mathbf{n}/\|\mathbf{n}\| \end{bmatrix}$$

where clearly $\mathbf{n}/\|\mathbf{n}\|$ is a unit vector; henceforth, this vector will be denoted by $\mathbf{f}$. Now let us take the limit of the above expression as P goes to infinity, i.e., when $\|\mathbf{p}\| \rightarrow \infty$, and consequently, as $\|\mathbf{n}\| \rightarrow \infty$. Thus,

$$\lim_{\|\mathbf{n}\| \rightarrow \infty} \boldsymbol{\kappa} = \left(\lim_{\|\mathbf{n}\| \rightarrow \infty} \|\mathbf{n}\| \right) \left(\lim_{\|\mathbf{n}\| \rightarrow \infty} \begin{bmatrix} \mathbf{e}/\|\mathbf{n}\| \\ \mathbf{f} \end{bmatrix} \right)$$

whence

$$\lim_{\|\mathbf{n}\| \rightarrow \infty} \boldsymbol{\kappa} = \left(\lim_{\|\mathbf{n}\| \rightarrow \infty} \|\mathbf{n}\| \right) \begin{bmatrix} \mathbf{0} \\ \mathbf{f} \end{bmatrix}$$

The 6-dimensional array appearing in the above equation is defined as the Plücker array of a line at infinity, $\boldsymbol{\kappa}_\infty$, namely,

$$\boldsymbol{\kappa}_\infty = \begin{bmatrix} \mathbf{0} \\ \mathbf{f} \end{bmatrix} \tag{3.35}$$

Note that a line at infinity of *unit moment* $\mathbf{f}$ can be thought of as being a line lying in a plane perpendicular to the unit vector $\mathbf{f}$, but otherwise with an indefinite location in the plane, except that it is an infinitely large distance from the origin. Thus, lines at infinity vary only in the orientation of the plane in which they lie.

3.2.3 The Pose of a Rigid Body

A possible form of describing a general rigid-body motion, then, is through a set of eight real numbers, namely, the six Plücker coordinates of its screw axis, its pitch, and its amplitude, i.e., its angle. Hence, *a rigid-body motion is fully described by six independent parameters.* Moreover, the pitch can attain values from $-\infty$ to $+\infty$. Alternatively, a rigid-body motion can be described by seven dependent parameters as follows: four invariants of the concomitant rotation—the linear invariants, the natural invariants, or the Euler-Rodrigues parameters, introduced in Section 2.3—and the three components of the displacement of an arbitrary point. Since those invariants are not independent, but subject to one constraint, this description consistently involves six independent parameters. Thus, let a particular point A of a rigid body undergoing a general motion from a reference configuration $\mathcal{C}_0$ of rotation $\mathbf{Q}$ have a position vector $\mathbf{a}$ and a displacement $\mathbf{d}_A$. The *pose array*, or simply the *pose*, $\mathbf{s}$ of the body in its current configuration $\mathcal{C}$, with respect to $\mathcal{C}_0$, can now be defined as a 7-dimensional array, namely,

$$\mathbf{s} \equiv \begin{bmatrix} \mathbf{q} \\ q_0 \\ \mathbf{d}_A \end{bmatrix} \tag{3.36}$$

where the 3-dimensional vector $\mathbf{q}$ and the scalar q_0 are *any* four invariants of $\mathbf{Q}$. For example, if these are the Euler-Rodrigues parameters, then

$$\mathbf{q} \equiv \sin(\frac{\phi}{2})\mathbf{e}, \quad q_0 \equiv \cos(\frac{\phi}{2})$$

If alternatively, we work with the linear invariants, then

$$\mathbf{q} \equiv (\sin\phi)\mathbf{e}, \quad q_0 \equiv \cos\phi$$

and of course, if we work instead with the natural invariants, then

$$\mathbf{q} \equiv \mathbf{e}, \quad q_0 \equiv \phi$$

In the first two cases, the constraint mentioned above is

$$\|\mathbf{q}\|^2 + q_0^2 = 1 \tag{3.37}$$

In the last case, the constraint is simply

$$\|\mathbf{e}\|^2 = 1 \tag{3.38}$$

An important problem in kinematics is the computation of the screw parameters, i.e., the components of $\mathbf{s}$, as given in eq.(3.36), from coordinate measurements over a certain finite set of points. From the foregoing discussion, it is clear that the computation of the attitude of a rigid body, given by matrix **Q** or its invariants, is crucial in solving this problem. Moreover, besides its theoretical importance, this problem, known as *pose estimation*, has also practical relevance. Shown in Fig. 3.2 is the *helmet-mounted display system* used in flight simulators. The helmet is supplied with a set of LEDs (light-emitting diodes) that emit infrared light signals at different frequencies each. These signals are then picked up by two cameras, from whose images the Cartesian coordinates of the LEDs centers are inferred. With these coordinates and knowledge of the LED pattern, the attitude of the pilot's head is determined from the rotation matrix **Q**. Moreover, with this information and that provided via sensors mounted on the lenses, the position of the center of the pupil of the pilot's eyes is then estimated. This position, then, indicates on which part of his or her visual field the pilot's eyes are focusing. In this way, a high-resolution graphics monitor synthesizes the image that the pilot would be viewing with a high level of detail. The rest of the visual field is rendered as a rather blurred image, in order to allocate computer resources where it really matters.

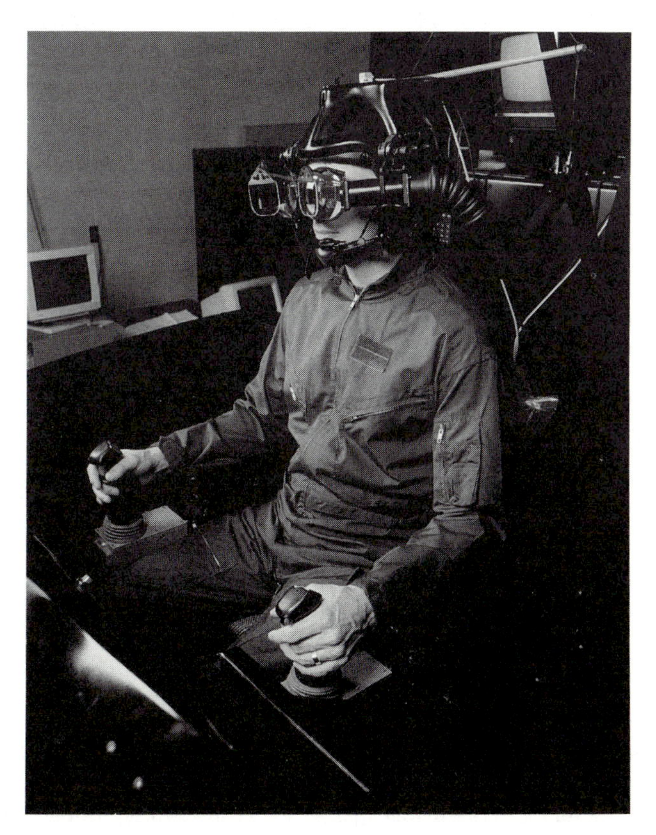

FIGURE 3.2. Helmet-mounted display system (courtesy of CAE Electronics Ltd., St.-Laurent, Quebec, Canada.)

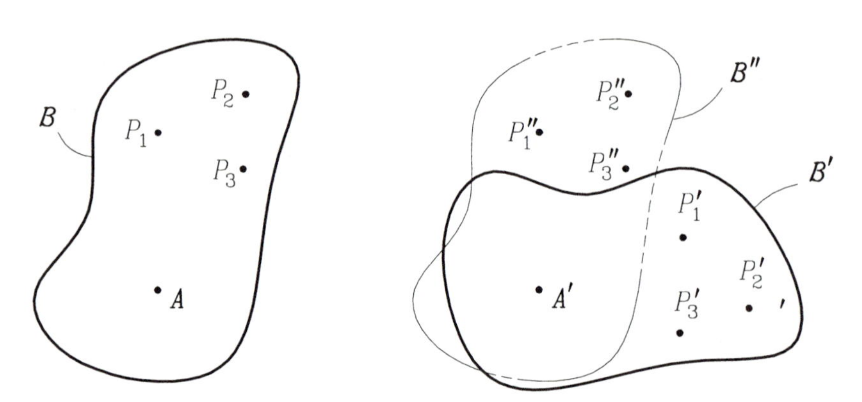

FIGURE 3.3. Decomposition of the displacement of a rigid body.

A straightforward method of computing the screw parameters consists of regarding the motion as follows: Choose a certain point A of the body, of position vector $\mathbf{a}$, and track it as the body moves to a displaced configuration, at which point A moves to A', of position vector $\mathbf{a}'$. Assume that the body reaches the displaced configuration $\mathcal{B}'$, passing through an intermediate one $\mathcal{B}''$, which is attained by pure translation. Next, configuration $\mathcal{B}'$ is reached by rotating the body about point A', as indicated in Fig. 3.3.

Matrix $\mathbf{Q}$ can now be readily determined. To do this, define three points of the body, P_1, P_2, and P_3, in such a way that the three vectors defined below are orthonormal and form a right-hand system:

$$\mathbf{e}_1 \equiv \overrightarrow{AP_1}, \quad \mathbf{e}_2 \equiv \overrightarrow{AP_2}, \quad \mathbf{e}_3 \equiv \overrightarrow{AP_3} \tag{3.39}$$

$$\mathbf{e}_i \cdot \mathbf{e}_j = \delta_{ij}, \quad i,j = 1,2,3, \quad \mathbf{e}_3 = \mathbf{e}_1 \times \mathbf{e}_2 \tag{3.40}$$

where δ_{ij} is the *Kronecker delta*, defined as 1 if $i = j$ and 0 otherwise. Now, let the set $\{\mathbf{e}_i\}_1^3$ be labelled $\{\mathbf{e}_i'\}_1^3$ and $\{\mathbf{e}_i''\}_1^3$ in configurations $\mathcal{B}'$ and $\mathcal{B}''$, respectively. Moreover, let q_{ij} denote the entries of the matrix representation of the rotation $\mathbf{Q}$ in a frame X, Y, Z with origin at A and such that the foregoing axes are parallel to vectors $\mathbf{e}_1$, $\mathbf{e}_2$, and $\mathbf{e}_3$, respectively. It is clear, from Definition 2.2.1, that

$$q_{ij} = \mathbf{e}_i \cdot \mathbf{e}_j' \tag{3.41}$$

i.e.,

$$[\mathbf{Q}] = \begin{bmatrix} \mathbf{e}_1 \cdot \mathbf{e}_1' & \mathbf{e}_1 \cdot \mathbf{e}_2' & \mathbf{e}_1 \cdot \mathbf{e}_3' \\ \mathbf{e}_2 \cdot \mathbf{e}_1' & \mathbf{e}_2 \cdot \mathbf{e}_2' & \mathbf{e}_2 \cdot \mathbf{e}_3' \\ \mathbf{e}_3 \cdot \mathbf{e}_1' & \mathbf{e}_3 \cdot \mathbf{e}_2' & \mathbf{e}_3 \cdot \mathbf{e}_3' \end{bmatrix} \tag{3.42}$$

Note that all $\mathbf{e}_i$ and $\mathbf{e}_i'$ appearing in eq.(3.42) must be represented in the same coordinate frame. Once $\mathbf{Q}$ is determined, computing the remaining screw parameters is straightforward. One can use, for example, eq.(3.19) to determine the point of the screw axis that lies closest to the origin, which would thus allow one to compute the Plücker coordinates of the screw axis.

3.3 Rotation of a Rigid Body About a Fixed Point

In this section, the motion of a rigid body having a point fixed is analyzed. This motion is fully described by a rotation matrix $\mathbf{Q}$ that is proper orthogonal. Now, $\mathbf{Q}$ will be assumed to be a smooth function of time, and hence, the position vector of a point P in an original configuration, denoted here by $\mathbf{p}_0$, is mapped smoothly into a new vector $\mathbf{p}(t)$, namely,

$$\mathbf{p}(t) = \mathbf{Q}(t)\mathbf{p}_0 \tag{3.43}$$

The velocity of P is computed by differentiating both sides of eq.(3.43) with respect to time, thus obtaining

$$\dot{\mathbf{p}}(t) = \dot{\mathbf{Q}}(t)\mathbf{p}_0 \tag{3.44}$$

which is not a very useful expression, because it requires knowledge of the original position of P. A more useful expression can be derived if eq.(3.43) is solved for $\mathbf{p}_0$ and the expression thus resulting is substituted into eq.(3.44), which yields

$$\dot{\mathbf{p}} = \dot{\mathbf{Q}}\mathbf{Q}^T\mathbf{p} \tag{3.45}$$

where the argument t has been dropped because all quantities are now time-varying, and hence, this argument is self-evident. The product $\dot{\mathbf{Q}}\mathbf{Q}^T$ is known as the *angular-velocity matrix* of the rigid-body motion and is denoted by $\mathbf{\Omega}$, i.e.,

$$\mathbf{\Omega} \equiv \dot{\mathbf{Q}}\mathbf{Q}^T \tag{3.46}$$

As a consequence of the orthogonality of $\mathbf{Q}$, one has a basic result, namely,

Theorem 3.3.1 *The angular-velocity matrix is skew symmetric.*

In order to derive the *angular-velocity vector* of a rigid-body motion, we recall the concept of *axial vector*, or simply *vector*, of a 3×3 matrix. Thus, the angular-velocity vector $\boldsymbol{\omega}$ of the rigid-body motion under study is defined as the vector of $\mathbf{\Omega}$, i.e.,

$$\boldsymbol{\omega} \equiv \mathrm{vect}(\mathbf{\Omega}) \tag{3.47}$$

and hence, eq.(3.45) can be written as

$$\dot{\mathbf{p}} = \mathbf{\Omega}\mathbf{p} = \boldsymbol{\omega} \times \mathbf{p} \tag{3.48}$$

from which it is apparent that *the velocity of any point P of a body moving with a point O fixed is perpendicular to line OP.*

3.4 General Instantaneous Motion of a Rigid Body

If a rigid body now undergoes the most general motion, none of its points remains fixed, and the position vector of any of these, P, in a displaced configuration is given by eq.(3.2). Let $\mathbf{a}_0$ and $\mathbf{p}_0$ denote the position vectors of points A and P of Section 3.2, respectively, in the reference configuration $\mathcal{C}_0$, $\mathbf{a}(t)$ and $\mathbf{p}(t)$ being the position vectors of the same points in the displaced configuration $\mathcal{C}$. Moreover, if $\mathbf{Q}(t)$ denotes the rotation matrix, then

$$\mathbf{p}(t) = \mathbf{a}(t) + \mathbf{Q}(t)(\mathbf{p}_0 - \mathbf{a}_0) \tag{3.49}$$

Now, the velocity of P is computed by differentiating both sides of eq.(3.49) with respect to time, thus obtaining

$$\dot{\mathbf{p}}(t) = \dot{\mathbf{a}}(t) + \dot{\mathbf{Q}}(t)(\mathbf{p}_0 - \mathbf{a}_0) \tag{3.50}$$

which again, as expression (3.50), is not very useful, for it requires the values of the position vectors of A and P in the original configuration. However, if eq.(3.49) is solved for $\mathbf{p}_0 - \mathbf{a}_0$ and the expression thus resulting is substituted into eq.(3.50), we obtain

$$\dot{\mathbf{p}} = \dot{\mathbf{a}} + \mathbf{\Omega}(\mathbf{p} - \mathbf{a}) \tag{3.51}$$

or in terms of the angular-velocity vector,

$$\dot{\mathbf{p}} = \dot{\mathbf{a}} + \boldsymbol{\omega} \times (\mathbf{p} - \mathbf{a}) \tag{3.52}$$

where the argument t has been dropped for brevity but is implicit, since all variables of the foregoing equation are now functions of time. Furthermore, from eq.(3.52), it is apparent that the result below holds:

$$(\dot{\mathbf{p}} - \dot{\mathbf{a}}) \cdot (\mathbf{p} - \mathbf{a}) = 0 \tag{3.53}$$

which can be summarized as

Theorem 3.4.1 *The relative velocity of two points of the same rigid body is perpendicular to the line joining them.*

Moreover, similar to the outcome of Theorem 3.2.1, one now has an additional result that is derived upon dot-multiplying both sides of eq.(3.52) by $\boldsymbol{\omega}$, namely,

$$\boldsymbol{\omega} \cdot \dot{\mathbf{p}} = \boldsymbol{\omega} \cdot \dot{\mathbf{a}} = \text{constant}$$

and hence,

Corollary 3.4.1 *The projections of the velocities of all the points of a rigid body onto the angular-velocity vector are identical.*

Furthermore, similar to the Mozzi-Chasles Theorem, we have now

Theorem 3.4.2 *Given a rigid body under general motion, a set of its points located on a line $\mathcal{L}'$ undergoes the identical minimum-magnitude velocity $\mathbf{v}_0$ parallel to the angular velocity.*

Definition 3.4.1 *The line containing the points of a rigid body undergoing minimum-magnitude velocities is called the* instant screw axis *(ISA) of the body under the given motion.*

3.4.1 The Instant Screw of a Rigid-Body Motion

From Theorem 3.4.2, the instantaneous motion of a body is equivalent to that of the bolt of a screw of axis $\mathcal{L}'$, the ISA. Clearly, as the body moves, the ISA changes, and the motion of the body is called an *instantaneous screw*. Moreover, since $\mathbf{v}_0$ is parallel to $\boldsymbol{\omega}$, it can be written in the form

$$\mathbf{v}_0 = v_0 \frac{\boldsymbol{\omega}}{\|\boldsymbol{\omega}\|} \tag{3.54}$$

where v_0 is a scalar quantity denoting the signed magnitude of $\mathbf{v}_0$ and bears the sign of $\mathbf{v}_0 \cdot \boldsymbol{\omega}$. Furthermore, the pitch of the instantaneous screw, p', is defined as

$$p' \equiv \frac{v_0}{\|\boldsymbol{\omega}\|} \equiv \frac{\dot{\mathbf{p}} \cdot \boldsymbol{\omega}}{\|\boldsymbol{\omega}\|^2} \quad \text{or} \quad p' \equiv \frac{2\pi v_0}{\|\boldsymbol{\omega}\|} \tag{3.55}$$

which thus bears units of m/rad or correspondingly, of m/turn.

Again, the ISA $\mathcal{L}'$ can be specified uniquely through its Plücker coordinates, stored in the $\mathbf{p}_{\mathcal{L}'}$ array defined as

$$\mathbf{p}_{\mathcal{L}'} \equiv \begin{bmatrix} \mathbf{e}' \\ \mathbf{n}' \end{bmatrix} \tag{3.56}$$

where $\mathbf{e}'$ and $\mathbf{n}'$ are, respectively, the unit vector defining the direction of $\mathcal{L}'$ and its moment about the origin, i.e.,

$$\mathbf{e}' \equiv \frac{\boldsymbol{\omega}}{\|\boldsymbol{\omega}\|}, \qquad \mathbf{n}' \equiv \mathbf{p} \times \mathbf{e}' \tag{3.57}$$

$\mathbf{p}$ being the position vector of any point of the ISA. Clearly, $\mathbf{e}'$ is defined uniquely but becomes trivial when the rigid body instantaneously undergoes a pure translation, i.e., a motion during which, instantaneously, $\boldsymbol{\omega} = \mathbf{0}$. In this case, $\mathbf{e}'$ is defined as the unit vector parallel to the associated displacement field. Thus, an instantaneous rigid-body motion is defined by a line $\mathcal{L}'$, a pitch p', and an amplitude $\|\boldsymbol{\omega}\|$. Such a motion is, then, fully determined by six independent parameters, namely, the four independent Plücker coordinates of $\mathcal{L}'$, its pitch, and its amplitude. A line supplied with a pitch is, in general, called a *screw*; a screw supplied with an amplitude representing the magnitude of an angular velocity provides the representation of an instantaneous rigid-body motion that is sometimes called the *twist*, an item that will be discussed more in detail below.

Hence, the instantaneous screw is fully defined by six independent real numbers. Moreover, such as in the case of the screw motion, the pitch of the instantaneous screw can attain values from $-\infty$ to $+\infty$.

The ISA can be alternatively described in terms of the position vector $\mathbf{p}_0'$ of its point lying closest to the origin. Expressions for $\mathbf{p}_0'$ in terms of the position and the velocity of an arbitrary body-point and the angular velocity are derived below. To this end, we decompose $\dot{\mathbf{p}}$ into two orthogonal components, $\dot{\mathbf{p}}_\parallel$ and $\dot{\mathbf{p}}_\perp$, along and transverse to the angular-velocity vector, respectively. To this end, $\dot{\mathbf{a}}$ is first decomposed into two such orthogonal components, $\dot{\mathbf{a}}_\parallel$ and $\dot{\mathbf{a}}_\perp$, the former being parallel, the latter normal to the ISA, i.e.,

$$\dot{\mathbf{a}} \equiv \dot{\mathbf{a}}_\parallel + \dot{\mathbf{a}}_\perp \tag{3.58}$$

These orthogonal components are given as

$$\dot{\mathbf{a}}_\parallel \equiv \dot{\mathbf{a}} \cdot \boldsymbol{\omega} \frac{\boldsymbol{\omega}}{\|\boldsymbol{\omega}\|^2} \equiv \frac{\boldsymbol{\omega}\boldsymbol{\omega}^T}{\|\boldsymbol{\omega}\|^2}\dot{\mathbf{a}}, \quad \dot{\mathbf{a}}_\perp \equiv (\mathbf{1} - \frac{\boldsymbol{\omega}\boldsymbol{\omega}^T}{\|\boldsymbol{\omega}\|^2})\dot{\mathbf{a}} \equiv -\frac{1}{\|\boldsymbol{\omega}\|^2}\boldsymbol{\Omega}^2\dot{\mathbf{a}} \tag{3.59}$$

In the derivation of eq.(3.59) we have used the identity introduced in eq.(2.39), namely,

$$\boldsymbol{\Omega}^2 \equiv \boldsymbol{\omega}\boldsymbol{\omega}^T - \|\boldsymbol{\omega}\|^2\mathbf{1} \tag{3.60}$$

Upon substitution of eq.(3.59) into eq.(3.52), we obtain

$$\dot{\mathbf{p}} = \underbrace{\frac{\boldsymbol{\omega}\boldsymbol{\omega}^T}{\|\boldsymbol{\omega}\|^2}\dot{\mathbf{a}}}_{\dot{\mathbf{p}}_\parallel} \underbrace{- \frac{1}{\|\boldsymbol{\omega}\|^2}\boldsymbol{\Omega}^2\dot{\mathbf{a}} + \boldsymbol{\Omega}(\mathbf{p} - \mathbf{a})}_{\dot{\mathbf{p}}_\perp} \tag{3.61}$$

Of the three components of $\dot{\mathbf{p}}$, the first, henceforth referred to as its *axial component*, is parallel, the last two being normal to $\boldsymbol{\omega}$. The sum of the last two components is referred to as the *normal component* of $\dot{\mathbf{p}}$. From eq.(3.61) it is apparent that the axial component is independent of $\mathbf{p}$, while the normal component is a linear function of $\mathbf{p}$. An obvious question now arises: *For an arbitrary motion, is it possible to find a certain point of position vector* $\mathbf{p}$ *whose velocity normal component vanishes?* The vanishing of the normal component obviously implies the minimization of the magnitude of $\dot{\mathbf{p}}$. The condition under which this happens can now be written as

$$\dot{\mathbf{p}}_\perp = \mathbf{0}$$

or

$$\boldsymbol{\Omega}(\mathbf{p} - \mathbf{a}) - \frac{1}{\|\boldsymbol{\omega}\|^2}\boldsymbol{\Omega}^2\,\dot{\mathbf{a}} = \mathbf{0} \tag{3.62}$$

which can be further expressed as a vector equation linear in $\mathbf{p}$, namely,

$$\mathbf{\Omega p} = \mathbf{\Omega}(\mathbf{a} + \frac{1}{\|\boldsymbol{\omega}\|^2}\mathbf{\Omega}\dot{\mathbf{a}}) \tag{3.63}$$

or

$$\mathbf{\Omega}(\mathbf{p} - \mathbf{r}) = \mathbf{0} \tag{3.64a}$$

with $\mathbf{r}$ defined as

$$\mathbf{r} \equiv \mathbf{a} + \frac{1}{\|\boldsymbol{\omega}\|^2}\mathbf{\Omega}\dot{\mathbf{a}} \tag{3.64b}$$

and hence, a possible solution of the foregoing problem is

$$\mathbf{p} = \mathbf{r} = \mathbf{a} + \frac{1}{\|\boldsymbol{\omega}\|^2}\mathbf{\Omega}\dot{\mathbf{a}} \tag{3.65}$$

However, this solution is not unique, for eq.(3.64a) does not require that $\mathbf{p}-\mathbf{r}$ be zero, only that this difference lie in the nullspace of $\mathbf{\Omega}$, i.e., that $\mathbf{p}-\mathbf{r}$ be linearly dependent with $\boldsymbol{\omega}$. In other words, if a vector $\alpha\boldsymbol{\omega}$ is added to $\mathbf{p}$, then the sum also satisfies eq.(3.63). It is then apparent that eq.(3.63) does not determine a single point whose normal velocity component vanishes but a set of points lying on the ISA, and thus, other solutions are possible. For example, we can find the point of the ISA lying closest to the origin. To this end, let $\mathbf{p}_0'$ be the position vector of that point. This vector is obviously perpendicular to $\boldsymbol{\omega}$, i.e.,

$$\boldsymbol{\omega}^T\mathbf{p}_0' = 0 \tag{3.66}$$

Next, eq.(3.63) is rewritten for $\mathbf{p}_0'$, and eq.(3.66) is adjoined to it, thereby deriving an expanded linear system of equations, namely,

$$\mathbf{A}\mathbf{p}_0' = \mathbf{b} \tag{3.67}$$

where $\mathbf{A}$ is a 4×3 matrix and $\mathbf{b}$ is a 4-dimensional vector, both of which are given below:

$$\mathbf{A} \equiv \begin{bmatrix} \mathbf{\Omega} \\ \boldsymbol{\omega}^T \end{bmatrix}, \quad \mathbf{b} \equiv \begin{bmatrix} \mathbf{\Omega a} + (1/\|\boldsymbol{\omega}\|^2)\mathbf{\Omega}^2\dot{\mathbf{a}} \\ 0 \end{bmatrix} \tag{3.68}$$

This system is of the same nature as that appearing in eq.(3.11), and hence, it can be solved for $\mathbf{p}_0'$ following the same procedure. Thus, both sides of eq.(3.67) are multiplied from the left by $\mathbf{A}^T$, thereby obtaining

$$\mathbf{A}^T\mathbf{A}\mathbf{p}_0' = \mathbf{A}^T\mathbf{b} \tag{3.69}$$

where

$$\mathbf{A}^T\mathbf{A} = \mathbf{\Omega}^T\mathbf{\Omega} + \boldsymbol{\omega}\boldsymbol{\omega}^T = -\mathbf{\Omega}^2 + \boldsymbol{\omega}\boldsymbol{\omega}^T \tag{3.70}$$

Moreover, from eq.(3.60), the rightmost side of the foregoing relation becomes $\|\boldsymbol{\omega}\|^2\mathbf{1}$, and hence, the matrix coefficient of the left-hand side of eq.(3.69) and the right-hand side of the same equation reduce, respectively, to

$$\mathbf{A}^T\mathbf{A} = \|\boldsymbol{\omega}\|^2\mathbf{1}, \quad \mathbf{A}^T\mathbf{b} = \boldsymbol{\Omega}(\dot{\mathbf{a}} - \boldsymbol{\Omega}\mathbf{a}) \tag{3.71}$$

Upon substitution of eq.(3.71) into eq.(3.69) and further solving for $\mathbf{p}_0'$, the desired expression is derived:

$$\mathbf{p}_0' = \frac{\boldsymbol{\Omega}(\dot{\mathbf{a}} - \boldsymbol{\Omega}\mathbf{a})}{\|\boldsymbol{\omega}\|^2} \equiv \frac{\boldsymbol{\omega} \times (\dot{\mathbf{a}} - \boldsymbol{\omega} \times \mathbf{a})}{\|\boldsymbol{\omega}\|^2} \tag{3.72}$$

Thus, the instantaneous screw is fully defined by an alternative set of six independent scalars, namely, the three components of its angular velocity $\boldsymbol{\omega}$ and the three components of the velocity of an arbitrary body point A, denoted by $\dot{\mathbf{a}}$. As in the case of the screw motion, we can also represent the instantaneous screw by a line and two additional parameters, as we explain below.

3.4.2 *The Twist of a Rigid Body*

A line, as we saw earlier, is fully defined by its 6-dimensional Plücker array, which contains only four independent components. Now, if a pitch p is added as a fifth feature to the line or correspondingly, to its Plücker array, we obtain a screw $\mathbf{s}$, namely,

$$\mathbf{s} \equiv \begin{bmatrix} \mathbf{e} \\ \mathbf{p} \times \mathbf{e} + p\mathbf{e} \end{bmatrix} \tag{3.73}$$

An *amplitude* is any scalar A multiplying the foregoing screw. The amplitude produces a twist or a *wrench*, to be discussed presently, depending on its units. The twist or the wrench thus defined can be regarded as an eight-parameter array. These eight parameters, of which only six are independent, are the amplitude, the pitch, and the six Plücker coordinates of the associated line. Clearly, a twist or a wrench is defined completely by six independent real numbers. More generally, a twist can be regarded as a 6-dimensional array defining completely the velocity field of a rigid body, and it comprises the three components of the angular velocity and the three components of the velocity of any of the points of the body.

Below we elaborate on the foregoing concepts. Upon multiplication of the screw appearing in eq.(3.73) by the amplitude A representing the magnitude of an angular velocity, we obtain a twist $\mathbf{t}$, namely,

$$\mathbf{t} \equiv \begin{bmatrix} A\mathbf{e} \\ \mathbf{p} \times (A\mathbf{e}) + p(A\mathbf{e}) \end{bmatrix}$$

where the product $A\mathbf{e}$ can be readily identified as the angular velocity $\boldsymbol{\omega}$ parallel to vector $\mathbf{e}$, of magnitude A. Moreover, the lower part of $\mathbf{t}$ can be readily identified with the velocity of a point of a rigid body. Indeed, if we regard the line $\mathcal{L}$ and point O as sets of points of a rigid body $\mathcal{B}$ moving with an angular velocity $\boldsymbol{\omega}$ and such that point P moves with a velocity $p\boldsymbol{\omega}$ parallel to the angular velocity, then the lower vector of $\mathbf{t}$, denoted by $\mathbf{v}$, represents the velocity of point O, i.e.,

$$\mathbf{v} = -\boldsymbol{\omega} \times \mathbf{p} + p\boldsymbol{\omega}$$

We can thus express the twist $\mathbf{t}$ as

$$\mathbf{t} \equiv \begin{bmatrix} \boldsymbol{\omega} \\ \mathbf{v} \end{bmatrix} \tag{3.74}$$

A special case of great interest in kinematics is the screw of infinitely large pitch. The form of this screw is derived, very informally, by taking the limit of expression (3.73) as $p \to \infty$, namely,

$$\lim_{p\to\infty} \begin{bmatrix} \mathbf{e} \\ \mathbf{p} \times \mathbf{e} + p\mathbf{e} \end{bmatrix} \equiv \lim_{p\to\infty} \left(p \begin{bmatrix} \mathbf{e}/p \\ (\mathbf{p} \times \mathbf{e})/p + \mathbf{e} \end{bmatrix} \right)$$

which readily leads to

$$\lim_{p\to\infty} \begin{bmatrix} \mathbf{e} \\ \mathbf{p} \times \mathbf{e} + p\mathbf{e} \end{bmatrix} = \left(\lim_{p\to\infty} p \right) \begin{bmatrix} \mathbf{0} \\ \mathbf{e} \end{bmatrix}$$

The *screw of infinite pitch* $\mathbf{s}_\infty$ is defined as the 6-dimensional array appearing in the above equation, namely,

$$\mathbf{s}_\infty \equiv \begin{bmatrix} \mathbf{0} \\ \mathbf{e} \end{bmatrix} \tag{3.75}$$

Note that this screw array is identical to the Plücker array of the line at infinity lying in a plane of unit normal $\mathbf{e}$.

The twist array, as defined in eq.(3.74), with $\boldsymbol{\omega}$ on top, represents the *ray coordinates* of the twist. An exchange of the order of the two Cartesian vectors of this array, in turn, gives rise to the *axis coordinates* of the twist.

The foregoing twist was also termed *motor* by Everett (1875). As Phillips (1990) points out, the word motor is an abbreviation of *mo*ment and vec*tor*. An extensive introduction into motor algebra was published by von Mises (1924), a work that is now available in English (von Mises, 1996). Roth (1984), in turn, provided a summary of these concepts, as applicable to robotics. The foregoing array goes also by other names, such as the German *Kinemate*.

The relationships between the angular-velocity vector and the time derivatives of the invariants of the associated rotation are linear. Indeed, let the three sets of four invariants of rotation, namely, the natural invariants, the linear invariants, and the Euler-Rodrigues parameters be grouped in the 4-dimensional arrays $\boldsymbol{\nu}$, $\boldsymbol{\lambda}$, and $\boldsymbol{\eta}$, respectively, i.e.,

$$\boldsymbol{\nu} \equiv \begin{bmatrix} \mathbf{e} \\ \phi \end{bmatrix}, \quad \boldsymbol{\lambda} \equiv \begin{bmatrix} (\sin\phi)\mathbf{e} \\ \cos\phi \end{bmatrix}, \quad \boldsymbol{\eta} \equiv \begin{bmatrix} [\sin(\phi/2)]\mathbf{e} \\ \cos(\phi/2) \end{bmatrix} \tag{3.76}$$

We then have the linear relations derived in full detail elsewhere (Angeles, 1988), and outlined in Appendix A for quick reference, namely,

$$\dot{\boldsymbol{\nu}} = \mathbf{N}\boldsymbol{\omega}, \quad \dot{\boldsymbol{\lambda}} = \mathbf{L}\boldsymbol{\omega}, \quad \dot{\boldsymbol{\eta}} = \mathbf{H}\boldsymbol{\omega} \tag{3.77a}$$

with $\mathbf{N}$, $\mathbf{L}$, and $\mathbf{H}$ defined as

$$\mathbf{N} \equiv \begin{bmatrix} [\sin\phi/(2(1-\cos\phi))](\mathbf{1}-\mathbf{e}\mathbf{e}^T) - (1/2)\mathbf{E} \\ \mathbf{e}^T \end{bmatrix}, \tag{3.77b}$$

$$\mathbf{L} \equiv \begin{bmatrix} (1/2)[\mathrm{tr}(\mathbf{Q})\mathbf{1} - \mathbf{Q}] \\ -(\sin\phi)\mathbf{e}^T \end{bmatrix}, \tag{3.77c}$$

$$\mathbf{H} \equiv \frac{1}{2}\begin{bmatrix} \cos(\phi/2)\mathbf{1} - \sin(\phi/2)\mathbf{E} \\ -\sin(\phi/2)\mathbf{e}^T \end{bmatrix} \tag{3.77d}$$

where, it is recalled, $\mathrm{tr}(\cdot)$ denotes the trace of its square matrix argument $(\cdot)$, i.e., the sum of the diagonal entries of that matrix.

The inverse relations of those shown in eqs.(3.77a) are to be derived by resorting to the approach introduced when solving eq.(3.67) for $\mathbf{p}_0'$, thereby obtaining

$$\boldsymbol{\omega} = \tilde{\mathbf{N}}\dot{\boldsymbol{\nu}} = \tilde{\mathbf{L}}\dot{\boldsymbol{\lambda}} = \tilde{\mathbf{H}}\dot{\boldsymbol{\eta}} \tag{3.78a}$$

the 3×4 matrices $\tilde{\mathbf{N}}$, $\tilde{\mathbf{L}}$, and $\tilde{\mathbf{H}}$ being defined below:

$$\tilde{\mathbf{N}} \equiv [\,(\sin\phi)\mathbf{1} + (1-\cos\phi)\mathbf{E} \quad \mathbf{e}\,], \tag{3.78b}$$

$$\tilde{\mathbf{L}} \equiv [\,\mathbf{1} + [(\sin\phi)/(1+\cos\phi)]\mathbf{E} \quad -[(\sin\phi)/(1+\cos\phi)]\mathbf{e}\,], \tag{3.78c}$$

$$\tilde{\mathbf{H}} \equiv 2\,[\,[\cos(\phi/2)]\mathbf{1} + [\sin(\phi/2)]\mathbf{E} \quad -[\sin(\phi/2)]\mathbf{e}\,] \tag{3.78d}$$

As a consequence, we have the following:

Caveat *The angular velocity vector is* **not** *a time-derivative, i.e., no Cartesian vector exists whose time-derivative is the angular-velocity vector.*

However, matrices $\mathbf{N}$, $\mathbf{L}$, and $\mathbf{H}$ of eqs.(3.77b–d) can be regarded as *integration factors* that yield time-derivatives.

Now we can write the relationship between the twist and the time-rate of change of the 7-dimensional pose array $\mathbf{s}$, namely,

$$\dot{\mathbf{s}} = \mathbf{T}\mathbf{t} \tag{3.79}$$

where

$$\mathbf{T} \equiv \begin{bmatrix} \mathbf{F} & \mathbf{O}_{43} \\ \mathbf{O} & \mathbf{1} \end{bmatrix} \tag{3.80}$$

in which $\mathbf{O}$ and $\mathbf{O}_{43}$ are the 3×3 and the 4×3 zero matrices, while $\mathbf{1}$ is the 3×3 identity matrix and $\mathbf{F}$ is, correspondingly, $\mathbf{N}$, $\mathbf{L}$, or $\mathbf{H}$, depending upon the invariant representation chosen for the rotation. The inverse relationship of eq.(3.79) takes the form

$$\mathbf{t} = \mathbf{S}\dot{\mathbf{s}} \tag{3.81a}$$

where

$$\mathbf{S} \equiv \begin{bmatrix} \widetilde{\mathbf{F}} & \mathbf{O} \\ \mathbf{O}_{34} & \mathbf{1} \end{bmatrix} \tag{3.81b}$$

in which $\mathbf{O}_{34}$ is the 3×4 zero matrix. Moreover, $\widetilde{\mathbf{F}}$ is one of $\widetilde{\mathbf{N}}$, $\widetilde{\mathbf{L}}$, or $\widetilde{\mathbf{H}}$, depending on the rotation representation adopted, namely, the natural invariants, the linear invariants, or the Euler-Rodrigues parameters, respectively.

A formula that relates the twist of the same rigid body at two different points is now derived. Let A and P be two arbitrary points of a rigid body. The twist at each of these points is defined as

$$\mathbf{t}_A = \begin{bmatrix} \boldsymbol{\omega} \\ \mathbf{v}_A \end{bmatrix}, \quad \mathbf{t}_P = \begin{bmatrix} \boldsymbol{\omega} \\ \mathbf{v}_P \end{bmatrix} \tag{3.82}$$

Moreover, eq.(3.52) can be rewritten as

$$\mathbf{v}_P = \mathbf{v}_A + (\mathbf{a} - \mathbf{p}) \times \boldsymbol{\omega} \tag{3.83}$$

Combining eq.(3.82) with eq.(3.83) yields

$$\mathbf{t}_P = \mathbf{U}\mathbf{t}_A \tag{3.84}$$

where

$$\mathbf{U} \equiv \begin{bmatrix} \mathbf{1} & \mathbf{O} \\ \mathbf{A} - \mathbf{P} & \mathbf{1} \end{bmatrix} \tag{3.85}$$

with the 6×6 matrix $\mathbf{U}$ defined as in eq.(3.31), while $\mathbf{A}$ and $\mathbf{P}$ denote the cross-product matrices of vectors $\mathbf{a}$ and $\mathbf{p}$, respectively. Thus, eq.(3.84) can be fairly called the *twist-transfer formula.*

3.5 Acceleration Analysis of Rigid-Body Motions

Upon differentiation of both sides of eq.(3.51) with respect to time, one obtains

$$\ddot{\mathbf{p}} = \ddot{\mathbf{a}} + \dot{\boldsymbol{\Omega}}(\mathbf{p} - \mathbf{a}) + \boldsymbol{\Omega}(\dot{\mathbf{p}} - \dot{\mathbf{a}}) \tag{3.86}$$

Now, eq.(3.51) is solved for $\dot{\mathbf{p}} - \dot{\mathbf{a}}$, and the expression thus resulting is substituted into eq.(3.86), thereby obtaining

$$\ddot{\mathbf{p}} = \ddot{\mathbf{a}} + (\dot{\boldsymbol{\Omega}} + \boldsymbol{\Omega}^2)(\mathbf{p} - \mathbf{a}) \tag{3.87}$$

where the matrix sum in parentheses is termed the *angular-acceleration matrix* of the rigid-body motion and is represented by $\mathbf{W}$, i.e.,

$$\mathbf{W} \equiv \dot{\boldsymbol{\Omega}} + \boldsymbol{\Omega}^2 \tag{3.88}$$

Clearly, the first term of the right-hand side of eq.(3.88) is skew-symmetric, whereas the second one is symmetric. Thus,

$$\text{vect}(\mathbf{W}) = \text{vect}(\dot{\mathbf{\Omega}}) = \dot{\boldsymbol{\omega}} \tag{3.89}$$

$\dot{\boldsymbol{\omega}}$ being termed the *angular-acceleration vector* of the rigid-body motion. We have now an interesting result, namely,

$$\begin{aligned}\text{tr}(\mathbf{W}) &= \text{tr}(\mathbf{\Omega}^2) = \text{tr}(-\|\boldsymbol{\omega}\|^2\mathbf{1} + \boldsymbol{\omega}\boldsymbol{\omega}^T) \\ &= -\|\boldsymbol{\omega}\|^2\text{tr}(\mathbf{1}) + \boldsymbol{\omega}\cdot\boldsymbol{\omega} = -2\|\boldsymbol{\omega}\|^2\end{aligned} \tag{3.90}$$

Moreover, eq.(3.87) can be written as

$$\ddot{\mathbf{p}} = \ddot{\mathbf{a}} + \dot{\boldsymbol{\omega}} \times (\mathbf{p} - \mathbf{a}) + \boldsymbol{\omega} \times [\boldsymbol{\omega} \times (\mathbf{p} - \mathbf{a})] \tag{3.91}$$

On the other hand, the time derivative of $\mathbf{t}$, henceforth referred to as the *twist rate*, is displayed below:

$$\dot{\mathbf{t}} \equiv \begin{bmatrix} \dot{\boldsymbol{\omega}} \\ \dot{\mathbf{v}} \end{bmatrix} \tag{3.92}$$

in which $\dot{\mathbf{v}}$ is the acceleration of a point of the body. The relationship between the twist rate and the second time derivative of the screw is derived by differentiation of both sides of eq.(3.79), which yields

$$\ddot{\mathbf{s}} = \mathbf{T}\dot{\mathbf{t}} + \dot{\mathbf{T}}\mathbf{t} \tag{3.93}$$

where

$$\dot{\mathbf{T}} \equiv \begin{bmatrix} \dot{\mathbf{F}} & \mathbf{O}_{43} \\ \mathbf{O} & \mathbf{O} \end{bmatrix} \tag{3.94}$$

and $\mathbf{F}$ is one of $\mathbf{N}$, $\mathbf{L}$, or $\mathbf{H}$, accordingly. The inverse relationship of eq.(3.93) is derived by differentiating both sides of eq.(3.81a) with respect to time, which yields

$$\dot{\mathbf{t}} = \mathbf{S}\ddot{\mathbf{s}} + \dot{\mathbf{S}}\dot{\mathbf{s}} \tag{3.95}$$

where

$$\dot{\mathbf{S}} = \begin{bmatrix} \dot{\overline{\mathbf{F}}} & \mathbf{O} \\ \mathbf{O}_{34} & \mathbf{O} \end{bmatrix} \tag{3.96}$$

with $\mathbf{O}$ and $\mathbf{O}_{34}$ already defined in eq.(3.81b) as the 3×3 and the 3×4 zero matrices, respectively, while $\dot{\overline{\mathbf{F}}}$ is one of $\dot{\overline{\mathbf{N}}}$, $\dot{\overline{\mathbf{L}}}$, or $\dot{\overline{\mathbf{H}}}$, according with the type of rotation representation at hand.

Before we take to differentiating the foregoing matrices, we introduce a few definitions: Let

$$\boldsymbol{\lambda} \equiv \begin{bmatrix} \mathbf{u} \\ u_0 \end{bmatrix}, \quad \boldsymbol{\eta} \equiv \begin{bmatrix} \mathbf{r} \\ r_0 \end{bmatrix} \tag{3.97a}$$

i.e.,

$$\mathbf{u} \equiv \sin\phi\mathbf{e}, \quad u_0 \equiv \cos\phi, \quad \mathbf{r} \equiv \sin(\frac{\phi}{2})\mathbf{e}, \quad r_0 \equiv \cos(\frac{\phi}{2}) \tag{3.97b}$$

Thus, the time derivatives sought take on the forms

$$\dot{\mathbf{N}} = \frac{1}{4(1-\cos\phi)}\begin{bmatrix}\mathbf{B}\\ \dot{\mathbf{e}}\end{bmatrix} \tag{3.98a}$$

$$\begin{aligned}\dot{\mathbf{L}} &= \begin{bmatrix}(1/2)[\mathbf{1}\mathrm{tr}(\dot{\mathbf{Q}}) - \dot{\mathbf{Q}}]\\ -(1/2)\boldsymbol{\omega}^T[\mathbf{1}\mathrm{tr}(\mathbf{Q}) - \mathbf{Q}^T]\end{bmatrix}\\ &= \begin{bmatrix}-(\boldsymbol{\omega}\cdot\mathbf{u})\mathbf{1} - (1/2)\boldsymbol{\Omega}\mathbf{Q}\\ -(1/2)\boldsymbol{\omega}^T[\mathbf{1}\mathrm{tr}(\mathbf{Q}) - \mathbf{Q}^T]\end{bmatrix}\end{aligned} \tag{3.98b}$$

$$\dot{\mathbf{H}} = \frac{1}{2}\begin{bmatrix}\dot{r}_0\mathbf{1} - \dot{\mathbf{R}}\\ -\dot{\mathbf{r}}^T\end{bmatrix} \tag{3.98c}$$

where we have used the identities below, which are derived in Appendix A.

$$\mathrm{tr}(\dot{\mathbf{Q}}) \equiv \mathrm{tr}(\boldsymbol{\Omega}\mathbf{Q}) \equiv -2\boldsymbol{\omega}^T\mathbf{u} \tag{3.98d}$$

Furthermore, $\mathbf{R}$ denotes the cross-product matrix of $\mathbf{r}$, and $\mathbf{B}$ is defined as

$$\begin{aligned}\mathbf{B} \equiv& -2(\mathbf{e}\cdot\boldsymbol{\omega})\mathbf{1} + 2(3-\cos\phi)(\mathbf{e}\cdot\boldsymbol{\omega})\mathbf{e}\mathbf{e}^T - 2(1+\sin\phi)\boldsymbol{\omega}\mathbf{e}^T\\ &-(2\cos\phi + \sin\phi)\mathbf{e}\boldsymbol{\omega}^T - (\sin\phi)[\boldsymbol{\Omega} - (\mathbf{e}\cdot\boldsymbol{\omega})\mathbf{E}]\end{aligned} \tag{3.98e}$$

Moreover,

$$\dot{\tilde{\mathbf{N}}} = [\,\dot{\phi}(\cos\phi)\mathbf{1} + \dot{\phi}(\sin\theta)\mathbf{E} \quad \dot{\mathbf{e}}\,] \tag{3.99a}$$

$$\dot{\tilde{\mathbf{L}}} = [\,\mathbf{V}/D \quad \dot{\mathbf{u}}\,] \tag{3.99b}$$

$$\dot{\tilde{\mathbf{H}}} = [\,\dot{r}_0\mathbf{1} + \dot{\mathbf{R}} \quad -\dot{\mathbf{r}}\,] \tag{3.99c}$$

where $\mathbf{U}$ denotes the cross-product matrix of $\mathbf{u}$, while $\mathbf{V}$ and D are defined below:

$$\mathbf{V} \equiv \dot{\mathbf{U}} - (\dot{\mathbf{u}}\mathbf{u}^T + \mathbf{u}\dot{\mathbf{u}}^T) - \frac{\dot{u}_0}{D}(\mathbf{U} - \mathbf{u}\mathbf{u}^T) \tag{3.99d}$$

$$D \equiv 1 + u_0 \tag{3.99e}$$

3.6 Rigid-Body Motion Referred to Moving Coordinate Axes

Although in kinematics no "preferred" coordinate system exists, in dynamics the governing equations of rigid-body motions are valid only in *inertial frames*. An inertial frame can be defined as a coordinate system that translates with uniform velocity and constant orientation with respect to the stars. Thus, it is important to refer vectors and matrices to inertial frames, but sometimes it is not possible to do so directly. For instance, a spacecraft can be supplied with instruments to measure the velocity and the acceleration of a satellite drifting in space, but the measurements taken from the spacecraft will be referred to a coordinate frame fixed to it, which is not inertial. If the motion of the spacecraft with respect to an inertial coordinate frame is recorded, e.g., from an Earth-based station, then the acceleration of the satellite with respect to an inertial frame can be computed using the foregoing information. How to do this is the subject of this section. In the realm of kinematics, it is not necessary to distinguish between inertial and noninertial coordinate frames, and hence, it will suffice to call the coordinate systems involved *fixed* and *moving*. Thus, consider the fixed coordinate frame X, Y, Z, which will be labeled $\mathcal{F}$, and the moving coordinate frame $\mathcal{X}$, $\mathcal{Y}$, and $\mathcal{Z}$, which will be labeled $\mathcal{M}$, both being depicted in Fig. 3.4. Moreover, let $\mathbf{Q}$ be the rotation matrix taking frame $\mathcal{F}$ into the orientation of $\mathcal{M}$, and $\mathbf{o}$ the position vector of the origin of $\mathcal{M}$ from the origin of $\mathcal{F}$. Further, let $\mathbf{p}$ be the position vector of point P from the origin of $\mathcal{F}$ and $\boldsymbol{\rho}$ the position vector of the same point from the origin of $\mathcal{M}$. From Fig. 3.4 one has

$$[\mathbf{p}]_{\mathcal{F}} = [\mathbf{o}]_{\mathcal{F}} + [\boldsymbol{\rho}]_{\mathcal{F}} \tag{3.100}$$

where it will be assumed that $\boldsymbol{\rho}$ is not available in frame $\mathcal{F}$, but in $\mathcal{M}$. Hence,

$$[\boldsymbol{\rho}]_{\mathcal{F}} = [\mathbf{Q}]_{\mathcal{F}}[\boldsymbol{\rho}]_{\mathcal{M}} \tag{3.101}$$

Substitution of eq.(3.101) into eq.(3.100) yields

$$[\mathbf{p}]_{\mathcal{F}} = [\mathbf{o}]_{\mathcal{F}} + [\mathbf{Q}]_{\mathcal{F}}[\boldsymbol{\rho}]_{\mathcal{M}} \tag{3.102}$$

Now, in order to compute the velocity of P, both sides of eq.(3.102) are differentiated with respect to time, which leads to

$$[\dot{\mathbf{p}}]_{\mathcal{F}} = [\dot{\mathbf{o}}]_{\mathcal{F}} + [\dot{\mathbf{Q}}]_{\mathcal{F}}[\boldsymbol{\rho}]_{\mathcal{M}} + [\mathbf{Q}]_{\mathcal{F}}[\dot{\boldsymbol{\rho}}]_{\mathcal{M}} \tag{3.103}$$

Furthermore, from the definition of $\boldsymbol{\Omega}$, eq.(3.46), we have

$$[\dot{\mathbf{Q}}]_{\mathcal{F}} = [\boldsymbol{\Omega}]_{\mathcal{F}}[\mathbf{Q}]_{\mathcal{F}} \tag{3.104}$$

Upon substitution of the foregoing relation into eq.(3.103), we obtain

$$[\dot{\mathbf{p}}]_{\mathcal{F}} = [\dot{\mathbf{o}}]_{\mathcal{F}} + [\boldsymbol{\Omega}]_{\mathcal{F}}[\mathbf{Q}]_{\mathcal{F}}[\boldsymbol{\rho}]_{\mathcal{M}} + [\mathbf{Q}]_{\mathcal{F}}[\dot{\boldsymbol{\rho}}]_{\mathcal{M}} \tag{3.105}$$

which is an expression for the velocity of P in $\mathcal{F}$ in terms of the velocity of P in $\mathcal{M}$ and the twist of $\mathcal{M}$ with respect to $\mathcal{F}$. Next, the acceleration of

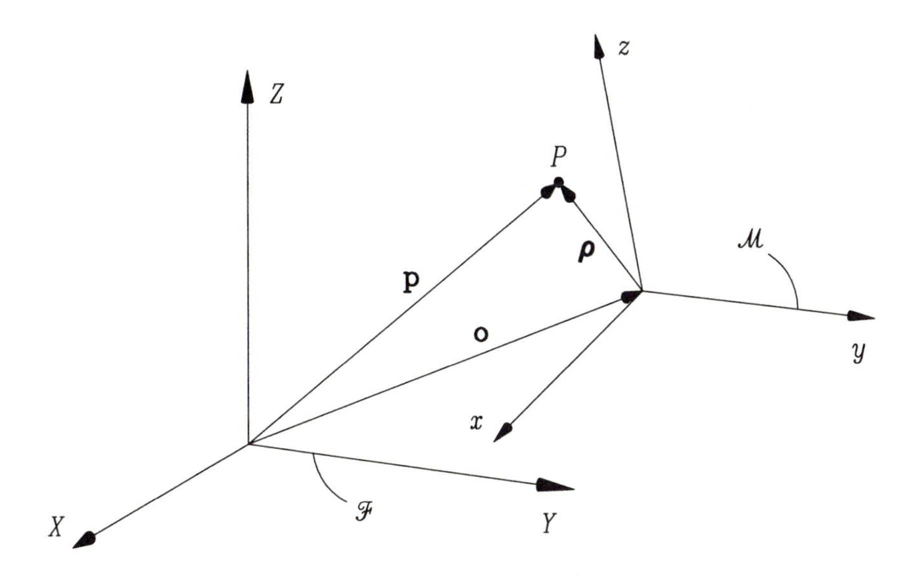

FIGURE 3.4. Fixed and moving coordinate frames.

P in $\mathcal{F}$ is derived by differentiation of both sides of eq.(3.105) with respect to time, which yields

$$\begin{aligned}[\ddot{\mathbf{p}}]_{\mathcal{F}} = [\ddot{\mathbf{o}}]_{\mathcal{F}} + [\dot{\mathbf{\Omega}}]_{\mathcal{F}}[\mathbf{Q}]_{\mathcal{F}}[\boldsymbol{\rho}]_{\mathcal{M}} + [\mathbf{\Omega}]_{\mathcal{F}}[\dot{\mathbf{Q}}]_{\mathcal{F}}[\boldsymbol{\rho}]_{\mathcal{M}} \\ +[\mathbf{\Omega}]_{\mathcal{F}}[\mathbf{Q}]_{\mathcal{F}}[\dot{\boldsymbol{\rho}}]_{\mathcal{M}} + [\dot{\mathbf{Q}}]_{\mathcal{F}}[\dot{\boldsymbol{\rho}}]_{\mathcal{M}} + [\mathbf{Q}]_{\mathcal{F}}[\ddot{\boldsymbol{\rho}}]_{\mathcal{M}} \qquad (3.106)\end{aligned}$$

Further, upon substitution of identity (3.104) into eq.(3.106), we obtain

$$\begin{aligned}[\ddot{\mathbf{p}}]_{\mathcal{F}} = [\ddot{\mathbf{o}}]_{\mathcal{F}} + ([\dot{\mathbf{\Omega}}]_{\mathcal{F}} + [\mathbf{\Omega}^2]_{\mathcal{F}})[\mathbf{Q}]_{\mathcal{F}}[\boldsymbol{\rho}]_{\mathcal{M}} \\ +2[\mathbf{\Omega}]_{\mathcal{F}}[\mathbf{Q}]_{\mathcal{F}}[\dot{\boldsymbol{\rho}}]_{\mathcal{M}} + [\mathbf{Q}]_{\mathcal{F}}[\ddot{\boldsymbol{\rho}}]_{\mathcal{M}} \qquad (3.107)\end{aligned}$$

Moreover, from the results of Section 3.5, it is clear that the first two terms of the right-hand side of eq.(3.107) represent the acceleration of P as a point of $\mathcal{M}$, whereas the fourth term is the acceleration of P measured from $\mathcal{M}$. The third term is what is called the *Coriolis acceleration*, as it was first pointed out by the French mathematician Gustave Gaspard Coriolis (1835).

3.7 Static Analysis of Rigid Bodies

Germane to the velocity analysis of rigid bodies is their force-and-moment analysis. In fact, striking similarities exist between the velocity relations associated with rigid bodies and the forces and moments acting on them.

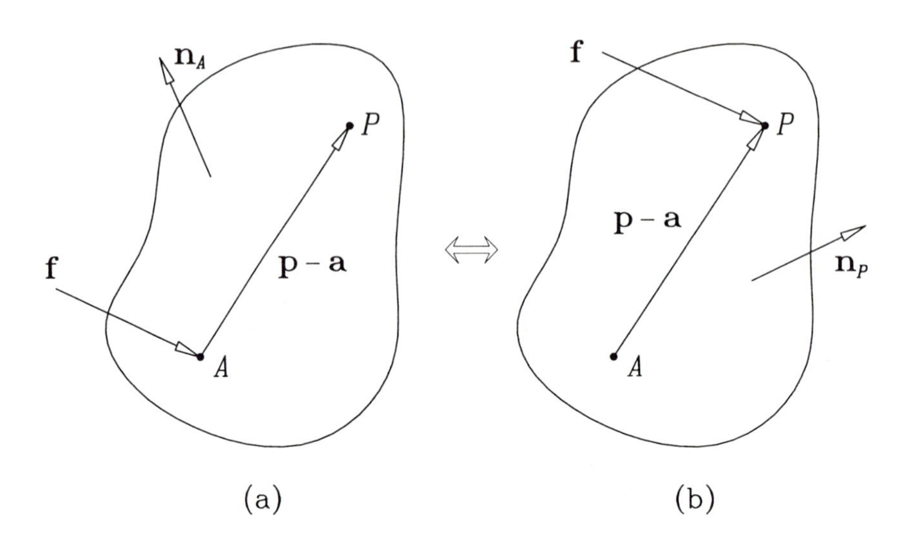

FIGURE 3.5. Equivalent systems of force and moment acting on a rigid body.

From elementary statics it is known that the resultant of all external actions, i.e., forces and moments, exerted on a rigid body can be reduced to a force $\mathbf{f}$ acting at a point, say A, and a moment $\mathbf{n}_A$. Alternatively, the aforementioned force $\mathbf{f}$ can be defined as acting at an arbitrary point P of the body, as depicted in Fig. 3.5, but then the resultant moment, $\mathbf{n}_P$, changes correspondingly.

In order to establish a relationship between $\mathbf{n}_A$ and $\mathbf{n}_P$, the moment of the first system of force and moment with respect to point P is equated to the moment about the same point of the second system, thus obtaining

$$\mathbf{n}_P = \mathbf{n}_A + (\mathbf{a} - \mathbf{p}) \times \mathbf{f} \tag{3.108}$$

which can be rewritten as

$$\mathbf{n}_P = \mathbf{n}_A + \mathbf{f} \times (\mathbf{p} - \mathbf{a}) \tag{3.109}$$

whence the analogy with eq.(3.52) is apparent. Indeed, $\mathbf{n}_P$ and $\mathbf{n}_A$ of eq.(3.109) play the role of the velocities of P and A, $\dot{\mathbf{p}}$ and $\dot{\mathbf{a}}$, respectively, whereas $\mathbf{f}$ of eq.(3.109) plays the role of $\boldsymbol{\omega}$ of eq.(3.52). Thus, similar to Theorem 3.4.2, one has

Theorem 3.7.1 *For a given system of forces and moments acting on a rigid body, if the resultant force is applied at any point of a particular line $\mathcal{L}''$, then the resultant moment is of minimum magnitude. Moreover, that minimum-magnitude moment is parallel to the resultant force.*

Hence, the resultant of the system of forces and moments is equivalent to a force $\mathbf{f}$ acting at a point of $\mathcal{L}''$ and a moment $\mathbf{n}$, with both $\mathbf{f}$ and $\mathbf{n}$ parallel to $\mathcal{L}''$. Paraphrasing the definition of the ISA, one defines line $\mathcal{L}''$ as the *axis of the wrench* acting on the body. Let $\mathbf{n}_0$ be the minimum-magnitude moment. Clearly, $\mathbf{n}_0$ can be expressed as $\mathbf{v}_0$ was in eq.(3.54), namely, as

$$\mathbf{n}_0 = n_0 \frac{\mathbf{f}}{\|\mathbf{f}\|}, \quad n_0 \equiv \frac{\mathbf{n}_P \cdot \mathbf{f}}{\|\mathbf{f}\|} \tag{3.110}$$

Moreover, the *pitch of the wrench*, p'', is defined as

$$p'' \equiv \frac{n_0}{\|\mathbf{f}\|} = \frac{\mathbf{n}_P \cdot \mathbf{f}}{\|\mathbf{f}\|^2} \quad \text{or} \quad p'' = \frac{2\pi \mathbf{n}_P \cdot \mathbf{f}}{\|\mathbf{f}\|^2} \tag{3.111}$$

which again has units of m/rad or correspondingly, of m/turn. Of course, the wrench axis can be defined by its Plücker array, $\mathbf{p}_{\mathcal{L}''}$, i.e.,

$$\mathbf{p}_{\mathcal{L}''} \equiv \begin{bmatrix} \mathbf{e}'' \\ \mathbf{n}'' \end{bmatrix}, \quad \mathbf{e}'' = \frac{\mathbf{f}}{\|\mathbf{f}\|}, \quad \mathbf{n}'' = \mathbf{p} \times \mathbf{e}'' \tag{3.112}$$

where $\mathbf{e}''$ is the unit vector parallel to $\mathcal{L}''$, $\mathbf{n}''$ is the moment of $\mathcal{L}''$ about the origin, and $\mathbf{p}$ is the position vector of any point on $\mathcal{L}''$.

The wrench axis is fully specified, then, by the direction of $\mathbf{f}$ and point P_0'' lying closest to the origin of position vector $\mathbf{p}_0''$, which can be derived by analogy with eq.(3.72), namely, as

$$\mathbf{p}_0'' = \frac{1}{\|\mathbf{f}\|^2} \mathbf{f} \times (\mathbf{n}_A - \mathbf{f} \times \mathbf{a}) \tag{3.113}$$

Similar to Theorem 3.4.1, one has

Theorem 3.7.2 *The projection of the resultant moment of a system of moments and forces acting on a rigid body that arises when the resultant force is applied at an arbitrary point of the body onto the wrench axis is constant.*

From the foregoing discussion, then, the wrench applied to a rigid body can be fully specified by the resultant force $\mathbf{f}$ acting at an arbitrary point P and the associated moment, $\mathbf{n}_P$. We shall derive presently the counterpart of the 6-dimensional array of the twist, namely, the wrench array. Upon multiplication of the screw of eq.(3.73) by an amplitude A with units of force, what we will obtain would be a wrench $\mathbf{w}$, i.e., a 6-dimensional array with its first three components having units of force and its last components units of moment. We would like to be able to obtain the power developed by the wrench on the body moving with the twist $\mathbf{t}$ by a simple inner product of the two arrays. However, because of the form the wrench $\mathbf{w}$ has taken, the inner product of these two arrays would be meaningless, for it would involve the sum of two scalar quantities with different units, and moreover, each of the two quantities is without an immediate physical meaning. In fact, the first scalar would have units of force by frequency (angular velocity by force), while the second would have units of moment of moment multiplied by frequency (velocity by moment), thereby leading to a physically meaningless result. This inconsistency can be resolved if we redefine the wrench not simply as the product of a screw by an amplitude, but as a linear transformation of that screw involving the 6×6 array $\mathbf{\Gamma}$ defined as

$$\mathbf{\Gamma} \equiv \begin{bmatrix} \mathbf{O} & \mathbf{1} \\ \mathbf{1} & \mathbf{O} \end{bmatrix} \tag{3.114}$$

where $\mathbf{O}$ and $\mathbf{1}$ denote, respectively, the 3×3 zero and identity matrices. Now we define the wrench as a linear transformation of the screw $\mathbf{s}$ defined in eq.(3.73). This transformation is obtained upon multiplying $\mathbf{s}$ by the product $A\mathbf{\Gamma}$, the amplitude A having units of force, i.e.,

$$\mathbf{w} \equiv A\mathbf{\Gamma}\mathbf{s} \equiv \begin{bmatrix} \mathbf{p} \times (A\mathbf{e}) + p(A\mathbf{e}) \\ A\mathbf{e} \end{bmatrix}$$

The foregoing wrench is said to be given in *axis coordinates*, as opposed to the twist, which was given in ray coordinates.

Now, the first three components of the foregoing array can be readily identified as the moment of a force of magnitude A acting along a line of action given by the Plücker array of eq.(3.112), with respect to a point P, to which a moment parallel to that line and of magnitude pA is added. Moreover, the last three components of that array pertain apparently to a force of magnitude A and parallel to the same line. We denote here the above-mentioned moment by $\mathbf{n}$ and the force by $\mathbf{f}$, i.e.,

$$\mathbf{f} \equiv A\mathbf{e}, \quad \mathbf{n} \equiv \mathbf{p} \times \mathbf{f} + p\mathbf{f}$$

The wrench $\mathbf{w}$ is then defined as

$$\mathbf{w} \equiv \begin{bmatrix} \mathbf{n} \\ \mathbf{f} \end{bmatrix} \tag{3.115}$$

which can thus be interpreted as a representation of a system of forces and moments acting on a rigid body, with the force acting at point P of the body $\mathcal{B}$ defined above and a moment $\mathbf{n}$. Under these circumstances, we say that $\mathbf{w}$ *acts* at point P of $\mathcal{B}$.

With the foregoing definitions it is now apparent that the wrench has been defined so that the *inner product* $\mathbf{t}^T\mathbf{w}$ will produce the power Π developed by $\mathbf{w}$ acting at P when $\mathcal{B}$ moves with a twist $\mathbf{t}$ defined at the same point, i.e.,

$$\Pi = \mathbf{t}^T\mathbf{w} \tag{3.116}$$

When a wrench $\mathbf{w}$ that acts on a rigid body moving with the twist $\mathbf{t}$ develops zero power onto the body, we say that the wrench and the twist are *reciprocal* to each other. By the same token, the screws associated with that wrench-twist pair are said to be *reciprocal*. More specifically, let the wrench and the twist be given in terms of their respective screws, $\mathbf{s}_w$ and $\mathbf{s}_t$, as

$$\mathbf{w} = W\mathbf{\Gamma}\mathbf{s}_w, \quad \mathbf{t} = T\mathbf{s}_t, \tag{3.117}$$

where W and T are the amplitudes of the wrench and the twist, respectively, while $\mathbf{\Gamma}$ is as defined in eq.(3.114). Thus, the two screws $\mathbf{s}_w$ and $\mathbf{s}_t$ are reciprocal if

$$(\mathbf{\Gamma}\mathbf{s}_w)^T\mathbf{s}_t \equiv \mathbf{s}_w^T\mathbf{\Gamma}^T\mathbf{s}_t = 0 \tag{3.118}$$

and by virtue of the symmetry of $\mathbf{\Gamma}$, the foregoing relation can be further expressed as

$$\mathbf{s}_w^T\mathbf{\Gamma}\mathbf{s}_t = 0 \quad \text{or} \quad \mathbf{s}_t^T\mathbf{\Gamma}\mathbf{s}_w = 0 \tag{3.119}$$

Now, if A and P are arbitrary points of a rigid body, we define the wrench at these points as

$$\mathbf{w}_A \equiv \begin{bmatrix} \mathbf{n}_A \\ \mathbf{f} \end{bmatrix}, \quad \mathbf{w}_P \equiv \begin{bmatrix} \mathbf{n}_P \\ \mathbf{f} \end{bmatrix} \tag{3.120}$$

Therefore, eq.(3.108) leads to

$$\mathbf{w}_P = \mathbf{V}\mathbf{w}_A \tag{3.121}$$

where

$$\mathbf{V} \equiv \begin{bmatrix} \mathbf{1} & \mathbf{A}-\mathbf{P} \\ \mathbf{0} & \mathbf{1} \end{bmatrix} \tag{3.122}$$

and $\mathbf{A}$ and $\mathbf{P}$ were defined in eq.(3.85) as the cross-product matrices of vectors $\mathbf{a}$ and $\mathbf{p}$, respectively. Thus, $\mathbf{w}_P$ is a linear transformation of $\mathbf{w}_A$. By analogy with the twist-transfer formula of eq.(3.84), eq.(3.121) is termed here the *wrench-transfer formula.*

Multiplying the transpose of eq.(3.84) by eq.(3.121) yields

$$\mathbf{t}_P^T\mathbf{w}_P = \mathbf{t}_A^T\mathbf{U}^T\mathbf{V}\mathbf{w}_A \tag{3.123}$$

where

$$\mathbf{U}^T\mathbf{V} = \begin{bmatrix} \mathbf{1} & -\mathbf{A}+\mathbf{P} \\ \mathbf{0} & \mathbf{1} \end{bmatrix} \begin{bmatrix} \mathbf{1} & \mathbf{A}-\mathbf{P} \\ \mathbf{0} & \mathbf{1} \end{bmatrix} = \mathbf{1}_{6\times 6} \tag{3.124}$$

with $\mathbf{1}_{6\times 6}$ denoting the 6×6 identity matrix. Thus, $\mathbf{t}_P^T\mathbf{w}_P = \mathbf{t}_A^T\mathbf{w}_A$, as expected, since the wrench develops the same amount of power, regardless of where the force is assumed to be applied. Also note that an interesting relation between $\mathbf{U}$ and $\mathbf{V}$ follows from eq.(3.124), namely,

$$\mathbf{V}^{-1} = \mathbf{U}^T \tag{3.125}$$

3.8 Dynamics of Rigid Bodies

The equations governing the motion of rigid bodies are recalled in this section and cast into a form suitable to multibody dynamics. To this end, a few definitions are introduced. If a rigid body has a mass density ρ, which need not be constant, then its mass m is defined as

$$m = \int_{\mathcal{B}} \rho d\mathcal{B} \tag{3.126}$$

where $\mathcal{B}$ denotes the region of the 3-dimensional space occupied by the body. Now, if $\mathbf{p}$ denotes the position vector of an arbitrary point of the body, from a previously defined origin O, the *mass first moment* of the body with respect to O, $\mathbf{q}_O$, is defined as

$$\mathbf{q}_O \equiv \int_{\mathcal{B}} \rho \mathbf{p} d\mathcal{B} \tag{3.127}$$

Furthermore, the *mass second moment* of the body with respect to O is defined as

$$\mathbf{I}_O \equiv \int_{\mathcal{B}} \rho[\,(\mathbf{p}\cdot\mathbf{p})\mathbf{1} - \mathbf{p}\mathbf{p}^T\,]d\mathcal{B} \tag{3.128}$$

which is clearly a symmetric matrix. This matrix is also called the moment-of-inertia matrix of the body under study with respect to O. One can readily prove the following result:

Theorem 3.8.1 *The moment of inertia of a rigid body with respect to a point O is positive definite.*

Proof: All we need to prove is that for any vector $\boldsymbol{\omega}$, the quadratic form $\boldsymbol{\omega}^T\mathbf{I}_O\boldsymbol{\omega}$ is positive. But this is so, because

$$\boldsymbol{\omega}^T\mathbf{I}_O\boldsymbol{\omega} = \int_{\mathcal{B}} \rho[\,\|\mathbf{p}\|^2\|\boldsymbol{\omega}\|^2 - (\mathbf{p}\cdot\boldsymbol{\omega})^2\,]d\mathcal{B} \tag{3.129}$$

Now, we recall that

$$\mathbf{p}\cdot\boldsymbol{\omega} = \|\mathbf{p}\|\|\boldsymbol{\omega}\|\cos(\mathbf{p},\boldsymbol{\omega}) \tag{3.130}$$

in which $(\mathbf{p},\boldsymbol{\omega})$ stands for the angle between the two vectors within the parentheses. Substitution of eq.(3.130) into eq.(3.131) leads to

$$\begin{aligned}\boldsymbol{\omega}^T\mathbf{I}_O\boldsymbol{\omega} &= \int_{\mathcal{B}} \rho\|\mathbf{p}\|^2\|\boldsymbol{\omega}\|^2[\,1-\cos^2(\mathbf{p},\boldsymbol{\omega})\,]d\mathcal{B} \\ &= \int_{\mathcal{B}} \rho\|\mathbf{p}\|^2\|\boldsymbol{\omega}\|^2\sin^2(\mathbf{p},\boldsymbol{\omega})d\mathcal{B}\end{aligned}$$

which is a positive quantity that vanishes only in the ideal case of a slender body having all its mass concentrated along a line passing through O and parallel to $\boldsymbol{\omega}$, which would thus render $\sin(\mathbf{p}, \boldsymbol{\omega}) = 0$ within the body, thereby completing the proof.

Alternatively, one can prove the positive definiteness of the mass moment of inertia based on physical arguments. Indeed, if vector $\boldsymbol{\omega}$ of the previous discussion is the angular velocity of the rigid body, then the quadratic form of eq.(3.129) turns out to be twice the kinetic energy of the body. Indeed, the said kinetic energy, denoted by T, is defined as

$$T \equiv \int_{\mathcal{B}} \frac{1}{2} \rho \|\dot{\mathbf{p}}\|^2 d\mathcal{B}$$

where $\dot{\mathbf{p}}$ is the velocity of any point P of the body. For the purposes of this discussion, it will be assumed that point O, about which the second moment is defined, is a point of the body that is instantaneously at rest. Thus, if this point is defined as the origin of the Euclidean space, the velocity of any point of the body, moving with an angular velocity $\boldsymbol{\omega}$, is given by

$$\dot{\mathbf{p}} = \boldsymbol{\omega} \times \mathbf{p}$$

which can be rewritten as

$$\dot{\mathbf{p}} = -\mathbf{P}\boldsymbol{\omega}$$

with $\mathbf{P}$ defined as the cross-product matrix of $\mathbf{p}$. Hence,

$$\|\dot{\mathbf{p}}\|^2 = (\mathbf{P}\boldsymbol{\omega})^T \mathbf{P}\boldsymbol{\omega} = \boldsymbol{\omega}^T \mathbf{P}^T \mathbf{P}\boldsymbol{\omega} = -\boldsymbol{\omega}^T \mathbf{P}^2 \boldsymbol{\omega}$$

Moreover, by virtue of eq.(2.39), the foregoing expression is readily reducible to

$$\|\dot{\mathbf{p}}\|^2 = \boldsymbol{\omega}^T (\|\mathbf{p}\|^2 \mathbf{1} - \mathbf{p}\mathbf{p}^T) \boldsymbol{\omega} \tag{3.131}$$

Therefore, the kinetic energy reduces to

$$T = \frac{1}{2} \int_{\mathcal{B}} \rho \boldsymbol{\omega}^T (\|\mathbf{p}\|^2 \mathbf{1} - \mathbf{p}\mathbf{p}^T) \boldsymbol{\omega} d\mathcal{B} \tag{3.132}$$

and since the angular velocity is constant throughout the body, it can be taken out of the integral sign, i.e.,

$$T = \frac{1}{2} \boldsymbol{\omega}^T \left[\int_{\mathcal{B}} \rho (\|\mathbf{p}\|^2 \mathbf{1} - \mathbf{p}\mathbf{p}^T) d\mathcal{B} \right] \boldsymbol{\omega} \tag{3.133}$$

The term inside the brackets of the latter equation is readily identified as $\mathbf{I}_O$, and hence, the kinetic energy can be written as

$$T = \frac{1}{2} \boldsymbol{\omega}^T \mathbf{I}_O \boldsymbol{\omega} \tag{3.134}$$

Now, since the kinetic energy is a positive-definite quantity, the quadratic form of eq.(3.134) is consequently positive-definite as well, thereby proving the positive-definiteness of the second moment.

The *mass center* of a rigid body, measured from O, is defined as a point C, not necessarily within the body—think of a homogeneous torus—of position vector $\mathbf{c}$ given by

$$\mathbf{c} \equiv \frac{\mathbf{q}_O}{m} \tag{3.135}$$

Naturally, the mass moment of inertia of the body with respect to its centroid is defined as

$$\mathbf{I}_C \equiv \int_{\mathcal{B}} \rho[\,\|\mathbf{r}\|^2\mathbf{1} - \mathbf{r}\mathbf{r}^T\,]d\mathcal{B} \tag{3.136}$$

where $\mathbf{r}$ is defined, in turn, as

$$\mathbf{r} \equiv \mathbf{p} - \mathbf{c} \tag{3.137}$$

Obviously, the mass moment of inertia of a rigid body about its mass center, also termed its *centroidal mass moment of inertia*, is positive-definite as well. In fact, the mass—or the volume, for that matter—moment of inertia of a rigid body *with respect to any point* is positive-definite. As a consequence, its three eigenvalues are positive and are referred to as the *principal moments of inertia* of the body. The eigenvectors of the inertia matrix are furthermore mutually orthogonal and define the *principal axes of inertia* of the body. These axes are parallel to the eigenvectors of that matrix and pass through the point about which the moment of inertia is taken. Note, however, that the principal moments and the principal axes of inertia of a rigid body depend on the point with respect to which the moment of inertia is defined. Moreover, let $\mathbf{I}_O$ and $\mathbf{I}_C$ be defined as in eqs.(3.128) and (3.136), with $\mathbf{r}$ defined as in eq.(3.137). It is possible to show that

$$\mathbf{I}_O = \mathbf{I}_C + m(\|\mathbf{c}\|^2\mathbf{1} - \mathbf{c}\mathbf{c}^T) \tag{3.138}$$

Furthermore, the smallest principal moment of inertia of a rigid body attains its minimum value at the mass center of the body. The relationship appearing in eq.(3.138) constitutes the *Theorem of Parallel Axes.*

Henceforth, we assume that $\mathbf{c}$ is the position vector of the mass center in an inertial frame. Now, we recall the Newton-Euler equations governing the motion of a rigid body. Let the body at hand be acted upon by a wrench of force $\mathbf{f}$ applied at its mass center, and a moment $\mathbf{n}_C$. The Newton equation then takes the form

$$\mathbf{f} = m\ddot{\mathbf{c}} \tag{3.139a}$$

whereas the Euler equation is

$$\mathbf{n}_C = \mathbf{I}_C\dot{\boldsymbol{\omega}} + \boldsymbol{\omega} \times \mathbf{I}_C\boldsymbol{\omega} \tag{3.139b}$$

The *momentum* $\mathbf{m}$ and the *angular momentum* $\mathbf{h}_C$ of a rigid body moving with an angular velocity $\boldsymbol{\omega}$ are defined below, the angular momentum being defined with respect to the mass center. These are

$$\mathbf{m} \equiv m\dot{\mathbf{c}}, \quad \mathbf{h}_C \equiv \mathbf{I}_C\boldsymbol{\omega} \tag{3.140}$$

Moreover, the time-derivatives of the foregoing quantities are readily computed as

$$\dot{\mathbf{m}} = m\ddot{\mathbf{c}}, \quad \dot{\mathbf{h}}_C = \mathbf{I}_C\dot{\boldsymbol{\omega}} + \boldsymbol{\omega} \times \mathbf{I}_C\boldsymbol{\omega} \tag{3.141}$$

and hence, eqs.(3.139a & b) take on the forms

$$\mathbf{f} = \dot{\mathbf{m}}, \quad \mathbf{n}_C = \dot{\mathbf{h}}_C \tag{3.142}$$

The set of equations (3.139a) and (3.139b) are known as the *Newton-Euler equations*. These can be written in a more compact form as we describe below. First, we introduce a 6×6 matrix $\mathbf{M}$ that following von Mises (1924), we term the *inertia dyad*, namely,

$$\mathbf{M} \equiv \begin{bmatrix} \mathbf{I}_C & \mathbf{O} \\ \mathbf{O} & m\mathbf{1} \end{bmatrix} \tag{3.143}$$

where $\mathbf{O}$ and $\mathbf{1}$ denote the 3×3 zero and identity matrices. A similar 6×6 matrix was defined by von Mises under the above name. However, von Mises's inertia dyad is full, while the matrix defined above is block-diagonal. Both matrices, nevertheless, denote the same physical property of a rigid body, i.e., its mass and moment of inertia. Now the Newton-Euler equations can be written as

$$\mathbf{M}\dot{\mathbf{t}} + \mathbf{W}\mathbf{M}\mathbf{t} = \mathbf{w} \tag{3.144}$$

in which matrix $\mathbf{W}$, which we shall term, by similarity with the inertia dyad, the *angular-velocity dyad*, is defined, in turn, as

$$\mathbf{W} \equiv \begin{bmatrix} \boldsymbol{\Omega} & \mathbf{O} \\ \mathbf{O} & \mathbf{O} \end{bmatrix} \tag{3.145}$$

with $\boldsymbol{\Omega}$ already defined as the angular-velocity matrix; it is, of course, the cross-product matrix of the angular-velocity vector $\boldsymbol{\omega}$. Note that the twist of a rigid body lies in the nullspace of its angular-velocity dyad, i.e.,

$$\mathbf{W}\mathbf{t} = \mathbf{0} \tag{3.146}$$

Further definitions are introduced below: The *momentum screw* of the rigid body about the mass center is the 6-dimensional vector $\boldsymbol{\mu}$ defined as

$$\boldsymbol{\mu} \equiv \begin{bmatrix} \mathbf{I}_C\boldsymbol{\omega} \\ m\dot{\mathbf{c}} \end{bmatrix} = \mathbf{M}\mathbf{t} \tag{3.147}$$

Furthermore, from eqs.(3.141) and definition (3.147), the time-derivative of $\boldsymbol{\mu}$ can be readily derived as

$$\dot{\boldsymbol{\mu}} = \mathbf{M}\dot{\mathbf{t}} + \mathbf{W}\boldsymbol{\mu} = \mathbf{M}\dot{\mathbf{t}} + \mathbf{W}\mathbf{M}\mathbf{t} \tag{3.148}$$

The kinetic energy of a rigid body undergoing a motion in which its mass center moves with velocity $\dot{\mathbf{c}}$ and rotates with an angular velocity $\boldsymbol{\omega}$ is given by

$$T = \frac{1}{2}m\|\dot{\mathbf{c}}\|^2 + \frac{1}{2}\boldsymbol{\omega}^T\mathbf{I}_C\boldsymbol{\omega} \tag{3.149}$$

From the foregoing definitions, then, the kinetic energy can be written in compact form as

$$T = \frac{1}{2}\mathbf{t}^T\mathbf{M}\mathbf{t} \tag{3.150}$$

Finally, the Newton-Euler equations can be written in an even more compact form as

$$\dot{\boldsymbol{\mu}} = \mathbf{w} \tag{3.151}$$

which is a 6-dimensional vector equation.

4

Kinetostatics of Simple Robotic Manipulators

4.1 Introduction

This chapter is devoted to the *kinetostatics* of robotic manipulators of the serial type, i.e., to the kinematics and statics of these systems. The study is general, but with regard to what is called the *inverse kinematics problem*, we limit the chapter to *decoupled manipulators*, to be defined below. The inverse displacement analysis of general six-axis manipulators is addressed in Chapter 8.

More specifically, we will define a serial, n-axis manipulator. In connection with this manipulator, additionally, we will (i) introduce the *Denavit-Hartenberg notation* for the definition of *link frames* that uniquely determine the *architecture* and the *configuration*, or *posture*, of the manipulator at hand; (ii) define the *Cartesian* and *joint coordinates* of this manipulator; and (iii) introduce its *Jacobian matrix*.

Moreover, with regard to six-axis manipulators, we will (i) define *decoupled* manipulators and provide a procedure for the solution of their displacement inverse kinematics; (ii) formulate and solve the *velocity-resolution* problem, give simplified solutions for decoupled manipulators, and identify their *singularities*; (iii) define the *workspace* of a three-axis positioning manipulator and provide means to display it; (iv) formulate and solve the *acceleration-resolution* problem and give simplified solutions for decoupled manipulators; and (v) analyze manipulators statically, while giving simplified analyses for decoupled manipulators. While doing this, we will devote special attention to *planar manipulators*.

4.2 The Denavit-Hartenberg Notation

One of the first tasks of a robotics engineer is the kinematic modeling of a robotic manipulator. This task consists of devising a model that can be *unambiguously* (*i*) described to a control unit through a database and (*ii*) interpreted by other robotics engineers. The purpose of this task is to give manipulating instructions to a robot, regardless of the dynamics of the manipulated load and the robot itself. The simplest way of kinematically modeling a robotic manipulator is by means of the concept of *kinematic chain*. A kinematic chain is a set of *rigid bodies*, also called *links*, coupled by *kinematic pairs*. A kinematic pair is, then, the coupling of two rigid bodies so as to constrain their relative motion. We distinguish two basic types of kinematic pairs, namely, *upper* and *lower* kinematic pairs. An upper kinematic pair arises between rigid bodies when contact takes place along a line or at a point. This type of coupling occurs in cam-and-follower mechanisms, gear trains, and roller bearings, for example. A lower kinematic pair occurs when contact takes place along a surface common to the two bodies. Six different types of lower kinematic pairs can be distinguished (Denavit and Hartenberg, 1964; Angeles, 1982), but all these can be produced from two basic types, namely, the *rotating pair*, denoted by R and also called *revolute*, and the *sliding pair*, represented by P and also called *prismatic*.

The common surface along which contact takes place in a revolute pair is a circular cylinder, a typical example of this pair being the coupling through journal bearings. Thus, two rigid bodies coupled by a revolute can rotate relative to each other about the axis of the common cylinder, which is thus referred to as the *axis of the revolute*, but are prevented from undergoing relative translations as well as rotations about axes other than the cylinder axis. On the other hand, the common surface of contact between two rigid bodies coupled by a prismatic pair is a prism of arbitrary cross section, and hence, the two bodies coupled in this way are prevented from undergoing any relative rotation and can move only in a pure-translation motion along a direction parallel to the axis of the prism. As an example of this kinematic pair, one can cite the dovetail coupling. Note that whereas the revolute axis is a totally defined line in three-dimensional space, the prismatic pair has no defined axis; this pair has only a direction. That is, the prismatic pair does not have a particular location in space. Bodies coupled by a revolute and a prismatic pair are shown in Fig. 4.1.

Serial manipulators will be considered in this chapter, their associated kinematic chains thus being of the *simple* type, i.e., each and every link is coupled to at most two other links. A *simple kinematic chain* can be either closed or open. It is closed if each and every link is coupled to two other links, the chain then being called a *linkage*; it is open if it contains exactly two links, the end ones, that are coupled to only one other link. Thus, simple kinematic chains studied in this chapter are open, and in the

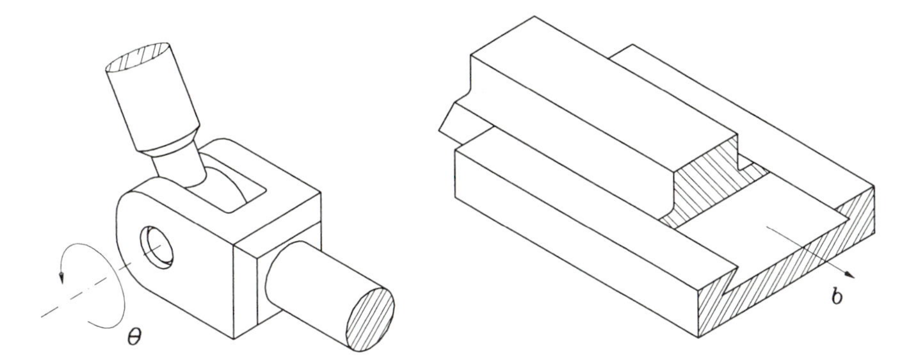

FIGURE 4.1. Revolute and prismatic pair.

particular robotics terminology, their first link is called the *manipulator base*, whereas their last link is termed the *end-effector (EE)*.

Thus, the kinematic chains associated with manipulators of the serial type are composed of *binary links*, the intermediate ones, and exactly two *simple links*, those at the ends. Hence, except for the end links, all links carry two kinematic pairs, and as a consequence, two pair axes—however, notice that a prismatic pair has a direction but no axis. In order to uniquely describe the *architecture* of a kinematic chain, i.e., the relative location and orientation of its neighboring pair axes, the Denavit-Hartenberg nomenclature (Denavit and Hartenberg, 1955) is introduced. To this end, links are numbered 0, 1, ..., n, the ith pair being defined as that coupling the $(i-1)$st link with the ith link. Hence, the manipulator is assumed to be composed of $n+1$ links and n pairs; each of the latter can be either R or P, where link 0 is the fixed base, while link n is the end-effector. Next, a coordinate frame $\mathcal{F}_i$ is defined with origin O_i and axes X_i, Y_i, Z_i. This frame is attached to the $(i-1)$st link—**not** to the ith link!—for $i = 1, \ldots, n+1$. For the first n frames, this is done following the rules given below:

1. Z_i is the axis of the ith pair. Notice that there are two possibilities of defining the positive direction of this axis, since each pair axis is only a line, not a directed segment. Moreover, the Z_i axis of a prismatic pair can be located arbitrarily, since only its direction is defined by the axis of this pair.

2. X_i is defined as the common perpendicular to Z_{i-1} and Z_i, directed from the former to the latter, as shown in Fig. 4.2a. Notice that if these two axes intersect, the positive direction of X_i is undefined and hence, can be freely assigned. Henceforth, we will follow the *right-hand* rule in this case. This means that if unit vectors $\mathbf{i}_i$, $\mathbf{k}_{i-1}$, and $\mathbf{k}_i$ are attached to axes X_i, Z_{i-1}, and Z_i, respectively, as indicated in Fig. 4.2b, then $\mathbf{i}_i$ is defined as $\mathbf{k}_{i-1} \times \mathbf{k}_i$. Moreover, if Z_{i-1} and Z_i are

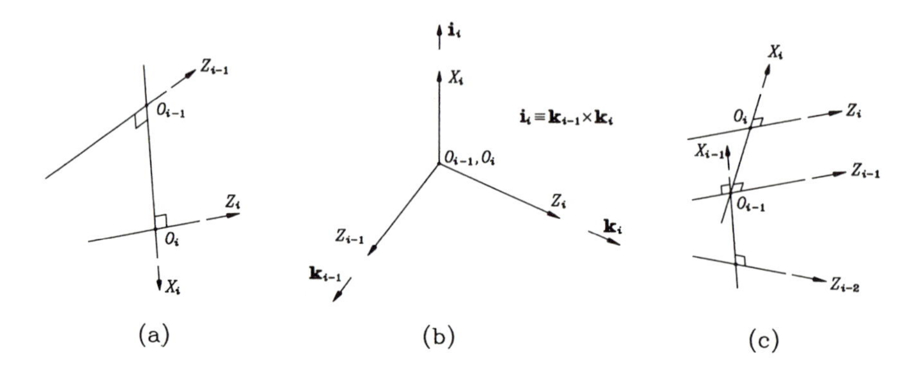

FIGURE 4.2. Definition of X_i when Z_{i-1} and Z_i: (a) are skew; (b) intersect; and (c) are parallel.

parallel, the location of X_i is undefined. In order to define it uniquely, we will specify X_i as passing through the origin of the $(i-1)$st frame, as shown in Fig. 4.2c.

3. The *distance* between Z_i and Z_{i+1} is defined as a_i, which is thus *nonnegative.*

4. The Z_i-coordinate of the intersection O_i' of Z_i with X_{i+1} is denoted by b_i. Note that this quantity is a coordinate and hence, can be either positive or negative. Its absolute value is the distance between X_i and X_{i+1}, also called the *offset* between successive common perpendiculars.

5. The angle between Z_i and Z_{i+1} is defined as α_i and is measured about the positive direction of X_{i+1}. This item is known as the *twist angle* between successive pair axes.

6. The angle between X_i and X_{i+1} is defined as θ_i and is measured about the positive direction of Z_i.

The $(n+1)$st coordinate frame is attached to the far end of the nth link. Since the manipulator has no $(n+1)$st link, the foregoing rules do not apply to the definition of the last frame. The analyst, thus, has the freedom to define this frame as it best suits the task at hand. Notice that $n+1$ frames, $\mathcal{F}_1$, $\mathcal{F}_2$, ..., $\mathcal{F}_{n+1}$, have been defined, whereas links are numbered from 0 to n. In summary, an n-axis manipulator is composed of $n+1$ links and $n+1$ coordinate frames. These rules are illustrated with an example below.

Consider the architecture depicted in Fig. 4.3, usually referred to as a *Puma robot*, which shows seven links, numbered from 0 to 6, and seven coordinate frames, numbered from 1 to 7. Note that the last frame is arbitrarily defined, but its origin is placed at a specific point of the EE, namely,

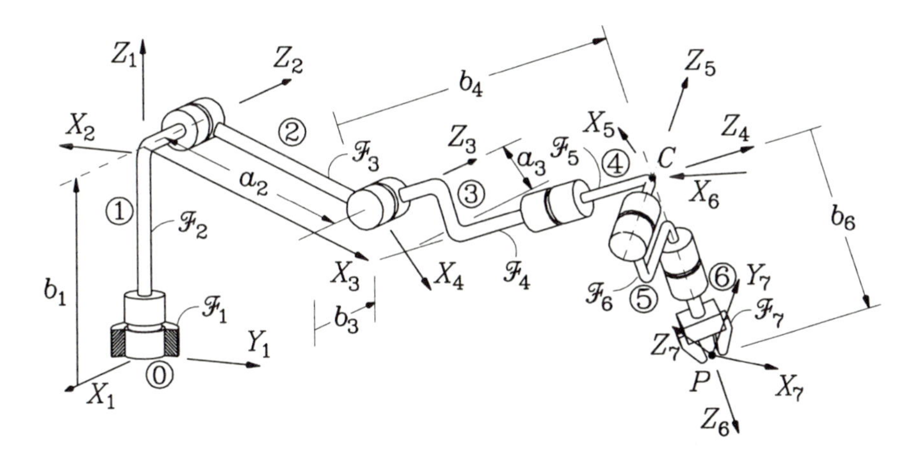

FIGURE 4.3. Coordinate frames of a Puma robot.

at the *operation point*, P, which is used to define the task at hand. Furthermore, three axes intersect at a point C, and hence, all points of the last three links move on concentric spheres with respect to C, for which reason the subchain comprising these three links is known as a *spherical wrist*, point C being its *center*. By the same token, the subchain composed of the first four links is called the *arm*. Thus, the wrist is *decoupled* from the arm, and is used for orientation purposes, the arm being used for the positioning of point C. The arm is sometimes called the *regional structure* and the wrist the *local structure*, the overall manipulator thus being of the *decoupled* type.

In the foregoing discussion, if the ith pair is R, then all quantities involved in those definitions are constant, except for θ_i, which is variable and is thus termed the *joint variable* of the ith pair. The other quantities, i.e., a_i, b_i, and α_i, are the *joint parameters* of the said pair. If, alternatively, the ith pair is P, then b_i is variable, and the other quantities are constant. In this case, the joint variable is b_i, and the joint parameters are a_i, α_i, and θ_i. Notice that associated with each joint there are exactly one joint variable and three constant parameters. Hence, an n-axis manipulator has n joint variables—which are henceforth grouped in the n-dimensional vector $\boldsymbol{\theta}$, regardless of whether the joint variables are angular or translational—and $3n$ constant parameters. The latter define the *architecture* of the manipulator, while the former determine its *configuration*, or *posture*.

Whereas the manipulator architecture is fully defined by its $3n$ *Denavit-Hartenberg (DH) parameters*, its posture is fully defined by its n joint variables, also called its *joint coordinates*, once the DH parameters are known. The relative position and orientation between links is fully specified, then, from the discussions of Chapter 2, by (i) the rotation matrix taking the X_i, Y_i, Z_i axes into a configuration in which they are parallel pairwise

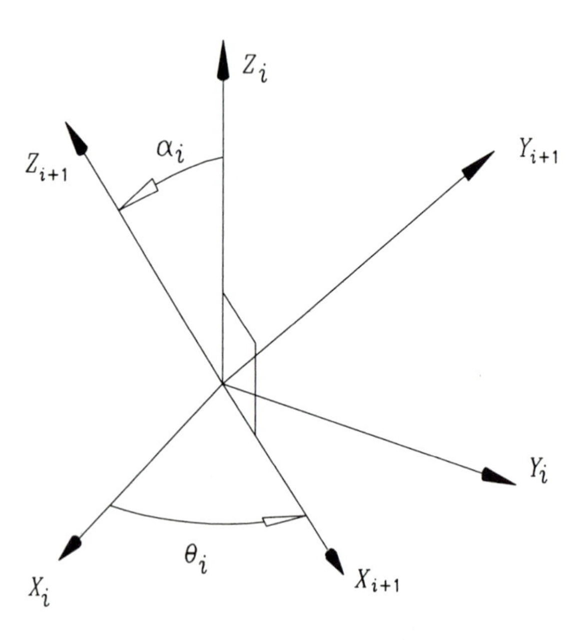

FIGURE 4.4. Relative orientation of the ith and $(i+1)$st coordinate frames.

to the X_{i+1}, Y_{i+1}, Z_{i+1} axes, and (*ii*) the position vector of the origin of the latter in the former. The representations of the foregoing items in coordinate frame $\mathcal{F}_i$ will be discussed presently. First, we obtain the matrix representation of the rotation $\mathbf{Q}_i$ carrying $\mathcal{F}_i$ into an orientation coincident with that of $\mathcal{F}_{i+1}$, assuming, without loss of generality because we are interested only in changes of orientation, that the two origins are coincident, as depicted in Fig. 4.4. This matrix is most easily derived if the rotation of interest is decomposed into two rotations, as indicated in Fig. 4.5. In that figure, X_i', Y_i', Z_i' is an intermediate coordinate frame $\mathcal{F}_i'$, obtained by rotating $\mathcal{F}_i$ about the Z_i axis through an angle θ_i. Then, the intermediate frame is rotated about $X_{i'}$ through an angle α_i, which takes it into a configuration coincident with $\mathcal{F}_{i+1}$. Let the foregoing rotations be denoted by $[\,\mathbf{C}_i\,]_i$ and $[\,\mathbf{\Lambda}_i\,]_{i'}$, respectively, which are readily derived for they are in the canonical forms (2.55c) and (2.55a), respectively.

Moreover, let

$$\lambda_i \equiv \cos\alpha_i, \quad \mu_i \equiv \sin\alpha_i \tag{4.1a}$$

One thus has, using subscripted brackets as introduced in Section 2.2,

$$[\,\mathbf{C}_i\,]_i = \begin{bmatrix} \cos\theta_i & -\sin\theta_i & 0 \\ \sin\theta_i & \cos\theta_i & 0 \\ 0 & 0 & 1 \end{bmatrix}, \quad [\,\mathbf{\Lambda}_i\,]_{i'} = \begin{bmatrix} 1 & 0 & 0 \\ 0 & \lambda_i & -\mu_i \\ 0 & \mu_i & \lambda_i \end{bmatrix}$$

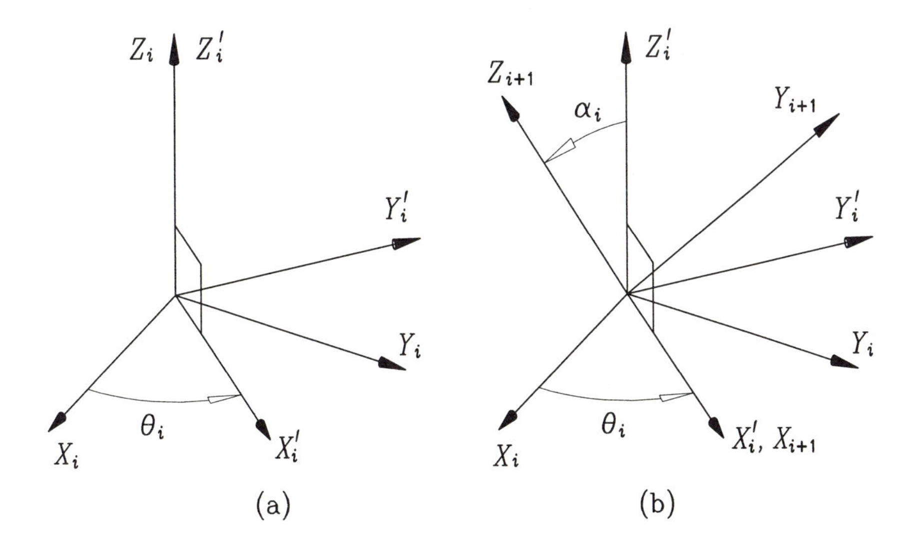

FIGURE 4.5. (a) Rotation about axis Z_i through an angle θ_i; and (b) relative orientation of the i'th and the $(i+1)$st coordinate frames.

and clearly, the matrix sought is computed simply as

$$[\,\mathbf{Q}_i\,]_i = [\,\mathbf{C}_i\,]_i\,[\,\mathbf{\Lambda}_i\,]_{i'} \tag{4.1b}$$

Henceforth, we will use the abbreviations introduced below:

$$\mathbf{Q}_i \equiv [\,\mathbf{Q}_i\,]_i, \quad \mathbf{C}_i \equiv [\,\mathbf{C}_i\,]_i, \quad \mathbf{\Lambda}_i \equiv [\,\mathbf{\Lambda}_i\,]_{i'} \tag{4.1c}$$

thereby doing away with brackets, when these are self-evident. Thus,

$$\mathbf{Q}_i \equiv [\,\mathbf{Q}_i\,]_i \equiv \begin{bmatrix} \cos\theta_i & -\lambda_i\sin\theta_i & \mu_i\sin\theta_i \\ \sin\theta_i & \lambda_i\cos\theta_i & -\mu_i\cos\theta_i \\ 0 & \mu_i & \lambda_i \end{bmatrix} \tag{4.1d}$$

One more factoring of matrix $\mathbf{Q}_i$, which finds applications in manipulator kinematics, is given below:

$$\mathbf{Q}_i = \mathbf{Z}_i\mathbf{X}_i \tag{4.2a}$$

with $\mathbf{X}_i$ and $\mathbf{Z}_i$ defined as two reflections, the former about the Y_iZ_i plane, the latter about the X_iY_i plane, namely,

$$\mathbf{X}_i \equiv \begin{bmatrix} 1 & 0 & 0 \\ 0 & -\lambda_i & \mu_i \\ 0 & \mu_i & \lambda_i \end{bmatrix}, \quad \mathbf{Z}_i \equiv \begin{bmatrix} \cos\theta_i & \sin\theta_i & 0 \\ \sin\theta_i & -\cos\theta_i & 0 \\ 0 & 0 & 1 \end{bmatrix} \tag{4.2b}$$

In order to derive an expression for the position vector $\mathbf{a}_i$ connecting the origin O_i of $\mathcal{F}_i$ with that of $\mathcal{F}_{i+1}$, O_{i+1}, reference is made to Fig. 4.6, showing the relative positions of the different origins and axes involved. From this figure, clearly,

$$\mathbf{a}_i \equiv \overrightarrow{O_iO}_{i+1} = \overrightarrow{O_iO}_{i'} + \overrightarrow{O_{i'}O}_{i+1} \tag{4.3a}$$

where obviously,

$$[\,\overrightarrow{O_iO}_{i'}\,]_i = \begin{bmatrix} 0 \\ 0 \\ b_i \end{bmatrix}, \quad [\,\overrightarrow{O_{i'}O}_{i+1}\,]_{i+1} = \begin{bmatrix} a_i \\ 0 \\ 0 \end{bmatrix}$$

Now, in order to compute the sum appearing in eq.(4.3a), the two foregoing vectors should be expressed in the same coordinate frame, namely, $\mathcal{F}_i$. Thus,

$$[\,\overrightarrow{O_{i'}O}_{i+1}\,]_i = [\,\mathbf{Q}_i\,]_i\,[\,\overrightarrow{O_{i'}O}_{i+1}\,]_{i+1} = \begin{bmatrix} a_i \cos\theta_i \\ a_i \sin\theta_i \\ 0 \end{bmatrix}$$

and hence,

$$[\,\mathbf{a}_i\,]_i = \begin{bmatrix} a_i \cos\theta_i \\ a_i \sin\theta_i \\ b_i \end{bmatrix} \tag{4.3b}$$

For brevity, we introduce the following definition:

$$\mathbf{a}_i \equiv [\,\mathbf{a}_i\,]_i \tag{4.3c}$$

Similar to the foregoing factoring of $\mathbf{Q}_i$, vector $\mathbf{a}_i$ admits the factoring

$$\mathbf{a}_i = \mathbf{Q}_i\mathbf{b}_i \tag{4.3d}$$

where $\mathbf{b}_i$ is given by

$$\mathbf{b}_i \equiv \begin{bmatrix} a_i \\ b_i\mu_i \\ b_i\lambda_i \end{bmatrix} \tag{4.3e}$$

with the definitions introduced in eq.(4.1a). Hence, vector $\mathbf{b}_i$ is constant for revolute pairs.

Matrices $\mathbf{Q}_i$ can also be regarded as coordinate transformations. Indeed, let $\mathbf{i}_i$, $\mathbf{j}_i$, and $\mathbf{k}_i$ be the unit vectors parallel to the X_i, Y_i, and Z_i axes, respectively, directed in the positive direction of these axes. From Fig. 4.6, it is apparent that

$$[\,\mathbf{i}_{i+1}\,]_i = \begin{bmatrix} \cos\theta_i \\ \sin\theta_i \\ 0 \end{bmatrix}, \quad [\,\mathbf{k}_{i+1}\,]_i = \begin{bmatrix} \mu_i \sin\theta_i \\ -\mu \cos\theta_i \\ \lambda_i \end{bmatrix}$$

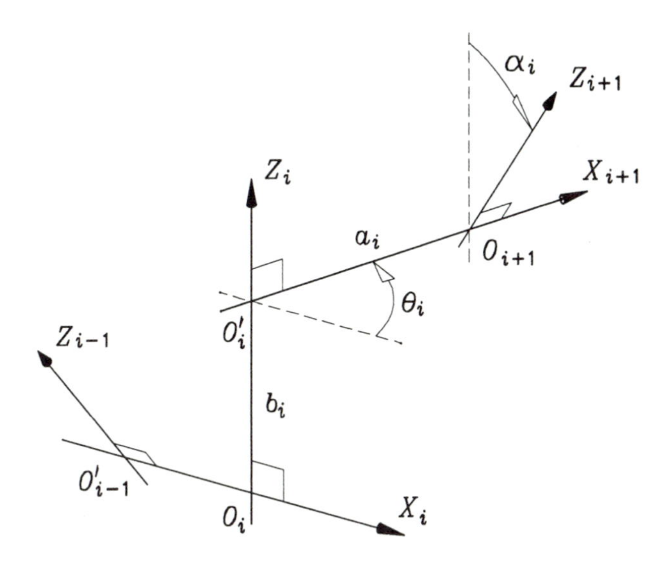

FIGURE 4.6. Layout of three successive coordinate frames.

whence

$$[\mathbf{j}_{i+1}]_i = [\mathbf{k}_{i+1} \times \mathbf{i}_{i+1}]_i = \begin{bmatrix} -\lambda_i \sin\theta_i \\ \lambda_i \cos\theta_i \\ \mu_i \end{bmatrix}$$

Therefore, the components of $\mathbf{i}_{i+1}$, $\mathbf{j}_{i+1}$, and $\mathbf{k}_{i+1}$ in $\mathcal{F}_i$ are nothing but the first, second, and third columns of $\mathbf{Q}_i$. In general, then, any vector $\mathbf{v}$ in $\mathcal{F}_{i+1}$ is transformed into $\mathcal{F}_i$ in the form

$$[\mathbf{v}]_i = [\mathbf{Q}_i]_i[\mathbf{v}]_{i+1}$$

which is a similarity transformation, as defined in eq.(2.117). Likewise, any matrix $\mathbf{M}$ in $\mathcal{F}_{i+1}$ is transformed into $\mathcal{F}_i$ by the corresponding similarity transformation, as given by eq.(2.128):

$$[\mathbf{M}]_i = [\mathbf{Q}_i]_i[\mathbf{M}]_{i+1}[\mathbf{Q}_i^T]_i$$

The inverse relations follow immediately in the form

$$[\mathbf{v}]_{i+1} = [\mathbf{Q}_i^T]_i[\mathbf{v}]_i, \quad [\mathbf{M}]_{i+1} = [\mathbf{Q}_i^T]_i[\mathbf{M}]_i[\mathbf{Q}_i]_i$$

or upon recalling the first of definitions (4.1c),

$$[\mathbf{v}]_i = \mathbf{Q}_i[\mathbf{v}]_{i+1}, \quad [\mathbf{M}]_i = \mathbf{Q}_i[\mathbf{M}]_{i+1}\mathbf{Q}_i^T \tag{4.4a}$$

$$[\mathbf{v}]_{i+1} = \mathbf{Q}_i^T[\mathbf{v}]_i, \quad [\mathbf{M}]_{i+1} = \mathbf{Q}_i^T[\mathbf{M}]_i\mathbf{Q}_i \tag{4.4b}$$

Moreover, if we have a chain of i frames, $\mathcal{F}_1, \mathcal{F}_2, \ldots, \mathcal{F}_i$, then the *inward* coordinate transformation from $\mathcal{F}_i$ to $\mathcal{F}_1$ is given by

$$[\mathbf{v}]_1 = \mathbf{Q}_1\mathbf{Q}_2\cdots\mathbf{Q}_{i-1}[\mathbf{v}]_i \tag{4.5a}$$

$$[\mathbf{M}]_1 = \mathbf{Q}_1\mathbf{Q}_2\cdots\mathbf{Q}_{i-1}[\mathbf{M}]_i(\mathbf{Q}_1\mathbf{Q}_2\cdots\mathbf{Q}_{i-1})^T \tag{4.5b}$$

Likewise, the outward coordinate transformation takes the form

$$[\mathbf{v}]_i = (\mathbf{Q}_1\mathbf{Q}_2\cdots\mathbf{Q}_{i-1})^T[\mathbf{v}]_1 \tag{4.6a}$$

$$[\mathbf{M}]_i = (\mathbf{Q}_1\mathbf{Q}_2\cdots\mathbf{Q}_{i-1})^T[\mathbf{M}]_1\mathbf{Q}_1\mathbf{Q}_2\cdots\mathbf{Q}_{i-1} \tag{4.6b}$$

4.3 The Kinematics of Six-Revolute Manipulators

The kinematics of serial manipulators comprises the study of the relations between *joint variables* and *Cartesian variables*. The former were defined in Section 4.2 as those determining the posture of a given manipulator, with one such variable per joint; a six-axis manipulator, like the one displayed in Fig. 4.7, thus has six joint variables, $\theta_1, \theta_2, \ldots, \theta_6$. The Cartesian variables of a manipulator, in turn, are those variables defining the pose of the EE; since six independent variables are needed to define the pose of a rigid body, the manipulator of Fig. 4.7 thus contains six Cartesian variables.

The study outlined above pertains to the *geometry* of the manipulator, for it involves one single pose of the EE. Besides geometry, the kinematics of manipulators comprises the study of the relations between the time-rates of change of the joint variables, referred to as the *joint rates*, and the twist of the EE. Additionally, the relations between the second time-derivatives of the joint variables, referred to as the *joint accelerations*, with the time-rate of change of the twist of the EE are also studied in this chapter.

In this section and in Section 4.4 we study the geometry of manipulators. In this regard, we distinguish two problems, commonly referred to as the *direct* and the *inverse* kinematic problems, or DKP and correspondingly, IKP, for brevity. In the DKP, the six joint variables of a given six-axis manipulator are assumed to be known, the problem consisting of finding the pose of the EE. In the IKP, on the contrary, the pose of the EE is given, while the six joint variables that produce this pose are to be found.

The DKP reduces to matrix multiplications, and as we shall show presently, poses no major problem. The IKP, however, is more challenging, for it involves intensive variable-elimination and nonlinear-equation solving. Indeed, in the most general case, the IKP amounts to eliminating five out of the six unknowns, with the aim of reducing the problem to a single monovariate polynomial of up to 16th degree. While finding the roots of a polynomial of this degree is no longer an insurmountable task, reducing the underlying system of nonlinear equations to a monovariate polynomial requires intensive computer-algebra work that must be very carefully planned

That is, if we regard $\mathbf{e}$ in the first of the foregoing relations as $[\,\mathbf{e}_{i+1}\,]_{i+1}$, and as $[\,\mathbf{e}_i\,]_i$ in the second relation, then, from the coordinate transformations of eqs.(4.4a & b),

$$\mathbf{u}_i = [\,\mathbf{e}_{i+1}\,]_i, \quad \text{and} \quad \mathbf{o}_i = [\,\mathbf{e}_i\,]_{i+1} \tag{4.15}$$

4.4 The IKP of Decoupled Manipulators

Industrial manipulators are frequently supplied with a special architecture that allows a decoupling of the positioning problem from the orientation problem. In fact, a determinant design criterion in this regard has been that the manipulator be kinematically invertible, i.e., that it lend itself to a closed-form inverse kinematics solution. Although the class of manipulators with this feature is quite broad, we will focus on a special kind, the most frequently encountered in commercial manipulators, that we will term *decoupled*. Decoupled manipulators were defined in Section 4.2 as those whose last three joints have intersecting axes. These joints, then, constitute the *wrist* of the manipulator, which is said to be *spherical*, because when the point of intersection of the three wrist axes, C, is kept fixed, all the points of the wrist move on spheres centered at C. In terms of the DH parameters of the manipulator, in a decoupled manipulator $a_4 = a_5 = b_5 = 0$, and thus, the origins of frames 4, 5, and 6 are coincident. All other DH parameters can assume arbitrary values. A general decoupled manipulator is shown in Fig. 4.8, where the wrist is represented as a concatenation of three revolutes with intersecting axes.

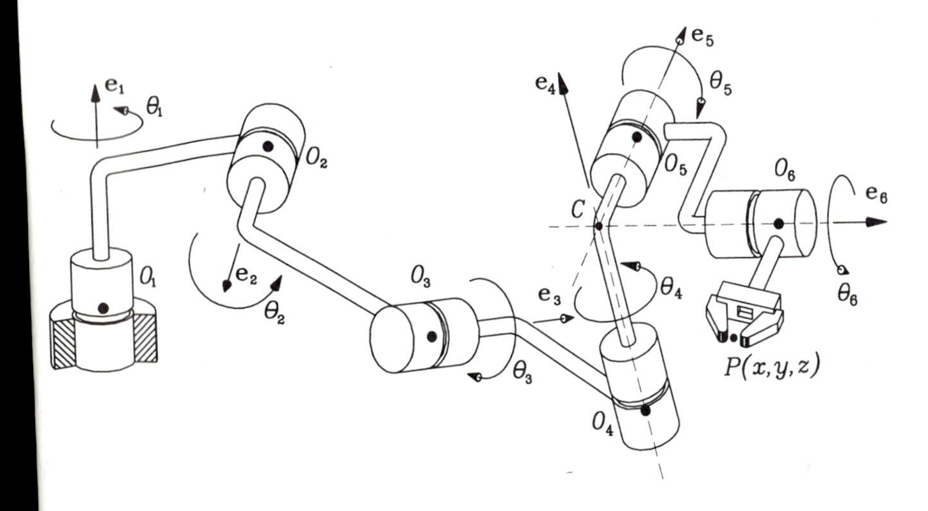

FIGURE 4.8. A general 6R manipulator with decoupled architecture.

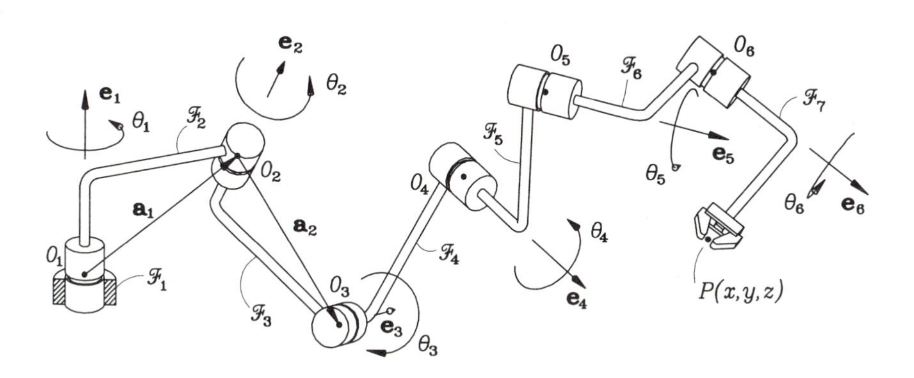

FIGURE 4.7. Serial six-axis manipulator.

to avoid the introduction of spurious roots and, with this, an increase in the degree of that polynomial. For this reason, we limit this chapter to the study of the geometric IKP of *decoupled* six-axis manipulators, to be defined presently. The IKP of the most general six-revolute serial manipulator is studied in Chapter 8.

In studying the DKP of six-axis manipulators, we need not limit ourselves to a particular architecture. We thus study here the DKP of general manipulators, such as the one sketched in Fig. 4.7. This manipulator consists of seven rigid bodies, or links, coupled by six revolute joints. Correspondingly, we have seven frames, $\mathcal{F}_1, \mathcal{F}_2, \ldots, \mathcal{F}_7$, the ith frame fixed to the $(i-1)$st link, $\mathcal{F}_1$ being termed the *base frame*, because it is fixed to the base of the manipulator. Manipulators with joints of the prismatic type are simpler to study and can be treated using correspondingly simpler procedures.

A line $\mathcal{L}_i$ is associated with the axis of the ith revolute joint, and a positive direction along this line is defined arbitrarily through a unit vector $\mathbf{e}_i$. For a prismatic pair, a line $\mathcal{L}_i$ can be also defined, as a line having the direction of the pair but whose location is undefined; the analyst, then, has the freedom to locate this axis conveniently. Thus, a rotation of the ith link with respect to the $(i-1)$st link or correspondingly, of $\mathcal{F}_{i+1}$ with respect to $\mathcal{F}_i$, is totally defined by the geometry of the said link, i.e., by the DH parameters a_i, b_i, and α_i, plus $\mathbf{e}_i$ and its associated joint variable θ_i. Then, the DH parameters and the joint variables define uniquely the posture of the manipulator. In particular, the relative position and orientation of $\mathcal{F}_{i+1}$ with respect to $\mathcal{F}_i$ is given by matrix $\mathbf{Q}_i$ and vector $\mathbf{a}_i$, respectively, which were defined in Section 4.2 and are displayed below for quick reference:

$$\mathbf{Q}_i = \begin{bmatrix} \cos\theta_i & -\lambda_i\sin\theta_i & \mu_i\sin\theta_i \\ \sin\theta_i & \lambda_i\cos\theta_i & -\mu_i\cos\theta_i \\ 0 & \mu_i & \lambda_i \end{bmatrix}, \quad \mathbf{a}_i = \begin{bmatrix} a_i\cos\theta_i \\ a_i\sin\theta_i \\ b_i \end{bmatrix} \tag{4.7}$$

Thus, $\mathbf{Q}_i$ and $\mathbf{a}_i$ denote, respectively, the matrix rotating $\mathcal{F}_i$ into an orientation coincident with that of $\mathcal{F}_{i+1}$ and the vector joining the origin of $\mathcal{F}_i$ with that of $\mathcal{F}_{i+1}$, directed from the former to the latter. Moreover, $\mathbf{Q}_i$ and $\mathbf{a}_i$, as given in eq.(4.7), are represented in $\mathcal{F}_i$ coordinates. The equations leading to the kinematic model under study are known as the *kinematic displacement equations.* It is noteworthy that the problem under study is equivalent to the *input-output analysis problem* of a seven-revolute linkage with one degree of freedom and one single kinematic loop (Duffy, 1980). Because of this equivalence with a *closed kinematic chain*, sometimes the displacement equations are also termed *closure equations.* These equations relate the orientation of the EE, as produced by the joint coordinates, with the prescribed orientation $\mathbf{Q}$ and the position vector $\mathbf{p}$ of the *operation point* P of the EE. That is, the orientation $\mathbf{Q}$ of the EE is obtained as a result of the six individual rotations $\{\mathbf{Q}_i\}_1^6$ about each revolute axis through an angle θ_i, in a *sequential order*, from 1 to 6. If, for example, the foregoing relations are expressed in $\mathcal{F}_1$, then

$$[\mathbf{Q}_6]_1[\mathbf{Q}_5]_1[\mathbf{Q}_4]_1[\mathbf{Q}_3]_1[\mathbf{Q}_2]_1[\mathbf{Q}_1]_1 = [\mathbf{Q}]_1 \qquad (4.8a)$$

$$[\mathbf{a}_1]_1 + [\mathbf{a}_2]_1 + [\mathbf{a}_3]_1 + [\mathbf{a}_4]_1 + [\mathbf{a}_5]_1 + [\mathbf{a}_6]_1 = [\mathbf{p}]_1 \qquad (4.8b)$$

Notice that the above equations require that all vectors and matrices involved be expressed *in the same coordinate frame*. However, we derived in Section 4.2 general expressions for $\mathbf{Q}_i$ and $\mathbf{a}_i$ in $\mathcal{F}_i$, eqs.(4.1d) and (4.3b), respectively. It is hence convenient to represent the foregoing relations in each individual frame, which can be readily done by means of similarity transformations. Indeed, if we apply the transformations (4.5a & b) to each of $[\mathbf{a}_i]_1$ and $[\mathbf{Q}_i]_1$, respectively, we obtain $\mathbf{a}_i$ or correspondingly, $\mathbf{Q}_i$, in $\mathcal{F}_i$. Therefore, eq.(4.8a) becomes

$$[\mathbf{Q}_1]_1[\mathbf{Q}_2]_2[\mathbf{Q}_3]_3[\mathbf{Q}_4]_4[\mathbf{Q}_5]_5[\mathbf{Q}_6]_6 = [\mathbf{Q}]_1$$

Now for compactness, let us represent $[\mathbf{Q}]_1$ simply by $\mathbf{Q}$ and let us recall the abbreviated notation introduced in eq.(4.1c), whereby $[\mathbf{Q}_i]_i$ is denoted simply by $\mathbf{Q}_i$, thereby obtaining

$$\mathbf{Q}_1\mathbf{Q}_2\mathbf{Q}_3\mathbf{Q}_4\mathbf{Q}_5\mathbf{Q}_6 = \mathbf{Q} \qquad (4.9a)$$

Likewise, eq.(4.8b) becomes

$$\mathbf{a}_1+\mathbf{Q}_1\mathbf{a}_2+\mathbf{Q}_1\mathbf{Q}_2\mathbf{a}_3+\mathbf{Q}_1\mathbf{Q}_2\mathbf{Q}_3\mathbf{a}_4+\mathbf{Q}_1\mathbf{Q}_2\mathbf{Q}_3\mathbf{Q}_4\mathbf{a}_5+\mathbf{Q}_1\mathbf{Q}_2\mathbf{Q}_3\mathbf{Q}_4\mathbf{Q}_5\mathbf{a}_6 = \mathbf{p} \qquad (4.9b)$$

in which both sides are given in base-frame coordinates. Equations (4.9a & b) above can be cast in a more compact form if homogeneous transformations, as defined in Section 2.5, are now introduced. Thus, if we let $\mathbf{T}_i \equiv \{\mathbf{T}_i\}_i$ be the 4×4 matrix transforming $\mathcal{F}_{i+1}$-coordinates into $\mathcal{F}_i$-coordinates, the foregoing equations can be written in 4×4 matrix form, namely,

$$\mathbf{T}_1\mathbf{T}_2\mathbf{T}_3\mathbf{T}_4\mathbf{T}_5\mathbf{T}_6 = \mathbf{T} \qquad (4.10)$$

with $\mathbf{T}$ denoting the transformation of coordinates from the end frame to the base frame. Thus, $\mathbf{T}$ contains the pose of the end-eff

In order to ease the discussion ahead, we introduce now a fev tions. A scalar, vector, or matrix expression is said to be *multi* a set of vectors $\{\mathbf{v}_i\}_1^N$ if those vectors appear only linearly in t expression. This does not prevent products of components of thos from occurring, as long as each product contains only one comp the same vector. Alternatively, we can say that the expression o is multilinear in the aforementioned set of vectors if and only if th derivative of that expression with respect to vector $\mathbf{v}_i$ is indepe $\mathbf{v}_i$, for $i = 1,\ldots,N$. For example, every matrix $\mathbf{Q}_i$ and every v defined in eqs.(4.1d) and (4.3b), respectively, is linear in vector $\mathbf{x}_i$ is defined as

$$\mathbf{x}_i \equiv \begin{bmatrix} \cos\theta_i \\ \sin\theta_i \end{bmatrix}$$

Moreover, the product $\mathbf{Q}_1\mathbf{Q}_2\mathbf{Q}_3\mathbf{Q}_4\mathbf{Q}_5\mathbf{Q}_6$ appearing in eq.(4.9a) i *ear*, or simply, *multilinear*, in vectors $\{\mathbf{x}_i\}_1^6$. Likewise, the sum in eq.(4.9b) is multilinear in the same set of vectors. By the sa a scalar, vector, or matrix expression is said to be *multiquadra* same set of vectors if those vectors appear at most quadratica said expression. That is, the expression of interest may contain of the components of all those vectors, as long as those product in turn, a maximum of two components of the same vector, incl same component squared. Qualifiers like *multicubic*, *multiquartic* similar meanings.

Further, we partition matrix $\mathbf{Q}_i$ rowwise and columnwise, nar

$$\mathbf{Q}_i \equiv \begin{bmatrix} \mathbf{m}_i^T \\ \mathbf{n}_i^T \\ \mathbf{o}_i^T \end{bmatrix} \equiv [\,\mathbf{p}_i \quad \mathbf{q}_i \quad \mathbf{u}_i\,]$$

It is pointed out that the third row of $\mathbf{Q}_i$, $\mathbf{o}_i^T$, is independent o that will be found useful in the forthcoming derivations. Further that according with the DH notation, the unit vector $\mathbf{e}_i$ in the d the ith joint axis in Fig. 4.7 has $\mathcal{F}_i$-components given by

$$[\mathbf{e}_i]_i = \begin{bmatrix} 0 \\ 0 \\ 1 \end{bmatrix} \equiv \mathbf{e}$$

Henceforth, $\mathbf{e}$ is used to represent a 3-dimensional array wi component equal to unity, its other components vanishing. Thu

$$\mathbf{Q}_i\mathbf{o}_i \equiv \mathbf{Q}_i^T\mathbf{u}_i = \mathbf{e}$$

or

$$\mathbf{u}_i = \mathbf{Q}_i\mathbf{e}, \quad \mathbf{o}_i = \mathbf{Q}_i^T\mathbf{e}$$

In the two subsections below, a procedure is derived for determining all the inverse kinematics solutions of decoupled manipulators. In view of the decoupled architecture of these manipulators, we study their inverse kinematics by decoupling the positioning problem from the orientation problem.

4.4.1 The Positioning Problem

The inverse kinematics of the robotic manipulators under study begins by *decoupling* the positioning and orientation problems. Moreover, we must solve first the positioning problem, which is done in this subsection.

Let C denote the intersection of axes 4, 5, and 6, i.e., the center of the spherical wrist, and let $\mathbf{c}$ denote the position vector of this point. Clearly, the position of C is independent of joint angles θ_4, θ_5, and θ_6; hence, only the first three joints are to be considered for this analysis. The arm structure depicted in Fig. 4.9 will then be analyzed. From that figure,

$$\mathbf{a}_1 + \mathbf{Q}_1\mathbf{a}_2 + \mathbf{Q}_1\mathbf{Q}_2\mathbf{a}_3 + \mathbf{Q}_1\mathbf{Q}_2\mathbf{Q}_3\mathbf{a}_4 = \mathbf{c} \tag{4.16}$$

where the two sides are expressed in $\mathcal{F}_1$-coordinates. This equation can be readily rewritten in the form

$$\mathbf{a}_2 + \mathbf{Q}_2\mathbf{a}_3 + \mathbf{Q}_2\mathbf{Q}_3\mathbf{a}_4 = \mathbf{Q}_1^T(\mathbf{c} - \mathbf{a}_1)$$

or if we recall eq.(4.3d),

$$\mathbf{Q}_2(\mathbf{b}_2 + \mathbf{Q}_3\mathbf{b}_3 + \mathbf{Q}_3\mathbf{Q}_4\mathbf{b}_4) = \mathbf{Q}_1^T\mathbf{c} - \mathbf{b}_1$$

However, since we are dealing with a decoupled manipulator, we have

$$\mathbf{a}_4 \equiv \mathbf{Q}_4\mathbf{b}_4 \equiv \begin{bmatrix} 0 \\ 0 \\ b_4 \end{bmatrix} \equiv b_4\mathbf{e}$$

which has been rewritten as the product of constant b_4 times the unit vector $\mathbf{e}$ defined in eq.(4.13).

Thus, the product $\mathbf{Q}_3\mathbf{Q}_4\mathbf{b}_4$ reduces to

$$\mathbf{Q}_3\mathbf{Q}_4\mathbf{b}_4 \equiv b_4\mathbf{Q}_3\mathbf{e} \equiv b_4\mathbf{u}_3$$

with $\mathbf{u}_i$ defined in eq.(4.14b). Hence, eq.(4.16) leads to

$$\mathbf{Q}_2(\mathbf{b}_2 + \mathbf{Q}_3\mathbf{b}_3 + b_4\mathbf{u}_3) = \mathbf{Q}_1^T\mathbf{c} - \mathbf{b}_1 \tag{4.17}$$

Further, an expression for $\mathbf{c}$ can be derived in terms of $\mathbf{p}$, the position vector of the operation point of the EE, and $\mathbf{Q}$, namely,

$$\mathbf{c} = \mathbf{p} - \mathbf{Q}_1\mathbf{Q}_2\mathbf{Q}_3\mathbf{Q}_4\mathbf{a}_5 - \mathbf{Q}_1\mathbf{Q}_2\mathbf{Q}_3\mathbf{Q}_4\mathbf{Q}_5\mathbf{a}_6 \tag{4.18a}$$

Now, since $a_5 = b_5 = 0$, we have that $\mathbf{a}_5 = \mathbf{0}$, eq.(4.18a) thus yielding

$$\mathbf{c} = \mathbf{p} - \mathbf{Q}\mathbf{Q}_6^T\mathbf{a}_6 \equiv \mathbf{p} - \mathbf{Q}\mathbf{b}_6 \tag{4.18b}$$

Moreover, the base coordinates of P and C, and hence, the $\mathcal{F}_1$-components of their position vectors $\mathbf{p}$ and $\mathbf{c}$, are defined as

$$[\mathbf{p}]_1 = \begin{bmatrix} x \\ y \\ z \end{bmatrix}, \quad [\mathbf{c}]_1 = \begin{bmatrix} x_C \\ y_C \\ z_C \end{bmatrix}$$

so that eq.(4.18b) can be expanded in the form

$$\begin{bmatrix} x_C \\ y_C \\ z_C \end{bmatrix} = \begin{bmatrix} x - (q_{11}a_6 + q_{12}b_6\mu_6 + q_{13}b_6\lambda_6) \\ y - (q_{21}a_6 + q_{22}b_6\mu_6 + q_{23}b_6\lambda_6) \\ z - (q_{31}a_6 + q_{32}b_6\mu_6 + q_{33}b_6\lambda_6) \end{bmatrix} \tag{4.18c}$$

where q_{ij} is the (i,j) entry of $[\mathbf{Q}]_1$, and the positioning problem now becomes one of finding the first three joint angles necessary to position point C at a point of base coordinates x_C, y_C, and z_C. We thus have three unknowns, but we also have three equations at our disposal, namely, the three scalar equations of eq.(4.17), and we should be able to solve the problem at hand.

In solving the foregoing system of equations, we first note that (i) the left-hand side of eq.(4.17) appears multiplied by $\mathbf{Q}_2$; and (ii) θ_2 does not appear in the right-hand side. This implies that (i) if the Euclidean norms of the two sides of that equation are equated, the resulting equation will

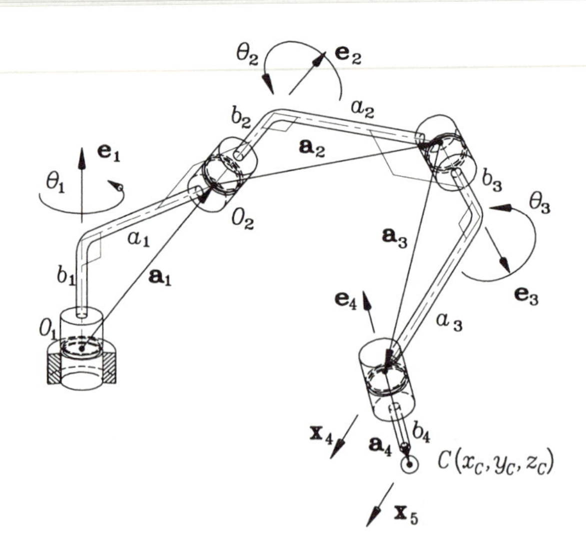

FIGURE 4.9. Three-axis, serial, positioning manipulator.

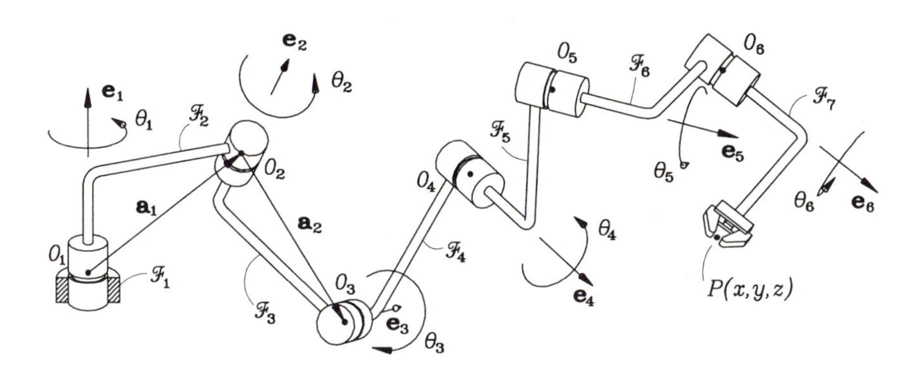

FIGURE 4.7. Serial six-axis manipulator.

to avoid the introduction of spurious roots and, with this, an increase in the degree of that polynomial. For this reason, we limit this chapter to the study of the geometric IKP of *decoupled* six-axis manipulators, to be defined presently. The IKP of the most general six-revolute serial manipulator is studied in Chapter 8.

In studying the DKP of six-axis manipulators, we need not limit ourselves to a particular architecture. We thus study here the DKP of general manipulators, such as the one sketched in Fig. 4.7. This manipulator consists of seven rigid bodies, or links, coupled by six revolute joints. Correspondingly, we have seven frames, $\mathcal{F}_1$, $\mathcal{F}_2$, ..., $\mathcal{F}_7$, the ith frame fixed to the $(i-1)$st link, $\mathcal{F}_1$ being termed the *base frame*, because it is fixed to the base of the manipulator. Manipulators with joints of the prismatic type are simpler to study and can be treated using correspondingly simpler procedures.

A line $\mathcal{L}_i$ is associated with the axis of the ith revolute joint, and a positive direction along this line is defined arbitrarily through a unit vector $\mathbf{e}_i$. For a prismatic pair, a line $\mathcal{L}_i$ can be also defined, as a line having the direction of the pair but whose location is undefined; the analyst, then, has the freedom to locate this axis conveniently. Thus, a rotation of the ith link with respect to the $(i-1)$st link or correspondingly, of $\mathcal{F}_{i+1}$ with respect to $\mathcal{F}_i$, is totally defined by the geometry of the said link, i.e., by the DH parameters a_i, b_i, and α_i, plus $\mathbf{e}_i$ and its associated joint variable θ_i. Then, the DH parameters and the joint variables define uniquely the posture of the manipulator. In particular, the relative position and orientation of $\mathcal{F}_{i+1}$ with respect to $\mathcal{F}_i$ is given by matrix $\mathbf{Q}_i$ and vector $\mathbf{a}_i$, respectively, which were defined in Section 4.2 and are displayed below for quick reference:

$$\mathbf{Q}_i = \begin{bmatrix} \cos\theta_i & -\lambda_i \sin\theta_i & \mu_i \sin\theta_i \\ \sin\theta_i & \lambda_i \cos\theta_i & -\mu_i \cos\theta_i \\ 0 & \mu_i & \lambda_i \end{bmatrix}, \quad \mathbf{a}_i = \begin{bmatrix} a_i \cos\theta_i \\ a_i \sin\theta_i \\ b_i \end{bmatrix} \tag{4.7}$$

Thus, $\mathbf{Q}_i$ and $\mathbf{a}_i$ denote, respectively, the matrix rotating $\mathcal{F}_i$ into an orientation coincident with that of $\mathcal{F}_{i+1}$ and the vector joining the origin of $\mathcal{F}_i$ with that of $\mathcal{F}_{i+1}$, directed from the former to the latter. Moreover, $\mathbf{Q}_i$ and $\mathbf{a}_i$, as given in eq.(4.7), are represented in $\mathcal{F}_i$ coordinates. The equations leading to the kinematic model under study are known as the *kinematic displacement equations.* It is noteworthy that the problem under study is equivalent to the *input-output analysis problem* of a seven-revolute linkage with one degree of freedom and one single kinematic loop (Duffy, 1980). Because of this equivalence with a *closed kinematic chain*, sometimes the displacement equations are also termed *closure equations.* These equations relate the orientation of the EE, as produced by the joint coordinates, with the prescribed orientation $\mathbf{Q}$ and the position vector $\mathbf{p}$ of the *operation point* P of the EE. That is, the orientation $\mathbf{Q}$ of the EE is obtained as a result of the six individual rotations $\{\mathbf{Q}_i\}_1^6$ about each revolute axis through an angle θ_i, in a *sequential order*, from 1 to 6. If, for example, the foregoing relations are expressed in $\mathcal{F}_1$, then

$$[\mathbf{Q}_6]_1[\mathbf{Q}_5]_1[\mathbf{Q}_4]_1[\mathbf{Q}_3]_1[\mathbf{Q}_2]_1[\mathbf{Q}_1]_1 = [\mathbf{Q}]_1 \tag{4.8a}$$

$$[\mathbf{a}_1]_1 + [\mathbf{a}_2]_1 + [\mathbf{a}_3]_1 + [\mathbf{a}_4]_1 + [\mathbf{a}_5]_1 + [\mathbf{a}_6]_1 = [\mathbf{p}]_1 \tag{4.8b}$$

Notice that the above equations require that all vectors and matrices involved be expressed *in the same coordinate frame.* However, we derived in Section 4.2 general expressions for $\mathbf{Q}_i$ and $\mathbf{a}_i$ in $\mathcal{F}_i$, eqs.(4.1d) and (4.3b), respectively. It is hence convenient to represent the foregoing relations in each individual frame, which can be readily done by means of similarity transformations. Indeed, if we apply the transformations (4.5a & b) to each of $[\mathbf{a}_i]_1$ and $[\mathbf{Q}_i]_1$, respectively, we obtain $\mathbf{a}_i$ or correspondingly, $\mathbf{Q}_i$, in $\mathcal{F}_i$. Therefore, eq.(4.8a) becomes

$$[\mathbf{Q}_1]_1[\mathbf{Q}_2]_2[\mathbf{Q}_3]_3[\mathbf{Q}_4]_4[\mathbf{Q}_5]_5[\mathbf{Q}_6]_6 = [\mathbf{Q}]_1$$

Now for compactness, let us represent $[\mathbf{Q}]_1$ simply by $\mathbf{Q}$ and let us recall the abbreviated notation introduced in eq.(4.1c), whereby $[\mathbf{Q}_i]_i$ is denoted simply by $\mathbf{Q}_i$, thereby obtaining

$$\mathbf{Q}_1\mathbf{Q}_2\mathbf{Q}_3\mathbf{Q}_4\mathbf{Q}_5\mathbf{Q}_6 = \mathbf{Q} \tag{4.9a}$$

Likewise, eq.(4.8b) becomes

$$\mathbf{a}_1+\mathbf{Q}_1\mathbf{a}_2+\mathbf{Q}_1\mathbf{Q}_2\mathbf{a}_3+\mathbf{Q}_1\mathbf{Q}_2\mathbf{Q}_3\mathbf{a}_4+\mathbf{Q}_1\mathbf{Q}_2\mathbf{Q}_3\mathbf{Q}_4\mathbf{a}_5+\mathbf{Q}_1\mathbf{Q}_2\mathbf{Q}_3\mathbf{Q}_4\mathbf{Q}_5\mathbf{a}_6 = \mathbf{p} \tag{4.9b}$$

in which both sides are given in base-frame coordinates. Equations (4.9a & b) above can be cast in a more compact form if homogeneous transformations, as defined in Section 2.5, are now introduced. Thus, if we let $\mathbf{T}_i \equiv \{\mathbf{T}_i\}_i$ be the 4×4 matrix transforming $\mathcal{F}_{i+1}$-coordinates into $\mathcal{F}_i$-coordinates, the foregoing equations can be written in 4×4 matrix form, namely,

$$\mathbf{T}_1\mathbf{T}_2\mathbf{T}_3\mathbf{T}_4\mathbf{T}_5\mathbf{T}_6 = \mathbf{T} \tag{4.10}$$

with $\mathbf{T}$ denoting the transformation of coordinates from the end-effector frame to the base frame. Thus, $\mathbf{T}$ contains the pose of the end-effector.

In order to ease the discussion ahead, we introduce now a few definitions. A scalar, vector, or matrix expression is said to be *multilinear* in a set of vectors $\{\mathbf{v}_i\}_1^N$ if those vectors appear only linearly in the same expression. This does not prevent products of components of those vectors from occurring, as long as each product contains only one component of the same vector. Alternatively, we can say that the expression of interest is multilinear in the aforementioned set of vectors if and only if the partial derivative of that expression with respect to vector $\mathbf{v}_i$ is independent of $\mathbf{v}_i$, for $i = 1, \ldots, N$. For example, every matrix $\mathbf{Q}_i$ and every vector $\mathbf{a}_i$, defined in eqs.(4.1d) and (4.3b), respectively, is linear in vector $\mathbf{x}_i$, where $\mathbf{x}_i$ is defined as

$$\mathbf{x}_i \equiv \begin{bmatrix} \cos\theta_i \\ \sin\theta_i \end{bmatrix} \tag{4.11}$$

Moreover, the product $\mathbf{Q}_1\mathbf{Q}_2\mathbf{Q}_3\mathbf{Q}_4\mathbf{Q}_5\mathbf{Q}_6$ appearing in eq.(4.9a) is *hexalinear*, or simply, *multilinear*, in vectors $\{\mathbf{x}_i\}_1^6$. Likewise, the sum appearing in eq.(4.9b) is multilinear in the same set of vectors. By the same token, a scalar, vector, or matrix expression is said to be *multiquadratic* in the same set of vectors if those vectors appear at most quadratically in the said expression. That is, the expression of interest may contain products of the components of all those vectors, as long as those products contain, in turn, a maximum of two components of the same vector, including the same component squared. Qualifiers like *multicubic*, *multiquartic*, etc. bear similar meanings.

Further, we partition matrix $\mathbf{Q}_i$ rowwise and columnwise, namely,

$$\mathbf{Q}_i \equiv \begin{bmatrix} \mathbf{m}_i^T \\ \mathbf{n}_i^T \\ \mathbf{o}_i^T \end{bmatrix} \equiv [\,\mathbf{p}_i \quad \mathbf{q}_i \quad \mathbf{u}_i\,] \tag{4.12}$$

It is pointed out that the third row of $\mathbf{Q}_i$, $\mathbf{o}_i^T$, is independent of θ_i, a fact that will be found useful in the forthcoming derivations. Furthermore, note that according with the DH notation, the unit vector $\mathbf{e}_i$ in the direction of the ith joint axis in Fig. 4.7 has $\mathcal{F}_i$-components given by

$$[\,\mathbf{e}_i\,]_i = \begin{bmatrix} 0 \\ 0 \\ 1 \end{bmatrix} \equiv \mathbf{e} \tag{4.13}$$

Henceforth, $\mathbf{e}$ is used to represent a 3-dimensional array with its last component equal to unity, its other components vanishing. Thus, we have

$$\mathbf{Q}_i\mathbf{o}_i \equiv \mathbf{Q}_i^T\mathbf{u}_i = \mathbf{e} \tag{4.14a}$$

or

$$\mathbf{u}_i = \mathbf{Q}_i\mathbf{e}, \quad \mathbf{o}_i = \mathbf{Q}_i^T\mathbf{e} \tag{4.14b}$$

That is, if we regard $\mathbf{e}$ in the first of the foregoing relations as $[\,\mathbf{e}_{i+1}\,]_{i+1}$, and as $[\,\mathbf{e}_i\,]_i$ in the second relation, then, from the coordinate transformations of eqs.(4.4a & b),

$$\mathbf{u}_i = [\,\mathbf{e}_{i+1}\,]_i, \quad \text{and} \quad \mathbf{o}_i = [\,\mathbf{e}_i\,]_{i+1} \tag{4.15}$$

4.4 The IKP of Decoupled Manipulators

Industrial manipulators are frequently supplied with a special architecture that allows a decoupling of the positioning problem from the orientation problem. In fact, a determinant design criterion in this regard has been that the manipulator be kinematically invertible, i.e., that it lend itself to a closed-form inverse kinematics solution. Although the class of manipulators with this feature is quite broad, we will focus on a special kind, the most frequently encountered in commercial manipulators, that we will term *decoupled.* Decoupled manipulators were defined in Section 4.2 as those whose last three joints have intersecting axes. These joints, then, constitute the *wrist* of the manipulator, which is said to be *spherical,* because when the point of intersection of the three wrist axes, C, is kept fixed, all the points of the wrist move on spheres centered at C. In terms of the DH parameters of the manipulator, in a decoupled manipulator $a_4 = a_5 = b_5 = 0$, and thus, the origins of frames 4, 5, and 6 are coincident. All other DH parameters can assume arbitrary values. A general decoupled manipulator is shown in Fig. 4.8, where the wrist is represented as a concatenation of three revolutes with intersecting axes.

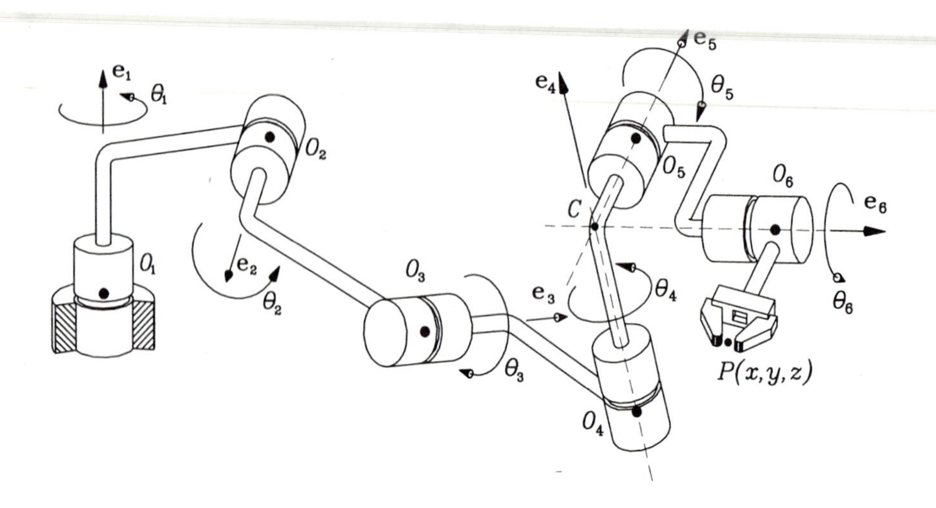

FIGURE 4.8. A general 6R manipulator with decoupled architecture.

In the two subsections below, a procedure is derived for determining all the inverse kinematics solutions of decoupled manipulators. In view of the decoupled architecture of these manipulators, we study their inverse kinematics by decoupling the positioning problem from the orientation problem.

4.4.1 The Positioning Problem

The inverse kinematics of the robotic manipulators under study begins by *decoupling* the positioning and orientation problems. Moreover, we must solve first the positioning problem, which is done in this subsection.

Let C denote the intersection of axes 4, 5, and 6, i.e., the center of the spherical wrist, and let $\mathbf{c}$ denote the position vector of this point. Clearly, the position of C is independent of joint angles θ_4, θ_5, and θ_6; hence, only the first three joints are to be considered for this analysis. The arm structure depicted in Fig. 4.9 will then be analyzed. From that figure,

$$\mathbf{a}_1 + \mathbf{Q}_1\mathbf{a}_2 + \mathbf{Q}_1\mathbf{Q}_2\mathbf{a}_3 + \mathbf{Q}_1\mathbf{Q}_2\mathbf{Q}_3\mathbf{a}_4 = \mathbf{c} \tag{4.16}$$

where the two sides are expressed in $\mathcal{F}_1$-coordinates. This equation can be readily rewritten in the form

$$\mathbf{a}_2 + \mathbf{Q}_2\mathbf{a}_3 + \mathbf{Q}_2\mathbf{Q}_3\mathbf{a}_4 = \mathbf{Q}_1^T(\mathbf{c} - \mathbf{a}_1)$$

or if we recall eq.(4.3d),

$$\mathbf{Q}_2(\mathbf{b}_2 + \mathbf{Q}_3\mathbf{b}_3 + \mathbf{Q}_3\mathbf{Q}_4\mathbf{b}_4) = \mathbf{Q}_1^T\mathbf{c} - \mathbf{b}_1$$

However, since we are dealing with a decoupled manipulator, we have

$$\mathbf{a}_4 \equiv \mathbf{Q}_4\mathbf{b}_4 \equiv \begin{bmatrix} 0 \\ 0 \\ b_4 \end{bmatrix} \equiv b_4\mathbf{e}$$

which has been rewritten as the product of constant b_4 times the unit vector $\mathbf{e}$ defined in eq.(4.13).

Thus, the product $\mathbf{Q}_3\mathbf{Q}_4\mathbf{b}_4$ reduces to

$$\mathbf{Q}_3\mathbf{Q}_4\mathbf{b}_4 \equiv b_4\mathbf{Q}_3\mathbf{e} \equiv b_4\mathbf{u}_3$$

with $\mathbf{u}_i$ defined in eq.(4.14b). Hence, eq.(4.16) leads to

$$\mathbf{Q}_2(\mathbf{b}_2 + \mathbf{Q}_3\mathbf{b}_3 + b_4\mathbf{u}_3) = \mathbf{Q}_1^T\mathbf{c} - \mathbf{b}_1 \tag{4.17}$$

Further, an expression for $\mathbf{c}$ can be derived in terms of $\mathbf{p}$, the position vector of the operation point of the EE, and $\mathbf{Q}$, namely,

$$\mathbf{c} = \mathbf{p} - \mathbf{Q}_1\mathbf{Q}_2\mathbf{Q}_3\mathbf{Q}_4\mathbf{a}_5 - \mathbf{Q}_1\mathbf{Q}_2\mathbf{Q}_3\mathbf{Q}_4\mathbf{Q}_5\mathbf{a}_6 \tag{4.18a}$$

Now, since $a_5 = b_5 = 0$, we have that $\mathbf{a}_5 = \mathbf{0}$, eq.(4.18a) thus yielding

$$\mathbf{c} = \mathbf{p} - \mathbf{Q}\mathbf{Q}_6^T\mathbf{a}_6 \equiv \mathbf{p} - \mathbf{Q}\mathbf{b}_6 \tag{4.18b}$$

Moreover, the base coordinates of P and C, and hence, the $\mathcal{F}_1$-components of their position vectors $\mathbf{p}$ and $\mathbf{c}$, are defined as

$$[\mathbf{p}]_1 = \begin{bmatrix} x \\ y \\ z \end{bmatrix}, \quad [\mathbf{c}]_1 = \begin{bmatrix} x_C \\ y_C \\ z_C \end{bmatrix}$$

so that eq.(4.18b) can be expanded in the form

$$\begin{bmatrix} x_C \\ y_C \\ z_C \end{bmatrix} = \begin{bmatrix} x - (q_{11}a_6 + q_{12}b_6\mu_6 + q_{13}b_6\lambda_6) \\ y - (q_{21}a_6 + q_{22}b_6\mu_6 + q_{23}b_6\lambda_6) \\ z - (q_{31}a_6 + q_{32}b_6\mu_6 + q_{33}b_6\lambda_6) \end{bmatrix} \tag{4.18c}$$

where q_{ij} is the (i,j) entry of $[\mathbf{Q}]_1$, and the positioning problem now becomes one of finding the first three joint angles necessary to position point C at a point of base coordinates x_C, y_C, and z_C. We thus have three unknowns, but we also have three equations at our disposal, namely, the three scalar equations of eq.(4.17), and we should be able to solve the problem at hand.

In solving the foregoing system of equations, we first note that (i) the left-hand side of eq.(4.17) appears multiplied by $\mathbf{Q}_2$; and (ii) θ_2 does not appear in the right-hand side. This implies that (i) if the Euclidean norms of the two sides of that equation are equated, the resulting equation will

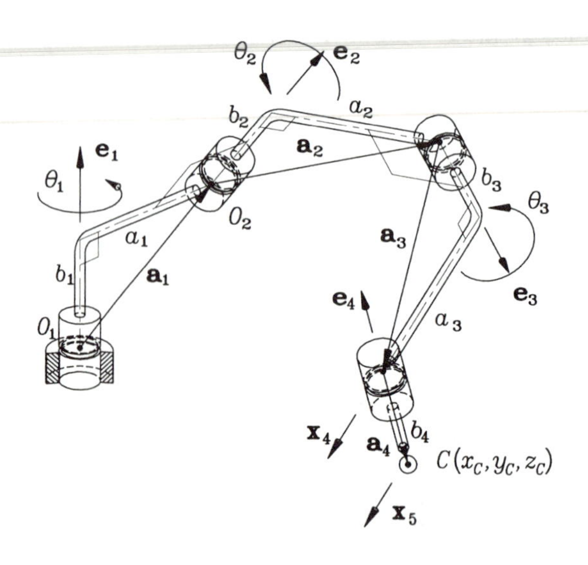

FIGURE 4.9. Three-axis, serial, positioning manipulator.

not contain θ_2; and (*ii*) the third scalar equation of the same equation is independent of θ_2, by virtue of the structure of the $\mathbf{Q}_i$ matrices displayed in eq.(4.1d). Thus, we have two equations free of θ_2, which allows us to calculate the two remaining unknowns θ_1 and θ_3.

Let the Euclidean norm of the left-hand side of eq.(4.17) be denoted by l, that of its right-hand side by r. We then have

$$l^2 \equiv b_2^2 + b_3^2 + b_4^2 + 2\mathbf{b}_2^T\mathbf{Q}_3\mathbf{b}_3 + 2b_4\mathbf{b}_2^T\mathbf{u}_3 + 2b_3$$
$$r^2 \equiv \|\mathbf{c}\|^2 + \|\mathbf{b}_1\|^2 - 2\mathbf{b}_1^T\mathbf{Q}_1^T\mathbf{c}$$

from which it is apparent that l^2 is linear in $\mathbf{x}_3$ and r^2 is linear in $\mathbf{x}_1$, for $\mathbf{x}_i$ defined in eq.(4.11). Upon equating l^2 with r^2, then, an equation linear in $\mathbf{x}_1$ and $\mathbf{x}_3$—not bilinear in these vectors—is readily derived, namely,

$$Ac_1 + Bs_1 + Cc_3 + Ds_3 + E = 0 \tag{4.19a}$$

whose coefficients do not contain any unknown, i.e.,

$$A = 2a_1x_C \tag{4.19b}$$
$$B = 2a_1y_C \tag{4.19c}$$
$$C = 2a_2a_3 - 2b_2b_4\mu_2\mu_3 \tag{4.19d}$$
$$D = 2a_3b_2\mu_2 + 2a_2b_4\mu_3 \tag{4.19e}$$
$$E = a_2^2 + a_3^2 + b_2^2 + b_3^2 + b_4^2 - a_1^2 - x_C^2 - y_C^2 - (z_C - b_1)^2 + 2b_2b_3\lambda_2 + 2b_2b_4\lambda_2\lambda_3 + 2b_3b_4\lambda_3 \tag{4.19f}$$

Moreover, the third scalar equation of eq.(4.17) takes on the form

$$Fc_1 + Gs_1 + Hc_3 + Is_3 + J = 0 \tag{4.20a}$$

whose coefficients, again, do not contain any unknown, as shown below:

$$F = y_C\mu_1 \tag{4.20b}$$
$$G = -x_C\mu_1 \tag{4.20c}$$
$$H = -b_4\mu_2\mu_3 \tag{4.20d}$$
$$I = a_3\mu_2 \tag{4.20e}$$
$$J = b_2 + b_3\lambda_2 + b_4\lambda_2\lambda_3 - (z_C - b_1)\lambda_1 \tag{4.20f}$$

Thus, we have derived two nonlinear equations in θ_1 and θ_3 that are linear in c_1, s_1, c_3, and s_3. Each of these equations thus defines a contour in the θ_1-θ_3 plane, their intersections determining all real solutions to the problem at hand.

Note that if c_i and s_i are substituted for their equivalents in terms of $\tan(\theta_i/2)$, for $i = 1, 3$, then two biquadratic polynomial equations in $\tan(\theta_1/2)$ and $\tan(\theta_3/2)$ are derived. Thus, one can eliminate one of these variables from the foregoing equations, thereby reducing the two equations to a single quartic polynomial equation in the other variable. The quartic equation thus resulting is called the *characteristic equation* of the problem at hand. Alternatively, the two above equations, eqs.(4.19a) and (4.20a), can be solved for, say, c_1 and s_1 in terms of the data and c_3 and s_3, namely,

$$c_1 = \frac{-G(Cc_3 + Ds_3 + E) + B(Hc_3 + Is_3 + J)}{\Delta_1} \tag{4.21a}$$

$$s_1 = \frac{F(Cc_3 + Ds_3 + E) - A(Hc_3 + Is_3 + J)}{\Delta_1} \tag{4.21b}$$

with Δ_1 defined as

$$\Delta_1 = AG - FB = -2a_1\mu_1(x_C^2 + y_C^2) \tag{4.21c}$$

Note that in trajectory planning, to be studied in Chapter 5, Δ_1 can be computed *off-line*, i.e., prior to setting the manipulator into operation, for it is a function solely of the manipulator parameters and the Cartesian coordinates of a point lying on the path to be tracked. Moreover, the above calculations are possible as long as Δ_1 does not vanish. Now, Δ_1 vanishes if and only if any of the factors a_1, μ_1, and $x_C^2 + y_C^2$ does. The first two conditions are architecture-dependent, whereas the third is position-dependent. The former occur frequently in industrial manipulators, although not both at the same time. If both parameters a_1 and μ_1 vanished, then the arm would be useless to position arbitrarily a point in space. The third condition, i.e., the vanishing of $x_C^2 + y_C^2$, means that point C lies on the Z_1 axis. Now, even if neither a_1 nor μ_1 vanishes, the manipulator can be positioned in a configuration at which point C lies on the Z_1 axis. Such a configuration is termed the *first singularity*. Note, however, that with point C being located on the Z_1 axis, any motion of the first joint, with the two other joints locked, does not change the location of C. For the moment, it will be assumed that Δ_1 does not vanish, the particular cases under which it does being studied later. Next, both sides of eqs.(4.21a & b) are squared, the squares thus obtained are then added, and the sum is equated to 1, which leads to a quadratic equation in $\mathbf{x}_3$, namely,

$$Kc_3^2 + Ls_3^2 + Mc_3s_3 + Nc_3 + Ps_3 + Q = 0 \tag{4.22}$$

whose coefficients, after simplification, are given below:

$$K = 4a_1^2H^2 + \mu_1^2C^2 \tag{4.23a}$$

$$L = 4a_1^2I^2 + \mu_1^2D^2 \tag{4.23b}$$

$$M = 2(4a_1^2HI + \mu_1^2CD) \tag{4.23c}$$

$$N = 2(4a_1^2HJ + \mu_1^2CE) \tag{4.23d}$$

$$P = 2(4a_1^2IJ + \mu_1^2DE) \tag{4.23e}$$

$$Q = 4a_1^2J^2 + \mu_1^2E^2 - 4a_1^2\mu_1^2\rho^2 \tag{4.23f}$$

with ρ^2 defined as

$$\rho^2 \equiv x_C^2 + y_C^2$$

Now, two well-known trigonometric identities are introduced, namely,

$$c_3 \equiv \frac{1-\tau_3^2}{1+\tau_3^2}, \quad s_3 \equiv \frac{2\tau_3}{1+\tau_3^2}, \quad \text{where } \tau_3 \equiv \tan\left(\frac{\theta_3}{2}\right) \tag{4.24}$$

Henceforth, the foregoing identities will be referred to as the *tan-half-angle identities.* We will be resorting to them throughout the book. Upon substitution of the foregoing identities into eq.(4.22), a quartic equation in τ_3 is obtained, i.e.,

$$R\tau_3^4 + S\tau_3^3 + T\tau_3^2 + U\tau_3 + V = 0 \tag{4.25}$$

whose coefficients are all computable from the data. After some simplifications, these coefficients take on the forms

$$R = 4a_1^2(J-H)^2 + \mu_1^2(E-C)^2 - 4\rho^2 a_1^2\mu_1^2 \tag{4.26a}$$
$$S = 4[4a_1^2 I(J-H) + \mu_1^2 D(E-C)] \tag{4.26b}$$
$$T = 2[4a_1^2(J^2 - H^2 + 2I^2) + \mu_1^2(E^2 - C^2 + 2D^2) - 4\rho^2 a_1^2\mu_1^2] \tag{4.26c}$$
$$U = 4[4a_1^2 I(H+J) + \mu_1^2 D(C+E)] \tag{4.26d}$$
$$V = 4a_1^2(J+H)^2 + \mu_1^2(E+C)^2 - 4\rho^2 a_1^2\mu_1^2 \tag{4.26e}$$

Furthermore, let $\{ (\tau_3)_i \}_1^4$ be the four roots of eq.(4.25). Thus, up to four possible values of θ_3 can be obtained, namely,

$$(\theta_3)_i = 2\arctan[(\tau_3)_i], \quad i = 1, 2, 3, 4 \tag{4.27}$$

Once the four values of θ_3 are available, each of these is substituted into eqs.(4.21a & b), which thus produce four different values of θ_1. For each value of θ_1 and θ_3, then, one value of θ_2 can be computed from the first two scalar equations of eq.(4.17), which are displayed below:

$$A_{11}\cos\theta_2 + A_{12}\sin\theta_2 = x_C\cos\theta_1 + y_C\sin\theta_1 - a_1 \tag{4.28a}$$
$$-A_{12}\cos\theta_2 + A_{11}\sin\theta_2 = -x_C\lambda_1\sin\theta_1 + y_C\lambda_1\cos\theta_1 + (z_C - b_1)\mu_1 \tag{4.28b}$$

where

$$A_{11} \equiv a_2 + a_3\cos\theta_3 + b_4\mu_3\sin\theta_3 \tag{4.28c}$$
$$A_{12} \equiv -a_3\lambda_2\sin\theta_3 + b_3\mu_2 + b_4\lambda_2\mu_3\cos\theta_3 + b_4\mu_2\lambda_3 \tag{4.28d}$$

Thus, if A_{11} and A_{12} do not vanish simultaneously, angle θ_2 is readily computed in terms of θ_1 and θ_3 from eqs.(4.28a & b) as

$$\cos\theta_2 = \frac{1}{\Delta_2}\{A_{11}(x_C\cos\theta_1 + y_C\sin\theta_1 - a_1)$$

$$-A_{12}[-x_C\lambda_1 \sin\theta_1 + y_C\lambda_1 \cos\theta_1 + (z_C - b_1)\mu_1]\} \tag{4.29a}$$

$$\sin\theta_2 = \frac{1}{\Delta_2}\{A_{12}(x_C \cos\theta_1 + y_C \sin\theta_1 - a_1) + A_{11}[-x_C\lambda_1 \sin\theta_1 + y_C\lambda_1 \cos\theta_1 + (z_C - b_1)\mu_1]\} \tag{4.29b}$$

where Δ_2 is defined as

$$\Delta_2 \equiv A_{11}^2 + A_{12}^2 \equiv a_2^2 + (\cos\theta_3^2 + \lambda_2^2 \sin\theta_3^2)a_3^2 + b_3^2\mu_2^2 + 2a_2a_3\cos\theta_3 - 2a_3b_3\lambda_2\mu_2 \sin\theta_3 \tag{4.29c}$$

the case in which $\Delta_2 = 0$, which leads to what is termed here the *second singularity*, being discussed presently.

Takano (1985) considered the solution of the positioning problem for all possible combinations of prismatic and revolute pairs in the regional structure of a manipulator, thereby finding that

1. In the case of arms containing either three revolutes, or two revolutes and one prismatic pair, with a general layout in all cases, a quartic equation in $\cos\theta_3$ was obtained;

2. in the case of one revolute and two prismatic pairs, the positioning problem was reduced to a single quadratic equation, the problem at hand thus admitting two solutions;

3. finally, for three prismatic pairs, a single linear equation was derived, the problem thus admitting a unique solution.

The Vanishing of Δ_1

In the above derivations we have assumed that neither μ_1 nor a_1 vanishes. However, if either $\mu_1 = 0$ or $a_1 = 0$, then one can readily show that eq.(4.25) reduces to a quadratic equation, and hence, this case differs essentially from the general one. Note that one of these conditions can occur, and the second occurs indeed frequently, but both together never occur, because their simultaneous occurrence would render the manipulator useless for a three-dimensional task. We thus have the two cases discussed below:

1. $\mu_1 = 0$, $a_1 \neq 0$. In this case, one has

$$A,\ B \neq 0, \quad F = G = 0$$

Under these conditions, eq.(4.20a) and the tan-half-angle identities given in eq.(4.24) yield

$$(J - H)\tau_3^2 + 2I\tau_3 + (J + H) = 0$$

which thus produces two values of τ_3, namely,

$$(\tau_3)_{1,2} = \frac{-I \pm \sqrt{I^2 - J^2 + H^2}}{J - H} \tag{4.30a}$$

Once two values of θ_3 have been determined according to the above equation, θ_1 can be found using eq.(4.19a) and the tan-half-angle identities, thereby deriving

$$(E' - A)\tau_1^2 + 2B\tau_1 + (E' + A) = 0$$

where

$$E' = Cc_3 + Ds_3 + E, \quad \tau_1 \equiv \tan\left(\frac{\theta_1}{2}\right)$$

whose roots are

$$(\tau_1)_{1,2} = \frac{-B \pm \sqrt{B^2 - E'^2 + A^2}}{E' - A} \tag{4.30b}$$

Thus, two values of θ_1 are found for each of the two values of θ_3, which results in four positioning solutions. Values of θ_2 are obtained using eqs.(4.29a & b).

2. $a_1 = 0$, $\mu_1 \neq 0$. In this case, one has an architecture similar to that of the robot of Fig. 4.3. We have now

$$A = B = 0, \quad F, G \neq 0$$

Under the present conditions, eq.(4.19a) reduces to

$$(E - C)\tau_3^2 + 2D\tau_3 + (E + C) = 0$$

which produces two values of τ_3, namely,

$$(\tau_3)_{1,2} = \frac{-D \pm \sqrt{D^2 - E^2 + C^2}}{E - C} \tag{4.31a}$$

With the two values of θ_3 obtained, θ_1 can be found using eq.(4.20a) and the tan-half-angle identities to produce

$$(J' - F)\tau_1^2 + 2G\tau_1 + (J' + F) = 0$$

where

$$J' = Hc_3 + Is_3 + J, \quad \tau_1 \equiv \tan(\frac{\theta_1}{2})$$

whose roots are

$$(\tau_1)_{1,2} = \frac{-G \pm \sqrt{G^2 - J'^2 + F^2}}{J' - F} \tag{4.31b}$$

Once again, the solution results in a cascade of two quadratic equations, one for θ_3 and one for θ_1, which yields four positioning solutions. As above, θ_2 is then determined using eqs.(4.29a & b). Note that for the special case of the manipulator of Fig. 4.3, we have

$$a_1 = b_2 = 0, \quad \alpha_1 = \alpha_3 = 90^\circ, \quad \alpha_2 = 0^\circ$$

and hence,

$$C = H = I = 0, \quad E = a_2^2 + a_3^2 + b_3^2 + b_4^2 - \left[x_C^2 + y_C^2 + (z_C - b_1)^2\right],$$

$$D = 2a_2 b_4, \quad F = y_C, \quad G = -x_C, \quad J = b_3$$

In this case, the foregoing solutions reduce to

$$(\tau_3)_{1,2} = \frac{-D \pm \sqrt{D^2 - E^2}}{E}, \quad (\tau_1)_{1,2} = \frac{x_C \pm \sqrt{x_C^2 + y_C^2 - b_3^2}}{b_3 - y_C}$$

A robot with the architecture studied here is the Puma, which is displayed in Fig. 4.10 in its four distinct postures for the same location of its wrist center. Notice that the orientation of the EE is kept constant in all four postures.

The Vanishing of Δ_2

In some instances, Δ_2, as defined in eq.(4.29c), may vanish, thereby preventing the calculation of θ_2 from eqs.(4.29a & b). This posture, termed the second singularity, occurs if both coefficients A_{11} and A_{12} of eqs.(4.28a & b) vanish. Note that from their definitions, eqs.(4.28c & d), these coefficients are not only position- but also architecture-dependent. Thus, an arbitrary manipulator cannot take on this configuration unless its geometric dimensions allow it. This type of singularity will be termed architecture-dependent, to distinguish it from others that are common to all robots, regardless of their particular architectures.

We can now give a geometric interpretation of the singularity at hand: First, note that the right-hand side of eq.(4.17), from which eqs.(4.28a & b) were derived, is identical to $\mathbf{Q}_1^T(\mathbf{c}-\mathbf{a}_1)$, which means that this expression is nothing but the $\mathcal{F}_2$-representation of the position vector of C. That is, the components of vector $\mathbf{Q}_1^T(\mathbf{c} - \mathbf{a}_1)$ are the $\mathcal{F}_2$-components of vector $\overrightarrow{O_2C}$. Therefore, the two sides of eqs.(4.28a & b) are, respectively, the X_2- and Y_2-components of vector $\overrightarrow{O_2C}$. Therefore, if $A_{11} = A_{12} = 0$, then the two foregoing components vanish and, as a consequence, point C lies on the Z_2 axis. Hence, the first singularity occurs when point C lies on the axis of the first revolute, while the second occurs when the same point lies on the axis of the second revolute.

Many industrial manipulators are designed with an orthogonal architecture, which means that the angles between neighbor axes are multiples

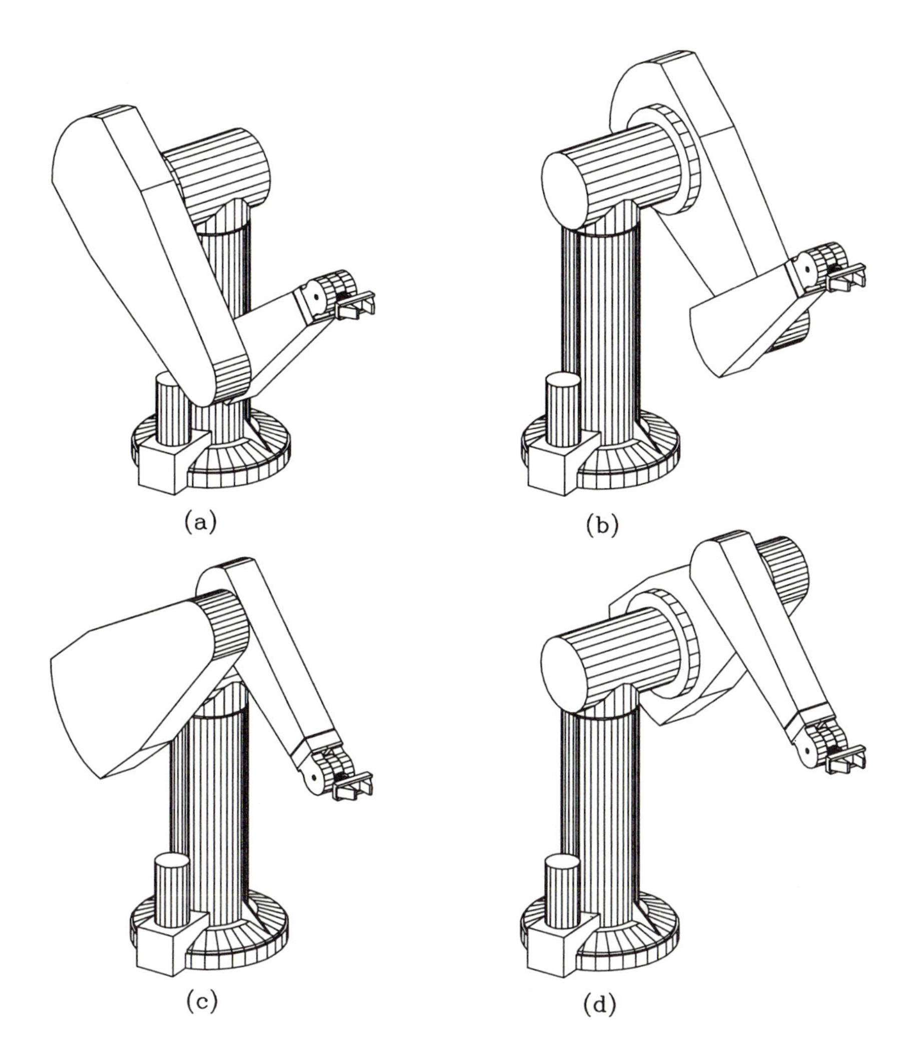

FIGURE 4.10. The four arm configurations for the positioning problem of the Puma robot: (a) and (b), elbow down; (a) and (c), shoulder fore; (c) and (d), elbow up; (b) and (d), shoulder aft.

of 90°. Moreover, with the purpose of maximizing their workspace, orthogonal manipulators are designed with their second and third links of equal lengths, thereby rendering them vulnerable to this type of singularity. An architecture common to many manipulators such as the Cincinnati-Milacron, ABB, Fanuc, and others, comprises a planar two-axis layout with equal link lengths, which is capable of turning about an axis orthogonal to these two axes. This layout allows for the architecture singularity under discussion, as shown in Fig. 4.11a. The well-known Puma manipulator is

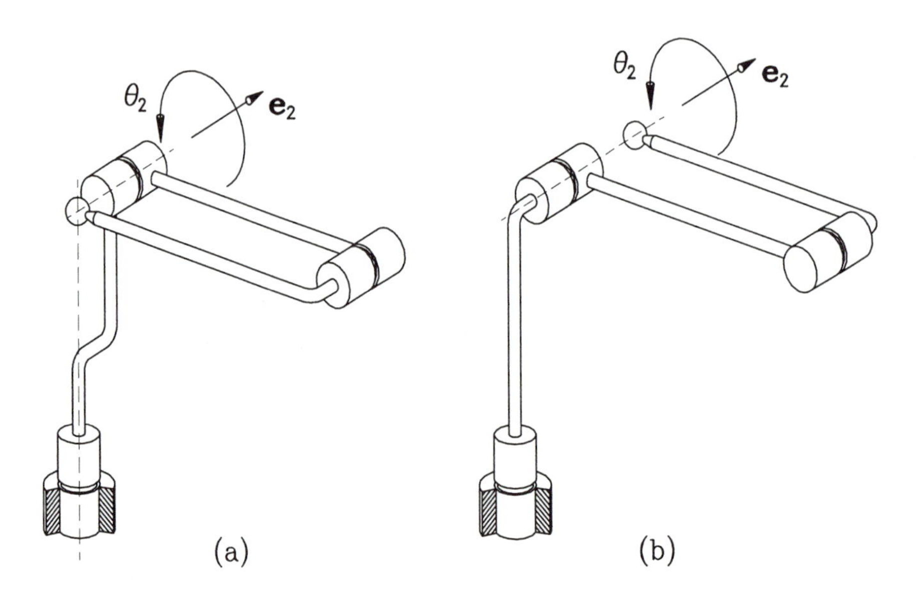

FIGURE 4.11. Architecture-dependent singularities of (a) the Cincinnati-Milacron and (b) the Puma robots.

similar to the aforementioned manipulators, except that it is supplied with what is called a shoulder offset b_3, as illustrated in Fig. 4.3. This offset, however, does not prevent the Puma from attaining the same singularity as depicted in Fig. 4.11b. Note that in the presence of this singularity, angle θ_2 is undetermined, but θ_1 and θ_3 are determined in the case of the Puma robot. However, in the presence of the singularity of Fig. 4.11a, neither θ_1 nor θ_2 are determined; only θ_3 of the arm structure is determined.

Example 4.4.1 *A manipulator with a common orthogonal architecture is displayed in Fig. 4.12 in an arbitrary configuration. The arm architecture of this manipulator has the DH parameters shown below:*

$$a_1 = 0, \quad b_1 = b_2 = b_3 = 0, \quad \alpha_1 = 90^\circ, \quad \alpha_2 = 0^\circ$$

Find its inverse kinematics solutions.

Solution: A common feature of this architecture is that it comprises $a_2 = b_4$. In the present discussion, however, the latter feature need not be included, and hence, the result that follows applies even in its absence. In this case, coefficients C, D, and E take on the forms

$$C = 2a_2a_3, \quad D = 0, \quad E = a_2^2 + a_3^2 - (x_C^2 + y_C^2 + z_C^2)$$

Hence,

$$E - C = (a_2 - a_3)^2 - (x_C^2 + y_C^2 + z_C^2), \quad E + C = (a_2 + a_3)^2 - (x_C^2 + y_C^2 + z_C^2)$$

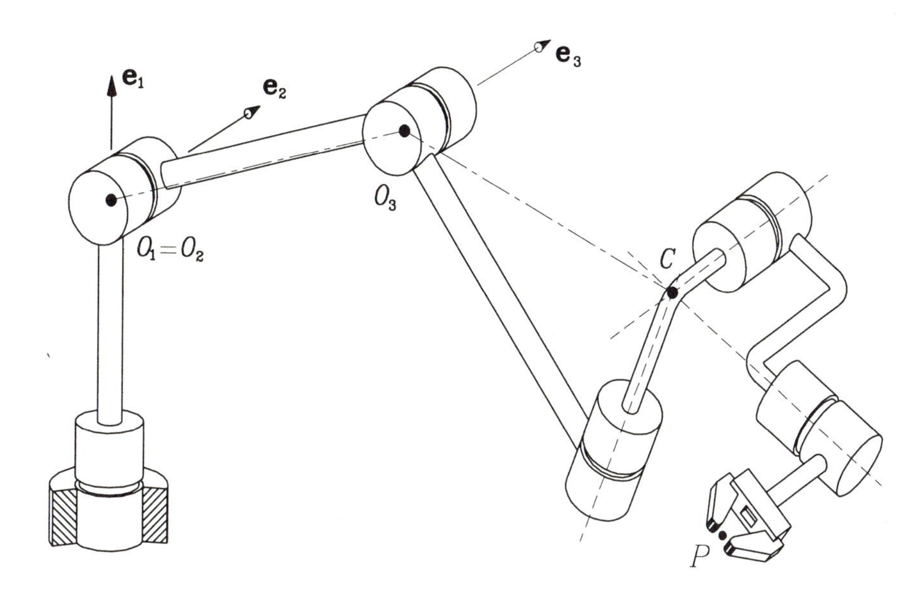

FIGURE 4.12. An orthogonal decoupled manipulator.

Moreover,

$$H = I = J = 0$$

and so

$$J' = 0, \quad F = y_C, \quad G = -x_C$$

The radical of eq.(4.31b) reduces to $x_C^2 + y_C^2$. Thus,

$$\tan(\frac{\theta_1}{2}) = \frac{x_C \pm \sqrt{x_C^2 + y_C^2}}{-y_C} \equiv \frac{-1 \pm \sqrt{1 + (y_C/x_C)^2}}{y_C/x_C} \qquad (4.32a)$$

Now we recall the relation between $\tan(\theta_1/2)$ and $\tan\theta_1$, namely,

$$\tan(\frac{\theta_1}{2}) \equiv \frac{-1 \pm \sqrt{1 + \tan^2\theta_1}}{\tan\theta_1} \qquad (4.32b)$$

Upon comparison of eqs.(4.32a) and (4.32b), it is apparent that

$$\theta_1 = \arctan\left(\frac{y_C}{x_C}\right)$$

a result that can be derived geometrically for this simple arm architecture. Given that the arctan(·) function is double-valued, its two values differing in 180°, we obtain here, again, two values for θ_1. On the other hand, θ_3 is calculated from eq.(4.31a) as

$$(\tau_3)_{1,2} = \pm\frac{\sqrt{C^2 - E^2}}{E - C}$$

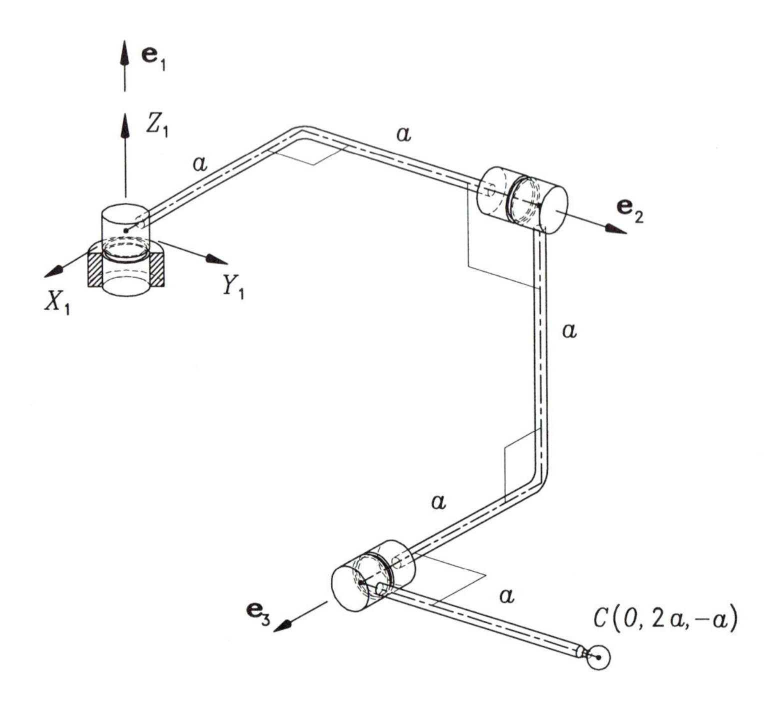

FIGURE 4.13. An orthogonal RRR manipulator.

thereby obtaining two values of θ_3. As a consequence, the inverse positioning problem of this arm architecture admits four solutions as well. These solutions give rise to two pairs of arm postures that are usually referred to as *elbow-up* and *elbow-down.*

Example 4.4.2 *Find all real inverse kinematic solutions of the manipulator shown in Fig. 4.13, when point* C *of its end-effector has the base coordinates* $C(0,\ 2a,\ -a)$.

Solution: The Denavit-Hartenberg parameters of this manipulator are derived from Fig. 4.14, where the coordinate frames involved are indicated. In defining the coordinate frames of that figure, the Denavit-Hartenberg notation was followed, with Z_4 defined, arbitrarily, as parallel to Z_3. From Fig. 4.14, then, we have

$$a_1 = a_2 = a_3 = b_2 = b_3 = a, \quad b_1 = b_4 = 0, \quad \alpha_1 = \alpha_2 = 90^\circ, \quad \alpha_3 = 0^\circ$$

One inverse kinematic solution can be readily inferred from the geometry of Fig. 4.14. For illustration purposes, and in order to find all other inverse kinematic solutions, we will use the procedure derived above. To this end, we first proceed to calculate the coefficients of the quartic polynomial equation, eq.(4.25), which are given, nevertheless, in terms of coefficients

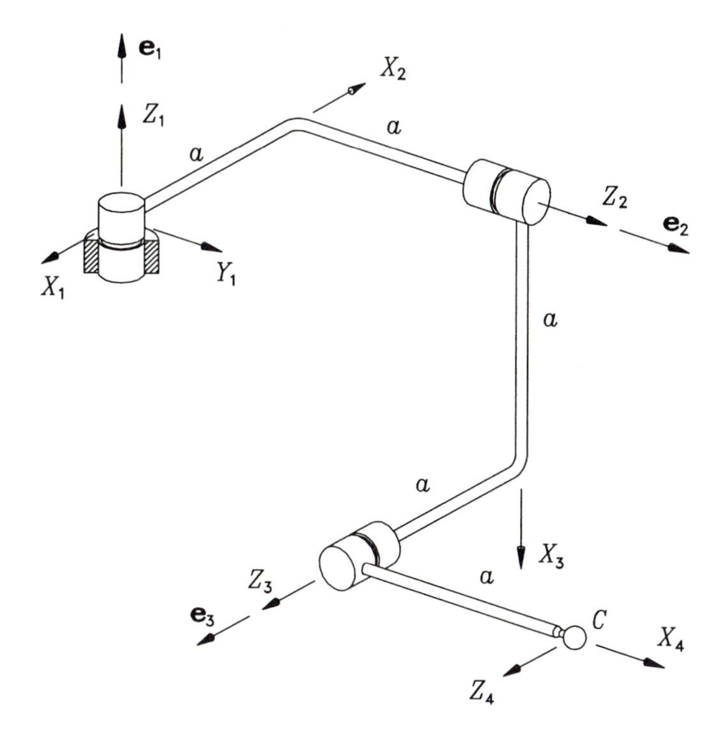

FIGURE 4.14. The coordinate frames of the orthogonal RRR manipulator.

K, ..., Q of eqs.(4.23a–f). These coefficients are given, in turn, in terms of coefficients A, ..., J of eqs.(4.19b–f) and (4.20b–f). We then proceed to calculate all the necessary coefficients in the proper order:

$$A = 0, \quad B = 4a^2, \quad C = D = -E = 2a^2$$
$$F = 2a, \quad G = H = 0, \quad I = J = a$$

Moreover,

$$\begin{aligned}
K &= C^2F^2 = 16a^2 \\
L &= B^2I^2 + D^2F^2 = 32a^6 \\
M &= 2F^2CD = 32a^6 \\
N &= 2F^2CE = -32a^6 \\
P &= 2(B^2IJ + DEF^2) = 0 \\
Q &= E^2F^2 + B^2J^2 - F^2B^2 = -32a^6
\end{aligned}$$

The set of coefficients sought thus reduces to

$$R = K - N + Q = 16a^6$$

$$S = 2(P - M) = -64a^6$$
$$T = 2(Q + 2L - K) = 32a^6$$
$$U = 2(M + P) = 64a^6$$
$$V = K + N + Q = -48a^6$$

which leads to the quartic equation given below:

$$\tau_3^4 - 4\tau_3^3 + 2\tau_3^2 + 4\tau_3 - 3 = 0$$

with four real roots, namely,

$$(\tau_3)_1 = (\tau_3)_2 = 1, \quad (\tau_3)_3 = -1, \quad (\tau_3)_4 = 3$$

These roots yield the θ_3 values that follow:

$$(\theta_3)_1 = (\theta_3)_2 = 90°, \quad (\theta_3)_3 = -90°, \quad (\theta_3)_4 = 143.13°$$

The quartic polynomial thus admits one double root, which means that at the configurations resulting from this root, two solutions meet, thereby producing a *singularity*, an issue that is discussed in Subsection 4.5.2. Below, we calculate the remaining angles for each solution: Angle θ_1 is computed from relations (4.21a–c), where $\Delta_1 = -8a^3$.

The first two roots, $(\theta_3)_1 = (\theta_3)_2 = 90°$, yield $c_3 = 0$ and $s_3 = 1$. Hence, eqs.(4.21a & b) lead to

$$c_1 = \frac{B(I+J)}{\Delta_1} = \frac{4a^2(a+a)}{-8a^3} = -1$$
$$s_1 = \frac{F(D+E)}{\Delta_1} = \frac{2a(2a^2 - 2a^2)}{-8a^3} = 0$$

and hence,

$$(\theta_1)_1 = (\theta_1)_2 = 180°$$

With θ_1 known, θ_2 is computed from the first two of eqs.(4.17), namely,

$$c_2 = 0, \quad s_2 = -1$$

and hence,

$$(\theta_2)_1 = (\theta_2)_2 = -90°$$

The remaining roots are treated likewise. These are readily calculated as shown below:

$$(\theta_1)_3 = -90°, \quad (\theta_2)_3 = 0, \quad (\theta_1)_4 = 143.13°, \quad (\theta_2)_4 = 0$$

It is noteworthy that the architecture of this manipulator does not allow for the second singularity, associated with $\Delta_2 = 0$.

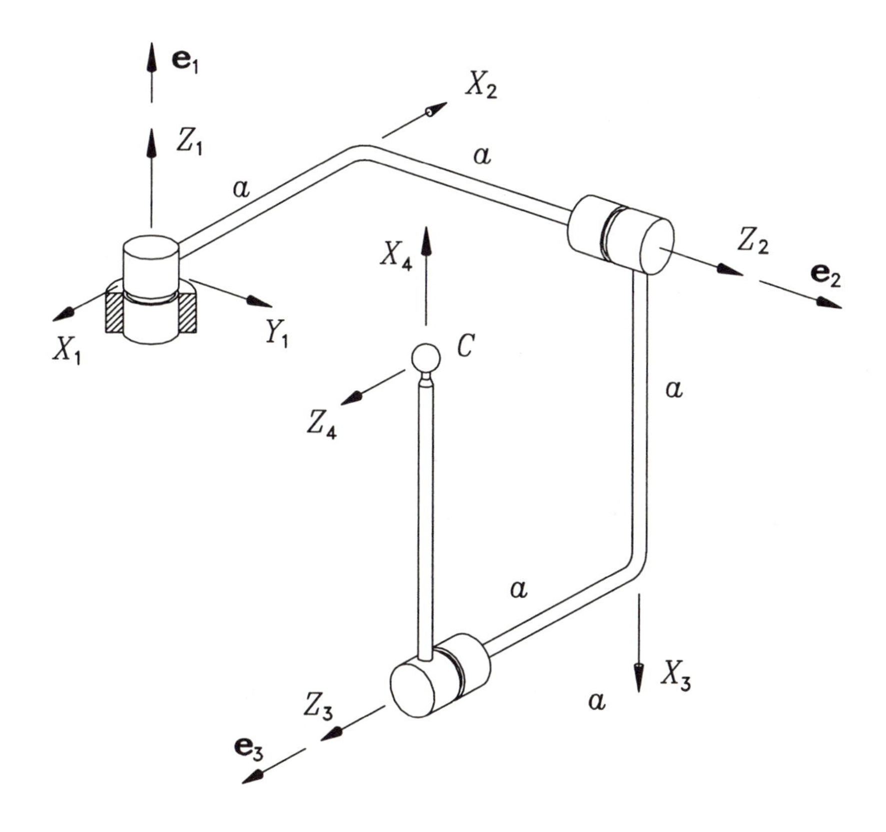

FIGURE 4.15. Manipulator configuration for $P(0,\, a,\, 0)$.

Example 4.4.3 *For the same manipulator of Example 4.4.2, find all real inverse kinematic solutions when point C of its end-effector has the base coordinates $C(0,\, a,\, 0)$, as displayed in Fig. 4.15.*

Solution: In this case, one obtains, successively,

$$\begin{aligned}
&A = 0, \quad B = C = D = E = 2a^2, \\
&F = a, \quad G = 0 \quad H = 0, \quad I = J = a \\
&K = 4a^6, L = M = N = 8a^6, \quad P = 16a^6, \quad Q = 4a^6 \\
&R = 0, \quad S = 16a^6, \quad T = 32a^6, \quad U = 48a^6, \quad V = 16a^6
\end{aligned}$$

Moreover, for this case, the quartic eq.(4.22) degenerates into a cubic equation, namely,

$$\tau_3^3 + 2\tau_3^2 + 3\tau_3 + 1 = 0$$

whose roots are readily found as

$$(\tau_3)_1 = -0.43016, \quad (\tau_3)_{2,3} = -0.78492 \pm j1.30714$$

where j is the imaginary unit, i.e., $j \equiv \sqrt{-1}$. That is, only one real solution is obtained, namely, $(\theta_3)_1 = -46.551°$. However, shown in Fig. 4.15 is a quite symmetric posture of this manipulator at the given position of point C of its end-effector, which does not correspond to the real solution obtained above. In fact, the solution yielding the posture of Fig. 4.15 disappeared because of the use of the quartic polynomial equation in $\tan(\theta_3/2)$. Note that if the two contours derived from eqs.(4.19a) and (4.20a) are plotted, as in Fig. 4.16, their intersections yield the two real roots, including the one leading to the posture of Fig. 4.15.

The explanation of how the fourth root of the quartic equation disappeared is given below: Let us write the quartic polynomial in full, with a "small" leading coefficient ϵ, namely,

$$\epsilon\tau_3^4 + \tau_3^3 + 2\tau_3^2 + 3\tau_3 + 1 = 0$$

Upon dividing both sides of the foregoing equation by τ_3^4, we obtain

$$\epsilon + \frac{1}{\tau_3} + \frac{2}{\tau_3^2} + \frac{3}{\tau_3^3} + \frac{1}{\tau_3^4} = 0$$

from which it is clear that the original equation is satisfied as $\epsilon \rightarrow 0$ if and only if $\tau_3 \rightarrow \pm\infty$, i.e, if $\theta_3 = 180°$. It is then apparent that the missing root is $\theta_3 = 180°$. The remaining angles are readily calculated as

$$(\theta_1)_1 = -105.9°, \quad (\theta_2)_1 = -149.35°, \quad (\theta_1)_4 = 180°, \quad (\theta_2)_4 = 180°$$

4.4.2 The Orientation Problem

Now the orientation inverse kinematic problem is addressed. This problem consists of determining the wrist angles that will produce a prescribed orientation of the end-effector. This orientation, in turn, is given in terms of the rotation matrix $\mathbf{Q}$ taking the end-effector from its home attitude to its current one. Alternatively, the orientation can be given by the natural invariants of the rotation matrix, vector $\mathbf{e}$ and angle ϕ. Moreover, since θ_1, θ_2, and θ_3 are available, $\mathbf{Q}_1$, $\mathbf{Q}_2$, and $\mathbf{Q}_3$ become data for this problem. One now has the general layout of Fig. 4.17, where angles $\{\,\theta_i\,\}_4^6$ are to be determined from the problem data, which are in this case the orientation of the end-effector and the architecture of the wrist; the latter is defined by angles α_4 and α_5, neither of which can be either 0 or π.

Now, since the orientation of the end-effector is given, we know the components of vector $\mathbf{e}_6$ in any coordinate frame. In particular, let

$$[\,\mathbf{e}_6\,]_4 = \begin{bmatrix} \xi \\ \eta \\ \zeta \end{bmatrix} \tag{4.33}$$

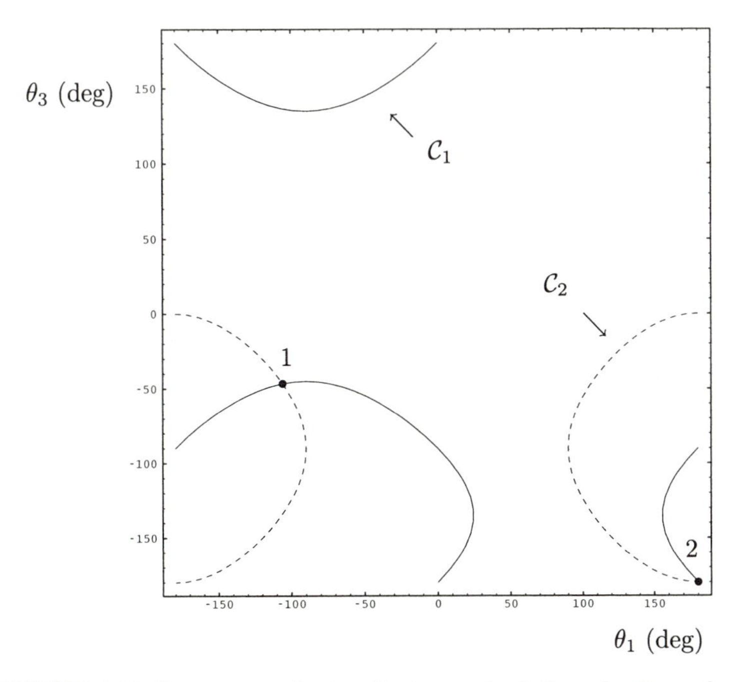

FIGURE 4.16. Contours producing the two real solutions for Example 4.3.

Moreover, the components of vector $\mathbf{e}_5$ in $\mathcal{F}_4$ are nothing but the entries of the third column of matrix $\mathbf{Q}_4$, i.e.,

$$[\,\mathbf{e}_5\,]_4 = \begin{bmatrix} \mu_4 \sin\theta_4 \\ -\mu_4 \cos\theta_4 \\ \lambda_4 \end{bmatrix} \tag{4.34}$$

Furthermore, vectors $\mathbf{e}_5$ and $\mathbf{e}_6$ make an angle α_5, and hence,

$$\mathbf{e}_6^T \mathbf{e}_5 = \lambda_5 \quad \text{or} \quad [\,\mathbf{e}_6\,]_4^T [\,\mathbf{e}_5\,]_4 = \lambda_5 \tag{4.35}$$

Upon substitution of eqs.(4.33) and (4.34) into eq.(4.35), we obtain

$$\xi\mu_4 \sin\theta_4 - \eta\mu_4 \cos\theta_4 + \zeta\lambda_4 = \lambda_5 \tag{4.36}$$

which can be readily transformed, with the aid of the tan-half-angle identities, into a quadratic equation in $\tau_4 \equiv \tan(\theta_4/2)$, namely,

$$(\lambda_5 - \eta\mu_4 - \zeta\lambda_4)\tau_4^2 - 2\xi\mu_4\tau_4 + (\lambda_5 + \eta\mu_4 - \zeta\lambda_4) = 0 \tag{4.37}$$

its two roots being given by

$$\tau_4 = \frac{\xi\mu_4 \pm \sqrt{(\xi^2 + \eta^2)\mu_4^2 - (\lambda_5 - \zeta\lambda_4)^2}}{\lambda_5 - \zeta\lambda_4 - \eta\mu_4} \tag{4.38}$$

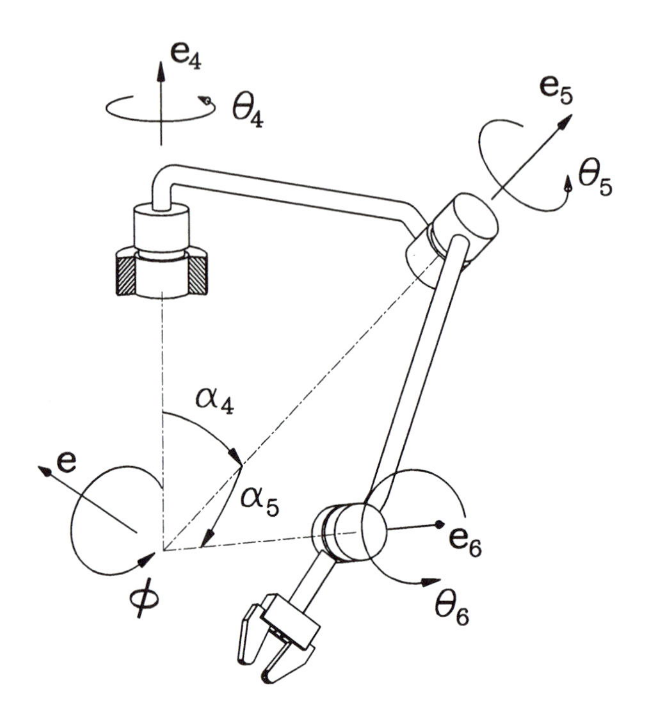

FIGURE 4.17. General architecture of a spherical wrist.

Note that the two foregoing roots are real as long as the radical is positive, the two roots merging into a single one when the radical vanishes. Thus, a negative radical means an attitude of the EE that is not feasible with the wrist. It is to be pointed out here that a three-revolute spherical wrist is kinematically equivalent to a spherical joint. However, the spherical wrist differs essentially from a spherical joint in that the latter has, kinematically, an unlimited workspace—a physical spherical joint, of course, has a limited workspace by virtue of its mechanical construction—and can orient a rigid body arbitrarily. Therefore, the workspace $\mathcal{W}$ of the wrist is not unlimited, but rather defined by the set of values of ξ, η, and ζ that satisfy the two relations shown below:

$$\xi^2 + \eta^2 + \zeta^2 = 1 \tag{4.39a}$$

$$f(\xi, \eta, \zeta) \equiv (\xi^2 + \eta^2)\mu_4^2 - (\lambda_5 - \zeta\lambda_4)^2 \geq 0 \tag{4.39b}$$

In view of condition (4.39a), however, relation (4.39b) simplifies to an inequality in ζ alone, namely,

$$F(\zeta) \equiv \zeta^2 - 2\lambda_4\lambda_5\zeta - (\mu_4^2 - \lambda_5^2) \leq 0 \tag{4.40}$$

As a consequence,

1. $\mathcal{W}$ is a region of the unit sphere $\mathcal{S}$ centered at the origin of the three-dimensional space;

2. $\mathcal{W}$ is bounded by the curve $F(\zeta) = 0$ on the sphere;

3. the wrist attains its singular configurations along the curve $F(\zeta) = 0$ lying on the surface of $\mathcal{S}$.

In order to gain more insight on the shape of the workspace $\mathcal{W}$, let us look at the boundary defined by $F(\zeta) = 0$. Upon setting $F(\zeta)$ to zero, we obtain a quadratic equation in ζ, whose two roots can be readily found to be

$$\zeta_{1,2} = \lambda_4\lambda_5 \pm |\mu_4\mu_5| \tag{4.41}$$

which thus defines two planes, Π_1 and Π_2, parallel to the ξ-η plane of the three-dimensional space, intersecting the ζ-axis at ζ_1 and ζ_2, respectively. Thus, the workspace $\mathcal{W}$ of the spherical wrist at hand is that region of the surface of the unit sphere $\mathcal{S}$ contained between Π_1 and Π_2. For example, a common wrist design involves an orthogonal architecture, i.e., $\alpha_4 = \alpha_5 = 90°$. For such wrists,

$$\zeta_{1,2} = \pm 1$$

and hence, orthogonal wrists become singular when $[\mathbf{e}_6]_4 = [0, 0, \pm 1]^T$, i.e., when the fourth and the sixth axes are aligned. Thus, the workspace of orthogonal spherical wrists is the whole surface of the unit sphere centered at the origin, the singularity curve thus degenerating into two points, namely, the two intersections of this sphere with the ζ-axis. If one views $\zeta = 0$ as the equatorial plane, then the two singularity points of the workspace are the poles.

An alternative design is the so-called *three-roll wrist* of some Cincinnati-Milacron robots, with $\alpha_4 = \alpha_5 = 120°$, thereby leading to $\lambda_4 = \lambda_5 = -1/2$ and $\mu_4 = \mu_5 = \sqrt{3}/2$. For this wrist, the two planes Π_1 and Π_2 are found below: First, we note that with the foregoing values,

$$\zeta_{1,2} = 1, -\frac{1}{2}$$

and hence, the workspace of this wrist is the part of the surface of the unit sphere $\mathcal{S}$ that lies between the planes Π_1 and Π_2 parallel to the ξ-η plane, intersecting the ζ-axis at $\zeta_1 = 1$ and $\zeta_2 = -1/2$, respectively. Hence, if $\zeta = 0$ is regarded as the equatorial plane, then the points of the sphere $\mathcal{S}$ that are outside of the workspace of this wrist are those lying at a latitude of less than $-30°$. The singularity points are thus the north pole and the parallel of latitude $-30°$.

Once θ_4 is calculated from the two foregoing values of τ_4, if these are real, angle θ_5 is obtained uniquely for each value of θ_4, as explained below:

First, eq.(4.9a) is rewritten in a form in which the data are collected in the right-hand side, which produces

$$\mathbf{Q}_4\mathbf{Q}_5\mathbf{Q}_6 = \mathbf{R} \tag{4.42a}$$

with $\mathbf{R}$ defined as

$$\mathbf{R} = \mathbf{Q}_3^T\mathbf{Q}_2^T\mathbf{Q}_1^T\mathbf{Q} \tag{4.42b}$$

Moreover, let the entries of $\mathbf{R}$ in the fourth coordinate frame be given as

$$[\mathbf{R}]_4 = \begin{bmatrix} r_{11} & r_{12} & r_{13} \\ r_{21} & r_{22} & r_{23} \\ r_{31} & r_{32} & r_{33} \end{bmatrix}$$

Expressions for θ_5 and θ_6 can be readily derived by solving first for $\mathbf{Q}_5$ from eq.(4.42a), namely,

$$\mathbf{Q}_5 = \mathbf{Q}_4^T\mathbf{R}\mathbf{Q}_6^T \tag{4.43}$$

Now, by virtue of the form of the $\mathbf{Q}_i$ matrices, as appearing in eq.(4.1d), it is apparent that the third row of $\mathbf{Q}_i$ does not contain θ_i. Hence, the third column of the matrix product of eq.(4.43) is independent of θ_6. Thus, two equations for θ_5 are obtained by equating the first two components of the third rows of that equation, thereby obtaining

$$\mu_5 s_5 = (\mu_6 r_{12} + \lambda_6 r_{13})c_4 + (\mu_6 r_{22} + \lambda_6 r_{23})s_4$$
$$-\mu_5 c_5 = -\lambda_4(\mu_6 r_{12} + \lambda_6 r_{13})s_4 + \lambda_4(\mu_6 r_{22} + \lambda_6 r_{23})c_4 + \mu_4(\mu_6 r_{32} + \lambda_6 r_{33})$$

which thus yield a unique value of θ_5 for every value of θ_4. Finally, with θ_4 and θ_5 known, it is a simple matter to calculate θ_6. This is done upon solving for $\mathbf{Q}_6$ from eq.(4.42a), i.e.,

$$\mathbf{Q}_6 = \mathbf{Q}_5^T\mathbf{Q}_4^T\mathbf{R}$$

and if the partitioning (4.12) of $\mathbf{Q}_i$ is now recalled, a useful vector equation is derived, namely,

$$\mathbf{p}_6 = \mathbf{Q}_5^T\mathbf{Q}_4^T\mathbf{r}_1 \tag{4.45}$$

where $\mathbf{r}_1$ is the first column of $\mathbf{R}$. Let $\mathbf{w}$ denote the product $\mathbf{Q}_4^T\mathbf{r}_1$, i.e.,

$$\mathbf{w} \equiv \mathbf{Q}_4^T\mathbf{r}_1 \equiv \begin{bmatrix} r_{11}c_4 + r_{21}s_4 \\ -\lambda_4(r_{11}s_4 - r_{21}c_4) + \mu_4 r_{31} \\ \mu_4(r_{11}s_4 - r_{21}c_4) + \lambda_4 r_{31} \end{bmatrix}$$

Hence,

$$\mathbf{Q}_5^T\mathbf{Q}_4^T\mathbf{r}_1 \equiv \begin{bmatrix} w_1c_5 + w_2s_5 \\ \lambda_5(-w_1s_5 + w_2c_5) + w_3\mu_5 \\ \mu_5(w_1s_5 - w_2c_5) + w_3\lambda_5 \end{bmatrix}$$

in which w_i denotes the ith component of $\mathbf{w}$. Hence, c_6 and s_6 are determined from the first two scalar equations of eq.(4.45), namely,

$$c_6 = w_1 c_5 + w_2 s_5$$
$$s_6 = -w_1 \lambda_5 s_5 + w_2 \lambda_5 c_5 + w_3 \mu_5$$

thereby deriving a unique value of θ_6 for every pair of values (θ_4, θ_5). In summary, then, two values of θ_4 have been determined, each value determining, in turn, one single corresponding set of θ_5 and θ_6 values. Therefore, there are two sets of solutions for the orientation problem under study, which lead to two corresponding wrist postures. The two distinct postures of an orthogonal three-revolute spherical wrist for a given orientation of its EE are displayed in Fig. 4.18.

When combined with the four postures of a decoupled manipulator leading to one and the same location of its wrist center—positioning problem—a maximum of eight possible combinations of joint angles for a single pose of the end-effector of a decoupled manipulator are found.

4.5 Velocity Analysis of Serial Manipulators

The relationships between the prescribed twist of the EE, also referred to as the *Cartesian velocity* of the manipulator, and the corresponding joint-rates are derived in this section. First, a serial n-axis manipulator containing only revolute pairs is considered. Then, relations associated with prismatic pairs are introduced, and finally, the joint rates of six-axis manipulators are calculated in terms of the EE twist. Particular attention is given to decoupled manipulators, for which simplified velocity relations are derived.

We consider here the manipulator of Fig. 4.19, in which a joint coordinate θ_i, a joint rate $\dot{\theta}_i$, and a unit vector $\mathbf{e}_i$ are associated with each revolute

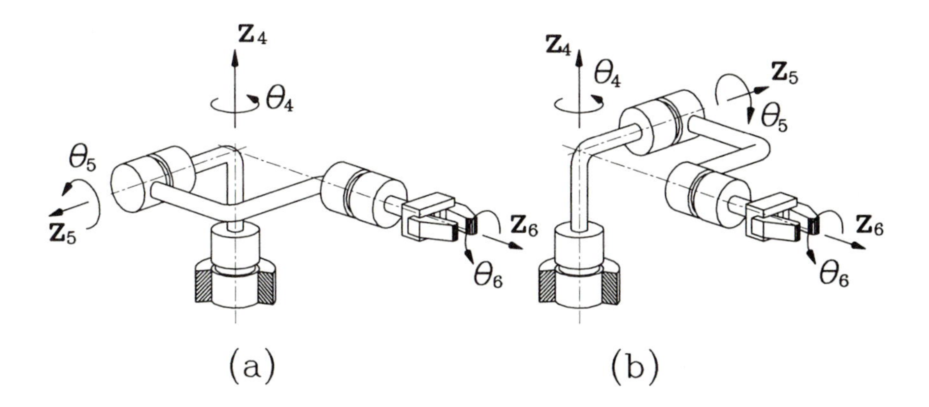

FIGURE 4.18. The two configurations of a three-axis spherical wrist.

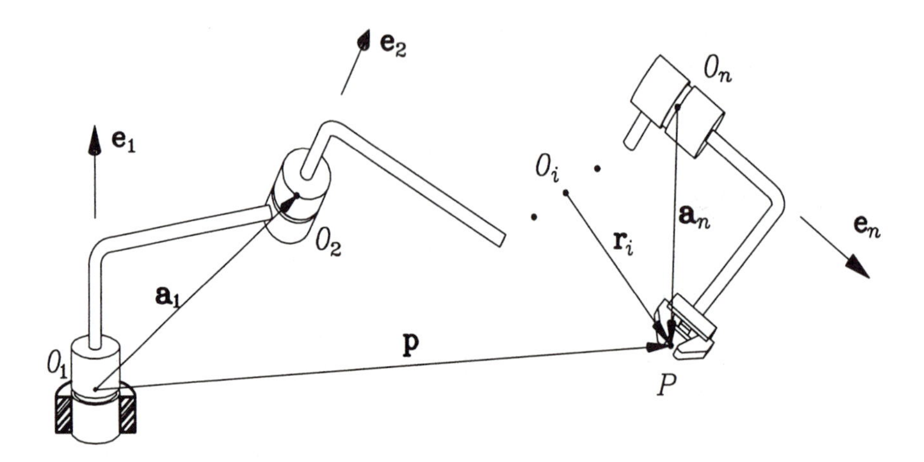

FIGURE 4.19. General n-axis manipulator.

axis. The X_i, Y_i, Z_i coordinate frame, attached to the $(i-1)$st link, is not shown, but its origin O_i is indicated. The relations that follow are apparent from that figure.

$$\begin{aligned}\boldsymbol{\omega}_0 &= \mathbf{0}\\ \boldsymbol{\omega}_1 &= \dot{\theta}_1\mathbf{e}_1\\ \boldsymbol{\omega}_2 &= \dot{\theta}_1\mathbf{e}_1 + \dot{\theta}_2\mathbf{e}_2\\ &\vdots\\ \boldsymbol{\omega}_n &= \dot{\theta}_1\mathbf{e}_1 + \dot{\theta}_2\mathbf{e}_2 + \cdots + \dot{\theta}_n\mathbf{e}_n\end{aligned} \tag{4.46}$$

and if the angular velocity of the EE is denoted by $\boldsymbol{\omega}$, then

$$\boldsymbol{\omega} \equiv \boldsymbol{\omega}_n = \dot{\theta}_1\mathbf{e}_1 + \dot{\theta}_2\mathbf{e}_2 + \cdots + \dot{\theta}_n\mathbf{e}_n = \sum_1^n \dot{\theta}_i\mathbf{e}_i$$

Likewise, from Fig. 4.19, one readily derives

$$\mathbf{p} = \mathbf{a}_1 + \mathbf{a}_2 + \cdots + \mathbf{a}_n \tag{4.47}$$

where $\mathbf{p}$ denotes the position vector of point P of the EE. Upon differentiating both sides of eq.(4.47), we have

$$\dot{\mathbf{p}} = \dot{\mathbf{a}}_1 + \dot{\mathbf{a}}_2 + \cdots + \dot{\mathbf{a}}_n \tag{4.48}$$

where

$$\dot{\mathbf{a}}_i = \boldsymbol{\omega}_i \times \mathbf{a}_i, \quad i = 1, 2, \ldots, n \tag{4.49}$$

Furthermore, substitution of eqs.(4.46) and (4.49) into eq.(4.48) yields

$$\begin{aligned}\dot{\mathbf{p}} = \dot{\theta}_1\mathbf{e}_1 \times \mathbf{a}_1 + (\dot{\theta}_1\mathbf{e}_1 + \dot{\theta}_2\mathbf{e}_2) \times \mathbf{a}_2 + \\ \vdots \\ +(\dot{\theta}_1\mathbf{e}_1 + \dot{\theta}_2\mathbf{e}_2 + \cdots + \dot{\theta}_n\mathbf{e}_n) \times \mathbf{a}_n\end{aligned} \tag{4.50}$$

which can be readily rearranged as

$$\begin{aligned}\dot{\mathbf{p}} = & \dot{\theta}_1\mathbf{e}_1 \times (\mathbf{a}_1 + \mathbf{a}_2 + \cdots + \mathbf{a}_n) \\ & +\dot{\theta}_2\mathbf{e}_2 \times (\mathbf{a}_2 + \mathbf{a}_3 + \cdots + \mathbf{a}_n) + \\ & \vdots \\ & +\dot{\theta}_n\mathbf{e}_n \times \mathbf{a}_n\end{aligned}$$

Now vector $\mathbf{r}_i$ is defined as that joining O_i with P, directed from the former to the latter, i.e.,

$$\mathbf{r}_i \equiv \mathbf{a}_i + \mathbf{a}_{i+1} + \cdots + \mathbf{a}_n \tag{4.51}$$

and hence, $\dot{\mathbf{p}}$ can be rewritten as

$$\dot{\mathbf{p}} = \sum_1^n \dot{\theta}_i\mathbf{e}_i \times \mathbf{r}_i$$

Further, let $\mathbf{A}$ and $\mathbf{B}$ denote the $3 \times n$ matrices defined as

$$\mathbf{A} \equiv [\,\mathbf{e}_1 \quad \mathbf{e}_2 \quad \cdots \quad \mathbf{e}_n\,] \tag{4.52a}$$
$$\mathbf{B} \equiv [\,\mathbf{e}_1 \times \mathbf{r}_1 \quad \mathbf{e}_2 \times \mathbf{r}_2 \quad \cdots \quad \mathbf{e}_n \times \mathbf{r}_n\,] \tag{4.52b}$$

Furthermore, the n-dimensional *joint-rate vector* $\dot{\boldsymbol{\theta}}$ is defined as

$$\dot{\boldsymbol{\theta}} \equiv [\,\dot{\theta}_1 \quad \dot{\theta}_2 \quad \cdots \quad \dot{\theta}_n\,]^T$$

Thus, $\boldsymbol{\omega}$ and $\dot{\mathbf{p}}$ can be expressed in a more compact form as

$$\boldsymbol{\omega} = \mathbf{A}\dot{\boldsymbol{\theta}}, \quad \dot{\mathbf{p}} = \mathbf{B}\dot{\boldsymbol{\theta}}$$

the twist of the EE being defined, in turn, as

$$\mathbf{t} \equiv \begin{bmatrix} \boldsymbol{\omega} \\ \dot{\mathbf{p}} \end{bmatrix} \tag{4.53}$$

The EE twist is thus related to the joint-rate vector $\dot{\boldsymbol{\theta}}$ in the form

$$\mathbf{J}\dot{\boldsymbol{\theta}} = \mathbf{t} \tag{4.54}$$

where $\mathbf{J}$ is the *Jacobian matrix*, or *Jacobian*, for brevity, of the manipulator under study, first introduced by Whitney (1972). The Jacobian is defined as the $6 \times n$ matrix shown below:

$$\mathbf{J} = \begin{bmatrix} \mathbf{A} \\ \mathbf{B} \end{bmatrix} \tag{4.55a}$$

or

$$\mathbf{J} = \begin{bmatrix} \mathbf{e}_1 & \mathbf{e}_2 & \cdots & \mathbf{e}_n \\ \mathbf{e}_1 \times \mathbf{r}_1 & \mathbf{e}_2 \times \mathbf{r}_2 & \cdots & \mathbf{e}_n \times \mathbf{r}_n \end{bmatrix} \tag{4.55b}$$

Clearly, an alternative definition of the foregoing Jacobian matrix can be given as

$$\mathbf{J} = \frac{\partial \mathbf{t}}{\partial \dot{\boldsymbol{\theta}}}$$

Moreover, if $\mathbf{j}_i$ denotes the ith column of $\mathbf{J}$, one has

$$\mathbf{j}_i = \begin{bmatrix} \mathbf{e}_i \\ \mathbf{e}_i \times \mathbf{r}_i \end{bmatrix}$$

It is important to note that if the axis of the ith revolute is denoted by $\mathcal{R}_i$, then $\mathbf{j}_i$ is nothing but the Plücker array of that line, with the moment of $\mathcal{R}_i$ being taken with respect to the operation point P of the EE.

On the other hand, if the ith pair is not rotational, but prismatic, then the $(i-1)$st and the ith links have the same angular velocity, for a prismatic pair does not allow any relative rotation. However, vector $\mathbf{a}_i$ joining the origins of the ith and $(i+1)$st frames is no longer of constant magnitude but undergoes a change of magnitude along the axis of the prismatic pair. That is,

$$\boldsymbol{\omega}_i = \boldsymbol{\omega}_{i-1}, \quad \dot{\mathbf{a}}_i = \boldsymbol{\omega}_{i-1} \times \mathbf{a}_i + \dot{b}_i \mathbf{e}_i$$

One can readily prove, in this case, that

$$\begin{aligned} \boldsymbol{\omega} &= \dot{\theta}_1 \mathbf{e}_1 + \dot{\theta}_2 \mathbf{e}_2 + \cdots + \dot{\theta}_{i-1} \mathbf{e}_{i-1} + \dot{\theta}_{i+1} \mathbf{e}_{i+1} + \cdots + \dot{\theta}_n \mathbf{e}_n \\ \dot{\mathbf{p}} &= \dot{\theta}_1 \mathbf{e}_1 \times \mathbf{r}_1 + \dot{\theta}_2 \mathbf{e}_2 \times \mathbf{r}_2 + \cdots + \dot{\theta}_{i-1} \mathbf{e}_{i-1} \times \mathbf{r}_{i-1} + \dot{b}_i \mathbf{e}_i \\ &\quad + \dot{\theta}_{i+1} \mathbf{e}_{i+1} \times \mathbf{r}_{i+1} + \cdots + \dot{\theta}_n \mathbf{e}_n \times \mathbf{a}_n \end{aligned}$$

from which it is apparent that the relation between the twist of the EE and the joint-rate vector is formally identical to that appearing in eq.(4.54) if vector $\dot{\boldsymbol{\theta}}$ is defined as

$$\dot{\boldsymbol{\theta}} \equiv [\dot{\theta}_1 \quad \dot{\theta}_2 \quad \cdots \quad \dot{\theta}_{i-1} \quad \dot{b}_i \quad \dot{\theta}_{i+1} \quad \cdots \quad \dot{\theta}_n]^T$$

and the ith column of $\mathbf{J}$ changes to

$$\mathbf{j}_i = \begin{bmatrix} \mathbf{0} \\ \mathbf{e}_i \end{bmatrix} \tag{4.57}$$

Note that the Plücker array of the axis of the ith joint, if prismatic, is that of a line at infinity lying in a plane normal to the unit vector $\mathbf{e}_i$, as defined in eq.(3.35).

In particular, for six-axis manipulators, $\mathbf{J}$ is a 6×6 matrix. Whenever this matrix is nonsingular, eq.(4.54) can be solved for $\dot{\boldsymbol{\theta}}$, namely,

$$\dot{\boldsymbol{\theta}} = \mathbf{J}^{-1}\mathbf{t} \tag{4.58}$$

Equation (4.58) is only symbolic, for the inverse of the Jacobian matrix need not be computed explicitly. Indeed, in the general case, matrix $\mathbf{J}$ cannot be inverted symbolically, and hence, $\dot{\boldsymbol{\theta}}$ is computed using a numerical procedure, the most suitable one being the *Gauss-elimination algorithm*, also known as *LU decomposition* (Golub and Van Loan, 1989). Gaussian elimination produces the solution by recognizing that a system of linear equations is most easily solved when it is in either upper- or lower-triangular form. To exploit this fact, matrix $\mathbf{J}$ is factored into the *unique* $\mathbf{L}$ and $\mathbf{U}$ factors in the form:

$$\mathbf{J} = \mathbf{L}\mathbf{U} \tag{4.59a}$$

where $\mathbf{L}$ is lower- and $\mathbf{U}$ is upper-triangular. Moreover, they have the forms

$$\mathbf{L} = \begin{bmatrix} 1 & 0 & \cdots & 0 \\ l_{21} & 1 & \cdots & 0 \\ \vdots & \vdots & \ddots & \vdots \\ l_{n1} & l_{n2} & \cdots & 1 \end{bmatrix} \tag{4.59b}$$

$$\mathbf{U} = \begin{bmatrix} u_{11} & u_{12} & \cdots & u_{1n} \\ 0 & u_{22} & \cdots & u_{2n} \\ \vdots & \vdots & \ddots & \vdots \\ 0 & 0 & \cdots & u_{nn} \end{bmatrix} \tag{4.59c}$$

where in the particular case at hand, $n = 6$. Thus, the unknown vector of joint rates can now be computed from two triangular systems, namely,

$$\mathbf{L}\mathbf{y} = \mathbf{t}, \quad \mathbf{U}\dot{\boldsymbol{\theta}} = \mathbf{y} \tag{4.60}$$

The latter equations are then solved, first for $\mathbf{y}$ and then for $\dot{\boldsymbol{\theta}}$, by application of only forward and backward substitutions, respectively. The LU decomposition of an $n \times n$ matrix requires M_n' multiplications and A_n' additions, whereas the forward substitution needed in solving the lower-triangular system of eq.(4.60) requires M_n'' multiplications and A_n'' additions. Moreover, the backward substitution needed in solving the upper-triangular system of eq.(4.60) requires M_n''' multiplications and A_n''' additions. These figures are (Dahlquist and Björck, 1974)

$$M_n' = \frac{n^3}{3} + \frac{n^2}{2} + \frac{n}{6}, \quad A_n' = \frac{n^3}{3} - \frac{n}{3} \tag{4.61a}$$

$$M_n'' = \frac{n(n-1)}{2}, \qquad A_n'' = \frac{n(n-1)}{2} \tag{4.61b}$$

$$M_n''' = \frac{n(n+1)}{2}, \qquad A_n''' = \frac{n(n-1)}{2} \tag{4.61c}$$

Thus, the solution of a system of n linear equations in n unknowns, using the LU-decomposition method, can be accomplished with M_n multiplications and A_n additions, as given below (Dahlquist and Björck, 1974):

$$M_n = \frac{n}{6}(2n^2 + 9n + 1), \quad A_n = \frac{n}{3}(n^2 + 3n - 4) \tag{4.62a}$$

Hence, the velocity resolution of a six-axis manipulator of *arbitrary architecture* requires M_6 multiplications and A_6 additions, as given below:

$$M_6 = 127, \quad A_6 = 100 \tag{4.62b}$$

Decoupled manipulators allow an even simpler velocity resolution. For manipulators with this type of architecture, it is more convenient to deal with the velocity of the center C of the wrist than with that of the operation point P. Thus, one has

$$\mathbf{t}_C = \mathbf{J}\dot{\boldsymbol{\theta}}$$

where $\mathbf{t}_C$ is defined as

$$\mathbf{t}_C = \begin{bmatrix} \boldsymbol{\omega} \\ \dot{\mathbf{c}} \end{bmatrix}$$

and can be obtained from $\mathbf{t}_P \equiv [\boldsymbol{\omega}^T, \dot{\mathbf{p}}^T]^T$ using the twist-transfer formula given by eqs.(3.84) and (3.85) as

$$\mathbf{t}_C = \begin{bmatrix} \mathbf{1} & \mathbf{0} \\ \mathbf{P} - \mathbf{C} & \mathbf{1} \end{bmatrix} \mathbf{t}_P$$

with $\mathbf{C}$ and $\mathbf{P}$ defined as the cross-product matrices of the position vectors $\mathbf{p}$ and $\mathbf{c}$, respectively.

If in general, $\mathbf{J}_A$ denotes the Jacobian defined for a point A of the EE and $\mathbf{J}_B$ that defined for another point B, then the relation between $\mathbf{J}_A$ and $\mathbf{J}_B$ is

$$\mathbf{J}_B = \mathbf{U}\mathbf{J}_A \tag{4.63a}$$

where the 6×6 matrix $\mathbf{U}$ is defined as

$$\mathbf{U} \equiv \begin{bmatrix} \mathbf{1} & \mathbf{O} \\ \mathbf{A} - \mathbf{B} & \mathbf{1} \end{bmatrix} \tag{4.63b}$$

while $\mathbf{A}$ and $\mathbf{B}$ are now the cross-product matrices of the position vectors $\mathbf{a}$ and $\mathbf{b}$ of points A and B, respectively. Moreover, this matrix $\mathbf{U}$ is identical to the matrix defined under the same name in eq.(3.31), and hence, it belongs to the 6×6 unimodular group, i.e., the group of 6×6 matrices whose determinant is unity. Thus,

$$\det(\mathbf{J}_B) = \det(\mathbf{J}_A) \tag{4.64}$$

We have then proven the result below:

Theorem 4.5.1 *The determinant of the Jacobian matrix of a six-axis manipulator is not affected under a change of* operation point *of the EE.*

Note, however, that the Jacobian matrix itself changes under a change of operation point. By analogy with the twist- and the wrench-transfer formulas, eq.(4.63a) can be called the *Jacobian-transfer formula.*

Since C is on the last three joint axes, its velocity is not affected by the motion of the last three joints, and we can write

$$\dot{\mathbf{c}} = \dot{\theta}_1 \mathbf{e}_1 \times \mathbf{r}_1 + \dot{\theta}_2 \mathbf{e}_2 \times \mathbf{r}_2 + \dot{\theta}_3 \mathbf{e}_3 \times \mathbf{r}_3$$

where in the case of a decoupled manipulator, vector $\mathbf{r}_i$ is defined as that directed from O_i to C. On the other hand, we have

$$\boldsymbol{\omega} = \dot{\theta}_1 \mathbf{e}_1 + \dot{\theta}_2 \mathbf{e}_2 + \dot{\theta}_3 \mathbf{e}_3 + \dot{\theta}_4 \mathbf{e}_4 + \dot{\theta}_5 \mathbf{e}_5 + \dot{\theta}_6 \mathbf{e}_6$$

and thus, the Jacobian takes on the following simple form

$$\mathbf{J} = \begin{bmatrix} \mathbf{J}_{11} & \mathbf{J}_{12} \\ \mathbf{J}_{21} & \mathbf{O} \end{bmatrix} \tag{4.65}$$

where $\mathbf{O}$ denotes the 3×3 zero matrix, the other 3×3 blocks being given below, *for manipulators with revolute pairs only*, as

$$\mathbf{J}_{11} = [\, \mathbf{e}_1 \quad \mathbf{e}_2 \quad \mathbf{e}_3 \,] \tag{4.66a}$$

$$\mathbf{J}_{12} = [\, \mathbf{e}_4 \quad \mathbf{e}_5 \quad \mathbf{e}_6 \,] \tag{4.66b}$$

$$\mathbf{J}_{21} = [\, \mathbf{e}_1 \times \mathbf{r}_1 \quad \mathbf{e}_2 \times \mathbf{r}_2 \quad \mathbf{e}_3 \times \mathbf{r}_3 \,] \tag{4.66c}$$

Further, vector $\dot{\boldsymbol{\theta}}$ is *partitioned* accordingly:

$$\dot{\boldsymbol{\theta}} \equiv \begin{bmatrix} \dot{\boldsymbol{\theta}}_a \\ \dot{\boldsymbol{\theta}}_w \end{bmatrix}$$

where

$$\dot{\boldsymbol{\theta}}_a \equiv \begin{bmatrix} \dot{\theta}_1 \\ \dot{\theta}_2 \\ \dot{\theta}_3 \end{bmatrix}, \quad \dot{\boldsymbol{\theta}}_w \equiv \begin{bmatrix} \dot{\theta}_4 \\ \dot{\theta}_5 \\ \dot{\theta}_6 \end{bmatrix}$$

Henceforth, the three components of $\dot{\boldsymbol{\theta}}_a$ will be referred to as the *arm rates*, whereas those of $\dot{\boldsymbol{\theta}}_w$ will be called the *wrist rates.* Now eqs.(4.54) can be written, for this particular case, as

$$\mathbf{J}_{11}\dot{\boldsymbol{\theta}}_a + \mathbf{J}_{12}\dot{\boldsymbol{\theta}}_w = \boldsymbol{\omega} \tag{4.67a}$$

$$\mathbf{J}_{21}\dot{\boldsymbol{\theta}}_a = \dot{\mathbf{c}} \tag{4.67b}$$

from which the solution is derived successively from the two systems of three equations and three unknowns that follow:

$$\mathbf{J}_{21}\dot{\boldsymbol{\theta}}_a = \dot{\mathbf{c}} \tag{4.68a}$$

$$\mathbf{J}_{12}\dot{\boldsymbol{\theta}}_w = \boldsymbol{\omega} - \mathbf{J}_{11}\dot{\boldsymbol{\theta}}_a \tag{4.68b}$$

From the general expressions (4.61), then, it is apparent that each of the foregoing systems can be solved with the numbers of operations shown below:

$$M_3 = 23, \quad A_3 = 14$$

Since the computation of the right-hand side of eq.(4.68b) requires, additionally, nine multiplications and nine additions, the total numbers of operations required to perform one joint-rate resolution of a decoupled manipulator, M_v multiplications and A_v additions, are given by

$$M_v = 55, \quad A_v = 37 \tag{4.69}$$

which are fairly low figures and can be performed in a matter of microseconds using a modern processor.

It is apparent from the foregoing kinematic relations that eq.(4.68a) should be first solved for $\dot{\boldsymbol{\theta}}_a$; with this value available, eq.(4.68b) can then be solved for $\dot{\boldsymbol{\theta}}_w$. We thus have, symbolically,

$$\dot{\boldsymbol{\theta}}_a = \mathbf{J}_{21}^{-1}\dot{\mathbf{c}} \tag{4.70}$$

$$\dot{\boldsymbol{\theta}}_w = \mathbf{J}_{12}^{-1}(\boldsymbol{\omega} - \mathbf{J}_{11}\dot{\boldsymbol{\theta}}_a) \tag{4.71}$$

Now, if we recall the concept of reciprocal bases introduced in Subsection 2.7.1, the above inverses can be represented explicitly. Indeed, let

$$\Delta_{21} \equiv \det(\mathbf{J}_{21}) = (\mathbf{e}_1 \times \mathbf{r}_1) \times (\mathbf{e}_2 \times \mathbf{r}_2) \cdot (\mathbf{e}_3 \times \mathbf{r}_3) \tag{4.72}$$

$$\Delta_{12} \equiv \det(\mathbf{J}_{12}) = \mathbf{e}_4 \times \mathbf{e}_5 \cdot \mathbf{e}_6 \tag{4.73}$$

Then

$$\mathbf{J}_{21}^{-1} = \frac{1}{\Delta_{21}} \begin{bmatrix} [(\mathbf{e}_2 \times \mathbf{r}_2) \times (\mathbf{e}_3 \times \mathbf{r}_3)]^T \\ [(\mathbf{e}_3 \times \mathbf{r}_3) \times (\mathbf{e}_1 \times \mathbf{r}_1)]^T \\ [(\mathbf{e}_1 \times \mathbf{r}_1) \times (\mathbf{e}_2 \times \mathbf{r}_2)]^T \end{bmatrix} \tag{4.74}$$

$$\mathbf{J}_{12}^{-1} = \frac{1}{\Delta_{21}} \begin{bmatrix} (\mathbf{e}_5 \times \mathbf{e}_6)^T \\ (\mathbf{e}_6 \times \mathbf{e}_4)^T \\ (\mathbf{e}_4 \times \mathbf{e}_5)^T \end{bmatrix} \tag{4.75}$$

Therefore,

$$\dot{\boldsymbol{\theta}}_a = \frac{1}{\Delta_{21}} \begin{bmatrix} (\mathbf{e}_2 \times \mathbf{r}_2) \times (\mathbf{e}_3 \times \mathbf{r}_3) \cdot \mathbf{c} \\ (\mathbf{e}_3 \times \mathbf{r}_3) \times (\mathbf{e}_1 \times \mathbf{r}_1) \cdot \mathbf{c} \\ (\mathbf{e}_1 \times \mathbf{r}_1) \times (\mathbf{e}_2 \times \mathbf{r}_2) \cdot \mathbf{c} \end{bmatrix} \tag{4.76a}$$

and if we let

$$\boldsymbol{\varpi} \equiv \boldsymbol{\omega} - \mathbf{J}_{11}\dot{\theta}_a \tag{4.76b}$$

where $\boldsymbol{\varpi}$ is read *varpi*, then

$$\dot{\boldsymbol{\theta}}_w = \frac{1}{\Delta_{12}} \begin{bmatrix} \mathbf{e}_5 \times \mathbf{e}_6 \cdot \boldsymbol{\varpi} \\ \mathbf{e}_6 \times \mathbf{e}_4 \cdot \boldsymbol{\varpi} \\ \mathbf{e}_4 \times \mathbf{e}_5 \cdot \boldsymbol{\varpi} \end{bmatrix} \tag{4.76c}$$

4.5.1 Jacobian Evaluation

The evaluation of the Jacobian matrix of a manipulator with n revolutes is discussed in this subsection, the presence of a prismatic pair leading to simplifications that will be outlined. Our aim here is to devise algorithms requiring a minimum number of operations, for these calculations are needed in real-time applications. We assume at the outset that all joint variables producing the desired EE pose are available. We divide this subsection into two subsubsections, one for the evaluation of the upper part of the Jacobian matrix and one for the evaluation of its lower part.

Evaluation of Submatrix **A**

The upper part **A** of the Jacobian matrix is composed of the set $\{ \mathbf{e}_i \}_1^n$, and hence, our aim here is the calculation of these unit vectors. Note, moreover, that vector $[\mathbf{e}_i]_1$ is nothing but the last column of $\mathbf{P}_{i-1} \equiv \mathbf{Q}_1 \cdots \mathbf{Q}_{i-1}$, our task then being the calculation of these matrix products. According to the DH nomenclature,

$$[\mathbf{e}_i]_i = [0 \quad 0 \quad 1]^T$$

Hence, $[\mathbf{e}_1]_1$ is available at no cost. However, each of the remaining $[\mathbf{e}_i]_1$ vectors, for $i = 2, \ldots, n$, is obtained as the last column of matrices $\mathbf{P}_{i-1}$. The *recursive calculation* of these matrices is described below:

$$\begin{aligned} \mathbf{P}_1 &\equiv \mathbf{Q}_1 \\ \mathbf{P}_2 &\equiv \mathbf{P}_1\mathbf{Q}_2 \\ &\vdots \\ \mathbf{P}_n &\equiv \mathbf{P}_{n-1}\mathbf{Q}_n \end{aligned}$$

and hence, a simple algorithm follows:

$\mathbf{P}_1 \leftarrow \mathbf{Q}_1$

```
For i = 2 to n do
```

$\quad \mathbf{P}_i \leftarrow \mathbf{P}_{i-1}\mathbf{Q}_i$

```
enddo
```

Now, since $\mathbf{P}_1$ is identical to $\mathbf{Q}_1$, the first product appearing in the do-loop, $\mathbf{P}_1\mathbf{Q}_2$, is identical to $\mathbf{Q}_1\mathbf{Q}_2$, whose two factors have a special structure. The computation of this product, then, requires special treatment, which warrants further discussion because of its particular features. From the structure of matrices $\mathbf{Q}_i$, as displayed in eq.(4.1d), we have

$$\mathbf{P}_2 \equiv \begin{bmatrix} \cos\theta_1 & -\lambda_1\sin\theta_1 & \mu_1\sin\theta_1 \\ \sin\theta_1 & \lambda_1\cos\theta_1 & -\mu_1\cos\theta_1 \\ 0 & \mu_1 & \lambda_1 \end{bmatrix} \begin{bmatrix} \cos\theta_2 & -\lambda_2\sin\theta_2 & \mu_2\sin\theta_2 \\ \sin\theta_2 & \lambda_2\cos\theta_2 & -\mu_2\cos\theta_2 \\ 0 & \mu_2 & \lambda_2 \end{bmatrix}$$

The foregoing product is calculated now by first computing the products $\lambda_1\lambda_2$, $\lambda_1\mu_2$, $\mu_1\mu_2$, and $\lambda_2\mu_1$, which involve only constant quantities, these terms thus being posture-independent. Thus, in tracking a prescribed Cartesian trajectory, the manipulator posture changes continuously, and hence, its joint variables also change. However, its DH parameters, those defining its architecture, remain constant. Therefore, the four above products remain constant and are computed prior to tracking a trajectory, i.e., *off-line*. In computing these products, we store them as

$$\lambda_{12} \equiv \lambda_1\lambda_2, \quad \mu_{21} \equiv \lambda_1\mu_2, \quad \mu_{12} \equiv \mu_1\mu_2, \quad \lambda_{21} \equiv \lambda_2\mu_1$$

Next, we perform the *on-line* computations. First, let[1]

$$\begin{aligned} \sigma &\leftarrow \lambda_1 \sin\theta_2 \\ \tau &\leftarrow \sin\theta_1\cos\theta_2 \\ \upsilon &\leftarrow \cos\theta_1\cos\theta_2 \\ u &\leftarrow \cos\theta_1\sin\theta_2 + \lambda_1\tau \\ v &\leftarrow \sin\theta_1\sin\theta_2 - \lambda_1\upsilon \end{aligned}$$

and hence,

$$\mathbf{P}_2 = \begin{bmatrix} \upsilon - \sigma\sin\theta_1 & -\lambda_2 u + \lambda_{12}\sin\theta_1 & \mu_2 u + \mu_{12}\sin\theta_1 \\ \tau + \sigma\cos\theta_1 & -\lambda_2 v - \lambda_{12}\cos\theta_1 & \mu_2 v - \mu_{12}\cos\theta_1 \\ \mu_1\sin\theta_2 & \lambda_{21}\cos\theta_2 + \mu_{21} & -\mu_{12}\cos\theta_2 + \lambda_{12} \end{bmatrix}$$

As the reader can verify, the foregoing calculations consume 20 multiplications and 10 additions. Now, we proceed to compute the remaining products in the foregoing do-loop.

Here, notice that the product $\mathbf{P}_{i-1}\mathbf{Q}_i$, for $3 \le i \le n$, can be computed *recursively*, as described below: Let $\mathbf{P}_{i-1}$ and $\mathbf{P}_i$ be given as

$$\mathbf{P}_{i-1} \equiv \begin{bmatrix} p_{11} & p_{12} & p_{13} \\ p_{21} & p_{22} & p_{23} \\ p_{31} & p_{32} & p_{33} \end{bmatrix}$$

[1]Although υ and v look similar, they should not be confused with each other, the former being the lowercase Greek letter *upsilon*. As a matter of fact, no confusion should arise, because upsilon is used only once, and does not appear further in the book.

$$\mathbf{P}_i \equiv \begin{bmatrix} p'_{11} & p'_{12} & p'_{13} \\ p'_{21} & p'_{22} & p'_{23} \\ p'_{31} & p'_{32} & p'_{33} \end{bmatrix}$$

Now matrix $\mathbf{P}_i$ is computed by first defining

$$\begin{aligned} u_i &= p_{11}\sin\theta_i - p_{12}\cos\theta_i \\ v_i &= p_{21}\sin\theta_i - p_{22}\cos\theta_i \\ w_i &= p_{31}\sin\theta_i - p_{32}\cos\theta_i \end{aligned} \tag{4.78a}$$

and

$$\begin{aligned} p'_{11} &= p_{11}\cos\theta_i + p_{12}\sin\theta_i \\ p'_{12} &= -u_i\lambda_i + p_{13}\mu_i \\ p'_{13} &= u_i\mu_i + p_{13}\lambda_i \\ p'_{21} &= p_{21}\cos\theta_i + p_{22}\sin\theta_i \\ p'_{22} &= -v_i\lambda_i + p_{23}\mu_i \\ p'_{23} &= v_i\mu_i + p_{23}\lambda_i \\ p'_{31} &= p_{31}\cos\theta_i + p_{32}\sin\theta_i \\ p'_{32} &= -w_i\lambda_i + p_{33}\mu_i \\ p'_{33} &= w_i\mu_i + p_{33}\lambda_i \end{aligned} \tag{4.78b}$$

Computing u_i, v_i, and w_i requires six multiplications and three additions, whereas each of the p'_{ij} entries requires two multiplications and one addition. Hence, the computation of each $\mathbf{P}_i$ matrix requires 24 multiplications and 12 additions, the total number of operations required to compute the $n-1$ products $\{\,\mathbf{P}_i\,\}_2^{n-1}$ thus being $24(n-2)+20 = 24n-28$ multiplications and $12(n-2)+10 = 12n-14$ additions, for $n \geq 2$. Moreover, $\mathbf{P}_1$, i.e., $\mathbf{Q}_1$, requires four multiplications and no additions, the total number of multiplications M_A and additions A_A required to compute matrix $\mathbf{A}$ thus being

$$M_A = 24n - 24, \quad A_A = 12n - 14 \tag{4.79}$$

Before concluding this section, a remark is in order: The reader may realize that $\mathbf{P}_n$ is nothing but $\mathbf{Q}$, and hence, the same reader may wonder whether we could not save some operations in the foregoing computations by stopping the above recursive algorithm at $n-1$, rather than at n. This is not a good idea, for the above equality holds if and only if the manipulator is capable of tracking *perfectly* a given trajectory. However, reality is quite different, and errors are always present when tracking. As a matter of fact, the mismatch between $\mathbf{P}_n$ and $\mathbf{Q}$ is very useful in estimating orientation errors, which are then used in a feedback-control scheme to synthesize the corrective signals that are meant to correct those errors.

Evaluation of Submatrix **B**

The computation of submatrix $\mathbf{B}$ of the Jacobian is studied here. This submatrix comprises the set of vectors $\{\mathbf{e}_i \times \mathbf{r}_i\}_1^n$. We thus proceed first to the computation of vectors $\mathbf{r}_i$, for $i = 1, \ldots, n$, which is most efficiently done using a recursive scheme, similar to that of Horner for polynomial evaluation (Henrici, 1964), namely,

$$[\mathbf{r}_6]_6 \leftarrow [\mathbf{a}_6]_6$$

For i = 5 to 1 do

$$[\mathbf{r}_i]_i \leftarrow [\mathbf{a}_i]_i + \mathbf{Q}_i[\mathbf{r}_{i+1}]_{i+1}$$

enddo

In the foregoing algorithm, a simple scheme is introduced to perform the product $\mathbf{Q}_i[\mathbf{r}_{i+1}]_{i+1}$, in order to economize operations: If we let $[\mathbf{r}_{i+1}]_{i+1} = [r_1,\, r_2,\, r_3]^T$, then

$$\mathbf{Q}_i[\mathbf{r}_{i+1}]_{i+1} = \begin{bmatrix} \cos\theta_i & -\lambda_i\sin\theta_i & \mu_i\sin\theta_i \\ \sin\theta_i & \lambda_i\cos\theta_i & -\mu_i\cos\theta_i \\ 0 & \mu_i & \lambda_i \end{bmatrix} \begin{bmatrix} r_1 \\ r_2 \\ r_3 \end{bmatrix} = \begin{bmatrix} r_1\cos\theta_i - u\sin\theta_i \\ r_1\sin\theta_i + u\cos\theta_i \\ r_2\mu_i + r_3\lambda_i \end{bmatrix} \tag{4.80a}$$

where

$$u \equiv r_2\lambda_i - r_3\mu_i \tag{4.80b}$$

Therefore, the product of matrix $\mathbf{Q}_i$ by an arbitrary vector consumes eight multiplications and four additions.

Furthermore, each vector $[\mathbf{a}_i]_i$, for $i = 1, \ldots, n$, requires 2 multiplications and no additions, as made apparent from their definitions in eq.(4.3b). Moreover, from the foregoing evaluation of $\mathbf{Q}_i[\mathbf{r}_{i+1}]_{i+1}$, it is apparent that each vector $\mathbf{r}_i$, in frame $\mathcal{F}_i$, is computed with 10 multiplications and seven additions—two more multiplications are needed to calculate each vector $[\mathbf{a}_i]_i$ and three more additions are required to add the latter to vector $\mathbf{Q}_i[\mathbf{r}_{i+1}]_{i+1}$—the whole set of vectors $\{\mathbf{r}_i\}_1^n$ thus being computed, in $\mathcal{F}_i$-coordinates, with $10(n-1) + 2 = 10n - 8$ multiplications and $7(n-1)$ additions, where one coordinate transformation, that of $\mathbf{r}_1$, is not counted, since this vector is computed directly in $\mathcal{F}_1$.

Now we turn to the transformation of the components of all the foregoing vectors into $\mathcal{F}_1$-coordinates. First, note that we can proceed now in two ways: in the first, we transform the individual vectors $\mathbf{e}_i$ and $\mathbf{r}_i$ from $\mathcal{F}_i$- into $\mathcal{F}_1$-coordinates and then compute their cross product; in the second, we first perform the cross products and then transform each of these products into $\mathcal{F}_1$-coordinates. It is apparent that the second approach is more efficient, which is why we choose it here.

In order to calculate the products $\mathbf{e}_i \times \mathbf{r}_i$ in $\mathcal{F}_i$-coordinates, we let $[\,\mathbf{r}_i\,]_i = [\,\rho_1,\, \rho_2,\, \rho_3\,]^T$. Moreover, $[\,\mathbf{e}_i\,]_i = [\,0,\,0,\,1\,]^T$, and hence,

$$[\,\mathbf{e}_i \times \mathbf{r}_i\,]_i = \begin{bmatrix} -\rho_2 \\ \rho_1 \\ 0 \end{bmatrix}$$

which is thus obtained at no cost. Now, the transformation from $\mathcal{F}_i$- into $\mathcal{F}_1$-coordinates is simply

$$[\,\mathbf{e}_i \times \mathbf{r}_i\,]_1 = \mathbf{P}_{i-1}[\,\mathbf{e}_i \times \mathbf{r}_i\,]_i \tag{4.81}$$

In particular, $[\,\mathbf{e}_1 \times \mathbf{r}_1\,]_1$ needs no transformation, for its two factors are given in $\mathcal{F}_1$-coordinates. The $\mathcal{F}_1$-components of the remaining cross products are computed using the general transformation of eq.(4.81). In the case at hand, this transformation requires, for each i, six multiplications and three additions, for this transformation involves the product of a full 3×3 matrix, $\mathbf{P}_{i-1}$, by a 3-dimensional vector, $\mathbf{e}_i \times \mathbf{r}_i$, whose third component vanishes. Thus, the computation of matrix $\mathbf{B}$ requires M_B multiplications and A_B additions, as given below:

$$M_B = 16n - 14, \quad A_B = 10(n-1) \tag{4.82}$$

In total, then, the evaluation of the complete Jacobian requires M_J multiplications and A_J additions, namely,

$$M_J = 40n - 38, \quad A_J = 22n - 24 \tag{4.83}$$

In particular, for a six-revolute manipulator, these figures are 202 multiplications and 108 additions.

Now, if the manipulator contains some prismatic pairs, the foregoing figures diminish correspondingly. Indeed, if the ith joint is prismatic, then the ith column of the Jacobian matrix changes as indicated in eq.(4.57). Hence, one cross-product calculation is spared, along with the associated coordinate transformation. As a matter of fact, as we saw above, the cross product is computed at no cost in local coordinates, and so each prismatic pair of the manipulator reduces the foregoing numbers of operations by only one coordinate transformation, i.e., by 10 multiplications and seven additions.

4.5.2 Singularity Analysis of Decoupled Manipulators

In performing the computation of the joint rates for a decoupled manipulator, it was assumed that neither $\mathbf{J}_{12}$ nor $\mathbf{J}_{21}$ was singular. If the latter is singular, then none of the joint rates can be evaluated, even if the former is nonsingular. However, if $\mathbf{J}_{21}$ is nonsingular, then eq.(4.67a) can be solved for the arm rates even if $\mathbf{J}_{12}$ is singular. Each of these sub-Jacobians is analyzed for singularities below.

We will start analyzing $\mathbf{J}_{21}$, whose singularity determines whether any joint-rate resolution is possible at all. First, we note from eq.(4.66c) that the columns of $\mathbf{J}_{21}$ are the three vectors $\mathbf{e}_1 \times \mathbf{r}_1$, $\mathbf{e}_2 \times \mathbf{r}_2$, and $\mathbf{e}_3 \times \mathbf{r}_3$. Hence, $\mathbf{J}_{21}$ becomes singular if either these three vectors become coplanar or at least one of them vanishes. Furthermore, neither the relative layout of these three vectors nor their magnitudes change if the manipulator undergoes a motion about the first revolute axis while keeping the second and the third revolute axes locked. This means that θ_1 does not affect the singularity of the manipulator, a result that can also be derived from invariance arguments—see Section 2.6—and by noticing that singularity is, indeed, an invariant property. Hence, whether a configuration is singular or not is independent of the viewpoint of the observer, a change in θ_1 being nothing but a change of viewpoint.

The singularity of a three-revolute arm for positioning tasks was analyzed by Burdick (1995), by recognizing that *i*) given three arbitrary lines in space, the three revolute axes in our case, it is always possible to find a set of lines that intersects all three, and *ii*) the moments of the three lines about any point on the intersecting line are all zero. As a matter of fact, the locus of those lines is a quadric ruled surface, namely, a one-sheet hyperboloid—see Exercise 3.4 of Appendix C. Therefore, if the endpoint of the third moving link lies in this quadric, the manipulator is in a singular posture, and velocities of C along the intersecting line cannot be produced. This means that the manipulator has lost, to a *first order*, one degree of freedom. Here we emphasize that this loss is meaningful only at a first order because, in fact, a motion along that intersecting line is still possible, provided that the full nonlinear relations of eq.(4.16) are considered. Even in this case, however, only a *unidirectional* motion is possible, i.e., in the direction in which the distance from C to the farthest axis decreases. Motions in the opposite direction are not feasible because of the rigidity of the links.

We will illustrate the foregoing concepts as they pertain to the most common types of industrial manipulators, i.e., those of the orthogonal type. In these cases, two consecutive axes either intersect at right angles or are parallel; most of the time, the first two axes intersect at right angles and the last two are parallel. Below we study each of these cases separately.

Case 1: Two consecutive axes intersect and C lies in their plane. Here, the ruled hyperboloid containing the lines that intersect all three axes degenerates into a plane, namely, that of the two intersecting axes. For conciseness, let us assume that the first two axes intersect, but the derivations are the same if the intersecting axes are the last two. Moreover, let O_{12} be the intersection of the first two axes, Π_{12} being the plane of these axes and $\mathbf{n}_{12}$ its normal. If we recall the notation adopted in Section 4.5, we have now that the vector directed from O_{12} to C can be regarded as both $\mathbf{r}_1$ and $\mathbf{r}_2$. Furthermore, $\mathbf{e}_1 \times \mathbf{r}_1$ and $\mathbf{e}_2 \times \mathbf{r}_2$ $(= \mathbf{e}_2 \times \mathbf{r}_1)$ are both parallel to $\mathbf{n}_{12}$. Hence, the first two axes can only produce velocities of C in the direction of $\mathbf{n}_{12}$. As a consequence, velocities of C in Π_{12} and perpendicular to $\mathbf{e}_3 \times \mathbf{r}_3$ cannot be produced in the presence of this singularity. The set of infeasible velocities, then, lies in a line whose direction is the geometric representation of the nullspace of $\mathbf{J}_{21}^T$. Likewise, the manipulator can withstand forces applied at C in the direction of the same line purely by reaction wrenches, i.e., without any motor torques. The last issue falls into the realm of manipulator statics, upon which we will elaborate in Section 4.7.

We illustrate this singularity, termed here *shoulder singularity*, in a manipulator with the architecture of Fig. 4.3, as postured in Fig. 4.20. In this figure, the line intersecting all three arm axes is not as obvious and needs further explanation. This line is indicated by $\mathcal{L}$ in that figure, and is parallel to the last two axes. It is apparent that this line intersects the first axis at right angles at a point I. Now, if we take into account that all parallel lines intersect at infinity, then it becomes apparent that $\mathcal{L}$ intersects the axis of the third revolute as well, and hence, $\mathcal{L}$ intersects all three axes.

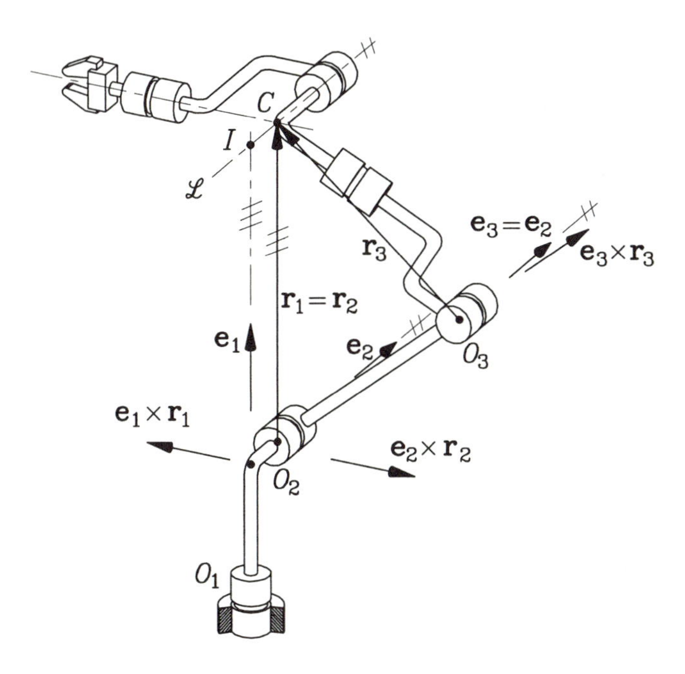

FIGURE 4.20. Shoulder singularity of the Puma robot.

Case 2: Two consecutive axes are parallel and C lies in their plane, as shown in Fig. 4.21. For conciseness, again, we assume that the parallel axes are now the last two, a rather common case in commercial manipulators, but the derivations below are the same if the parallel axes are the first two. We now let Π_{23} be the plane of the last two axes and $\mathbf{n}_{23}$ its normal. Furthermore, $\mathbf{e}_3 = \mathbf{e}_2$, $\mathbf{r}_2 = \mathbf{r}_1$, and $\mathbf{e}_2 \times \mathbf{r}_3 = \alpha(\mathbf{e}_2 \times \mathbf{r}_2)$, where $\alpha = b_4/(a_2 + b_4)$ in terms of the Denavit-Hartenberg notation, thereby making apparent that the last two columns of $\mathbf{J}_{21}$ are linearly dependent. Moreover, $\mathbf{e}_2 \times \mathbf{r}_2$ and consequently, $\mathbf{e}_3 \times \mathbf{r}_3$, are parallel to $\mathbf{n}_{23}$, the last two axes being capable of producing velocities of C only in the direction of $\mathbf{n}_{23}$. Hence, velocities of C in Π_{23} that are normal to $\mathbf{e}_1 \times \mathbf{r}_1$ cannot be produced in this configuration, and the manipulator loses, again, to a first-order approximation, one degree of freedom. The set of infeasible velocities, then, is parallel to the line $\mathcal{L}$ of Fig. 4.21, whose direction is the geometric representation of the nullspace of $\mathbf{J}_{21}^T$. The singularity displayed in the foregoing figure, termed here the *elbow singularity*, pertains also to a manipulator with the architecture of Fig. 4.3.

With regard to the wrist singularities, these were already studied when solving the orientation problem for the inverse kinematics of decoupled manipulators. Here, we study the same in light of the sub-Jacobian $\mathbf{J}_{12}$ of eq.(4.66b). This sub-Jacobian obviously vanishes when the wrist is so configured that its three revolute axes are coplanar, which thus leads to

$$\mathbf{e}_4 \times \mathbf{e}_5 \cdot \mathbf{e}_6 = 0$$

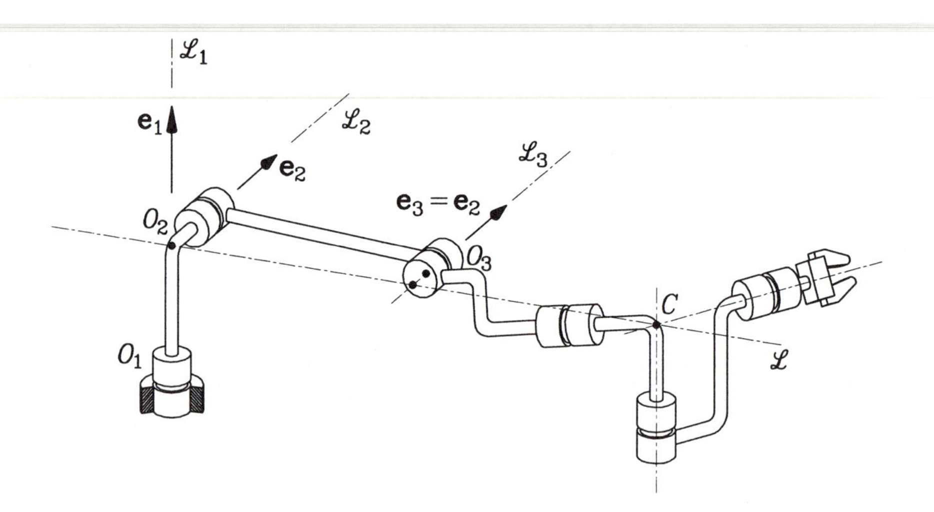

FIGURE 4.21. Elbow singularity of the Puma robot.

Note that when studying the orientation problem of decoupled manipulators, we found that orthogonal wrists are singular when the sixth and fourth axes are aligned, in full agreement with the foregoing condition. Indeed, if these two axes are aligned, then $\mathbf{e}_4 = -\mathbf{e}_6$, and the above equation holds.

4.5.3 Manipulator Workspace

The workspace of spherical wrists for orientation tasks was discussed in Subsection 4.4.2. Here we focus on the workspaces of three-axis positioning manipulators in light of their singularities.

In order to gain insight into the problem, we study first the workspace of manipulators with the architecture of Fig. 4.3. Figures 4.20 and 4.21 show such a manipulator with point C at the limit of its positioning capabilities in one direction, i.e., at the boundary of its workspace. Moreover, with regard to the posture of Fig. 4.20, it is apparent that the first singularity is preserved if (i) point C moves on a line parallel to the first axis and intersecting the second axis; and (ii) with the second and third joints locked, the first joint goes through a full turn. Under the second motion, the line of the first motion sweeps a circular cylinder whose axis is the first manipulator axis and with radius equal to b_3, the shoulder offset. This cylinder constitutes a part of the workspace boundary, the other part consisting of a toroidal surface. Indeed, the second singularity is preserved if (i) with point C in the plane of the second and third axes, the second joint makes a full turn, thereby tracing a circle with center on $\mathcal{L}_2$, a distance b_3 from the first axis, and radius $a_2 + b_4$; and (ii) with point C still in the plane of the second and third joints, the first joint makes a full turn. Under the second motion, the circle generated by the first motion describes a toroid whose major axis is the first manipulator axis. Moreover, since the circle generating the toroid lies in a plane outside this axis, the cross section of the toroid is an ellipse of semiaxes $a_2 + b_4$ and $\sqrt{b_3^2 + (a_2 + b_4)^2} - b_3$, as shown in Fig. 4.22.

The determination of the workspace boundaries of more general manipulators requires, obviously, more general approaches, like that proposed by Ceccarelli (1996). By means of an alternative approach, Ranjbaran, Angeles, and Patel (1992) found the workspace boundary with the aid of the general characteristic equation of a three-revolute manipulator. This equation is a quartic polynomial, as displayed in eq.(4.25). From the discussion of Subsection 4.4.1, it is apparent that at singularities, two distinct roots of the IKP merge into a single one. This happens at points where the plot of the characteristic polynomial of eq.(4.25) is tangent to the τ_3 axis, which occurs in turn at points where the derivative of this polynomial with respect to τ_3 vanishes. The condition for θ_3 to correspond to a point C on the boundary of the workspace is, then, that both the characteristic polynomial and its derivative with respect to τ_3 vanish concurrently. These two

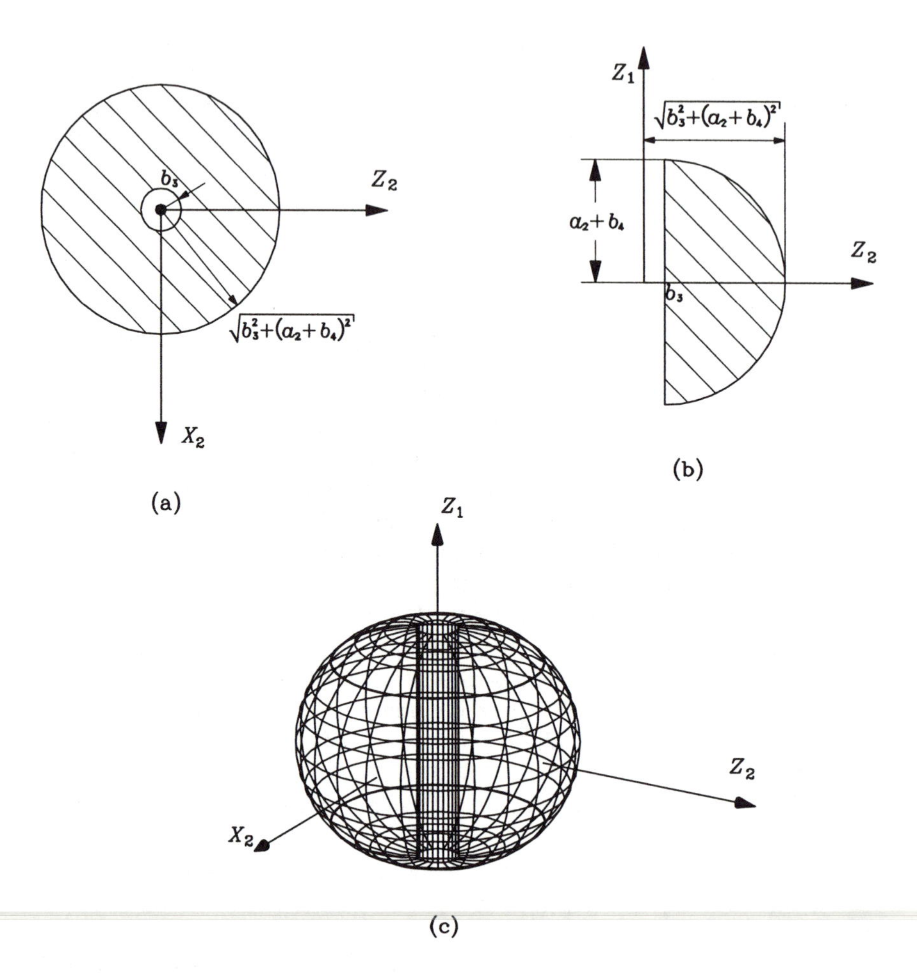

FIGURE 4.22. Workspace of a Puma manipulator (a) top view; (b) side view; (c) perspective.

polynomials are displayed below:

$$P(\tau_3) \equiv R\tau_3^4 + S\tau_3^3 + T\tau_3^2 + U\tau_3 + V = 0 \tag{4.84a}$$

$$P'(\tau_3) \equiv 4R\tau_3^3 + 3S\tau_3^2 + 2T\tau_3 + U = 0 \tag{4.84b}$$

with coefficients R, S, T, U, and V defined in eqs.(4.26a–e). From these equations and eqs.(4.19d–f) and (4.20d–f), it is apparent that the foregoing coefficients are solely functions of the manipulator architecture and the Cartesian coordinates of point C. Moreover, from the same equations, it is clear that the above coefficients are all *quadratic* in $\rho^2 \equiv x_C^2 + y_C^2$ and *quartic* in z_C. Thus, since the Cartesian coordinates x_C and y_C do not appear in the foregoing coefficients explicitly, the workspace is symmetric about the Z_1 axis, a result to be expected by virtue of the independence of singularities from angle θ_1. Hence, the workspace boundary is given by a function $f(\rho^2, z_C) = 0$ that can be derived by eliminating τ_3 from eqs.(4.84a & b). This can be readily done by resorting to any elimination procedure, the simplest one being *dialytic elimination*, as discussed below.

In order to eliminate τ_3 from the above two equations, we proceed in two steps: In the first step, six additional polynomial equations are derived from eqs.(4.84a & b) by multiplying the two sides of each of these equations by τ_3, τ_3^2, and τ_3^3, thereby obtaining a total of eight polynomial equations in τ_3, namely,

$$
\begin{aligned}
R\tau_3^7 + S\tau_3^6 + T\tau_3^5 + U\tau_3^4 + V\tau_3^3 &= 0\\
4R\tau_3^6 + 3S\tau_3^5 + 2T\tau_3^4 + U\tau_3^3 &= 0\\
R\tau_3^6 + S\tau_3^5 + T\tau_3^4 + U\tau_3^3 + V\tau_3^2 &= 0\\
4R\tau_3^5 + 3S\tau_3^4 + 2T\tau_3^3 + U\tau_3^2 &= 0\\
R\tau_3^5 + S\tau_3^4 + T\tau_3^3 + U\tau_3^2 + V\tau_3 &= 0\\
4R\tau_3^4 + 3S\tau_3^3 + 2T\tau_3^2 + U\tau_3 &= 0\\
R\tau_3^4 + S\tau_3^3 + T\tau_3^2 + U\tau_3 + V &= 0\\
4R\tau_3^3 + 3S\tau_3^2 + 2T\tau_3 + U &= 0
\end{aligned}
$$

In the second elimination step we write the above eight equations in *linear homogeneous form*, namely,

$$\mathbf{M}\boldsymbol{\tau}_3 = \mathbf{0} \tag{4.85a}$$

with the 8×8 matrix $\mathbf{M}$ and the 8-dimensional vector $\boldsymbol{\tau}_3$ defined as

$$
\mathbf{M} \equiv \begin{bmatrix}
R & S & T & U & V & 0 & 0 & 0\\
0 & 4R & 3S & 2T & U & 0 & 0 & 0\\
0 & R & S & T & U & V & 0 & 0\\
0 & 0 & 4R & 3S & 2T & U & 0 & 0\\
0 & 0 & R & S & T & U & V & 0\\
0 & 0 & 0 & 4R & 3S & 2T & U & 0\\
0 & 0 & 0 & R & S & T & U & V\\
0 & 0 & 0 & 0 & 4R & 3S & 2T & U
\end{bmatrix}, \quad
\boldsymbol{\tau}_3 = \begin{bmatrix}
\tau_3^7\\ \tau_3^6\\ \tau_3^5\\ \tau_3^4\\ \tau_3^3\\ \tau_3^2\\ \tau_3\\ 1
\end{bmatrix} \tag{4.85b}
$$

It is now apparent that any feasible solution of eq.(4.85a) must be nontrivial, and hence, $\mathbf{M}$ must be singular. The desired boundary equation is then derived from the singularity condition on $\mathbf{M}$, i.e.,

$$f(\rho^2, z_C) \equiv \det(\mathbf{M}) = 0 \tag{4.86}$$

Note that all entries of matrix $\mathbf{M}$ are linear in the coefficients R, S, ..., V, which are, in turn, quadratic in ρ^2 and quartic in z_C. Therefore, the workspace boundary is a surface of 16th degree in ρ^2 and of 32nd degree in z_C.

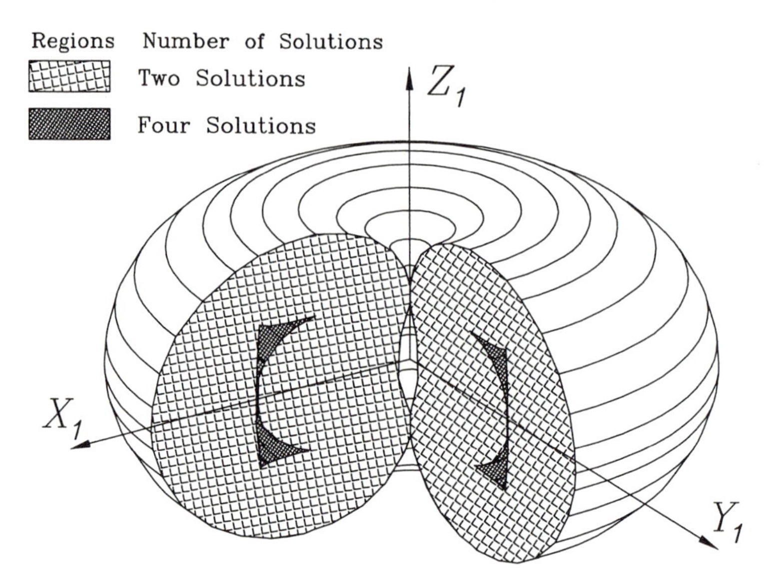

FIGURE 4.23. The workspace of the manipulator of Figs. 4.17–19.

We used the foregoing procedure, with the help of symbolic computations, to obtain a rendering of the workspace boundary of the manipulator of Figs. 4.13–4.15, the workspace thus obtained being displayed in Fig. 4.23.

4.6 Acceleration Analysis of Serial Manipulators

The subject of this section is the computation of vector $\ddot{\boldsymbol{\theta}}$ of second joint-variable derivatives, also called the *joint accelerations*. This vector is computed from Cartesian position, velocity, and acceleration data. To this end, both sides of eq.(4.54) are differentiated with respect to time, thus obtaining

$$\mathbf{J}\ddot{\boldsymbol{\theta}} = \dot{\mathbf{t}} - \dot{\mathbf{J}}\dot{\boldsymbol{\theta}} \tag{4.87}$$

and hence,

$$\ddot{\boldsymbol{\theta}} = \mathbf{J}^{-1}(\dot{\mathbf{t}} - \dot{\mathbf{J}}\dot{\boldsymbol{\theta}}) \tag{4.88}$$

From eq.(4.87), it is clear that the joint-acceleration vector is computed in exactly the same way as the joint-rate vector. In fact, the LU decomposition of $\mathbf{J}$ is the same in this case and hence, need not be recomputed. All that is needed is the solution of a lower- and an upper-triangular system, namely,

$$\mathbf{L}\mathbf{z} = \dot{\mathbf{t}} - \dot{\mathbf{J}}\dot{\boldsymbol{\theta}}, \quad \mathbf{U}\ddot{\boldsymbol{\theta}} = \mathbf{z}$$

The two foregoing systems are solved first for $\mathbf{z}$ and then for $\ddot{\boldsymbol{\theta}}$ by forward and backward substitution, respectively. The first of the foregoing systems is solved with M_n'' multiplications and A_n'' additions; the second with M_n'''

multiplications and A_n''' additions. These figures appear in eqs.(4.63b & c). Thus, the total numbers of multiplications, M_t, and additions, A_t, that the forward and backward solutions of the aforementioned systems require are

$$M_t = n^2, \quad A_t = n(n-1) \tag{4.89}$$

In eq.(4.87), the right-hand side comprises two terms, the first being the specified time-rate of change of the twist of the EE, or twist-rate, for brevity, which is readily available. The second term is not available and must be computed. This term involves the product of the time-derivative of $\mathbf{J}$ times the previously computed joint-rate vector. Hence, in order to evaluate the right-hand side of that equation, all that is further required is $\dot{\mathbf{J}}$. From eq.(4.55a), one has

$$\dot{\mathbf{J}} = \begin{bmatrix} \dot{\mathbf{A}} \\ \dot{\mathbf{B}} \end{bmatrix}$$

where, from eqs.(4.52a & b),

$$\dot{\mathbf{A}} = [\,\dot{\mathbf{e}}_1 \quad \dot{\mathbf{e}}_2 \quad \cdots \quad \dot{\mathbf{e}}_n\,] \tag{4.90a}$$

$$\dot{\mathbf{B}} = [\,\dot{\mathbf{u}}_1 \quad \dot{\mathbf{u}}_2 \quad \cdots \quad \dot{\mathbf{u}}_n\,] \tag{4.90b}$$

and $\mathbf{u}_i$ denotes $\mathbf{e}_i \times \mathbf{r}_i$, for $i = 1, 2, \ldots, n$. Moreover,

$$\dot{\mathbf{e}}_1 = \boldsymbol{\omega}_0 \times \mathbf{e}_1 = \mathbf{0} \tag{4.91a}$$

$$\dot{\mathbf{e}}_i = \boldsymbol{\omega}_{i-1} \times \mathbf{e}_i \equiv \boldsymbol{\omega}_i \times \mathbf{e}_i, \quad i = 2, 3, \ldots, n \tag{4.91b}$$

and

$$\dot{\mathbf{u}}_i = \dot{\mathbf{e}}_i \times \mathbf{r}_i + \mathbf{e}_i \times \dot{\mathbf{r}}_i, \quad i = 1, 2, \ldots, n \tag{4.91c}$$

Next, an expression for $\dot{\mathbf{r}}_i$ is derived by time-differentiating both sides of eq.(4.51), which produces

$$\dot{\mathbf{r}}_i = \dot{\mathbf{a}}_i + \dot{\mathbf{a}}_{i+1} + \cdots + \dot{\mathbf{a}}_n, \quad i = 1, 2, \ldots\ n$$

Recalling eq.(4.49), the above equation reduces to

$$\dot{\mathbf{r}}_i = \boldsymbol{\omega}_i \times \mathbf{a}_i + \boldsymbol{\omega}_{i+1} \times \mathbf{a}_{i+1} + \cdots + \boldsymbol{\omega}_n \times \mathbf{a}_n \tag{4.92}$$

Substitution of eqs.(4.91) and (4.92) into eqs.(4.90a & b) leads to

$$\dot{\mathbf{A}} = [\,\mathbf{0} \quad \boldsymbol{\omega}_1 \times \mathbf{e}_2 \quad \cdots \quad \boldsymbol{\omega}_{n-1} \times \mathbf{e}_n\,]$$

$$\dot{\mathbf{B}} = [\,\mathbf{e}_1 \times \dot{\mathbf{r}}_1 \quad \boldsymbol{\omega}_{12} \times \mathbf{r}_2 + \mathbf{e}_2 \times \dot{\mathbf{r}}_2 \quad \cdots \quad \boldsymbol{\omega}_{n-1,n} \times \mathbf{r}_n + \mathbf{e}_n \times \dot{\mathbf{r}}_n\,]$$

with $\dot{\mathbf{r}}_k$ and $\boldsymbol{\omega}_{k,k+1}$ defined as

$$\dot{\mathbf{r}}_k \equiv \sum_k^n \boldsymbol{\omega}_i \times \mathbf{a}_i, \quad k = 1, \ldots, n \tag{4.93a}$$

$$\boldsymbol{\omega}_{k,k+1} \equiv \boldsymbol{\omega}_k \times \mathbf{e}_{k+1}, \quad k = 1, \ldots, n-1 \tag{4.93b}$$

The foregoing expressions are invariant and hence, valid in any coordinate frame. However, they are going to be incorporated into matrix $\dot{\mathbf{J}}$, and then the latter is to be multiplied by vector $\dot{\boldsymbol{\theta}}$, as indicated in eq.(4.87). Thus, eventually all columns of both $\dot{\mathbf{A}}$ and $\dot{\mathbf{B}}$ will have to be represented in the same coordinate frame. Hence, coordinate transformations will have to be introduced in the foregoing matrix columns in order to have all of these represented in the same coordinate frame, say, the first one. We then have the expansion below:

$$\dot{\mathbf{J}}\dot{\boldsymbol{\theta}} = \dot{\theta}_1 \begin{bmatrix} \mathbf{0} \\ \dot{\mathbf{u}}_1 \end{bmatrix} + \dot{\theta}_2 \begin{bmatrix} \dot{\mathbf{e}}_2 \\ \dot{\mathbf{u}}_2 \end{bmatrix} + \cdots + \dot{\theta}_n \begin{bmatrix} \dot{\mathbf{e}}_n \\ \dot{\mathbf{u}}_n \end{bmatrix} \tag{4.94}$$

The right-hand side of eq.(4.94) is computed recursively as described below in five steps, the number of operations required being included at the end of each step.

1. Compute $\{[\boldsymbol{\omega}_i]_i\}_1^n$:

$$[\boldsymbol{\omega}_1]_1 \leftarrow \dot{\theta}_1[\mathbf{e}_1]_1$$

`For i = 1 to n − 1 do`

$$[\boldsymbol{\omega}_{i+1}]_{i+1} \leftarrow \dot{\theta}_{i+1}[\mathbf{e}_{i+1}]_{i+1} + \mathbf{Q}_i^T[\boldsymbol{\omega}_i]_i$$

`enddo` $\qquad 8(n-1)\,M$ & $5(n-1)\,A$

2. Compute $\{[\dot{\mathbf{e}}_i]_i\}_1^n$:

$$[\dot{\mathbf{e}}_1]_1 \leftarrow [\mathbf{0}]_1$$

`For i = 2 to n do`

$$[\dot{\mathbf{e}}_i]_i \leftarrow [\boldsymbol{\omega}_i \times \mathbf{e}_i]_i$$

`enddo` $\qquad 0M$ & $0A$

3. Compute $\{[\dot{\mathbf{r}}_i]_i\}_1^n$:

$$[\dot{\mathbf{r}}_n]_n \leftarrow [\boldsymbol{\omega}_n \times \mathbf{a}_n]_n$$

`For i = n − 1 to 1 do`

$$[\dot{\mathbf{r}}_i]_i \leftarrow [\boldsymbol{\omega}_i \times \mathbf{a}_i]_i + \mathbf{Q}_i[\dot{\mathbf{r}}_{i+1}]_{i+1}$$

`enddo` $\qquad (14n-8)M$ & $(10n-7)A$

4. Compute $\{[\dot{\mathbf{u}}_i]_i\}_1^n$ using the expression appearing in eq.(4.91c):

$$[\dot{\mathbf{u}}_1]_1 \leftarrow [\mathbf{e}_1 \times \dot{\mathbf{r}}_1]_1$$

`For i = 2 to n do`

$$[\dot{\mathbf{u}}_i]_i \leftarrow [\dot{\mathbf{e}}_i \times \mathbf{r}_i + \mathbf{e}_i \times \dot{\mathbf{r}}_i]_i$$

`enddo` $\qquad 4(n-1)\,M$ & $3(n-1)\,A$

5. Compute $\dot{\mathbf{J}}\dot{\boldsymbol{\theta}}$:

   Let $\mathbf{v} \equiv \dot{\mathbf{J}}\dot{\boldsymbol{\theta}}$, which is a 6-dimensional vector. A coordinate transformation of its two 3-dimensional vector components will be implemented using the 6×6 matrices $\mathbf{U}_i$, which are defined as

$$\mathbf{U}_i \equiv \begin{bmatrix} \mathbf{Q}_i & \mathbf{O} \\ \mathbf{O} & \mathbf{Q}_i \end{bmatrix}$$

   where $\mathbf{O}$ stands for the 3×3 zero matrix. Thus, the foregoing 6×6 matrices are block-diagonal, their diagonal blocks being simply matrices $\mathbf{Q}_i$. One then has the algorithm below:

$$[\,\mathbf{v}\,]_n \leftarrow \dot{\theta}_n \begin{bmatrix} \dot{\mathbf{e}}_n \\ \dot{\mathbf{u}}_n \end{bmatrix}_n$$

   For i = n − 1 to 1 do

$$[\,\mathbf{v}\,]_i \leftarrow \dot{\theta}_i \begin{bmatrix} \dot{\mathbf{e}}_i \\ \dot{\mathbf{u}}_i \end{bmatrix}_i + \mathbf{U}_i[\,\mathbf{v}\,]_{i+1}$$

   enddo

$$\dot{\mathbf{J}}\dot{\boldsymbol{\theta}} \leftarrow [\,\mathbf{v}\,]_1 \qquad\qquad 19(n-1) + 3\,M \;\&\; 12(n-1)\,A$$

   thereby completing the computation of $\dot{\mathbf{J}}\dot{\boldsymbol{\theta}}$.

The figures given above for the floating-point operations involved were obtained based on a few facts, namely,

1. It is recalled that $[\,\mathbf{e}_i\,]_i = [\,0,\,0,\,1\,]^T$. Moreover, if we let $[\,\mathbf{w}\,]_i = [\,w_x,\,w_y,\,w_z\,]^T$ be an arbitrary 3-dimensional vector, then

$$[\,\mathbf{e}_i \times \mathbf{w}\,]_i = \begin{bmatrix} -w_y \\ w_x \\ 0 \end{bmatrix}$$

   and hence, this product requires zero multiplications and zero additions.

2. $[\,\dot{\mathbf{e}}_i\,]_i$, computed as in eq.(4.91b), takes on the form $[\,\omega_y,\,-\omega_x,\,0\,]^T$, where ω_x and ω_y are the X_i and Y_i components of $\boldsymbol{\omega}_i$. Moreover, let $[\,\mathbf{r}_i\,]_i = [\,x,\,y,\,z\,]^T$. Then

$$[\,\dot{\mathbf{e}}_i \times \mathbf{r}_i\,]_i = \begin{bmatrix} -z\omega_x \\ -z\omega_y \\ x\omega_x + y\omega_y \end{bmatrix}$$

   and this product is computed with four multiplications and one addition.

3. As found in Subsection 4.5.1, any coordinate transformation from $\mathcal{F}_i$ to $\mathcal{F}_{i+1}$, or vice versa, of any 3-dimensional vector is computed with eight multiplications and four additions.

Thus, the total numbers of multiplications and additions required to compute $\dot{\mathbf{J}}\dot{\boldsymbol{\theta}}$ in frame $\mathcal{F}_1$, denoted by M_J and A_J, respectively, are as shown below:

$$M_J = 45n - 36, \quad A_J = 30n - 27$$

Since the right-hand side of eq.(4.87) involves the algebraic sum of two 6-dimensional vectors, then, the total numbers of multiplications and additions needed to compute the aforementioned right-hand side, denoted by M_r and A_r, are

$$M_r = 45n - 36, \quad A_r = 30n - 21$$

These figures yield 234 multiplications and 159 additions for a six-revolute manipulator of arbitrary architecture. Finally, if the latter figures are added to those of eq.(4.89), one obtains the numbers of multiplications and additions required for an acceleration resolution of a six-revolute manipulator of arbitrary architecture as

$$M_a = 270, \quad A_a = 189$$

Furthermore, for six-axis, decoupled manipulators, the operation counts of steps 1 and 2 above do not change. However, step 3 is reduced by 42 multiplications and 30 additions, whereas step 4 by 12 multiplications and 9 additions. Moreover, step 5 is reduced by 57 multiplications and 36 additions. With regard to the solution of eq.(4.87) for $\ddot{\boldsymbol{\theta}}$, an additional reduction of *floating-point operations*, or flops, is obtained, for now we need only 18 multiplications and 12 additions to solve two systems of three equations with three unknowns, thereby saving 18 multiplications and 18 additions. Thus, the corresponding figures for such a manipulator, M_a' and A_a', respectively, are

$$M_a' = 141, \quad A_a' = 96$$

4.7 Static Analysis of Serial Manipulators

In this section, the static analysis of a serial n-axis manipulator is undertaken, particular attention being given to six-axis, decoupled manipulators. Let τ_i be the torque acting at the ith revolute or the force acting at the ith prismatic pair. Moreover, let $\boldsymbol{\tau}$ be the n-dimensional vector of joint forces and torques, whose ith component is τ_i, whereas $\mathbf{w} = [\,\mathbf{n}^T, \mathbf{f}^T\,]^T$ denotes the wrench acting on the EE, with $\mathbf{n}$ denoting the resultant moment and $\mathbf{f}$ the resultant force applied at point P of the end-effector of the manipulator of Fig. 4.19. Then the power exerted on the manipulator by all forces and moments acting on the end-effector is

$$\Pi_E = \mathbf{w}^T\mathbf{t} = \mathbf{n}^T\boldsymbol{\omega} + \mathbf{f}^T\dot{\mathbf{p}}$$

whereas the power exerted on the manipulator by all joint motors, Π_J, is

$$\Pi_J = \boldsymbol{\tau}^T\dot{\boldsymbol{\theta}} \tag{4.95}$$

Under static, conservative conditions, there is neither power dissipation nor change in the kinetic energy of the manipulator, and hence, the two foregoing powers are equal, which is just a restatement of the *First Law of Thermodynamics* or equivalently, the *Principle of Virtual Work*, i.e.,

$$\mathbf{w}^T\mathbf{t} = \boldsymbol{\tau}^T\dot{\boldsymbol{\theta}} \tag{4.96a}$$

Upon substitution of eq.(4.54) into eq.(4.96a), we obtain

$$\mathbf{w}^T\mathbf{J}\dot{\boldsymbol{\theta}} = \boldsymbol{\tau}^T\dot{\boldsymbol{\theta}} \tag{4.96b}$$

which is a relation valid for arbitrary $\dot{\boldsymbol{\theta}}$. Under these conditions, if $\mathbf{J}$ is not singular, eq.(4.96b) leads to

$$\mathbf{J}^T\mathbf{w} = \boldsymbol{\tau} \tag{4.97}$$

This equation relates the wrench acting on the EE with the joint forces and torques exerted by the actuators. Therefore, this equation finds applications in the *sensing* of the wrench $\mathbf{w}$ acting on the EE by means of torque sensors located at the revolute axes. These sensors measure the motor-supplied torques via the current flowing through the motor armatures, the sensor readouts being the joint torques—or forces, in the case of prismatic joints—$\{\,\tau_k\,\}_1^n$, grouped into vector $\boldsymbol{\tau}$.

For a six-axis manipulator, in the absence of singularities, the foregoing equation can be readily solved for $\mathbf{w}$ in the form

$$\mathbf{w} = \mathbf{J}^{-T}\boldsymbol{\tau}$$

where $\mathbf{J}^{-T}$ stands for the inverse of $\mathbf{J}^T$. Thus, using the figures recorded in eq.(4.62b), $\mathbf{w}$ can be computed from eq.(4.97) with 127 multiplications and 100 additions for a manipulator of arbitrary architecture. However, if the manipulator is of the decoupled type, the Jacobian takes on the form appearing in eq.(4.65), and hence, the foregoing computation can be performed in two steps, namely,

$$\mathbf{J}_{12}^T\mathbf{n}_w = \boldsymbol{\tau}_w$$
$$\mathbf{J}_{21}^T\mathbf{f} = \boldsymbol{\tau}_a - \mathbf{J}_{11}^T\mathbf{n}_w$$

where $\mathbf{n}_w$ is the resultant moment acting on the end-effector when $\mathbf{f}$ is applied at the center of the wrist, while $\boldsymbol{\tau}$ has been partitioned as

$$\boldsymbol{\tau} \equiv \begin{bmatrix} \boldsymbol{\tau}_a \\ \boldsymbol{\tau}_w \end{bmatrix}$$

with $\boldsymbol{\tau}_a$ and $\boldsymbol{\tau}_w$ defined as the wrist- and the arm torques, respectively. These two vectors are given, in turn, as

$$\boldsymbol{\tau}_a = \begin{bmatrix} \tau_1 \\ \tau_2 \\ \tau_3 \end{bmatrix}, \quad \boldsymbol{\tau}_w = \begin{bmatrix} \tau_4 \\ \tau_5 \\ \tau_6 \end{bmatrix}$$

Hence, the foregoing calculations, as pertaining to a six-axis, decoupled manipulator, are performed with 55 multiplications and 37 additions, which follows from a result that was derived in Section 4.5 and is summarized in eq.(4.69).

In solving for the wrench acting on the EE from the above relations, the wrist equilibrium equation is first solved for $\mathbf{n}_w$, thus obtaining

$$\mathbf{n}_w = \mathbf{J}_{12}^{-T}\boldsymbol{\tau}_w \tag{4.98}$$

where $\mathbf{J}_{12}^{-T}$ stands for the inverse of $\mathbf{J}_{12}^{T}$, and is available in eq.(4.75). Therefore,

$$\begin{aligned}\mathbf{n}_w &= \frac{1}{\Delta_{21}}\left[\,(\mathbf{e}_5 \times \mathbf{e}_6) \quad (\mathbf{e}_6 \times \mathbf{e}_4) \quad (\mathbf{e}_4 \times \mathbf{e}_5)\,\right]\boldsymbol{\tau}_w \\ &= \frac{1}{\Delta_{21}}\left[\,\tau_4(\mathbf{e}_5 \times \mathbf{e}_6) + \tau_5(\mathbf{e}_6 \times \mathbf{e}_4) + \tau_6(\mathbf{e}_4 \times \mathbf{e}_5)\,\right]\end{aligned} \tag{4.99}$$

Now, if we let

$$\overline{\boldsymbol{\tau}}_a \equiv \boldsymbol{\tau}_a - \mathbf{J}_{11}^{T}\mathbf{n}_w \tag{4.100}$$

we have, from eq.(4.74),

$$\mathbf{f} = \left[\,\mathbf{u}_2 \times \mathbf{u}_3 \quad \mathbf{u}_3 \times \mathbf{u}_1 \quad \mathbf{u}_1 \times \mathbf{u}_2\,\right]\frac{\overline{\boldsymbol{\tau}}_a}{\Delta_{21}}$$

where

$$\mathbf{u}_i \equiv \mathbf{e}_i \times \mathbf{r}_i$$

or

$$\mathbf{f} = \frac{1}{\Delta_{21}}\left[\,\overline{\tau}_1(\mathbf{u}_2 \times \mathbf{u}_3) + \overline{\tau}_2(\mathbf{u}_3 \times \mathbf{u}_1) + \overline{\tau}_3(\mathbf{u}_1 \times \mathbf{u}_2)\,\right] \tag{4.101}$$

4.8 Planar Manipulators

Shown in Fig. 4.24 is a three-axis planar manipulator. Note that in this case, the DH parameters b_i and α_i vanish, for $i = 1, 2, 3$, the nonvanishing parameters a_i being indicated in the same figure. Below we proceed with the displacement, velocity, acceleration, and static analyses of this manipulator. Here, we recall a few relations of planar mechanics that will be found useful in the discussion below.

A 2×2 matrix $\mathbf{A}$ can be partitioned either columnwise or rowwise, as shown below:

$$\mathbf{A} \equiv [\mathbf{a} \quad \mathbf{b}] \equiv \begin{bmatrix} \mathbf{c}^T \\ \mathbf{d}^T \end{bmatrix}$$

where $\mathbf{a}$, $\mathbf{b}$, $\mathbf{c}$, and $\mathbf{d}$ are all 2-dimensional column vectors. Furthermore, let $\mathbf{E}$ be defined as an orthogonal matrix rotating 2-dimensional vectors through an angle of 90° counterclockwise. Hence,

$$\mathbf{E} \equiv \begin{bmatrix} 0 & -1 \\ 1 & 0 \end{bmatrix} \tag{4.102}$$

We thus have

Fact 4.8.1

$$\mathbf{E}^{-1} = \mathbf{E}^T = -\mathbf{E}$$

and hence,

Fact 4.8.2

$$\mathbf{E}^2 = -\mathbf{1}$$

where $\mathbf{1}$ *is the* 2×2 *identity matrix.*

Moreover,

Fact 4.8.3

$$\begin{aligned} \det(\mathbf{A}) &= -\mathbf{a}^T\mathbf{E}\mathbf{b} = \mathbf{b}^T\mathbf{E}\mathbf{a} \\ &= -\mathbf{c}^T\mathbf{E}\mathbf{d} = \mathbf{d}^T\mathbf{E}\mathbf{c} \end{aligned}$$

and

Fact 4.8.4

$$\begin{aligned} \mathbf{A}^{-1} &= \frac{1}{\det(\mathbf{A})} \begin{bmatrix} \mathbf{b}^T \\ -\mathbf{a}^T \end{bmatrix} \mathbf{E} \\ &= \frac{1}{\det(\mathbf{A})} \mathbf{E} [-\mathbf{d} \quad \mathbf{c}] \end{aligned}$$

4.8.1 Displacement Analysis

The inverse kinematics of the manipulator at hand now consists of determining the values of angles θ_i, for $i = 1, 2, 3$, that will place the end-effector

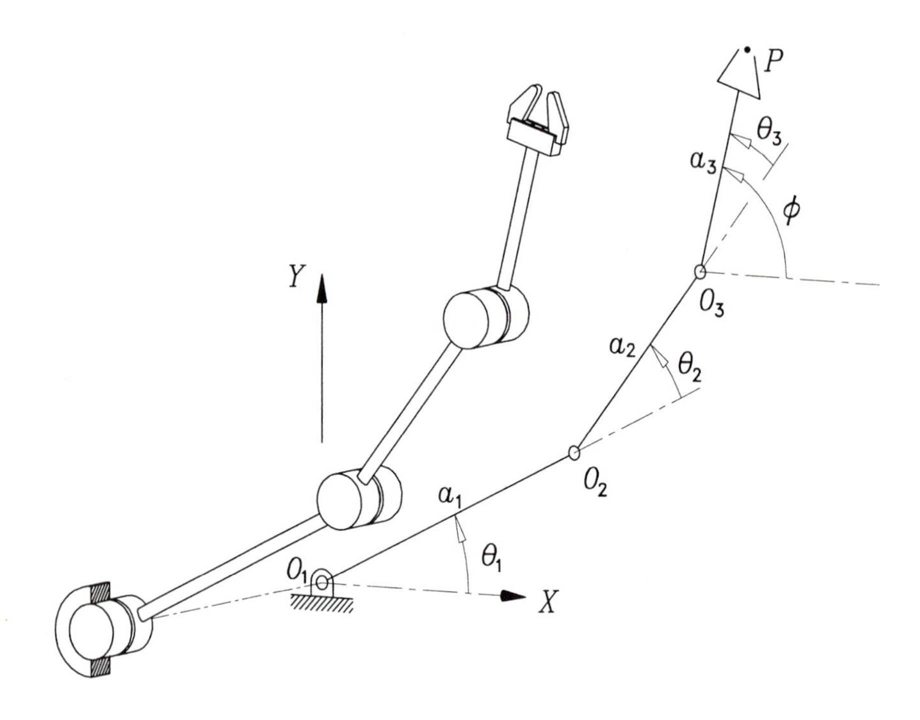

FIGURE 4.24. Three-axis planar manipulator.

so that its operation point P will be positioned at the prescribed Cartesian coordinates x, y and be oriented at a given angle ϕ with the X axis of Fig. 4.24. Note that this manipulator can be considered as decoupled, for the end-effector can be placed at the desired pose by first positioning point O_3 with the aid of the first two joints and then orienting it with the third joint only. We then solve for the joint angles in two steps, one for positioning and one for orienting.

We now have, from the geometry of Fig. 4.24,

$$a_1 c_1 + a_2 c_{12} = x$$
$$a_1 s_1 + a_2 s_{12} = y$$

where x and y denote the Cartesian coordinates of point O_3, while c_{12} and s_{12} stand for $\cos(\theta_1+\theta_2)$ and $\sin(\theta_1+\theta_2)$, respectively. We have thus derived two equations for the two unknown angles, from which we can determine these angles in various ways. For example, we can solve the problem using a semigraphical approach similar to that of Subsection 8.2.1. Indeed, from the two foregoing equations we can eliminate both c_{12} and s_{12} by solving for the second terms of the left-hand sides of those equations, namely,

$$a_2 c_{12} = x - a_1 c_1 \tag{4.103a}$$
$$a_2 s_{12} = y - a_1 s_1 \tag{4.103b}$$

If both sides of the above two equations are now squared, then added, and the ensuing sum is equated to a_2^2, we obtain, after simplification, a linear equation in c_1 and s_1 that represents a line $\mathcal{L}$ in the c_1-s_1 plane:

$$\mathcal{L}: \qquad -a_1^2 + a_2^2 + 2a_1 x c_1 + 2a_1 y s_1 - (x^2 + y^2) = 0 \tag{4.104}$$

Clearly, the two foregoing variables are constrained by a quadratic equation defining a circle $\mathcal{C}$ in the same plane:

$$\mathcal{C}: \qquad c_1^2 + s_1^2 = 1$$

which is a circle $\mathcal{C}$ of unit radius centered at the origin of the aforementioned plane. The real roots of interest are then obtained as the intersections of $\mathcal{L}$ and $\mathcal{C}$. Thus, the problem can admit (*i*) two real and distinct roots, if the line and the circle intersect; (*ii*) one repeated root if the line is tangent to the circle; and (*iii*) no real root if the line does not intersect the circle. The two real roots constitute the so-called *elbow-up* and *elbow-down* arm postures, depicted in Fig. 4.25. The same result could have been obtained if we had transformed eq.(4.104) into a quadratic equation by resorting to the usual identities relating the cosine and the sine functions with the tangent of half the corresponding angle.

With c_1 and s_1 known, angle θ_1 is fully determined. Note that the two real intersections of $\mathcal{L}$ with $\mathcal{C}$ provide each one value of θ_1, as depicted in Fig. 4.26.

Once θ_1 is available, both c_{12} and s_{12} are computed uniquely from eqs.(4.103a & b), and hence, each value of θ_1 yields a unique value of

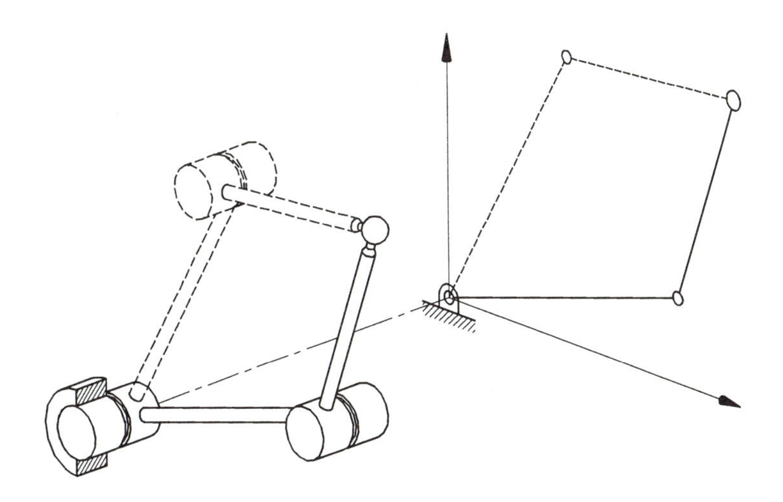

FIGURE 4.25. The two real solutions of a planar manipulator.

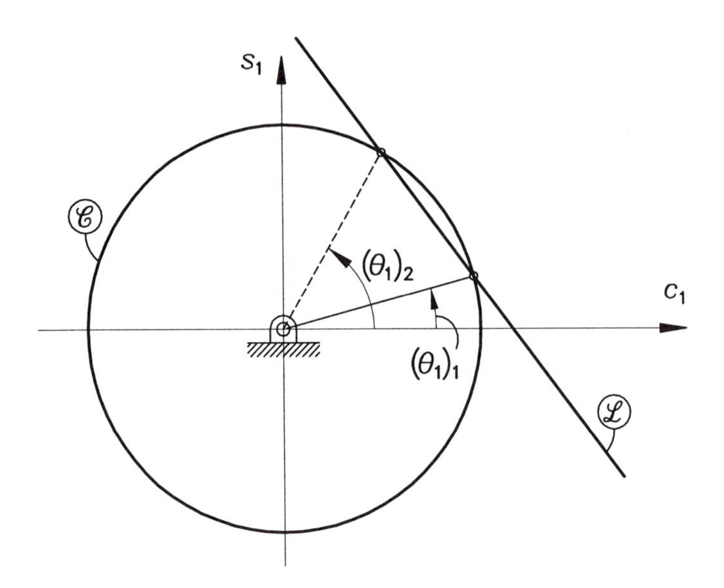

FIGURE 4.26. The two real values of θ_1 depicted in Fig. 4.25.

$\theta_1+\theta_2$. Moreover, from this value, one unique value of θ_2 is readily derived for each value of θ_1, namely,

$$(\theta_2)_i = \tan^{-1}\left(\frac{y - a_1(s_1)_i}{x - a_1(c_1)_i}\right) - (\theta_1)_i, \quad \text{for} \quad i = 1, 2$$

Once θ_1 and θ_2 are available, θ_3 is readily derived from the geometry of Fig. 4.24, namely,

$$\theta_3 = \phi - (\theta_1 + \theta_2)$$

and hence, each pair of (θ_1, θ_2) values yields one single value for θ_3. Since we have two such pairs, the problem admits two real solutions.

4.8.2 Velocity Analysis

Velocity analysis is most easily accomplished if the general velocity relations derived in Section 4.5 are recalled and adapted to planar manipulators. Thus we have, as in eq.(4.54),

$$\mathbf{J}\dot{\boldsymbol{\theta}} = \mathbf{t} \tag{4.105a}$$

where now,

$$\mathbf{J} \equiv \begin{bmatrix} \mathbf{e}_1 & \mathbf{e}_2 & \mathbf{e}_3 \\ \mathbf{e}_1 \times \mathbf{r}_1 & \mathbf{e}_2 \times \mathbf{r}_2 & \mathbf{e}_3 \times \mathbf{r}_3 \end{bmatrix}, \quad \dot{\boldsymbol{\theta}} \equiv \begin{bmatrix} \dot{\theta}_1 \\ \dot{\theta}_2 \\ \dot{\theta}_3 \end{bmatrix}, \quad \mathbf{t} \equiv \begin{bmatrix} \boldsymbol{\omega} \\ \dot{\mathbf{p}} \end{bmatrix} \tag{4.105b}$$

and $\{\mathbf{r}_i\}_1^3$ are defined as in Subsection 4.8.2, i.e., as the vectors directed from O_i to P. As in the previous subsection, we assume here that the manipulator moves in the X-Y plane, and hence, all revolute axes are parallel to the Z axis, vectors $\mathbf{e}_i$ and $\mathbf{r}_i$, for $i = 1, 2, 3$, thus taking on the forms

$$\mathbf{e}_1 = \mathbf{e}_2 = \mathbf{e}_3 = \mathbf{e} \equiv \begin{bmatrix} 0 \\ 0 \\ 1 \end{bmatrix}, \quad \mathbf{r}_i = \begin{bmatrix} x_i \\ y_i \\ 0 \end{bmatrix}$$

with $\mathbf{t}$ reducing to

$$\mathbf{t} = \begin{bmatrix} 0 & 0 & \dot{\phi} & \dot{x}_P & \dot{y}_P & 0 \end{bmatrix}^T \tag{4.105c}$$

in which $\dot{x}_P$ and $\dot{y}_P$ denote the components of the velocity of P. Thus,

$$\mathbf{e}_i \times \mathbf{r}_i = \begin{bmatrix} -y_i \\ x_i \\ 0 \end{bmatrix}$$

and hence, the foregoing cross product can be expressed as

$$\mathbf{e}_i \times \mathbf{r}_i = \begin{bmatrix} \mathbf{E}\mathbf{s}_i \\ 0 \end{bmatrix}$$

where $\mathbf{E}$ was defined in eq.(4.102) and $\mathbf{s}_i$ is the 2-dimensional projection of $\mathbf{r}_i$ onto the x-y plane of motion, i.e., $\mathbf{s}_i \equiv [\, x_i \quad y_i \,]^T$. Equation (4.105a) thus reduces to

$$\begin{bmatrix} \mathbf{0} & \mathbf{0} & \mathbf{0} \\ 1 & 1 & 1 \\ \mathbf{E}\mathbf{s}_1 & \mathbf{E}\mathbf{s}_2 & \mathbf{E}\mathbf{s}_3 \\ 0 & 0 & 0 \end{bmatrix} \dot{\boldsymbol{\theta}} = \begin{bmatrix} \mathbf{0} \\ \dot{\phi} \\ \dot{\mathbf{p}} \\ 0 \end{bmatrix} \tag{4.106}$$

where $\mathbf{0}$ is the 2-dimensional zero vector and $\dot{\mathbf{p}}$ is now reduced to $\dot{\mathbf{p}} \equiv [\, \dot{x}, \ \dot{y} \,]^T$. In summary, then, by working only with the three nontrivial equations of eq.(4.106), we can represent the velocity relation using a 3×3 Jacobian in eq.(4.105a). To this end, we redefine $\mathbf{J}$ and $\mathbf{t}$ as

$$\mathbf{J} \equiv \begin{bmatrix} 1 & 1 & 1 \\ \mathbf{E}\mathbf{s}_1 & \mathbf{E}\mathbf{s}_2 & \mathbf{E}\mathbf{s}_3 \end{bmatrix}, \quad \mathbf{t} \equiv \begin{bmatrix} \dot{\phi} \\ \dot{\mathbf{p}} \end{bmatrix} \tag{4.107}$$

The velocity resolution of this manipulator thus reduces to solving for the three joint rates from eq.(4.105a), with $\mathbf{J}$ and $\mathbf{t}$ defined as in eq.(4.107), which thus leads to the system below:

$$\begin{bmatrix} 1 & 1 & 1 \\ \mathbf{E}\mathbf{s}_1 & \mathbf{E}\mathbf{s}_2 & \mathbf{E}\mathbf{s}_3 \end{bmatrix} \begin{bmatrix} \dot{\theta}_1 \\ \dot{\theta}_2 \\ \dot{\theta}_3 \end{bmatrix} = \begin{bmatrix} \dot{\phi} \\ \dot{\mathbf{p}} \end{bmatrix} \tag{4.108}$$

Solving for $\{\dot{\theta}_i\}_1^3$ is readily done by first reducing the system of equations appearing in eq.(4.105a) to one of two equations in two unknowns by resorting to Gaussian elimination. Indeed, if the first scalar equation of eq.(4.108) is multiplied by $\mathbf{E}\mathbf{s}_1$ and the product is subtracted from the 2-dimensional vector equation, we obtain

$$\begin{bmatrix} 1 & 1 & 1 \\ \mathbf{0} & \mathbf{E}(\mathbf{s}_2 - \mathbf{s}_1) & \mathbf{E}(\mathbf{s}_3 - \mathbf{s}_1) \end{bmatrix} \begin{bmatrix} \dot{\theta}_1 \\ \dot{\theta}_2 \\ \dot{\theta}_3 \end{bmatrix} = \begin{bmatrix} \dot{\phi} \\ \dot{\mathbf{p}} - \dot{\phi}\mathbf{E}\mathbf{s}_1 \end{bmatrix} \tag{4.109}$$

from which a reduced system of two equations in two unknowns is readily obtained, namely,

$$[\,\mathbf{E}(\mathbf{s}_2 - \mathbf{s}_1) \quad \mathbf{E}(\mathbf{s}_3 - \mathbf{s}_1)\,] \begin{bmatrix} \dot{\theta}_2 \\ \dot{\theta}_3 \end{bmatrix} = \dot{\mathbf{p}} - \dot{\phi}\mathbf{E}\mathbf{s}_1 \tag{4.110}$$

The system of equations (4.110) can be readily solved if Fact 4.8.4 is recalled, namely,

$$\begin{bmatrix} \dot{\theta}_2 \\ \dot{\theta}_3 \end{bmatrix} = \frac{1}{\Delta} \begin{bmatrix} -(\mathbf{s}_3 - \mathbf{s}_1)^T\mathbf{E} \\ (\mathbf{s}_2 - \mathbf{s}_1)^T\mathbf{E} \end{bmatrix} \mathbf{E}(\dot{\mathbf{p}} - \dot{\phi}\mathbf{E}\mathbf{s}_1)$$

$$= \frac{1}{\Delta} \begin{bmatrix} (\mathbf{s}_3 - \mathbf{s}_1)^T(\dot{\mathbf{p}} - \dot{\phi}\mathbf{E}\mathbf{s}_1) \\ -(\mathbf{s}_2 - \mathbf{s}_1)^T(\dot{\mathbf{p}} - \dot{\phi}\mathbf{E}\mathbf{s}_1) \end{bmatrix}$$

where Δ is the determinant of the 2×2 matrix involved, i.e.,

$$\Delta \equiv \det([\,\mathbf{E}(\mathbf{s}_2 - \mathbf{s}_1) \quad \mathbf{E}(\mathbf{s}_3 - \mathbf{s}_1)\,]) \equiv -(\mathbf{s}_2 - \mathbf{s}_1)^T\mathbf{E}(\mathbf{s}_3 - \mathbf{s}_1) \tag{4.111}$$

We thus have

$$\dot{\theta}_2 = -\frac{(\mathbf{s}_3 - \mathbf{s}_1)^T(\dot{\mathbf{p}} - \dot{\phi}\mathbf{E}\mathbf{s}_1)}{(\mathbf{s}_2 - \mathbf{s}_1)^T\mathbf{E}(\mathbf{s}_3 - \mathbf{s}_1)} \tag{4.112a}$$

$$\dot{\theta}_3 = \frac{(\mathbf{s}_2 - \mathbf{s}_1)^T(\dot{\mathbf{p}} - \dot{\phi}\mathbf{E}\mathbf{s}_1)}{(\mathbf{s}_2 - \mathbf{s}_1)^T\mathbf{E}(\mathbf{s}_3 - \mathbf{s}_1)} \tag{4.112b}$$

Further, $\dot{\theta}_1$ is computed from the first scalar equation of eq.(4.108), i.e.,

$$\dot{\theta}_1 = \dot{\phi} - (\dot{\theta}_2 + \dot{\theta}_3) \tag{4.112c}$$

thereby completing the velocity analysis.

The foregoing calculations are summarized below in algorithmic form, with the numbers of multiplications and additions indicated at each stage. In those numbers, we have taken into account that a multiplication of $\mathbf{E}$ by any 2-dimensional vector incurs no computational cost, but rather a simple rearrangement of the entries of this vector, with a reversal of one sign.

1. $\mathbf{d}_{21} \leftarrow \mathbf{s}_2 - \mathbf{s}_1$ $\qquad 0M + 2A$

2. $\mathbf{d}_{31} \leftarrow \mathbf{s}_3 - \mathbf{s}_1$ $\qquad 0M + 2A$

3. $\Delta \leftarrow \mathbf{d}_{31}^T \mathbf{E} \mathbf{d}_{21}$ $\qquad 2M + 1A$

4. $\mathbf{u} \leftarrow \dot{\mathbf{p}} - \dot{\phi} \mathbf{E} \mathbf{s}_1$ $\qquad 2M + 2A$

5. $\mathbf{u} \leftarrow \mathbf{u}/\Delta$ $\qquad 2M + 0A$

6. $\dot{\theta}_2 \leftarrow \mathbf{u}^T \mathbf{d}_{31}$ $\qquad 2M + 1A$

7. $\dot{\theta}_3 \leftarrow -\mathbf{u}^T \mathbf{d}_{21}$ $\qquad 2M + 1A$

8. $\dot{\theta}_1 \leftarrow \dot{\phi} - \dot{\theta}_2 - \dot{\theta}_3$ $\qquad 0M + 2A$

The complete calculation of joint rates thus consumes only $10M$ and $11A$, which represents a savings of about 67% of the computations involved if Gaussian elimination is applied without regarding the algebraic structure of the Jacobian $\mathbf{J}$ and its kinematic and geometric significance. In fact, the solution of an arbitrary system of three equations in three unknowns requires, from eq.(4.62a), 28 additions and 23 multiplications. If the cost of calculating the right-hand side is added, namely, $4A$ and $6M$, a total of $32A$ and $29M$ is required to solve for the joint rates if straightforward Gaussian elimination is used.

4.8.3 Acceleration Analysis

The calculation of the joint accelerations needed to produce a given twist rate of the EE is readily accomplished by differentiating both sides of eq.(4.105a), with definitions (4.107), i.e.,

$$\mathbf{J}\ddot{\boldsymbol{\theta}} + \dot{\mathbf{J}}\dot{\boldsymbol{\theta}} = \dot{\mathbf{t}}$$

from which we readily derive a system of equations similar to eq.(4.105a) with $\ddot{\boldsymbol{\theta}}$ as unknown, namely,

$$\mathbf{J}\ddot{\boldsymbol{\theta}} = \dot{\mathbf{t}} - \dot{\mathbf{J}}\dot{\boldsymbol{\theta}}$$

where

$$\dot{\mathbf{J}} = \begin{bmatrix} 0 & 0 & 0 \\ \mathbf{E}\dot{\mathbf{s}}_1 & \mathbf{E}\dot{\mathbf{s}}_2 & \mathbf{E}\dot{\mathbf{s}}_3 \end{bmatrix}, \quad \ddot{\boldsymbol{\theta}} \equiv \begin{bmatrix} \ddot{\theta}_1 \\ \ddot{\theta}_2 \\ \ddot{\theta}_3 \end{bmatrix}, \quad \dot{\mathbf{t}} \equiv \begin{bmatrix} \ddot{\phi} \\ \ddot{\mathbf{p}} \end{bmatrix}$$

and

$$\begin{aligned}
\dot{\mathbf{s}}_3 &= (\dot{\theta}_1 + \dot{\theta}_2 + \dot{\theta}_3)\mathbf{E}\mathbf{a}_3 \\
\dot{\mathbf{s}}_2 &= \dot{\mathbf{a}}_2 + \dot{\mathbf{s}}_3 = (\dot{\theta}_1 + \dot{\theta}_2)\mathbf{E}\mathbf{a}_2 + \dot{\mathbf{s}}_3 \\
\dot{\mathbf{s}}_1 &= \dot{\mathbf{a}}_1 + \dot{\mathbf{s}}_2 = \dot{\theta}_1\mathbf{E}\mathbf{s}_1 + \dot{\mathbf{s}}_2
\end{aligned}$$

Now we can proceed by Gaussian elimination to solve for the joint accelerations in exactly the same manner as in Subsection 4.8.2, thereby obtaining the counterpart of eq.(4.110), namely,

$$[\mathbf{E}(\mathbf{s}_2 - \mathbf{s}_1) \quad \mathbf{E}(\mathbf{s}_3 - \mathbf{s}_1)] \begin{bmatrix} \ddot{\theta}_2 \\ \ddot{\theta}_3 \end{bmatrix} = \mathbf{w} \tag{4.113a}$$

with $\mathbf{w}$ defined as

$$\mathbf{w} \equiv \ddot{\mathbf{p}} - \mathbf{E}(\dot{\theta}_1 \dot{\mathbf{s}}_1 + \dot{\theta}_2 \dot{\mathbf{s}}_2 + \dot{\theta}_3 \dot{\mathbf{s}}_3 + \ddot{\phi} \mathbf{s}_1) \tag{4.113b}$$

and hence, similar to eqs.(4.112a–c), one has

$$\ddot{\theta}_2 = \frac{(\mathbf{s}_3 - \mathbf{s}_1)^T \mathbf{w}}{\Delta} \tag{4.114a}$$

$$\ddot{\theta}_3 = -\frac{(\mathbf{s}_2 - \mathbf{s}_1)^T \mathbf{w}}{\Delta} \tag{4.114b}$$

$$\ddot{\theta}_1 = \ddot{\phi} - (\ddot{\theta}_2 + \ddot{\theta}_3) \tag{4.114c}$$

Below we summarize the foregoing calculations in algorithmic form, indicating the numbers of operations required at each stage.

1. $\dot{\mathbf{s}}_3 \leftarrow (\dot{\theta}_1 + \dot{\theta}_2 + \dot{\theta}_3)\mathbf{E}\mathbf{a}_3$ $\quad 2M$ & $2A$
2. $\dot{\mathbf{s}}_2 \leftarrow (\dot{\theta}_1 + \dot{\theta}_2)\mathbf{E}\mathbf{a}_2 + \dot{\mathbf{s}}_3$ $\quad 2M$ & $3A$
3. $\dot{\mathbf{s}}_1 \leftarrow \dot{\theta}_1 \mathbf{E}\mathbf{s}_1 + \dot{\mathbf{s}}_2$ $\quad 2M$ & $2A$
4. $\mathbf{w} \equiv \ddot{\mathbf{p}} - \mathbf{E}(\dot{\theta}_1 \dot{\mathbf{s}}_1 + \dot{\theta}_2 \dot{\mathbf{s}}_2 + \dot{\theta}_3 \dot{\mathbf{s}}_3 + \ddot{\phi} \mathbf{s}_1)$ $\quad 8M$ & $8A$
5. $\mathbf{w} \leftarrow \mathbf{w}/\Delta$ $\quad 2M + 0A$
6. $\ddot{\theta}_2 \leftarrow \mathbf{w}^T \mathbf{d}_{31}$ $\quad 2M + 1A$
7. $\ddot{\theta}_3 \leftarrow -\mathbf{w}^T \mathbf{d}_{21}$ $\quad 2M + 1A$
8. $\ddot{\theta}_1 \leftarrow \ddot{\phi} - (\ddot{\theta}_2 + \ddot{\theta}_3)$ $\quad 0M + 2A$

where $\mathbf{d}_{21}$ and $\mathbf{d}_{31}$ are available from velocity calculations. The joint accelerations thus require a total of 20 multiplications and 19 additions. These figures represent substantial savings when compared with the numbers of operations required if plain Gaussian elimination were used, namely, 33 multiplications and 35 additions.

It is noteworthy that in the foregoing algorithm, we have replaced neither the sum $\dot{\theta}_1 + \dot{\theta}_2 + \dot{\theta}_3$ nor $\dot{\theta}_1 \mathbf{E}(\mathbf{s}_1 + \mathbf{s}_2 + \mathbf{s}_3)$ by ω and correspondingly, by $\dot{\mathbf{p}}$, because in path tracking, there is no perfect match between joint and Cartesian variables. In fact, joint-rate and joint-acceleration calculations are needed in feedback control schemes to estimate the position, velocity, and acceleration errors by proper corrective actions.

4.8.4 Static Analysis

Here we assume that a planar wrench acts at the end-effector of the manipulator appearing in Fig. 4.24. In accordance with the definition of the planar twist in Subsection 4.8.2, eq.(4.107), the planar wrench is now defined as

$$\mathbf{w} \equiv \begin{bmatrix} n \\ \mathbf{f} \end{bmatrix} \tag{4.115}$$

where n is the scalar couple acting on the end-effector and $\mathbf{f}$ is the 2-dimensional force acting at the operation point P of the end-effector. If additionally, we denote by $\boldsymbol{\tau}$ the 3-dimensional vector of joint torques, the planar counterpart of eq.(4.97) follows, i.e.,

$$\mathbf{J}^T\mathbf{w} = \boldsymbol{\tau} \tag{4.116}$$

where

$$\mathbf{J}^T = \begin{bmatrix} 1 & (\mathbf{E}\mathbf{s}_1)^T \\ 1 & (\mathbf{E}\mathbf{s}_2)^T \\ 1 & (\mathbf{E}\mathbf{s}_3)^T \end{bmatrix}$$

Now, in order to solve for the wrench $\mathbf{w}$ acting on the end-effector, given the joint torques $\boldsymbol{\tau}$ and the posture of the manipulator, we can still apply our compact Gaussian-elimination scheme, as introduced in Subsection 4.8.2. To this end, we subtract the first scalar equation from the second and the third scalar equations of eq.(4.116), which renders the foregoing system in the form

$$\begin{bmatrix} 1 & (\mathbf{E}\mathbf{s}_1)^T \\ 0 & [\mathbf{E}(\mathbf{s}_2 - \mathbf{s}_1)]^T \\ 0 & [\mathbf{E}(\mathbf{s}_3 - \mathbf{s}_1)]^T \end{bmatrix} \begin{bmatrix} n \\ \mathbf{f} \end{bmatrix} = \begin{bmatrix} \tau_1 \\ \tau_2 - \tau_1 \\ \tau_3 - \tau_1 \end{bmatrix}$$

Thus, the last two equations have been decoupled from the first one, which allows us to solve them separately, i.e., we have reduced the system to one of two equations in two unknowns, namely,

$$\begin{bmatrix} [\mathbf{E}(\mathbf{s}_2 - \mathbf{s}_1)]^T \\ [\mathbf{E}(\mathbf{s}_3 - \mathbf{s}_1)]^T \end{bmatrix} \mathbf{f} = \begin{bmatrix} \tau_2 - \tau_1 \\ \tau_3 - \tau_1 \end{bmatrix} \tag{4.117}$$

from which we readily obtain

$$\mathbf{f} = \begin{bmatrix} [\mathbf{E}(\mathbf{s}_2 - \mathbf{s}_1)]^T \\ [\mathbf{E}(\mathbf{s}_3 - \mathbf{s}_1)]^T \end{bmatrix}^{-1} \begin{bmatrix} \tau_2 - \tau_1 \\ \tau_3 - \tau_1 \end{bmatrix} \tag{4.118}$$

and hence, upon expansion of the above inverse,

$$\mathbf{f} = \frac{1}{\Delta}[(\tau_2 - \tau_1)(\mathbf{s}_3 - \mathbf{s}_1) - (\tau_3 - \tau_1)(\mathbf{s}_2 - \mathbf{s}_1)] \tag{4.119}$$

where Δ is exactly as defined in eq.(4.111). Finally, the resultant moment n acting on the end-effector is readily calculated from the first scalar equation of eq.(4.116), namely, as

$$n = \tau_1 + \mathbf{s}_1^T\mathbf{E}\mathbf{f}$$

thereby completing the static analysis of the manipulator under study. A quick analysis of computational costs shows that the foregoing solution needs $8M$ and $6A$, or a savings of about 70% if straightforward Gaussian elimination is applied.

4.9 Kinetostatic Performance Indices

The balance of Part I of the book does not depend on this section, which can thus be skipped. We have included it here because *(i)* it is a simple matter to render the section self-contained, while introducing the concept of *condition number* and its relevance in robotics; *(ii)* kinetostatic performance can be studied with the background of the material included up to this section; and *(iii)* kinetostatic performance is becoming increasingly relevant as a design criterion and as a figure of merit in robot control.

A *kinetostatic performance index* of a robotic mechanical system is a scalar quantity that measures how well the system behaves with regard to force and motion transmission, the latter being understood in the differential sense, i.e., at the velocity level. Now, a kinetostatic performance index, or kinetostatic index for brevity, may be needed to assess the performance of a robot at the design stage, in which case we need a posture-independent index. In this case, the index becomes a function of the robot architecture only. If, on the other hand, we want to assess the performance of a *given* robot while performing a task, what we need is a posture-dependent index. In many instances, this difference is not mentioned in the robotics literature, although it is extremely important. Moreover, while performance indices can be defined for all kinds of robotic mechanical systems, we focus here on those associated with serial manipulators, which are the ones studied most intensively.

Among the various performance indices that have been proposed, one can cite the concept of *service angle*, first introduced by Vinogradov et al. (1971), and the *conditioning* of robotic manipulators, as proposed by Yang and Lai (1985). Yoshikawa (1985), in turn, introduced the concept of *manipulability*, which is defined as the square root of the determinant of the product of the manipulator Jacobian by its transpose. Paul and Stevenson (1983) used the absolute value of the determinant of the Jacobian to assess the kinematic performance of spherical wrists. Note that for square Jacobians, Yoshikawa's manipulability is identical to the absolute value of the determinant of the Jacobian, and hence, the latter coincides with Paul and Stevenson's performance index. It should be pointed out that these indices were defined for control purposes and hence, are posture-dependent. Germane to these concepts is that of *dextrous workspace*, introduced by Kumar and Waldron (1981), and used for geometric optimization by Vijaykumar et al. (1986). Although the concepts of service angle and manipulability are clearly different, they touch upon a common underlying issue, namely, the kinematic, or alternatively, the static performance of a manipulator from an accuracy viewpoint.

What is at stake when discussing the manipulability of a robotic manipulator is a measure of the *invertibility* of the associated Jacobian matrix, since this is required for velocity and force-feedback control. One further performance index is based on the *condition number* of the Jacobian, which was first used by Salisbury and Craig (1982) to design mechanical fingers. Here, we shall call such an index the *conditioning* of the manipulator. For the sake of brevity, we devote the discussion below to only two indices, namely, manipulability and conditioning. Prior to discussing these indices, we recall a few facts from linear algebra.

Although the concepts discussed here are equally applicable to square and rectangular matrices, we shall focus on the former. First, we give a geometric interpretation of the mapping induced by an $n \times n$ matrix $\mathbf{A}$. Here, we do not assume any particular structure of $\mathbf{A}$, which can thus be totally arbitrary. However, by invoking the *polar-decomposition theorem* (Strang, 1988), we can factor $\mathbf{A}$ as

$$\mathbf{A} \equiv \mathbf{RU} \equiv \mathbf{VR} \tag{4.120}$$

where $\mathbf{R}$ is orthogonal, although not necessarily proper, while $\mathbf{U}$ and $\mathbf{V}$ are both at least positive-semidefinite. Moreover, if $\mathbf{A}$ is nonsingular, then $\mathbf{U}$ and $\mathbf{V}$ are both positive-definite, and $\mathbf{R}$ is unique. Clearly, $\mathbf{U}$ can be readily determined as the positive-semidefinite or correspondingly, positive-definite *square root* of the product $\mathbf{A}^T\mathbf{A}$, which is necessarily positive-semidefinite; it is, in fact, positive-definite if $\mathbf{A}$ is nonsingular. We recall here that the square root of arbitrary matrices was briefly discussed in Subsection 2.3.6. The square root of a positive-semidefinite matrix can be most easily understood if that matrix is assumed to be in diagonal form, which is possible because such a matrix is necessarily symmetric, and every symmetric matrix is diagonalizable. The matrix at hand being positive-semidefinite, its eigenvalues are nonnegative, and hence, their square roots are all real. The positive-semidefinite square root of interest is, then, readily obtained as the diagonal matrix whose nontrivial entries are the nonnegative square roots of the aforementioned eigenvalues. With $\mathbf{U}$ determined, $\mathbf{R}$ can be found uniquely only if $\mathbf{A}$ is nonsingular, in which case $\mathbf{U}$ is positive-definite. If this is the case, then we have

$$\mathbf{R} = \mathbf{U}^{-1}\mathbf{A} \tag{4.121a}$$

It is a simple matter to show that $\mathbf{V}$ can be found, in turn, as a similarity transformation of $\mathbf{U}$, namely, as

$$\mathbf{V} = \mathbf{RUR}^T \tag{4.121b}$$

Now, let vector $\mathbf{x}$ be mapped by $\mathbf{A}$ into $\mathbf{z}$, i.e.,

$$\mathbf{z} = \mathbf{Ax} \equiv \mathbf{RUx} \tag{4.122a}$$

Moreover, let

$$\mathbf{y} \equiv \mathbf{U}\mathbf{x} \tag{4.122b}$$

and hence, we have a concatenation of mappings, namely, $\mathbf{U}$ maps $\mathbf{x}$ into $\mathbf{y}$, while $\mathbf{R}$ maps $\mathbf{y}$ into $\mathbf{z}$. Thus, by virtue of the nature of matrices $\mathbf{R}$ and $\mathbf{U}$, the latter maps the unit n-dimensional ball into an n-axis ellipsoid whose semiaxis lengths bear the ratios of the eigenvalues of $\mathbf{U}$. Moreover, $\mathbf{R}$ maps this ellipsoid into another one with identical semiaxes, except that it is rotated about its center or reflected, depending upon whether $\mathbf{R}$ is proper or improper orthogonal. In fact, the eigenvalues of $\mathbf{U}$ or for that matter, those of $\mathbf{V}$, are nothing but the *singular values* of $\mathbf{A}$. Yoshikawa (1985) explained the foregoing relations resorting to the *singular-value decomposition theorem*. We prefer to invoke the polar-decomposition theorem instead, because of the geometric nature of the latter, as opposed to the former, which is of an algebraic nature—it is based on a diagonalization of either $\mathbf{U}$ or $\mathbf{V}$, which is really not needed.

We illustrate the two mappings $\mathbf{U}$ and $\mathbf{R}$ in Fig. 4.27, where we orient the X, Y, and Z axes along the three eigenvectors of $\mathbf{U}$. Therefore, the semiaxes of the ellipsoid are oriented as the eigenvectors of $\mathbf{U}$ as well. If $\mathbf{A}$ is singular, then the ellipsoid degenerates into one with at least one vanishing semiaxis. On the other hand, if matrix $\mathbf{A}$ is *isotropic*, i.e., if all its singular values are identical, then it maps the unit ball into another ball, either enlarged or shrunken.

For our purposes, we can regard the Jacobian of a serial manipulator as mapping the unit ball in the space of joint rates into a rotated or reflected ellipsoid in the space of Cartesian velocities, or twists. Now, let us assume that the polar decomposition of $\mathbf{J}$ is given by $\mathbf{R}$ and $\mathbf{U}$, the manipulability μ of the robot under study thus becoming

$$\mu \equiv |\det(\mathbf{J})| \equiv |\det(\mathbf{R})||\det(\mathbf{U})| \tag{4.123a}$$

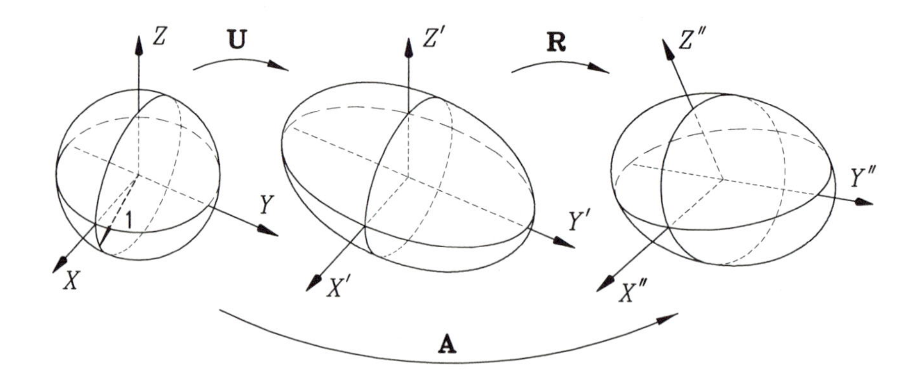

FIGURE 4.27. Geometric representation of mapping induced by matrix $\mathbf{A}$.

Since $\mathbf{R}$ is orthogonal, the absolute value of its determinant is unity. Additionally, the determinant of $\mathbf{U}$ is nonnegative, and hence,

$$\mu = \det(\mathbf{U}) \tag{4.123b}$$

which shows that the manipulability is the product of the eigenvalues of $\mathbf{U}$ or equivalently, of the singular values of $\mathbf{J}$. Now, the product of those singular values, in the geometric interpretation of the mapping induced by $\mathbf{J}$, is proportional to the volume of the ellipsoid at hand, and hence, μ can be interpreted as a measure of the volume of that ellipsoid. It is apparent that the manipulability defined in eq.(4.123b) is posture-dependent. For example, if $\mathbf{J}$ is singular, at least one of the semiaxes of the ellipsoid vanishes, and so does its volume. Manipulators at singular configurations thus have a manipulability of zero.

Now, if we want to use the concept of manipulability to define a posture-independent kinetostatic index, we have to define this index in a *global sense*. This can be done in the same way as magnitudes of vectors are defined as the sum of the squares of their components, i.e., as the integral of a certain power of the manipulability over the whole workspace of the manipulator, which would amount to defining the index as a *norm* of the manipulability in a space of functions. For example, we can use the maximum manipulability attained over the whole workspace, thereby ending up with what would be like a Chebyshev norm; alternatively, we can use the *root-mean square* (rms) value of the manipulability, thereby ending up with a measure similar to the Euclidean norm.

Furthermore, if we have a Jacobian $\mathbf{J}$ *whose entries all have the same units*, then we can define its condition number $\kappa(\mathbf{J})$ as the ratio of the largest singular value σ_l of $\mathbf{J}$ to the smallest one, σ_s, i.e.,

$$\kappa(\mathbf{J}) \equiv \frac{\sigma_l}{\sigma_s} \tag{4.124}$$

Note that $\kappa(\mathbf{J})$ can attain values from 1 to infinity. Clearly, the condition number attains its minimum value of unity for matrices with identical singular values; such matrices map the unit ball into another ball, although of a different size, and are, thus, called *isotropic*. By extension, we shall call manipulators whose Jacobian matrix can attain isotropic values *isotropic* as well. On the other side of the spectrum, singular matrices have a smallest singular value that vanishes, and hence, their condition number is infinity. The condition number of $\mathbf{J}$ can be thought of as indicating the *distortion* of the unit ball in the space of joint-variables. The larger this distortion, the greater the condition number, the worst-conditioned Jacobians being those that are singular. For these, one of the semiaxes of the ellipsoid vanishes and the ellipsoid degenerates into what would amount to an elliptical disk in the 3-dimensional space.

The condition number of a square matrix can also be understood as a measure of the relative roundoff-error amplification of the computed results upon solving a linear system of equations associated with that matrix, with respect to the relative roundoff error of the data (Dahlquist and Björck, 1974). Based on the condition number of the Jacobian, a posture-independent *kinematic conditioning index* of robotic manipulators can now be defined as a global measure of the condition number, or its reciprocal for that matter, which is better behaved because it is bounded between 0 and unity.

Now, if the entries of $\mathbf{J}$ have different units, the foregoing definition of $\kappa(\mathbf{J})$ cannot be applied, for we would face a problem of ordering singular values of different units from largest to smallest. We resolve this inconsistency by defining a *characteristic length*, by which we divide the Jacobian entries that have units of length, thereby producing a new Jacobian that is dimensionally homogeneous. We shall therefore divide our study into (*i*) manipulators for only positioning tasks, (*ii*) manipulators for only orientation tasks, and (*iii*) manipulators for both positioning and orientation tasks. The characteristic length will be introduced when studying the last of these.

In the sequel, we will need an interesting property of isotropic matrices that is recalled below. First note that given the polar decomposition of a square matrix $\mathbf{A}$ of eq.(4.120), its singular values are simply the—nonnegative—eigenvalues of matrix $\mathbf{U}$, or those of $\mathbf{V}$, for both matrices have identical eigenvalues. Moreover, if $\mathbf{A}$ is isotropic, all the foregoing eigenvalues are identical, say equal to σ, and hence, matrices $\mathbf{U}$ and $\mathbf{V}$ are proportional to the $n \times n$ identity matrix, i.e.,

$$\mathbf{U} = \mathbf{V} = \sigma \mathbf{1} \tag{4.125}$$

In this case, then,

$$\mathbf{A}^T \mathbf{A} = \sigma^2 \mathbf{1} \tag{4.126}$$

Given an arbitrary manipulator of the serial type with a Jacobian matrix whose entries all have the same units, we can calculate its condition number and use a global measure of this to define a posture-independent kinetostatic index. Let κ_m be the minimum value attained by the condition number of the dimensionally-homogeneous Jacobian over the whole workspace. Note that $1/\kappa_m$ can be regarded as a Chebyshev norm of the reciprocal of $\kappa(\mathbf{J})$, because now $1/\kappa_m$ represents the maximum value of this reciprocal in the whole workspace. We then introduce a posture-independent performance index, the *kinematic conditioning index*, or KCI for brevity, defined as

$$\mathrm{KCI} = \frac{1}{\kappa_m} \times 100 \tag{4.127}$$

Notice that since the condition number is bounded from below, the KCI is bounded by a value of 100%. Manipulators with a KCI of 100% are those identified above as isotropic because their Jacobians have, at the configuration of minimum condition number, all their singular values identical and different from zero.

4.9.1 *Positioning Manipulators*

Here, again, we shall distinguish between planar and spatial manipulators. These are studied separately.

Planar Manipulators

If the manipulator of Fig. 4.24 is limited to positioning tasks, we can dispense with its third axis, the manipulator thus reducing to the one shown in Fig. 4.25; its Jacobian reduces correspondingly to

$$\mathbf{J} = [\,\mathbf{E}\mathbf{s}_1 \quad \mathbf{E}\mathbf{s}_2\,]$$

If we want to design this manipulator for maximum manipulability, we need first to determine its manipulability as given by eq.(4.123a) or correspondingly, as $\mu = |\det(\mathbf{J})|$. Now, note that

$$\det(\mathbf{J}) = \det(\mathbf{E}\,[\,\mathbf{s}_1 \quad \mathbf{s}_2\,]) = \det(\mathbf{E})\det([\,\mathbf{s}_1 \quad \mathbf{s}_2\,])$$

and since matrix $\mathbf{E}$ is orthogonal, its determinant equals unity. Thus, the determinant of interest is now calculated using Fact 4.8.3 of Section 4.8, namely,

$$\det(\mathbf{J}) = -\mathbf{s}_1^T\mathbf{E}\mathbf{s}_2 \tag{4.128}$$

Therefore,

$$\mu = |\mathbf{s}_1^T\mathbf{E}\mathbf{s}_2| \equiv \|\mathbf{s}_1\|\|\mathbf{s}_2\||\sin(\mathbf{s}_1,\, \mathbf{s}_2)|$$

where $(\mathbf{s}_1,\, \mathbf{s}_2)$ stands for the angle between the two vectors inside the parentheses. Now let us denote the manipulator reach with R, i.e., $R = a_1 + a_2$, and let $a_k = R\rho_k$, where ρ_k, for $k = 1, 2$, is a dimensionless number. Hence,

$$\mu = R^2\rho_1\rho_2|\sin(\mathbf{s}_1,\, \mathbf{s}_2)| \tag{4.129}$$

with ρ_1 and ρ_2 subjected to

$$\rho_1 + \rho_2 = 1 \tag{4.130}$$

The design problem at hand, then, can be formulated as an optimization problem aimed at maximizing μ as given in eq.(4.129) over ρ_1 and ρ_2, subject to the constraint (4.130). This optimization problem can be readily solved using, for example, Lagrange multipliers, thereby obtaining

$$\rho_1 = \rho_2 = \frac{1}{2}$$

the absolute value of the sine of the angle between the vectors $\mathbf{s}_1$ and $\mathbf{s}_2$ attaining its maximum value when these vectors make an angle of 90°. The maximum manipulability thus becomes

$$\mu_{\max} = \frac{R^2}{4} \tag{4.131}$$

Incidentally, the equal-length condition maximizes the workspace volume as well.

On the other hand, if we want to minimize the condition number of $\mathbf{J}$, we should aim at rendering it isotropic, which means that the product $\mathbf{J}^T\mathbf{J}$ should be proportional to the identity matrix, and so,

$$\begin{bmatrix} \mathbf{s}_1^T\mathbf{s}_1 & \mathbf{s}_1^T\mathbf{s}_2 \\ \mathbf{s}_1^T\mathbf{s}_2 & \mathbf{s}_2^T\mathbf{s}_2 \end{bmatrix} = \begin{bmatrix} \sigma^2 & 0 \\ 0 & \sigma^2 \end{bmatrix}$$

where σ is the repeated singular value of $\mathbf{J}$. Hence, for $\mathbf{J}$ to be isotropic, all we need is that the two vectors $\mathbf{s}_1$ and $\mathbf{s}_2$ have the same norm and that they lie at right angles. The solution is a manipulator with link lengths observing a ratio of $\sqrt{2}/2$, i.e., with $a_1/a_2 = \sqrt{2}/2$, and the two link axes at an angle of 135°, as depicted in Fig. 4.28. Manipulators of the above type, used as mechanical fingers, were investigated by Salisburg and Craig (1982), who found that these manipulators can be rendered isotropic if given the foregoing dimensions and configured as shown in Fig. 4.28.

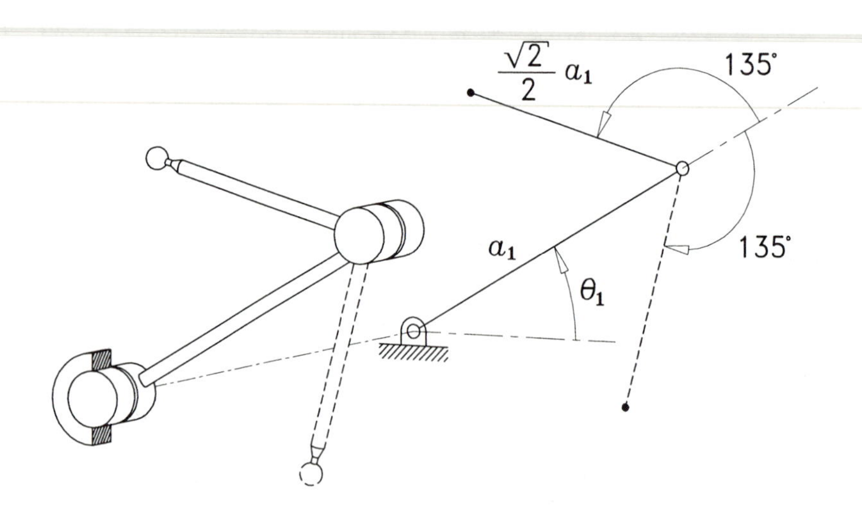

FIGURE 4.28. A two-axis isotropic manipulator.

Spatial Manipulators

Now we have a manipulator like that depicted in Fig. 4.9, its Jacobian matrix taking on the form

$$\mathbf{J} = [\mathbf{e}_1 \times \mathbf{r}_1 \quad \mathbf{e}_2 \times \mathbf{r}_2 \quad \mathbf{e}_3 \times \mathbf{r}_3] \tag{4.132}$$

The condition for isotropy of this kind of manipulator takes on the form of eq.(4.126), which thus leads to

$$\sum_1^3 (\mathbf{e}_k \times \mathbf{r}_k)(\mathbf{e}_k \times \mathbf{r}_k)^T = \sigma^2 \mathbf{1} \tag{4.133}$$

This condition can be attained by various designs, one example being the manipulator of Fig. 4.15. Another isotropic manipulator for 3-dimensional positioning tasks is displayed in Fig. 4.29.

Note that the manipulator of Fig. 4.29 has an orthogonal architecture, the ratio of its last link length to the length of the intermediate link being, as in the 2-dimensional case, $\sqrt{2}/2$. Since the first axis does not affect singularities, neither does it affect isotropy, and hence, not only does one location of the operation point exist that renders the manipulator isotropic, but a whole locus, namely, the circle known as the *isotropy circle*, indicated in the same figure. By the same token, the manipulator of Fig. 4.28 has an isotropy circle centered at the center of the first joint, with a radius of $(\sqrt{2}/2)a_1$.

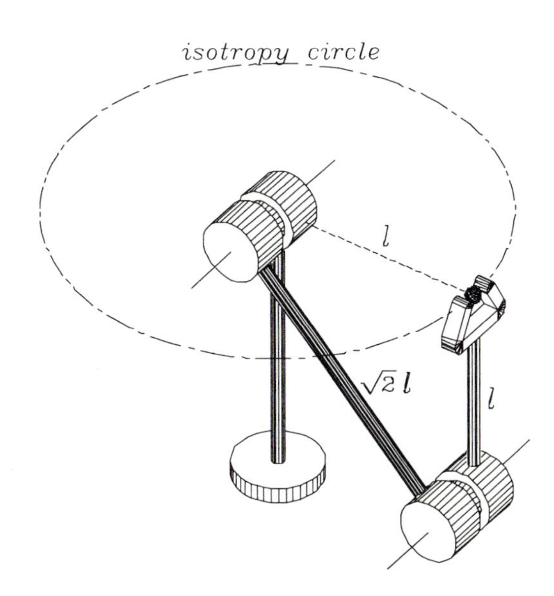

FIGURE 4.29. An isotropic manipulator for 3-dimensional positioning tasks.

4.9.2 Orienting Manipulators

We now have a three-revolute manipulator like that depicted in Fig. 4.17, its Jacobian taking on the simple form

$$\mathbf{J} = [\mathbf{e}_1 \quad \mathbf{e}_2 \quad \mathbf{e}_3] \tag{4.134}$$

and hence, the isotropy condition of eq. (4.126) leads to

$$\sum_1^3 \mathbf{e}_k \mathbf{e}_k^T = \sigma^2 \mathbf{1} \tag{4.135}$$

What the foregoing condition states is that a spherical wrist for orienting tasks is isotropic if its three unit vectors $\{\mathbf{e}_k\}_1^3$ are so laid out that the three products $\mathbf{e}_k \mathbf{e}_k^T$, for $k = 1, 2, 3$, add up to a multiple of the 3×3 identity matrix. This is the case if the three foregoing unit vectors are orthonormal, which occurs in orthogonal wrists when the two planes defined by the corresponding pairs of neighboring axes are at right angles. Moreover, the value of σ in this case can be readily found if we take the trace of both sides of the above equation, which yields

$$\sum_1^3 \mathbf{e}_k \cdot \mathbf{e}_k = 3\sigma^2 \tag{4.136}$$

and hence, $\sigma = 1$, because all three vectors on the left-hand side are of unit magnitude. In summary, then, orthogonal wrists, which are rather frequent among industrial manipulators, are isotropic. Here we have an example of engineering insight leading to an optimal design, for such wrists existed long before isotropy was introduced as a design criterion for manipulators. Moreover, notice that from the results of Subsection 4.4.2, spherical manipulators with an orthogonal architecture have a maximum workspace volume. That is, isotropic manipulators of the spherical type have two optimality properties: they have both a maximum workspace volume and a maximum KCI. Apparently, the manipulability of orthogonal spherical wrists is also optimal, as the reader is invited to verify, when the wrist is postured so that its three axes are mutually orthogonal. In this posture, the manipulability of the wrist is unity.

4.9.3 Positioning and Orienting Manipulators

We saw already in Subsubsection 4.9.3 that the optimization of the two indices studied here—the condition number of the Jacobian matrix and the manipulability—leads to different manipulators. In fact, the two indices entail even deeper differences, as we shall see presently. First and foremost, as we shall prove for both planar and spatial manipulators, the manipulability μ is independent of the operation point P of the end-effector, while the condition number is not. One more fundamental difference is that while calculating the manipulability of manipulators meant for both positioning and orienting tasks poses no problem, the condition number cannot be calculated, at least directly, for this kind of manipulator. Indeed, in order to determine the condition number of the Jacobian matrix, we must order its singular values from largest to smallest. However, in the presence of positioning and orienting tasks, three of these singular values, namely, those associated with orientation, are dimensionless, while those associated with positioning have units of length, thereby making impossible such an ordering. We resolve this dimensional inhomogeneity by introducing a normalizing *characteristic length*. Upon dividing the three *positioning* rows, i.e., the bottom rows, of the Jacobian by this length, a nondimensional Jacobian is obtained whose singular values are nondimensional as well. The characteristic length is then defined as the normalizing length that renders the condition number of the Jacobian matrix a minimum. Below we shall determine the characteristic length for isotropic manipulators; determining the same for nonisotropic manipulators requires solving a minimization problem that calls for numerical techniques, as illustrated with an example.

Planar Manipulators

In the ensuing development, we will need the planar counterpart of the twist-transfer formula of Subsection 3.4.2. First, we denote the 3-dimensional twist of a rigid body undergoing planar motion, defined at a point A, by $\mathbf{t}_A$; when defined at point B, the corresponding twist is denoted by $\mathbf{t}_B$, i.e.,

$$\mathbf{t}_A \equiv \begin{bmatrix} \omega \\ \dot{\mathbf{a}} \end{bmatrix}, \quad \mathbf{t}_B \equiv \begin{bmatrix} \omega \\ \dot{\mathbf{b}} \end{bmatrix} \tag{4.137}$$

The relation between the two twists, or the *planar twist-transfer formula*, is given by a linear transformation $\mathbf{U}$ as

$$\mathbf{t}_B = \mathbf{U}\mathbf{t}_A \tag{4.138a}$$

where $\mathbf{U}$ is now defined as

$$\mathbf{U} = \begin{bmatrix} 1 & \mathbf{0}^T \\ \mathbf{E}(\mathbf{b} - \mathbf{a}) & \mathbf{1}_2 \end{bmatrix} \tag{4.138b}$$

with $\mathbf{a}$ and $\mathbf{b}$ representing the position vectors of points A and B, and $\mathbf{1}_2$ stands for the 2×2 identity matrix. Moreover, $\mathbf{U}$ is, not surprisingly, a member of the 3×3 unimodular group, i.e.,

$$\det(\mathbf{U}) = 1$$

Because of the planar twist-transfer formula, the Jacobian defined at an operation point B is related to that defined at an operation point A of the same end-effector by the same linear transformation $\mathbf{U}$, i.e., if we denote the two Jacobians by $\mathbf{J}_A$ and $\mathbf{J}_B$, then

$$\mathbf{J}_B = \mathbf{U}\mathbf{J}_A \tag{4.139}$$

and if we denote by μ_A and μ_B the manipulability calculated at points A and B, respectively, then

$$\mu_B = |\det(\mathbf{J}_B)| = |\det(\mathbf{U})||\det(\mathbf{J}_A)| = |\det(\mathbf{J}_A)| = \mu_A \tag{4.140}$$

thereby proving that the manipulability is insensitive to a change of operation point, or to a change of end-effector, for that matter. Note that a similar analysis for the condition number cannot be completed at this stage because as pointed out earlier, the condition number of these Jacobian matrices cannot even be calculated directly.

In order to resolve the foregoing dimensional inhomogeneity, we introduce the characteristic length L, which will be defined as that rendering the Jacobian dimensionally homogeneous and optimally conditioned, i.e., with a minimum condition number. We thus redefine the Jacobian matrix of interest as

$$\mathbf{J} \equiv \begin{bmatrix} 1 & 1 & 1 \\ \frac{1}{L}\mathbf{E}\mathbf{r}_1 & \frac{1}{L}\mathbf{E}\mathbf{r}_2 & \frac{1}{L}\mathbf{E}\mathbf{r}_3 \end{bmatrix} \tag{4.141}$$

Now, if we want to size the manipulator at hand by properly choosing its geometric parameters so as to render it isotropic, we must observe the isotropy condition, eq.(4.126), which readily leads to

$$\begin{bmatrix} 3 & \frac{1}{L}\mathbf{E}\sum_1^3 \mathbf{r}_k \\ \frac{1}{L}\sum_1^3 \mathbf{r}_k^T\mathbf{E}^T & \frac{1}{L^2}\mathbf{E}[\sum_1^3(\mathbf{r}_k\mathbf{r}_k^T)]\mathbf{E}^T \end{bmatrix} = \begin{bmatrix} \sigma^2 & 0 & 0 \\ 0 & \sigma^2 & 0 \\ 0 & 0 & \sigma^2 \end{bmatrix} \tag{4.142}$$

and hence,

$$\sigma^2 = 3 \tag{4.143a}$$

$$\sum_1^3 \mathbf{r}_k = \mathbf{0} \tag{4.143b}$$

$$\frac{1}{L^2}\mathbf{E}\left(\sum_1^3(\mathbf{r}_k\mathbf{r}_k^T)\right)\mathbf{E}^T = \sigma^2\mathbf{1}_2 \tag{4.143c}$$

What eq.(4.143a) states is simply that the triple singular value of the isotropic $\mathbf{J}$ is $\sqrt{3}$; eq.(4.143b) states, in turn, that the operation point is the centroid of the centers of all manipulator joints if its Jacobian matrix is isotropic. Now, in order to gain more insight into eq.(4.143c), we note that since $\mathbf{E}$ is orthogonal and $\sigma^2 = 3$, this equation can be rewritten in a simpler form, namely,

$$\frac{1}{L^2}\left(\sum_1^3(\mathbf{r}_k\mathbf{r}_k^T)\right) = (3)\mathbf{1}_2 \tag{4.144}$$

Further, if we recall the definition of the moment of inertia of a rigid body, we can immediately realize that the moment of inertia $\mathbf{I}_P$ of a set of particles of unit mass located at the centers of the manipulator joints, with respect to the operation point P, is given by

$$\mathbf{I}_P \equiv \sum_1^3 \left(\|\mathbf{r}_k\|^2 \mathbf{1}_2 - \mathbf{r}_k \mathbf{r}_k^T \right) \tag{4.145}$$

from which it is apparent that the moment of inertia of the set comprises two parts, the first being isotropic—it is a multiple of the 2×2 identity matrix—the second not necessarily so. However, the second part has the form of the left-hand side of eq.(4.144). Hence, eq.(4.144) states that if the manipulator under study is isotropic, then its joint centers are located, at the isotropic configuration, at the corners of a triangle that has *circular inertial symmetry*. What we mean by this is that the 2×2 moment of inertia of the set of particles, with entries I_{xx}, I_{xy}, and I_{yy}, is similar to that of a circle, i.e., with $I_{xx} = I_{yy}$ and $I_{xy} = 0$. An obvious candidate for such a triangle is, obviously, an equilateral triangle, the operation point thus coinciding with the center of the triangle. Since the corners of an equilateral triangle are at equal distances d from the center, and these distances are nothing but $\|\mathbf{r}_k\|$, then the condition below is readily derived for isotropy:

$$\|\mathbf{r}_k\|^2 = d^2 \tag{4.146}$$

In order to compute the characteristic length of the manipulator under study, let us take the trace of both sides of eq.(4.144), thereby obtaining

$$\frac{1}{L^2} \sum_1^3 \|\mathbf{r}_k\|^2 = 6$$

and hence, upon substituting eq.(4.146) into the foregoing relation, an expression for the characteristic length, *as pertaining to planar isotropic manipulators*, is readily derived, namely,

$$L = \frac{\sqrt{2}}{2} d \tag{4.147}$$

It is now a simple matter to show that the three link lengths of this isotropic manipulator are $a_1 = a_2 = \sqrt{3}d$ and $a_3 = d$. Such a manipulator is sketched in an isotropic configuration in Fig. 4.30.

Spatial Manipulators

The entries of the Jacobian of a six-axis manipulator meant for both positioning and orienting tasks are dimensionally inhomogeneous as well. Indeed, as discussed in Section 4.5, the ith column of $\mathbf{J}$ is composed of the

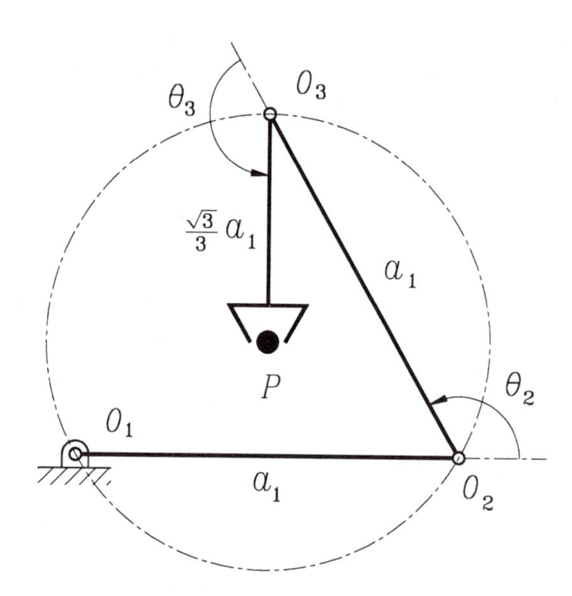

FIGURE 4.30. The planar 3-R isotropic manipulator.

Plücker coordinates of the ith axis of the manipulator, namely,

$$\mathbf{J} = \begin{bmatrix} \mathbf{e}_1 & \mathbf{e}_2 & \mathbf{e}_3 & \mathbf{e}_4 & \mathbf{e}_5 & \mathbf{e}_6 \\ \mathbf{e}_1 \times \mathbf{r}_1 & \mathbf{e}_2 \times \mathbf{r}_2 & \mathbf{e}_3 \times \mathbf{r}_3 & \mathbf{e}_4 \times \mathbf{r}_4 & \mathbf{e}_5 \times \mathbf{r}_5 & \mathbf{e}_6 \times \mathbf{r}_6 \end{bmatrix} \tag{4.148}$$

Now it is apparent that the first three rows of $\mathbf{J}$ are dimensionless, whereas the remaining three, corresponding to the *moments* of the axes with respect to the operation point of the end-effector, have units of length. This dimensional inhomogeneity is resolved in the same way as in the case of planar manipulators for both positioning and orienting tasks, i.e., by means of a characteristic length. This length is defined as the one that minimizes the condition number of the dimensionless Jacobian thus obtained. We then redefine the Jacobian as

$$\mathbf{J} \equiv \begin{bmatrix} \mathbf{e}_1 & \mathbf{e}_2 & \mathbf{e}_3 & \mathbf{e}_4 & \mathbf{e}_5 & \mathbf{e}_6 \\ \frac{1}{L}\mathbf{e}_1 \times \mathbf{r}_1 & \frac{1}{L}\mathbf{e}_2 \times \mathbf{r}_2 & \frac{1}{L}\mathbf{e}_3 \times \mathbf{r}_3 & \frac{1}{L}\mathbf{e}_4 \times \mathbf{r}_4 & \frac{1}{L}\mathbf{e}_5 \times \mathbf{r}_5 & \frac{1}{L}\mathbf{e}_6 \times \mathbf{r}_6 \end{bmatrix} \tag{4.149}$$

and hence, the isotropy condition of eq.(4.126) leads to

$$\sum_1^6 \mathbf{e}_k \mathbf{e}_k^T = \sigma^2 \mathbf{1} \tag{4.150a}$$

$$\sum_1^6 \mathbf{e}_k (\mathbf{e}_k \times \mathbf{r}_k)^T = \mathbf{O} \tag{4.150b}$$

$$\frac{1}{L^2}\sum_1^6(\mathbf{e}_k\times\mathbf{r}_k)(\mathbf{e}_k\times\mathbf{r}_k)^T=\sigma^2\mathbf{1} \tag{4.150c}$$

where $\mathbf{1}$ is the 3×3 identity matrix, and $\mathbf{O}$ is the 3×3 zero matrix. Now, if we take the trace of both sides of eq.(4.150a), we obtain

$$\sigma^2=2 \quad \text{or} \quad \sigma=\sqrt{2}$$

Furthermore, we take the trace of both sides of eq.(4.150c), which yields

$$\frac{1}{L^2}\sum_1^6\|\mathbf{e}_k\times\mathbf{r}_k\|^2=3\sigma^2$$

But $\|\mathbf{e}_k\times\mathbf{r}_k\|^2$ is nothing but the square of the distance d_k of the kth revolute axis to the operation point, the foregoing equation thus yielding

$$L=\sqrt{\frac{1}{6}\sum_1^6 d_k^2}$$

i.e., *the characteristic length of a spatial six-revolute isotropic manipulator is the root-mean square of the distances of the revolute axes to the operation point when the robot finds itself at the posture of minimum condition number.*

Furthermore, eq.(4.150a) states that if $\{\mathbf{e}_k\}_1^6$ is regarded as the set of position vectors of points $\{P_k\}_1^6$ on the surface of the unit sphere, then the moment-of-inertia matrix of the set of equal masses located at these points has *spherical symmetry.* What the latter means is that any direction of the 3-dimensional space is a principal axis of inertia of the foregoing set. Likewise, eq.(4.150c) states that if $\{\mathbf{e}_k\times\mathbf{r}_k\}_1^6$ is regarded as the set of position vectors of points $\{Q_k\}$ in the 3-dimensional Euclidean space, then the moment-of-inertia matrix of the set of equal masses located at these points has spherical symmetry.

Now, in order to gain insight into eq.(4.150b), let us take the axial vector of both sides of that equation, thus obtaining

$$\sum_1^6\mathbf{e}_k\times(\mathbf{e}_k\times\mathbf{r}_k)=\mathbf{0} \tag{4.151}$$

with $\mathbf{0}$ denoting the 3-dimensional zero vector. Furthermore, let us denote by $\mathbf{E}_k$ the cross-product matrix of $\mathbf{e}_k$, the foregoing equation thus taking on the form

$$\sum_1^6\mathbf{E}_k^2\mathbf{r}_k=\mathbf{0}$$

However,

$$\mathbf{E}_k^2 = -\mathbf{1} + \mathbf{e}_k\mathbf{e}_k^T$$

for every k, and hence, eq.(4.151) leads to

$$\sum_1^6 (\mathbf{1} - \mathbf{e}_k\mathbf{e}_k^T)\mathbf{r}_k = \mathbf{0}$$

Moreover, $(\mathbf{1} - \mathbf{e}_k\mathbf{e}_k^T)\mathbf{r}_k$ is nothing but the normal component of $\mathbf{r}_k$ with respect to $\mathbf{e}_k$, as defined in Section 2.2. Let us denote this component by $\mathbf{r}_k^{\perp}$, thereby obtaining an alternative expression for the foregoing equation, namely,

$$\sum_1^6 \mathbf{r}_k^{\perp} = \mathbf{0} \tag{4.152}$$

The geometric interpretation of the foregoing equation is readily derived: To this end, let O_k' be the foot of the perpendicular to the kth revolute axis from the operation point P; then, $\mathbf{r}_k$ is the vector directed from O_k' to P. Therefore, *the operation point of an isotropic manipulator, configured at the isotropic posture is the centroid of the set* $\{ O_k' \}_1^6$ *of perpendicular feet from the operation point.*

A six-axis manipulator designed with an isotropic architecture, DIESTRO, is displayed in Fig. 4.31. The Denavit-Hartenberg parameters of this manipulator are given in Table 4.1. DIESTRO is characterized by identical link lengths a and offsets identical with this common link length, besides twist angles of 90° between all pairs of neighboring axes. Not surprisingly, the characteristic length of this manipulator is a.

Example 4.9.1 *Find the KCI and the characteristic length of the Fanuc Arc Mate robot whose DH parameters are given in Table 4.2.*

Solution: Apparently, what we need is the minimum value $\kappa_{\min}$ that the condition number of the manipulator Jacobian can attain, in order to calculate its KCI as indicated in eq.(4.127). Now, the Fanuc Arc Mate robot is a six-revolute manipulator for positioning and orienting tasks. Hence,

TABLE 4.1. DH Parameters of DIESTRO

i	a_i (mm)	b_i (mm)	α_i	θ_i
1	50	50	90°	θ_1
2	50	50	−90°	θ_2
3	50	50	90°	θ_3
4	50	50	−90°	θ_4
5	50	50	90°	θ_5
6	50	50	−90°	θ_6

FIGURE 4.31. DIESTRO, a six-axis isotropic manipulator.

TABLE 4.2. DH Parameters of the Fanuc Arc Mate Manipulator

i	a_i (mm)	b_i (mm)	α_i	θ_i
1	200	810	90°	θ_1
2	600	0	0°	θ_2
3	130	30	90°	θ_3
4	0	550	90°	θ_4
5	0	100	90°	θ_5
6	0	100	0°	θ_6

its Jacobian matrix has to be first recast in nondimensional form, as in eq.(4.149). Next, we find L, along with the joint variables that determine the posture of minimum condition number via an optimization procedure. Prior to the formulation of the underlying optimization problem, however, we must realize that the first joint, accounting for motions of the manipulator as a single rigid body, does not affect its Jacobian condition number. We thus define the *design vector* $\mathbf{x}$ of the optimization problem at hand as

$$\mathbf{x} \equiv [\theta_2 \quad \theta_3 \quad \theta_4 \quad \theta_5 \quad \theta_6 \quad L]$$

and set up the optimization problem as

$$\min_{\mathbf{x}} \kappa(\mathbf{J})$$

The condition number having been defined as the ratio of the largest to the smallest singular values of the Jacobian matrix at hand, the gradient of the above objective function, $\partial\kappa/\partial\mathbf{x}$, is apparently elusive to calculate. Thus, we use a direct-search method, i.e., a method not requiring any partial derivatives, but rather, only objective-function evaluations, to solve the above optimization problem. There are various methods of this kind at our disposal; the one we chose is the *simplex method*, as implemented in Matlab. The results reported are displayed below:

$$\mathbf{x}_{\text{opt}} = [\,26.82^\circ \quad -56.06^\circ \quad 15.79^\circ \quad -73.59^\circ \quad -17.83^\circ \quad 0.3573\,]$$

where the last entry, the characteristic length of the robot, is in meters, i.e.,

$$L = 357.3 \text{ mm}$$

Furthermore, the minimum condition number attained at the foregoing posture, with the characteristic length found above, is

$$\kappa_m = 2.589$$

Therefore, the KCI of the Fanuc Arc Mate is

$$\text{KCI} = 38.625\%$$

and so this robot is apparently far from being kinematically isotropic. To be sure, the KCI of this manipulator can still be improved dramatically by noting that the condition number is highly dependent on the location of the operation point of the end-effector. As reported by Tandirci, Angeles, and Ranjbaran (1992), an optimum selection of the operation point for the robot at hand yields a minimum condition number of 1.591, which thus leads to a KCI of 62.85%. The point of the EE that yields the foregoing minimum is thus termed the *characteristic point* of the manipulator in the foregoing reference. Its location in the EE is given by the DH parameters a_6 and b_6, namely,

$$a_6 = 223.6 \text{ mm}, \quad b_6 = 274.2 \text{ mm}$$

5

Trajectory Planning: Pick-and-Place Operations

5.1 Introduction

The motions undergone by robotic mechanical systems should be, as a rule, as smooth as possible; i.e., abrupt changes in position, velocity, and acceleration should be avoided. Indeed, abrupt motions require unlimited amounts of power to be implemented, which the motors cannot supply because of their physical limitations. On the other hand, abrupt motion changes arise when the robot collides with an object, a situation that should also be avoided. While smooth motions can be planned with simple techniques, as described below, these are no guarantees that no abrupt motion changes will occur. In fact, if the work environment is cluttered with objects, whether stationary or mobile, collisions may occur. Under ideal conditions, a flexible manufacturing cell is a work environment in which all objects, machines and workpieces alike, move with preprogrammed motions that by their nature, can be predicted at any instant. Actual situations, however, are far from being ideal, and system failures are unavoidable. Unpredictable situations should thus be accounted for when designing a robotic system, which can be done by supplying the system with sensors for the automatic detection of unexpected events or by providing for human monitoring. Nevertheless, robotic systems find applications not only in the well-structured environments of flexible manufacturing cells, but also in unstructured environments such as exploration of unknown terrains and systems in which humans are present. The planning of robot motions in the latter case is obviously much more challenging than in the former. Robot motion planning in unstructured environments calls for techniques beyond the scope of those studied in this book, involving such areas as pattern recognition and artificial intelligence. For this reason, we have devoted this book to the planning of robot motions in structured environments only.

Two typical tasks call for trajectory planning techniques, namely,

- pick-and-place operations (PPO), and
- continuous paths (CP).

We will study PPO in this chapter, with Chapter 9 devoted to CP. Moreover, we will focus on simple robotic manipulators of the serial type, although these techniques can be directly applied to other, more advanced, robotic mechanical systems.

5.2 Background on PPO

In PPO, a robotic manipulator is meant to take a workpiece from a given *initial pose*, specified by the position of one of its points and its orientation with respect to a certain coordinate frame, to a *final pose*, specified likewise. However, how the object moves from its initial to its final pose is immaterial, as long as the motion is smooth and no collisions occur. Pick-and-place operations are executed in elementary manufacturing operations such as loading and unloading of belt conveyors, tool changes in machine tools, and simple assembly operations such as putting roller bearings on a shaft. The common denominator of these tasks is *material handling*, which usually requires the presence of conventional machines whose motion is very simple and is usually characterized by a uniform velocity. In some instances, such as in *packing operations*, a set of workpieces, e.g., in a magazine, is to be relocated in a prescribed pattern in a container, which constitutes an operation known as *palletizing*. Although palletizing is a more elaborate operation than simple pick-and-place, it can be readily decomposed into a sequence of the latter operations.

It should be noted that although the initial and the final poses in a PPO are prescribed in the Cartesian space, robot motions are implemented in the joint space. Hence, the planning of PPO will be conducted in the latter space, which brings about the need of mapping the motion thus planned into the Cartesian space, in order to ensure that the robot will not collide with other objects in its surroundings. The latter task is far from being that simple, since it involves the rendering of the motion of all the moving links of the robot, each of which has a particular geometry. An approach to path planning first proposed by Lozano-Pérez (1981) consists of mapping the obstacles in the joint space, thus producing obstacles in the joint space in the form of regions that the joint-space trajectory should avoid. The

FIGURE 5.1. Still image of the animation of a palletizing operation.

idea can be readily implemented for simple planar motions and simple geometries of the obstacles. However, for general 3-D motions and arbitrary geometries, the computational requirements make the procedure impractical. A more pragmatic approach would consist of two steps, namely, (*i*) planning a preliminary trajectory in the joint space, disregarding the obstacles, and (*ii*) visually verifying if collisions occur with the aid of a graphics system rendering the animation of the robot motion in the presence of obstacles. The availability of powerful graphics hardware enables the fast animation of robot motions within a highly realistic environment. Shown in Fig. 5.1 is a still image of the animation produced by RVS, the McGill University *Robot-Visualization System*, of the motion of a robot performing a palletizing operation. Commercial software for robot-motion rendering is available.

By inspection of the kinematic closure equations of robotic manipulators—see eqs.(4.5a & b)—it is apparent that in the absence of singularities, the mapping of joint to Cartesian variables, and vice versa, is continuous. Hence, a smooth trajectory planned in the joint space is guaranteed to be smooth in the Cartesian space, and the other way around, as long as the trajectory does not encounter a singularity.

In order to proceed to synthesize the joint trajectory, we must then start by mapping the initial and final poses of the workpiece, which is assumed to be rigidly attached to the EE of the manipulator, into manipulator configurations described in the joint space. This is readily done with the methods described in Chapter 4. Let the vector of joint variables at the initial and final robot configurations be denoted by $\boldsymbol{\theta}_I$ and $\boldsymbol{\theta}_F$, respectively. Moreover, the initial pose in the Cartesian space is defined by the position vector $\mathbf{p}_I$ of the operation point P of the EE and a rotation matrix $\mathbf{Q}_I$. Likewise, the final pose in the Cartesian space is defined by the position vector $\mathbf{p}_F$ of P and the rotation matrix $\mathbf{Q}_F$. Moreover, let $\dot{\mathbf{p}}_I$ and $\ddot{\mathbf{p}}_I$ denote the velocity and acceleration of P, while $\boldsymbol{\omega}_I$ and $\dot{\boldsymbol{\omega}}_I$ denote the angular velocity and angular acceleration of the workpiece, all of these at the initial pose. These variables at the final pose are denoted likewise, with the subscript I changed to F. Furthermore, we assume that time is counted from the initial pose, i.e., at this pose, $t = 0$. If the operation takes place in time T, then at the final pose, $t = T$. We have thus the set of conditions that define a smooth motion between the initial and the final poses, namely,

$$\mathbf{p}(0) = \mathbf{p}_I \qquad \dot{\mathbf{p}}(0) = \mathbf{0} \qquad \ddot{\mathbf{p}}(0) = \mathbf{0}, \tag{5.1a}$$
$$\mathbf{Q}(0) = \mathbf{Q}_I \qquad \boldsymbol{\omega}(0) = \mathbf{0} \qquad \dot{\boldsymbol{\omega}}(0) = \mathbf{0} \tag{5.1b}$$
$$\mathbf{p}(T) = \mathbf{p}_F \qquad \dot{\mathbf{p}}(T) = \mathbf{0} \qquad \ddot{\mathbf{p}}(T) = \mathbf{0} \tag{5.1c}$$
$$\mathbf{Q}(T) = \mathbf{Q}_F \qquad \boldsymbol{\omega}(T) = \mathbf{0} \qquad \dot{\boldsymbol{\omega}}(T) = \mathbf{0} \tag{5.1d}$$

In the absence of singularities, then, the conditions of zero velocity and acceleration imply zero joint velocity and acceleration, and hence,

$$\boldsymbol{\theta}(0) = \boldsymbol{\theta}_I \qquad \dot{\boldsymbol{\theta}}(0) = \mathbf{0} \qquad \ddot{\boldsymbol{\theta}}(0) = \mathbf{0} \tag{5.2a}$$
$$\boldsymbol{\theta}(T) = \boldsymbol{\theta}_F \qquad \dot{\boldsymbol{\theta}}(T) = \mathbf{0} \qquad \ddot{\boldsymbol{\theta}}(T) = \mathbf{0} \tag{5.2b}$$

5.3 Polynomial Interpolation

A simple inspection of conditions (5.2a) and (5.2b) reveals that a linear interpolation between initial and final configurations will not work here, and neither will a quadratic interpolation, for its slope vanishes only at a single point. Hence, a higher-order interpolation is needed. On the other hand, these conditions imply, in turn, six conditions for every joint trajectory, which means that if a polynomial is to be employed to represent the motion of every joint, then this polynomial should be at least of the fifth degree. We thus start by studying trajectory planning with the aid of a 5th-degree polynomial.

5.3.1 A 3-4-5 Interpolating Polynomial

In order to represent each joint motion, we use here a fifth-order polynomial $s(\tau)$, namely,

$$s(\tau) = a\tau^5 + b\tau^4 + c\tau^3 + d\tau^2 + e\tau + f \tag{5.3}$$

such that

$$0 \le s \le 1, \qquad 0 \le \tau \le 1 \tag{5.4}$$

and

$$\tau = \frac{t}{T} \tag{5.5}$$

We will thus aim at a *normal polynomial* that, upon scaling both its argument and the polynomial itself, will allow us to represent each of the joint variables θ_j throughout its range of motion, so that

$$\theta_j(t) = \theta_j^I + (\theta_j^F - \theta_j^I)s(\tau) \tag{5.6a}$$

where θ_j^I and θ_j^F are the given initial and final values of the jth joint variable. In vector form, eq.(5.6a) becomes

$$\boldsymbol{\theta}(t) = \boldsymbol{\theta}_I + (\boldsymbol{\theta}_F - \boldsymbol{\theta}_I)s(\tau) \tag{5.6b}$$

and hence,

$$\dot{\boldsymbol{\theta}}(t) = (\boldsymbol{\theta}_F - \boldsymbol{\theta}_I)s'(\tau)\dot{\tau}(t) = (\boldsymbol{\theta}_F - \boldsymbol{\theta}_I)\frac{1}{T}s'(\tau) \tag{5.6c}$$

Likewise,

$$\ddot{\boldsymbol{\theta}}(t) = \frac{1}{T^2}(\boldsymbol{\theta}_F - \boldsymbol{\theta}_I)s''(\tau) \tag{5.6d}$$

and

$$\dddot{\boldsymbol{\theta}}(t) = \frac{1}{T^3}(\boldsymbol{\theta}_F - \boldsymbol{\theta}_I)s'''(\tau) \tag{5.6e}$$

What we now need are the values of the coefficients of $s(\tau)$ that appear in eq.(5.3). These are readily found by recalling conditions (5.2a & b), upon consideration of eqs.(5.6b–d). We thus obtain the end conditions for $s(\tau)$, namely,

$$s(0) = 0, \quad s'(0) = 0, \quad s''(0) = 0, \quad s(1) = 1, \quad s'(1) = 0, \quad s''(1) = 0 \tag{5.7}$$

The derivatives of $s(\tau)$ appearing above are readily derived from eq.(5.3), i.e.,

$$s'(\tau) = 5a\tau^4 + 4b\tau^3 + 3c\tau^2 + 2d\tau + e \tag{5.8}$$

and

$$s''(\tau) = 20a\tau^3 + 12b\tau^2 + 6c\tau + 2d \tag{5.9}$$

Thus, the first three conditions of eq.(5.7) lead to

$$f = e = d = 0 \tag{5.10}$$

while the last three conditions yield three linear equations in a, b, and c, namely,

$$a + b + c = 1 \tag{5.11a}$$

$$5a + 4b + 3c = 0 \tag{5.11b}$$

$$20a + 12b + 6c = 0 \tag{5.11c}$$

Upon solving the three foregoing equations for the three aforementioned unknowns, we obtain

$$a = 6, \qquad b = -15, \qquad c = 10 \tag{5.12}$$

and hence, the normal polynomial sought is

$$s(\tau) = 6\tau^5 - 15\tau^4 + 10\tau^3 \tag{5.13}$$

which is called a *3-4-5 polynomial.*

This polynomial and its first three derivatives, all normalized to fall within the $(-1, 1)$ range, are shown in Fig. 5.2. Note that the smoothness conditions imposed at the outset are respected and that the curve thus obtained is a monotonically growing function of τ, a rather convenient property for the problem at hand.

It is thus possible to determine the evolution of each joint variable if we know both its end values and the time T required to complete the motion. If no extra conditions are imposed, we then have the freedom to perform the desired motion in as short a time T as possible. Note, however, that this time cannot be given an arbitrarily small value, for we must respect the motor specifications on maximum velocity and maximum torque, the latter being the subject of Chapter 6. In order to ease the discussion, we limit ourselves to specifications of maximum joint velocity and acceleration rather than maximum torque. From the form of function $\theta_j(t)$ of eq.(5.6a), it is apparent that this function takes on extreme values at points corresponding to those at which the normal polynomial attains its extrema. In order to find the values of τ at which the first and second derivatives of $s(\tau)$ attain maximum values, we need to zero its second and third derivatives. These derivatives are displayed below:

$$s'(\tau) = 30\tau^4 - 60\tau^3 + 30\tau^2 \tag{5.14a}$$

$$s''(\tau) = 120\tau^3 - 180\tau^2 + 60\tau \tag{5.14b}$$

$$s'''(\tau) = 360\tau^2 - 360\tau + 60 \tag{5.14c}$$

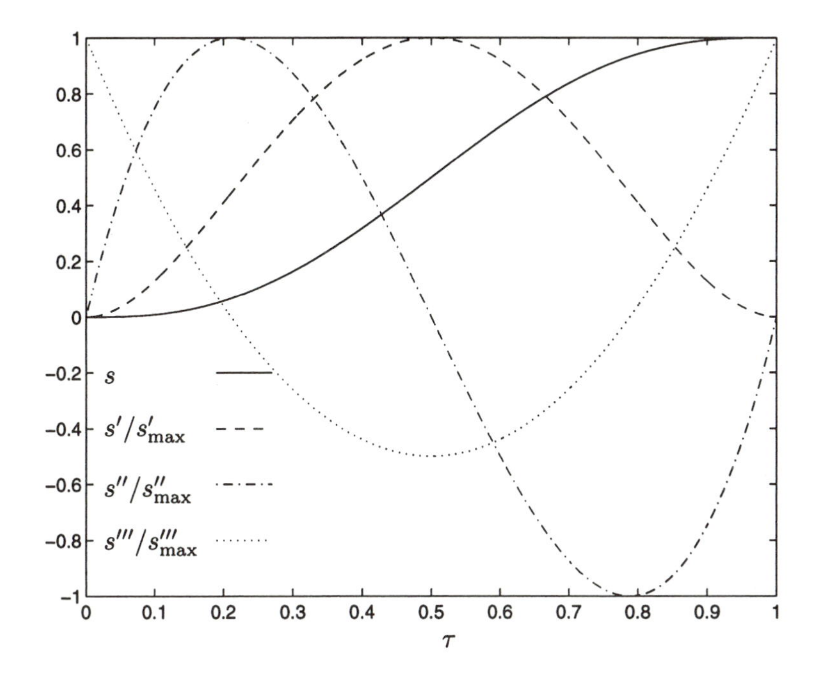

FIGURE 5.2. 3-4-5 interpolation polynomial and its derivatives.

from which it is apparent that the second derivative vanishes at the two ends of the interval $0 \leq \tau \leq 1$. Additionally, the same derivative vanishes at the midpoint of the same interval, i.e., at $\tau = 1/2$. Hence, the maximum value of $s'(\tau)$, s'_{max}, is readily found as

$$s'_{max} = s'\left(\frac{1}{2}\right) = \frac{15}{8} \tag{5.15}$$

and hence, the maximum value of the jth joint rate takes on the value

$$(\dot{\theta}_j)_{max} = \frac{15(\theta_j^F - \theta_j^I)}{8T} \tag{5.16}$$

which becomes negative, and hence, a local minimum, if the difference in the numerator is negative. The values of τ at which the second derivative attains its extreme values are likewise determined. The third derivative vanishes at two intermediate points τ_1 and τ_2 of the interval $0 \leq \tau \leq 1$, namely, at

$$\tau_{1,2} = \frac{1}{2} \pm \frac{\sqrt{3}}{6} \tag{5.17}$$

and hence, the maximum value of $s''(\tau)$ is readily found as

$$s''_{max} = s''\left(\frac{1}{2} - \frac{\sqrt{3}}{6}\right) = \frac{10\sqrt{3}}{3} \tag{5.18}$$

while the minimum is given as

$$s''_{\min} = s''\left(\frac{1}{2} + \frac{\sqrt{3}}{6}\right) = -\frac{10\sqrt{3}}{3} \tag{5.19}$$

Therefore, the maximum value of the joint acceleration is as shown below:

$$(\ddot{\theta}_j)_{\max} = \frac{10\sqrt{3}}{3}\frac{(\theta_j^F - \theta_j^I)}{T^2} \tag{5.20}$$

Likewise,

$$s'''_{\max} = s'''(0) = s'''(1) = 60$$

and hence,

$$(\dddot{\theta}_j)_{\max} = 60\frac{\theta_J^F - \theta_j^I}{T^3} \tag{5.21}$$

Thus, eqs.(5.16) and (5.20) allow us to determine T for each joint so that the joint rates and accelerations lie within the allowed limits. Obviously, since the motors of different joints are different, the minimum values of T allowed by the joints will be, in general, different. Of those various values of T, we will, of course, choose the largest one.

5.3.2 A 4-5-6-7 Interpolating Polynomial

Now, from eq.(5.14c), it is apparent that the third derivative of the normal polynomial does not vanish at the end points of the interval of interest. This implies that the third time derivative of $\theta_j(t)$, also known as the joint *jerk*, does not vanish at those ends either. It is desirable to have this derivative as smooth as the first two, but this requires us to increase the order of the normal polynomial. In order to attain the desired smoothness, we will then impose two more conditions, namely,

$$s'''(0) = 0, \qquad s'''(1) = 0 \tag{5.22}$$

We now have eight conditions on the normal polynomial, which means that the polynomial degree should be increased to seven, namely,

$$s(\tau) = a\tau^7 + b\tau^6 + c\tau^5 + d\tau^4 + e\tau^3 + f\tau^2 + g\tau + h \tag{5.23a}$$

whose derivatives are readily determined as shown below:

$$s'(\tau) = 7a\tau^6 + 6b\tau^5 + 5c\tau^4 + 4d\tau^3 + 3e\tau^2 + 2f\tau + g \tag{5.23b}$$
$$s''(\tau) = 42a\tau^5 + 30b\tau^4 + 20c\tau^3 + 12d\tau^2 + 6e\tau + 2f \tag{5.23c}$$
$$s'''(\tau) = 210a\tau^4 + 120b\tau^3 + 60c\tau^2 + 24d\tau + 6e \tag{5.23d}$$

The first three conditions of eq.(5.7) and the first condition of eq.(5.22) readily lead to

$$e = f = g = h = 0 \tag{5.24}$$

Furthermore, the last three conditions of eq.(5.7) and the second condition of eq.(5.22) lead to four linear equations in four unknowns, namely,

$$a + b + c + d = 1 \tag{5.25a}$$

$$7a + 6b + 5c + 4d = 0 \tag{5.25b}$$

$$42a + 30b + 20c + 12d = 0 \tag{5.25c}$$

$$210a + 120b + 60c + 24d = 0 \tag{5.25d}$$

and hence, we obtain the solution

$$a = -20, \qquad b = 70, \qquad c = -84, \qquad d = 35 \tag{5.26}$$

the desired polynomial thus being

$$s(\tau) = -20\tau^7 + 70\tau^6 - 84\tau^5 + 35\tau^4 \tag{5.27}$$

which is a *4-5-6-7 polynomial.* This polynomial and its first three derivatives, normalized to fall within the range $(-1, 1)$, are plotted in Fig. 5.3. Note that the 4-5-6-7 polynomial is similar to that of Fig. 5.2, except that the third derivative of the former vanishes at the extremes of the interval of interest. As we will presently show, this smoothness has been obtained at the expense of higher maximum values of the first and second derivatives.

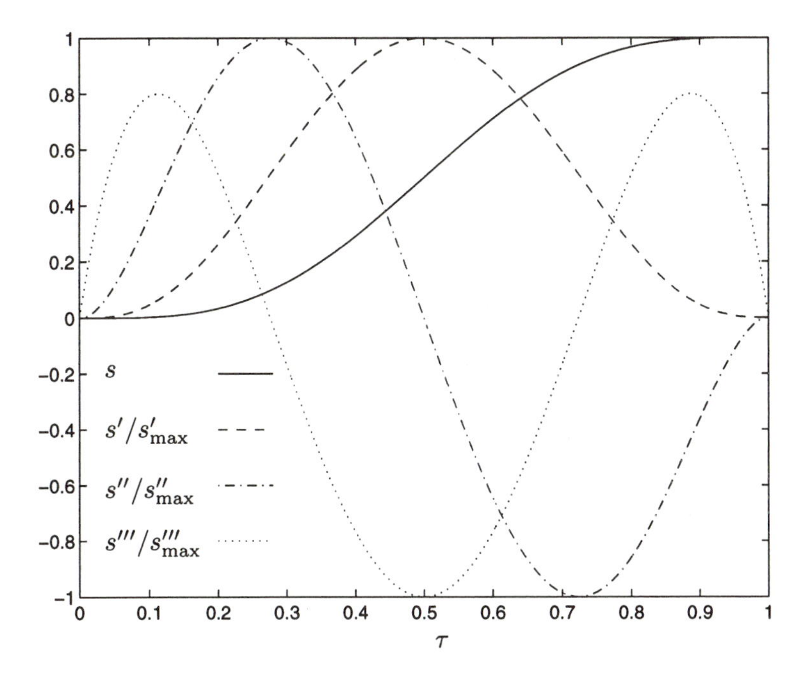

FIGURE 5.3. 4-5-6-7 interpolating polynomial and its derivatives.

We now determine the maximum values of the velocity and acceleration produced with this motion. To this end, we display below the first three derivatives, namely,

$$s'(\tau) = -140\tau^6 + 420\tau^5 - 420\tau^4 + 140\tau_3 \tag{5.28a}$$

$$s''(\tau) = -840\tau^5 + 2100\tau^4 - 1680\tau^3 + 420\tau^2 \tag{5.28b}$$

$$s'''(\tau) = -4200\tau^4 + 8400\tau^3 - 5040\tau^2 + 840\tau \tag{5.28c}$$

The first derivative attains its extreme values at points where the second derivative vanishes. Upon zeroing the latter, we obtain

$$\tau^2(-2\tau^3 + 5\tau^2 - 4\tau + 1) = 0 \tag{5.29}$$

which clearly contains a double root at $\tau = 0$. Moreover, the cubic polynomial in the parentheses above admits one real root, namely, $\tau = 1/2$, which yields the maximum value of $s'(\tau)$, i.e.,

$$s'_{\max} = s'\left(\frac{1}{2}\right) = \frac{35}{16} \tag{5.30}$$

whence the maximum value of the jth joint rate is found as

$$(\dot{\theta}_j)_{\max} = \frac{35(\theta_j^F - \theta_j^I)}{16T} \tag{5.31}$$

Likewise, the points of maximum joint acceleration are found upon zeroing the third derivative of $s(\tau)$, namely,

$$s'''(\tau) = -4200\tau^4 + 8400\tau^3 - 5040\tau^2 + 840\tau = 0 \tag{5.32}$$

or

$$\tau(\tau - 1)(5\tau^2 - 5\tau + 1) = 0 \tag{5.33}$$

which yields, in addition to the two end points, two intermediate extreme points, namely,

$$\tau_{1,2} = \frac{1}{2} \pm \frac{\sqrt{5}}{10} \tag{5.34}$$

and hence, the maximum value of acceleration is found to be

$$s''_{\max} = s''(\tau_1) = \frac{84\sqrt{5}}{25} \tag{5.35}$$

the minimum occurring at $\tau = \tau_2$, with $s''_{\min} = -s''_{\max}$. The maximum value of the jth joint acceleration is thus

$$(\ddot{\theta}_j)_{\max} = \frac{84\sqrt{5}}{25}\left(\frac{\theta_j^F - \theta_j^I}{T^2}\right) \tag{5.36}$$

which becomes a minimum if the difference in the numerator is negative. Likewise, the zeroing of the fourth derivative leads to

$$-20\tau^3 + 30\tau^2 - 12\tau + 1 = 0$$

whose three roots are

$$\tau_1 = \frac{1 - \sqrt{3/5}}{2}, \quad \tau_2 = \frac{1}{2}, \quad \tau_1 = \frac{1 + \sqrt{3/5}}{2}$$

and hence,

$$s'''_{\max} = s'''\left(\frac{1 \pm \sqrt{3/5}}{2}\right) = 42, \quad s'''_{\min} = s'''(0.5) = -\frac{105}{2}$$

i.e.,

$$\max_{\tau}\{|s'''(\tau)|\} = \frac{105}{2} \equiv s'''_M \tag{5.37}$$

As in the case of the fifth-order polynomial, it is possible to use the foregoing relations to determine the minimum time T during which it is possible to perform a given PPO while observing the physical limitations of the motors.

5.4 Cycloidal Motion

An alternative motion that produces zero velocity and acceleration at the ends of a finite interval is the *cycloidal motion*. In normal form, this motion is given by

$$s(\tau) = \tau - \frac{1}{2\pi}\sin 2\pi\tau \tag{5.38a}$$

its derivatives being readily derived as

$$s'(\tau) = 1 - \cos 2\pi\tau \tag{5.38b}$$

$$s''(\tau) = 2\pi \sin 2\pi\tau \tag{5.38c}$$

$$s'''(\tau) = 4\pi^2 \cos 2\pi\tau \tag{5.38d}$$

The cycloidal motion and its first three time-derivatives, normalized to fall within the range $(-1, 1)$, are shown in Fig. 5.4. Note that while this motion, indeed, has zero velocity and acceleration at the ends of the interval $0 \le \tau \le 1$, its jerk is nonzero at these points and hence, exhibits jump discontinuities at the ends of that interval.

When implementing the cycloidal motion in PPO, we have, for the jth joint,

$$\theta_j(t) = \theta_j^I + (\theta_j^F - \theta_j^I)s(\tau) \tag{5.39a}$$

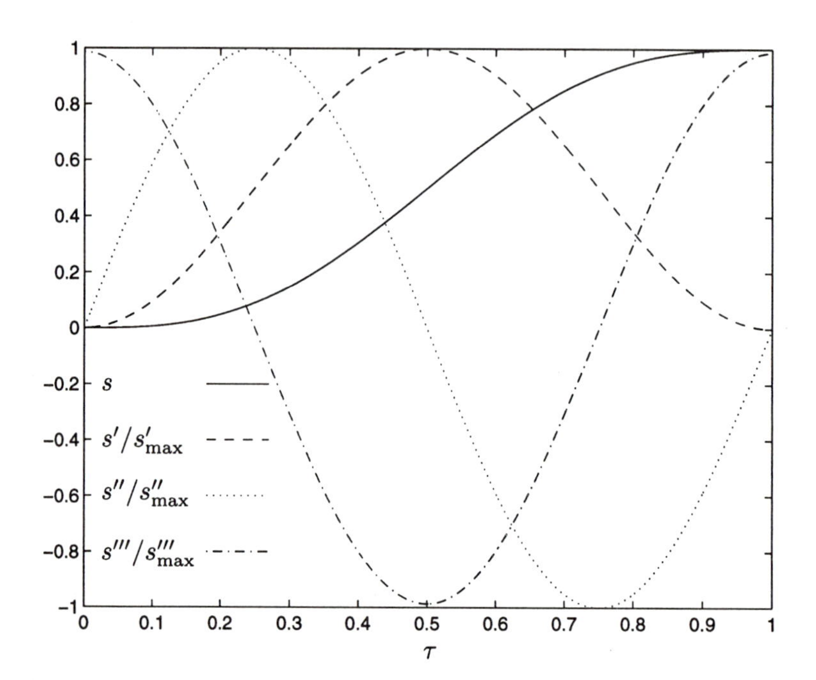

FIGURE 5.4. The normal cycloidal motion and its time derivatives.

$$\dot{\theta}_j(t) = \frac{\theta_j^F - \theta_j^I}{T} s'(\tau) \tag{5.39b}$$

$$\ddot{\theta}_j(t) = \frac{\theta_j^F - \theta_j^I}{T^2} s''(\tau) \tag{5.39c}$$

Moreover, as the reader can readily verify, under the assumption that $\theta_j^F > \theta_j^I$, this motion attains its maximum velocity at the center of the interval, i.e., at $\tau = 0.5$, the maximum being

$$s'_{\max} = s'(0.5) = 2$$

and hence,

$$(\dot{\theta}_j)_{\max} = \frac{2}{T}(\theta_j^F - \theta_j^I) \tag{5.40a}$$

Likewise, the jth joint acceleration attains its maximum and minimum values at $\tau = 0.25$ and $\tau = 0.75$, respectively, i.e.,

$$s''_{\max} = s''(0.25) = s''(0.75) = 2\pi \tag{5.40b}$$

and hence,

$$(\ddot{\theta}_j)_{\max} = \frac{2\pi}{T^2}(\theta_j^F - \theta_j^I), \quad (\ddot{\theta}_j)_{\min} = -\frac{2\pi}{T^2}(\theta_j^F - \theta_j^I) \tag{5.40c}$$

Moreover, $s'''(\tau)$ attains its extrema at the ends of the interval, i.e.,

$$s'''_{\max} = s'''(0) = s'''(1) = 4\pi^2 \tag{5.41}$$

and hence,

$$(\dddot{\theta}_j)_{\max} = \frac{4\pi^2}{T^3}(\theta_j^F - \theta_j^I) \tag{5.42}$$

Thus, if motion is constrained by the maximum speed delivered by the motors, the minimum time T_j for the jth joint to produce the given PPO can be readily determined from eq.(5.40a) as

$$T_j = \frac{2(\theta_j^F - \theta_j^I)}{(\dot{\theta}_j)_{\max}} \tag{5.43}$$

and hence, the minimum time in which the operation can take place can be readily found as

$$T_{\min} = 2 \max_j \left\{ \frac{\theta_j^F - \theta_j^I}{(\dot{\theta}_j)_{\max}} \right\} \tag{5.44}$$

If joint-acceleration constraints are imposed, then a similar procedure can be followed to find the minimum time in which the operation can be realized. As a matter of fact, rather than maximum joint accelerations, maximum joint torques are to be respected. How to determine these torques is studied in detail in Chapter 6.

5.5 Trajectories with Via Poses

The polynomial trajectories discussed above do not allow the specification of intermediate Cartesian poses of the EE. All they guarantee is that the Cartesian trajectories prescribed at the initial and final instants are met. One way of verifying the feasibility of the Cartesian trajectories thus synthesized was outlined above and consists of using a graphics system, preferably with animation capabilities, to produce an animated rendering of the robot motion, thereby allowing for verification of collisions. If the latter occur, we can either try alternative branches of the inverse kinematics solutions computed at the end poses or modify the trajectory so as to eliminate collisions. We discuss below the second approach. This is done with what are called *via poses*, i.e., poses of the EE in the Cartesian space that lie between the initial and the final poses, and are determined so as to avoid collisions. For example, if upon approaching the final pose of the PPO, the manipulator is detected to interfere with the surface on which the workpiece is to be placed, a via pose is selected close to the final point so that at this pose, the workpiece is far enough from the surface. From inverse kinematics, values of the joint variables can be determined that correspond to the aforementioned via poses. These values can now be regarded as points on the joint-space trajectory and are hence called *via points*. Obviously, upon plotting each joint variable vs. time, via points appear as points on those plots as well.

The introduction of via points in the joint-space trajectories amounts to an increase in the number of conditions to be satisfied by the desired trajectory. For example, in the case of the polynomial trajectory synthesized for continuity up to second derivatives, we can introduce two via points by requiring that

$$s(\tau_1) = s_1, \qquad s(\tau_2) = s_2 \tag{5.45}$$

where τ_1, τ_2, s_1, and s_2 depend on the via poses prescribed and the instants at which these poses are desired to occur. Hence, s_1 and s_2 differ from joint to joint, although the occurrence instants τ_1 and τ_2 are the same for all joints. Thus, we will have to determine one normal polynomial for each joint. Furthermore, the ordinate values s_1 and s_2 of the normal polynomial at via points are determined from the corresponding values of the joint variable determined, in turn, from given via poses through inverse kinematics. Once the via values of the joint variables are known, the ordinate values of the via points of the normal polynomial are found from eq.(5.6a). Since we have now eight conditions to satisfy, namely, the six conditions (5.7) plus the two conditions (5.45), we need a seventh-order polynomial, i.e.,

$$s(\tau) = a\tau^7 + b\tau^6 + c\tau^5 + d\tau^4 + e\tau^3 + f\tau^2 + g\tau + h \tag{5.46}$$

Again, the first three conditions of eq.(5.7) lead to the vanishing of the last three coefficients, i.e.,

$$f = g = h = 0 \tag{5.47}$$

Further, the five remaining conditions are now introduced, which leads to a system of five linear equations in five unknowns, namely,

$$a + b + c + d + e = 1 \tag{5.48a}$$

$$7a + 6b + 5c + 4d + 3e = 0 \tag{5.48b}$$

$$42a + 30b + 20c + 12d + 6e = 0 \tag{5.48c}$$

$$\tau_1^7 a + \tau_1^6 b + \tau_1^5 c + \tau_1^4 d + \tau_1^3 e = s_1 \tag{5.48d}$$

$$\tau_2^7 a + \tau_2^6 b + \tau_2^5 c + \tau_2^4 d + \tau_2^3 e = s_2 \tag{5.48e}$$

where τ_1, τ_2, s_1, and s_2 are all data. For example, if the via poses occur at 10% and 90% of T, we have

$$\tau_1 = 1/10, \qquad \tau_2 = 9/10 \tag{5.48f}$$

the polynomial coefficients being found as

$$a = 100(12286 + 12500s_1 - 12500s_2)/729 \tag{5.49a}$$
$$b = 100(-38001 - 48750s_1 + 38750s_2)/729 \tag{5.49b}$$
$$c = (1344358 + 2375000s_1 - 1375000s_2)/243 \tag{5.49c}$$
$$d = (-1582435 - 4625000s_1 + 1625000s_2)/729 \tag{5.49d}$$
$$e = 10(12159 + 112500s_1 - 12500s_2)/729 \tag{5.49e}$$

The shape of each joint trajectory thus depends on the values of s_1 and s_2 found from eq.(5.6a) for that trajectory.

5.6 Synthesis of PPO Using Cubic Splines

When the number of via poses increases, the foregoing approach may become impractical, or even unreliable. Indeed, forcing a trajectory to pass through a number of via points and meet endpoint conditions is equivalent to interpolation. We have seen that an increase in the number of conditions to be met by the normal polynomial amounts to an increase in the degree of this polynomial. Now, finding the coefficients of the interpolating polynomial requires solving a system of linear equations. As we saw in Section 4.9, the computed solution, when solving a system of linear equations, is corrupted with a relative roundoff error that is roughly equal to the relative roundoff error of the data multiplied by an amplification factor that is known as the *condition number* of the system matrix. As we increase the order of the interpolating polynomial, the associated condition number rapidly increases, a fact that numerical analysts discovered some time ago (Kahaner, Moler, and Nash, 1989). In order to cope with this problem, *orthogonal polynomials*, such as those bearing the names of *Chebyshev, Laguerre, Legendre*, and so on, have been proposed. While orthogonal polynomials alleviate the problem of a large condition number, they do this only up to a certain extent. As an alternative to higher-order polynomials, *spline functions* have been found to offer more robust interpolation schemes (Dierckx, 1993). Spline functions, or *splines*, for brevity, are piecewise polynomials with continuity properties imposed at the *supporting points*. The latter are those points at which two neighboring polynomials join.

The attractive feature of splines is that they are defined as a set of rather lower-degree polynomials joined at a number of supporting points. Moreover, the matrices that arise from an interpolation problem associated with a spline function are such that their condition number is only slightly dependent on the number of supporting points, and hence, splines offer the possibility of interpolating over a virtually unlimited number of points without producing serious numerical conditioning problems.

Below we expand on periodic cubic splines, for these will be shown to be specially suited for path planning in robotics.

A cubic spline function $s(x)$ connecting N points P_k $(x_k,\ y_k)$, for $k = 1, 2, \ldots, N$, is a *function* defined *piecewise* by $N - 1$ cubic polynomials joined at the points P_k, such that $s(x_k) = y_k$. Furthermore, the spline function thus defined is twice differentiable everywhere in $x_1 \le x \le x_N$. Hence, cubic splines are said to be C^2 functions, i.e., to have continuous derivatives up to the second order.

Cubic splines are optimal in the sense that they minimize a *functional*, i.e., an integral defined as

$$F = \int_0^T s''^2(x)\, dx$$

subject to the constraints

$$s(x_k) = y_k, \qquad k = 1, \ldots, N$$

where x_k and y_k are given. The aforementioned optimality property has a simple kinematic interpretation: Among all functions defining a motion so that the plot of this function passes through a set of points $P_1(x_1,\ s_1)$, $P_2(x_2,\ s_2)$, $\ldots$, $P_N(x_N,\ s_N)$ in the x-s plane, the cubic spline is the one containing the minimum *acceleration magnitude*. In fact, F, as given above, is the square of the *Euclidean norm* (Halmos, 1974) of $s''(x)$, i.e., F turns out to be a measure of the *magnitude* of the acceleration of a displacement program given by $s(x)$, if we interpret s as displacement and x as time.

Let $P_k(x_k,\ y_k)$ and $P_{k+1}(x_{k+1},\ y_{k+1})$ be two consecutive supporting points. The kth cubic polynomial $s_k(x)$ between those points is assumed to be given by

$$s_k(x) = A_k\,(x - x_k)^3 + B_k\,(x - x_k)^2 + C_k\,(x - x_k) + D_k \tag{5.50a}$$

for $x_k \le x \le x_{k+1}$. Thus, for the spline $s(x)$, $4(N - 1)$ coefficients A_k, B_k, C_k, D_k, for $k = 1, \ldots, N - 1$, are to be determined. These coefficients will be computed presently in terms of the given function values $\{s_k\}_1^N$ and the second derivatives of the spline at the supporting points, $\{s_k''(x_k)\}_1^N$, as explained below:

We will need the first and second derivatives of $s_k(x)$ as given above, namely,

$$s_k'(x) = 3A_k(x - x_k)^2 + 2B_k(x - x_k) + C_k \tag{5.50b}$$

$$s_k''(x) = 6A_k(x - x_k) + 2B_k \tag{5.50c}$$

whence the relations below follow immediately:

$$B_k = \frac{1}{2} s_k'' \tag{5.51a}$$

$$C_k = s_k' \tag{5.51b}$$

$$D_k = s_k \tag{5.51c}$$

where we have used the abbreviations

$$s_k \equiv s(x_k), \quad s'_k \equiv s'(x_k), \quad s''_k \equiv s''(x_k) \tag{5.52}$$

Furthermore, let

$$\Delta x_k \equiv x_{k+1} - x_k \tag{5.53}$$

From the above relations, we have expressions for coefficients B_k and D_k in terms of s''_k and s_k, respectively, but the expression for C_k is given in terms of s'_k. What we would like to have are similar expressions for A_k and C_k, i.e., in terms of s_k and s''_k. The relations sought will be found by imposing the continuity conditions on the spline function and its first and second derivatives with respect to x at the supporting points. These conditions are, then, for $k = 1, 2, \ldots, N-1$,

$$s_k(x_{k+1}) = s_{k+1} \tag{5.54a}$$

$$s'_k(x_{k+1}) = s'_{k+1} \tag{5.54b}$$

$$s''_k(x_{k+1}) = s''_{k+1} \tag{5.54c}$$

Upon substituting $s''_k(x_{k+1})$, as given by eq.(5.50c), into eq.(5.54c), we obtain

$$6A_k\Delta x_k + 2B_k = 2B_{k+1}$$

but from eq.(5.51a), we have already an expression for B_k, and hence, one for B_{k+1} as well. Substituting these two expressions in the above equation, we obtain an expression for A_k, namely,

$$A_k = \frac{1}{6\,\Delta x_k}\,(s''_{k+1} - s''_k) \tag{5.54d}$$

Furthermore, if we substitute $s_k(x_{k+1})$, as given by eq.(5.50a), into eq.(5.54a), we obtain

$$A_k(\Delta x_k)^3 + B_k(\Delta x_k)^2 + C_k\Delta x_k = s_{k+1}$$

But we already have values for A_k and B_k from eqs.(5.54d) and (5.51a), respectively. Upon substituting these values in the foregoing equation, we obtain the desired expression for C_k in terms of function and second-derivative values, i.e.,

$$C_k = \frac{\Delta s_k}{\Delta x_k} - \frac{1}{6}\,\Delta x_k\,(s''_{k+1} + 2s''_k) \tag{5.54e}$$

In summary, then, we now have expressions for all four coefficients of the kth polynomial in terms of function and second-derivative values at the supporting points, namely,

$$A_k = \frac{1}{6\,\Delta x_k}\,(s''_{k+1} - s''_k) \tag{5.55a}$$

$$B_k = \frac{1}{2}\,s''_k \tag{5.55b}$$

$$C_k = \frac{\Delta s_k}{\Delta x_k} - \frac{1}{6}\Delta x_k\,(s''_{k+1} + 2s''_k) \tag{5.55c}$$

$$D_k = s_k \tag{5.55d}$$

with

$$\Delta s_k \equiv s_{k+1} - s_k \tag{5.55e}$$

Furthermore, from the requirement of continuity in the first derivative, eq.(5.54b), after substitution of eq.(5.50b), one obtains

$$3A_k(\Delta x_k)^2 + 2B_k\Delta x_k + C_k = C_{k+1}$$

or if we shift to the previous polynomials,

$$3A_{k-1}(\Delta x_{k-1})^2 + 2B_{k-1}\Delta x_{k-1} + C_{k-1} = C_k$$

Now, if we substitute expressions (5.55a–c) in the above equation, a linear system of $N-2$ simultaneous equations for the N unknowns $\{s''_k\}_1^N$ is obtained, namely,

$$(\Delta x_k)s''_{k+1} + 2(\Delta x_{k-1} + \Delta x_k)s''_k + (\Delta x_{k-1})s''_{k-1} = 6(\frac{\Delta s_k}{\Delta x_k} - \frac{\Delta s_{k-1}}{\Delta x_{k-1}}) \tag{5.56}$$

for $k = 2, \ldots, N-1$.

Further, let $\mathbf{s}$ be the N-dimensional vector whose kth component is s_k, with vector $\mathbf{s}''$ being defined likewise, i.e.,

$$\mathbf{s} = [\,s_1, \cdots, s_N\,]^T, \qquad \mathbf{s}'' = [\,s''_1, \cdots, s''_N\,]^T \tag{5.57}$$

The relationship between $\mathbf{s}$ and $\mathbf{s}''$ of eq.(5.56) can then be written in vector form as

$$\mathbf{A}\,\mathbf{s}'' = 6\,\mathbf{C}\,\mathbf{s} \tag{5.58a}$$

where $\mathbf{A}$ and $\mathbf{C}$ are $(N-2)\times N$ matrices defined as:

$$\mathbf{A} = \begin{bmatrix} \alpha_1 & 2\alpha_{1,2} & \alpha_2 & 0 & \cdots & 0 & 0 \\ 0 & \alpha_2 & 2\alpha_{2,3} & \alpha_3 & \cdots & 0 & 0 \\ \vdots & \vdots & \ddots & \ddots & \ddots & \vdots & \vdots \\ 0 & 0 & \cdots & \alpha_{N'''} & 2\alpha_{N''',N''} & \alpha_{N''} & 0 \\ 0 & 0 & 0 & \cdots & \alpha_{N''} & 2\alpha_{N'',N'} & \alpha_{N'} \end{bmatrix} \tag{5.58b}$$

and

$$\mathbf{C} = \begin{bmatrix} \beta_1 & -\beta_{1,2} & \beta_2 & 0 & \cdots & 0 & 0 \\ 0 & \beta_2 & -\beta_{2,3} & \beta_3 & \cdots & 0 & 0 \\ \vdots & \vdots & \ddots & \ddots & \ddots & \vdots & \vdots \\ 0 & 0 & \cdots & \beta_{N'''} & -\beta_{N''',N''} & \beta_{N''} & 0 \\ 0 & 0 & 0 & \cdots & \beta_{N''} & -\beta_{N'',N'} & \beta_{N'} \end{bmatrix} \tag{5.58c}$$

while for $i, j, k = 1, \ldots, N-1$,

$$\alpha_k \equiv \Delta x_k, \quad \alpha_{i,j} \equiv \alpha_i + \alpha_j, \tag{5.58d}$$
$$\beta_k \equiv 1/\alpha_k, \quad \beta_{i,j} \equiv \beta_i + \beta_j \tag{5.58e}$$

and

$$N' \equiv N-1, \qquad N'' \equiv N-2, \qquad N''' \equiv N-3 \tag{5.58f}$$

Thus, two additional equations are needed to render eq.(5.58a) a determined system. The additional equations are derived, in turn, depending upon the class of functions one is dealing with, which thus gives rise to various types of splines. For example, if s_1'' and s_N'' are defined as zero, then one obtains *natural cubic splines*, the name arising by an analogy with beam analysis. Indeed, in beam theory, the boundary conditions of a simply-supported beam establish the vanishing of the bending moments at the ends. From beam theory, moreover, the bending moment is proportional to the second derivative of the *elastica*, or *neutral axis*, of the beam with respect to the abscissa along the beam axis in the undeformed configuration. In this case, vector $\mathbf{s}''$ becomes of dimension $N-2$, and hence, matrix $\mathbf{A}$ becomes, correspondingly, of $(N-2) \times (N-2)$, namely,

$$\mathbf{A} = \begin{bmatrix} 2\alpha_{1,2} & \alpha_2 & 0 & \cdots & 0 \\ \alpha_2 & 2\alpha_{2,3} & \alpha_3 & \cdots & 0 \\ \vdots & \ddots & \ddots & \ddots & \vdots \\ 0 & \cdots & \alpha_{N'''} & 2\alpha_{N''',N''} & \alpha_{N''} \\ 0 & 0 & \cdots & \alpha_{N''} & 2\alpha_{N'',N'} \end{bmatrix} \tag{5.59}$$

On the other hand, if one is interested in periodic functions, which is often the case when synthesizing pick-and-place motions, then the conditions $s_1 = s_N$, $s_1' = s_N'$, $s_1'' = s_N''$ are imposed, thereby producing *periodic cubic splines*. The last of these conditions is used to eliminate one unknown in eq.(5.58a), while the second condition, namely the continuity of the first derivative, is used to add an equation. We have, then,

$$s_1' = s_N' \tag{5.60}$$

which can be written, using eq.(5.54b), as

$$s_1' = s_{N-1}'(x_N) \tag{5.61}$$

Upon substituting $s_{N-1}'(x_N)$, as given by eq.(5.50b), into the above equation, we obtain

$$s_1' = 3A_{N-1}\Delta x_{N-1}^2 + 2B_{N-1}\Delta x_{N-1} + C_{N-1} \tag{5.62}$$

Now we use eqs.(5.55a–c) and simplify the expression thus resulting, which leads to

$$2(\Delta x_1 + \Delta x_{N-1})s_1'' + \Delta x_1 s_2'' + \Delta x_{N-1} s_{N-1}'' = 6\left(\frac{\Delta s_1}{\Delta x_1} - \frac{\Delta s_{N-1}}{\Delta x_{N-1}}\right) \quad (5.63)$$

thereby obtaining the last equation required to solve the system of equations given by eqs.(5.58a–c). We thus have $(N-1)$ independent equations to solve for $(N-1)$ unknowns, namely, s_k'', for $k = 1, \ldots, N-1$, s_N'' being equal to s_1''. Expressions for matrices $\mathbf{A}$ and $\mathbf{C}$, as applicable to periodic cubic splines, are given in eqs.(9.59a & b).

While we focused in the above discussion on cubic splines, other types of splines could have been used. For example, Thompson and Patel (1987) used B-splines in robotics trajectory planning.

Example 5.6.1 (Approximation of a 4-5-6-7 polynomial with a cubic spline) *Find the cubic spline that interpolates the 4-5-6-7 polynomial of Fig. 5.3 with $N+1$ equally-spaced supporting points and plot the interpolation error for $N = 3$ and $N = 10$.*

Solution: Let us use a natural spline, in which case the second derivative at the end points vanishes, with vector $\mathbf{s}''$ thus losing two components. That is, we now have only $N-1$ unknowns $\{ s_k'' \}_1^{N-1}$ to determine. Correspondingly, matrix $\mathbf{A}$ then loses its first and last columns and hence, becomes a square $(N-1) \times (N-1)$ matrix. Moreover,

$$\Delta x_k = \frac{1}{N}, \quad k = 1, \ldots, N$$

and matrices $\mathbf{A}$ and $\mathbf{C}$ become, correspondingly,

$$\mathbf{A} = \frac{1}{N}\begin{bmatrix} 4 & 1 & 0 & \cdots & 0 \\ 1 & 4 & 1 & \cdots & 0 \\ \vdots & \ddots & \ddots & \ddots & \vdots \\ 0 & \cdots & 1 & 4 & 1 \\ 0 & 0 & \cdots & 1 & 4 \end{bmatrix}$$

and

$$\mathbf{C} = N\begin{bmatrix} 1 & -2 & 1 & 0 & \cdots & 0 & 0 \\ 0 & 1 & -2 & 1 & \cdots & 0 & 0 \\ \vdots & \vdots & \ddots & \ddots & \ddots & \vdots & \vdots \\ 0 & 0 & \cdots & 1 & -2 & 1 & 0 \\ 0 & 0 & 0 & \cdots & 1 & -2 & 1 \end{bmatrix}$$

the vector of second derivatives at the supporting points, $\mathbf{s}''$, then being readily obtained as

$$\mathbf{s}'' = 6\mathbf{A}^{-1}\mathbf{C}\mathbf{s}$$

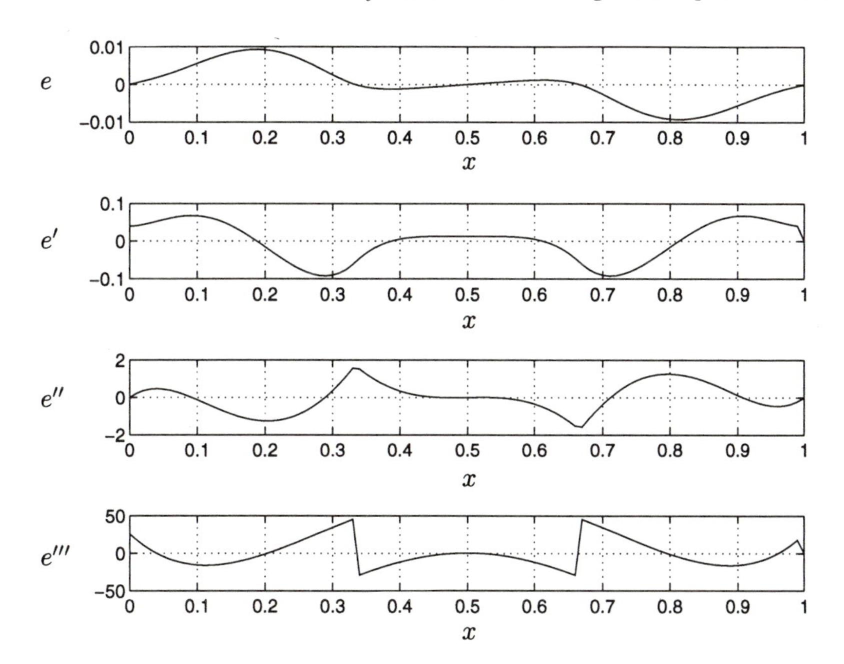

FIGURE 5.5. Errors in the approximation of a 4-5-6-7 polynomial with a natural cubic spline, using four supporting points.

With the values of the second derivatives at the supporting points known, the calculation of the spline coefficients A_k, B_k, C_k, and D_k, for $k = 1, \ldots, N$, is now straightforward. Let the interpolation error, $e(x)$, be defined as $e(x) \equiv s(x) - p(x)$, where $s(x)$ is the interpolating spline and $p(x)$ is the given polynomial. This error and its derivatives $e'(x)$, $e''(x)$, and $e'''(x)$ are plotted in Figs. 5.5 and 5.6 for $N = 3$ and $N = 10$, respectively. What we observe is an increase of more than one order of magnitude in the error as we increase the order of the derivative by one. Thus, the order of magnitude of acceleration errors is usually higher than two orders of magnitude above the displacement errors, a fact that should not be overlooked in applications.

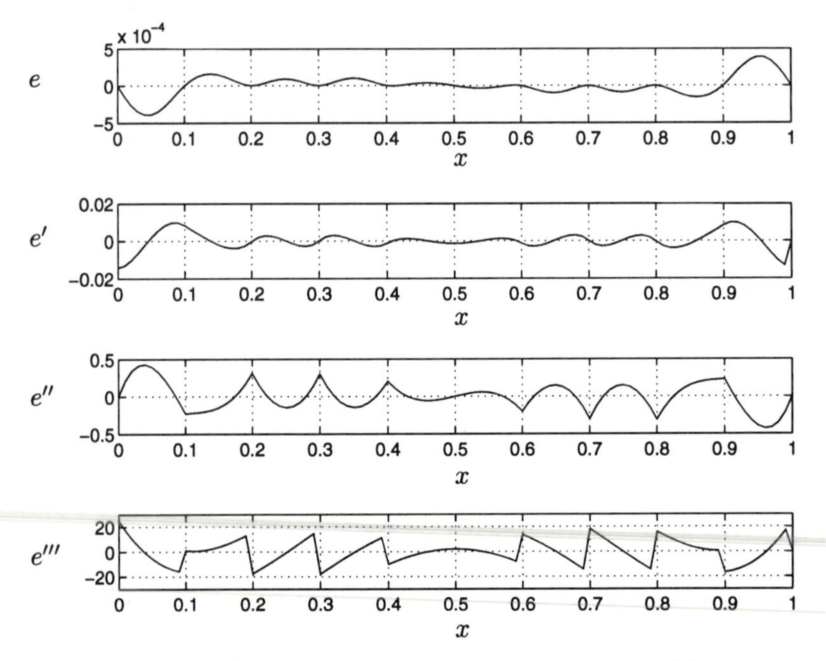

FIGURE 5.6. Errors in the approximation of a 4-5-6-7 polynomial with a natural cubic spline, using eleven supporting points.

6
Dynamics of Serial Robotic Manipulators

6.1 Introduction

The main objectives of this chapter are (i) to devise an algorithm for the real-time *computed torque control* and (ii) to derive the system of second-order ordinary differential equations (ODE) governing the motion of an n-axis manipulator. We will focus on serial manipulators, the dynamics of a much broader class of robotic mechanical systems, namely, parallel manipulators and mobile robots, being the subject of Chapter 10. Moreover, we will study mechanical systems with rigid links and rigid joints and will put aside systems with flexible elements, which pertain to a more specialized realm.

6.2 Inverse vs. Forward Dynamics

The two basic problems associated with the dynamics of robotic mechanical systems, namely, the *inverse* and the *forward* problems, are thoroughly discussed in this chapter. The relevance of these problems cannot be overstated: the former is essential for the computed-torque control of robotic manipulators, while the latter is required for the simulation and the real-time feedback control of the same systems. Because the inverse problem is purely algebraic, it is conceptually simpler to grasp than the forward problem, and hence, the inverse problem will be discussed first. Moreover, the inverse problem is also computationally simpler than the forward problem. In the inverse problem, a time-history of either the Cartesian or the joint coordinates is given, and from knowledge of these histories and the architecture and inertial parameters of the system at hand, the torque or force requirements at the different actuated joints are determined as time-histories as well. In the forward problem, current values of the joint coordinates and their first time-derivatives are known at a given instant, the time-histories of the applied torques or forces being also known, along with the architecture and the inertial parameters of the manipulator at hand. With the aforementioned data, the values of the joint coordinates and their time-derivatives are computed at a later sampling instant by integration of the underlying system of nonlinear ordinary differential equations.

The study of the dynamics of systems of multiple rigid bodies is classical, but up until the advent of the computer, it was limited only to theoretical results and a reduced number of bodies. First Uicker (1965) and then Kahn (1969) produced a method based on the Euler-Lagrange equations of mechanical systems of rigid bodies that they used to simulate the dynamical behavior of such systems. A breakthrough in the development of algorithms for dynamics computations was reported by Luh, Walker, and Paul (1980), who proposed a recursive formulation of multibody dynamics that is applicable to systems with serial kinematic chains. This formulation, based on the Newton-Euler equations of rigid bodies, allowed the calculation of the joint torques of a six-revolute manipulator with only 800 multiplications and 595 additions, a tremendous gain if we consider that the straightforward calculation of the Euler-Lagrange equations for the same type of manipulator involves 66,271 multiplications and 51,548 additions, as pointed out by Hollerbach (1980). In the aforementioned reference, a recursive derivation of the Euler-Lagrange equations was proposed whereby the computational complexity was reduced to only 2,195 multiplications and 1,719 additions.

The aforementioned results provoked a discussion on the merits and demerits of each of the Euler-Lagrange and the Newton-Euler formulations. Silver (1982) pointed out that since both formulations are equivalent, they should lead to the same computational complexity. In fact, Silver showed how to derive the Euler-Lagrange equations from the Newton-Euler formulation by following an approach first introduced by Kane (1961) in connection with nonholonomic systems. Kane and Levinson (1983) then showed how Kane's equations can be applied to particular robotic manipulators and arrived at lower computational complexities. They applied the said equations to the Stanford Arm (Paul, 1981) and computed its inverse dynamics with 646 multiplications and 394 additions. Thereafter, Khalil, Kleinfinger, and Gautier (1986) proposed a condensed recursive Newton-Euler method that reduced the computational complexity to 538 multiplications and 478 additions, for *arbitrary architectures*. Further developments in this area, summarized by Balafoutis and Patel (1991), have shown that the underlying computational complexity can be reduced to 489 multiplications and 420 additions for the most general case of a six-revolute manipulator, i.e., without exploiting particular features of the manipulator geometry. Balafoutis and Patel based their algorithm on tensor analysis, whereby tensor identities are exploited to their fullest extent in order to reduce the number of operations involved.

In this chapter, the inverse dynamics problem is solved with the well-known recursive Newton-Euler algorithm, while the direct dynamics problem is handled with a novel approach, based on the reciprocity relations between the *constraint wrenches* and the *feasible twists* of a manipulator. This technique is developed with the aid of a modeling tool known as *natural orthogonal complement*, thoroughly discussed in Section 6.5.

Throughout the chapter, we will follow a multibody system approach, which requires a review of the underlying fundamentals.

6.3 Fundamentals of Multibody System Dynamics

6.3.1 On Nomenclature and Basic Definitions

We consider here a mechanical system composed of r rigid bodies and denote by $\mathbf{M}_i$ the 6×6 *inertia dyad*—see Section 3.8—of the ith body. Moreover, we let $\mathbf{W}_i$, already introduced in eq.(3.145), be the *angular-velocity dyad* of the same body. As pertaining to the case at hand, the said matrices are displayed below:

$$\mathbf{M}_i \equiv \begin{bmatrix} \mathbf{I}_i & \mathbf{O} \\ \mathbf{O} & m_i\mathbf{1} \end{bmatrix}, \quad \mathbf{W}_i \equiv \begin{bmatrix} \mathbf{\Omega}_i & \mathbf{O} \\ \mathbf{O} & \mathbf{O} \end{bmatrix}, \quad i = 1, \ldots, r \tag{6.1}$$

where $\mathbf{1}$ and $\mathbf{O}$ denote the 3×3 identity and zero matrices, respectively, while $\mathbf{\Omega}_i$ and $\mathbf{I}_i$ are the angular-velocity and the inertia matrices of the ith body, this last being defined with respect to the mass center C_i of this body. Moreover, the mass of this body is denoted by m_i, whereas $\mathbf{c}_i$ and $\dot{\mathbf{c}}_i$ denote the position and the velocity vectors of C_i. Furthermore, let $\mathbf{t}_i$ denote the twist of the same body, the latter being defined in terms of the angular velocity vector $\boldsymbol{\omega}_i$, the vector of $\mathbf{\Omega}_i$, and the velocity of C_i. The 6-dimensional *momentum screw*, $\boldsymbol{\mu}_i$, is defined likewise. Furthermore, $\mathbf{w}_i^W$ and $\mathbf{w}_i^C$ are defined as the *working wrench* and the *nonworking constraint wrench* exerted on the ith body by its neighbors, in which forces are assumed to be applied at C_i. We thus have, for $i = 1, \ldots, r$,

$$\mathbf{t}_i = \begin{bmatrix} \boldsymbol{\omega}_i \\ \dot{\mathbf{c}}_i \end{bmatrix}, \quad \boldsymbol{\mu}_i = \begin{bmatrix} \mathbf{I}_i\boldsymbol{\omega}_i \\ m_i\dot{\mathbf{c}}_i \end{bmatrix}, \quad \mathbf{w}_i^W = \begin{bmatrix} \mathbf{n}_i^W \\ \mathbf{f}_i^W \end{bmatrix}, \quad \mathbf{w}_i^C = \begin{bmatrix} \mathbf{n}_i^C \\ \mathbf{f}_i^C \end{bmatrix} \tag{6.2}$$

where superscripted $\mathbf{n}_i$ and $\mathbf{f}_i$ stand, respectively, for the moment and the force acting on the ith body, the force being applied at the mass center C_i. Thus, whereas $\mathbf{w}_i^W$ accounts for forces and moments exerted by both the environment and the actuators, including driving forces as well as dissipative effects, $\mathbf{w}_i^C$, whose sole function is to keep the links together, accounts for those forces and moments exerted by the neighboring links, which do not produce any mechanical work. Therefore, friction wrenches applied by the $(i-1)$st and the $(i+1)$st links onto the ith link are not included in $\mathbf{w}_i^C$; rather, they are included in $\mathbf{w}_i^W$.

Clearly, from the definitions of $\mathbf{M}_i$, $\boldsymbol{\mu}_i$, and $\mathbf{t}_i$, we have

$$\boldsymbol{\mu}_i = \mathbf{M}_i \mathbf{t}_i \tag{6.3}$$

Moreover, from eq.(3.173),

$$\dot{\boldsymbol{\mu}}_i = \mathbf{M}_i \dot{\mathbf{t}}_i + \mathbf{W}_i \boldsymbol{\mu}_i = \mathbf{M}_i \dot{\mathbf{t}}_i + \mathbf{W}_i \mathbf{M}_i \mathbf{t}_i \tag{6.4}$$

We now recall the Newton-Euler equations for a rigid body, namely,

$$\mathbf{I}_i \dot{\boldsymbol{\omega}}_i = -\boldsymbol{\omega}_i \times \mathbf{I}_i \boldsymbol{\omega}_i + \mathbf{n}_i^W + \mathbf{n}_i^C \tag{6.5a}$$

$$m_i \ddot{\mathbf{c}}_i = \mathbf{f}_i^W + \mathbf{f}_i^C \tag{6.5b}$$

which can be written in compact form using the foregoing 6-dimensional twist and wrench arrays as well as the 6×6 inertia and angular-velocity dyads. We thus obtain the Newton-Euler equations of the ith body in the form

$$\mathbf{M}_i \dot{\mathbf{t}}_i = -\mathbf{W}_i \mathbf{M}_i \mathbf{t}_i + \mathbf{w}_i^W + \mathbf{w}_i^C \tag{6.5c}$$

6.3.2 *The Euler-Lagrange Equations of Serial Manipulators*

The Euler-Lagrange dynamical equations of a mechanical system are now recalled, and apply them to serial manipulators. Thus, the mechanical system at hand has n degrees of freedom, its n independent generalized coordinates being the n joint variables, which are stored in the n-dimensional vector $\boldsymbol{\theta}$. We thus have

$$\frac{d}{dt}\left(\frac{\partial T}{\partial \dot{\boldsymbol{\theta}}}\right) - \frac{\partial T}{\partial \boldsymbol{\theta}} = \boldsymbol{\phi} \tag{6.6}$$

where T is a scalar function denoting the *kinetic energy* of the system and $\boldsymbol{\phi}$ is the n-dimensional vector of *generalized force*. If some forces on the right-hand side stem from a potential V, we can, then decompose $\boldsymbol{\phi}$ into two parts, $\boldsymbol{\phi}_p$ and $\boldsymbol{\phi}_n$, the former arising from V and termed the *conservative force* of the system; the latter is the *nonconservative force* $\boldsymbol{\phi}_n$. That is,

$$\boldsymbol{\phi}_p \equiv -\frac{\partial V}{\partial \boldsymbol{\theta}} \tag{6.7}$$

the above Euler-Lagrange equations thus becoming

$$\frac{d}{dt}\left(\frac{\partial L}{\partial \dot{\boldsymbol{\theta}}}\right) - \frac{\partial L}{\partial \boldsymbol{\theta}} = \boldsymbol{\phi}_n \tag{6.8}$$

where L is the *Lagrangian* of the system, defined as

$$L \equiv T - V \tag{6.9}$$

Moreover, the kinetic energy of the system is simply the sum of the kinetic energies of all the r links. Recalling eq.(3.150), which gives the kinetic energy of a rigid body in terms of 6-dimensional arrays, one has

$$T = \sum_1^r T_i = \sum_1^r \frac{1}{2}\mathbf{t}_i^T \mathbf{M}_i \mathbf{t}_i \tag{6.10}$$

whereas the vector of nonconservative generalized forces is given by

$$\boldsymbol{\phi}_n \equiv \frac{\partial \Pi^A}{\partial \dot{\boldsymbol{\theta}}} - \frac{\partial \Delta}{\partial \dot{\boldsymbol{\theta}}} \tag{6.11}$$

in which Π^A and Δ denote the power supplied to the system and the *Rayleigh dissipation function*, or for brevity, the *dissipation function* of the system. The first of these items is discussed below; the latter is only outlined in this section but is discussed extensively in Section 6.8. First, the wrench $\mathbf{w}_i^W$ is decomposed into two parts, $\mathbf{w}_i^A$ and $\mathbf{w}_i^D$, the former being the wrench supplied by the actuators and the latter being the wrench that arises from viscous and Coulomb friction, the gravity wrench being not needed here because gravity effects are considered in the potential $V(\boldsymbol{\theta})$. We thus call $\mathbf{w}_i^A$ the *active wrench* and $\mathbf{w}_i^D$ the dissipative wrench. Here, the wrenches supplied by the actuators are assumed to be prescribed functions of time. Moreover, these wrenches are supplied by single-dof actuators in the form of forces along a line of action or moments in a given direction, both line and direction being fixed to the two bodies that are coupled by an active joint. Hence, the actuator-supplied wrenches are dependent on the posture of the manipulator as well, but not on its twist. That is, the actuator wrenches are functions of both the vector of generalized coordinates, or joint variables, and time, but not of the generalized speeds, or joint-rates. Forces dependent on the latter to be considered here are assumed to be all *dissipative*. As a consequence, they can be readily incorporated into the mathematical model at hand via the dissipation function, to be discussed in Section 6.8. Note that feedback control schemes require actuator forces that are functions not only of the generalized coordinates, but also of the generalized speeds. These forces or moments are most easily incorporated into the underlying mathematical model, once this model is derived in the state-variable space, i.e., in the space of generalized coordinates and generalized speeds.

Thus, the power supplied to the ith link, Π_i^A, is readily computed as

$$\Pi_i^A = (\mathbf{w}_i^A)^T \mathbf{t}_i \tag{6.12a}$$

Similar to the kinetic energy, then, the power supplied to the overall system is simply the sum of the individual powers supplied to each link, and expressed as in eq.(6.12a), i.e.,

$$\Pi^A \equiv \sum_1^r \Pi_i^A \tag{6.12b}$$

Further definitions are now introduced. These are the $6n$-dimensional vectors of *manipulator twist*, $\mathbf{t}$; *manipulator momentum*, $\boldsymbol{\mu}$; *manipulator constraint wrench*, $\mathbf{w}^C$; *manipulator active wrench*, $\mathbf{w}^A$; and *manipulator dissipative wrench*, $\mathbf{w}^D$. Additionally, the $6n \times 6n$ matrices of *manipulator mass*, $\mathbf{M}$, and *manipulator angular velocity*, $\mathbf{W}$, are also introduced below:

$$\mathbf{t} = \begin{bmatrix} \mathbf{t}_1 \\ \vdots \\ \mathbf{t}_n \end{bmatrix}, \quad \mathbf{w}^C = \begin{bmatrix} \mathbf{w}_1^C \\ \vdots \\ \mathbf{w}_n^C \end{bmatrix}, \quad \mathbf{w}^A = \begin{bmatrix} \mathbf{w}_1^A \\ \vdots \\ \mathbf{w}_n^A \end{bmatrix}, \quad \mathbf{w}^D = \begin{bmatrix} \mathbf{w}_1^D \\ \vdots \\ \mathbf{w}_n^D \end{bmatrix} \tag{6.13a}$$

$$\mathbf{M} = \text{diag}\,(\,\mathbf{M}_1, \,\ldots,\, \mathbf{M}_n\,), \qquad \mathbf{W} = \text{diag}\,(\,\mathbf{W}_1, \,\ldots,\, \mathbf{W}_n\,) \tag{6.13b}$$

It is now apparent that, from definitions (6.13a & 6.13b) and relation (6.3), we have

$$\boldsymbol{\mu} = \mathbf{M}\mathbf{t} \tag{6.14}$$

Moreover, from definitions (6.1),

$$\dot{\boldsymbol{\mu}} = \mathbf{M}\dot{\mathbf{t}} + \mathbf{W}\mathbf{M}\mathbf{t} \tag{6.15}$$

With the foregoing definitions, then, the kinetic energy of the manipulator takes on a simple form, namely,

$$T = \frac{1}{2}\mathbf{t}^T\mathbf{M}\mathbf{t} \equiv \frac{1}{2}\mathbf{t}^T\boldsymbol{\mu} \tag{6.16}$$

which is a quadratic form in the system twist. Moreover, the twists are linear expressions in the velocity variables $\dot{\mathbf{c}}_i$ and $\boldsymbol{\omega}_i$. As a consequence, the twists are bound to be linear in the vector of generalized speeds, or joint-rates, $\dot{\boldsymbol{\theta}}$. Therefore, the kinetic energy is bound to be a quadratic form in the joint-rates, i.e., a symmetric, positive-definite $n \times n$ matrix $\mathbf{I}$ exists, independent of the joint rates, although in general, dependent on the generalized coordinates, or joint variables, $\boldsymbol{\theta}$, such that

$$T = \frac{1}{2}\dot{\boldsymbol{\theta}}^T\mathbf{I}(\boldsymbol{\theta})\dot{\boldsymbol{\theta}} + \mathbf{h}(\boldsymbol{\theta}, t)^T\dot{\boldsymbol{\theta}} + T_0(t) \tag{6.17}$$

the $n \times n$ matrix $\mathbf{I}$ being known as the *generalized inertia matrix.* We will study the inertia matrix in greater detail when deriving the Euler-Lagrange equations of a manipulator using the natural orthogonal complement in Section 6.6. What we can readily conclude is that the kinetic energy being *intrinsically positive*, the generalized inertia matrix is *positive-definite.* Obviously, $\mathbf{I}$ is symmetric as well. Moreover, note that one term linear in the joint rates and one independent of these are included in the above form of T, the latter accounting for motions of the robotic system that are not affected by the dynamics of the system. These motions would occur, for example, if a robot were mounted on a space station rotating at a uniform speed about an axis fixed in an inertial frame. This rotation would be intended to provide an acceleration field emulating gravity. We will consider in this chapter mainly manipulators mounted on an inertial base, which are characterized by a *homogeneous* kinetic energy, i.e., by $\mathbf{h} = \mathbf{0}$ and $T_0 = 0$ in eq.(6.17). Hence, the kinetic energy will be assumed of the form

$$T = \frac{1}{2}\dot{\boldsymbol{\theta}}^T \mathbf{I}(\boldsymbol{\theta})\dot{\boldsymbol{\theta}} \tag{6.18}$$

whence it is apparent that

$$\mathbf{I}(\boldsymbol{\theta}) = \frac{\partial^2}{\partial \dot{\boldsymbol{\theta}}^2}(T) \tag{6.19}$$

which means that the $n \times n$ generalized inertia matrix is the *Hessian* matrix of the kinetic energy with respect to the vector of generalized speed.

Furthermore, the Euler-Lagrange equations can be written in the form

$$\frac{d}{dt}\left(\frac{\partial T}{\partial \dot{\boldsymbol{\theta}}}\right) - \frac{\partial T}{\partial \boldsymbol{\theta}} + \frac{\partial V}{\partial \boldsymbol{\theta}} = \boldsymbol{\phi}_n \tag{6.20a}$$

Now, from the form of T given in eq.(6.18), the partial derivatives appearing in the foregoing equation take the forms derived below:

$$\frac{\partial T}{\partial \dot{\boldsymbol{\theta}}} = \mathbf{I}(\boldsymbol{\theta})\dot{\boldsymbol{\theta}}$$

and hence,

$$\frac{d}{dt}\left(\frac{\partial T}{\partial \dot{\boldsymbol{\theta}}}\right) = \mathbf{I}(\boldsymbol{\theta})\ddot{\boldsymbol{\theta}} + \dot{\mathbf{I}}(\boldsymbol{\theta}, \dot{\boldsymbol{\theta}})\dot{\boldsymbol{\theta}} \tag{6.20b}$$

Moreover, in order to calculate the second term of the left-hand side of eq.(6.20a), we express the kinetic energy in the form

$$T = \frac{1}{2}\mathbf{p}(\boldsymbol{\theta}, \dot{\boldsymbol{\theta}})^T \dot{\boldsymbol{\theta}} \tag{6.20c}$$

where $\mathbf{p}(\boldsymbol{\theta}, \dot{\boldsymbol{\theta}})$ is the *generalized momentum* of the manipulator, defined as

$$\mathbf{p}(\boldsymbol{\theta}, \dot{\boldsymbol{\theta}}) \equiv \mathbf{I}(\boldsymbol{\theta})\dot{\boldsymbol{\theta}} \tag{6.20d}$$

Hence,

$$\frac{\partial T}{\partial \boldsymbol{\theta}} = \frac{1}{2}\left(\frac{\partial \mathbf{p}}{\partial \boldsymbol{\theta}}\right)^T \dot{\boldsymbol{\theta}} \tag{6.20e}$$

or

$$\frac{\partial T}{\partial \boldsymbol{\theta}} = \frac{1}{2}\left[\frac{\partial (\mathbf{I}\dot{\boldsymbol{\theta}})}{\partial \boldsymbol{\theta}}\right]^T \dot{\boldsymbol{\theta}} \tag{6.20f}$$

the Euler-Lagrange equations thus taking on the alternative form

$$\mathbf{I}(\boldsymbol{\theta})\ddot{\boldsymbol{\theta}} + \dot{\mathbf{I}}(\boldsymbol{\theta}, \dot{\boldsymbol{\theta}})\dot{\boldsymbol{\theta}} - \frac{1}{2}\left[\frac{\partial (\mathbf{I}\dot{\boldsymbol{\theta}})}{\partial \boldsymbol{\theta}}\right]^T \dot{\boldsymbol{\theta}} + \frac{\partial V}{\partial \boldsymbol{\theta}} = \boldsymbol{\phi}_n \tag{6.21}$$

Example 6.3.1 (Euler-Lagrange equations of a planar manipulator) *Consider the manipulator of Fig. 6.1, with links designed so that their mass centers, C_1, C_2, and C_3, are located at the midpoints of segments O_1O_2, O_2O_3, and O_3P, respectively. Moreover, the ith link has a mass m_i and a centroidal moment of inertia in a direction normal to the plane of motion I_i, while the joints are actuated by motors delivering torques τ_1, τ_2, and τ_3, the lubricant of the joints producing dissipative torques that we will neglect in this model. Under the assumption that gravity acts in the direction of $-Y$, find the associated Euler-Lagrange equations.*

Solution: Here we recall the kinematic analysis of Section 4.8 and the definitions introduced therein for the analysis of planar motion. In this light, all vectors introduced below are 2-dimensional, the scalar angular velocities of the links, ω_i, for $i = 1, 2, 3$, being

$$\omega_1 = \dot{\theta}_1, \quad \omega_2 = \dot{\theta}_1 + \dot{\theta}_2, \quad \omega_3 = \dot{\theta}_1 + \dot{\theta}_2 + \dot{\theta}_3$$

Moreover, the velocities of the mass centers are

$$\dot{\mathbf{c}}_1 = \frac{1}{2}\dot{\theta}_1 \mathbf{E}\mathbf{a}_1$$

$$\dot{\mathbf{c}}_2 = \dot{\theta}_1 \mathbf{E}\mathbf{a}_1 + \frac{1}{2}(\dot{\theta}_1 + \dot{\theta}_2)\mathbf{E}\mathbf{a}_2$$

$$\dot{\mathbf{c}}_3 = \dot{\theta}_1 \mathbf{E}\mathbf{a}_1 + (\dot{\theta}_1 + \dot{\theta}_2)\mathbf{E}\mathbf{a}_2 + \frac{1}{2}(\dot{\theta}_1 + \dot{\theta}_2 + \dot{\theta}_3)\mathbf{E}\mathbf{a}_3$$

the kinetic energy then becoming

$$T = \frac{1}{2}\sum_1^3 (m_i \|\dot{\mathbf{c}}_i\|^2 + I_i \omega_i^2)$$

The squared magnitudes of the mass-center velocities are now computed using the expressions derived above. After simplifications, these yield

$$\|\dot{\mathbf{c}}_1\|^2 = \frac{1}{4}a_1^2\dot{\theta}_1^2$$

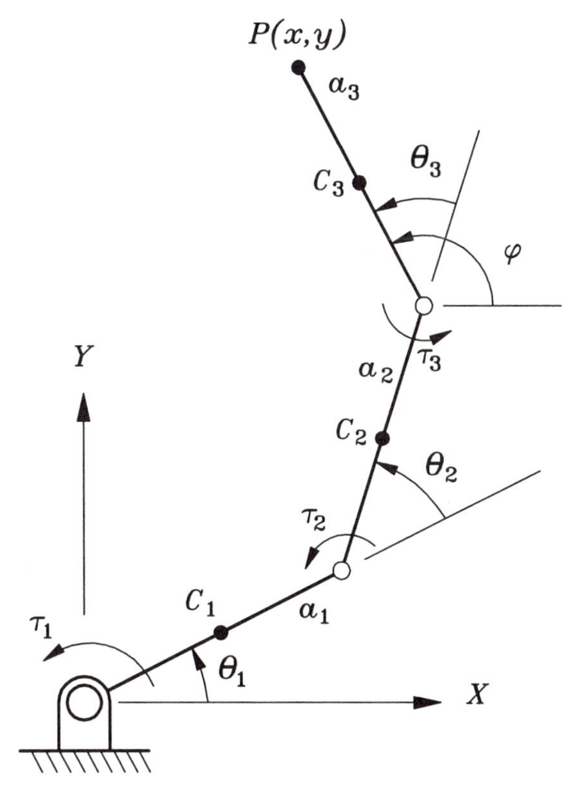

FIGURE 6.1. A planar manipulator.

$$\begin{aligned}
\|\dot{\mathbf{c}}_2\|^2 &= a_1^2\dot\theta_1^2 + \frac{1}{4}a_2^2(\dot\theta_1^2 + 2\dot\theta_1\dot\theta_2 + \dot\theta_2^2) + a_1a_2\cos\theta_2(\dot\theta_1^2 + \dot\theta_1\dot\theta_2) \\
\|\dot{\mathbf{c}}_3\|^2 &= a_1^2\dot\theta_1^2 + a_2^2(\dot\theta_1^2 + 2\dot\theta_1\dot\theta_2 + \dot\theta_2^2) \\
&\quad + \frac{1}{4}a_3^2(\dot\theta_1^2 + \dot\theta_2^2 + \dot\theta_3^2 + 2\dot\theta_1\dot\theta_2 + 2\dot\theta_1\dot\theta_3 + 2\dot\theta_2\dot\theta_3) \\
&\quad + 2a_1a_2\cos\theta_2(\dot\theta_1^2 + \dot\theta_1\dot\theta_2) + a_1a_3\cos(\theta_2+\theta_3)(\dot\theta_1^2 + \dot\theta_1\dot\theta_2 + \dot\theta_1\dot\theta_3) \\
&\quad + a_2a_3\cos\theta_3(\dot\theta_1^2 + \dot\theta_2^2 + 2\dot\theta_1\dot\theta_2 + \dot\theta_1\dot\theta_3 + \dot\theta_2\dot\theta_3)
\end{aligned}$$

The kinetic energy of the whole manipulator thus becomes

$$T = \frac{1}{2}(I_{11}\dot\theta_1^2 + 2I_{12}\dot\theta_1\dot\theta_2 + 2I_{23}\dot\theta_2\dot\theta_3 + I_{22}\dot\theta_2^2 + 2I_{13}\dot\theta_1\dot\theta_3 + I_{33}\dot\theta_3^2)$$

with coefficients I_{ij}, for $i = 1, 2, 3$, and $j = i$ to 3 being the distinct entries of the 3×3 matrix of generalized inertia of the system. These entries are given below:

$$I_{11} \equiv I_1 + I_2 + I_3 + \frac{1}{4}m_1a_1^2 + m_2(a_1^2 + \frac{1}{4}a_2^2 + a_1a_2c_2)$$

$$+m_3(a_1^2 + a_2^2 + \frac{1}{4}a_3^2 + 2a_1a_2c_2 + a_1a_3c_{23} + a_2a_3c_3)$$

$$I_{12} \equiv I_2 + I_3 + \frac{1}{2}\left[m_2\left(\frac{1}{2}a_2^2 + a_1a_2c_2\right)\right.$$

$$\left. + m_3\left(2a_2^2 + \frac{1}{2}a_3^2 + 2a_1a_2c_2 + a_1a_3c_{23} + 2a_2a_3c_3\right)\right]$$

$$I_{13} \equiv I_3 + \frac{1}{2}\left(\frac{1}{2}a_3^2 + a_1a_3c_{23} + a_2a_3c_3\right)$$

$$I_{22} \equiv I_2 + I_3 + \frac{1}{4}m_2a_2^2 + m_3\left(a_2^2 + \frac{1}{4}a_3^2 + a_2a_3c_3\right)$$

$$I_{23} \equiv I_3 + \frac{1}{2}m_3\left(\frac{1}{2}a_3^2 + a_2a_3c_3\right)$$

$$I_{33} \equiv I_3 + \frac{1}{4}m_3a_3^2$$

where c_i and c_{ij} stand for $\cos\theta_i$ and $\cos(\theta_i + \theta_j)$, respectively. From the foregoing expressions, it is apparent that the generalized inertia matrix is not a function of θ_1, which is only natural, for if the second and third joints are locked while leaving the first one free, the whole manipulator becomes a single rigid body pivoting about point O_1. Now, the polar moment of inertia of a rigid body in planar motion about a fixed point is constant, and hence, the first joint variable should not affect the generalized inertia matrix.

Furthermore, the potential energy of the manipulator is computed as the sum of the individual link potential energies, i.e.,

$$V = \frac{1}{2}m_1ga_1\sin\theta_1 + m_2g\left[a_1\sin\theta_1 + \frac{1}{2}a_2\sin(\theta_1+\theta_2)\right]$$

$$+m_3g\left[a_1\sin\theta_1 + a_2\sin(\theta_1+\theta_2) + \frac{1}{2}a_3\sin(\theta_1+\theta_2+\theta_3)\right]$$

while the total power delivered to the manipulator takes the form

$$\Pi = \tau_1\dot\theta_1 + \tau_2\dot\theta_2 + \tau_3\dot\theta_3$$

We now proceed to compute the various terms in eq.(6.21). We already have $\mathbf{I}(\boldsymbol{\theta})$, but we do not have, as yet, its time-derivative. However, the entries of $\dot{\mathbf{I}}$ are merely the time-derivatives of the entries of $\mathbf{I}$. From the above expressions for these entries, their time-rates of change are readily calculated, namely,

$$\dot I_{11} = -m_2a_1a_2s_2\dot\theta_2 - m_3[2a_1a_2s_2\dot\theta_2 + a_1a_3s_{23}(\dot\theta_2+\dot\theta_3) + a_2a_3s_3\dot\theta_3]$$

$$\dot I_{12} = \frac{1}{2}\{-m_2a_1a_2s_2\dot\theta_2 - m_3[2a_1a_2s_2\dot\theta_2 + a_1a_3s_{23}(\dot\theta_2+\dot\theta_3) + 2a_2a_3s_3\dot\theta_3]\}$$

$$\dot I_{13} = -\frac{1}{2}m_3[a_1a_3s_{23}(\dot\theta_2+\dot\theta_3) + a_2a_3s_3\dot\theta_3]$$

$$\dot{I}_{22} = -m_3 a_2 a_3 s_3 \dot{\theta}_3$$
$$\dot{I}_{23} = -\frac{1}{2} m_3 a_2 a_3 s_3 \dot{\theta}_3$$
$$\dot{I}_{33} = 0$$

with s_{ij} defined as $\sin(\theta_i + \theta_j)$. It should now be apparent that the time-rate of change of the generalized inertia matrix is independent of $\dot{\theta}_1$, as one should have expected, for this matrix is independent of θ_1. That is, if all joints but the first one are frozen, no matter how fast the first joint rotates, the manipulator moves as a single rigid body whose polar moment of inertia about O_1, the center of the first joint, is constant. As a matter of fact, I_{33} is constant for the same reason and $\dot{I}_{33}$ hence vanishes. We have, then,[1]

$$\dot{\mathbf{I}}\dot{\boldsymbol{\theta}} \equiv \boldsymbol{\iota} = \begin{bmatrix} \dot{I}_{11}\dot{\theta}_1 + \dot{I}_{12}\dot{\theta}_2 + \dot{I}_{13}\dot{\theta}_3 \\ \dot{I}_{12}\dot{\theta}_1 + \dot{I}_{22}\dot{\theta}_2 + \dot{I}_{23}\dot{\theta}_3 \\ \dot{I}_{13}\dot{\theta}_1 + \dot{I}_{23}\dot{\theta}_2 + \dot{I}_{33}\dot{\theta}_3 \end{bmatrix}$$

whose components, ι_i, for $i = 1, 2, 3$, are readily calculated as

$$\begin{aligned}
\iota_1 = &-[m_2 a_1 a_2 s_2 + m_3 a_1(2a_2 s_2 + a_3 s_{23})]\dot{\theta}_1\dot{\theta}_2 - m_3 a_3(a_1 s_{23} + a_2 s_3)\dot{\theta}_1\dot{\theta}_3 \\
&-\frac{1}{2}[m_2 a_1 a_2 s_2 + m_3 a_1(2a_2 s_2 + a_3 s_{23})]\dot{\theta}_2^2 - m_3 a_3(a_1 s_{23} + a_2 s_3)\dot{\theta}_2\dot{\theta}_3 \\
&-\frac{1}{2} m_3 a_3(a_1 s_{23} + a_2 s_3)\dot{\theta}_3^2 \\
\iota_2 = &-\frac{1}{2}[m_2 a_1 a_2 s_2 + m_3 a_1(2a_2 s_2 + a_3 s_{23})]\dot{\theta}_1\dot{\theta}_2 \\
&-\frac{1}{2} m_3 a_3(a_1 s_{23} + a_2 s_3)\dot{\theta}_1\dot{\theta}_3 - m_3 a_2 a_3 s_3 \dot{\theta}_2\dot{\theta}_3 - \frac{1}{2} m_3 a_2 a_3 s_3 \dot{\theta}_3^2 \\
\iota_3 = &-\frac{1}{2} m_3 a_1 a_3 s_{23}\dot{\theta}_1\dot{\theta}_3 - \frac{1}{2} m_3 a_3(a_1 s_{23} + a_2 s_3)\dot{\theta}_1\dot{\theta}_3 - \frac{1}{2} m_3 a_2 a_3 s_3 \dot{\theta}_2\dot{\theta}_3
\end{aligned}$$

The next term in the right-hand side of eq.(6.21) now requires the calculation of the partial derivatives of vector $\mathbf{I}\dot{\boldsymbol{\theta}}$ with respect to the joint variables, which are computed below. Let

$$\frac{\partial(\mathbf{I}\dot{\boldsymbol{\theta}})}{\partial\boldsymbol{\theta}} \equiv \mathbf{I}'$$

its entries being denoted by I'_{ij}. This matrix, in component form, is given by

$$\mathbf{I}' = \begin{bmatrix} 0 & I_{11,2}\dot{\theta}_1 + I_{12,2}\dot{\theta}_2 + I_{13,2}\dot{\theta}_3 & I_{11,3}\dot{\theta}_1 + I_{12,3}\dot{\theta}_2 + I_{13,3}\dot{\theta}_3 \\ 0 & I_{12,2}\dot{\theta}_1 + I_{22,2}\dot{\theta}_2 + I_{23,2}\dot{\theta}_3 & I_{12,3}\dot{\theta}_1 + I_{22,3}\dot{\theta}_2 + I_{23,3}\dot{\theta}_3 \\ 0 & I_{13,2}\dot{\theta}_1 + I_{23,2}\dot{\theta}_2 + I_{33,2}\dot{\theta}_3 & I_{13,3}\dot{\theta}_1 + I_{23,3}\dot{\theta}_2 + I_{33,3}\dot{\theta}_3 \end{bmatrix}$$

[1] ι is the Greek letter *iota* and denotes a vector; according to our notation, its components are ι_1, ι_2, and ι_3.

with the shorthand notation $I_{ij,k}$ indicating the partial derivative of I_{ij} with respect to θ_k. As the reader can verify, these entries are given as

$$\begin{aligned}
I'_{11} &= 0 \\
I'_{12} &= -[m_2a_1a_2s_2 + m_3(2a_1a_2s_2 + a_1a_3s_{23})]\dot{\theta}_1 \\
&\quad -\frac{1}{2}[m_2a_1a_2s_2 + m_3(2a_1a_2s_2 + a_1a_3s_{23})]\dot{\theta}_2 \\
&\quad -\frac{1}{2}m_3a_1a_3s_{23}\dot{\theta}_3 \\
I'_{13} &= -m_3(a_1a_3s_{23} + a_2a_3s_3)\dot{\theta}_1 - \frac{1}{2}m_3(a_1a_3s_{23} + 2a_2a_3s_3)\dot{\theta}_2 \\
&\quad -\frac{1}{2}m_3(a_1a_3s_{23} + a_2a_3s_3)\dot{\theta}_3 \\
I'_{21} &= 0 \\
I'_{22} &= -\frac{1}{2}[m_2a_1a_2s_2 + m_3(2a_1a_2s_2 + a_1a_3s_{23})]\dot{\theta}_1 \\
I'_{23} &= -\frac{1}{2}m_3(a_1a_3s_{23} + 2a_2a_3s_3)\dot{\theta}_1 - m_3a_2a_3s_3\dot{\theta}_2 - \frac{1}{2}m_3a_2a_3s_3\dot{\theta}_3 \\
I'_{31} &= 0 \\
I'_{32} &= -\frac{1}{2}m_3a_1a_3s_{23}\dot{\theta}_1 \\
I'_{33} &= -\frac{1}{2}m_3(a_1a_3s_{23} + a_2a_3s_3)\dot{\theta}_1 - \frac{1}{2}m_3a_2a_3s_3\dot{\theta}_2
\end{aligned}$$

Now, we define the 3-dimensional vector γ below:

$$\gamma \equiv \left[\frac{\partial(\mathbf{I}\dot{\boldsymbol{\theta}})}{\partial\boldsymbol{\theta}}\right]^T \dot{\boldsymbol{\theta}}$$

its three components, γ_i, for $i = 1, 2, 3$, being

$$\begin{aligned}
\gamma_1 &= 0 \\
\gamma_2 &= -[m_2a_1a_2s_2 + m_3(2a_1a_2s_2 + a_1a_3s_{23})]\dot{\theta}_1^2 \\
&\quad -[m_2a_1a_2s_2 + m_3(2a_1a_2s_2 + a_1a_3s_{23})]\dot{\theta}_1\dot{\theta}_2 \\
&\quad -m_3a_1a_3s_{23}\dot{\theta}_1\dot{\theta}_3 \\
\gamma_3 &= -m_3(a_1a_3s_{23} + a_2a_3s_3)\dot{\theta}_1^2 - m_3(a_1a_3s_{23} + 2a_2a_3s_3)\dot{\theta}_1\dot{\theta}_2 \\
&\quad -m_3(a_1a_3s_{23} + a_2a_3s_3)\dot{\theta}_1\dot{\theta}_3 - m_3a_2a_3s_3\dot{\theta}_3^2 - m_3a_2a_3s_3\dot{\theta}_2\dot{\theta}_3
\end{aligned}$$

We now turn to the computation of the partial derivatives of the potential energy:

$$\begin{aligned}
\frac{\partial V}{\partial\theta_1} &= \frac{1}{2}m_1ga_1c_1 + m_2g(a_1c_1 + \frac{1}{2}a_2c_{12}) + m_3g(a_1c_1 + a_2c_{12} + \frac{1}{2}a_3c_{123}) \\
\frac{\partial V}{\partial\theta_2} &= \frac{1}{2}m_2ga_2 + m_3g(a_2c_{12} + \frac{1}{2}a_3c_{123})
\end{aligned}$$

$$\frac{\partial V}{\partial \theta_3} = \frac{1}{2} m_3 g a_3 c_{123}$$

The Euler-Lagrange equations thus reduce to

$$I_{11}\ddot{\theta}_1 + I_{12}\ddot{\theta}_2 + I_{13}\ddot{\theta}_3 + \iota_1 - \frac{1}{2}\gamma_1 + \frac{1}{2}m_1 g a_1 c_1 + m_2 g(a_1 c_1 + \frac{1}{2}a_2 c_{12}) + m_3 g(a_1 c_1 + a_2 c_{12} + \frac{1}{2}a_3 c_{123}) = \tau_1$$

$$I_{12}\ddot{\theta}_1 + I_{22}\ddot{\theta}_2 + I_{23}\ddot{\theta}_3 + \iota_2 - \frac{1}{2}\gamma_2 + \frac{1}{2}m_2 g a_2 c_{12} + m_3 g(a_2 c_{12} + \frac{1}{2}a_3 c_{123}) = \tau_2$$

$$I_{13}\ddot{\theta}_1 + I_{23}\ddot{\theta}_2 + I_{33}\ddot{\theta}_3 + \iota_3 - \frac{1}{2}\gamma_3 + \frac{1}{2}m_3 g a_3 c_{123} = \tau_3$$

With this example, it becomes apparent that a straightforward differentiation procedure to derive the Euler-Lagrange equations of a robotic manipulator, or for that matter, of a mechanical system at large, is not practical. For example, these equations do not seem to lend themselves to symbolic manipulations for a six-axis manipulator of arbitrary architecture, given that they become quite cumbersome even for a three-axis planar manipulator with an architecture that is not so general. For this reason, procedures have been devised that lend themselves to an algorithmic treatment. We will study a procedure based on the *natural orthogonal complement* whereby the underlying equations are derived using matrix-times-vector multiplications.

6.4 Recursive Inverse Dynamics

The inverse dynamics problem associated with serial manipulators is studied here. We assume at the outset that the manipulator under study is of the serial type with $n+1$ links including the base link and n joints of either the revolute or the prismatic type.

The underlying algorithm consists of two steps: (i) *kinematic computations*, required to determine the twists of all the links and their time derivatives in terms of $\boldsymbol{\theta}$, $\dot{\boldsymbol{\theta}}$, and $\ddot{\boldsymbol{\theta}}$; and ($ii$) *dynamic computations*, required to determine both the constraint and the external wrenches. Each of these steps is described below, the aim here being to calculate the desired variables with as few computations as possible, for one purpose of inverse dynamics is to permit the real-time model-based control of the manipulator. Real-time performance requires, obviously, a low number of computations. For the sake of simplicity, we decided against discussing the algorithms with the lowest computational cost, mainly because these algorithms, fully discussed by Balafoutis and Patel (1991), rely heavily on tensor calculus, which we have not studied here. Henceforth, revolute joints are referred to as R, prismatic joints as P.

6.4.1 Kinematics Computations: Outward Recursions

We will use the Denavit-Hartenberg (DH) notation introduced in Section 4.2 and hence will refer to Fig. 4.7 for the basic notation required for the kinematic analysis to be described first. Note that the calculation of each $\mathbf{Q}_i$ matrix, as given by eq.(4.1d), requires four multiplications and zero additions.

Moreover, *every 3-dimensional vector-component transfer from the $\mathcal{F}_i$ frame to the $\mathcal{F}_{i+1}$ frame requires a multiplication by $\mathbf{Q}_i^T$. Likewise, every component transfer from the $\mathcal{F}_{i+1}$ frame to the $\mathcal{F}_i$ frame requires a multiplication by $\mathbf{Q}_i$.* Therefore, we will need to account for the aforementioned component transfers, which we will generically term *coordinate transformations* between successive coordinate frames. We derive below the number of operations required for such transformations. If we have $[\mathbf{r}]_i \equiv [r_1, r_2, r_3]^T$ and we need $[\mathbf{r}]_{i+1}$, then we proceed as follows:

$$[\mathbf{r}]_{i+1} = \mathbf{Q}_i^T[\mathbf{r}]_i \tag{6.22}$$

and if we recall the form of $\mathbf{Q}_i$ from eq.(4.1d), we then have

$$[\mathbf{r}]_{i+1} = \begin{bmatrix} \cos\theta_i & \sin\theta_i & 0 \\ -\lambda_i\sin\theta_i & \lambda_i\cos\theta_i & \mu_i \\ \mu_i\sin\theta_i & -\mu_i\cos\theta_i & \lambda_i \end{bmatrix} \begin{bmatrix} r_1 \\ r_2 \\ r_3 \end{bmatrix} = \begin{bmatrix} r_1\cos\theta_i + r_2\sin\theta_i \\ -\lambda_i r + \mu_i r_3 \\ \mu_i r + \lambda_i r_3 \end{bmatrix} \tag{6.23a}$$

where $\lambda_i \equiv \cos\alpha_i$ and $\mu_i \equiv \sin\alpha_i$, while

$$r \equiv r_1\sin\theta_i - r_2\cos\theta_i \tag{6.23b}$$

Likewise, if we have $[\mathbf{v}]_{i+1} \equiv [v_1, v_2, v_3]^T$ and we need $[\mathbf{v}]_i$, we use the component transformation given below:

$$[\mathbf{v}]_i = \begin{bmatrix} \cos\theta_i & -\lambda_i\sin\theta_i & \mu_i\sin\theta_i \\ \sin\theta_i & \lambda_i\cos\theta_i & -\mu_i\cos\theta_i \\ 0 & \mu_i & \lambda_i \end{bmatrix} \begin{bmatrix} v_1 \\ v_2 \\ v_3 \end{bmatrix} = \begin{bmatrix} v_1\cos\theta_i - v\sin\theta_i \\ v_1\sin\theta_i + v\cos\theta_i \\ v_2\mu_i + v_3\lambda_i \end{bmatrix} \tag{6.24a}$$

where

$$v \equiv v_2\lambda_i - v_3\mu_i \tag{6.24b}$$

It is now apparent that *every coordinate transformation between successive frames, whether forward or backward, requires eight multiplications and four additions.* Here, as in Chapter 4, we indicate the units of multiplications and additions with M and A, respectively.

The angular velocity and acceleration of the ith link are computed recursively as follows:

$$\boldsymbol{\omega}_i = \begin{cases} \boldsymbol{\omega}_{i-1} + \dot{\theta}_i \mathbf{e}_i, & \text{if the } i\text{th joint is } R \\ \boldsymbol{\omega}_{i-1}, & \text{if the } i\text{th joint is } P \end{cases} \tag{6.25a}$$

$$\dot{\boldsymbol{\omega}}_i = \begin{cases} \dot{\boldsymbol{\omega}}_{i-1} + \boldsymbol{\omega}_{i-1} \times \dot{\theta}_i \mathbf{e}_i + \ddot{\theta}_i \mathbf{e}_i, & \text{if the } i\text{th joint is } R \\ \dot{\boldsymbol{\omega}}_{i-1}, & \text{if the } i\text{th joint is } P \end{cases} \tag{6.25b}$$

for $i = 1, 2, \ldots, n$, where $\boldsymbol{\omega}_0$ and $\dot{\boldsymbol{\omega}}_0$ are the angular velocity and angular acceleration of the base link. Note that eqs.(6.25a & b) are frame-invariant; i.e., they are valid in *any* coordinate frame, as long as the same frame is used to represent all quantities involved. Below we derive the equivalent relations applicable when taking into account that quantities with a subscript i are available in $\mathcal{F}_{i+1}$-coordinates. Hence, operations involving quantities with different subscripts require a change of coordinates, which is taken care of by the corresponding rotation matrices.

In order to reduce the numerical complexity of the algorithm developed here, all vector and matrix quantities of the ith link will be expressed in $\mathcal{F}_{i+1}$. Note, however, that the two vectors $\mathbf{e}_i$ and $\mathbf{e}_{i+1}$ are fixed to the ith link, which is a potential source of confusion. Now, since $\mathbf{e}_i$ has very simple components in $\mathcal{F}_i$, namely, $[\,0,\ 0,\ 1\,]^T$, this will be regarded as a vector of the $(i-1)$st link. Therefore, this vector, or multiples of it, will be added to vectors bearing the $(i-1)$st subscript without any coordinate transformation. Moreover, subscripted brackets, as introduced in Section 2.2, can be avoided if all vector and matrix quantities subscripted with i, except for vector $\mathbf{e}_i$, are assumed to be expressed in $\mathcal{F}_{i+1}$. Furthermore, in view of the serial type of the underlying kinematic chain, only additions of quantities with two successive subscripts will appear in the relations below.

Quantities given in two successive frames can be added if both are expressed in the same frame, the obvious frame of choice being the frame of one of the two quantities. Hence, all we need to add two quantities with successive subscripts is to multiply one of these by a suitable orthogonal matrix. Additionally, in view of the *outwards* recursive nature of the kinematic relations above, it is apparent that a transfer from $\mathcal{F}_i$- to $\mathcal{F}_{i+1}$-coordinates is needed, which can be accomplished by multiplying either $\mathbf{e}_i$ or any other vector with the $(i-1)$ subscript by matrix $\mathbf{Q}_i^T$. Hence, the angular velocities and accelerations are computed recursively, as indicated below:

$$\boldsymbol{\omega}_i = \begin{cases} \mathbf{Q}_i^T(\boldsymbol{\omega}_{i-1} + \dot{\theta}_i \mathbf{e}_i), & \text{if the } i\text{th joint is } R \\ \mathbf{Q}_i^T \boldsymbol{\omega}_{i-1}, & \text{if the } i\text{th joint is } P \end{cases} \tag{6.26a}$$

$$\dot{\boldsymbol{\omega}}_i = \begin{cases} \mathbf{Q}_i^T(\dot{\boldsymbol{\omega}}_{i-1} + \boldsymbol{\omega}_{i-1} \times \dot{\theta}_i \mathbf{e}_i + \ddot{\theta}_i \mathbf{e}_i), & \text{if the } i\text{th joint is } R \\ \mathbf{Q}_i^T \dot{\boldsymbol{\omega}}_{i-1}, & \text{if the } i\text{th joint is } P \end{cases} \tag{6.26b}$$

If the base link is an inertial frame, then

$$\boldsymbol{\omega}_0 = \mathbf{0}, \qquad \dot{\boldsymbol{\omega}}_0 = \mathbf{0} \tag{6.27}$$

Thus, calculating each $\boldsymbol{\omega}_i$ vector in $\mathcal{F}_{i+1}$ when $\boldsymbol{\omega}_{i-1}$ is given in $\mathcal{F}_i$ requires $8M$ and $5A$ if the ith joint is R; if it is P, the said calculation reduces to $8M$ and $4A$. Here, note that $\dot{\theta}_i \mathbf{e}_i = [\,0,\, 0,\, \dot{\theta}_i\,]^T$ in $\mathcal{F}_i$-coordinates, and hence, the vector addition of the upper right-hand side of eq.(6.26a) requires only $1A$. Furthermore, in order to determine the number of operations required to calculate $\dot{\boldsymbol{\omega}}_i$ in $\mathcal{F}_{i+1}$ when $\dot{\boldsymbol{\omega}}_{i-1}$ is available in $\mathcal{F}_i$, we note that

$$[\,\mathbf{e}_i\,]_i = \begin{bmatrix} 0 \\ 0 \\ 1 \end{bmatrix} \tag{6.28}$$

Moreover, we let

$$[\,\boldsymbol{\omega}_{i-1}\,]_i = \begin{bmatrix} \omega_x \\ \omega_y \\ \omega_z \end{bmatrix} \tag{6.29}$$

Hence,

$$[\,\boldsymbol{\omega}_{i-1} \times \dot{\theta}_i \mathbf{e}_i\,]_i = \begin{bmatrix} \dot{\theta}_i\, \omega_y \\ -\dot{\theta}_i\, \omega_x \\ 0 \end{bmatrix} \tag{6.30}$$

Furthermore, we note that

$$[\,\ddot{\theta}_i \mathbf{e}_i\,]_i = \begin{bmatrix} 0 \\ 0 \\ \ddot{\theta}_i \end{bmatrix} \tag{6.31}$$

and hence, the calculation of $\dot{\boldsymbol{\omega}}_i$ in $\mathcal{F}_{i+1}$ when $\dot{\boldsymbol{\omega}}_{i-1}$ is given in $\mathcal{F}_i$ requires $10M$ and $7A$ if the ith joint is R; if it is P, the same calculation requires $8M$ and $4A$.

Furthermore, let $\mathbf{c}_i$ be the position vector of C_i, the mass center of the ith link, $\boldsymbol{\rho}_i$ being the vector directed from O_i to C_i, as shown in Figs. 6.2 and 6.3. The position vectors of two successive mass centers thus observe the relationships

(*i*) if the ith joint is R,

$$\boldsymbol{\delta}_{i-1} \equiv \mathbf{a}_{i-1} - \boldsymbol{\rho}_{i-1} \tag{6.32a}$$

$$\mathbf{c}_i = \mathbf{c}_{i-1} + \boldsymbol{\delta}_{i-1} + \boldsymbol{\rho}_i \tag{6.32b}$$

(*ii*) if the ith joint is P,

$$\boldsymbol{\delta}_{i-1} \equiv \mathbf{d}_{i-1} - \boldsymbol{\rho}_{i-1} \tag{6.32c}$$

$$\mathbf{c}_i = \mathbf{c}_{i-1} + \boldsymbol{\delta}_{i-1} + b_i \mathbf{e}_i + \boldsymbol{\rho}_i \tag{6.32d}$$

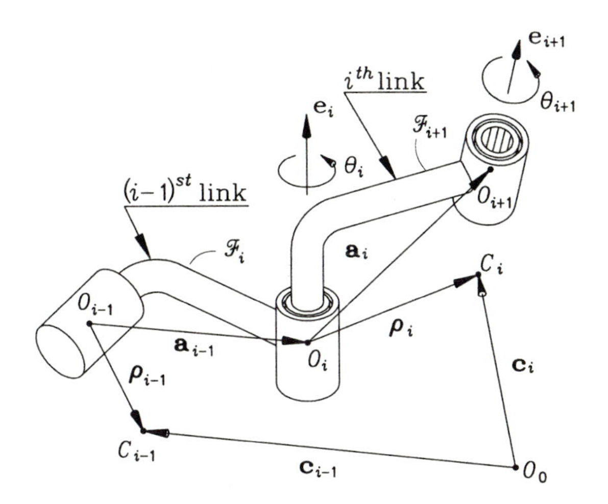

FIGURE 6.2. A revolute joint.

where point O_i, in this case, is a point of the $(i-1)$st link conveniently defined, as dictated by the particular geometry of the manipulator at hand. The foregoing freedom in the choice of O_i is a consequence of prismatic pairs having only a defined direction but no axis, properly speaking.

Notice that in the presence of a revolute pair at the ith joint, the difference $\mathbf{a}_{i-1} - \boldsymbol{\rho}i - 1$ is constant in $\mathcal{F}_i$. Likewise, in the presence of a prismatic pair at the same joint, the difference $\mathbf{d}_{i-1} - \boldsymbol{\rho}i - 1$ is constant in $\mathcal{F}_i$. Therefore, these differences are computed off-line, their evaluation not counting toward the computational complexity of the algorithm.

Upon differentiation of both sides of eqs.(6.32b & d) with respect to time, we derive the corresponding relations between the velocities and accelerations of the mass centers of links $i-1$ and i, namely,

(i) if the ith joint is R,

$$\dot{\mathbf{c}}_i = \dot{\mathbf{c}}_{i-1} + \boldsymbol{\omega}_{i-1} \times \boldsymbol{\delta}_{i-1} + \boldsymbol{\omega}_i \times \boldsymbol{\rho}_i \tag{6.33a}$$

$$\ddot{\mathbf{c}}_i = \ddot{\mathbf{c}}_{i-1} + \dot{\boldsymbol{\omega}}_{i-1} \times \boldsymbol{\delta}_{i-1} + \boldsymbol{\omega}_{i-1} \times (\boldsymbol{\omega}_{i-1} \times \boldsymbol{\delta}_{i-1}) + \dot{\boldsymbol{\omega}}_i \times \boldsymbol{\rho}_i + \boldsymbol{\omega}_i \times (\boldsymbol{\omega}_i \times \boldsymbol{\rho}_i) \tag{6.33b}$$

(ii) if the ith joint is P,

$$\boldsymbol{\omega}_i = \boldsymbol{\omega}_{i-1} \tag{6.34a}$$

$$\dot{\boldsymbol{\omega}}_i = \dot{\boldsymbol{\omega}}_{i-1} \tag{6.34b}$$

$$\mathbf{u}_i \equiv \boldsymbol{\delta}_{i-1} + \boldsymbol{\rho}_i + b_i \mathbf{e}_i \tag{6.34c}$$

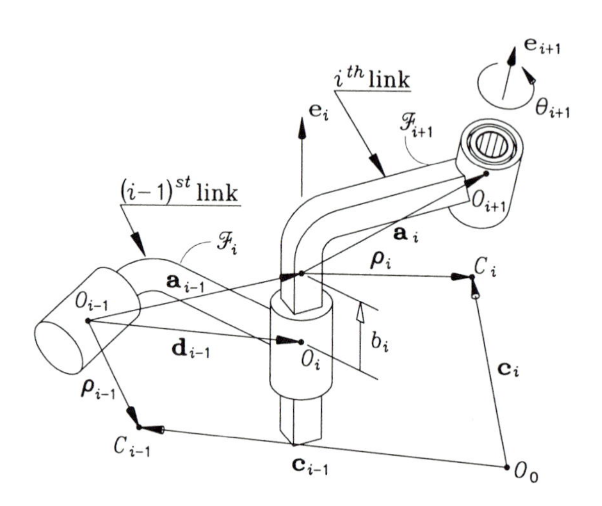

FIGURE 6.3. A prismatic joint.

$$\mathbf{v}_i \equiv \boldsymbol{\omega}_i \times \mathbf{u}_i \tag{6.34d}$$

$$\dot{\mathbf{c}}_i = \dot{\mathbf{c}}_{i-1} + \mathbf{v}_i + \dot{b}_i \mathbf{e}_i \tag{6.34e}$$

$$\ddot{\mathbf{c}}_i = \ddot{\mathbf{c}}_{i-1} + \dot{\boldsymbol{\omega}}_i \times \mathbf{u}_i + \boldsymbol{\omega}_i \times (\mathbf{v}_i + 2\dot{b}_i \mathbf{e}_i) + \ddot{b}_i \mathbf{e}_i \tag{6.34f}$$

for $i = 1, 2, \ldots, n$, where $\dot{\mathbf{c}}_0$ and $\ddot{\mathbf{c}}_0$ are the velocity and acceleration of the mass center of the base link. If the latter is an inertial frame, then

$$\boldsymbol{\omega}_0 = \mathbf{0}, \quad \dot{\boldsymbol{\omega}}_0 = \mathbf{0}. \quad \dot{\mathbf{c}}_0 = \mathbf{0}, \quad \ddot{\mathbf{c}}_0 = \mathbf{0} \tag{6.35}$$

Expressions (6.32b) to (6.34f) are *invariant*, i.e., they hold in *any* coordinate frame, as long as all vectors involved are expressed in that frame. However, we have vectors that are naturally expressed in the $\mathcal{F}_i$ frame added to vectors expressed in the $\mathcal{F}_{i+1}$ frame, and hence, a coordinate transformation is needed. This coordinate transformation is taken into account in Algorithm 6.4.1, whereby the logical variable `R` is `true` if the ith joint is R; otherwise it is `false`.

In performing the foregoing calculations, we need the cross product of a vector $\mathbf{w}$ times $\mathbf{e}_i$ in $\mathcal{F}_i$ coordinates, the latter being simply $[\mathbf{e}_i]_i = [0, 0, 1]^T$, and hence, this cross product reduces to $[w_2, -w_1, 0]^T$, whereby w_k, for $k = 1, 2, 3$, are the x, y, and z $\mathcal{F}_i$-components of $\mathbf{w}$. This cross product, then, requires no multiplications and no additions. Likewise, vectors $b_i\mathbf{e}_i$, $\dot{b}_i\mathbf{e}_i$, and $\ddot{b}_i\mathbf{e}_i$ take on the simple forms $[0, 0, b_i]^T$, $[0, 0, \dot{b}_i]^T$, and $[0, 0, \ddot{b}_i]^T$ in $\mathcal{F}_i$. Adding any of these vectors to any other vector in $\mathcal{F}_i$ then requires one single addition. If, moreover, we take into account that the cross product of two arbitrary vectors requires $6M$ and $3A$, we then have the operation counts given below:

Algorithm 6.4.1 (Outward Recursions):

read $\{\mathbf{Q}_i\}_0^{n-1}$, $\mathbf{c}_0$, $\boldsymbol{\omega}_0$, $\dot{\mathbf{c}}_0$, $\dot{\boldsymbol{\omega}}_0$, $\ddot{\mathbf{c}}_0$, $\{\boldsymbol{\rho}_i\}_1^n$, $\{\boldsymbol{\delta}_i\}_0^{n-1}$

For i = 1 to n step 1 do

update $\mathbf{Q}_i$

if R then

$$\mathbf{c}_i \leftarrow \mathbf{Q}_i^T(\mathbf{c}_{i-1} + \boldsymbol{\delta}_{i-1}) + \boldsymbol{\rho}_i$$
$$\boldsymbol{\omega}_i \leftarrow \mathbf{Q}_i^T(\boldsymbol{\omega}_{i-1} + \dot{\theta}_i \mathbf{e}_i)$$
$$\mathbf{u}_{i-1} \leftarrow \boldsymbol{\omega}_{i-1} \times \boldsymbol{\delta}_{i-1}$$
$$\mathbf{v}_i \leftarrow \boldsymbol{\omega}_i \times \boldsymbol{\rho}_i$$
$$\dot{\mathbf{c}}_i \leftarrow \mathbf{Q}_i^T(\dot{\mathbf{c}}_{i-1} + \mathbf{u}_{i-1}) + \mathbf{v}_i$$
$$\dot{\boldsymbol{\omega}}_i \leftarrow \mathbf{Q}_i^T(\dot{\boldsymbol{\omega}}_{i-1} + \boldsymbol{\omega}_{i-1} \times \dot{\theta}_i \mathbf{e}_i + \ddot{\theta}_i \mathbf{e}_i)$$
$$\ddot{\mathbf{c}}_i \leftarrow \mathbf{Q}_i^T(\ddot{\mathbf{c}}_{i-1} + \dot{\boldsymbol{\omega}}_{i-1} \times \boldsymbol{\delta}_{i-1} + \boldsymbol{\omega}_{i-1} \times \mathbf{u}_{i-1}) + \dot{\boldsymbol{\omega}}_i \times \boldsymbol{\rho}_i + \boldsymbol{\omega}_i \times \mathbf{v}_i$$

else

$$\mathbf{u}_i \leftarrow \mathbf{Q}_i^T \boldsymbol{\delta}_{i-1} + \boldsymbol{\rho}_i + b_i \mathbf{e}_i$$
$$\mathbf{c}_i \leftarrow \mathbf{Q}_i^T \mathbf{c}_{i-1} + \mathbf{u}_i$$
$$\boldsymbol{\omega}_i \leftarrow \mathbf{Q}_i^T \boldsymbol{\omega}_{i-1}$$
$$\mathbf{v}_i \leftarrow \boldsymbol{\omega}_i \times \mathbf{u}_i$$
$$\mathbf{w}_i \leftarrow \dot{b}_i \mathbf{e}_i$$
$$\dot{\mathbf{c}}_i \leftarrow \mathbf{Q}_i^T \dot{\mathbf{c}}_{i-1} + \mathbf{v}_i + \mathbf{w}_i$$
$$\dot{\boldsymbol{\omega}}_i \leftarrow \mathbf{Q}_i^T \dot{\boldsymbol{\omega}}_{i-1}$$
$$\ddot{\mathbf{c}}_i \leftarrow \mathbf{Q}_i^T \ddot{\mathbf{c}}_{i-1} + \dot{\boldsymbol{\omega}}_i \times \mathbf{u}_i + \boldsymbol{\omega}_i \times (\mathbf{v}_i + \mathbf{w}_i + \mathbf{w}_i) + \ddot{b}_i \mathbf{e}_i$$

endif

enddo

(*i*) If the ith joint is R,
$\mathbf{Q}_i$ requires $4M$ and $0A$
$\mathbf{c}_i$ requires $8M$ and $10A$
$\boldsymbol{\omega}_i$ requires $8M$ and $5A$
$\dot{\mathbf{c}}_i$ requires $20M$ and $16A$
$\dot{\boldsymbol{\omega}}_i$ requires $10M$ and $7A$
$\ddot{\mathbf{c}}_i$ requires $32M$ and $28A$

(*ii*) If the ith joint is P,
$\mathbf{Q}_i$ requires $4M$ and $0A$
$\mathbf{c}_i$ requires $16M$ and $15A$
$\boldsymbol{\omega}_i$ requires $8M$ and $4A$
$\dot{\mathbf{c}}_i$ requires $14M$ and $11A$
$\dot{\boldsymbol{\omega}}_i$ requires $8M$ and $4A$
$\ddot{\mathbf{c}}_i$ requires $20M$ and $19A$

TABLE 6.1. Complexity of the Kinematics Computations

Item	M	A
$\{\mathbf{Q}_i\}_1^n$	$4n$	0
$\{\mathbf{c}_i\}_1^n$	$8n$	$10n$
$\{\boldsymbol{\omega}_i\}_1^n$	$8n$	$5n$
$\{\dot{\mathbf{c}}_i\}_1^n$	$20n$	$16n$
$\{\dot{\boldsymbol{\omega}}_i\}_1^n$	$10n$	$7n$
$\{\ddot{\mathbf{c}}_i\}_1^n$	$32n$	$28n$
Total	$82n$	$66n$

The computational complexity for the forward recursions of the kinematics calculations for an n-revolute manipulator, as pertaining to various algorithms, are summarized in Table 6.1. Note that if some joints are P, then these figures become lower.

6.4.2 Dynamics Computations: Inward Recursions

A few additional definitions are first introduced. Let $\mathbf{w}_i^P$ denote the wrench exerted on the ith link by the $(i-1)$st link through contact at the ith kinematic pair. The moment and the force of this wrench are correspondingly defined as $\mathbf{n}_i^P$ and $\mathbf{f}_i^P$, the latter being applied at point O_i of the ith axis. A free-body diagram of the ith link is included in Fig. 6.4.

Moreover, a free-body diagram of the end-effector, or nth link, appears in Fig. 6.5. Note that this link is acted upon by a nonworking constraint

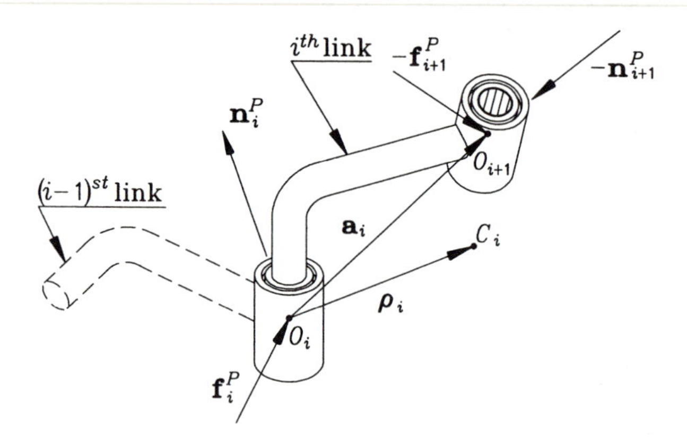

FIGURE 6.4. Free-body diagram of the ith link.

wrench, exerted through the nth pair, and a working wrench; the latter involves both active and dissipative forces and moments. Although dissipative forces and moments are difficult to model because of dry friction and striction, they can be readily incorporated into the dynamics model, once a suitable constitutive model for these items is available. Since these forces and moments depend only on joint variables and joint rates, they can be calculated once the kinematic variables are known. For the sake of simplicity, dissipative wrenches are not included here, their discussion being the subject of Section 6.8. Hence, the force and the moment that the $(i-1)$st link exerts on the ith link through the ith joint only produce nonworking constraint and active wrenches. That is, for a revolute pair, one has

$$\mathbf{n}_i^P = \begin{bmatrix} n_i^x \\ n_i^y \\ \tau_i \end{bmatrix}, \quad \mathbf{f}_i^P = \begin{bmatrix} f_i^x \\ f_i^y \\ f_i^z \end{bmatrix} \tag{6.36}$$

in which n_i^x and n_i^y are the nonzero $\mathcal{F}_i$-components of the nonworking constraint moment exerted by the $(i-1)$st link on the ith link; obviously, this moment lies in a plane perpendicular to Z_i, whereas τ_i is the active torque applied by the motor at the said joint. Vector $\mathbf{f}_i^P$ contains only nonworking constraint forces.

For a prismatic pair, one has

$$\mathbf{n}^P = \begin{bmatrix} n_i^x \\ n_i^y \\ n_i^z \end{bmatrix}, \quad \mathbf{f}^P = \begin{bmatrix} f_i^x \\ f_i^y \\ \tau_i \end{bmatrix} \tag{6.37}$$

where vector $\mathbf{n}_i^P$ contains only nonworking constraint torques, while τ_i is now the active force exerted by the ith motor in the Z_i direction, f_i^x and f_i^y being the nonzero $\mathcal{F}_i$-components of the nonworking constraint force exerted by the ith joint on the ith link, which is perpendicular to the Z_i axis.

In the algorithm below, the driving torques or forces $\{\tau_i\}_1^n$, are computed via vectors $\mathbf{n}_i^P$ and $\mathbf{f}_i^P$. In fact, in the case of a revolute pair, τ_i is simply the third component of $\mathbf{n}_i^P$; in the case of a prismatic pair, τ_i is, accordingly, the third component of $\mathbf{f}_i^P$. From Fig. 6.5, the Newton-Euler equations of the end-effector are

$$\mathbf{f}_n^P = m_n\ddot{\mathbf{c}}_n - \mathbf{f} \tag{6.38a}$$

$$\mathbf{n}_n^P = \mathbf{I}_n\dot{\boldsymbol{\omega}}_n + \boldsymbol{\omega}_n \times \mathbf{I}_n\boldsymbol{\omega}_n - \mathbf{n} + \boldsymbol{\rho}_n \times \mathbf{f}_n^P \tag{6.38b}$$

where $\mathbf{f}$ and $\mathbf{n}$ are the external force and moment, the former being applied at the mass center of the end-effector. The Newton-Euler equations for the remaining links are derived based on the free-body diagram of Fig. 6.4, namely,

$$\mathbf{f}_i^P = m_i\ddot{\mathbf{c}}_i + \mathbf{f}_{i+1}^P \tag{6.38c}$$

$$\mathbf{n}_i^P = \mathbf{I}_i\dot{\boldsymbol{\omega}}_i + \boldsymbol{\omega}_i \times \mathbf{I}_i\boldsymbol{\omega}_i + \mathbf{n}_{i+1}^P + \boldsymbol{\delta}_i \times \mathbf{f}_{i+1}^P + \boldsymbol{\rho}_i \times \mathbf{f}_i^P \tag{6.38d}$$

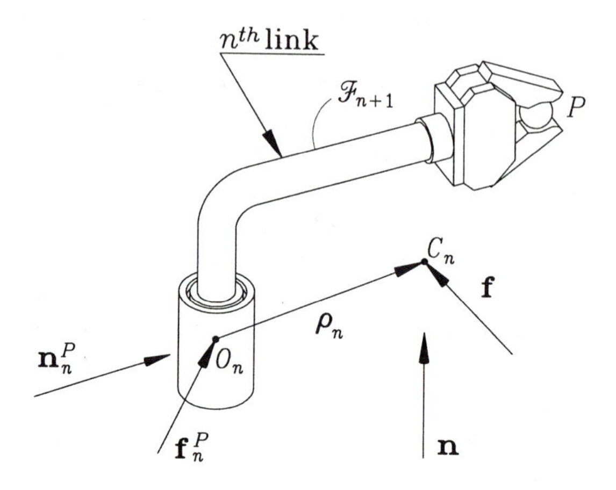

FIGURE 6.5. Free-body diagram of the end-effector.

with $\boldsymbol{\delta}_i$ defined as the difference $\mathbf{a}_i - \boldsymbol{\rho}_i$ in eqs.(6.32a & c).

Once the $\mathbf{n}_i^P$ and $\mathbf{f}_i^P$ vectors are available, the actuator torques and forces, denoted by τ_i, are readily computed. In fact, if the ith joint is a revolute, then

$$\tau_i = \mathbf{e}_i^T \mathbf{n}_i^P \tag{6.39}$$

which does not require any further operations, for τ_i reduces, in this case, to the Z_i component of vector $\mathbf{n}_i^P$. Similarly, if the ith joint is prismatic, then the corresponding actuator force reduces to

$$\tau_i = \mathbf{e}_i^T \mathbf{f}_i^P \tag{6.40}$$

Again, the foregoing relations are written in invariant form. In order to perform the computations involved, transformations that transfer coordinates between two successive frames are required. Here, we have to keep in mind that the components of a vector expressed in the $(i+1)$st frame can be transferred to the ith frame by multiplying the vector array in $(i+1)$st coordinates by matrix $\mathbf{Q}_i$. In taking these coordinate transformations into account, we derive the Newton-Euler algorithm from the above equations, namely,

Algorithm 6.4.2 (Inward Recursions):

$$
\begin{aligned}
\mathbf{f}_n^P &\leftarrow m_n\ddot{\mathbf{c}}_n - \mathbf{f} \\
\mathbf{n}_n^P &\leftarrow \mathbf{I}_n\dot{\boldsymbol{\omega}}_n + \boldsymbol{\omega}_n \times \mathbf{I}_n\boldsymbol{\omega}_n - \mathbf{n} + \boldsymbol{\rho}_n \times \mathbf{f}_n^P \\
\tau_n &\leftarrow (\mathbf{n}_n^P)_z
\end{aligned}
$$

For i = n − 1 to 1 step −1 do

$$
\begin{aligned}
\boldsymbol{\phi}_{i+1} &\leftarrow \mathbf{Q}_i\mathbf{f}_{i+1}^P \\
\mathbf{f}_i^P &\leftarrow m_i\ddot{\mathbf{c}}_i + \boldsymbol{\phi}_{i+1} \\
\mathbf{n}_i^P &\leftarrow \mathbf{I}_i\dot{\boldsymbol{\omega}}_i + \boldsymbol{\omega}_i \times \mathbf{I}_i\boldsymbol{\omega}_i + \boldsymbol{\rho}_i \times \mathbf{f}_i^P + \mathbf{Q}_i\mathbf{n}_{i+1}^P + \boldsymbol{\delta}_i \times \boldsymbol{\phi}_{i+1} \\
\tau_i &\leftarrow (\mathbf{n}_i^P)_z
\end{aligned}
$$

enddo

Note that, within the do-loop of the foregoing algorithm, the vectors to the left of the arrow are expressed in the ith frame, while $\mathbf{f}_{i+1}^P$ and $\mathbf{n}_{i+1}^P$, to the right of the arrow, are expressed in the $(i+1)$st frame. Moreover, if either the nth or the ith pair is prismatic, then the last statement both outside and inside the do-loop must be changed accordingly; one has, in that case,

$$\tau_n \leftarrow (\mathbf{f}_n^P)_z \tag{6.41a}$$

$$\tau_i \leftarrow (\mathbf{f}_i^P)_z \tag{6.41b}$$

In calculating the computational complexity of this algorithm, note that the $\mathbf{a}_i - \boldsymbol{\rho}_i$ term is constant in the $(i+1)$st frame, and hence, it is computed *off-line*. Thus, its computation need not be accounted for. A summary of computational costs is given in Table 6.2 for an n-revolute manipulator, with the row number indicating the step in Algorithm 6.4.2.

The total numbers of multiplications M_d and additions A_d required by the foregoing algorithm are readily obtained, with the result shown below:

$$M_d = 55n - 22, \quad A_d = 44n - 14 \tag{6.42}$$

TABLE 6.2. Complexity of Dynamics Computations

Row #	M	A
1	3	3
2	30	27
5	$8(n-1)$	$4(n-1)$
6	$3(n-1)$	$3(n-1)$
7	$44(n-1)$	$37(n-1)$
Total	$55n - 22$	$44n - 14$

In particular, for a six-revolute manipulator, one has

$$n = 6, \quad M_d = 308, \quad A_d = 250 \tag{6.43}$$

If the kinematics computations are accounted for, then the Newton-Euler algorithm given above for the inverse dynamics of n-revolute manipulators requires M multiplications and A additions, as given below:

$$M = 137n - 22, \quad A = 110n - 14 \tag{6.44}$$

The foregoing number of multiplications is identical to that reported by Walker and Orin (1982); however, the number of additions is slightly higher than Walker and Orin's figure, namely, $101n - 11$.

Thus, the inverse dynamics of a six-revolute manipulator requires 800 multiplications and 646 additions. These computations can be performed in a few microseconds using a modern processor. Clearly, if the aforementioned algorithms are tailored to suit particular architectures, then they can be further simplified. Note that, in the presence of a prismatic pair in the jth joint, the foregoing complexity is reduced. In fact, if this is the case, the Newton-Euler equations for the jth link remain as in eqs.(6.38c & d) for the ith link, the only difference appearing in the implementing algorithm, which is simplified, in light of the results derived in discussing the kinematics calculations.

The incorporation of gravity in the Newton-Euler algorithm is done most economically by following the idea proposed by Luh, Walker, and Paul (1980), namely, by declaring that the inertial base undergoes an acceleration $-\mathbf{g}$, where $\mathbf{g}$ denotes the acceleration of gravity. That is

$$\dot{\mathbf{c}}_0 = -\mathbf{g} \tag{6.45}$$

the gravitational accelerations thus propagating forward to the EE. A comparison of various algorithms with regard to their computational complexity is displayed in Table 6.3 for an n-revolute manipulator. For $n = 6$, the corresponding figures appear in Table 6.4.

6.5 The Natural Orthogonal Complement in Robot Dynamics

In simulation studies, we need to integrate the system of ordinary differential equations (ODE) describing the dynamics of a robotic mechanical system. This system of ODE is known as the *mathematical model* of the system at hand. Note that the Newton-Euler equations derived above for a serial manipulator do not constitute the mathematical model because we cannot use the recursive relations derived therein to set up the underlying

TABLE 6.3. Complexity of Different Algorithms for Inverse Dynamics

Author(s)	Methods	Multiplications	Additions
Hollerbach (1980)	E-L	$412n - 277$	$320n - 201$
Luh et al. (1980)	N-E	$150n - 48$	$131n - 48$
Walker & Orin (1982)	N-E	$137n - 22$	$101n - 11$
Khalil et al. (1986)	N-E	$105n - 92$	$94n - 86$
Angeles, Ma & Rojas (1989)	Kane	$105n - 109$	$90n - 105$
Balafoutis & Patel (1991)	tensor	$93n - 69$	$81n - 65$

TABLE 6.4. Complexity of Different Algorithms for Inverse Dynamics, for $n = 6$

Author(s)	Methods	Multiplications ($n = 6$)	Additions ($n = 6$)
Hollerbach (1980)	E-L	2195	1719
Luh et al. (1980)	N-E	852	738
Walker & Orin (1982)	N-E	800	595
Hollerbach and Sahar (1983)	N-E	688	558
Kane & Levinson (1983)	Kane	646	394
Khalil et al. (1986)	N-E	538	478
Angeles, Ma & Rojas (1989)	Kane	521	435
Balafoutis & Patel (1991)	tensor	489	420

ODE *directly.* What we need is a model relating the *state* of the system with its external generalized forces of the form

$$\dot{\mathbf{x}} = \mathbf{f}(\mathbf{x}, \mathbf{u}), \quad \mathbf{x}(t_0) = \mathbf{x}_0 \tag{6.46}$$

where $\mathbf{x}$ is the *state vector,* $\mathbf{u}$ is the *input* or *control vector,* $\mathbf{x}_0$ is the state vector at a certain time t_0, and $\mathbf{f}(\mathbf{x}, \mathbf{u})$ is a nonlinear function of $\mathbf{x}$ and $\mathbf{u}$, derived from the dynamics of the system. The state of a dynamical system is defined, in turn, as *the set of variables that separate the past from the future of the system* (Bryson and Ho, 1975). Thus, if we take t_0 as the present time, we can predict from eqs.(6.46) the future states of the system upon integration of the initial-value problem at hand, even if we do not know the complete past history of the system in full detail. Now, if we regard the vector $\boldsymbol{\theta}$ of independent joint variables and its time-rate of change, $\dot{\boldsymbol{\theta}}$, as the vectors of generalized coordinates and generalized speeds, then an obvious definition of $\mathbf{x}$ is

$$\mathbf{x} \equiv \begin{bmatrix} \boldsymbol{\theta}^T & \dot{\boldsymbol{\theta}}^T \end{bmatrix}^T \tag{6.47}$$

The n generalized coordinates, then, define the configuration of the system, while their time-derivatives determine its generalized momentum, an item defined in eq.(6.20d). Hence, knowing $\boldsymbol{\theta}$ and $\dot{\boldsymbol{\theta}}$, we can predict the future values of these variables with the aid of eqs.(6.46).

Below we will derive the mathematical model, eq.(6.46), explicitly, as pertaining to serial manipulators, in terms of the kinematic structure of the system and its inertial properties, i.e., the mass, mass-center coordinates, and inertia matrix of each of its bodies. To this end, we first write the underlying system of uncoupled Newton-Euler equations for each link. We have $n+1$ links numbered from 0 to n, which are coupled by n kinematic pairs. Moreover, the base link 0 need not be an inertial frame; if it is noninertial, then the force and moment exerted by the environment upon it must be known. For ease of presentation, we will assume in this section that the base frame is inertial, the modifications needed to handle a noninertial base frame to be introduced in Subsection 6.5.2.

We now recall the Newton-Euler equations of the ith body in 6-dimensional form, eqs.(6.5b), which we reproduce below for quick reference:

$$\mathbf{M}_i\dot{\mathbf{t}}_i = -\mathbf{W}_i\mathbf{M}_i\mathbf{t}_i + \mathbf{w}_i^W + \mathbf{w}_i^C, \quad i = 1, \ldots, n \tag{6.48}$$

Furthermore, the definitions of eqs.(6.13a) and (6.13b) are recalled. Apparently, $\mathbf{M}$ and $\mathbf{W}$ are now $6n \times 6n$ matrices, while $\mathbf{t}$, $\mathbf{w}^C$, $\mathbf{w}^A$, and $\mathbf{w}^D$ are all $6n$-dimensional vectors. Then the foregoing $6n$ scalar equations for the n moving links take on the simple form

$$\mathbf{M}\dot{\mathbf{t}} = -\mathbf{W}\mathbf{M}\mathbf{t} + \mathbf{w}^A + \mathbf{w}^G + \mathbf{w}^D + \mathbf{w}^C \tag{6.49}$$

in which $\mathbf{w}^W$ has been decomposed into its active, gravitational, and dissipative parts $\mathbf{w}^A$, $\mathbf{w}^G$, and $\mathbf{w}^D$, respectively. Now, since gravity acts at the mass center of a body, the gravity wrench $\mathbf{w}_i^G$ acting on the ith link takes the form

$$\mathbf{w}_i^G = \begin{bmatrix} \mathbf{0} \\ m_i\mathbf{g} \end{bmatrix} \tag{6.50}$$

The mathematical model displayed in eq.(6.49) represents the *uncoupled* Newton-Euler equations of the overall manipulator. The following step of this derivation consists of representing the coupling between every two consecutive links as a *linear homogeneous system* of algebraic equations on the link twists. Moreover, we note that all kinematic pairs allow a relative one-degree-of-freedom motion between the coupled bodies. We can then express the kinematic constraints of the system in *linear homogeneous form* in the $6n$-dimensional vector of manipulator twist, namely,

$$\mathbf{K}\mathbf{t} = \mathbf{0} \tag{6.51}$$

with $\mathbf{K}$ being a $6n \times 6n$ matrix, to be derived in Subsection 6.5.1. What is important to note at the moment is that the *kinematic constraint equations*, or *constraint equations*, for brevity, eqs.(6.51), consist of a system of $6n$ scalar equations, i.e., six scalar equations for each joint, for the manipulator at hand has n joints. Moreover, when the system is in motion, $\mathbf{t}$ is different from zero, and hence, matrix $\mathbf{K}$ is singular. In fact, the dimension of the nullspace of $\mathbf{K}$, termed its *nullity*, is exactly equal to n, the degree of freedom of the manipulator. Furthermore, since the nonworking constraint wrench $\mathbf{w}^C$ produces no work on the manipulator, its sole function being to keep the links together, the power developed by this wrench on $\mathbf{t}$, for any possible motion of the manipulator, is zero, i.e.,

$$\mathbf{t}^T\mathbf{w}^C = 0 \tag{6.52}$$

On the other hand, if the two sides of eq.(6.51) are transposed and then multiplied by a $6n$-dimensional vector $\boldsymbol{\lambda}$, one has

$$\mathbf{t}^T\mathbf{K}^T\boldsymbol{\lambda} = 0 \tag{6.53}$$

Upon comparing eqs.(6.52) and (6.53), it is apparent that $\mathbf{w}^C$ is of the form

$$\mathbf{w}^C = \mathbf{K}^T\boldsymbol{\lambda} \tag{6.54}$$

More formally, the inner product of $\mathbf{w}^C$ and $\mathbf{t}$, as stated by eq.(6.52), vanishes, and hence, $\mathbf{t}$ lies in the nullspace of $\mathbf{K}$, as stated by eq.(6.51). This means that $\mathbf{w}^C$ lies in the range of $\mathbf{K}^T$, as stated in eq.(6.54). The following step will be to represent $\mathbf{t}$ as a linear transformation of the independent generalized speeds, i.e., as

$$\mathbf{t} = \mathbf{T}\dot{\boldsymbol{\theta}} \tag{6.55}$$

with $\mathbf{T}$ defined as a $6n \times n$ matrix that can be fairly termed the *twist-shaping matrix*. Moreover, the above mapping will be referred to as the *twist-shape relations*. The derivation of expressions for matrices $\mathbf{K}$ and $\mathbf{T}$ will be described in detail in Subsection 6.5.1 below. Now, upon substitution of eq.(6.55) into eq.(6.51), we obtain

$$\mathbf{K}\mathbf{T}\dot{\boldsymbol{\theta}} = \mathbf{0} \tag{6.56a}$$

Furthermore, since the degree of freedom of the manipulator is n, the n generalized speeds $\{\,\dot{\theta}_i\,\}_1^n$ can be assigned arbitrarily. However, while doing this, eq.(6.56a) has to hold. Thus, the only possibility for this to happen is that the product $\mathbf{KT}$ vanish, i.e.,

$$\mathbf{K}\mathbf{T} = \mathbf{O} \tag{6.56b}$$

where $\mathbf{O}$ denotes the $6n\times n$ zero matrix. The above equation states that $\mathbf{T}$ is an *orthogonal complement* of $\mathbf{K}$. Because of the particular form of choosing this complement—see eq.(6.55)—we refer to $\mathbf{T}$ as the *natural orthogonal complement* of $\mathbf{K}$ (Angeles and Lee, 1988).

In the final step of this method, $\dot{\mathbf{t}}$ of eq.(6.49) is obtained from eq.(6.55), namely,

$$\dot{\mathbf{t}} = \mathbf{T}\ddot{\boldsymbol{\theta}} + \dot{\mathbf{T}}\dot{\boldsymbol{\theta}} \tag{6.57}$$

Furthermore, the uncoupled equations, eqs.(6.49), are multiplied on the left by $\mathbf{T}^T$, thereby eliminating $\mathbf{w}^C$ from those equations and reducing these to a system of only n independent equations, free of nonworking constraint wrenches. These are nothing but the Euler-Lagrange equations of the manipulator, namely,

$$\mathbf{I}\ddot{\boldsymbol{\theta}} = -\mathbf{T}^T(\mathbf{M}\dot{\mathbf{T}} + \mathbf{W}\mathbf{M}\mathbf{T})\dot{\boldsymbol{\theta}} + \mathbf{T}^T(\mathbf{w}^A + \mathbf{w}^D) \tag{6.58}$$

where $\mathbf{I}$ is the positive definite $n \times n$ *generalized inertia matrix* of the manipulator and is defined as

$$\mathbf{I} \equiv \mathbf{T}^T\mathbf{M}\mathbf{T} \tag{6.59}$$

which is identical to the inertia matrix derived using the Euler-Lagrange equations, with $\boldsymbol{\theta}$ as the vector of generalized coordinates. Now, we let $\boldsymbol{\tau}$ and $\boldsymbol{\delta}$ denote the n-dimensional vectors of active and dissipative generalized force. Moreover, we let $\mathbf{C}(\boldsymbol{\theta}, \dot{\boldsymbol{\theta}})\dot{\boldsymbol{\theta}}$ be the n-dimensional vector of *quadratic* terms of inertia force. The aforementioned items are defined as

$$\boldsymbol{\tau} \equiv \mathbf{T}^T\mathbf{w}^A, \quad \boldsymbol{\delta} \equiv \mathbf{T}^T\mathbf{w}^D, \quad \mathbf{C}(\boldsymbol{\theta}, \dot{\boldsymbol{\theta}}) \equiv \mathbf{T}^T\mathbf{M}\dot{\mathbf{T}} + \mathbf{T}^T\mathbf{W}\mathbf{M}\mathbf{T} \tag{6.60}$$

Clearly, the sum $\boldsymbol{\tau} + \boldsymbol{\delta}$ produces $\boldsymbol{\phi}$, the generalized force defined in eq.(6.11). Thus, the Euler-Lagrange equations of the system take on the form

$$\mathbf{I}\ddot{\boldsymbol{\theta}} = -\mathbf{C}\dot{\boldsymbol{\theta}} + \boldsymbol{\tau} + \boldsymbol{\delta} \tag{6.61}$$

As a matter of fact, $\boldsymbol{\delta}$ is defined in eq.(6.60) only for conceptual reasons. In practice, this term is most easily calculated once a dissipation function in terms of the generalized coordinates and generalized speeds is available, as described in Section 6.8. Thus, $\boldsymbol{\delta}$ is computed as

$$\boldsymbol{\delta} = -\frac{\partial \Delta}{\partial \dot{\boldsymbol{\theta}}} \tag{6.62}$$

It is pointed out that the first term of the right-hand side of eq.(6.61) is *quadratic* in $\dot{\boldsymbol{\theta}}$ because matrix $\mathbf{C}$, defined in eq.(6.60), is linear in $\dot{\boldsymbol{\theta}}$. In fact, the first term of that expression is linear in a factor $\dot{\mathbf{T}}$ that is, in turn, linear in $\dot{\boldsymbol{\theta}}$. Moreover, the second term of the same expression is linear in $\mathbf{W}$, which is linear in $\dot{\boldsymbol{\theta}}$ as well. However, $\mathbf{C}$ is *nonlinear* in $\boldsymbol{\theta}$. Because of the quadratic nature of that term, it is popularly known as the vector of *Coriolis and centrifugal forces*, whereas the left-hand side of that equation is given the name of vector of *inertia forces*. Properly speaking, both the left-hand side and the first term of the right-hand side of eq.(6.61) arise from inertia forces.

Example 6.5.1 (A minimum-time trajectory) *A pick-and-place operation is to be performed with an n-axis manipulator in the shortest possible time. Moreover, the maneuver is defined so that the n-dimensional vector of joint variables is given by a common shape function $s(x)$, with $0 \le x \le 1$ and $0 \le s \le 1$, which is prescribed. Thus, for a fixed n-dimensional vector $\boldsymbol{\theta}_0$, the time-history of the joint-variable vector, $\boldsymbol{\theta}(t)$, is given by*

$$\boldsymbol{\theta}(t) = \boldsymbol{\theta}_0 + s\left(\frac{t}{T}\right)\Delta\boldsymbol{\theta}, \quad 0 \le t \le T$$

with T defined as the time taken by the maneuver, while $\boldsymbol{\theta}_0$ and $\boldsymbol{\theta}_0 + \Delta\boldsymbol{\theta}$ are the values of the joint-variable vector at the pick- and the place-postures of the manipulator, respectively. These vectors are computed from inverse kinematics, as explained in Chapter 4. Furthermore, the load-carrying capacity of the manipulator is specified in terms of the maximum torques delivered by the motors, namely,

$$|\tau_i| \le \overline{\tau}_i, \quad \text{for} \quad i = 1, \ldots, n$$

where the constant values $\overline{\tau}_i$ are supplied by the manufacturer. In order to keep the analysis simple, we neglect power loses in this example. Find the minimum time in which the maneuver can take place.

Solution: Let us first calculate the vector of joint-rates and its time-derivative:

$$\dot{\boldsymbol{\theta}}(t) = \frac{1}{T}s'(x)\Delta\boldsymbol{\theta}, \quad \ddot{\boldsymbol{\theta}}(t) = \frac{1}{T^2}s''(x)\Delta\boldsymbol{\theta}, \quad x \equiv \frac{t}{T}$$

Now we substitute the aforementioned values into the mathematical model of eq.(6.61), with $\boldsymbol{\delta}(t) = \mathbf{0}$, thereby obtaining

$$\begin{aligned}
\boldsymbol{\tau} &= \mathbf{I}(\boldsymbol{\theta})\ddot{\boldsymbol{\theta}} + \mathbf{C}(\boldsymbol{\theta}, \dot{\boldsymbol{\theta}})\dot{\boldsymbol{\theta}} \\
&= \frac{1}{T^2}s''(x)\mathbf{I}(x)\Delta\boldsymbol{\theta} + \frac{1}{T^2}s'^2(x)\mathbf{C}(x)\Delta\boldsymbol{\theta} \\
&\equiv \frac{1}{T^2}\mathbf{f}(x)
\end{aligned}$$

with $\mathbf{f}(x)$ defined, of course, as

$$\mathbf{f}(x) \equiv [\mathbf{I}(x)s''(x) + \mathbf{C}(x)s'^2(x)]\Delta\boldsymbol{\theta}$$

the $1/T^2$ factor in the term of Coriolis and centrifugal forces stemming from the quadratic nature of the $\mathbf{C}(\boldsymbol{\theta}, \dot{\boldsymbol{\theta}})\dot{\boldsymbol{\theta}}$ term. What we now have is the vector of motor torques, $\boldsymbol{\tau}$, expressed as a function of the scalar argument x. Now, let $f_i(x)$ be the ith component of vector $\mathbf{f}(x)$, and

$$F_i \equiv \max_x\{|f_i(x)\|\}, \quad \text{for} \quad i = 1, \ldots, n$$

We would then like to have each value F_i produce the maximum available torque $\overline{\tau}_i$, namely,

$$\overline{\tau}_i = \frac{F_i}{T^2}, \quad i = 1, \ldots n$$

and hence, for each joint we have a value T_i of T given by

$$T_i^2 \equiv \frac{F_i}{\overline{\tau}_i}, \quad i = 1, \ldots n$$

Obviously, the minimum value sought, $T_{\min}$, is nothing but the maximum of the foregoing values, i.e.,

$$T_{\min} = \max_i \{T_i\}_1^n$$

thereby completing the solution.

6.5.1 *Derivation of Constraint Equations and Twist-Shape Relations*

In order to illustrate the general ideas behind the method of the natural orthogonal complement, we derive below the underlying kinematic constraint equations and the twist-shape relations. We first note, from eq.(6.25a), that the relative angular velocity of the ith link with respect to the $(i-1)$st link, $\boldsymbol{\omega}_i - \boldsymbol{\omega}_{i-1}$, is $\dot{\theta}_i \mathbf{e}_i$. Thus, if matrix $\mathbf{E}_i$ is defined as the cross-product matrix of vector $\mathbf{e}_i$, then, the angular velocities of two successive links obey a simple relation, namely,

$$\mathbf{E}_i(\boldsymbol{\omega}_i - \boldsymbol{\omega}_{i-1}) = \mathbf{0} \tag{6.63}$$

Furthermore, we rewrite now eq.(6.33a) in the form

$$\dot{\mathbf{c}}_i - \dot{\mathbf{c}}_{i-1} + \mathbf{R}_i\boldsymbol{\omega}_i + \mathbf{D}_{i-1}\boldsymbol{\omega}_{i-1} = \mathbf{0} \tag{6.64}$$

where $\mathbf{D}_i$ and $\mathbf{R}_i$ are defined as the cross-product matrices of vectors $\boldsymbol{\delta}_i$, defined in Subsection 6.4.1 as $\mathbf{a}_i - \boldsymbol{\rho}_i$, and $\boldsymbol{\rho}_i$, respectively. In particular, when the first link is inertial, eqs.(6.63 & b), as pertaining to the first link, reduce to

$$\mathbf{E}_1\boldsymbol{\omega}_1 = \mathbf{0} \tag{6.65a}$$

$$\dot{\mathbf{c}}_1 + \mathbf{R}_1\boldsymbol{\omega}_1 = \mathbf{0} \tag{6.65b}$$

Now, eqs.(6.63) and (6.64), as well as their counterparts for $i = 1$, eqs.(6.65a & b), are further expressed in terms of the link twists, thereby producing the constraints below:

$$\mathbf{K}_{11}\mathbf{t}_1 = \mathbf{0} \tag{6.66a}$$

$$\mathbf{K}_{i,i-1}\mathbf{t}_{i-1} + \mathbf{K}_{ii}\mathbf{t}_i = \mathbf{0}, \quad i = 1, \ldots, n \tag{6.66b}$$

with $\mathbf{K}_{11}$ and $\mathbf{K}_{ij}$, for $i = 2, \ldots, n$ and $j = i-1, i$, defined as

$$\mathbf{K}_{11} \equiv \begin{bmatrix} \mathbf{E}_1 & \mathbf{O} \\ \mathbf{R}_1 & \mathbf{1} \end{bmatrix} \tag{6.67a}$$

$$\mathbf{K}_{i,i-1} \equiv \begin{bmatrix} -\mathbf{E}_i & \mathbf{O} \\ \mathbf{D}_{i-1} & -\mathbf{1} \end{bmatrix} \tag{6.67b}$$

$$\mathbf{K}_{ii} \equiv \begin{bmatrix} \mathbf{E}_i & \mathbf{O} \\ \mathbf{R}_i & \mathbf{1} \end{bmatrix} \tag{6.67c}$$

where $\mathbf{1}$ and $\mathbf{O}$ denote the 3×3 identity and zero matrices, respectively. Furthermore, from eqs.(6.66a & b) and (6.67a–c), it is apparent that matrix $\mathbf{K}$ appearing in eq.(6.56b) takes on the form

$$\mathbf{K} = \begin{bmatrix} \mathbf{K}_{11} & \mathbf{O}_6 & \mathbf{O}_6 & \cdots & \mathbf{O}_6 & \mathbf{O}_6 \\ \mathbf{K}_{21} & \mathbf{K}_{22} & \mathbf{O}_6 & \cdots & \mathbf{O}_6 & \mathbf{O}_6 \\ \vdots & \vdots & \vdots & \ddots & \vdots & \vdots \\ \mathbf{O}_6 & \mathbf{O}_6 & \mathbf{O}_6 & \cdots & \mathbf{K}_{n-1,n-1} & \mathbf{O}_6 \\ \mathbf{O}_6 & \mathbf{O}_6 & \mathbf{O}_6 & \cdots & \mathbf{K}_{n,n-1} & \mathbf{K}_{nn} \end{bmatrix} \tag{6.68}$$

with $\mathbf{O}_6$ denoting the 6×6 zero matrix.

Further, the link-twists are expressed as linear combinations of the joint-rate vector $\dot{\boldsymbol{\theta}}$. To this end, we define the $6 \times n$ *partial Jacobian* $\mathbf{J}_i$ as the matrix mapping the joint-rate vector $\dot{\boldsymbol{\theta}}$ into the twist $\mathbf{t}_i$ of that link, i.e.,

$$\mathbf{J}_i \dot{\boldsymbol{\theta}} = \mathbf{t}_i \tag{6.69}$$

whose jth column, $\mathbf{t}_{ij}$, is given, for $i, j = 1, 2, \ldots, n$, by

$$\mathbf{t}_{ij} = \begin{cases} \begin{bmatrix} \mathbf{e}_j \\ \mathbf{e}_j \times \mathbf{r}_{ij} \end{bmatrix}, & \text{if } j \le i; \\ \begin{bmatrix} \mathbf{0} \\ \mathbf{0} \end{bmatrix}, & \text{otherwise.} \end{cases} \tag{6.70}$$

with $\mathbf{r}_{ij}$ illustrated in Fig. 6.6 and defined, for $i, j = 1, \ldots, n$, as

$$\mathbf{r}_{ij} \equiv \begin{cases} \mathbf{a}_j + \mathbf{a}_{j+1} + \cdots + \mathbf{a}_{i-1} + \boldsymbol{\rho}_i, & \text{if } j < i; \\ \boldsymbol{\rho}_i, & \text{if } j = i; \\ \mathbf{0}, & \text{otherwise.} \end{cases} \tag{6.71}$$

We can thus readily express the twist $\mathbf{t}_i$ of the ith link as a linear combination of the first i joint rates, namely,

$$\mathbf{t}_i = \dot{\theta}_1 \mathbf{t}_{i1} + \dot{\theta}_2 \mathbf{t}_{i2} + \cdots + \dot{\theta}_i \mathbf{t}_{ii}, \quad i = 1, \ldots, n \tag{6.72}$$

and hence, matrix $\mathbf{T}$ of eq.(6.55) takes the form

$$\mathbf{T} \equiv \begin{bmatrix} \mathbf{t}_{11} & \mathbf{0} & \cdots & \mathbf{0} \\ \mathbf{t}_{21} & \mathbf{t}_{22} & \cdots & \mathbf{0} \\ \vdots & \vdots & \ddots & \vdots \\ \mathbf{t}_{n1} & \mathbf{t}_{n2} & \cdots & \mathbf{t}_{nn} \end{bmatrix} \tag{6.73}$$

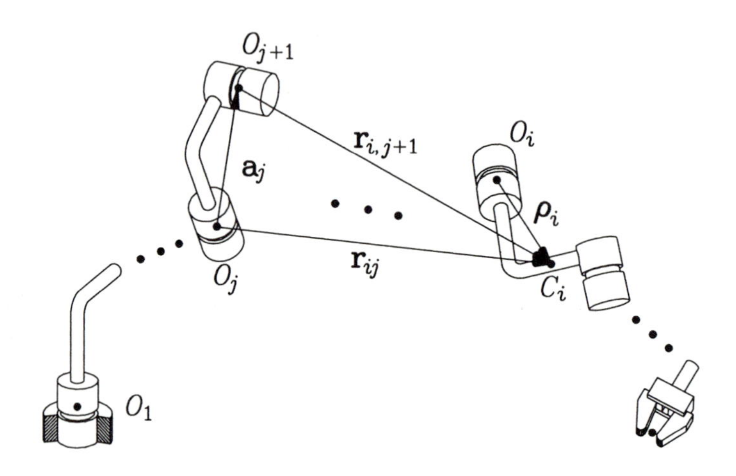

FIGURE 6.6. Kinematic subchain comprising links j, $j+1\ldots$, i.

As a matter of verification, one can readily prove that the product of matrix $\mathbf{T}$, as given by eq.(6.73), by matrix $\mathbf{K}$, as given by eq.(6.68), vanishes, and hence, relation (6.56b) holds.

The kinematic constraint equations on the twists, for the case in which the ith joint is prismatic, are derived likewise. In this case, we use eqs.(6.34a & e), with the latter rewritten more conveniently for our purposes, namely,

$$\boldsymbol{\omega}_i = \boldsymbol{\omega}_{i-1} \tag{6.74a}$$

$$\dot{\mathbf{c}}_i = \dot{\mathbf{c}}_{i-1} + \boldsymbol{\omega}_{i-1} \times (\boldsymbol{\delta}_{i-1} + \boldsymbol{\rho}_i + b_i\mathbf{e}_i) + \dot{b}_i\mathbf{e}_i \tag{6.74b}$$

We now introduce one further definition:

$$\mathbf{R}'_i \equiv \mathbf{D}'_{i-1} + \mathbf{R}_i \tag{6.75}$$

where $\mathbf{D}'_{i-1}$ is the cross-product matrix of vector $\boldsymbol{\delta}_{i-1}$, defined in Subsection 6.4.1 as $\mathbf{d}_{i-1} - \boldsymbol{\rho}_{i-1}$, while $\mathbf{R}_i$ is the cross-product matrix of $\boldsymbol{\rho}_i + b_i\mathbf{e}_i$. Hence, eq.(6.74b) can be rewritten as

$$\dot{\mathbf{c}}_i - \dot{\mathbf{c}}_{i-1} + \mathbf{R}'_i\boldsymbol{\omega}_i - \dot{b}_i\mathbf{e}_i = \mathbf{0} \tag{6.76}$$

Upon multiplication of both sides of eq.(6.76) by $\mathbf{E}_i$, the term in $\dot{b}_i$ cancels, and we obtain

$$\mathbf{E}_i(\dot{\mathbf{c}}_i - \dot{\mathbf{c}}_{i-1} + \mathbf{R}'_i\boldsymbol{\omega}_i) = \mathbf{0} \tag{6.77}$$

Hence, eqs.(6.74a) and (6.77) can now be regrouped in a single 6-dimensional linear homogeneous equation in the twists, namely,

$$\mathbf{K}'_{i,i-1}\mathbf{t}_{i-1} + \mathbf{K}'_{ii}\mathbf{t}_i = \mathbf{0} \tag{6.78}$$

the associated matrices being defined below:

$$\mathbf{K}'_{i,i-1} \equiv \begin{bmatrix} -\mathbf{1} & \mathbf{O} \\ \mathbf{O} & -\mathbf{E}_i \end{bmatrix} \tag{6.79a}$$

$$\mathbf{K}'_{ii} \equiv \begin{bmatrix} \mathbf{1} & \mathbf{O} \\ \mathbf{E}_i\mathbf{R}'_i & \mathbf{E}_i \end{bmatrix} \tag{6.79b}$$

with $\mathbf{1}$ and $\mathbf{O}$ defined already as the 3×3 identity and zero matrices, respectively. If the first joint is prismatic, then the corresponding constraint equation takes on the form

$$\mathbf{K}'_{11}\mathbf{t}_1 = \mathbf{0} \tag{6.80}$$

with $\mathbf{K}'_{11}$ defined as

$$\mathbf{K}'_{11} \equiv \begin{bmatrix} \mathbf{1} & \mathbf{O} \\ \mathbf{O} & \mathbf{E}_1 \end{bmatrix} \tag{6.81}$$

Furthermore, if the kth pair is prismatic and $1 \leq k \leq i$, then the twist $\mathbf{t}_i$ of the ith link changes to

$$\mathbf{t}_i = \dot{\theta}_1\mathbf{t}_{i1} + \cdots + \dot{b}_k\mathbf{t}'_{ik} + \cdots + \dot{\theta}_i\mathbf{t}_{ii}, \quad i = 1, \ldots, n \tag{6.82}$$

where $\mathbf{t}'_{ik}$ is defined as

$$\mathbf{t}'_{ik} \equiv \begin{bmatrix} \mathbf{0} \\ \mathbf{e}_k \end{bmatrix} \tag{6.83}$$

In order to set up eq.(6.61), then all we now need is $\dot{\mathbf{T}}$, which is computed below. Two cases will be distinguished again, namely, whether the joint at hand is a revolute or a prismatic pair. In the first case, from eq.(6.70) one readily derives, for $i, j = 1, 2, \ldots, n$,

$$\dot{\mathbf{t}}_{ij} = \begin{cases} \begin{bmatrix} \boldsymbol{\omega}_j \times \mathbf{e}_j \\ (\boldsymbol{\omega}_j \times \mathbf{e}_j) \times \mathbf{r}_{ij} + \mathbf{e}_j \times \dot{\mathbf{r}}_{ij} \end{bmatrix}, & \text{if } j \leq i; \\ \begin{bmatrix} \mathbf{0} \\ \mathbf{0} \end{bmatrix}, & \text{otherwise} \end{cases} \tag{6.84}$$

where, from eq.(6.71),

$$\dot{\mathbf{r}}_{ij} = \boldsymbol{\omega}_j \times \mathbf{a}_j + \cdots + \boldsymbol{\omega}_{i-1} \times \mathbf{a}_{i-1} + \boldsymbol{\omega}_i \times \boldsymbol{\rho}_i \tag{6.85}$$

On the other hand, if the kth pair is prismatic and $1 \leq k \leq i$, then from eq.(6.83), the time-rate of change of $\mathbf{t}'_{ik}$ becomes

$$\dot{\mathbf{t}}'_{ik} = \begin{bmatrix} \mathbf{0} \\ \boldsymbol{\omega}_k \times \mathbf{e}_k \end{bmatrix} \tag{6.86}$$

thereby completing the desired derivations.

Note that the natural orthogonal complement can also be used for the inverse dynamics calculations. In this case, if the manipulator is subjected to a gravity field, then the twist-rate of the first link will have to be modified by adding a nonhomogeneous term to it, thereby accounting for the gravity-acceleration terms. This issue is discussed in Section 6.7.

6.5.2 Noninertial Base Link

Noninertial bases occur in space applications, e.g., in the case of a manipulator mounted on a space platform or on the space shuttle. A noninertial base can be readily handled with the use of the natural orthogonal complement, as discussed in this subsection. Since the base is free of attachments to an inertial frame, we have to add its six degrees of freedom (dof) to the n dof of the rest of the manipulator. Correspondingly, $\mathbf{t}$, $\mathbf{w}^C$, $\mathbf{w}^A$, and $\mathbf{w}^D$ now become $6(n+1)$-dimensional vectors. In particular, $\mathbf{t}$ takes the form

$$\mathbf{t} = [\,\mathbf{t}_0^T \quad \mathbf{t}_1^T \quad \ldots \quad \mathbf{t}_n^T\,]^T \tag{6.87}$$

with $\mathbf{t}_0$ defined as the twist of the base. Furthermore, the vector of independent generalized speeds, $\dot{\boldsymbol{\theta}}$, is now of dimension $n+6$, its first six components being those of $\mathbf{t}_0$, the other n remaining as in the previous case. Thus, $\dot{\boldsymbol{\theta}}$ has the components shown below:

$$\dot{\boldsymbol{\theta}} \equiv [\,\mathbf{t}_0^T \quad \dot{\theta}_1 \quad \ldots \quad \dot{\theta}_n\,]^T \tag{6.88}$$

Correspondingly, $\mathbf{T}$ becomes a $6(n+1) \times (n+6)$ matrix, namely,

$$\mathbf{T} \equiv \begin{bmatrix} \mathbf{1} & \mathbf{O} \\ \mathbf{O}' & \mathbf{T}' \end{bmatrix} \tag{6.89}$$

where $\mathbf{1}$ is the 6×6 identity matrix, $\mathbf{O}$ denotes the $6 \times n$ zero matrix, $\mathbf{O}'$ represents the $6n \times 6$ zero matrix, and $\mathbf{T}'$ is the $6n \times n$ matrix defined in eq.(6.73) as $\mathbf{T}$. Otherwise, the model remains as in the case of an inertial base.

A word of caution is in order here. Because of the presence of the twist vector $\mathbf{t}_0$ in the definition of the vector of generalized speeds above, the latter cannot, properly speaking, be regarded as a time-derivative. Indeed, as studied in Chapter 3, the angular velocity appearing in the twist vector is not a time-derivative. Hence, the vector of independent generalized speeds defined in eq.(6.88) is represented instead by $\mathbf{v}$, which does not imply a time-derivative, namely,

$$\mathbf{v} = [\,\mathbf{t}_0^T \quad \dot{\theta}_1 \quad \cdots \quad \dot{\theta}_n\,]^T \tag{6.90}$$

6.6 Manipulator Forward Dynamics

Forward dynamics is needed either for purposes of simulation or for the model-based control of manipulators (Craig, 1989), and hence, a fast calculation of the joint-variable time-histories $\boldsymbol{\theta}(t)$ is needed. These time-histories are calculated from the model displayed in eq.(6.61), reproduced below for quick reference, in terms of vector $\boldsymbol{\theta}(t)$, i.e.,

$$\mathbf{I}\ddot{\boldsymbol{\theta}} = -\mathbf{C}(\boldsymbol{\theta}, \dot{\boldsymbol{\theta}})\dot{\boldsymbol{\theta}} + \boldsymbol{\tau} + \boldsymbol{\delta}(\boldsymbol{\theta}, \dot{\boldsymbol{\theta}}) \tag{6.91}$$

Clearly, what is at stake here is the calculation of $\ddot{\boldsymbol{\theta}}$ from the foregoing model. Indeed, the right-hand side of eq.(6.91) can be calculated with the aid of the Newton-Euler recursive algorithm, as we will describe below, and needs no further discussion for the time being. Now, the calculation of $\ddot{\boldsymbol{\theta}}$ from eq.(6.91) is similar to the calculation of $\dot{\boldsymbol{\theta}}$ from the relation between the joint-rates and the twist, derived in Section 4.5. From the discussion in that section, such calculations take a number of floating-point operations, or *flops*, that is proportional to n^3, and are thus said to have a complexity of $O(n^3)$—read "order n^3". In real-time calculations, we would like to have a computational scheme of $O(n)$. In attempting to derive such schemes, Walker and Orin (1982) proposed a procedure that they called the *composite rigid-body method*, whereby the number of flops is minimized by cleverly calculating $\mathbf{I}(\boldsymbol{\theta})$ and the right-hand side of eq.(6.91) by means of the recursive Newton-Euler algorithm. In their effort, they produced an $O(n^2)$ algorithm to calculate $\ddot{\boldsymbol{\theta}}$. Thereafter, Featherstone (1983) proposed an $O(n)$ algorithm that is based, however, on the assumption that Coriolis and centrifugal forces are negligible. The same author reported an improvement to the aforementioned algorithm, namely, the *articulated-body method*, that takes into account Coriolis and centrifugal forces (Featherstone, 1987.) The outcome, for an n-revolute manipulator, is an algorithm requiring $300n - 267$ multiplications and $279n - 259$ additions. For $n = 6$, these figures yield 1,533 multiplications and 1,415 additions.

In this subsection, we illustrate the application of the method of the natural orthogonal complement to the modeling of an n-axis serial manipulator for purposes of simulation. While this algorithm gives an $O(n^3)$ complexity, its derivation is straightforward and gives, for a six-axis manipulator, a computational cost similar to that of Featherstone's, namely, 1,596 multiplications and 1,263 additions. Moreover, a clever definition of coordinate frames leads to even lower figures, i.e., 1,353 multiplications and 1,165 additions, as reported by Angeles and Ma (1988).

The manipulator at hand is assumed to be constituted by n moving links coupled by n kinematic pairs of the revolute or prismatic types. Again, for brevity, the base link is assumed to be inertial, noninertial bases being readily incorporated as described in Subsection 6.5.2. For the sake of conciseness, we will henceforth consider only manipulators mounted on an inertial base. Moreover, we assume that the generalized coordinates $\boldsymbol{\theta}$ and the generalized speeds $\dot{\boldsymbol{\theta}}$ are known at an instant t_k, along with the driving torque $\boldsymbol{\tau}(t)$, for $t \geq t_k$, and of course, the DH and the inertial parameters of the manipulator are assumed to be known as well. Based on the aforementioned information, then, $\ddot{\boldsymbol{\theta}}$ is evaluated at t_k and with a suitable integration scheme, the values of $\boldsymbol{\theta}$ and $\dot{\boldsymbol{\theta}}$ are determined at instant t_{k+1}. Obviously, the governing equation, eq.(6.61), enables us to solve for $\ddot{\boldsymbol{\theta}}(t_k)$. This requires, of course, the *inversion* of the $n \times n$ matrix of generalized inertia $\mathbf{I}$. Since the said matrix is positive-definite, solving for $\ddot{\boldsymbol{\theta}}$ from eq.(6.61) can be done economically using the *Cholesky-decomposition* algorithm (Dahlquist and Björck, 1974). The sole remaining task is, then, the computation of $\mathbf{I}$, the quadratic inertia term $\mathbf{C}\dot{\boldsymbol{\theta}}$, and the dissipative torque $\boldsymbol{\delta}$. The last of these is dependent on the manipulator and the constitutive model adopted for the representation of viscous and Coulomb friction forces and will not be considered at this stage. Models for dissipative forces will be studied in Section 6.8. Thus, the discussion below will focus on the computation of $\mathbf{I}$ and $\mathbf{C}\dot{\boldsymbol{\theta}}$ appearing in the mathematical model of eq.(6.91).

Furthermore, we will take into account that the end-effector is normally acted upon by a working wrench $\mathbf{w}^W$ that is exerted by the environment. For example, in deburring operations, this force is that exerted by the workpiece onto the grinding wheel. This wrench is static, in that it does not stem from inertia effects. In order to incorporate this wrench into the above model, we calculate the joint torque $\boldsymbol{\tau}^W$ that the motors must supply in order to balance that wrench. We do this by invoking the *First Law of Thermodynamics*, also known as the *Principle of Virtual Work*, already invoked when deriving eq.(4.96a). To this end, we equate the power developed by $\mathbf{w}^W$ with that developed by $\boldsymbol{\tau}^W$, namely,

$$\mathbf{t}^T\mathbf{w}^W = \dot{\boldsymbol{\theta}}^T\boldsymbol{\tau}^W \tag{6.92}$$

But, from manipulator kinematics, eq.(4.54), $\mathbf{t}$ is nothing but $\mathbf{J}\dot{\boldsymbol{\theta}}$, and hence,

$$\dot{\boldsymbol{\theta}}^T\mathbf{J}^T\mathbf{w}^W = \dot{\boldsymbol{\theta}}^T\boldsymbol{\tau}^W$$

a relation valid for arbitrary $\dot{\boldsymbol{\theta}}$. Therefore,

$$\mathbf{J}^T\mathbf{w}^W = \boldsymbol{\tau}^W \tag{6.93}$$

The mathematical model of the manipulator now takes the form

$$\mathbf{I}\ddot{\boldsymbol{\theta}} = -\mathbf{C}\dot{\boldsymbol{\theta}} + \boldsymbol{\tau} - \boldsymbol{\delta} + \mathbf{J}^T\mathbf{w}^W \tag{6.94}$$

with $\mathbf{I}$ defined already in eq.(6.59). Next, the $6n \times 6n$ matrix $\mathbf{M}$ is factored as

$$\mathbf{M} = \mathbf{H}^T\mathbf{H} \tag{6.95}$$

which is possible because $\mathbf{M}$ is at least positive-semidefinite. In particular, for manipulators of the type at hand, $\mathbf{M}$ is positive-definite if no link-mass is neglected. Moreover, due to the diagonal-block structure of this matrix, its factoring is straightforward. In fact, $\mathbf{H}$ is given simply by

$$\mathbf{H} = \mathrm{diag}(\mathbf{H}_1, \ldots, \mathbf{H}_n) \tag{6.96}$$

each 6×6 block $\mathbf{H}_i$ of eq.(6.96) being given, in turn, as

$$\mathbf{H}_i = \begin{bmatrix} \mathbf{N}_i & \mathbf{O} \\ \mathbf{O} & n_i\mathbf{1} \end{bmatrix} \tag{6.97}$$

with $\mathbf{1}$ and $\mathbf{O}$ defined as the 3×3 identity and zero matrices, respectively. We thus have

$$\mathbf{M}_i = \mathbf{H}_i^T \mathbf{H}_i \tag{6.98}$$

Furthermore, $\mathbf{N}_i$ can be obtained from the Cholesky decomposition of $\mathbf{I}_i$, while n_i is the *positive* square root of m_i, i.e.,

$$\mathbf{I}_i = \mathbf{N}_i^T \mathbf{N}_i, \quad m_i = n_i^2 \tag{6.99}$$

Now, since each 6×6 $\mathbf{M}_i$ block is constant, the above factoring can be done off-line. From the foregoing definitions, then, the $n \times n$ matrix of generalized inertia $\mathbf{I}$ can now be expressed as

$$\mathbf{I} = \mathbf{P}^T \mathbf{P} \tag{6.100}$$

where $\mathbf{P}$ is defined, in turn, as the $6n \times n$ matrix given below:

$$\mathbf{P} \equiv \mathbf{HT} \tag{6.101}$$

The computation of $\mathbf{P}$ is now discussed. If we recall the structure of $\mathbf{T}$ from eq.(6.73) and that of $\mathbf{H}$ from eq.(6.96), along with the definition of $\mathbf{P}$, eq.(6.101), we readily obtain

$$\mathbf{P} = \begin{bmatrix} \mathbf{H}_1\mathbf{t}_{11} & \mathbf{0} & \cdots & \mathbf{0} \\ \mathbf{H}_2\mathbf{t}_{21} & \mathbf{H}_2\mathbf{t}_{22} & \cdots & \mathbf{0} \\ \vdots & \vdots & \ddots & \vdots \\ \mathbf{H}_n\mathbf{t}_{n1} & \mathbf{H}_n\mathbf{t}_{n2} & \cdots & \mathbf{H}_n\mathbf{t}_{nn} \end{bmatrix} = \begin{bmatrix} \mathbf{p}_{11} & \mathbf{0} & \cdots & \mathbf{0} \\ \mathbf{p}_{21} & \mathbf{p}_{22} & \cdots & \mathbf{0} \\ \vdots & \vdots & \ddots & \vdots \\ \mathbf{p}_{n1} & \mathbf{p}_{n2} & \cdots & \mathbf{p}_{nn} \end{bmatrix} \tag{6.102}$$

with $\mathbf{0}$ denoting the 6-dimensional zero vector. Moreover, each of the above nontrivial 6-dimensional arrays $\mathbf{p}_{ij}$ is given as

$$\mathbf{p}_{ij} \equiv \mathbf{H}_i\mathbf{t}_{ij} = \begin{cases} \begin{bmatrix} \mathbf{N}_i\mathbf{e}_j \\ n_i\mathbf{e}_j \times \mathbf{r}_{ij} \end{bmatrix} & \text{if the } j\text{th joint is } R; \\ \begin{bmatrix} \mathbf{0} \\ n_i\mathbf{e}_j \end{bmatrix} & \text{if the } j\text{th joint is } P \end{cases} \tag{6.103}$$

Thus, the (i, j) entry of $\mathbf{I}$ is computed as the sum of the inner products of the (k, i) and the (k, j) blocks of $\mathbf{P}$, for $k = j, \ldots, n$, i.e.,

$$I_{ij} = I_{ji} = \sum_{k=j}^{n} \mathbf{p}_{ki}^T \mathbf{p}_{kj} \tag{6.104}$$

with both $\mathbf{p}_{ki}$ and $\mathbf{p}_{kj}$ expressed in $\mathcal{F}_{k+1}$-coordinates, i.e., in kth-link coordinates. Now, the Cholesky decomposition of $\mathbf{I}$ can be expressed as

$$\mathbf{I} = \mathbf{L}^T \mathbf{L} \tag{6.105}$$

where $\mathbf{L}$ is an $n \times n$ lower-triangular matrix with positive diagonal entries. Moreover, eq.(6.94) is now rewritten as

$$\mathbf{L}^T\mathbf{L}\ddot{\boldsymbol{\theta}} = -(\mathbf{C}\dot{\boldsymbol{\theta}} - \mathbf{J}^T\mathbf{w}^W) + \boldsymbol{\delta} + \boldsymbol{\tau} \tag{6.106}$$

From eq.(6.94), it is apparent that the term inside the parentheses in the right-hand side of the above equation is nothing but the torque required to produce the motion prescribed by the current values of $\boldsymbol{\theta}$ and $\dot{\boldsymbol{\theta}}$, in the absence of dissipative wrenches and with zero joint accelerations, when the manipulator is acted upon by a static wrench $\mathbf{w}^W$. That is,

$$\mathbf{C}\dot{\boldsymbol{\theta}} - \mathbf{J}^T\mathbf{w}^W = \boldsymbol{\tau}\big|_{\mathbf{w}^D=\mathbf{0},\ddot{\boldsymbol{\theta}}=\mathbf{0}} \equiv \overline{\boldsymbol{\tau}} \tag{6.107}$$

which can be clearly computed from inverse dynamics. Now eq.(6.105) is solved for $\ddot{\boldsymbol{\theta}}$ in two steps, namely,

$$\mathbf{L}^T\mathbf{x} = -\overline{\boldsymbol{\tau}} + \boldsymbol{\tau} + \boldsymbol{\delta} \tag{6.108a}$$

$$\mathbf{L}\ddot{\boldsymbol{\theta}} = \mathbf{x} \tag{6.108b}$$

In the above equations, then, $\mathbf{x}$ is first computed from eq.(6.108a) by backward substitution. With $\mathbf{x}$ known, $\ddot{\boldsymbol{\theta}}$ is computed from eq.(6.108b) by forward substitution, thereby completing the computation of $\ddot{\boldsymbol{\theta}}$. The complexity of the foregoing algorithm is discussed in Subsection 6.6.2.

Alternatively, $\ddot{\boldsymbol{\theta}}$ can be calculated in two steps from two linear systems of equations, the first one underdetermined, the second overdetermined. Indeed, if we let the product $\mathbf{P}\ddot{\boldsymbol{\theta}}$ be denoted by $\mathbf{y}$, then the dynamics model of the manipulator, eq.(6.61), along with the factoring of eq.(6.100), leads to

$$\mathbf{P}^T\mathbf{y} = -\overline{\boldsymbol{\tau}} + \boldsymbol{\tau} + \boldsymbol{\delta} \tag{6.109a}$$

$$\mathbf{P}\ddot{\boldsymbol{\theta}} = \mathbf{y} \tag{6.109b}$$

Thus, in the above equations, $\mathbf{y}$ is calculated first as the *minimum-norm* solution of eq.(6.109a); then, the desired value of $\ddot{\boldsymbol{\theta}}$ is calculated as the *least-square approximation* of eq.(6.109b). These two solutions are computed most efficiently using an orthogonalization algorithm that reduces matrix $\mathbf{P}$ to upper-triangular form (Golub and Van Loan, 1989). A straightforward calculation based on the explicit calculation of the generalized inverses involved is not recommended, because of the frequent numerical ill-conditioning incurred. Two orthogonalization procedures, one based on *Householder reflections*, the other on the Gram-Schmidt procedure, for the computation of both the least-square approximation of an overdetermined system of equations and the minimum-norm solution of its underdetermined counterpart are outlined in Appendix B.

The complexity of the foregoing calculations is discussed in Subsection 6.6.2, based on the Cholesky decomposition of the generalized inertia matrix, details on the alternative approach being available elsewhere (Angeles and Ma, 1988).

6.6.1 Planar Manipulators

The application of the natural orthogonal complement to planar manipulators is straightforward. Here, we assume that the manipulator at hand is composed of n links coupled by n joints of the revolute or the prismatic type. Moreover, for conciseness, we assume that the first link, labeled the base, is fixed to an inertial frame. We now adopt the planar representation of the twists and wrenches introduced in Section 4.8; that is, we define the twist of the ith link and the wrench acting on it as 3-dimensional arrays, namely,

$$\mathbf{t}_i \equiv \begin{bmatrix} \omega_i \\ \dot{\mathbf{c}}_i \end{bmatrix}, \quad \mathbf{w}_i \equiv \begin{bmatrix} n_i \\ \mathbf{f}_i \end{bmatrix} \tag{6.110}$$

where ω_i is the scalar angular velocity of this link; $\dot{\mathbf{c}}_i$ is the 2-dimensional velocity of its mass center, C_i; n_i is the scalar moment acting on the link; and $\mathbf{f}_i$ is the 2-dimensional force acting at C_i. Moreover, the inertia dyad is now a 3×3 matrix, i.e.,

$$\mathbf{M}_i \equiv \begin{bmatrix} I_i & \mathbf{0}^T \\ \mathbf{0} & m_i\mathbf{1} \end{bmatrix} \tag{6.111}$$

with I_i defined as the scalar moment of inertia of the ith link about an axis passing through its center of mass, in the direction normal to the plane of motion, while $\mathbf{0}$ is the 2-dimensional zero vector and $\mathbf{1}$ is the 2×2 identity matrix.

Furthermore, the Newton-Euler equations of the ith link take on the forms

$$n_i = I_i\dot{\omega}_i \tag{6.112a}$$

$$\mathbf{f}_i = m_i\ddot{\mathbf{c}}_i \tag{6.112b}$$

and so, these equations can now be cast in the form

$$\mathbf{M}_i\dot{\mathbf{t}}_i = \mathbf{w}_i^W + \mathbf{w}_i^C, \quad i = 1, \ldots, n \tag{6.113}$$

where we have decomposed the total wrench acting on the ith link into its *working* component $\mathbf{w}_i^W$, supplied by the environment and accounting for motor and joint dissipative torques, and $\mathbf{w}_i^C$, the nonworking constraint wrench, supplied by the neighboring links via the coupling joints. The latter, it is recalled, develop no power, their sole role being to keep the links together. An essential difference from the general 6-dimensional counterpart of the foregoing equation, namely, eq.(6.49), is the lack of a quadratic term in ω_i in eq.(6.112a) and consequently, the lack of a $\mathbf{W}_i\mathbf{M}_i\mathbf{t}_i$ term in eq.(6.113).

Upon assembling the foregoing $3n$ equations of motion, we obtain a system of $3n$ uncoupled equations in the form

$$\mathbf{M}\dot{\mathbf{t}} = \mathbf{w}^W + \mathbf{w}^C$$

Now, the wrench $\mathbf{w}^W$ accounts for active forces and moments exerted on the manipulator, and so we can decompose this wrench into an actuator-supplied wrench $\mathbf{w}^A$ and a gravity wrench $\mathbf{w}^G$.

In the next step of the formulation, we set up the kinematic constraints in linear homogeneous form, as in eq.(6.51), with the difference that now, in the presence of n kinematic pairs of the revolute or the prismatic type, $\mathbf{K}$ is a $3n \times n$ matrix. Moreover, we set up the twist-shape relations in the form of eq.(6.57), except that now, $\mathbf{T}$ is a $3n \times n$ matrix. The derivation of the Euler-Lagrange equations for planar motion using the natural orthogonal complement, then, parallels that of general 3-dimensional motion, the model sought taking the form

$$\mathbf{I}(\boldsymbol{\theta})\ddot{\boldsymbol{\theta}} + \mathbf{C}(\boldsymbol{\theta}, \dot{\boldsymbol{\theta}})\dot{\boldsymbol{\theta}} = \boldsymbol{\tau} + \boldsymbol{\gamma} + \boldsymbol{\delta} \tag{6.114a}$$

with the definitions

$$\mathbf{I}(\boldsymbol{\theta}) \equiv \mathbf{T}^T\mathbf{M}\mathbf{T}, \quad \mathbf{C}(\boldsymbol{\theta}, \dot{\boldsymbol{\theta}}) \equiv \mathbf{T}^T\mathbf{M}\dot{\mathbf{T}}, \tag{6.114b}$$

$$\boldsymbol{\tau} \equiv \mathbf{T}^T\mathbf{w}^A, \quad \boldsymbol{\gamma} \equiv \mathbf{T}^T\mathbf{w}^G, \quad \boldsymbol{\delta} \equiv \mathbf{T}^T\mathbf{w}^D \tag{6.114c}$$

We can illustrate best this formulation with the aid of the example below.

Example 6.6.1 (Dynamics of a planar three-revolute manipulator) *Derive the model of the manipulator of Fig. 4.24, under the assumptions of Example 6.3.1, but now using the natural orthogonal complement.*

Solution: We start by deriving all kinematics-related variables, and thus,

$$\omega_1 = \dot{\theta}_1, \quad \omega_2 = \dot{\theta}_1 + \dot{\theta}_2, \quad \omega_3 = \dot{\theta}_1 + \dot{\theta}_2 + \dot{\theta}_3$$

Furthermore,

$$\begin{aligned}\mathbf{t}_1 &= \dot{\theta}_1\mathbf{t}_{11}\\ \mathbf{t}_2 &= \dot{\theta}_1\mathbf{t}_{12} + \dot{\theta}_2\mathbf{t}_{22}\\ \mathbf{t}_3 &= \dot{\theta}_1\mathbf{t}_{13} + \dot{\theta}_2\mathbf{t}_{23} + \dot{\theta}_3\mathbf{t}_{33}\end{aligned}$$

where

$$\mathbf{t}_{11} = \begin{bmatrix} 1 \\ \mathbf{E}\mathbf{r}_{11} \end{bmatrix} = \begin{bmatrix} 1 \\ \mathbf{E}\boldsymbol{\rho}_1 \end{bmatrix} = \begin{bmatrix} 1 \\ (1/2)\mathbf{E}\mathbf{a}_1 \end{bmatrix}$$

$$\mathbf{t}_{21} = \begin{bmatrix} 1 \\ \mathbf{E}\mathbf{r}_{12} \end{bmatrix} = \begin{bmatrix} 1 \\ \mathbf{E}(\mathbf{a}_1 + \boldsymbol{\rho}_2) \end{bmatrix} = \begin{bmatrix} 1 \\ \mathbf{E}(\mathbf{a}_1 + (1/2)\mathbf{a}_2) \end{bmatrix}$$

$$\mathbf{t}_{22} = \begin{bmatrix} 1 \\ \mathbf{E}\mathbf{r}_{22} \end{bmatrix} = \begin{bmatrix} 1 \\ \mathbf{E}\boldsymbol{\rho}_2 \end{bmatrix} = \begin{bmatrix} 1 \\ (1/2)\mathbf{E}\mathbf{a}_2 \end{bmatrix}$$

$$\mathbf{t}_{31} = \begin{bmatrix} 1 \\ \mathbf{E}\mathbf{r}_{13} \end{bmatrix} = \begin{bmatrix} 1 \\ \mathbf{E}(\mathbf{a}_1 + \mathbf{a}_2 + \boldsymbol{\rho}_3) \end{bmatrix} = \begin{bmatrix} 1 \\ \mathbf{E}(\mathbf{a}_1 + \mathbf{a}_2 + (1/2)\mathbf{a}_3) \end{bmatrix}$$

$$\mathbf{t}_{32} = \begin{bmatrix} 1 \\ \mathbf{E}\mathbf{r}_{23} \end{bmatrix} = \begin{bmatrix} 1 \\ \mathbf{E}(\mathbf{a}_2 + \boldsymbol{\rho}_3) \end{bmatrix} = \begin{bmatrix} 1 \\ \mathbf{E}(\mathbf{a}_2 + (1/2)\mathbf{a}_3) \end{bmatrix}$$

$$\mathbf{t}_{33} = \begin{bmatrix} 1 \\ \mathbf{E}\boldsymbol{\rho}_3 \end{bmatrix} = \begin{bmatrix} 1 \\ (1/2)\mathbf{E}\mathbf{a}_3 \end{bmatrix}$$

and hence, the 9×3 twist-shaping matrix $\mathbf{T}$ becomes

$$\mathbf{T} = \begin{bmatrix} 1 & 0 & 0 \\ (1/2)\mathbf{E}\mathbf{a}_1 & \mathbf{0} & \mathbf{0} \\ 1 & 1 & 0 \\ \mathbf{E}(\mathbf{a}_1 + (1/2)\mathbf{a}_2) & (1/2)\mathbf{E}\mathbf{a}_2 & \mathbf{0} \\ 1 & 1 & 1 \\ \mathbf{E}(\mathbf{a}_1 + \mathbf{a}_2 + (1/2)\mathbf{a}_3) & \mathbf{E}(\mathbf{a}_2 + (1/2)\mathbf{a}_3) & (1/2)\mathbf{E}\mathbf{a}_3 \end{bmatrix}$$

The 9×9 matrix of inertia dyads of this manipulator now takes the form

$$\mathbf{M} = \text{diag}(\mathbf{M}_1, \mathbf{M}_2, \mathbf{M}_3)$$

with each 3×3 $\mathbf{M}_i$ matrix defined as

$$\mathbf{M}_i \equiv \begin{bmatrix} I_i & \mathbf{0}^T \\ \mathbf{0} & m_i\mathbf{1} \end{bmatrix}$$

Now, the 3×3 generalized inertia matrix is readily derived as

$$\mathbf{I} \equiv \mathbf{T}^T\mathbf{M}\mathbf{T}$$

whose entries are given below:

$$\begin{aligned} I_{11} &= \mathbf{t}_{11}^T\mathbf{M}_1\mathbf{t}_{11} + \mathbf{t}_{21}^T\mathbf{M}_2\mathbf{t}_{21} + \mathbf{t}_{31}^T\mathbf{M}_3\mathbf{t}_{31} \\ I_{12} &= \mathbf{t}_{21}^T\mathbf{M}_2\mathbf{t}_{22} + \mathbf{t}_{31}^T\mathbf{M}_3\mathbf{t}_{32} = I_{21} \\ I_{13} &= \mathbf{t}_{31}^T\mathbf{M}_3\mathbf{t}_{33} = I_{31} \\ I_{22} &= \mathbf{t}_{22}^T\mathbf{M}_2\mathbf{t}_{22} + \mathbf{t}_{32}^T\mathbf{M}_3\mathbf{t}_{32} \\ I_{23} &= \mathbf{t}_{32}^T\mathbf{M}_3\mathbf{t}_{33} = I_{32} \\ I_{33} &= \mathbf{t}_{33}^T\mathbf{M}_3\mathbf{t}_{33} \end{aligned}$$

Upon expansion, the above entries result in exactly the same expressions as those derived in Example 6.3.1, thereby confirming the correctness of the two derivations. Furthermore, the next term in the Euler-Lagrange equations is derived below. Here, we will need $\dot{\mathbf{T}}$, which is readily derived from the above expression for $\mathbf{T}$. In deriving this time-derivative, we note that in general, for $i = 1, 2, 3$,

$$\dot{\mathbf{a}}_i = \omega_i\mathbf{E}\mathbf{a}_i, \quad \mathbf{E}^2\mathbf{a}_i = -\mathbf{a}_i$$

and hence,

$$\dot{\mathbf{T}} = -\begin{bmatrix} 0 & 0 & 0 \\ (1/2)\dot{\theta}_1\mathbf{a}_1 & \mathbf{0} & \mathbf{0} \\ 0 & 0 & 0 \\ \dot{\theta}_1\mathbf{a}_1 + (1/2)\dot{\theta}_{12}\mathbf{a}_2 & (1/2)\dot{\theta}_{12}\mathbf{a}_2 & \mathbf{0} \\ 0 & 0 & 0 \\ \dot{\theta}_1\mathbf{a}_1 + \dot{\theta}_{12}\mathbf{a}_2 + (1/2)\dot{\theta}_{123}\mathbf{a}_3 & \dot{\theta}_{12}\mathbf{a}_2 + (1/2)\dot{\theta}_{123}\mathbf{a}_3 & (1/2)\dot{\theta}_{123}\mathbf{a}_3 \end{bmatrix}$$

where $\dot{\theta}_{12}$ and $\dot{\theta}_{123}$ stand for $\dot{\theta}_1 + \dot{\theta}_2$ and $\dot{\theta}_1 + \dot{\theta}_2 + \dot{\theta}_3$, respectively.

We now can perform the product $\mathbf{T}^T\mathbf{M}\dot{\mathbf{T}}$, whose (i, j) entry will be represented as μ_{ij}. Below we display the expressions for these entries:

$$
\begin{aligned}
\mu_{11} &= -\frac{1}{2}[m_2a_1a_2s_2 + m_3(2a_1a_2s_2 + a_1a_3s_{23})]\dot{\theta}_2 - \frac{1}{2}(a_1a_3s_{23} + a_2a_3s_3)\dot{\theta}_3 \\
\mu_{12} &= -\frac{1}{2}[m_2a_1a_2s_2 + m_3(2a_1a_2s_2 + a_1a_3s_{23})]\dot{\theta}_1 \\
&\quad -\frac{1}{2}[m_2a_1a_2s_2 + m_3(2a_1a_2s_2 + a_1a_3s_{23})]\dot{\theta}_2 \\
&\quad -\frac{1}{2}m_3(a_1a_3s_{23} + a_2a_3s_3)\dot{\theta}_3 \\
\mu_{13} &= -\frac{1}{2}m_3(a_1a_3s_{23} + a_2a_3s_3)(\dot{\theta}_1 + \dot{\theta}_2 + \dot{\theta}_3) \\
\mu_{21} &= \frac{1}{2}[m_2a_1a_2s_2 + m_3(2a_1a_2s_2 + a_1a_3s_{23})]\dot{\theta}_1 - \frac{1}{2}m_3a_2a_3s_3\dot{\theta}_3 \\
\mu_{22} &= -\frac{1}{2}m_3a_2a_3s_3\dot{\theta}_3 \\
\mu_{23} &= -\frac{1}{2}m_3a_2a_3s_3(\dot{\theta}_1 + \dot{\theta}_2 + \dot{\theta}_3) \\
\mu_{31} &= \frac{1}{2}[m_3(a_1a_3s_{23} + a_2a_3s_3)\dot{\theta}_1 + a_2a_3s_3\dot{\theta}_2] \\
\mu_{32} &= \frac{1}{2}m_3a_2a_3s_3(\dot{\theta}_1 + \dot{\theta}_2) \\
\mu_{33} &= 0
\end{aligned}
$$

Now, let us define

$$\boldsymbol{\nu} \equiv \mathbf{T}^T\mathbf{M}\dot{\mathbf{T}}\dot{\boldsymbol{\theta}}$$

whose three components are given below:

$$
\begin{aligned}
\nu_1 &= -[m_2a_1a_2s_2 + m_3(2a_1a_2s_2 + a_1a_3s_{23})]\dot{\theta}_1\dot{\theta}_2 \\
&\quad -m_3(a_1a_3s_{23} + a_2a_3s_3)\dot{\theta}_1\dot{\theta}_3 \\
&\quad -\frac{1}{2}[m_2a_1a_2s_2 + m_3(2a_1a_2s_2 + a_1a_3s_{23})]\dot{\theta}_2^2 \\
&\quad -m_3(a_1a_3s_{23} + a_2a_3s_3)\dot{\theta}_2\dot{\theta}_3 - \frac{1}{2}m_3(a_1a_3s_{23} + a_2a_3s_3)\dot{\theta}_3^2 \\
\nu_2 &= \frac{1}{2}[m_2a_1a_2s_2 + m_3(2a_1a_2s_2 + a_1a_3s_{23})]\dot{\theta}_1^2 - m_3a_2a_3s_3\dot{\theta}_1\dot{\theta}_3 \\
&\quad -m_3a_2a_3s_3\dot{\theta}_2\dot{\theta}_3 - \frac{1}{2}m_3a_2a_3s_3\dot{\theta}_3^2 \\
\nu_3 &= \frac{1}{2}m_3(a_1a_3s_{23} + a_2a_3s_3)\dot{\theta}_1^2 + m_3a_2a_3s_3\dot{\theta}_1\dot{\theta}_2 + \frac{1}{2}m_3a_2a_3s_3\dot{\theta}_3^2
\end{aligned}
$$

The mathematical model sought, thus, takes the form

$$\mathbf{I}(\boldsymbol{\theta})\ddot{\boldsymbol{\theta}} + \boldsymbol{\nu}(\boldsymbol{\theta}, \dot{\boldsymbol{\theta}}) = \boldsymbol{\tau} + \boldsymbol{\gamma}$$

where $\boldsymbol{\delta} = \mathbf{0}$ because we have not included dissipation. Moreover, γ is derived as described below: Let $\mathbf{w}_i^G$ be the gravity wrench acting on the ith link, $\mathbf{w}^G$ then being

$$\mathbf{w}^G = \begin{bmatrix} \mathbf{w}_1^G \\ \mathbf{w}_2^G \\ \mathbf{w}_3^G \end{bmatrix}$$

and

$$\mathbf{w}_1^G = \begin{bmatrix} 0 \\ -m_1 g\mathbf{j} \end{bmatrix}, \quad \mathbf{w}_2^G = \begin{bmatrix} 0 \\ -m_2 g\mathbf{j} \end{bmatrix}, \quad \mathbf{w}_3^G = \begin{bmatrix} 0 \\ -m_3 g\mathbf{j} \end{bmatrix}$$

Therefore,

$$\begin{aligned} \gamma &= \mathbf{T}^T \mathbf{w}^G \\ &= \frac{g}{2} \begin{bmatrix} m_1 \mathbf{a}_1^T \mathbf{E}\mathbf{j} + m_2(2\mathbf{a}_1 + \mathbf{a}_2)^T \mathbf{E}\mathbf{j} + m_3[2(\mathbf{a}_1 + \mathbf{a}_2) + \mathbf{a}_3)^T \mathbf{E}\mathbf{j} \\ m_2 \mathbf{a}_1^T \mathbf{E}\mathbf{j} + m_3(2\mathbf{a}_2 + \mathbf{a}_3)^T \mathbf{E}\mathbf{j} \\ m_3 \mathbf{a}_3^T \mathbf{E}\mathbf{j} \end{bmatrix} \end{aligned}$$

But

$$\begin{aligned} \mathbf{a}_1^T \mathbf{E}\mathbf{j} &= -\mathbf{a}_1^T \mathbf{i} = -a_1 \cos\theta_1 \\ \mathbf{a}_2^T \mathbf{E}\mathbf{j} &= -\mathbf{a}_2^T \mathbf{i} = -a_2 \cos(\theta_1 + \theta_2) \\ \mathbf{a}_3^T \mathbf{E}\mathbf{j} &= -\mathbf{a}_3^T \mathbf{i} = -a_3 \cos(\theta_1 + \theta_2 + \theta_3) \end{aligned}$$

Hence,

$$\gamma = \frac{g}{2} \begin{bmatrix} -m_1 a_1 c_1 - 2m_2(a_1 c_1 + a_2 c_{12}) - 2m_3(a_1 c_1 + a_2 c_{12} + a_3 c_{123}) \\ -m_2 a_2 c_{12} - 2m_3(a_2 c_{12} + a_3 c_{123}) \\ -m_3 a_3 c_{123} \end{bmatrix}$$

with the definitions for c_1, c_{12}, and c_{123} introduced in Example 6.3.1. As the reader can verify, the foregoing model is identical to the model derived with the Euler-Lagrange equations in that example.

Example 6.6.2 (Dynamics of a spatial 3-revolute manipulator) *The manipulator of Fig. 4.15 is reproduced in Fig. 6.7, in a form that is kinematically equivalent to the sketch of that figure, but more suitable for the purposes of this example. For this manipulator, (i) find its inertia matrix at the configuration depicted in that figure; (ii) find the time-rate of change of the inertia matrix under a maneuver whereby* $\dot\theta_1 = \dot\theta_2 = \dot\theta_3 = p\ \mathrm{s}^{-1}$ *and* $\ddot\theta_1 = \ddot\theta_2 = \ddot\theta_3 = 0$*; and (iii) under the same maneuver, find the centrifugal and Coriolis terms of its governing equation. Furthermore, assume that all links are identical and* dynamically isotropic. *What we mean by "dynamically isotropic" is that the moment of inertia of all three links about their mass centers are proportional to the* 3×3 *identity matrix, the proportionality factor being* I*. Moreover, all three links are designed so that the mass center of each is located as shown in Fig. 6.7.*

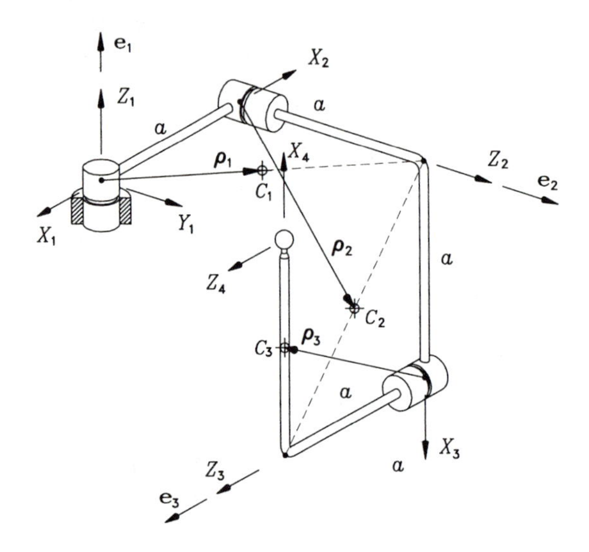

FIGURE 6.7. Mass-center locations of the manipulator of Fig. 4.19.

Solution:

(*i*) Henceforth, we represent all vectors and matrices with respect to the $\mathcal{F}_1$-frame of Fig. 6.7, while denoting by $\mathbf{i}, \mathbf{j}$, and $\mathbf{k}$ the unit vectors parallel to the X_1, Y_1, and Z_1 axes, respectively. Under these conditions, we have, for the unit vectors parallel to the revolute axes,

$$\mathbf{e}_1 = \mathbf{k}, \quad \mathbf{e}_2 = \mathbf{j}, \quad \mathbf{e}_3 = \mathbf{i}$$

while vector $\mathbf{a}_i$ is directed from the origin of $\mathcal{F}_i$ to that of $\mathcal{F}_{i+1}$, for $i = 1, 2, 3$. Hence,

$$\mathbf{a}_1 = -a\mathbf{i}, \quad \mathbf{a}_2 = a(\mathbf{j} - \mathbf{k}), \quad \mathbf{a}_3 = a(\mathbf{i} + \mathbf{k})$$

Likewise, the position vectors of the mass centers, $\boldsymbol{\rho}_i$, for $i = 1$, 2, and 3, with respect to the origins of their respective frames, are given by

$$\boldsymbol{\rho}_1 = \frac{1}{2}a(-\mathbf{i} + \mathbf{j})$$
$$\boldsymbol{\rho}_2 = \frac{1}{2}a(\mathbf{i} + 2\mathbf{j} - \mathbf{k})$$
$$\boldsymbol{\rho}_3 = \frac{1}{2}a(2\mathbf{i} + \mathbf{k})$$

We can now calculate the various 6-dimensional arrays $\mathbf{t}_{ij}$, for $i = 1$, 2, 3, and $j = 1$ till i, i.e.,

$$\mathbf{t}_{11} = \begin{bmatrix} \mathbf{e}_1 \\ \mathbf{e}_1 \times \boldsymbol{\rho}_1 \end{bmatrix} = \begin{bmatrix} \mathbf{k} \\ -(a/2)(\mathbf{i} + \mathbf{j}) \end{bmatrix}$$

$$\mathbf{t}_{21} = \begin{bmatrix} \mathbf{e}_1 \\ \mathbf{e}_1 \times (\mathbf{a}_1 + \boldsymbol{\rho}_2) \end{bmatrix} = \begin{bmatrix} \mathbf{k} \\ -(a/2)(2\mathbf{i} + \mathbf{j}) \end{bmatrix}$$

$$\mathbf{t}_{22} = \begin{bmatrix} \mathbf{e}_2 \\ \mathbf{e}_2 \times \boldsymbol{\rho}_2 \end{bmatrix} = \begin{bmatrix} \mathbf{j} \\ -(a/2)(\mathbf{i} + \mathbf{k}) \end{bmatrix}$$

$$\mathbf{t}_{31} = \begin{bmatrix} \mathbf{e}_1 \\ \mathbf{e}_1 \times (\mathbf{a}_1 + \mathbf{a}_2 + \boldsymbol{\rho}_3) \end{bmatrix} = \begin{bmatrix} \mathbf{k} \\ -a\mathbf{i} \end{bmatrix}$$

$$\mathbf{t}_{32} = \begin{bmatrix} \mathbf{e}_2 \\ \mathbf{e}_2 \times (\mathbf{a}_2 + \boldsymbol{\rho}_3) \end{bmatrix} = \begin{bmatrix} \mathbf{j} \\ -(a/2)(\mathbf{i} + 2\mathbf{k}) \end{bmatrix}$$

$$\mathbf{t}_{33} = \begin{bmatrix} \mathbf{e}_3 \\ \mathbf{e}_3 \times \boldsymbol{\rho}_3 \end{bmatrix} = \begin{bmatrix} \mathbf{i} \\ -(a/2)\mathbf{j} \end{bmatrix}$$

and so, the 18×3 matrix $\mathbf{T}$ is given by

$$\mathbf{T} = \begin{bmatrix} \mathbf{k} & \mathbf{0} & \mathbf{0} \\ -(a/2)(\mathbf{i} + \mathbf{j}) & \mathbf{0} & \mathbf{0} \\ \mathbf{k} & \mathbf{j} & \mathbf{0} \\ -(a/2)(2\mathbf{i} + \mathbf{j}) & -(a/2)(\mathbf{i} + \mathbf{k}) & \mathbf{0} \\ \mathbf{k} & \mathbf{j} & \mathbf{i} \\ -a\mathbf{i} & -(a/2)(\mathbf{i} + 2\mathbf{k}) & -(a/2)\mathbf{j} \end{bmatrix}$$

Moreover, the 6×6 inertia dyad of the ith link takes the form

$$\mathbf{M}_i = \begin{bmatrix} I\mathbf{1} & \mathbf{O} \\ \mathbf{O} & m\mathbf{1} \end{bmatrix}, \quad i = 1, 2, 3$$

with $\mathbf{1}$ and $\mathbf{O}$ denoting the 3×3 identity and zero matrices, respectively. Thus, the 18×18 system mass matrix is given as

$$\mathbf{M} = \mathrm{diag}(\mathbf{M}_1, \mathbf{M}_2, \mathbf{M}_3)$$

and the 3×3 generalized inertia matrix $\mathbf{I}$ of the manipulator is

$$\mathbf{I} = \mathbf{T}^T \mathbf{M} \mathbf{T}$$

whose entries are given by

$$\begin{aligned} I_{11} &= \mathbf{t}_{11}^T \mathbf{M}_1 \mathbf{t}_{11} + \mathbf{t}_{21}^T \mathbf{M}_2 \mathbf{t}_{21} + \mathbf{t}_{31}^T \mathbf{M}_3 \mathbf{t}_{31} \\ I_{12} &= \mathbf{t}_{21}^T \mathbf{M}_2 \mathbf{t}_{22} + \mathbf{t}_{31}^T \mathbf{M}_3 \mathbf{t}_{32} = I_{21} \\ I_{13} &= \mathbf{t}_{31}^T \mathbf{M}_3 \mathbf{t}_{33} = I_{31} \\ I_{22} &= \mathbf{t}_{22}^T \mathbf{M}_2 \mathbf{t}_{22} + \mathbf{t}_{32}^T \mathbf{M}_3 \mathbf{t}_{32} \\ I_{23} &= \mathbf{t}_{32}^T \mathbf{M}_3 \mathbf{t}_{33} = I_{32} \\ I_{33} &= \mathbf{t}_{33}^T \mathbf{M}_3 \mathbf{t}_{33} \end{aligned}$$

Upon expansion, the foregoing expressions yield

$$\mathbf{I} = \frac{1}{4} m a^2 \begin{bmatrix} 11 & 4 & 0 \\ 4 & 7 & 0 \\ 0 & 0 & 1 \end{bmatrix} + I \begin{bmatrix} 3 & 0 & 0 \\ 0 & 2 & 0 \\ 0 & 0 & 1 \end{bmatrix}$$

(*ii*) Now, the time-rate of change of $\mathbf{I}$, $\dot{\mathbf{I}}$, is calculated as

$$\dot{\mathbf{I}} = \mathbf{T}^T\mathbf{M}\dot{\mathbf{T}} + \dot{\mathbf{T}}^T\mathbf{M}\mathbf{T}^T + \mathbf{T}^T(\mathbf{W}\mathbf{M} - \mathbf{M}\mathbf{W})\mathbf{T}$$

We proceed first to compute $\dot{\mathbf{T}}$. This time-derivative is nothing but the 18×3 matrix whose entries are the time-derivatives of the entries of $\mathbf{T}$, namely, $\dot{\mathbf{t}}_{ij}$, as given in eq.(6.84), which is reproduced below for quick reference:

$$\dot{\mathbf{t}}_{ij} = \begin{bmatrix} \boldsymbol{\omega}_j \times \mathbf{e}_j \\ (\boldsymbol{\omega}_j \times \mathbf{e}_j) \times \mathbf{r}_{ij} + \mathbf{e}_j \times \dot{\mathbf{r}}_{ij} \end{bmatrix}$$

where $\dot{\mathbf{r}}_{ij}$ is given, in turn, by

$$\dot{\mathbf{r}}_{ij} = \boldsymbol{\omega}_j \times \mathbf{a}_j + \ldots + \boldsymbol{\omega}_{i-1} \times \mathbf{a}_{i-1} + \boldsymbol{\omega}_i \times \boldsymbol{\rho}_i$$

Hence, we will need vectors $\boldsymbol{\omega}_i$, for $i = 1$, 2, and 3. These are calculated below:

$$\begin{aligned}
\boldsymbol{\omega}_1 &= \dot{\theta}_1\mathbf{e}_1 = p\mathbf{k} \\
\boldsymbol{\omega}_2 &= \dot{\theta}_1\mathbf{e}_1 + \dot{\theta}_2\mathbf{e}_2 = p(\mathbf{j} + \mathbf{k}) \\
\boldsymbol{\omega}_3 &= \dot{\theta}_1\mathbf{e}_1 + \dot{\theta}_2\mathbf{e}_2 + \dot{\theta}_3\mathbf{e}_3 = p(\mathbf{i} + \mathbf{j} + \mathbf{k})
\end{aligned}$$

We have, therefore,

$$\dot{\mathbf{t}}_{11} = \begin{bmatrix} \dot{\mathbf{e}}_1 \\ \dot{\mathbf{e}}_1 \times \boldsymbol{\rho}_1 + \mathbf{e}_1 \times \dot{\boldsymbol{\rho}}_1 \end{bmatrix} = \begin{bmatrix} \mathbf{0} \\ \mathbf{e}_1 \times (\boldsymbol{\omega}_1 \times \boldsymbol{\rho}_1) \end{bmatrix} = p\begin{bmatrix} \mathbf{0} \\ (1/2)a(\mathbf{i} - \mathbf{j}) \end{bmatrix}$$

$$\begin{aligned}
\dot{\mathbf{t}}_{21} &= \begin{bmatrix} \dot{\mathbf{e}}_1 \\ \dot{\mathbf{e}}_1 \times (\mathbf{a}_1 + \boldsymbol{\rho}_2) + \mathbf{e}_1 \times (\dot{\mathbf{a}}_1 + \dot{\boldsymbol{\rho}}_2) \end{bmatrix} \\
&= \begin{bmatrix} \mathbf{0} \\ \mathbf{e}_1 \times (\boldsymbol{\omega}_1 \times \mathbf{a}_1 + \boldsymbol{\omega}_2 \times \boldsymbol{\rho}_2) \end{bmatrix} = p\begin{bmatrix} \mathbf{0} \\ (1/2)a\mathbf{j} \end{bmatrix}
\end{aligned}$$

$$\begin{aligned}
\dot{\mathbf{t}}_{22} &= \begin{bmatrix} \dot{\mathbf{e}}_2 \\ \dot{\mathbf{e}}_2 \times \boldsymbol{\rho}_2 + \mathbf{e}_2 \times \dot{\boldsymbol{\rho}}_2 \end{bmatrix} \\
&= \begin{bmatrix} p\mathbf{e}_1 \times \mathbf{e}_2 \\ (p\mathbf{e}_1 \times \mathbf{e}_2) \times \boldsymbol{\rho}_2 + \mathbf{e}_2 \times [p(\mathbf{e}_1 + \mathbf{e}_2) \times \boldsymbol{\rho}_2] \end{bmatrix} \\
&= p\begin{bmatrix} -\mathbf{i} \\ -(1/2)a(\mathbf{i} + \mathbf{j} - \mathbf{k}) \end{bmatrix}
\end{aligned}$$

$$\begin{aligned}
\dot{\mathbf{t}}_{31} &= \begin{bmatrix} \dot{\mathbf{e}}_1 \\ \dot{\mathbf{e}}_1 \times (\mathbf{a}_1 + \mathbf{a}_2 + \boldsymbol{\rho}_3) + \mathbf{e}_1 \times (\dot{\mathbf{a}}_1 + \dot{\mathbf{a}}_2 + \dot{\boldsymbol{\rho}}_3) \end{bmatrix} \\
&= \begin{bmatrix} \mathbf{0} \\ \mathbf{e}_1 \times (\boldsymbol{\omega}_1 \times \mathbf{a}_1 + \boldsymbol{\omega}_2 \times \mathbf{a}_2 + \boldsymbol{\omega}_3 \times \boldsymbol{\rho}_3) \end{bmatrix} \\
&= \begin{bmatrix} \mathbf{0} \\ \mathbf{e}_1 \times [p\mathbf{e}_1 \times \mathbf{a}_1 + p(\mathbf{e}_1 + \mathbf{e}_2) \times \mathbf{a}_2 + p(\mathbf{e}_1 + \mathbf{e}_2 + \mathbf{e}_3) \times \boldsymbol{\rho}_3] \end{bmatrix} \\
&= p\begin{bmatrix} \mathbf{0} \\ -a\mathbf{j} \end{bmatrix}
\end{aligned}$$

$$
\begin{aligned}
\dot{\mathbf{t}}_{32} &= \begin{bmatrix} \dot{\mathbf{e}}_2 \\ \dot{\mathbf{e}}_2 \times (\mathbf{a}_2 + \boldsymbol{\rho}_3) + \mathbf{e}_2 \times (\dot{\mathbf{a}}_2 + \dot{\boldsymbol{\rho}}_3) \end{bmatrix} \\
&= \begin{bmatrix} p\mathbf{e}_1 \times \mathbf{e}_2 \\ (p\mathbf{e}_1 \times \mathbf{e}_2) \times (\mathbf{a}_2 + \boldsymbol{\rho}_3) + p\mathbf{e}_2 \times [(\mathbf{e}_1 + \mathbf{e}_2) \times (\mathbf{a}_2 + \boldsymbol{\rho}_3)] \end{bmatrix} \\
&= p \begin{bmatrix} -\mathbf{i} \\ -(1/2)a(2\mathbf{i} + \mathbf{j} - \mathbf{k}) \end{bmatrix} \\
\dot{\mathbf{t}}_{33} &= \begin{bmatrix} \dot{\mathbf{e}}_3 \\ \dot{\mathbf{e}}_3 \times \boldsymbol{\rho}_3 + \mathbf{e}_3 \times \dot{\boldsymbol{\rho}}_3 \end{bmatrix} = \begin{bmatrix} \boldsymbol{\omega}_2 \times \mathbf{e}_3 \\ (\boldsymbol{\omega}_2 \times \mathbf{e}_3) \times \boldsymbol{\rho}_3 + \mathbf{e}_3 \times (\boldsymbol{\omega}_3 \times \boldsymbol{\rho}_3) \end{bmatrix} \\
&= \begin{bmatrix} p(\mathbf{e}_1 + \mathbf{e}_2) \times \mathbf{e}_3 \\ p[(\mathbf{e}_1 + \mathbf{e}_2) \times \mathbf{e}_3] \times \boldsymbol{\rho}_3 + p\mathbf{e}_3 \times [(\mathbf{e}_1 + \mathbf{e}_2 + \mathbf{e}_3) \times \boldsymbol{\rho}_3] \end{bmatrix} \\
&= \begin{bmatrix} p(\mathbf{e}_2 - \mathbf{e}_1) \\ p(\mathbf{e}_2 - \mathbf{e}_1) \times \boldsymbol{\rho}_3 + p[(\mathbf{e}_3 \cdot \boldsymbol{\rho}_3)(\mathbf{e}_1 + \mathbf{e}_2 + \mathbf{e}_3) - \boldsymbol{\rho}_3] \end{bmatrix} \\
&= p \begin{bmatrix} \mathbf{j} - \mathbf{k} \\ (1/2)a(\mathbf{i} - \mathbf{k}) \end{bmatrix}
\end{aligned}
$$

Now, let

$$\mathbf{P} \equiv \mathbf{T}^T \mathbf{M} \dot{\mathbf{T}}$$

whose entries are displayed below:

$$
\begin{aligned}
p_{11} &= \mathbf{t}_{11}^T \mathbf{M}_1 \dot{\mathbf{t}}_{11} + \mathbf{t}_{21}^T \mathbf{M}_2 \dot{\mathbf{t}}_{21} + \mathbf{t}_{31}^T \mathbf{M}_3 \dot{\mathbf{t}}_{31} \\
p_{12} &= \mathbf{t}_{21}^T \mathbf{M}_2 \dot{\mathbf{t}}_{22} + \mathbf{t}_{31}^T \mathbf{M}_3 \dot{\mathbf{t}}_{32} \\
p_{13} &= \mathbf{t}_{31}^T \mathbf{M}_3 \dot{\mathbf{t}}_{33} \\
p_{21} &= \mathbf{t}_{22}^T \mathbf{M}_2 \dot{\mathbf{t}}_{21} + \mathbf{t}_{32}^T \mathbf{M}_3 \dot{\mathbf{t}}_{31} \\
p_{22} &= \mathbf{t}_{22}^T \mathbf{M}_2 \dot{\mathbf{t}}_{22} + \mathbf{t}_{32}^T \mathbf{M}_3 \dot{\mathbf{t}}_{32} \\
p_{23} &= \mathbf{t}_{32}^T \mathbf{M}_3 \dot{\mathbf{t}}_{33} \\
p_{31} &= \mathbf{t}_{33}^T \mathbf{M}_3 \dot{\mathbf{t}}_{31} \\
p_{32} &= \mathbf{t}_{33}^T \mathbf{M}_3 \dot{\mathbf{t}}_{32} \\
p_{33} &= \mathbf{t}_{33}^T \mathbf{M}_3 \dot{\mathbf{t}}_{33}
\end{aligned}
$$

Upon performing the foregoing operations, we end up with

$$
\mathbf{T}^T \mathbf{M} \dot{\mathbf{T}} = p \begin{bmatrix} -(1/4)a^2 m & (7/4)a^2 m & -(1/2)a^2 m - I \\ -(1/2)a^2 m & 0 & (1/4)a^2 m + I \\ (1/2)a^2 m & (1/4)a^2 m - I & 0 \end{bmatrix} \equiv \mathbf{P}
$$

the second term of the above expression for $\dot{\mathbf{I}}$ simply being $\mathbf{P}^T$. In order to compute the third term, we need the products $\mathbf{WM}$ and $\mathbf{MW}$. However, it is apparent that the latter is the negative of the transpose of the former, and so, all we need is one of the two terms.

Furthermore, note that since both matrices $\mathbf{M}$ and $\mathbf{W}$ are block-diagonal, their product is block-diagonal as well, namely,

$$\mathbf{WM} = \mathrm{diag}(\mathbf{W}_1\mathbf{M}_1,\ \mathbf{W}_2\mathbf{M}_2,\ \mathbf{W}_3\mathbf{M}_3)$$

where for $i = 1$, 2, and 3,

$$\mathbf{W}_i = \begin{bmatrix} \mathbf{\Omega}_i & \mathbf{O} \\ \mathbf{O} & \mathbf{O} \end{bmatrix}$$

with $\mathbf{O}$ denoting the 3×3 zero matrix, while $\mathbf{\Omega}_i$ is the cross-product matrix of vector $\boldsymbol{\omega}_i$. Moreover,

$$\mathbf{W}_i\mathbf{M}_i = \begin{bmatrix} I\mathbf{\Omega}_i & \mathbf{O} \\ \mathbf{O} & \mathbf{O} \end{bmatrix}$$

Therefore, $\mathbf{W}_i\mathbf{M}_i$ is skew-symmetric; as a consequence, $\mathbf{WM}$ is also skew-symmetric, and the difference $\mathbf{WM} - \mathbf{MW}$ vanishes. Hence, in this particular case, $\dot{\mathbf{I}}$ reduces to

$$\dot{\mathbf{I}} = \mathbf{P} + \mathbf{P}^T$$

That is,

$$\dot{\mathbf{I}} = p \begin{bmatrix} -(1/2)a^2m & (5/4)a^2m & -I \\ (5/4)a^2m & 0 & a^2m + I \\ -I & (1/2)a^2m & 0 \end{bmatrix}$$

(*iii*) Now, the term of Coriolis and centrifugal forces can be computed in two ways, namely, (a) as $(\mathbf{T}^T\mathbf{M}\dot{\mathbf{T}} + \mathbf{T}^T\mathbf{WMT})\dot{\boldsymbol{\theta}}$, and (b) by using the Newton-Euler algorithm with $\ddot{\theta}_i = 0$, for $i = 1$, 2, and 3. We proceed in these two ways in order to verify the correctness of our results.

In proceeding with the first alternative, we already have the first term in the foregoing parentheses; the second term is now computed. First, we note that

$$\mathbf{WMT} = \begin{bmatrix} \mathbf{W}_1\mathbf{M}_1\mathbf{t}_{11} & \mathbf{0} & \mathbf{0} \\ \mathbf{W}_2\mathbf{M}_2\mathbf{t}_{21} & \mathbf{W}_2\mathbf{M}_2\mathbf{t}_{22} & \mathbf{0} \\ \mathbf{W}_3\mathbf{M}_3\mathbf{t}_{31} & \mathbf{W}_3\mathbf{M}_3\mathbf{t}_{32} & \mathbf{W}_3\mathbf{M}_3\mathbf{t}_{33} \end{bmatrix}$$

with $\mathbf{0}$ defined as the 6-dimensional zero vector. The foregoing nontrivial 6-dimensional arrays are computed below:

$$\mathbf{W}_1\mathbf{M}_1\mathbf{t}_{11} = \begin{bmatrix} I\mathbf{\Omega}_1 & \mathbf{O} \\ \mathbf{O} & \mathbf{O} \end{bmatrix} \begin{bmatrix} \mathbf{k} \\ -(a/2)(\mathbf{i}+\mathbf{j}) \end{bmatrix} = \begin{bmatrix} I\mathbf{\Omega}_1\mathbf{k} \\ \mathbf{0} \end{bmatrix} = \begin{bmatrix} \mathbf{0} \\ \mathbf{0} \end{bmatrix}$$

$$\mathbf{W}_2\mathbf{M}_2\mathbf{t}_{21} = \begin{bmatrix} I\mathbf{\Omega}_2 & \mathbf{O} \\ \mathbf{O} & \mathbf{O} \end{bmatrix} \begin{bmatrix} \mathbf{k} \\ -(a/2)(2\mathbf{i}+\mathbf{j}) \end{bmatrix} = \begin{bmatrix} I\mathbf{\Omega}_2\mathbf{k} \\ \mathbf{0} \end{bmatrix}$$

$$= \begin{bmatrix} pI(\mathbf{j}+\mathbf{k})\times\mathbf{k} \\ \mathbf{0} \end{bmatrix} = pI \begin{bmatrix} \mathbf{i} \\ \mathbf{0} \end{bmatrix}$$

$$\mathbf{W}_2\mathbf{M}_2\mathbf{t}_{22} = \begin{bmatrix} I\mathbf{\Omega}_2 & \mathbf{O} \\ \mathbf{O} & \mathbf{O} \end{bmatrix} \begin{bmatrix} \mathbf{j} \\ -(a/2)(\mathbf{i}+\mathbf{k}) \end{bmatrix} = \begin{bmatrix} I\mathbf{\Omega}_2\mathbf{j} \\ \mathbf{0} \end{bmatrix}$$

$$= \begin{bmatrix} pI(\mathbf{j}+\mathbf{k})\times\mathbf{j} \\ \mathbf{0} \end{bmatrix} = pI \begin{bmatrix} -\mathbf{i} \\ \mathbf{0} \end{bmatrix}$$

$$\mathbf{W}_3\mathbf{M}_3\mathbf{t}_{31} = \begin{bmatrix} I\mathbf{\Omega}_3 & \mathbf{O} \\ \mathbf{O} & \mathbf{O} \end{bmatrix} \begin{bmatrix} \mathbf{k} \\ -a\mathbf{i} \end{bmatrix} = \begin{bmatrix} I\mathbf{\Omega}_3\mathbf{k} \\ \mathbf{0} \end{bmatrix}$$

$$= \begin{bmatrix} pI(\mathbf{i}+\mathbf{j}+\mathbf{k})\times\mathbf{k} \\ \mathbf{0} \end{bmatrix} = pI \begin{bmatrix} \mathbf{i}-\mathbf{j} \\ \mathbf{0} \end{bmatrix}$$

$$\mathbf{W}_3\mathbf{M}_3\mathbf{t}_{32} = \begin{bmatrix} I\mathbf{\Omega}_3 & \mathbf{O} \\ \mathbf{O} & \mathbf{O} \end{bmatrix} \begin{bmatrix} \mathbf{j} \\ -(a/2)(\mathbf{i}+2\mathbf{k}) \end{bmatrix} = \begin{bmatrix} I\mathbf{\Omega}_3\mathbf{j} \\ \mathbf{0} \end{bmatrix}$$

$$= \begin{bmatrix} pI(\mathbf{i}+\mathbf{j}+\mathbf{k})\times\mathbf{j} \\ \mathbf{0} \end{bmatrix} = pI \begin{bmatrix} -\mathbf{i}+\mathbf{k} \\ \mathbf{0} \end{bmatrix}$$

$$\mathbf{W}_3\mathbf{M}_3\mathbf{t}_{33} = \begin{bmatrix} I\mathbf{\Omega}_3 & \mathbf{O} \\ \mathbf{O} & \mathbf{O} \end{bmatrix} \begin{bmatrix} \mathbf{i} \\ -(a/2)\mathbf{j} \end{bmatrix} = \begin{bmatrix} I\mathbf{\Omega}_3\mathbf{i} \\ \mathbf{0} \end{bmatrix}$$

$$= \begin{bmatrix} pI(\mathbf{i}+\mathbf{j}+\mathbf{k})\times\mathbf{i} \\ \mathbf{0} \end{bmatrix} = pI \begin{bmatrix} \mathbf{j}-\mathbf{k} \\ \mathbf{0} \end{bmatrix}$$

where $\mathbf{0}$ now denotes the 3-dimensional zero vector. Therefore,

$$\mathbf{WMT} = pI \begin{bmatrix} \mathbf{0} & \mathbf{0} & \mathbf{0} \\ \mathbf{0} & \mathbf{0} & \mathbf{0} \\ \mathbf{i} & -\mathbf{i} & \mathbf{0} \\ \mathbf{0} & \mathbf{0} & \mathbf{0} \\ \mathbf{i}-\mathbf{j} & -\mathbf{i}+\mathbf{k} & \mathbf{j}-\mathbf{k} \\ \mathbf{0} & \mathbf{0} & \mathbf{0} \end{bmatrix}$$

and hence,

$$\mathbf{T}^T\mathbf{WMT} = pI \begin{bmatrix} 0 & 1 & -1 \\ -1 & 0 & 1 \\ 1 & -1 & 0 \end{bmatrix}$$

which turns out to be skew-symmetric. Notice, however, that this will not always be the case. The reason why the above product turned out to be skew-symmetric in this example is that the individual matrices $\mathbf{W}_i$ and $\mathbf{M}_i$ commute, a consequence of the assumed inertial isotropy, which leads to the isotropy of matrices $\mathbf{I}_i$, for $i = 1$, 2, and 3. Now, we have

$$\mathbf{T}^T\mathbf{M}\dot{\mathbf{T}} + \mathbf{T}^T\mathbf{WMT} = p\mathbf{A}$$

with $\mathbf{A}$ defined as

$$\mathbf{A} \equiv \begin{bmatrix} -(3/4)a^2m & (7/4)a^2m + I & -(1/2)a^2m - 2I \\ -(1/2)a^2m - I & 0 & (1/4)a^2m + 2I \\ (3/4)a^2m + I & (1/4)a^2m - 2I & 0 \end{bmatrix}$$

Hence, the term of Coriolis and centrifugal forces is

$$(\mathbf{T}^T\mathbf{M}\dot{\mathbf{T}} + \mathbf{T}^T\mathbf{W}\mathbf{M}\mathbf{T})\dot{\boldsymbol{\theta}} = p^2 \begin{bmatrix} (1/2)a^2m - I \\ -(1/4)a^2m + I \\ a^2m - I \end{bmatrix}$$

thereby completing the desired calculations.

Now, in order to verify the correctness of the above results, we will compute the same term using the Newton-Euler algorithm. To this end, we set $\ddot{\theta}_i = 0$, for $i = 1$, 2, and 3, in that algorithm, and calculate the desired expression as the torque required to produce the joint rates given above.

Since we have already calculated the angular velocities, we will skip these calculations here and limit ourselves to the mass-center velocities, angular accelerations, and mass-center accelerations. We thus have

$$\dot{\mathbf{c}}_1 = \boldsymbol{\omega}_1 \times \boldsymbol{\rho}_1 = p\mathbf{k} \times \left(-\frac{1}{2}a\right)(\mathbf{i} - \mathbf{j}) = -\frac{1}{2}ap(\mathbf{i} + \mathbf{j})$$

$$\begin{aligned}\dot{\mathbf{c}}_2 &= \dot{\mathbf{c}}_1 + \boldsymbol{\omega}_1 \times (\mathbf{a}_1 - \boldsymbol{\rho}_1) + \boldsymbol{\omega}_2 \times \boldsymbol{\rho}_2 \\ &= \frac{1}{2}ap[-\mathbf{i} - \mathbf{j} - \mathbf{k} \times (\mathbf{i} + \mathbf{j}) + (\mathbf{j} + \mathbf{k}) \times (\mathbf{i} + \mathbf{j} - \mathbf{k})] = -\frac{1}{2}ap(3\mathbf{i} + \mathbf{j} + \mathbf{k})\end{aligned}$$

$$\begin{aligned}\dot{\mathbf{c}}_3 &= \dot{\mathbf{c}}_2 + \boldsymbol{\omega}_2 \times (\mathbf{a}_2 - \boldsymbol{\rho}_2) + \boldsymbol{\omega}_3 \times \boldsymbol{\rho}_3 \\ &= -\frac{1}{2}ap[3\mathbf{i} + \mathbf{j} + \mathbf{k} + (\mathbf{j} + \mathbf{k}) \times (\mathbf{i} + \mathbf{k}) - (\mathbf{i} + \mathbf{j} + \mathbf{k}) \times (2\mathbf{i} + \mathbf{k})] \\ &= -\frac{1}{2}ap(3\mathbf{i} + \mathbf{j} + 2\mathbf{k})\end{aligned}$$

Now, the acceleration calculations are implemented recursively, which yields

$$\dot{\boldsymbol{\omega}}_1 = \ddot{\theta}_1\mathbf{e}_1 = \mathbf{0}$$

$$\dot{\boldsymbol{\omega}}_2 = \dot{\boldsymbol{\omega}}_1 + \boldsymbol{\omega}_1 \times \dot{\theta}_2\mathbf{e}_2 = p^2\mathbf{k} \times \mathbf{j} = -p^2\mathbf{i}$$

$$\dot{\boldsymbol{\omega}}_3 = \dot{\boldsymbol{\omega}}_2 + \boldsymbol{\omega}_2 \times \dot{\theta}_3\mathbf{e}_3 = -p^2\mathbf{i} + p^2(\mathbf{j} + \mathbf{k}) \times \mathbf{i} = -p^2(\mathbf{i} - \mathbf{j} + \mathbf{k})$$

$$\ddot{\mathbf{c}}_1 = \dot{\boldsymbol{\omega}}_1 \times \boldsymbol{\rho}_1 + \boldsymbol{\omega}_1 \times (\boldsymbol{\omega}_1 \times \boldsymbol{\rho}_1) = ap^2\mathbf{k} \times \left[\mathbf{k} \times \frac{1}{2}(-\mathbf{i} + \mathbf{j})\right] = \frac{1}{2}ap^2(\mathbf{i} - \mathbf{j})$$

$$\begin{aligned}\ddot{\mathbf{c}}_2 &= \ddot{\mathbf{c}}_1 + \dot{\boldsymbol{\omega}}_1 \times (\mathbf{a}_1 - \boldsymbol{\rho}_1) + \boldsymbol{\omega}_1 \times [\boldsymbol{\omega}_1 \times (\mathbf{a}_1 - \boldsymbol{\rho}_1)] + \dot{\boldsymbol{\omega}}_2 \times \boldsymbol{\rho}_2 \\ &\quad + \boldsymbol{\omega}_2 \times (\boldsymbol{\omega}_2 \times \boldsymbol{\rho}_2) = \frac{1}{2}ap^2(\mathbf{i} - \mathbf{j}) + \mathbf{0} + \frac{1}{2}ap^2(\mathbf{i} + \mathbf{j}) \\ &\quad - \frac{1}{2}ap^2(\mathbf{j} + 2\mathbf{k}) + \frac{1}{2}ap^2(-2\mathbf{i} - 3\mathbf{j} + 3\mathbf{k}) \\ &= \frac{1}{2}ap^2(-4\mathbf{j} + \mathbf{k})\end{aligned}$$

$$\begin{aligned}\ddot{\mathbf{c}}_3 &= \ddot{\mathbf{c}}_2 + \dot{\boldsymbol{\omega}}_2 \times (\mathbf{a}_2 - \boldsymbol{\rho}_2) + \boldsymbol{\omega}_2 \times [\boldsymbol{\omega}_2 \times (\mathbf{a}_2 - \boldsymbol{\rho}_2)] + \dot{\boldsymbol{\omega}}_3 \times \boldsymbol{\rho}_3 \\ &\quad + \boldsymbol{\omega}_3 \times (\boldsymbol{\omega}_3 \times \boldsymbol{\rho}_3) = \frac{1}{2}ap^2(-4\mathbf{j} + \mathbf{k}) - \frac{1}{2}ap^2\mathbf{j} + \frac{1}{2}ap^2(2\mathbf{i} - \mathbf{j} + \mathbf{k})\end{aligned}$$

$$
\begin{aligned}
&+\frac{1}{2}ap^2(\mathbf{i}-\mathbf{j}-2\mathbf{k})+\frac{1}{2}ap^2(-3\mathbf{i}+3\mathbf{j}) \\
&=-2ap^2\mathbf{j}
\end{aligned}
$$

With the foregoing values, we can now implement the inward Newton-Euler recursions, namely,

$$
\begin{aligned}
\mathbf{f}_3^P &= m_3\ddot{\mathbf{c}}_3-\mathbf{f}=-m(2ap^2\mathbf{j})-\mathbf{0}=-2amp^2\mathbf{j} \\
\mathbf{n}_3^P &= \mathbf{I}_3\dot{\boldsymbol{\omega}}_3+\boldsymbol{\omega}_3\times\mathbf{I}_3\boldsymbol{\omega}_3-\mathbf{n}+\boldsymbol{\rho}_3\times\mathbf{f}_3^P \\
&= -Ip^2(\mathbf{i}-\mathbf{j}+\mathbf{k})+\mathbf{0}-\mathbf{0}-a^2mp^2(-\mathbf{i}+2\mathbf{k}) \\
&= -Ip^2(\mathbf{i}-\mathbf{j}+\mathbf{k})+a^2mp^2(\mathbf{i}-2\mathbf{k}) \\
\mathbf{f}_2^P &= m_2\ddot{\mathbf{c}}_2+\mathbf{f}_3^P=\frac{1}{2}amp^2(-4\mathbf{j}+\mathbf{k})-amp^2\mathbf{j}=\frac{1}{2}amp^2(-6\mathbf{j}+\mathbf{k}) \\
\mathbf{n}_2^P &= \mathbf{I}_2\dot{\boldsymbol{\omega}}_2+\boldsymbol{\omega}_2\times\mathbf{I}_2\boldsymbol{\omega}_2+\mathbf{n}_3^P+(\mathbf{a}_2-\boldsymbol{\rho}_2)\times\mathbf{f}_3^P+\boldsymbol{\rho}_2\times\mathbf{f}_2^P \\
&= -p^2I\mathbf{i}+\mathbf{0}-Ip^2(\mathbf{i}-\mathbf{j}+\mathbf{k})+\frac{1}{2}a^2mp^2(\mathbf{i}-2\mathbf{k})+a^2mp^2\mathbf{i} \\
&\quad +\frac{1}{4}a^2mp^2(-4\mathbf{i}-\mathbf{j}-6\mathbf{k}) \\
&= -Ip^2(2\mathbf{i}-\mathbf{j}+\mathbf{k})+\frac{1}{4}a^2mp^2(2\mathbf{i}-\mathbf{j}-10\mathbf{k}) \\
\mathbf{f}_1^P &= m_1\ddot{\mathbf{c}}_1+\mathbf{f}_2^P=\frac{1}{2}amp^2(\mathbf{i}-\mathbf{j})+\frac{1}{2}amp^2(-6\mathbf{j}+\mathbf{k}) \\
&= \frac{1}{2}amp^2(\mathbf{i}-7\mathbf{j}+\mathbf{k}) \\
\mathbf{n}_1^P &= \mathbf{I}_1\dot{\boldsymbol{\omega}}_1+\boldsymbol{\omega}_1\times\mathbf{I}_1\boldsymbol{\omega}_1+\mathbf{n}_2^P+(\mathbf{a}_1-\boldsymbol{\rho}_1)\times\mathbf{f}_2^P+\boldsymbol{\rho}_1\times\mathbf{f}_1^P \\
&= \mathbf{0}+\mathbf{0}-p^2I(2\mathbf{i}-\mathbf{j}+\mathbf{k})+\frac{1}{4}a^2mp^2(2\mathbf{i}-\mathbf{j}-10\mathbf{k}) \\
&\quad -\frac{1}{4}a^2mp^2(\mathbf{i}-\mathbf{j}-6\mathbf{k})+\frac{1}{4}a^2mp^2(\mathbf{i}+\mathbf{j}-6\mathbf{k}) \\
&= -Ip^2(2\mathbf{i}-\mathbf{j}+\mathbf{k})+\frac{1}{4}a^2mp^2(2\mathbf{i}+\mathbf{j}+2\mathbf{k})
\end{aligned}
$$

and hence,

$$
\begin{aligned}
\tau_3 &= \mathbf{n}_3^P\cdot\mathbf{e}_3=-Ip^2+a^2mp^2 \\
\tau_2 &= \mathbf{n}_2^P\cdot\mathbf{e}_2=Ip^2-\frac{1}{4}a^2mp^2 \\
\tau_1 &= \mathbf{n}_1^P\cdot\mathbf{e}_1=-Ip^2+\frac{1}{2}a^2mp^2
\end{aligned}
$$

thereby completing the calculation of the term containing Coriolis and centrifugal forces, i.e.,

$$
\mathbf{C}(\boldsymbol{\theta},\dot{\boldsymbol{\theta}})\dot{\boldsymbol{\theta}}=\begin{bmatrix} -Ip^2+a^2mp^2 \\ Ip^2-(1/4)a^2mp^2 \\ -Ip^2+(1/2)a^2mp^2 \end{bmatrix}
$$

As the reader can verify, the natural orthogonal complement and the Newton-Euler algorithm produce the same result. In the process, the reader may have realized that when performing calculations by hand, the Newton-Euler algorithm is more prone to errors than the natural orthogonal complement, which is more systematic, for it is based on matrix-times-vector multiplications.

6.6.2 Algorithm Complexity

The complexity of this algorithm is analyzed with regard to the three items involved, namely, (*i*) the evaluation of $\mathbf{L}$, (*ii*) the solution of systems (6.108a & b), and (*iii*) the computation of $\overline{\boldsymbol{\tau}}$.

The evaluation of $\mathbf{L}$ involves, in turn, the three following steps: (a) the computation of $\mathbf{P}$; (b) the computation of $\mathbf{I}$; and (c) the Cholesky decomposition of $\mathbf{I}$ into the product $\mathbf{L}^T\mathbf{L}$.

(i.a) In the computation of $\mathbf{P}$, it is recalled that $\mathbf{H}_i$, $\mathbf{a}_i$, and $\boldsymbol{\rho}_i$, and consequently, $\boldsymbol{\delta}_i \equiv \mathbf{a}_i - \boldsymbol{\rho}_i$, are constant in $\mathcal{F}_{i+1}$, which is the frame fixed to the ith link. Moreover, at each step of the algorithm, both revolute and prismatic pairs are considered. If the jth joint is a revolute, then the logical variable `R` is `true`; if this joint is prismatic, then `R` is `false`. Additionally, it is recalled that $\mathbf{e}_{i+1}$, in $\mathcal{F}_i$-coordinates, is simply the last column of $\mathbf{Q}_i$. The columnwise evaluation of $\mathbf{P}$, with each $\mathbf{p}_{ij}$ array in $\mathcal{F}_{i+1}$-coordinates, is described in Algorithm 6.6.1. Note that in this algorithm, $\mathbf{r}_{ij}$ is calculated recursively from $\mathbf{r}_{i-1,j}$. To do this, we use the relation between these two vectors, as displayed in Fig. 6.8.

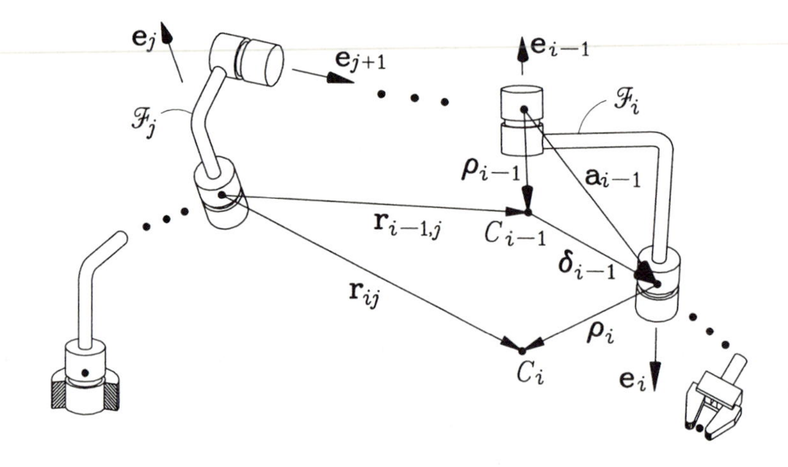

FIGURE 6.8. Recursive calculation of vectors $\mathbf{r}_{ij}$.

Algorithm 6.6.1:

For j = 1 to n step 1 do

$$\mathbf{r}_{jj} \leftarrow [\boldsymbol{\rho}_j]_{j+1}$$

$$\mathbf{p}_{jj} \leftarrow \begin{bmatrix} \mathbf{N}_j\mathbf{e}_j \\ n_j\mathbf{e}_j \times \mathbf{r}_{jj} \end{bmatrix}_{j+1}$$

For i = j + 1 to n step 1 do

$$\mathbf{e}_j \leftarrow \mathbf{Q}_i^T[\mathbf{e}_j]_i$$

if R then

$$\mathbf{r}_{ij} \leftarrow \mathbf{Q}_i^T[\mathbf{r}_{i-1,j} + \boldsymbol{\delta}_{i-1}]_i + [\boldsymbol{\rho}_i]_{i+1}$$

$$\mathbf{p}_{ij} \leftarrow \begin{bmatrix} \mathbf{N}_i\mathbf{e}_j \\ n_i\mathbf{e}_j \times \mathbf{r}_{ij} \end{bmatrix}_{i+1}$$

else

$$\mathbf{p}_{ij} \leftarrow \begin{bmatrix} \mathbf{0} \\ n_i\mathbf{e}_j \end{bmatrix}_{i+1}$$

endif

enddo

enddo

Algorithm 6.6.2:

For j = 1 to n step 1 do

$$I_{jj} \leftarrow \sum_{k=j}^{n}[\mathbf{p}_{kj}^T\mathbf{p}_{kj}]_{k+1}$$

For i = j + 1 to n step 1 do

$$I_{ij} \leftarrow I_{ji} \leftarrow \sum_{k=i}^{n}[\mathbf{p}_{ki}^T\mathbf{p}_{kj}]_{k+1}$$

enddo

enddo

(i.b) Now we go on to the computation of $\mathbf{I}$, as described in Algorithm 6.6.2. In that algorithm, the subscripted brackets indicate that the vectors inside these brackets are represented in $\mathcal{F}_{k+1}$ coordinates.

(i.c) Because the Cholesky decomposition of a positive-definite matrix is a standard item, it is not discussed here. This step completes the computation of $\mathbf{L}$.

(ii) The solution of systems (6.108a & 6.108b) is a standard issue as well, and hence, needs no further discussion.

(iii) The term $\overline{\boldsymbol{\tau}}$ is computed using the recursive Newton-Euler formulation, as discussed in Section 6.4. To do this, we calculate the aforementioned term by setting $\ddot{\boldsymbol{\theta}} = \mathbf{0}$ in that procedure, which introduces a slight simplification of the complexity of the inverse-dynamics algorithm.

Below we determine the computational complexity of each of the foregoing steps.

(i.a) This step includes Algorithm 6.6.1, which involves two nested do-loops. The first statement of the outermost loop involves no floating-point operations; the second statement involves (a) one multiplication of a matrix by a vector, (b) one cross product, and (c) one multiplication of a scalar by a vector. Of the last three items, (a) is done off-line, for the matrix and the vector factors are both constant in $\mathcal{F}_{j+1}$-coordinates, and so, this operation is not counted. Moreover, item (b) is nothing but the cross product of vector $[\,\mathbf{e}_j\,]_{j+1} \equiv [\,0,\,0,\,1\,]^T$ by vector $\mathbf{r}_{jj}$. A similar operation was already discussed in connection with Algorithm 4.1 and was found to involve zero floating-point operations, for the result is, simply, $[\,\mathbf{e}_j \times \mathbf{r}_{jj}\,]_{j+1} = [\,-y,\,x,\,0\,]^T$, with x and y denoting the X_{j+1} and Y_{j+1} components of $\mathbf{r}_{jj}$. Hence, item (b) requires no floating-point operations, while item (c) requires $2n$ multiplications and zero additions.

The innermost do-loop, as pertaining to revolute manipulators, involves two coordinate transformations between *two consecutive coordinate frames*, from $\mathcal{F}_i$- to $\mathcal{F}_{i+1}$-coordinates, plus two vector sums, which consumes $16(n-i)$ multiplications and $14(n-i)$ additions; this loop also consumes one matrix-times-vector multiplication, one cross product and one scalar-times-vector multiplication, which requires $18(n-i)$ multiplications and $12(n-i)$ additions. Thus, the total numbers of operations required by this step, for an n-revolute manipulator, are M_{ia} multiplications and A_{ia} additions, as given below:

$$M_{ia} = 2n + \sum_{i=1}^{n} 34(n-i) = 17n^2 - 15n \qquad (6.115a)$$

$$A_{ia} = \sum_{i=1}^{n} 26(n-i) = 13n^2 - 13n \qquad (6.115b)$$

the presence of prismatic pairs reducing the above figures.

(i.b) This step, summarized in Algorithm 6.6.2, is also composed of two do-loops, each containing the inner product of two 6-dimensional arrays, and hence, requires six multiplications and five additions. Moreover, in the outermost do-loop, this operation is performed n times, whereas in the innermost loop, $\sum_{i=1}^{n}(n-i)$ times, i.e., $n(n-1)/2$ times. Thus, the step requires M_{ib} multiplications and A_{ib} additions, as given below:

$$M_{ib} = 3n^2 + 3n, \quad A_{ib} = \frac{5}{2}n^2 + \frac{5}{2}n \tag{6.116}$$

(i.c) This step performs the Cholesky decomposition of an $n \times n$ symmetric and positive-definite matrix, a standard operation that requires M_{ic} multiplications and A_{ic} additions (Dahlquist and Björck, 1974), namely,

$$M_{ic} = \frac{1}{6}n^3 + \frac{1}{2}n^2 + \frac{1}{3}n, \quad A_{ic} = \frac{1}{6}n^3 + \frac{1}{2}n^2 + \frac{1}{3}n \tag{6.117}$$

(ii) In this step, the two triangular systems of equations, eqs.(6.108a & b), are solved first for $\mathbf{x}$ and then for $\ddot{\boldsymbol{\theta}}$. The numbers of operations it takes to solve each of the two systems, as derived by Dahlquist and Björck (1975), are repeated below for quick reference; these are labelled M_{ii} and A_{ii}, respectively, i.e.,

$$M_{ii} = n^2, \quad A_{ii} = n^2 - n \tag{6.118}$$

(iii) In this step, $\overline{\boldsymbol{\tau}}$ is computed from inverse dynamics, with $\mathbf{w}^D = \mathbf{0}$ and $\ddot{\boldsymbol{\theta}} = \mathbf{0}$. If this calculation is done with the Newton-Euler formulation, we then have the computational costs given in eq.(6.44), and reproduced below for quick reference:

$$M_{iii} = 137n - 22, \quad A_{iii} = 110n - 14 \tag{6.119}$$

Because of the simplifications introduced by setting the joint accelerations equal to zero, the aforementioned figures are, in fact, slightly lower than those required by the general recursive Newton-Euler algorithm.

Thus, the total numbers of multiplications and additions required for the forward dynamics of an n-revolute, serial manipulator are

$$M_f = \frac{1}{6}n^3 + \frac{43}{2}n^2 + \frac{376}{3}n - 22, \quad A_f = \frac{1}{6}n^3 + 17n^2 + \frac{593}{6}n - 14 \tag{6.120}$$

In particular, for a six-revolute manipulator, one obtains the figures shown below:

$$M_f = 1,540, \quad A_f = 1,227 \tag{6.121}$$

We have reduced the foregoing figures even more by introducing a modified Denavit-Hartenberg labeling of coordinate frames and very careful management of the computations involved. Indeed, in (Angeles and Ma, 1988), the complexity of the algorithm for a six-revolute manipulator of arbitrary architecture is reduced to 1,353 multiplications and 1,165 additions. Since the details of this simplification lie beyond the scope of the book, we do not elaborate on this item here.

6.6.3 Simulation

The purpose of the algorithm introduced in the foregoing subsection is to enable us to predict the behavior of a given manipulator under given initial conditions, applied, torques, and applied loads. The ability of predicting this behavior is important for several reasons: for example, in design, we want to know whether with a given selection of motors, the manipulator will be able to perform a certain typical task in a given time frame; in devising feedback control schemes, where stability is a major concern, the control engineer cannot risk a valuable piece of equipment by exposing it to untested control strategies. Hence, a facility capable of predicting the behavior of a robotic manipulator, or of a system at large, for that matter, becomes imperative.

The procedure whereby the motion of the manipulator is determined from initial conditions and applied torques and loads is known as *simulation*. Since we start with a second-order n-dimensional system of ordinary differential equations (ODE) in the joint variables of the manipulator, we have to integrate this system in order to determine the time-histories of all joint variables, which are grouped in vector $\boldsymbol{\theta}$. With current software available, this task has become routine work, the user being freed from the quite demanding task of writing code for integrating systems of ODE. Below we discuss a few issues pertaining to the implementation of the simulation-related algorithms available in commercial software packages.

As a rule, simulation code requires that the user supply a state-variable model of the form of eq.(6.46), with the state-variable vector, or state-vector for brevity, $\mathbf{x}$, and the *input* or *control vector* $\mathbf{u}$ defined as

$$\mathbf{x} \equiv \begin{bmatrix} \boldsymbol{\theta} \\ \dot{\boldsymbol{\theta}} \end{bmatrix} \equiv \begin{bmatrix} \boldsymbol{\theta} \\ \boldsymbol{\psi} \end{bmatrix}, \quad \mathbf{u}(t) = \boldsymbol{\tau}(t) \tag{6.122}$$

With the above definitions, then we can write the state-variable equations, or state equations for brevity, in the form of eq.(6.46), with $\mathbf{f}(\mathbf{x}, \boldsymbol{\tau})$ given by

$$\mathbf{f}(\mathbf{x}, \boldsymbol{\tau}) \equiv \begin{bmatrix} \boldsymbol{\psi} \\ -\mathbf{I}(\boldsymbol{\theta})^{-1}[\mathbf{C}(\boldsymbol{\theta}, \boldsymbol{\psi})\boldsymbol{\psi} - \boldsymbol{\delta}(\boldsymbol{\theta}, \boldsymbol{\psi})] + \mathbf{I}(\boldsymbol{\theta})^{-1}\boldsymbol{\tau}(t) \end{bmatrix} \tag{6.123}$$

thereby obtaining a system of $2n$ first-order ODE in the state-variable vector $\mathbf{x}$ defined above. Various methods are available to solve the ensuing initial-value problem, all of them being based on a discretization of the time variable. That is, if the behavior of the system is desired in the interval $t_0 \leq t \leq t_F$, then the software implementing these methods provides *approximations* $\{\mathbf{y}_k\}_1^N$ to the state-variable vector at a discrete set of instants, $\{t_k\}_0^N$, with $t_N \equiv t_F$.

The variety of methods available to solve the underlying initial-value problem can be classified into two main categories, *explicit methods* and *implicit methods*. The former provide $\mathbf{y}_{k+1}$ *explicitly* in terms of previously computed values. On the contrary, implicit methods provide $\mathbf{y}_{k+1}$ in terms of previously computed values $\mathbf{y}_k$, $\mathbf{y}_{k-1}$, ..., etc., *and* $\mathbf{y}_{k+1}$ itself. For example, in the simplest of implicit methods, namely, the *backward Euler method*, we can approximate the integral of $\mathbf{f}$ in the interval $t_k \leq t \leq t_{k+1}$ by resorting to the *trapezoidal rule* (Kahaner, Moler, and Nash, 1989), which leads to the expression

$$\mathbf{y}_{k+1} = \mathbf{y}_k + h_k \mathbf{f}(t_{k+1}, \mathbf{y}_{k+1}) \tag{6.124}$$

In eq.(6.124), h_k is the *current* time-step $t_{k+1} - t_k$ and $\mathbf{f}(t_{k+1}, \mathbf{y}_{k+1})$ can be an arbitrary function of $\mathbf{y}_{k+1}$. If this function is nonlinear in the said variable, then, a *direct*—as opposed to *iterative*—computation of $\mathbf{y}_{k+1}$ is very unlikely. Hence, most likely an iterative scheme must be implemented at every integration stage of an implicit method. While this feature might render implicit schemes unattractive, they offer interesting advantages. Indeed, the aforementioned iterative procedure requires a tolerance to decide when and whether the procedure has converged. The convergence criterion imposed thus brings about a self-correcting effect that helps keep the unavoidable *truncation error* under control. This error is incurred when approximating both the time derivative $\dot{\mathbf{x}}$ and the integral of $\mathbf{f}$ by floating-point operations.

Current software provides routines for both implicit and explicit methods, the user having to decide which method to invoke. Of the explicit methods in use, by far the most common ones are the *Runge-Kutta methods.* Of these, there are several versions, depending on the number of evaluations of the function $\mathbf{f}(t_i, \mathbf{y}_i)$, for various values of i, that they require. A two-stage Runge-Kutta method, for example, requires two function evaluations, while a four-stage Runge-Kutta method requires four. The self-correcting feature of implicit methods, not present in Runge-Kutta methods—to be sure, *implicit Runge-Kutta methods* also exist (Gear, 1971), but these are less common than their explicit counterparts—is compensated for by a very clever strategy that consists of computing $\mathbf{y}_{k+1}$ using two Runge-Kutta schemes of different numbers of stages. What is at stake here is the magnitude of the *local error* in computing $\mathbf{y}_{k+1}$, under the assumption that $\mathbf{y}_k$ is error-free. Here, the magnitude of the error is of order h^p, where p is the *order* of the method in use. In Runge-Kutta methods, the order of the method is identical to its number of stages. In general, a method is said to be of order p if it is capable of computing *exactly* the integral of an ordinary differential equation, provided that the solution is known to be a pth-degree polynomial. Now, upon computing $\mathbf{y}_{k+1}$ using two Runge-Kutta schemes with N and $N+1$ stages, we can compare the two computed values reported by each method, namely, $\mathbf{y}_{k+1}^N$ and $\mathbf{y}_{k+1}^{N+1}$. If a *norm* of the difference of these two values is smaller than a user-prescribed tolerance, then the step size in use is acceptable. If not, then the step size is halved, and the process is repeated until the foregoing norm is within the said tolerance. The most common Runge-Kutta methods are those combining two and three stages and those combining four and five.

A drawback of Runge-Kutta methods is their inability to deal with what are known as *stiff systems*, first identified by Gear (1971). As defined by Shampine and Gear (1979), a system of ordinary differential equations is said to be stiff if it is not unstable and its linear part—i.e., the linear part of the series expansion of $\mathbf{f}$, evaluated at the current instant—comprises a coefficient matrix that has an eigenvalue with a negative real part whose absolute value is much greater than that of the other eigenvalues. In other words, stiff systems of ODE are stable systems with very different time scales. Thus, stiff systems are not inherently difficult to integrate, but they require a special treatment. Gear's method, which is implicit, provides exactly the means to handle stiff systems. However, methods like Runge-Kutta's, with excellent performance for nonstiff systems, perform rather poorly for stiff systems, and the other way around. The mathematical models that arise in robotic mechanical systems are likely to be stiff because of the various orders of magnitude of the physical parameters involved. For example, robotic manipulators are provided, usually, with links close to the base that are very robust and hence, heavy and with links far from the base that are very light. As a consequence, when simulating robotic mechanical systems, a provision must be made for numerical stiffness.

Commercial software for scientific computations offers Runge-Kutta methods of various orders, with combinations thereof. For example, IMSL offers excellent FORTRAN routines, like `IVPRK`, for the implementation of Runge-Kutta methods, while Matlab's Simulink toolbox offers the C functions `rk23` and `rk45` for the implementation of second-and-third and fourth-and-fifth-order Runge-Kutta methods. With regard to stiff systems, IMSL offers a subroutine, `IVPAG`, implementing both Adams's and Gear's methods, while Simulink offers the `adams` and `gear` functions for the implementation of either of these. Since Matlab is written in C, communication between Matlab and FORTRAN programs is not as direct as when using IMSL, which may be disappointing to FORTRAN users. Details on linking FORTRAN code with Matlab and other related issues are discussed in the pertinent literature (Etter, 1993). Moreover, the FORTRAN `SDRIV2` subroutine (Kahaner, Moler, and Nash, 1989) comprises features that allow it to handle both stiff and nonstiff systems.

6.7 Incorporation of Gravity into the Dynamics Equations

Manipulators subjected to gravity fields have been discussed in Section 6.4 in connection with the Newton-Euler algorithm and with Kane's equations. As found in that section, gravitational forces can be incorporated into the underlying models without introducing any major modifications that would increase the computational load if the method of Luh, Walker, and Paul

(1980) is adopted. Within this approach, gravitational forces are taken into account by defining the acceleration of the mass center of the 0th link, the base link, as equal to $-\mathbf{g}$, the negative of the gravity-acceleration vector. The effect of this approach is to propagate the gravity effect into all the links composing the manipulator. Thus, the kinematics algorithm of Section 6.4 need not be modified in order to include gravitational forces, for all that is needed is to declare

$$[\ddot{\mathbf{c}}_0]_1 \quad \leftarrow \quad [-\mathbf{g}]_1 \tag{6.125}$$

If inverse dynamics is computed with the natural orthogonal complement, then the twist-rate of the first link will have to be modified by adding a nonhomogeneous term to it, thereby accounting for the gravity-acceleration terms. That is,

$$\dot{\mathbf{t}}_1 \quad \leftarrow \quad \ddot{\theta}_1\mathbf{t}_{11} + \dot{\theta}_1\dot{\mathbf{t}}_{11} + \begin{bmatrix} \mathbf{0} \\ -\mathbf{g} \end{bmatrix} \tag{6.126}$$

Otherwise, the foregoing algorithms require no modifications. Furthermore, with regard to simulation, it is pointed out that the $\overline{\boldsymbol{\tau}}$ term defined in eq.(6.107), and appearing in the right-hand side of eq.(6.108a), is computed from inverse dynamics with zero frictional forces and zero joint accelerations.

6.8 The Modeling of Dissipative Forces

Broadly speaking, frictional forces are of two basic types, namely, (*i*) viscous forces and (*ii*) *Coulomb*, or dry-friction, forces. The latter occur when contact between two solids takes place directly, the former when contact between the solids takes place via a viscous fluid, e.g., a lubricant. In the analysis of viscous fluids, a basic assumption is that the relative velocity between the fluid and the solid vanishes at the fluid-solid interface, i.e., at the solid boundary confining the fluid. Hence, a *velocity gradient* appears within the fluid, which is responsible for the power dissipation inside it. In fact, not all the velocity gradient within the fluid, but only its *symmetric part*, is responsible for power dissipation; the *skew-symmetric part* of the velocity gradient accounts for a rigid-body rotation of a small fluid element. Thus, if a velocity field $\mathbf{v}(\mathbf{r}, t)$ is defined within a region $\mathcal{R}$ occupied by a viscous fluid, for a point of the fluid of position vector $\mathbf{r}$ at a time t, then, the velocity gradient $\text{grad}(\mathbf{v}) \equiv \partial\mathbf{v}/\partial\mathbf{r}$, can be decomposed as

$$\text{grad}(\mathbf{v}) = \mathbf{D} + \mathbf{W} \tag{6.127}$$

where $\mathbf{D}$ and $\mathbf{W}$ are the symmetric and the skew-symmetric parts of the velocity gradient, i.e.,

$$\mathbf{D} \equiv \frac{1}{2}[\,\text{grad}(\mathbf{v}) + \text{grad}^T(\mathbf{v})\,], \quad \mathbf{W} \equiv \frac{1}{2}[\,\text{grad}(\mathbf{v}) - \text{grad}^T(\mathbf{v})\,] \tag{6.128}$$

The kinematic interpretation of $\mathbf{D}$ and $\mathbf{W}$ is given below: The former accounts for a *distorsion* of an infinitesimally small spherical element of fluid into a three-axis ellipsoid, the ratios of the time *rates* of change of the lengths of the three axes being identical to the ratios of the real eigenvalues of $\mathbf{D}$; the latter accounts for the angular velocity of the ellipsoid as a rigid-body. Clearly, both $\mathbf{D}$ and $\mathbf{W}$ change from point to point within the fluid and also from time to time, i.e.,

$$\mathbf{D} = \mathbf{D}(\mathbf{r}, t), \quad \mathbf{W} = \mathbf{W}(\mathbf{r}, t) \tag{6.129}$$

Since the skew-symmetric matrix $\mathbf{W}$ accounts only for the rotation of a differential element of fluid as a rigid body, it cannot be responsible for any energy dissipation, and hence, the only part that is responsible for this is $\mathbf{D}$. In fact, for a *linearly viscous, incompressible* fluid of viscosity coefficient μ, the power dissipated within $\mathcal{R}$ is given by

$$\Pi^D = \int_{\mathcal{R}} \mu \operatorname{tr}(\mathbf{D}^2) d\mathcal{R} \tag{6.130}$$

Now, if the motion of the lubricant separating the two cylindrical surfaces of a revolute pair is modeled as a purely tangential velocity field (Currie, 1993), which assumes that the two cylinders remain concentric, then the foregoing expression for Π^D leads to the dissipation function

$$\Delta = \frac{1}{2}\beta\dot{\theta}^2 \tag{6.131}$$

where $\dot{\theta}$ is the relative angular speed between the two cylinders and the coefficient β is a function of the lubricant viscosity and the geometry of the kinematic pair at hand. If the kinematic pair under study is prismatic, then we can model the motion of the lubricant between the two prismatic surfaces as a *Couette flow* between a pair of parallel surfaces of the sides of the prism. Under these conditions, then, the associated dissipation function Δ takes on the same form of that given for a revolute pair in eq.(6.131), in which the sole difference is that $\dot{\theta}$ changes to $\dot{b}$, the time rate of change of the associated joint variable. Of course, $\dot{b}$ is the relative speed between the two prismatic surfaces. Thus in any event, the dissipation function of the ith joint due to linearly viscous effects can be written as

$$\Delta_i = \frac{1}{2}\beta_i\dot{\theta}_i^2 \tag{6.132}$$

where $\dot{\theta}_i$ changes to $\dot{b}_i$ if the ith pair is prismatic. The dissipation function thus arising then reduces to

$$\Delta = \sum_1^n \Delta_i = \frac{1}{2}\dot{\boldsymbol{\theta}}^T \mathbf{B}\dot{\boldsymbol{\theta}} \tag{6.133}$$

where the constant $n \times n$ matrix $\mathbf{B}$ is given by

$$\mathbf{B} = \text{diag}(\beta_1, \beta_2 \ldots, \beta_n) \tag{6.134}$$

and hence, the generalized force $\boldsymbol{\delta}^V$ associated with linearly viscous effects is *linear* in the vector of joint rates, $\dot{\boldsymbol{\theta}}$, i.e.,

$$\boldsymbol{\delta}^V \equiv -\frac{\partial \Delta}{\partial \dot{\boldsymbol{\theta}}} = -\mathbf{B}\dot{\boldsymbol{\theta}} \tag{6.135}$$

and so, $\Delta = -(1/2)\Pi^D$, which was introduced in eqs.(6.11) and (6.12a & b).

Coulomb, or dry friction, is much more difficult to model. If δ_i^C denotes either the dissipative torque produced by Coulomb friction at a revolute or the dissipative force produced by Coulomb friction at a prismatic joint, and $\dot{\theta}_i$ the associated joint rate, then, the simplest model for the resulting generalized Coulomb-frictional force is

$$\delta_i^C = -\tau_i^C \text{sgn}(\dot{\theta}_i) \tag{6.136}$$

where sgn$(\cdot)$ denotes the *signum function*, which is defined as $+1$ or -1, depending on whether its argument is positive or negative, and τ_i^C is a positive constant representing a torque for revolute joints or a force for prismatic joints. The numerical value of this constant is to be determined experimentally. The foregoing model leads to a simple expression for the associated dissipation function, namely,

$$\Delta_i^C = \tau_i^C |\dot{\theta}_i| \tag{6.137}$$

The Coulomb dissipation function for the overall manipulator is, then,

$$\Delta^C = \sum_1^n \tau_i^C |\dot{\theta}_i| \tag{6.138}$$

The foregoing simplified model of Coulomb frictional forces is applicable when the relative speed between the two surfaces in contact is high. However, at low relative speed, that model becomes inaccurate. In robotics applications, where typical end-effector maximum speeds are of the order of 1 m/s, relative speeds are obviously low, and hence, a more accurate model should be introduced. Such a model should account for the empirical observation that Coulomb frictional forces are higher at low relative speeds and become constant at very high relative speeds. A model taking this fact into account has the form

$$\delta_i^C = -(\tau_i^C + \epsilon_i e^{-\gamma_i |\dot{\theta}_i|}) \text{sgn}(\dot{\theta}_i) \tag{6.139}$$

where α_i, γ_i, and ϵ_i are constants associated with the ith joint and are to be determined experimentally. The foregoing expression readily leads to the dissipation function associated with the same joint, namely,

$$\Delta_i^C = \tau_i^C |\dot{\theta}_i| + \frac{\epsilon_i}{\gamma_i}(1 - e^{-\gamma_i |\dot{\theta}_i|}) \tag{6.140}$$

and hence, the Coulomb dissipation function of the overall manipulator becomes

$$\Delta^C = \sum_1^n \left[\tau_i^C |\dot{\theta}_i| + \frac{\epsilon_i}{\gamma_i}(1 - e^{-\gamma_i |\dot{\theta}_i|}) \right] \tag{6.141}$$

Dissipation functions are very useful. On the one hand, they allow us to obtain associated generalized frictional forces when these are difficult, if not impossible, to express in formula form. On the other hand, since dissipation functions represent nonrecoverable forms of power, their integrals over time yield the dissipated energy. Moreover, the energy dissipated into unrecoverable heat can be estimated from an energy balance, and hence, the parameters associated with that dissipation function can be estimated with suitable identification techniques, once a suitable model for a dissipation function is available. Furthermore, the said parameters appear in the generalized frictional forces as well. For this reason, knowing these parameters is essential for the modeling of the corresponding generalized frictional forces.

7
Special Topics in Rigid-Body Kinematics

7.1 Introduction

The motivation for this chapter is twofold. On the one hand, the determination of the angular velocity and angular acceleration of a rigid body from point-velocity measurements is a fundamental problem in kinematics. On the other hand, the solution of this problem is becoming increasingly relevant in the kinematics of parallel manipulators, to be studied in Chapter 8. Moreover, the estimation of the attitude of a rigid body from knowledge of the Cartesian coordinates of some of its points is sometimes accomplished by time-integration of the velocity data. Likewise, the use of accelerometers in the area of motion control readily leads to estimates of the acceleration of a sample of points of a rigid body, which can be used to estimate the angular acceleration of the body, and hence, to better control its motion.

In order to keep the discussion at the level of fundamentals, we assume throughout this chapter that the information available on point velocity and point acceleration is error-free, a rather daring assumption, but useful for understanding the underlying concepts at this level. Once the fundamentals are well understood, devising algorithms that yield the best estimates of angular velocity and acceleration in the presence of noisy measurements becomes an easier task. For the sake of conciseness, the problem of motion estimation will not be discussed in this book.

7.2 Computation of Angular Velocity from Point-Velocity Data

The twist of a rigid body, as introduced in eq.(3.74), defines completely the velocity field of a rigid body under arbitrary motion. Notice that the twist involves two vector quantities, the angular velocity and the velocity of a point of the rigid body. Since we are assuming that point-velocity data are available, the only item to be computed is the angular velocity of the body under study, which is the subject of this section. Once the angular velocity is known and the velocities of a set of body points are available, other relevant motion parameters, like the location of the ISA, can be readily determined.

If the twist of a rigid body is known, the computation of the velocity of an arbitrary point of the body, of a given position vector, is straightforward. However, the inverse problem, namely, the computation of the twist of the motion under study given the velocities of a set of points of known position vectors, is a more difficult task. A solution to this problem is now outlined.

First and foremost, we acknowledge that the velocities of a minimum of three noncollinear points are needed in order to determine the angular velocity of the rigid body under study. Indeed, if the velocity of a single body point is known, we have no information on the angular motion of the body; if the velocities of two points are known, we can calculate two components of the angular-velocity vector of the body, namely, those that are orthogonal to the line joining the two given points, thereby leaving one component indeterminate, the one along that line. Therefore, in order to know the angular velocity of a rigid body in motion, we need at least the velocities of three noncollinear points of the body—obviously, knowing only the velocities of any number of points along one line yields no more information than knowing only the velocities of two points along that line. We thus assume henceforth that we have three noncollinear points and that we know *perfectly* their velocities.

Let the three noncollinear points of the body under study be denoted by $\{ P_i \}_1^3$ and let $\{ \mathbf{p}_i \}_1^3$ be their corresponding position vectors. The centroid C of the foregoing set has a position vector $\mathbf{c}$ that is the mean value of the three given position vectors, namely,

$$\mathbf{c} \equiv \frac{1}{3} \sum_1^3 \mathbf{p}_i \tag{7.1}$$

Likewise, if the velocities of the three points are denoted by $\dot{\mathbf{p}}_i$, and that of their centroid by $\dot{\mathbf{c}}$, one has

$$\dot{\mathbf{c}} \equiv \frac{1}{3} \sum_1^3 \dot{\mathbf{p}}_i \tag{7.2}$$

From eq.(3.51), the velocity of the three given points can be expressed as

$$\dot{\mathbf{p}}_i = \dot{\mathbf{c}} + \mathbf{\Omega}(\mathbf{p}_i - \mathbf{c}), \quad i = 1, 2, 3 \tag{7.3a}$$

or

$$\dot{\mathbf{p}}_i - \dot{\mathbf{c}} = \mathbf{\Omega}(\mathbf{p}_i - \mathbf{c}), \quad i = 1, 2, 3 \tag{7.3b}$$

Now, we define a 3×3 matrix $\mathbf{P}$ as

$$\mathbf{P} \equiv [\,\mathbf{p}_1 - \mathbf{c} \quad \mathbf{p}_2 - \mathbf{c} \quad \mathbf{p}_3 - \mathbf{c}\,] \tag{7.4}$$

Upon differentiation of both sides of eq.(7.4) with respect to time, one has

$$\dot{\mathbf{P}} \equiv [\,\dot{\mathbf{p}}_1 - \dot{\mathbf{c}} \quad \dot{\mathbf{p}}_2 - \dot{\mathbf{c}} \quad \dot{\mathbf{p}}_3 - \dot{\mathbf{c}}\,] \tag{7.5}$$

Thus, eqs.(7.3b) can be written in matrix form as

$$\dot{\mathbf{P}} = \mathbf{\Omega}\mathbf{P} \tag{7.6}$$

from which we want to solve for $\mathbf{\Omega}$, or equivalently, for $\boldsymbol{\omega}$. This cannot be done by simply multiplying by the inverse of $\mathbf{P}$, because the latter is a singular matrix. In fact, as the reader can readily verify, any vector having three identical components lies in the nullspace of $\mathbf{P}$, thereby showing that $\mathbf{P}$ is singular, its nullspace being spanned by that vector. Furthermore, notice that from eq.(7.3b), it is apparent that

$$(\dot{\mathbf{p}}_i - \dot{\mathbf{c}})^T \boldsymbol{\omega} = 0, \quad i = 1, 2, 3 \tag{7.7a}$$

Upon assembling all three scalar equations above in one single vector equation, we obtain

$$\dot{\mathbf{P}}^T \boldsymbol{\omega} = \mathbf{0} \tag{7.7b}$$

a result that is summarized below:

Theorem 7.2.1 *The angular-velocity vector lies in the nullspace of matrix* $\dot{\mathbf{P}}^T$, *with* $\dot{\mathbf{P}}$ *defined as in eq.(7.5).*

In order to find the desired expression for $\boldsymbol{\omega}$ from the above equation, we recall here a result which is proven in Appendix A: Let $\mathbf{S}$ be a skew-symmetric 3×3 matrix and $\mathbf{A}$ be an arbitrary 3×3 matrix. Then,

$$\operatorname{vect}(\mathbf{SA}) = \frac{1}{2}\left[\operatorname{tr}(\mathbf{A})\mathbf{1} - \mathbf{A}\right] \operatorname{vect}(\mathbf{S}) \tag{7.8}$$

Upon application of the foregoing result, eq.(7.6) leads to

$$\mathbf{D}\boldsymbol{\omega} = \operatorname{vect}(\dot{\mathbf{P}}) \tag{7.9}$$

where $\mathbf{D}$ is defined below and $\operatorname{vect}(\mathbf{\Omega})$ is nothing but $\boldsymbol{\omega}$:

$$\mathbf{D} \equiv \frac{1}{2}[\mathrm{tr}(\mathbf{P})\mathbf{1} - \mathbf{P}] \tag{7.10}$$

Thus, eq.(7.9) can be solved for $\boldsymbol{\omega}$ as long as $\mathbf{D}$ is invertible. But since the three points are noncollinear, $\mathbf{D}$ is invertible if and only if $\mathrm{tr}(\mathbf{P})$ does not vanish. Indeed, if $\mathrm{tr}(\mathbf{P})$ vanishes, $\mathbf{D}$ becomes just one-half the negative of $\mathbf{P}$, which as we saw above, is singular. Moreover, as we saw in Example 2.6.1, matrix $\mathbf{P}$ is not frame-invariant in the sense of eq.(7.4). Thus, $\mathrm{tr}(\mathbf{P})$ is not frame-invariant either. However, if the three given points are noncollinear, then it is always possible to find a coordinate frame in which the trace of $\mathbf{P}$ does not vanish. Furthermore, under the assumption that a suitable coordinate frame has been chosen, the inverse of $\mathbf{D}$ can be proven to be

$$\mathbf{D}^{-1} = \alpha\mathbf{1} - \beta\mathbf{P}^2 \tag{7.11}$$

where coefficients α and β are given below:

$$\alpha \equiv \frac{2}{\mathrm{tr}(\mathbf{P})}, \quad \beta \equiv \frac{4}{\mathrm{tr}(\mathbf{P})[\mathrm{tr}(\mathbf{P}^2) - \mathrm{tr}^2(\mathbf{P})]} \tag{7.12}$$

By looking at expressions (7.12), one might think that $\mathbf{D}$ fails to be invertible not only when $\mathrm{tr}(\mathbf{P})$ vanishes, but also when the term in the brackets in the denominator of β does. It is left as an exercise to the reader to prove that the foregoing term vanishes if and only if the three given points are collinear.

From the foregoing discussion, it is clear that given the velocities and the position vectors of three noncollinear points of a rigid body, the angular velocity of the body can always be determined. However, the data, i.e., the velocities of the three given points, cannot be arbitrary, for they must conform to eq.(7.6) or to Theorem 7.2.1. Equation (7.6) states that the columns of matrix $\dot{\mathbf{P}}$ must lie in the range of $\boldsymbol{\Omega}$, while Theorem 7.2.1 states that $\boldsymbol{\omega}$ lies in the nullspace of $\dot{\mathbf{P}}$. However, prior to the computation of $\boldsymbol{\omega}$, or equivalently, of $\boldsymbol{\Omega}$, it is not possible to verify this condition. An alternative approach to verify the compatibility of the data follows: Since lines P_iC belong to a rigid body, vectors $\mathbf{p}_i - \mathbf{c}$ must remain of the same magnitude throughout a rigid-body motion. Moreover, the angles between any two of the said lines must be preserved throughout the motion as well. This means that the conditions below must hold:

$$(\mathbf{p}_i - \mathbf{c})^T(\mathbf{p}_j - \mathbf{c}) = c_{ij}, \quad i, j = 1, 2, 3 \tag{7.13}$$

or in compact form,

$$\mathbf{P}^T\mathbf{P} = \mathbf{C} \tag{7.14}$$

where the (i, j) entry of the constant matrix $\mathbf{C}$ is c_{ij}, as defined in eq.(7.13) above. Upon differentiation of both sides of eq.(7.14) with respect to time, we obtain

Theorem 7.2.2 (Velocity Compatibility) *The velocities of three points of a rigid body satisfy the compatibility condition given below:*

$$\dot{\mathbf{P}}^T\mathbf{P} + \mathbf{P}^T\dot{\mathbf{P}} = \mathbf{O} \tag{7.15}$$

with matrices $\mathbf{P}$ *and* $\dot{\mathbf{P}}$ *defined in eqs.(7.4) and (7.5) and* $\mathbf{O}$ *denoting the* 3×3 *zero matrix.*

The above equation, then, states that for the given velocities of three points of a rigid body to be compatible, the product $\mathbf{P}^T\dot{\mathbf{P}}$ must be skew-symmetric. Note that the above matrix compatibility equation represents six independent scalar equations that the data of the problem at hand must satisfy. There is a tendency to neglect the foregoing six independent scalar compatibility conditions and to focus only on the three scalar conditions drawn from the diagonal entries of the above matrix equation. This is, however, a mistake, for these three conditions do not suffice to guarantee data compatibility in this context; all these three conditions guarantee is that the distance between any pair of points of the set remains constant, but they say nothing about the angles between the pairs of lines formed by each pair of points.

Note, on the other hand, that the product $\mathbf{P}\mathbf{P}^T$ has no direct geometric interpretation, although the difference $\text{tr}(\mathbf{P}\mathbf{P}^T)\mathbf{1} - \mathbf{P}\mathbf{P}^T$ does, as discussed in Exercise C.7.8. Furthermore, while Theorem 7.2.2 states that matrix $\mathbf{P}^T\dot{\mathbf{P}}$ is skew-symmetric, it says nothing about the product $\mathbf{P}\dot{\mathbf{P}}^T$. All we can say about this product is stated in the result below:

Theorem 7.2.3 *With matrices* $\mathbf{P}$ *and* $\dot{\mathbf{P}}$ *defined in eqs.(7.4) and (7.5), the product* $\mathbf{P}\dot{\mathbf{P}}^T$ *obeys the constraint*

$$\text{tr}(\mathbf{P}\dot{\mathbf{P}}^T) = 0 \tag{7.16}$$

If $m \times n$ matrices are regarded as forming a vector space, then an *inner product* of two such matrices $\mathbf{A}$ and $\mathbf{B}$, denoted by $(\mathbf{A}, \mathbf{B})$, can be defined as

$$(\mathbf{A}, \mathbf{B}) \equiv \text{tr}(\mathbf{A}\mathbf{B}^T) \tag{7.17}$$

two matrices being said to be *orthogonal* when the foregoing inner product vanishes. We thus have that Theorem 7.2.3 states that matrices $\dot{\mathbf{P}}$ and $\mathbf{P}$ are orthogonal, a result that parallels that about the orthogonality of the relative velocity of two points and the line joining them, as stated in eq.(3.53) and summarized in the ensuing theorem. The proof of Theorem 7.2.3 is left as an exercise.

Example 7.2.1 *The rigid cube shown in Fig. 7.1 moves in such a way that vertices* P_1, P_2, *and* P_3 *undergo the velocities shown in that figure, for three different possible motions. The length of the sides of the cube is 1, and*

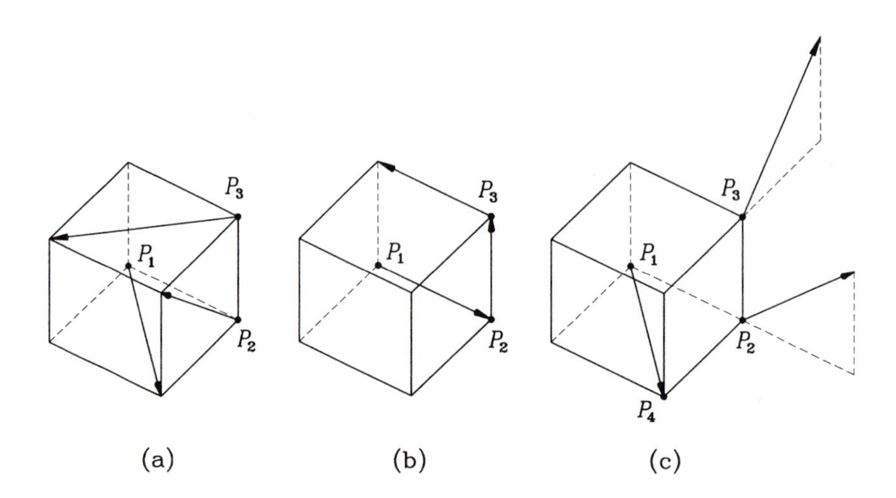

FIGURE 7.1. A rigid cube undergoing a motion determined by the velocities of three of its points.

the velocities all have magnitude $\sqrt{2}$ in Figs. 7.1a and c; these velocities are of unit magnitude in Fig. 7.1b. Furthermore, in the motion depicted in Fig. 7.1c, the velocity of P_3 is parallel to line P_4P_3, whereas that of P_2 is parallel to line P_1P_3. Out of the three different motions, it is known that at least one is compatible. Identify it. Moreover, compute the angular velocity of the compatible motion.

Solution: Let $\dot{\mathbf{p}}_i$ denote the velocity of P_i, of position vector $\mathbf{p}_i$. Each proposed motion is then analyzed: (a) The projection of $\dot{\mathbf{p}}_1$ onto P_1P_2 is 1, but that of $\dot{\mathbf{p}}_2$ onto the same line is 0, and hence, this motion is incompatible; (b) Again, the projection of $\dot{\mathbf{p}}_1$ onto P_1P_2 is 1, but that of $\dot{\mathbf{p}}_2$ onto the same line vanishes, and hence, this motion is also incompatible. Thus, the only possibility is (c), which is now analyzed more formally: Use a dextrous—right-handed—rectangular coordinate frame with origin at P_1, axis Y along P_1P_2, and axis Z parallel to P_2P_3. All vectors and matrices are now represented in this coordinate frame, and hence,

$$\mathbf{p}_1 = \begin{bmatrix} 0 \\ 0 \\ 0 \end{bmatrix}, \quad \mathbf{p}_2 = \begin{bmatrix} 0 \\ 1 \\ 0 \end{bmatrix}, \quad \mathbf{p}_3 = \begin{bmatrix} 0 \\ 1 \\ 1 \end{bmatrix}$$

$$\dot{\mathbf{p}}_1 = \begin{bmatrix} 1 \\ 1 \\ 0 \end{bmatrix}, \quad \dot{\mathbf{p}}_2 = \begin{bmatrix} 0 \\ 1 \\ 1 \end{bmatrix}, \quad \dot{\mathbf{p}}_3 = \begin{bmatrix} -1 \\ 0 \\ 1 \end{bmatrix}$$

Thus,

$$\mathbf{c} = \frac{1}{3}\begin{bmatrix} 0 \\ 2 \\ 1 \end{bmatrix}, \quad \dot{\mathbf{c}} = \frac{1}{3}\begin{bmatrix} 0 \\ 2 \\ 2 \end{bmatrix}$$

Now matrices $\mathbf{P}$ and $\dot{\mathbf{P}}$ are constructed:

$$\mathbf{P} = \frac{1}{3}\begin{bmatrix} 0 & 0 & 0 \\ -2 & 1 & 1 \\ -1 & -1 & 2 \end{bmatrix}, \quad \dot{\mathbf{P}} = \frac{1}{3}\begin{bmatrix} 3 & 0 & -3 \\ 1 & 1 & -2 \\ -2 & 1 & 1 \end{bmatrix}$$

Furthermore,

$$\mathbf{P}^T\dot{\mathbf{P}} = \frac{1}{9}\begin{bmatrix} 0 & -3 & 3 \\ 3 & 0 & -3 \\ -3 & 3 & 0 \end{bmatrix}$$

which is skew-symmetric, and hence, the motion is compatible. Now, matrix $\mathbf{D}$ is computed:

$$\mathbf{D} \equiv \frac{1}{2}[\,\mathbf{1}\mathrm{tr}(\mathbf{P}) - \mathbf{P}] = \frac{1}{6}\begin{bmatrix} 3 & 0 & 0 \\ 2 & 2 & -1 \\ 1 & 1 & 1 \end{bmatrix}$$

The angular velocity $\boldsymbol{\omega}$ is computed as the solution to

$$\mathbf{D}\boldsymbol{\omega} = \mathrm{vect}(\dot{\mathbf{P}})$$

where

$$\mathrm{vect}(\dot{\mathbf{P}}) = \frac{1}{6}\begin{bmatrix} 3 \\ -1 \\ 1 \end{bmatrix}$$

Equations (7.2) are thus

$$\begin{aligned} 3\omega_1 &= 3 \\ 2\omega_1 + 2\omega_2 - \omega_3 &= -1 \\ \omega_1 + \omega_2 + \omega_3 &= 1 \end{aligned}$$

The first of the foregoing equations leads to

$$\omega_1 = 1$$

whereas the second and the third lead to

$$\begin{aligned} 2\omega_2 - \omega_3 &= -3 \\ \omega_2 + \omega_3 &= 0 \end{aligned}$$

and hence,

$$\omega_2 = -1, \quad \omega_3 = 1$$

Now, as a verification, $\boldsymbol{\omega}$ should be normal to the three columns of $\dot{\mathbf{P}}$ as defined in eq.(7.15); in other words, $\boldsymbol{\omega}$ should lie in the nullspace of $\dot{\mathbf{P}}^T$. But this is so, because

$$\dot{\mathbf{P}}^T\boldsymbol{\omega} = \frac{1}{3}\begin{bmatrix} 3 & 1 & -2 \\ 0 & 1 & 1 \\ -3 & -2 & 1 \end{bmatrix}\begin{bmatrix} 1 \\ -1 \\ 1 \end{bmatrix} = \frac{1}{3}\begin{bmatrix} 0 \\ 0 \\ 0 \end{bmatrix}$$

thereby verifying that $\boldsymbol{\omega}$ lies, in fact, in the nullspace of $\dot{\mathbf{P}}^{\mathbf{T}}$.

7.3 Computation of Angular Acceleration from Point-Acceleration Data

The angular acceleration of a rigid body under general motion is determined in this section from knowledge of the position, velocity, and acceleration vectors of three noncollinear points of the body. The underlying procedure parallel that of Section 7.2. Indeed, recalling the notation introduced in that section, and letting vectors $\ddot{\mathbf{p}}_i$, for $i = 1, 2, 3$, denote the acceleration of the given points, one can rewrite eq.(3.87) for each point in the form

$$\ddot{\mathbf{p}}_i = \ddot{\mathbf{c}} + (\dot{\boldsymbol{\Omega}} + \boldsymbol{\Omega}^2)(\mathbf{p}_i - \mathbf{c}), \quad i = 1, 2, 3 \tag{7.18a}$$

or

$$\ddot{\mathbf{p}}_i - \ddot{\mathbf{c}} = (\dot{\boldsymbol{\Omega}} + \boldsymbol{\Omega}^2)(\mathbf{p}_i - \mathbf{c}), \quad i = 1, 2, 3 \tag{7.18b}$$

where $\mathbf{c}$ was defined in eq.(7.1), and $\ddot{\mathbf{c}}$ is the acceleration of the centroid, i.e.,

$$\ddot{\mathbf{c}} \equiv \frac{1}{3}\sum_{1}^{3} \ddot{\mathbf{p}}_i \tag{7.18c}$$

Furthermore, matrix $\ddot{\mathbf{P}}$ is defined as

$$\ddot{\mathbf{P}} \equiv [\,\ddot{\mathbf{p}}_1 - \ddot{\mathbf{c}} \quad \ddot{\mathbf{p}}_2 - \ddot{\mathbf{c}} \quad \ddot{\mathbf{p}}_3 - \ddot{\mathbf{c}}\,] \tag{7.19}$$

Thus, eqs.(7.18b) can be written in compact form as

$$\ddot{\mathbf{P}} = (\dot{\boldsymbol{\Omega}} + \boldsymbol{\Omega}^2)\mathbf{P} \tag{7.20}$$

from which one is interested in computing $\dot{\boldsymbol{\Omega}}$, or correspondingly, $\dot{\boldsymbol{\omega}}$. To this end, eq.(7.20) is rewritten as

$$\dot{\boldsymbol{\Omega}}\mathbf{P} = \mathbf{W} \tag{7.21a}$$

with matrix $\mathbf{W}$ defined as

$$\mathbf{W} \equiv \ddot{\mathbf{P}} - \boldsymbol{\Omega}^2\dot{\mathbf{P}} \tag{7.21b}$$

The counterpart of Theorem 7.2.1 is now derived from eqs.(7.18b). First, these equations are cast in the form

$$\ddot{\mathbf{p}}_i - \ddot{\mathbf{c}} - \boldsymbol{\Omega}^2(\mathbf{p}_i - \mathbf{c}) = \dot{\boldsymbol{\omega}} \times (\mathbf{p}_i - \mathbf{c}), \quad i = 1, 2, 3$$

It is now apparent that if we dot-multiply the above equations by $\dot{\boldsymbol{\omega}}$, we obtain

$$[\ddot{\mathbf{p}}_i - \ddot{\mathbf{c}} - \boldsymbol{\Omega}^2(\mathbf{p}_i - \mathbf{c})] \cdot \dot{\boldsymbol{\omega}} = 0, \quad i = 1, 2, 3 \tag{7.22a}$$

Upon assembling the three foregoing equations in one single vector equation, we derive the counterpart of eq.(7.7b), namely,

$$(\ddot{\mathbf{P}} - \boldsymbol{\Omega}^2\dot{\mathbf{P}})^T\dot{\boldsymbol{\omega}} = \mathbf{0} \tag{7.22b}$$

a result that is summarized below in theorem form:

Theorem 7.3.1 *The angular-acceleration vector $\dot{\boldsymbol{\omega}}$ lies in the nullspace of matrix $\mathbf{W}^T$, with $\mathbf{W}$ defined in eq.(7.21b).*

Just as we did in Section 7.2 when solving for $\boldsymbol{\omega}$ from eq.(7.9), we apply the result already invoked in connection with eq.(7.9), thereby deriving an alternative form of eq.(7.21a), namely,

$$\mathbf{D}\dot{\boldsymbol{\omega}} = \operatorname{vect}(\ddot{\mathbf{P}} - \boldsymbol{\Omega}^2\mathbf{P}) \tag{7.23}$$

where $\mathbf{D}$ is defined as in eq.(7.10). Thus,

$$\dot{\boldsymbol{\omega}} = \mathbf{D}^{-1}\operatorname{vect}(\ddot{\mathbf{P}} - \boldsymbol{\Omega}^2\mathbf{P}) \tag{7.24}$$

with $\mathbf{D}^{-1}$ given as in eqs.(7.11) and (7.12). As in Section 7.2, then, given the position, velocity, and acceleration vectors of three noncollinear points of a rigid body, it is always possible to compute the associated angular acceleration. However, as discussed in that section, the data cannot be given arbitrarily, for they must comply with eq.(7.21a), or correspondingly, with eq.(7.22b). The former implies that the three columns of matrix $\mathbf{W}$ lie in the range of matrix $\dot{\boldsymbol{\Omega}}$; alternatively, eq.(7.22b) implies that $\dot{\boldsymbol{\Omega}}$ lies in the nullspace of $\mathbf{W}^T$. Again, prior to the determination of $\dot{\boldsymbol{\Omega}}$, it is impossible to verify this condition, for which reason an alternative approach is taken to verify compatibility. The obvious one is to differentiate both sides of eq.(7.15), which produces

$$\ddot{\mathbf{P}}^T\mathbf{P} + 2\dot{\mathbf{P}}^T\dot{\mathbf{P}} + \mathbf{P}^T\ddot{\mathbf{P}} = \mathbf{0} \tag{7.25}$$

thereby deriving the *compatibility conditions* that the acceleration measurements should satisfy.

Finally, upon differentiation of both sides of eq.(7.16) with respect to time, and while doing this, resorting to Lemma A.2 of Appendix A, we have

$$\operatorname{tr}(\ddot{\mathbf{P}}\mathbf{P}^T + \dot{\mathbf{P}}\dot{\mathbf{P}}^T) = 0 \tag{7.26}$$

which is the counterpart of eq.(7.16).

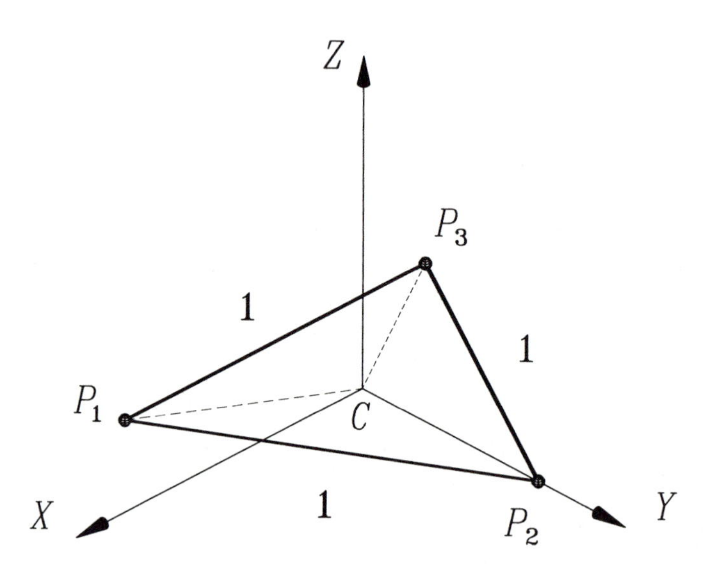

FIGURE 7.2. A rigid triangular plate undergoing a motion given by the velocity and acceleration of its vertices.

Example 7.3.1 *The three vertices of the equilateral triangular plate of Fig. 7.2, which lies in the X-Y plane, are labeled* P_1, P_2, *and* P_3, *their position vectors being* $\mathbf{p}_1$, $\mathbf{p}_2$, *and* $\mathbf{p}_3$. *Moreover, the velocities of the foregoing points are denoted by* $\dot{\mathbf{p}}_i$, *for* $i = 1, 2, 3$. *The origin of the coordinate frame* X, Y, Z *lies at the centroid* C *of the triangle, the velocities of the vertices, in this coordinate frame, being given as*

$$\dot{\mathbf{p}}_1 = \frac{4-\sqrt{2}}{4}\begin{bmatrix}0\\0\\1\end{bmatrix}, \quad \dot{\mathbf{p}}_2 = \frac{4-\sqrt{3}}{4}\begin{bmatrix}0\\0\\1\end{bmatrix}, \quad \dot{\mathbf{p}}_3 = \frac{4+\sqrt{2}}{4}\begin{bmatrix}0\\0\\1\end{bmatrix}$$

Likewise, $\ddot{\mathbf{p}}_1$, $\ddot{\mathbf{p}}_2$, *and* $\ddot{\mathbf{p}}_3$ *denote the accelerations of the three vertices of the plate, given below in the same coordinate frame:*

$$\ddot{\mathbf{p}}_1 = \frac{1}{24}\begin{bmatrix}-6+4\sqrt{3}\\12-3\sqrt{2}\\0\end{bmatrix}, \quad \ddot{\mathbf{p}}_2 = -\frac{1}{24}\begin{bmatrix}8\sqrt{3}+3\sqrt{6}\\3\sqrt{3}\\0\end{bmatrix},$$

$$\ddot{\mathbf{p}}_3 = \frac{1}{24}\begin{bmatrix}6+4\sqrt{3}\\-12+3\sqrt{2}\\0\end{bmatrix}$$

With the foregoing information,

(a) *show that the three given velocities are compatible;*

(b) *compute the angular velocity of the plate;*

(c) *determine the set of points of the plate that undergo a velocity of minimum magnitude;*

(d) *show that the given accelerations are compatible;*

(e) *compute the angular acceleration of the plate.*

Solution:

(a) Since the centroid of the triangle coincides with that of the three given points, we have $\mathbf{c} = \mathbf{0}$. Moreover,

$$\mathbf{p}_1 = \begin{bmatrix} 1/2 \\ -\sqrt{3}/6 \\ 0 \end{bmatrix}, \quad \mathbf{p}_2 = \begin{bmatrix} 0 \\ \sqrt{3}/3 \\ 0 \end{bmatrix}, \quad \mathbf{p}_3 = \begin{bmatrix} -1/2 \\ -\sqrt{3}/6 \\ 0 \end{bmatrix}$$

Thus,

$$\mathbf{P} = \frac{1}{6} \begin{bmatrix} 3 & 0 & -3 \\ -\sqrt{3} & 2\sqrt{3} & -\sqrt{3} \\ 0 & 0 & 0 \end{bmatrix}$$

Furthermore,

$$\dot{\mathbf{c}} = \begin{bmatrix} 0 \\ 0 \\ (12 - \sqrt{3})/12 \end{bmatrix}$$

and hence,

$$\dot{\mathbf{P}} = \frac{1}{12} \begin{bmatrix} 0 & 0 & 0 \\ 0 & 0 & 0 \\ \sqrt{3} - 3\sqrt{2} & -2\sqrt{3} & \sqrt{3} + 3\sqrt{2} \end{bmatrix}$$

We can readily show from the above results that

$$\mathbf{P}^T \dot{\mathbf{P}} = \mathbf{O}$$

with $\mathbf{O}$ denoting the 3×3 zero matrix. Hence, matrix $\mathbf{P}^T\dot{\mathbf{P}}$ is skew-symmetric and the velocities are compatible

(b) Next, we have

$$\mathbf{D} \equiv \frac{1}{2}[\mathrm{tr}(\mathbf{P})\mathbf{1} - \mathbf{P}] \equiv \frac{1}{12} \begin{bmatrix} 2\sqrt{3} & 0 & 3 \\ \sqrt{3} & 3 & \sqrt{3} \\ 0 & 0 & 3 + 2\sqrt{3} \end{bmatrix}$$

and

$$\mathrm{vect}(\dot{\mathbf{P}}) = \frac{1}{24} \begin{bmatrix} -2\sqrt{3} \\ -\sqrt{3} + 3\sqrt{2} \\ 0 \end{bmatrix}$$

Hence, if the components of $\boldsymbol{\omega}$ in the given coordinate frame are denoted by ω_i, for $i = 1, 2, 3$, then we obtain

$$2\sqrt{3}\omega_1 + 3\omega_3 = -\sqrt{3}$$
$$\sqrt{3}\omega_1 + 3\omega_2 + \sqrt{3}\omega_3 = \frac{-\sqrt{3} + 3\sqrt{2}}{2}$$
$$(3 + 2\sqrt{3})\omega_3 = 0$$

From the third equation,

$$\omega_3 = 0$$

Substitution of the foregoing value into the first of the above equations yields $\omega_1 = -1/2$. Further, upon substitution of the values of ω_1 and ω_3 into the second of the above equations, we obtain $\omega_2 = \sqrt{2}/2$ and hence,

$$\boldsymbol{\omega} = \frac{1}{2}\begin{bmatrix} -1 \\ \sqrt{2} \\ 0 \end{bmatrix}$$

(c) Let $\mathbf{p}'_0$ be the position vector of the point P'_0 on the instantaneous screw axis lying closest to the origin. Now, in order to find $\mathbf{p}'_0$, we can resort to eq.(3.72), using point C as a reference, i.e., with $\mathbf{c}$ and $\dot{\mathbf{c}}$ playing the roles of $\mathbf{a}$ and $\dot{\mathbf{a}}$ in that equation. Moreover, since $\mathbf{c} = \mathbf{0}$, the expression for $\mathbf{p}'_0$ reduces to

$$\mathbf{p}'_0 = \frac{1}{\|\boldsymbol{\omega}\|^2}\boldsymbol{\Omega}\dot{\mathbf{c}}$$

where from item (b),

$$\|\boldsymbol{\omega}\|^2 = \frac{3}{4}$$

while

$$\boldsymbol{\Omega}\dot{\mathbf{c}} = \frac{12 - \sqrt{3}}{24}\begin{bmatrix} \sqrt{2} \\ 1 \\ 0 \end{bmatrix}$$

and hence,

$$\mathbf{p}'_0 = \frac{12 - \sqrt{3}}{18}\begin{bmatrix} \sqrt{2} \\ 1 \\ 0 \end{bmatrix}$$

As a verification, $\mathbf{p}'_0$ should be perpendicular to the ISA, as it is, for the product $\boldsymbol{\omega}^T\mathbf{p}'_0$ to vanish. Next, the vector representing the direction of the screw axis is obtained simply as

$$\mathbf{e} = \frac{\boldsymbol{\omega}}{\|\boldsymbol{\omega}\|} = \frac{\sqrt{3}}{3}[-1 \quad \sqrt{2} \quad 0]^T$$

thereby defining completely the instant screw axis.

(d) The acceleration of the centroid of the three given points is given as follows:

$$\ddot{\mathbf{c}} = [-\frac{\sqrt{6}}{24}, -\frac{\sqrt{3}}{24}, 0]^T$$

Then, matrices $\ddot{\mathbf{P}}$, $\mathbf{P}^T\ddot{\mathbf{P}}$, $\ddot{\mathbf{P}}^T\mathbf{P}$, and $\dot{\mathbf{P}}^T\dot{\mathbf{P}}$ are readily computed as

$$\ddot{\mathbf{P}} = \frac{1}{24}\begin{bmatrix} -6+4\sqrt{3}+\sqrt{6} & -8\sqrt{3}-2\sqrt{6} & 6+4\sqrt{3}+\sqrt{6} \\ 12-3\sqrt{2}+\sqrt{3} & -2\sqrt{3} & -12+3\sqrt{2}+\sqrt{3} \\ 0 & 0 & 0 \end{bmatrix}$$

$$\mathbf{P}^T\ddot{\mathbf{P}} = \frac{1}{144}\begin{bmatrix} -21+6\sqrt{6} & 6-24\sqrt{3}-6\sqrt{6} & 15+24\sqrt{3} \\ 6+24\sqrt{3}-6\sqrt{6} & -12 & 6-24\sqrt{3}+6\sqrt{6} \\ 15-24\sqrt{3} & 6+24\sqrt{3}+6\sqrt{6} & -21-6\sqrt{6} \end{bmatrix}$$

$$\ddot{\mathbf{P}}^T\mathbf{P} = \frac{1}{144}\begin{bmatrix} -21+6\sqrt{6} & 6+24\sqrt{3}-6\sqrt{6} & 15-24\sqrt{3} \\ 6-24\sqrt{3}-6\sqrt{6} & -12 & 6+24\sqrt{3}+6\sqrt{6} \\ 15+24\sqrt{3} & 6-24\sqrt{3}+6\sqrt{6} & -21-6\sqrt{6} \end{bmatrix}$$

$$\dot{\mathbf{P}}^T\dot{\mathbf{P}} = \frac{1}{144}\begin{bmatrix} 21-6\sqrt{6} & -6+6\sqrt{6} & -15 \\ -6+6\sqrt{6} & 12 & -6-6\sqrt{6} \\ -15 & -6-6\sqrt{6} & 21+6\sqrt{6} \end{bmatrix}$$

Now, it is a simple matter to verify that

$$\ddot{\mathbf{P}}^T\mathbf{P} + 2\dot{\mathbf{P}}^T\dot{\mathbf{P}} + \mathbf{P}^T\ddot{\mathbf{P}} = \mathbf{O}$$

and hence, the given accelerations are compatible.

(e) $\boldsymbol{\Omega}$ is defined as the unique skew-symmetric matrix whose vector is $\boldsymbol{\omega}$, the latter having been computed in item (b). Thus,

$$\boldsymbol{\Omega} = \frac{1}{2}\begin{bmatrix} 0 & 0 & \sqrt{2} \\ 0 & 0 & 1 \\ -\sqrt{2} & -1 & 0 \end{bmatrix}, \quad \boldsymbol{\Omega}^2 = \frac{1}{4}\begin{bmatrix} -2 & -\sqrt{2} & 0 \\ -\sqrt{2} & -1 & 0 \\ 0 & 0 & -3 \end{bmatrix},$$

$$\boldsymbol{\Omega}^2\mathbf{P} = \frac{1}{24}\begin{bmatrix} -6+\sqrt{6} & -2\sqrt{6} & 6+\sqrt{6} \\ -3\sqrt{2}+\sqrt{3} & -2\sqrt{3} & 3\sqrt{2}+\sqrt{3} \\ 0 & 0 & 0 \end{bmatrix}$$

Hence,

$$\ddot{\mathbf{P}} - \boldsymbol{\Omega}^2\mathbf{P} = \frac{1}{24}\begin{bmatrix} 4\sqrt{3} & -8\sqrt{3} & 4\sqrt{3} \\ 12 & 0 & -12 \\ 0 & 0 & 0 \end{bmatrix}$$

The angular-acceleration vector is thus computed from

$$\mathbf{D}\dot{\boldsymbol{\omega}} = \operatorname{vect}(\ddot{\mathbf{P}} - \boldsymbol{\Omega}^2\mathbf{P})$$

where $\mathbf{D}$ was computed in item (b), while

$$\text{vect}(\ddot{\mathbf{P}} - \mathbf{\Omega}^2\mathbf{P}) = \frac{1}{12}\begin{bmatrix} 3 \\ \sqrt{3} \\ 3 + 2\sqrt{3} \end{bmatrix}$$

and hence, letting $\dot{\omega}_i$ denote the ith component of $\dot{\boldsymbol{\omega}}$ in the given coordinate frame, we obtain

$$\frac{1}{12}(2\sqrt{3}\dot{\omega}_1 + 3\dot{\omega}_3) = \frac{1}{4}$$
$$\frac{1}{12}(\sqrt{3}\dot{\omega}_1 + 3\dot{\omega}_2 + \sqrt{3}\dot{\omega}_3) = \frac{\sqrt{3}}{12}$$
$$\frac{1}{12}(3 + 2\sqrt{3})\dot{\omega}_3 = \frac{3 + 2\sqrt{3}}{12}$$

which yields

$$\dot{\boldsymbol{\omega}} = \begin{bmatrix} 0 \\ 0 \\ 1 \end{bmatrix}$$

thereby completing the solution. Note that $\dot{\boldsymbol{\omega}}$ lies, in fact, in the nullspace of matrix $(\ddot{\mathbf{P}} - \mathbf{\Omega}^2\dot{\mathbf{P}})^T$.

8

Kinematics of Complex Robotic Mechanical Systems

8.1 Introduction

Current robotic mechanical systems, encountered not only in research laboratories but also in production or construction environments, include robotic mechanical systems with features that deserve a chapter apart. Generically, we will call here *complex robotic mechanical systems* all such systems that do not fall in the category of those studied in Chapter 4. Thus, complex robotic mechanical systems are understood here as those systems of this kind that are either of the serial type but that do not allow a decoupling of the positioning and the orientation problems, or of type other than serial. Examples of the latter are parallel manipulators, dextrous hands, walking machines, and rolling robots. While redundant manipulators of the serial type fall within this category as well, we will leave these aside, for their redundancy resolution calls for a more specialized background than what we have either assumed or given here.

A special feature of serial manipulators of the kind studied here is that they can admit up to sixteen inverse kinematics solutions. Such manipulators are now in operation in industry, an example of which is the TELBOT System, shown in Fig. 8.1, which features all its six motors on its base, the motion and force transmission taking place via concentric tubes and bevel

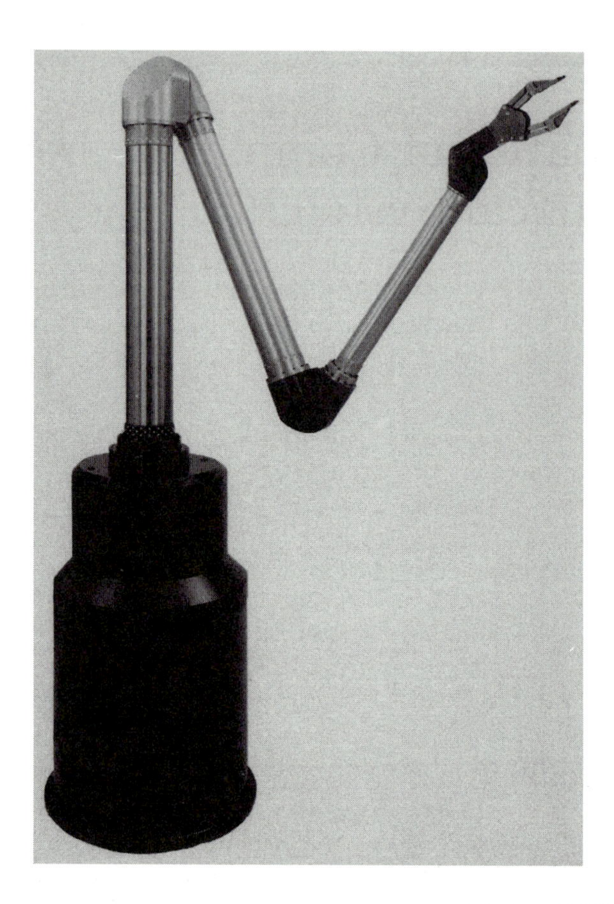

FIGURE 8.1. The TELBOT System (Courtesy of Wälischmiller GmbH, Meersburg, Germany.)

gears. This special feature allows TELBOT to have unlimited angular displacements at its joints, no cables traveling through its structure and no deadload on its links by virtue of the motors (Wälischmiller and Li, 1996).

8.2 The IKP of General Six-Revolute Manipulators

As shown in Chapter 4, the IKP of six-revolute manipulators of the most general type leads to a system of six independent equations in six unknowns. This is a highly nonlinear algebraic system whose solution posed a challenge to kinematicians for about two decades and that was not considered essentially solved until the late eighties. Below we give a short historical account of this problem.

Pieper (1968) reported what is probably the earliest attempt to formulate the inverse kinematic problem of six-axis serial manipulators in a univariate polynomial form. He showed that decoupled manipulators, studied in Section 4.4, and a few others, allow a closed-form solution of their inverse kinematics. However, apart from the simple architectures identified by Pieper, and others that have been identified more recently (Mavroidis and Roth, 1992), a six-axis manipulator does not admit a closed-form solution. Attempts to derive the *minimal characteristic polynomial* for this manipulator were reported by Duffy and Derby (1979), Duffy and Crane (1980), Albala (1982), and Alizade, Duffy, and Hatiyev (1983), who derived a 32nd-degree polynomial, but suspected that this polynomial was not minimal, in the sense that the manipulator at hand might not be able to admit up to 32 postures for a given set of joint variables. Tsai and Morgan (1985) used a technique known as *polynomial continuation* (Morgan, 1987) to solve *numerically* the nonlinear displacement equations, cast in the form of a system of quadratic equations. These researchers found that no more than 16 solutions were to be expected. Briefly stated, polynomial continuation consists basically of two stages, namely, reducing first the given problem to a system of polynomial equations; in the second stage, a continuous path, also known as a *homotopy* in mathematics, is defined with a real parameter t that can be regarded as time. The continuous path takes the system of equations from a given initial *state* to a final one. The initial state is so chosen that all solutions to the nonlinear system in this state are either apparent or much easier to find numerically than those of the originally proposed system. The final state of the system is the actual system to be solved. The initial system is thus deformed continuously into the final state upon varying its set of parameters. At each continuation step, a set of initial guesses for each of the solutions already exists, for it is simply the solution to the previous continuation step. Moreover, finding the solutions at the current continuation step is done using a standard Newton method (Dahlquist and Björck, 1974).

Primrose (1986) proved conclusively that the problem under discussion admits at most 16 solutions, while Lee and Liang (1988) showed that the same problem leads to a 16th-degree univariate polynomial. Using different elimination procedures, as described in Subsection 8.2.4 below, Li (1990) and Raghavan and Roth (1990, 1993) provided different approaches for the computation of the coefficients of the univariate polynomial. While the inverse kinematics problem can be considered basically solved, research on finding all its solutions *safely* and *quickly* still continues (Angeles, Hommel and Kovács, 1993). Below we describe two approaches to solving this problem: first, the semigraphical approach introduced in (Angeles and Etemadi Zanganeh, 1992) is described; then, we outline the methods of Raghavan and Roth (1990, 1993) and of Li (1990), aimed at reducing the kinematic relations to a single monovariate polynomial.

8.2.1 A Semigraphical Approach to the Solution of the IKP

In this subsection, we introduce a semigraphical solution to the general IKP. Unlike well-established procedures aiming at the reduction of all involved equations to one single univariate polynomial equation, we will proceed to a reduction of those equations to a system of two nonlinear bivariate equations. Each of these equations defines a contour in the plane of the two unknowns, the intersection of the two contours thus yielding all real solutions. Our aim will be to avoid the introduction of spurious solutions, which would increase the degree of the resulting equations. On the other hand, from the numerical viewpoint, it is convenient to avoid increasing the conditioning of the problem (Dahlquist and Björck, 1974; Golub and Van Loan, 1989), which calls for a judicious elimination procedure. However, it will become apparent that two contours, while producing all real solutions of the IKP under study, yield *spurious* solutions as well. These can be readily eliminated by introducing a third contour, i.e., a *discriminating contour* that will allow us to tell the actual from the spurious solutions.

Before introducing the solution procedure, we will give some background. Moreover, the notation we will use throughout is taken from Tsai and Morgan (1985) and Raghavan and Roth (1990, 1993).

Background

We start by recalling the definition of vector $\mathbf{x}_i$ of eq.(4.11), which is reproduced below for quick reference:

$$\mathbf{x}_i \equiv \begin{bmatrix} \cos\theta_i \\ \sin\theta_i \end{bmatrix}$$

Likewise, we recall the definitions of Subsection 4.4.1 pertaining to bilinear forms, with similar definitions for biquadratic, bicubic, trilinear, and multilinear forms. Now we have

Lemma 8.2.1 *Let matrix* $\mathbf{A}$ *be skew-symmetric and* $\mathbf{B}$ *be defined as the similarity transformation of* $\mathbf{A}$ *given below:*

$$\mathbf{B} \equiv \mathbf{Q}_i \mathbf{A} \mathbf{Q}_i^T \tag{8.1}$$

where $\mathbf{Q}_i$ *was defined in eq.(4.1d) and* $\mathbf{A}$ *is assumed to be independent of* θ_i*. Then,* $\mathbf{B}$ *is linear in* $\mathbf{x}_i$*.*

Proof: This result follows from relation (2.138). Indeed, as the reader can readily verify, $\mathbf{B}$ is skew-symmetric as well, and the product $\mathbf{Bv}$, for any 3-dimensional vector $\mathbf{v}$, can be expressed in terms of $\mathbf{b}$, defined as the vector of $\mathbf{B}$. That is,

$$\mathbf{Bv} = \mathbf{b} \times \mathbf{v}$$

But if $\mathbf{a}$ denotes the vector of $\mathbf{A}$, then $\mathbf{a}$ and $\mathbf{b}$, by virtue of eq.(8.1) and the results of Section 2.6, obey the relation

$$\mathbf{b} = \mathbf{Q}_i \mathbf{a}$$

Hence,

$$\mathbf{B}\mathbf{v} = \mathbf{Q}_i(\mathbf{a} \times \mathbf{v})$$

thereby showing that the resulting product is linear in $\mathbf{x}_i$, q.e.d.

Moreover, let

$$\tau_i \equiv \tan(\frac{\theta_i}{2}) \tag{8.2a}$$

which allows us to write the identities below, as suggested by Li (1990):

$$s_i - \tau_i c_i \equiv \tau_i, \quad \tau_i s_i + c_i \equiv 1 \tag{8.2b}$$

Next, we write eqs.(4.9a & b) in the form

$$\mathbf{Q}_3\mathbf{Q}_4\mathbf{Q}_5 = \mathbf{Q}_2^T\mathbf{Q}_1^T\mathbf{Q}\mathbf{Q}_6^T \tag{8.3a}$$

$$\mathbf{Q}_3(\mathbf{b}_3 + \mathbf{Q}_4\mathbf{b}_4 + \mathbf{Q}_4\mathbf{Q}_5\mathbf{b}_5) = \mathbf{Q}_2^T\mathbf{Q}_1^T(\mathbf{p} - \mathbf{Q}\mathbf{b}_6) - (\mathbf{b}_2 + \mathbf{Q}_2^T\mathbf{b}_1) \tag{8.3b}$$

Two more definitions are introduced below:

$$\boldsymbol{\sigma} \equiv \mathbf{Q}\mathbf{o}_6 \tag{8.4a}$$

$$\boldsymbol{\rho} \equiv \mathbf{p} - \mathbf{Q}\mathbf{b}_6 = \mathbf{p} - \mathbf{Q}\mathbf{Q}_6^T\mathbf{a}_6 \tag{8.4b}$$

Thus, $\mathbf{o}_6$ is $\mathbf{e}_6$ in $\mathcal{F}_7$-coordinates, and hence, $\boldsymbol{\sigma}$ represents $\mathbf{e}_6$ in $\mathcal{F}_1$-coordinates. Likewise, $\boldsymbol{\rho}$ is the vector directed from O_1 to O_6, as depicted in Fig. 8.2, in $\mathcal{F}_0$-coordinates. Furthermore, in the above definitions, vectors $\boldsymbol{\sigma}$ and $\boldsymbol{\rho}$ are independent of $\boldsymbol{\theta}$ because so are $\mathbf{Q}$, $\mathbf{o}_6$, $\mathbf{p}$, and $\mathbf{b}_6$.

Note that the matrix equation (8.3a) represents three vector equations, one for each column of the two sides of that equation. If we now separate the third of those vector equations from the others, introduce definition (8.4a) in the equation thus resulting, and rewrite eq.(8.3b) using definition (8.4b), we derive six scalar equations free of θ_6, namely,

$$\mathbf{Q}_3\mathbf{Q}_4\mathbf{u}_5 = \mathbf{Q}_2^T\mathbf{Q}_1^T\boldsymbol{\sigma} \tag{8.5a}$$

$$\mathbf{Q}_3(\mathbf{b}_3 + \mathbf{Q}_4\mathbf{b}_4 + \mathbf{Q}_4\mathbf{Q}_5\mathbf{b}_5) = \mathbf{Q}_2^T\mathbf{Q}_1^T\boldsymbol{\rho} - \mathbf{b}_2 - \mathbf{Q}_2^T\mathbf{b}_1 \tag{8.5b}$$

Moreover, the six foregoing scalar equations are multilinear in $\mathbf{x}_i$, for $i = 1, \ldots, 5$. In fact, their left-hand sides are trilinear in $\mathbf{x}_3$, $\mathbf{x}_4$, and $\mathbf{x}_5$, their right-hand sides being bilinear in $\mathbf{x}_1$ and $\mathbf{x}_2$. In the elimination procedure, we will try to keep this multilinearity inasmuch as it is possible. For brevity, we introduce further definitions below:

$$\mathbf{f} \equiv \mathbf{Q}_3(\mathbf{b}_3 + \mathbf{Q}_4\mathbf{b}_4 + \mathbf{Q}_4\mathbf{Q}_5\mathbf{b}_5) \tag{8.6a}$$

$$\mathbf{g} \equiv \mathbf{Q}_2^T\mathbf{Q}_1^T\boldsymbol{\rho} - (\mathbf{b}_2 + \mathbf{Q}_2^T\mathbf{b}_1) \tag{8.6b}$$

$$\mathbf{h} \equiv \mathbf{Q}_3\mathbf{Q}_4\mathbf{u}_5 \tag{8.6c}$$

$$\mathbf{i} \equiv \mathbf{Q}_2^T\mathbf{Q}_1^T\boldsymbol{\sigma} \tag{8.6d}$$

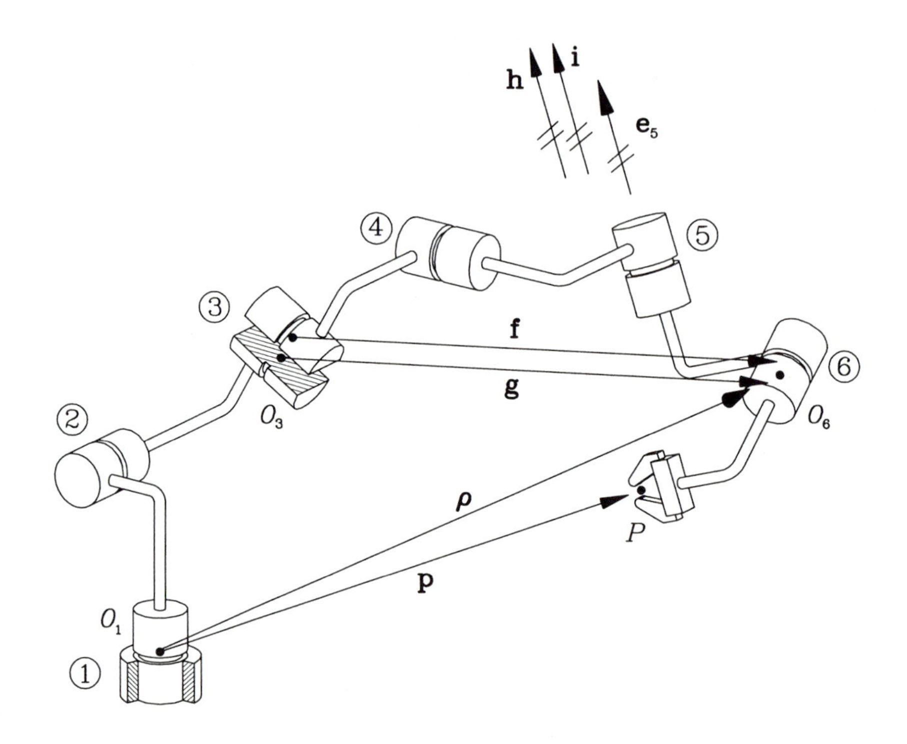

FIGURE 8.2. Partitioning of the manipulator loop into two subloops.

The geometric meaning of the foregoing definitions is illustrated in Fig. 8.2. In that figure, $\mathbf{f}$ and $\mathbf{g}$ represent the same vector directed from O_3 to O_6; the former denotes this vector in the form $\mathbf{a}_3+\mathbf{a}_4+\mathbf{a}_5$, the latter as $\boldsymbol{\rho}-(\mathbf{a}_1+\mathbf{a}_2)$. Moreover, both $\mathbf{f}$ and $\mathbf{g}$ are expressed in $\mathcal{F}_3$-coordinates. Likewise, $\mathbf{h}$ and $\mathbf{i}$ represent vector $\mathbf{e}_6$ in $\mathcal{F}_3$-coordinates.

The two vector equations of interest, i.e., those providing six scalar equations free of θ_6, are thus obtained upon equating the right-hand side of eq.(8.6a) with that of eq.(8.6b), and the right-hand side of eq.(8.6c) with that of eq.(8.6d).

Note that the foregoing equations are not independent, for the two sides of the second of those equations are subjected to the same *quadratic* constraint, i.e.,

$$\mathbf{h}\cdot\mathbf{h}=1, \quad \mathbf{i}\cdot\mathbf{i}=1$$

Hence, out of those six scalar equations, only five are independent, but these suffice to determine the five unknowns contained therein. Raghavan and Roth (1990, 1993), as well as Li, Woernle and Hiller (1991), proposed independent procedures to eliminate four of the five unknowns, thereby deriving a 16th-degree polynomial equation in the tangent of half the fifth unknown angle. We will outline in Subsection 8.2.4 these two procedures.

In this subsection, however, we follow a slightly different approach that will lead to a semigraphical solution to the problem at hand. We do this for several reasons, namely, (*i*) high-degree polynomials are prone to numerical ill-conditioning (Dahlquist and Björck, 1974); (*ii*) in the presence of a root yielding an angle of π, the polynomial degenerates into one of a lower degree, while, in the vicinity of π, one of the roots is extremely large, thereby producing numerical instabilities; (*iii*) a graphical or semigraphical procedure is more appealing to engineers than a purely numerical approach; (*iv*) by resorting to a semigraphical approach, we make use of very powerful tools in terms of software and hardware, that are available nowadays; and (*v*) the graphical solution provides information on the *numerical conditioning*—to be explained later—of the solutions, while purely numerical methods do not. Thus, our procedure aims at deriving not one single 16th-degree univariate polynomial, but rather two bivariate equations in the sines and cosines of two of the unknown angles.

Our goal is then to eliminate three of the five unknowns of eqs.(8.7a & b). In order to do this, we will proceed, as Raghavan and Roth (1990, 1993) did, to derive 14 equations out of the four vectors $\mathbf{f}$, $\mathbf{g}$, $\mathbf{h}$, and $\mathbf{i}$ given above. On the other hand, Li (1990)[1] derived 20 equations comprising the same 14 used by Raghavan and Roth, plus six supplementary equations, all of which are multilinear in $\{\mathbf{x}_i\}_1^5$. The basic 14 equations consist of 12 scalar equations derived from four 3-dimensional vector equations and two additional scalar equations. The first three of these vector equations are listed below:

$$\mathbf{f} = \mathbf{g} \tag{8.7a}$$

$$\mathbf{h} = \mathbf{i} \tag{8.7b}$$

$$\mathbf{f} \times \mathbf{h} = \mathbf{g} \times \mathbf{i} \tag{8.7c}$$

The fourth vector equation of interest is derived by first equating the *reflection*[2] of vector $\mathbf{h}$ onto a plane normal to $\mathbf{f}$ with its counterpart, namely, the reflection of vector $\mathbf{i}$ onto a plane normal to $\mathbf{g}$. We now define two unit vectors $\boldsymbol{\mu}$ and $\boldsymbol{\nu}$, namely,

$$\boldsymbol{\mu} \equiv \frac{\mathbf{f}}{\|\mathbf{f}\|}, \qquad \boldsymbol{\nu} \equiv \frac{\mathbf{g}}{\|\mathbf{g}\|}$$

Further, two pure reflections, as introduced in Section 2.2, are defined as

$$\mathbf{R}_\mu \equiv \mathbf{1} - 2\boldsymbol{\mu}\boldsymbol{\mu}^T, \quad \mathbf{R}_\nu \equiv \mathbf{1} - 2\boldsymbol{\nu}\boldsymbol{\nu}^T$$

---

[1]N. B. Lee and Li of the references in this chapter are one and the same person, namely, Dr.-Ing. Hongyou Lee (a.k.a. Dr.-Ing. Hongyou Li).

[2]Surprisingly, neither Li nor Raghavan and Roth realized the geometric significance of this fourth equation, first proposed by Lee.

Upon equating $\mathbf{R}_\mu \mathbf{h}$ with $\mathbf{R}_\nu \mathbf{i}$, the desired equation is obtained in the form

$$\mathbf{h} - 2(\boldsymbol{\mu} \cdot \mathbf{h})\boldsymbol{\mu} = \mathbf{i} - 2(\boldsymbol{\nu} \cdot \mathbf{i})\boldsymbol{\nu}$$

or

$$\mathbf{h} - 2\frac{\mathbf{f} \cdot \mathbf{h}}{\| \mathbf{f} \|^2}\mathbf{f} = \mathbf{i} - 2\frac{\mathbf{g} \cdot \mathbf{i}}{\| \mathbf{g} \|^2}\mathbf{g}$$

But, since $\mathbf{f} = \mathbf{g}$, we also have $\|\mathbf{f}\| = \|\mathbf{g}\|$, the above equation thus leading to

$$\|\mathbf{f}\|^2\mathbf{h} - 2(\mathbf{f} \cdot \mathbf{h})\mathbf{f} = \|\mathbf{g}\|^2\mathbf{i} - 2(\mathbf{g} \cdot \mathbf{i})\mathbf{g} \tag{8.7d}$$

Note that each side of the above equation is, in fact, an *elongated*, or correspondingly, a *contracted* reflection of a unit vector. Finally, the two scalar equations are derived upon equating the inner products $\mathbf{f} \cdot \mathbf{f}$ and $\mathbf{f} \cdot \mathbf{h}$ with their counterparts $\mathbf{g} \cdot \mathbf{g}$ and $\mathbf{g} \cdot \mathbf{i}$, respectively, i.e.,

$$\mathbf{f} \cdot \mathbf{f} = \mathbf{g} \cdot \mathbf{g} \tag{8.7e}$$
$$\mathbf{f} \cdot \mathbf{h} = \mathbf{g} \cdot \mathbf{i} \tag{8.7f}$$

In connection with the foregoing 14 scalar equations, eqs.(8.7a–f), we have a few facts that are proven below:

Fact 8.2.1 *The inner products* $\mathbf{f} \cdot \mathbf{f}$ *and* $\mathbf{f} \cdot \mathbf{h}$ *are bilinear in* $\{ \mathbf{x}_i \}_4^5$, *while their counterparts* $\mathbf{g} \cdot \mathbf{g}$ *and* $\mathbf{g} \cdot \mathbf{i}$ *are linear in* $\mathbf{x}_1$.

Proof:

$$\begin{aligned}
\mathbf{f} \cdot \mathbf{f} &\equiv \|\mathbf{Q}_3(\mathbf{b}_3 + \mathbf{Q}_4\mathbf{b}_4 + \mathbf{Q}_4\mathbf{Q}_5\mathbf{b}_5)\|^2 \\
&\equiv \|\mathbf{b}_3 + \mathbf{Q}_4\mathbf{b}_4 + \mathbf{Q}_4\mathbf{Q}_5\mathbf{b}_5\|^2 \\
&\equiv \sum_3^5 \|\mathbf{b}_i\|^2 + 2\mathbf{b}_3^T\mathbf{Q}_4(\mathbf{b}_4 + \mathbf{Q}_5\mathbf{b}_5) + 2\mathbf{b}_4^T\mathbf{Q}_5\mathbf{b}_5
\end{aligned}$$

whose rightmost-hand side is clearly free of $\mathbf{x}_3$ and is bilinear in $\{ \mathbf{x}_i \}_4^5$. Similarly,

$$\begin{aligned}
\mathbf{f} \cdot \mathbf{h} &\equiv (\mathbf{b}_3 + \mathbf{Q}_4\mathbf{b}_4 + \mathbf{Q}_4\mathbf{Q}_5\mathbf{b}_5)^T\mathbf{Q}_3^T\mathbf{Q}_3\mathbf{Q}_4\mathbf{u}_5 \\
&\equiv \mathbf{b}_3^T\mathbf{Q}_4\mathbf{u}_5 + \mathbf{b}_4^T\mathbf{u}_5 + \mathbf{b}_5^T\mathbf{Q}_5^T\mathbf{u}_5
\end{aligned}$$

whose rightmost-hand side is clearly bilinear in $\mathbf{x}_4$ and $\mathbf{x}_5$, except for the last term, which contains two terms that are linear in $\mathbf{x}_5$, and hence, can be suspected to be quadratic. However, $\mathbf{Q}_5\mathbf{b}_5$ is, in fact, $\mathbf{a}_5$, while $\mathbf{u}_5$ is the last column of $\mathbf{Q}_5$, the suspicious term thus reducing to a constant, namely, $b_5 \cos\alpha_5$. Similar proofs hold for $\mathbf{g} \cdot \mathbf{g}$ and $\mathbf{g} \cdot \mathbf{i}$. Moreover,

Fact 8.2.2 *Vector* $\mathbf{f} \times \mathbf{h}$ *is trilinear in* $\{ \mathbf{x}_i \}_3^5$, *while its counterpart,* $\mathbf{g} \times \mathbf{i}$, *is bilinear in* $\{ \mathbf{x}_i \}_1^2$.

Proof: If we want the cross product of two vectors in frame $\mathcal{A}$ but have these vectors in frame $\mathcal{B}$, then we can proceed in two ways: either (*i*) transform each of the two vectors into $\mathcal{A}$-coordinates and perform the cross product of the two transformed vectors; or (*ii*) perform the product of the two vectors in $\mathcal{B}$-coordinates and transform the product vector into $\mathcal{A}$-coordinates. Obviously, the two products will be the same, which allows us to write

$$\begin{aligned}\mathbf{f} \times \mathbf{h} &\equiv \mathbf{Q}_3\left[\mathbf{b}_3 \times (\mathbf{Q}_4\mathbf{u}_5) + (\mathbf{Q}_4\mathbf{b}_4) \times (\mathbf{Q}_4\mathbf{u}_5) + (\mathbf{Q}_4\mathbf{Q}_5\mathbf{b}_5) \times (\mathbf{Q}_4\mathbf{u}_5)\right] \\ &\equiv \mathbf{Q}_3\{\mathbf{b}_3 \times (\mathbf{Q}_4\mathbf{u}_5) + \mathbf{Q}_4(\mathbf{b}_4 \times \mathbf{u}_5) + \mathbf{Q}_4\left[(\mathbf{Q}_5\mathbf{b}_5) \times \mathbf{u}_5)\right]\}\end{aligned}$$

whose rightmost-hand side is apparently trilinear in $\{\mathbf{x}_i\}_3^5$, except for the term in brackets, which looks quadratic in $\mathbf{x}_5$. A quick calculation, however, reveals that this term is, in fact, linear in $\mathbf{x}_5$. Indeed, from the definitions given in eqs.(4.3d & d) and (4.12) we have

$$(\mathbf{Q}_5\mathbf{b}_5) \times \mathbf{u}_5 \equiv \mathbf{a}_5 \times \mathbf{u}_5 \equiv \begin{bmatrix} a_5\lambda_5 s_5 + b_5\mu_5 c_5 \\ -a_5\lambda_5 c_5 + b_5\mu_5 s_5 \\ -a_5\mu_5 \end{bmatrix}$$

which is obviously linear in $\mathbf{x}_5$. The proof for the counterpart product, $\mathbf{g}\times\mathbf{i}$, parallels the foregoing proof.

Fact 8.2.3 *Vector* $(\mathbf{f}\cdot\mathbf{f})\mathbf{h} - 2(\mathbf{f}\cdot\mathbf{h})\mathbf{f}$ *is trilinear in* $\{\mathbf{x}_i\}_3^5$, *its counterpart,* $(\mathbf{g}\cdot\mathbf{g})\mathbf{i} - 2(\mathbf{g}\cdot\mathbf{i})\mathbf{g}$, *being bilinear in* $\{\mathbf{x}_i\}_1^2$.

Proof: First, we write the aforementioned (elongated or contracted) reflection of vector $\mathbf{h}$ in the form

$$(\mathbf{f}\cdot\mathbf{f})\mathbf{h} - 2(\mathbf{f}\cdot\mathbf{h})\mathbf{f} \equiv \mathbf{Q}_3\mathbf{v}$$

where

$$\begin{aligned}\mathbf{v} \equiv (\sum_3^5 \|\mathbf{b}_i\|^2)\mathbf{Q}_4\mathbf{u}_5 - 2[&(\mathbf{u}_5^T\mathbf{Q}_4\mathbf{b}_3)\mathbf{b}_3 + (\mathbf{u}_5^T\mathbf{b}_4)\mathbf{b}_3 + (\mathbf{u}_5^T\mathbf{b}_4)\mathbf{Q}_4\mathbf{b}_4 \\ &+(\mathbf{u}_5^T\mathbf{Q}_5\mathbf{b}_5)\mathbf{b}_3 + (\mathbf{u}_5^T\mathbf{Q}_5\mathbf{b}_5)\mathbf{Q}_4\mathbf{b}_4 + (\mathbf{u}_5^T\mathbf{Q}_5\mathbf{b}_5)\mathbf{Q}_4\mathbf{Q}_5\mathbf{b}_5] + 2\mathbf{w}\end{aligned}$$

with all terms on the right-hand side, except for $\mathbf{w}$, which will be defined presently, clearly bilinear in $\mathbf{x}_4$ and $\mathbf{x}_5$. Vector $\mathbf{w}$ is defined as

$$\mathbf{w} \equiv [\ ]_1 + [\ ]_2 + [\ ]_3$$

each of the foregoing brackets being expanded below:

$$\begin{aligned}[\ ]_1 &\equiv \left[(\mathbf{b}_3^T\mathbf{Q}_4\mathbf{b}_4)\mathbf{Q}_4\mathbf{u}_5 - (\mathbf{u}_5^T\mathbf{Q}_4^T\mathbf{b}_3)\mathbf{Q}_4\mathbf{b}_4\right] \\ &\equiv \mathbf{Q}_4(\mathbf{u}_5\mathbf{b}_4^T\mathbf{Q}_4^T - \mathbf{b}_4\mathbf{u}_5^T\mathbf{Q}_4^T)\mathbf{b}_3 \\ &\equiv \mathbf{Q}_4(\mathbf{u}_5\mathbf{b}_4^T - \mathbf{b}_4\mathbf{u}_5^T)\mathbf{Q}_4^T\mathbf{b}_3\end{aligned}$$

which thus reduces to a product including the factor $\mathbf{Q}_i\mathbf{A}\mathbf{Q}_i^T$, with $\mathbf{A}$ being the term in parentheses in the rightmost-hand side of the last equation. This is obviously a skew-symmetric matrix, and Lemma 8.2.1 applies, i.e., the rightmost-hand side of the last equation is linear in $\mathbf{x}_4$. This term is, hence, bilinear in $\mathbf{x}_4$ and $\mathbf{x}_5$. Furthermore,

$$\begin{aligned}[\ ]_2 &\equiv \left[(\mathbf{b}_4^T\mathbf{Q}_5\mathbf{b}_5)\mathbf{Q}_4\mathbf{u}_5 - (\mathbf{u}_5^T\mathbf{b}_4)\mathbf{Q}_4\mathbf{Q}_5\mathbf{b}_5\right] \\ &\equiv \mathbf{Q}_4\left[(\mathbf{b}_5^T\mathbf{Q}_5^T\mathbf{b}_4)\mathbf{u}_5 - (\mathbf{u}_5^T\mathbf{b}_4)\mathbf{Q}_5\mathbf{b}_5\right] \\ &\equiv \mathbf{Q}_4(\mathbf{u}_5\mathbf{b}_5^T\mathbf{Q}_5^T - \mathbf{Q}_5\mathbf{b}_5\mathbf{u}_5^T)\mathbf{b}_4\end{aligned}$$

which is clearly linear in $\mathbf{x}_4$, but it is not obvious that it is also linear in $\mathbf{x}_5$. To show that the latter linearity also holds, we can proceed in two ways. First, note that the term in parentheses is the skew-symmetric matrix $\mathbf{u}_5\mathbf{a}_5^T - \mathbf{a}_5\mathbf{u}_5^T$, whose vector, $\mathbf{a}_5 \times \mathbf{u}_5$, was already proven to be linear in $\mathbf{x}_5$. Since the vector of a skew-symmetric matrix fully defines that matrix—see Section 2.3—the linearity of the foregoing term in $\mathbf{x}_5$ follows immediately. Alternatively, we can expand the aforementioned difference, thereby deriving

$$\mathbf{u}_5\mathbf{a}_5^T - \mathbf{a}_5\mathbf{u}_5^T = \begin{bmatrix} 0 & a_5\mu_5 & -a_5\lambda_5 c_5 + b_5\mu_5 s_5 \\ -a_5\mu_5 & 0 & -a_5\lambda_5 s_5 - b_5\mu_5 c_5 \\ a_5\lambda_5 c_5 - b_5\mu_5 s_5 & a_5\lambda_5 s_5 + b_5\mu_5 c_5 & 0 \end{bmatrix}$$

which is clearly linear in $\mathbf{x}_5$. Moreover, its vector can be readily identified as $\mathbf{a}_5 \times \mathbf{u}_5$, as calculated above. Finally,

$$\begin{aligned}[\ ]_3 &\equiv \left[(\mathbf{b}_3^T\mathbf{Q}_4\mathbf{Q}_5\mathbf{b}_5)\mathbf{Q}_4\mathbf{u}_5 - (\mathbf{u}_5^T\mathbf{Q}_4^T\mathbf{b}_3)\mathbf{Q}_4\mathbf{Q}_5\mathbf{b}_5\right] \\ &\equiv \mathbf{Q}_4(\mathbf{u}_5\mathbf{b}_5^T\mathbf{Q}_5^T - \mathbf{Q}_5\mathbf{b}_5\mathbf{u}_5^T)\mathbf{Q}_4^T\mathbf{b}_3 \\ &\equiv \mathbf{Q}_4(\mathbf{u}_5\mathbf{a}_5^T - \mathbf{a}_5\mathbf{u}_5^T)\mathbf{Q}_4^T\mathbf{b}_3\end{aligned}$$

this bracket thus reducing to a product including the factor $\mathbf{Q}_i\mathbf{A}\mathbf{Q}_i^T$, with $\mathbf{A}$ skew-symmetric. Hence, the foregoing expression is linear in $\mathbf{x}_4$, according to Lemma 8.2.1. Moreover, the matrix in parentheses was already proven to be linear in $\mathbf{x}_5$, thereby completing the proof for vector $(\mathbf{f}\cdot\mathbf{f})\mathbf{h} - 2(\mathbf{f}\cdot\mathbf{h})\mathbf{f}$. The proof for vector $(\mathbf{g}\cdot\mathbf{g})\mathbf{i} - 2(\mathbf{g}\cdot\mathbf{i})\mathbf{g}$ parallels the foregoing proof and hence, need not be included here. Finally, we have one more result that will be used below:

Fact 8.2.4 *If a scalar, vector, or matrix equation is linear in $\mathbf{x}_i$, then upon substitution of c_i and s_i by their equivalent forms in terms of $\tau_i \equiv \tan(\theta_i/2)$, the foregoing equation becomes quadratic in τ_i.*

Proof: We shall show that this result holds for a scalar equation, with the extension to vector and matrix equations following directly. The scalar equation under discussion takes on the general form

$$Ac_i + Bs_i + C = 0$$

where the coefficients A, B, and C do not contain θ_i. Upon performing the trigonometric substitutions of c_i and s_i in terms of $\tau_i \equiv \tan(\theta_i/2)$ and multiplying both sides of that equation by $1+\tau_i^2$, we obtain

$$A(1-\tau_i^2) + 2B\tau_i + C(1+\tau_i^2) = 0$$

which is clearly quadratic in τ_i. The same proof follows immediately for vector and matrix equations. Moreover, in general, if a scalar, vector, or matrix equation is of degree k in $\mathbf{x}_i$, upon introducing the same trigonometric substitution, the said equation becomes of degree $2k$ in τ_i.

Elimination Procedure

So far, we have derived four 3-dimensional vector equations—eqs.(8.7a–d)—and two scalar equations, eqs.(8.7e & f), which thus amount to a total of 14 scalar equations. Moreover, from the form of vectors $\mathbf{f}$ and $\mathbf{h}$, it is apparent that the left-hand sides of those four vector equations all have the form $\mathbf{Q}_3\mathbf{v}$, while their right-hand sides take on the form $\mathbf{Q}_2^T\mathbf{w} + \mathbf{k}$, where $\mathbf{v}$ is a 3-dimensional vector independent of θ_3; $\mathbf{w}$ is, in turn, a 3-dimensional vector independent of θ_2; and $\mathbf{k}$ is a constant 3-dimensional vector. Moreover, if we recall that the third row of $\mathbf{Q}_i$ is independent of θ_i, it is apparent that the third component of the right-hand sides of those four vector equations are independent of θ_3. Furthermore, the left-hand sides of eqs.(8.7e & f) are also independent of θ_3, while their right-hand sides are, in turn, independent of θ_2. We thus have produced two sets of equations: (i) the first set consists of the first two scalar equations of the four vector equations (8.7a–d), and is thus composed of eight scalar equations; we shall term this set Group 1; and (ii) the set of third scalar equations of the four vector equations (8.7a–d), along with the last two scalar equations (8.7e & f). Correspondingly, the second set will be termed Group 2.

The eight equations of Group 1 are now cast in the form

$$\mathbf{A}\mathbf{x}_{12} = \mathbf{b} \tag{8.8a}$$

where $\mathbf{A}$ is an 8×6 matrix whose entries are all functions of the data only and do not depend on the unknowns, while $\mathbf{b}$ is an 8-dimensional vector whose components are trilinear in $\{\mathbf{x}_i\}_3^5$, and $\mathbf{x}_{12}$ is the 6-dimensional vector defined below:

$$\mathbf{x}_{12} \equiv \begin{bmatrix} s_1s_2 & s_1c_2 & c_1s_2 & c_1c_2 & s_2 & c_2 \end{bmatrix}^T \tag{8.8b}$$

Moreover, the six equations in Group 2 are written in the form

$$\mathbf{C}\mathbf{x}_1 = \mathbf{d} \tag{8.9}$$

in which $\mathbf{C}$ is a 6×2 matrix whose entries are, like those of $\mathbf{A}$, functions of the data only, and do not depend on the unknowns, while $\mathbf{d}$ is a 6-dimensional vector that is trilinear in $\{\mathbf{x}_i\}_3^5$. We now pick out any two of the six scalar equations in eq.(8.9) and solve for $\mathbf{x}_1$ from them. To this end, we partition matrix $\mathbf{C}$ into the 2×2 block $\mathbf{C}_U$ and the 4×2 block $\mathbf{C}_L$. Likewise, vector $\mathbf{d}$ is partitioned correspondingly into an upper 2-dimensional part $\mathbf{d}_U$ and a lower 4-dimensional part $\mathbf{d}_L$, namely,

$$\mathbf{C} \equiv \begin{bmatrix} \mathbf{C}_U \\ \mathbf{C}_L \end{bmatrix}, \quad \mathbf{d} \equiv \begin{bmatrix} \mathbf{d}_U \\ \mathbf{d}_L \end{bmatrix} \tag{8.10}$$

In the above partitioning, $\mathbf{C}_U$ is chosen so that it is nonsingular, which may require a reordering of the equations. Under this condition, then, we have two subsystems of equations in $\mathbf{x}_1$, i.e.,

$$\mathbf{C}_U\mathbf{x}_1 = \mathbf{d}_U \tag{8.11a}$$

$$\mathbf{C}_L\mathbf{x}_1 = \mathbf{d}_L \tag{8.11b}$$

Upon solving for $\mathbf{x}_1$ from eq.(8.11a) and substituting the expression thus resulting into eq.(8.11b), we obtain four equations free of $\mathbf{x}_1$, namely,

$$\mathbf{C}_L\mathbf{C}_U^{-1}\mathbf{d}_U - \mathbf{d}_L = \mathbf{0}_4 \tag{8.12}$$

where $\mathbf{0}_4$ is the 4-dimensional zero vector. Note that the foregoing equations are linear in $\mathbf{d}$, and hence, trilinear in $\{\mathbf{x}_i\}_3^5$. We can then express these equations as a system of four equations that are linear in $\mathbf{x}_3$, or in homogeneous form, as

$$\mathbf{D}_1\mathbf{y}_3 = \mathbf{0}_4 \tag{8.13a}$$

where $\mathbf{D}_1$ is a 4×3 matrix that is bilinear in $\mathbf{x}_4$ and $\mathbf{x}_5$, while $\mathbf{y}_3$ is defined as

$$\mathbf{y}_3 \equiv \begin{bmatrix} \cos\theta_3 \\ \sin\theta_3 \\ 1 \end{bmatrix} \tag{8.13b}$$

Furthermore, we pick up any six equations of Group 1 and solve for $\mathbf{x}_{12}$ from them. To this end, we partition $\mathbf{A}$ into an upper 6×6 block $\mathbf{A}_U$ and a lower 2×6 block $\mathbf{A}_L$, with a corresponding partitioning of $\mathbf{b}$ into an upper 6-dimensional part $\mathbf{b}_U$ and a lower 2-dimensional part $\mathbf{b}_L$, namely,

$$\mathbf{A} \equiv \begin{bmatrix} \mathbf{A}_U \\ \mathbf{A}_L \end{bmatrix}, \quad \mathbf{b} \equiv \begin{bmatrix} \mathbf{b}_U \\ \mathbf{b}_L \end{bmatrix} \tag{8.14}$$

where block $\mathbf{A}_U$ is chosen so that it is nonsingular, which again may require a renumbering of the equations. We have now two subsystems of equation in $\mathbf{x}_{12}$, i.e.,

$$\mathbf{A}_U\mathbf{x}_{12} = \mathbf{b}_U \tag{8.15a}$$

$$\mathbf{A}_L\mathbf{x}_{12} = \mathbf{b}_L \tag{8.15b}$$

Upon solving for $\mathbf{x}_{12}$ from eq.(8.15a) and substituting the expression thus obtained into eq.(8.15b), we obtain a system of two equations free of θ_1 and θ_2, as shown below:

$$\mathbf{A}_L \mathbf{A}_U^{-1} \mathbf{b}_U - \mathbf{b}_L = \mathbf{0}_2$$

with $\mathbf{0}_2$ denoting the 2-dimensional zero vector. Note that this equation is linear in $\mathbf{b}$, and hence, trilinear in $\{\mathbf{x}_i\}_3^5$. This system can thus be cast in the form of two linear equations in $\mathbf{x}_3$, or correspondingly, in linear homogeneous form in $\mathbf{y}_3$, namely,

$$\mathbf{D}_2 \mathbf{y}_3 = \mathbf{0}_2 \tag{8.16}$$

with $\mathbf{D}_2$ being a 2×3 matrix whose entries are bilinear in $\mathbf{x}_4$ and $\mathbf{x}_5$. Further, we assemble eqs.(8.13a) and (8.16) in the form of six linear homogeneous equations in $\mathbf{y}_3$, namely,

$$\mathbf{D} \mathbf{y}_3 = \mathbf{0}_6 \tag{8.17}$$

where $\mathbf{D}$ is a 6×3 matrix whose entries are bilinear in $\mathbf{x}_4$ and $\mathbf{x}_5$. In the next step, we partition $\mathbf{D}$ into two 3×3 blocks, $\mathbf{D}_U$ and $\mathbf{D}_L$, thereby obtaining two systems of three equations each, that are linear homogeneous in $\mathbf{y}_3$, namely,

$$\mathbf{D}_U \mathbf{y}_3 = \mathbf{0}_3 \quad \text{and} \quad \mathbf{D}_L \mathbf{y}_3 = \mathbf{0}_3 \tag{8.18}$$

with $\mathbf{0}_3$ defined as the 3-dimensional zero vector. It is apparent that both $\mathbf{D}_U$ and $\mathbf{D}_L$ are bilinear in $\mathbf{x}_4$ and $\mathbf{x}_5$. Now, for the two systems of equations appearing in eq.(8.18) to yield acceptable solutions $\mathbf{y}_3$, the two associated matrices, $\mathbf{D}_U$ and $\mathbf{D}_L$ must be singular, for $\mathbf{y}_3$ is apparently nonzero from definition (8.13b), and hence, we must have

$$f_1(\theta_4, \theta_5) \equiv \det(\mathbf{D}_U) = 0 \quad \text{and} \quad f_2(\theta_4, \theta_5) \equiv \det(\mathbf{D}_L) = 0 \tag{8.19}$$

However, nothing so far guarantees that the nontrivial solution of the first of the two foregoing equations is identical to the solution of the second equation. In order to guarantee this, we impose the condition that each solution $\mathbf{y}_3$ must satisfy

$$\mathbf{D}_U \mathbf{y}_3 - \mathbf{D}_L \mathbf{y}_3 = \mathbf{0}_3$$

or

$$(\mathbf{D}_U - \mathbf{D}_L) \mathbf{y}_3 = \mathbf{0}_3$$

and hence, we must impose a third condition, namely,

$$f_3(\theta_4, \theta_5) \equiv \det(\mathbf{D}_U - \mathbf{D}_L) = 0 \tag{8.20}$$

Now, the set $\{f_i(\theta_4, \theta_5)\}_1^3$ defines three contours $\{\mathcal{C}_i\}_1^3$. The points where all three contours intersect define the solutions of the problem at hand. Intersections of only two contours are thus disregarded, for these are *spurious solutions*.

Once all common intersections of the three foregoing contours have been determined, we have already two of the unknowns, θ_4 and θ_5, the remaining four unknowns being calculated uniquely as described presently. First, θ_3 can be computed from eq.(8.17), which can be rewritten in the form

$$\mathbf{H}\mathbf{x}_3 = \boldsymbol{\tau} \tag{8.21a}$$

where the 6×2 matrix $\mathbf{H}$ and the 6-dimensional vector $\boldsymbol{\tau}$ are both bilinear in $\mathbf{x}_4$ and $\mathbf{x}_5$ and hence, known. Although any two of the six equations (8.21a) suffice, in principle, to determine $\mathbf{x}_3$, we should not forget that these computations will most likely be performed with finite precision, and hence, roundoff-error amplification is bound to occur. In order to keep roundoff errors as low as possible, we recommend to use all foregoing six equations and calculate $\mathbf{x}_3$ as the *least-square approximation* of the overdetermined system (8.21a). This approximation will be, in fact, the solution of the given system because all six equations are compatible. Moreover, this approximation can be expressed *symbolically* in the form

$$\mathbf{x}_3 = (\mathbf{H}^T\mathbf{H})^{-1}\mathbf{H}^T\boldsymbol{\tau} \tag{8.21b}$$

In practice, the foregoing least-square approximation is computed using an orthogonalization procedure (Golub and Van Loan, 1989), the explicit or the numerical inversion of the product $\mathbf{H}^T\mathbf{H}$ being advised against because of its frequent ill-conditioning. Appendix B outlines a robust numerical computation of the least-square approximation of an overdetermined system of equations using orthogonalization procedures. What is relevant to our discussion is that eq.(8.21b) determines θ_3 *uniquely* for given values of θ_4 and θ_5. Likewise, θ_1 can be readily found from eq.(8.9) using the least-square approximation of the overdetermined system of eq.(8.9) that contains six equations and two unknowns. Again, this approximation is, in fact, the solution of the said system, and can be expressed symbolically as

$$\mathbf{x}_1 = (\mathbf{C}^T\mathbf{C})^{-1}\mathbf{C}^T\mathbf{d} \tag{8.22}$$

With θ_1, θ_3, θ_4, and θ_5 available, vector $\mathbf{x}_2$ can be determined from eq.(8.8a) by rewriting it in the form

$$\tilde{\mathbf{A}}\mathbf{x}_2 = \mathbf{b} \tag{8.23a}$$

where the 8×2 matrix $\tilde{\mathbf{A}}$ is linear in $\mathbf{x}_1$. Thus, $\mathbf{x}_2$ can be determined uniquely from eq.(8.23a) as its least-square approximation, namely,

$$\mathbf{x}_2 = (\tilde{\mathbf{A}}^T\tilde{\mathbf{A}})^{-1}\tilde{\mathbf{A}}^T\mathbf{b} \tag{8.23b}$$

Finally, θ_6 is readily determined from eq.(4.9a). In fact, the first of the three vector equations represented by this matrix equation yields

$$\mathbf{Q}_1\mathbf{Q}_2\mathbf{Q}_3\mathbf{Q}_4\mathbf{Q}_5\mathbf{p}_6 = \mathbf{q} \tag{8.24a}$$

where $\mathbf{q}$ denotes the first column of $\mathbf{Q}$, while according to eq.(4.12), $\mathbf{p}_6$ denotes the first column of matrix $\mathbf{Q}_6$, i.e.,

$$\mathbf{p}_6 \equiv \begin{bmatrix} \cos\theta_6 \\ \sin\theta_6 \\ 0 \end{bmatrix}, \quad \mathbf{q} \equiv \begin{bmatrix} q_{11} \\ q_{21} \\ q_{31} \end{bmatrix} \tag{8.24b}$$

Thus, eq.(8.24a) can be readily solved for $\mathbf{p}_6$, thereby obtaining

$$\mathbf{p}_6 = \mathbf{Q}_5^T \mathbf{Q}_4^T \mathbf{Q}_3^T \mathbf{Q}_2^T \mathbf{Q}_1^T \mathbf{q} \tag{8.25}$$

which thus provides a unique value of θ_6 for every set of values of $\{ \theta_k \}_1^5$, thus completing the solution of the IKP under study.

8.2.2 Numerical Conditioning of the Solutions

In this subsection, we recall the concept of *condition number* of a square matrix (Golub and Van Loan, 1989), as introduced in Section 4.9. In this subsection, however, we stress the relevance of the concept in connection with the *numerical reliability* of the computed solutions of the general IKP.

The concept of condition number of a square matrix is of the utmost importance because it measures the roundoff-error amplification upon solving a system of linear equations having that matrix as coefficient. The condition number of a matrix was defined in Section 4.9 as the ratio of the largest to the smallest singular values of the matrix. While this is but one of the many possible ways in which the condition number of a matrix can be defined (Dahlquist and Björck, 1974), it is what best suits our purposes.

In the context of the foregoing contour-intersection method, we can intuitively argue that the accuracy in the computation of a solution is dictated by the angle at which the two contours giving a solution intersect. Thus, the numerically most reliable solutions are those determined by contours intersecting at right angles, the least reliable being those obtained by tangent contours. We shall formalize this observation in the discussion below.

Here, we distinguish between the condition number of a matrix and the conditioning of a solution of a nonlinear system of equations. We define the latter as the condition number of the Jacobian matrix of the system, evaluated at that particular solution.

For concreteness, let

$$f_1(x_1, x_2) = 0$$
$$f_2(x_1, x_2) = 0$$

be a system of two nonlinear equations in the two unknowns x_1 and x_2. Moreover, the Jacobian matrix of this system is defined as

$$\mathbf{F} \equiv \begin{bmatrix} (\nabla f_1)^T \\ (\nabla f_2)^T \end{bmatrix}, \quad \nabla f_k \equiv \begin{bmatrix} \partial f_k / \partial x_1 \\ \partial f_k / \partial x_2 \end{bmatrix} \tag{8.26}$$

where ∇f_k denotes the gradient of function $f_k(x_1, x_2)$, defined as

$$\nabla f_k \equiv \begin{bmatrix} \partial f_k/\partial x_1 \\ \partial f_k/\partial x_2 \end{bmatrix} \tag{8.27}$$

It is to be noted that multiplying each of the two given equations by a scalar other than zero does not affect its solutions, each Jacobian column being, then, correspondingly multiplied by the same scaling factor. To ease matters, we will assume henceforth that each of the above equations has been properly scaled so as to render its gradient a unit vector in the plane of the two unknowns. In order to calculate the condition number of $\mathbf{F}$, which determines the conditioning of the solutions, we calculate first its singular values as the positive square roots of the eigenvalues of $\mathbf{F}\mathbf{F}^T$. This matrix is given as

$$\mathbf{F}\mathbf{F}^T = \begin{bmatrix} 1 & \nabla f_1 \cdot \nabla f_2 \\ \nabla f_1 \cdot \nabla f_2 & 1 \end{bmatrix} \equiv \begin{bmatrix} 1 & \cos\gamma \\ \cos\gamma & 1 \end{bmatrix}$$

where γ is the angle at which the contours intersect. The eigenvalues λ_1 and λ_2 of the product $\mathbf{F}\mathbf{F}^T$ are thus given by

$$\lambda_1 = 1 - \cos\gamma, \quad \lambda_2 = 1 + \cos\gamma \tag{8.28}$$

and hence, the condition number κ of $\mathbf{F}$ can be readily computed as

$$\kappa = \sqrt{\frac{\lambda_2}{\lambda_1}} = \frac{1}{\tan(\gamma/2)}, \quad 0 \le \gamma \le \pi \tag{8.29}$$

which means that for the best possible solutions from the numerical conditioning viewpoint, the two contours cross each other at right angles, whereas at singular configurations, they are tangent to each other. The reader may have experienced that, when solving a system of two linear equations in two unknowns with the aid of drafting instruments, the solution becomes less reliable as the two lines representing those equations become closer and closer to parallel.

8.2.3 Examples

Example 8.2.1 *In this first example, all inverse kinematic solutions of the Fanuc Arc Mate manipulator are found for the pose of the end-effector given below:*

$$\mathbf{Q} = \begin{bmatrix} 0 & 1 & 0 \\ 0 & 0 & 1 \\ 1 & 0 & 0 \end{bmatrix} \qquad \mathbf{p} = \begin{bmatrix} 130 \\ 850 \\ 1540 \end{bmatrix}$$

in which $\mathbf{p}$ *is given in mm and the DH parameters of the manipulator are given in Table 4.2.*

TABLE 8.1. Inverse Kinematics Solutions of the Fanuc Arc Mate Manipulator

Sol'n No.	θ_1	θ_2	θ_3	θ_4	θ_5	θ_6
1 & 2	90°	90°	0°	180°	−180°	0°
3	90°	16.010°	153.403°	180°	100.588°	0°
4	75.157°	15.325°	150.851°	15.266°	−103.353°	176.393°

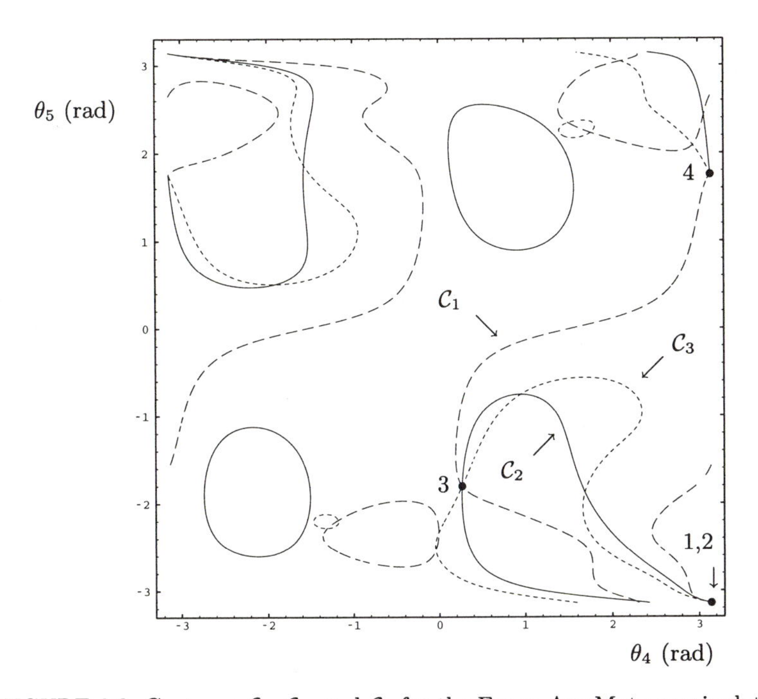

FIGURE 8.3. Contours C_1, C_2, and C_3 for the Fanuc Arc Mate manipulator.

Solution: The solutions are obtained from the intersections of the three contours C_1, C_2, and C_3, as shown in Fig. 8.3.

Four intersection points can be detected in this figure, which are numbered 1, 2, 3, and 4. Moreover, at points 1 and 2 the three contours are tangent to each other. Tangency indicates the existence of a multiple root at that point, and hence, a *singularity*, as discussed in Subsection 4.5.2 in connection with decoupled manipulators. The numerical values of the joint angles of the four solutions are given in Table 8.1.

Example 8.2.2 *In this example, we discuss the IKP of DIESTRO, the* isotropic *six-axis orthogonal manipulator shown in Fig. 4.31 (Williams, Angeles, and Bulca, 1993). For a meaning of kinematic isotropy, we refer*

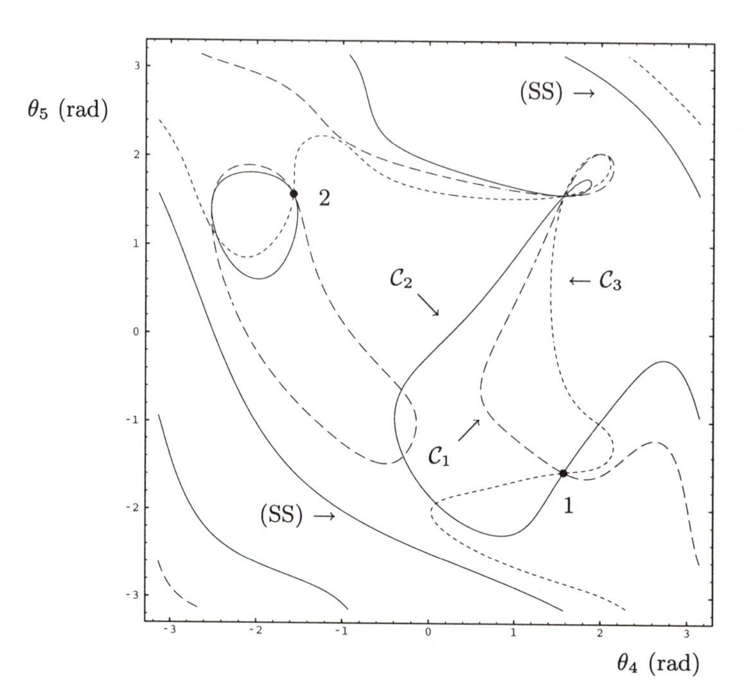

FIGURE 8.4. Contours C_1, C_2, and C_3 for the DIESTRO Manipulator.

the reader to Section 4.9. This manipulator has the DH parameters given in Table 4.1.

The isotropic *pose of the end-effector is defined by the orthogonal matrix* $\mathbf{Q}$ *and the position vector* $\mathbf{p}$ *displayed below:*

$$\mathbf{Q} = \begin{bmatrix} 0 & -1 & 0 \\ 0 & 0 & -1 \\ 1 & 0 & 0 \end{bmatrix} \qquad \mathbf{p} = \begin{bmatrix} 0 \\ -50 \\ 50 \end{bmatrix}$$

with $\mathbf{p}$ *given in mm.*

Solution: At the given pose, the manipulator admits two isolated solutions, those labeled 1 and 2, as well as an infinite number of singular solutions (SS), as shown in Fig. 8.4.

In this figure, the overlapping parts of the three contours C_1, C_2, and C_3 represent a manifold of singular solutions, which means that this manipulator admits a set of self-motions, i.e., joint motions leaving the end-effector stationary. These self-motions can be readily explained if one realizes that when the end-effector is located at the isotropic pose and the manipulator is postured at joint-variable values determined by any point on the overlapping part SS, the six links describe a Bricard mechanism (Bricard, 1927), which is exceptional in that its degree of freedom cannot be determined from the Chebyshev-Grübler-Kutzbach formula (Angeles, 1982). Here, the

TABLE 8.2. Inverse Kinematics Solutions of the DIESTRO Manipulator

Solution No.	θ_1	θ_2	θ_3	θ_4	θ_5	θ_6
1	$0°$	$90°$	$-90°$	$90°$	$-90°$	$180°$
2	$180°$	$-90°$	$90°$	$-90°$	$90°$	$0°$

TABLE 8.3. DH Parameters of Lee's Manipulator

i	a_i (m)	b_i (m)	α_i	θ_i
1	0.12	0	$-57°$	θ_1
2	1.76	0.89	$35°$	θ_2
3	0.07	0.25	$95°$	θ_3
4	0.88	-0.43	$79°$	θ_4
5	0.39	0.5	$-75°$	θ_5
6	0.93	-1.34	$-90°$	θ_6

one-dof motion of the mechanism occurs because the axes determine two intersecting triads. Moreover, the two contours $\mathcal{C}_1$ and $\mathcal{C}_2$ intersect at right angles at solution 1, which corresponds to the isotropic posture of the robot. The numerical values of the joint variables for the isolated solutions are given in Table 8.2

Example 8.2.3 *Here we include an example of a manipulator admitting sixteen real inverse kinematics solutions. This manipulator was proposed by Li (1990), its Denavit-Hartenberg parameters appearing in Table 8.3.*

Solution: The foregoing procedure was applied to this manipulator for an end-effector pose given as

$$\mathbf{Q} = \begin{bmatrix} -0.357279 & -0.850000 & 0.387106 \\ 0.915644 & -0.237000 & 0.324694 \\ -0.184246 & 0.470458 & 0.862973 \end{bmatrix}, \quad \mathbf{p} = \begin{bmatrix} 0.798811 \\ -0.000331 \\ 1.200658 \end{bmatrix}$$

where $\mathbf{p}$ is given in meters.

The contours for this manipulator at the given EE pose are shown in Fig. 8.5, the numerical values of the sixteen solutions being given in Table 8.4.

8.2.4 The Univariate Polynomial Approach to the Solution of the IKP

Alternatively, a univariate 16th-degree polynomial equation can be derived for six-revolute manipulators of general geometry. This polynomial is termed the "characteristic polynomial" of the problem at hand. Here we derive this polynomial using two systematic procedures, namely, those proposed by Raghavan and Roth (1990, 1993) and by Li (1990).

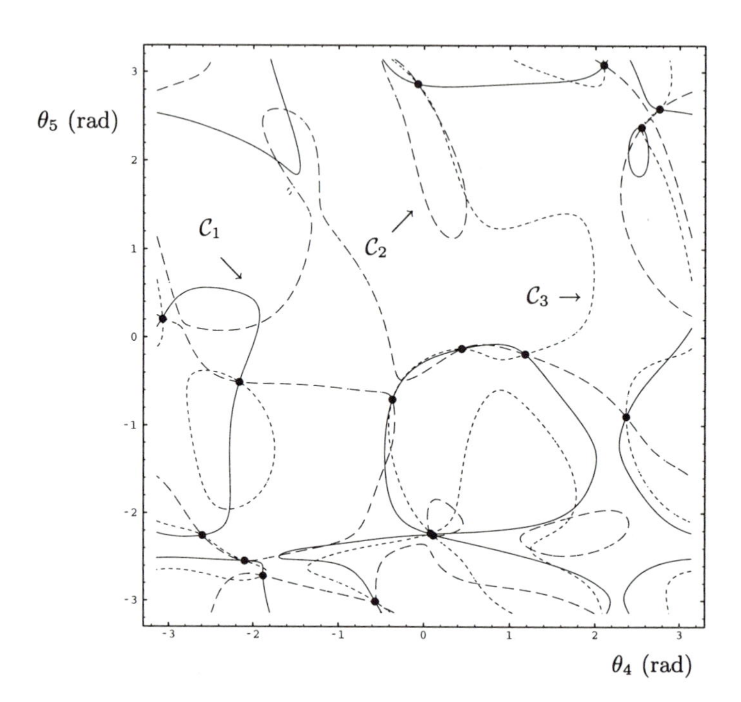

FIGURE 8.5. Contours $\mathcal{C}_1$, $\mathcal{C}_2$, and $\mathcal{C}_3$ for Lee's manipulator.

Raghavan and Roth's Procedure

In the discussion below, we outline Raghavan and Roth's procedure to derive the characteristic polynomial sought. To this end, we use their notation to a large extent. In fact, Raghavan and Roth's procedure to derive a set of six linear homogeneous equations of the form of eq.(8.17) from eqs.(8.7a–f) is different from the one we outlined above. Their procedure is based on simplifications that rely on the factoring of the rotation matrices $\mathbf{Q}_i$ into two reflections, as introduced in eqs.(4.2a). However, for purposes of the forthcoming derivations, this difference is immaterial, and we can start from eq.(8.17), written in the form

$$\boldsymbol{\Sigma}\mathbf{x}_{45} = \mathbf{0} \tag{8.30a}$$

where $\boldsymbol{\Sigma}$ is a 6×9 matrix whose entries are linear in $\mathbf{x}_3$, while $\mathbf{x}_{45}$ is defined as

$$\mathbf{x}_{45} \equiv \begin{bmatrix} s_4 s_5 & s_4 c_5 & c_4 s_5 & c_4 c_5 & s_4 & c_4 & s_5 & c_5 & 1 \end{bmatrix}^T$$

Now, the usual trigonometric identities relating s_i and c_i with $\tau_i \equiv \tan(\theta_i/2)$, for $i = 4, 5$, are substituted into eq.(8.30a). Upon multiplying the two sides of the equation thus resulting by $(1 + \tau_4^2)(1 + \tau_5^2)$, Raghavan and Roth obtain

$$\boldsymbol{\Sigma}'\mathbf{x}'_{45} = \mathbf{0} \tag{8.30b}$$

TABLE 8.4. Joint Angles of Lee's Manipulator

Sol'n No.	θ_1	θ_2	θ_3
1	174.083°	−163.302°	−164.791°
2	−159.859°	−159.324°	−111.347°
3	164.800°	−154.290°	-85.341°
4	−148.749°	−179.740°	−78.505°
5	−16.480°	−10.747°	−58.894°
6	−46.014°	−19.256°	−46.988°
7	−22.260°	−22.431°	−32.024°
8	−53.176°	26.165°	9.103°
9	−173.928°	150.697°	47.811°
10	−41.684°	−29.130°	52.360°
11	−137.195°	−156.920°	68.306°
12	−139.059°	128.112°	96.052°
13	−22.696°	29.214°	98.631°
14	−83.094°	57.022°	130.976°
15	1.227°	−7.353°	142.697°
16	177.538°	−148.178°	159.429°

Sol'n No.	θ_4	θ_5	θ_6
1	−107.818°	−155.738°	141.281°
2	120.250°	176.596°	21.654°
3	4.779°	−127.809°	−101.359°
4	158.091°	148.266°	55.719°
5	−4.164°	164.079°	5.677°
6	−120.218°	−145.864°	−114.768°
7	−32.411°	−172.616°	−17.155°
8	145.868°	136.351°	127.978°
9	−21.000°	−40.438°	−92.284°
10	6.559°	−129.124°	25.091°
11	135.685°	−51.347°	147.446°
12	25.440°	−7.345°	−119.837°
13	−176.071°	11.573°	170.303°
14	67.570°	−10.827°	−110.981°
15	−123.878°	−29.214°	149.208°
16	−148.647°	−129.278°	110.984°

where $\mathbf{\Sigma}'$ is a 6×9 matrix that is linear in $\mathbf{x}_3$, while $\mathbf{x}'_{45}$ is defined as

$$\mathbf{x}'_{45} \equiv [\tau_4^2\tau_5^2 \quad \tau_4^2\tau_5 \quad \tau_4^2 \quad \tau_4\tau_5^2 \quad \tau_4\tau_5 \quad \tau_4 \quad \tau_5^2 \quad \tau_5 \quad 1]^T$$

If the same trigonometric identities, for $i = 3$, are now substituted into eq.(8.30b) and then the first four scalar equations of this set are multiplied by $(1 + \tau_3^2)$ to clear denominators, the equation thus resulting then takes the form

$$\mathbf{\Sigma}''\mathbf{x}'_{45} = \mathbf{0} \tag{8.30c}$$

In the above equations, $\mathbf{\Sigma}''$ is a 6×9 matrix whose first four rows are quadratic in τ_3 while its last two rows are clearly rational functions of τ_3. However, as reported by Raghavan and Roth, the determinant of any 6×6 submatrix of $\mathbf{\Sigma}''$ is, in fact, an 8th-degree polynomial in τ_3 and not a rational function of the same. Moreover, in order to eliminate τ_4 and τ_5, they resort to *dialytic elimination* (Salmon, 1964). This is done by first multiplying the system appearing in eq.(8.30c) by τ_4; then the new system of equations thus obtained is adjoined to the original system of eq.(8.30c), thereby deriving a system of 12 linear homogeneous equations in $\tilde{\mathbf{x}}_{45}$, namely,

$$\mathbf{S}\tilde{\mathbf{x}}_{45} = \mathbf{0}_{12} \tag{8.30d}$$

where $\mathbf{0}_{12}$ is the 12-dimensional zero vector, while the 12-dimensional vector $\tilde{\mathbf{x}}_{45}$ is defined as

$$\tilde{\mathbf{x}}_{45} \equiv \begin{bmatrix} \tau_4^3\tau_5^2 & \tau_4^3\tau_5 & \tau_4^3 & \tau_4^2\tau_5^2 & \tau_4^2\tau_5 & \tau_4^2 & \tau_4\tau_5^2 & \tau_4\tau_5 & \tau_4 & \tau_5^2 & \tau_5 & 1 \end{bmatrix}^T$$

Furthermore, the 12×12 matrix $\mathbf{S}$ is defined as

$$\mathbf{S} \equiv \begin{bmatrix} \mathbf{G} \\ \mathbf{K} \end{bmatrix}$$

its 6×12 blocks $\mathbf{G}$ and $\mathbf{K}$ taking on the forms

$$\mathbf{G} \equiv [\,\mathbf{\Sigma}'' \quad \mathbf{O}_{63}\,], \qquad \mathbf{K} \equiv [\,\mathbf{O}_{63} \quad \mathbf{\Sigma}''\,]$$

with $\mathbf{O}_{63}$ defined as the 6×3 zero matrix.

Now, in order for eq.(8.30d) to admit a nontrivial solution, the determinant of its coefficient matrix must vanish, i.e.,

$$\det(\mathbf{S}) = 0 \tag{8.31}$$

The determinant of $\mathbf{S}$ is the characteristic equation sought. This determinant turns out to be a 16th-degree polynomial in τ_3. Moreover, the roots of this polynomial give the values of τ_3 corresponding to the 16 solutions of the IKP. It should be noted that using the same procedure, one can also derive this polynomial in terms of either τ_4 or τ_5 if the associated vector in eq.(8.30d) is written as $\mathbf{x}_{35}$ or $\mathbf{x}_{34}$, respectively. Consequently, the entries of matrix $\mathbf{\Sigma}$ would be linear in either $\mathbf{x}_4$ or $\mathbf{x}_5$.

Lee's Procedure

At the outset, the factoring of $\mathbf{Q}_i$ given in eq.(4.1b) and the identities first used by Li (1990), namely, eqs.(8.2b), are recalled. Additionally, Lee defines a matrix $\mathbf{T}_i$ as

$$\mathbf{T}_i \equiv \begin{bmatrix} -\tau_i & 1 & 0 \\ 1 & \tau_i & 0 \\ 0 & 0 & 1 \end{bmatrix}$$

Hence,

$$\mathbf{T}_i\mathbf{C}_i \equiv \mathbf{U}_i = \begin{bmatrix} \tau_i & 1 & 0 \\ 1 & -\tau_i & 0 \\ 0 & 0 & 1 \end{bmatrix}$$

Furthermore, we note that the left-hand sides of the four vector equations (8.7a–d) are of the form $\mathbf{Q}_3\mathbf{v}$, where $\mathbf{v}$ is a 3-dimensional vector independent of θ_3. Upon multiplication of the above-mentioned equations from the left by matrix $\mathbf{T}_3$, Lee obtains a new set of equations, namely,

$$\mathbf{U}_3\overline{\mathbf{f}} = \mathbf{T}_3\mathbf{g} \tag{8.32a}$$

$$\mathbf{U}_3\overline{\mathbf{r}} = \mathbf{T}_3\mathbf{i} \tag{8.32b}$$

$$\mathbf{U}_3(\overline{\mathbf{f}} \times \overline{\mathbf{r}}) = \mathbf{T}_3(\mathbf{g} \times \mathbf{i}) \tag{8.32c}$$

$$\mathbf{U}_3\left[(\mathbf{f}\cdot\mathbf{f})\overline{\mathbf{r}} - 2(\mathbf{f}\cdot\mathbf{h})\overline{\mathbf{f}}\right] = \mathbf{T}_3\left[(\mathbf{g}\cdot\mathbf{g})\mathbf{i} - 2(\mathbf{g}\cdot\mathbf{i})\mathbf{g}\right] \tag{8.32d}$$

where $\overline{\mathbf{f}}$ and $\overline{\mathbf{r}}$ are defined as

$$\overline{\mathbf{f}} \equiv \mathbf{\Lambda}_3(\mathbf{b}_3 + \mathbf{Q}_4\mathbf{b}_4 + \mathbf{Q}_4\mathbf{Q}_5\mathbf{b}_5) \tag{8.33}$$

$$\overline{\mathbf{r}} \equiv \mathbf{\Lambda}_3(\mathbf{Q}_4\mathbf{u}_5) \tag{8.34}$$

with $\mathbf{\Lambda}_i$ defined, in turn, in eq.(4.1b).

Because of the form of matrices $\mathbf{T}_3$ and $\mathbf{U}_3$, the third of each of the four vector equations (8.32a–d) is identical to its counterpart appearing in eqs.(8.7a–d). That is, if we denote by either v_i or $(\mathbf{v})_i$ the ith component of any 3-dimensional vector $\mathbf{v}$, the unchanged equations are

$$\overline{f}_3 = g_3 \tag{8.35a}$$

$$\overline{r}_3 = i_3 \tag{8.35b}$$

$$(\overline{\mathbf{f}} \times \overline{\mathbf{r}})_3 = (\mathbf{g} \times \mathbf{i})_3 \tag{8.35c}$$

$$(\mathbf{f}\cdot\mathbf{f})\overline{r}_3 - 2(\mathbf{f}\cdot\mathbf{h})\overline{f}_3 = (\mathbf{g}\cdot\mathbf{g})i_3 - 2(\mathbf{g}\cdot\mathbf{i})g_3 \tag{8.35d}$$

all of which are free of θ_3. Furthermore, six additional equations linear in τ_3 will be derived by multiplying both sides of eqs.(8.35a–d) and of (8.7e & f) by τ_3, i.e.,

$$\tau_3\overline{f}_3 = \tau_3 g_3 \tag{8.36a}$$

$$\tau_3\overline{r}_3 = \tau_3 i_3 \tag{8.36b}$$

$$\tau_3(\overline{\mathbf{f}} \times \overline{\mathbf{r}})_3 = \tau_3(\mathbf{g} \times \mathbf{i})_3 \tag{8.36c}$$

$$\tau_3[(\mathbf{f}\cdot\mathbf{f})\overline{r}_3 - 2(\mathbf{f}\cdot\mathbf{h})\overline{f}_3] = \tau_3[(\mathbf{g}\cdot\mathbf{g})i_3 - 2(\mathbf{g}\cdot\mathbf{i})g_3] \tag{8.36d}$$

$$\tau_3(\mathbf{f}\cdot\mathbf{f}) = \tau_3(\mathbf{g}\cdot\mathbf{g}) \tag{8.36e}$$

$$\tau_3(\mathbf{f}\cdot\mathbf{h}) = \tau_3(\mathbf{g}\cdot\mathbf{i}) \tag{8.36f}$$

We have now 20 scalar equations that are linear in τ_3, namely, the 12 eqs.(8.32a–d), plus the six equations (8.36a–f) and the two scalar equations

(8.7e & f). Moreover, the left-hand sides of the foregoing 20 equations are trilinear in τ_3, $\mathbf{x}_4$, and $\mathbf{x}_5$, while their right-hand sides are trilinear in τ_3, $\mathbf{x}_1$, and $\mathbf{x}_2$. These 20 equations can thus be written in the form

$$\mathbf{A}\mathbf{x} = \boldsymbol{\beta} \tag{8.37a}$$

where the 20×16 matrix $\mathbf{A}$ is a function of the data only, while the 20-dimensional vector $\boldsymbol{\beta}$ is trilinear in τ_3, $\mathbf{x}_1$, and $\mathbf{x}_2$, the 16-dimensional vector $\mathbf{x}$ being defined, in turn, as

$$\begin{aligned}\mathbf{x} \equiv [\tau_3 c_4 c_5 \quad \tau_3 c_4 s_5 \quad \tau_3 s_4 c_5 \quad \tau_3 s_4 s_5 \quad \tau_3 c_4 \quad \tau_3 s_4 \tau_3 c_5 \quad \tau_3 s_5 \\ c_4 c_5 \quad c_4 s_5 \quad s_4 c_5 \quad s_4 s_5 \quad c_4 \quad s_4 \quad c_5 \quad s_5]^T\end{aligned} \tag{8.37b}$$

Next, matrix $\mathbf{A}$ and vector $\boldsymbol{\beta}$ are partitioned as

$$\mathbf{A} \equiv \begin{bmatrix} \mathbf{A}_U \\ \mathbf{A}_L \end{bmatrix}, \quad \boldsymbol{\beta} \equiv \begin{bmatrix} \boldsymbol{\beta}_U \\ \boldsymbol{\beta}_L \end{bmatrix} \tag{8.37c}$$

where $\mathbf{A}_U$ is a nonsingular 16×16 matrix, $\mathbf{A}_L$ is a 4×16 matrix, vector $\boldsymbol{\beta}_U$ is 16-dimensional, and vector $\boldsymbol{\beta}_L$ is 4-dimensional. Moreover, the two foregoing matrices are functions of the data only. Thus, we can solve for $\mathbf{x}$ from the first sixteen equations of eq.(8.37a) in the form

$$\mathbf{x} = \mathbf{A}_U^{-1}\boldsymbol{\beta}_U$$

Upon substituting the foregoing value of $\mathbf{x}$ in the remaining four equations of eq.(8.37a), we derive an equation free of $\mathbf{x}$, namely,

$$\mathbf{A}_L \mathbf{A}_U^{-1}\boldsymbol{\beta}_U = \boldsymbol{\beta}_L \tag{8.38}$$

In eq.(8.38) the two matrices involved are functions of the data only, while the two vectors are trilinear in τ_3, $\mathbf{x}_1$, and $\mathbf{x}_2$. These equations are now cast in the form

$$(A_i c_2 + B_i s_2 + C_i)\tau_3 + D_i c_2 + E_i s_2 + F_i = 0, \quad i = 1, 2, 3, 4 \tag{8.39a}$$

where all coefficients $A_i, \ldots, F_i$ are linear in $\mathbf{x}_1$. Next, we substitute c_2 and s_2 in the foregoing equations by their equivalents in terms of $\tau_2 \equiv \tan(\theta_2/2)$, thereby obtaining, for $i = 1, 2, 3, 4$,

$$C_{ii}\tau_2^2\tau_3 + 2B_i\tau_2\tau_3 + A_{ii}\tau_3 + F_{ii}\tau_2^2 + 2E_i\tau_2 + D_i = 0 \tag{8.39b}$$

with the definitions

$$A_{ii} \equiv A_i + C_i \tag{8.39c}$$

$$C_{ii} \equiv C_i - A_i \tag{8.39d}$$

$$F_{ii} \equiv F_i - D_i \tag{8.39e}$$

(8.39f)

Further, both sides of all the four equations (8.39b) are multiplied by τ_2, which yields

$$C_{ii}\tau_2^3\tau_3 + 2B_i\tau_2^2\tau_3 + A_{ii}\tau_2\tau_3 + F_{ii}\tau_2^3 + 2E_i\tau_2^2 + D_i\tau_2 = 0 \tag{8.39g}$$

We have now eight equations that are linear homogeneous in the 8-dimensional nonzero vector $\mathbf{z}$ defined as

$$\mathbf{z} \equiv \begin{bmatrix} \tau_2^3\tau_3 & \tau_2^2\tau_3 & \tau_2^3 & \tau_2\tau_3 & \tau_2^2 & \tau_3 & \tau_2 & 1 \end{bmatrix}^T \tag{8.39h}$$

and hence, the foregoing 8-dimensional system of equations takes on the form

$$\mathbf{M}\mathbf{z} = \mathbf{0} \tag{8.40}$$

where the 8×8 matrix $\mathbf{M}$ is simply

$$\mathbf{M} \equiv \begin{bmatrix} 0 & C_{11} & 0 & 2B_1 & F_{11} & A_{11} & 2E_1 & D_1 \\ 0 & C_{22} & 0 & 2B_2 & F_{22} & A_{22} & 2E_2 & D_2 \\ 0 & C_{33} & 0 & 2B_3 & F_{33} & A_{33} & 2E_3 & D_3 \\ 0 & C_{44} & 0 & 2B_4 & F_{44} & A_{44} & 2E_4 & D_4 \\ C_{11} & 2B_1 & F_{11} & A_{11} & 2E_1 & 0 & D_1 & 0 \\ C_{22} & 2B_2 & F_{22} & A_{22} & 2E_2 & 0 & D_2 & 0 \\ C_{33} & 2B_3 & F_{33} & A_{33} & 2E_3 & 0 & D_3 & 0 \\ C_{44} & 2B_4 & F_{44} & A_{44} & 2E_4 & 0 & D_4 & 0 \end{bmatrix}$$

Now, since $\mathbf{z}$ is necessarily nonzero, eq.(8.40) should admit nontrivial solutions, and hence, matrix $\mathbf{M}$ should be singular, which leads to the condition below:

$$\det(\mathbf{M}) = 0 \tag{8.41}$$

Thus, considering that all entries of $\mathbf{M}$ are linear in $\mathbf{x}_1$, $\det(\mathbf{M})$ is *octic* in $\mathbf{x}_1$, and hence, eq.(8.41) is equally octic in $\mathbf{x}_1$. By virtue of Fact 4, then, eq.(8.41) is of 16th degree in τ_1, i.e., it takes on the form

$$\sum_0^{16} a_k\tau_1^k = 0 \tag{8.42}$$

which is the characteristic equation sought, whose roots provide up to 16 real values of θ_1 for the IKP at hand. Once θ_1 is available, the remaining angles are computed as indicated below:

Equations (8.40) are first rearranged in nonhomogeneous form, i.e., as

$$\mathbf{N}\mathbf{z}' = \mathbf{n} \tag{8.43}$$

with the 8×7 matrix $\mathbf{N}$ and the 7- and 8-dimensional vectors $\mathbf{z}'$ and $\mathbf{n}$ defined as

$$\mathbf{N} \equiv \begin{bmatrix} 0 & C_{11} & 0 & 2B_1 & F_{11} & A_{11} & 2E_1 \\ 0 & C_{22} & 0 & 2B_2 & F_{22} & A_{22} & 2E_2 \\ 0 & C_{33} & 0 & 2B_3 & F_{33} & A_{33} & 2E_3 \\ 0 & C_{44} & 0 & 2B_4 & F_{44} & A_{44} & 2E_4 \\ C_{11} & 2B_1 & F_{11} & A_{11} & 2E_1 & 0 & D_1 \\ C_{22} & 2B_2 & F_{22} & A_{22} & 2E_2 & 0 & D_2 \\ C_{33} & 2B_3 & F_{33} & A_{33} & 2E_3 & 0 & D_3 \\ C_{44} & 2B_4 & F_{44} & A_{44} & 2E_4 & 0 & D_4 \end{bmatrix}$$

and

$$\mathbf{z}' \equiv \begin{bmatrix} \tau_2^3\tau_3 \\ \tau_2^2\tau_3 \\ \tau_2^3 \\ \tau_2\tau_3 \\ \tau_2^2 \\ \tau_3 \\ \tau_2 \end{bmatrix}, \qquad \mathbf{n} \equiv \begin{bmatrix} D_1 \\ D_2 \\ D_3 \\ D_4 \\ 0 \\ 0 \\ 0 \\ 0 \end{bmatrix}$$

Now, eq.(8.43) represents an overdetermined linear algebraic system of eight equations but only seven unknowns. While we can solve, in principle, for the seven unknowns from any subset of seven equations of the given eight, we prefer not to do so, but rather use all information available, i.e., use all eight equations. This is recommended because of the unavoidable arbitrariness in the choice of the equation to be deleted from the system (8.43). This arbitrariness would lead unnecessarily to ill-conditioning, which can be readily avoided if the whole system of eight equations is solved for the 7-dimensional unknown. This is readily done using a least-square approach and an orthogonalization procedure, as outlined before. *Symbolically*, the solution thus obtained can be expressed, as usual, in the form

$$\mathbf{z}' = (\mathbf{N}^T\mathbf{N})^{-1}\mathbf{N}^T\mathbf{n}$$

We have emphasized here that the above expression is meant only to indicate the solution to the above underdetermined system of algebraic equations because, numerically, the least-square approximation at hand is not computed explicitly as displayed above. That formula was used verbatim in Subsection 3.4.1 to compute the position vector of the point of the instant screw axis lying closest to the origin. Although the formula produced excellent results in that case, the reader should not be misled to think that this will be always the case. Indeed, the reason why this formula worked in that subsection is that the matrix $\mathbf{A}$ to which it was applied happened to have all its columns of the same Euclidean norm and mutually orthogonal. In practice, this is seldom the case, and matrix $\mathbf{N}$ above may be far from having the properties of $\mathbf{A}$ in Subsection 3.4.1. Moreover, if the Polar-Decomposition Theorem, introduced in Section 4.9, is invoked and applied to rectangular matrices, something that is possible but on which we have not elaborated here, it may become apparent that the singular values of $\mathbf{N}^T\mathbf{N}$ are exactly the squares of the singular values of $\mathbf{N}$. What this means is that the condition number of the former, based on the definition adopted here, is exactly the square of the condition number of the latter. As a consequence, even if $\mathbf{N}$ is moderately ill-conditioned, the product $\mathbf{N}^T\mathbf{N}$ will turn out to be catastrophically ill-conditioned. The reader is advised to look at Appendix B for an outline on the robust computation of the least-square approximation of an overdetermined system of linear equations.

With $\mathbf{z}'$ known, both τ_2 and τ_3, and hence, θ_2 and θ_3, are known uniquely. Further, with θ_1, θ_2, and θ_3 known, the right-hand side of eq.(8.37a), $\boldsymbol{\beta}$, is known. Since the coefficient matrix $\mathbf{A}$ of that equation is independent of the joint angles, it is known, and that equation can be solved for vector $\mathbf{x}$ *uniquely*. Once $\mathbf{x}$ is known, the two angles θ_4 and θ_5 are uniquely determined, with θ_6 the sole remaining unknown, which can be readily determined, also uniquely, as discussed in Subsection 8.2.1.

8.3 Kinematics of Parallel Manipulators

Contrary to serial manipulators, parallel manipulators are composed of kinematic chains with closed subchains. A very general manipulator of this kind of machine is shown in Fig. 8.6, in which one can distinguish two platforms, one fixed to the ground, $\mathcal{B}$, and one capable of moving arbitrarily within its workspace, $\mathcal{M}$. The moving platform is connected to the fixed platform through six *legs*, each being regarded as a serial manipulator, the leg thus constituting a six-axis serial manipulator whose base is $\mathcal{B}$ and whose end-effector is $\mathcal{M}$. The whole leg is composed of six links coupled through six revolutes.

The manipulator shown in Fig. 8.6 is, in fact, too general, and of little use as such. A photograph of a simpler and more practical manipulator of this kind, which is used as a flight simulator, is shown in Fig. 1.5, its kinematic structure being depicted in Fig. 8.7a. In this figure, the fixed platform $\mathcal{B}$ is a regular hexagon, while the moving platform $\mathcal{M}$ is an equilateral triangle, as depicted in Fig. 8.7b. Moreover, $\mathcal{B}$ is connected to $\mathcal{M}$ by five revolutes and one prismatic pair. Three of the revolutes constitute a spherical joint, similar to the wrists studied in Section 4.4, while two more constitute a *universal joint*, i.e., the concatenation of two revolutes with intersecting axes. Of the foregoing six joints, only one, the prismatic pair, is actuated.

It is to be noted that although each leg of the manipulator of Fig. 8.7a has a spherical joint at only one end and a universal joint at the other end, we represent each leg in that figure with a spherical joint at each

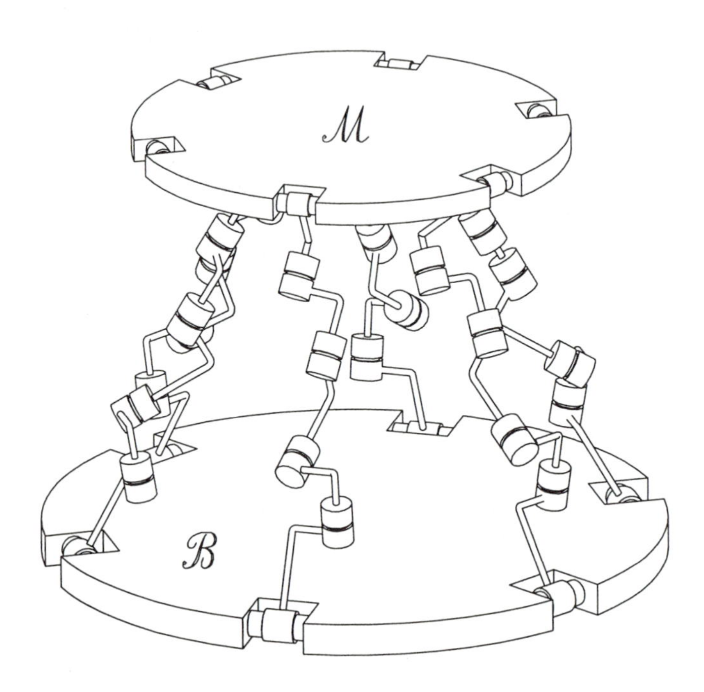

FIGURE 8.6. A general six-dof parallel manipulator.

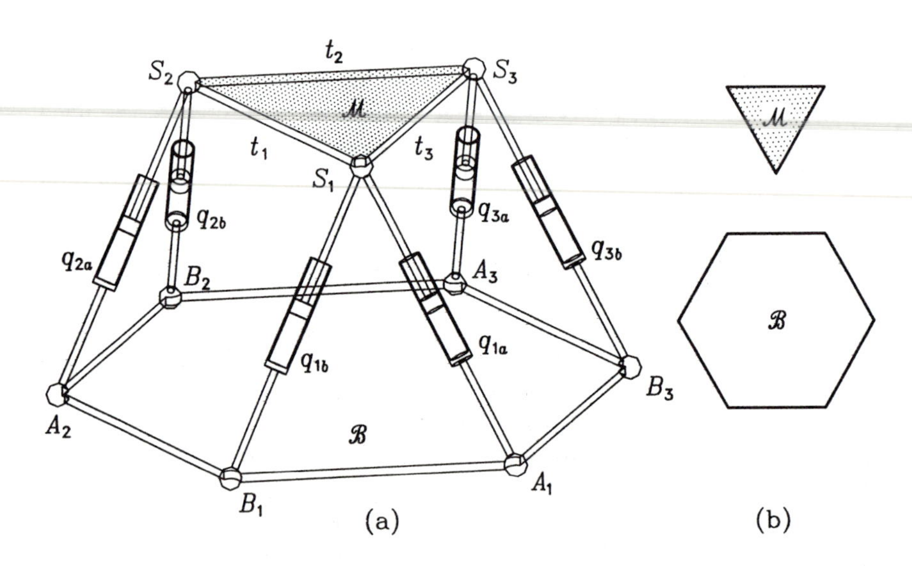

FIGURE 8.7. A six-dof flight simulator: (a) general layout; (b) geometry of its two platforms.

end. Kinematically, the leg depicted in Fig. 8.7a is equivalent to the actual one, the only difference being that the former appears to have a redundant joint. We use the model of Fig. 8.7a only to make the drawing simpler. A more accurate display of the leg architecture of this manipulator appears in Fig. 8.8.

The kinematics and statics of parallel manipulators at large being beyond the scope of this book, we will limit the discussion to parallel manipulators of the simplest type.

With regard to the manipulators under study, we can also distinguish between the inverse and the direct kinematics problems in exactly the same way as these problems were defined for serial manipulators. The inverse kinematics of the general manipulator of Fig. 8.6 is identical to that of the general serial manipulator studied in Section 8.2. In fact, each leg can be studied separately for this purpose, the problem thus becoming the same as that of the aforementioned section. For the particular architecture of the manipulator of Fig. 8.7a, in which the actuated joint variables are displacements measured along the leg axes, the inverse kinematics simplifies substantially and allows for a simple closed-form solution. However, the direct kinematics of the same manipulator is as challenging as that of the general serial manipulator of Section 8.2. With regard to the direct kinematics of manipulators of the type depicted in Fig. 8.7a, Charentus and Renaud (1989) and Nanua, Waldron, and Murthy (1990) showed independently that like the inverse kinematics of general six-axis serial manipulators, the direct kinematics of this manipulator reduces to a 16th-degree polynomial. Note, however, that the direct kinematics of a manipulator similar to that of Fig. 8.7a, but with arbitrary locations of the attachment points of each leg to the moving and fixed platforms, termed the *general platform manipulator*, has been the subject of intensive research (Merlet, 1991). A breakthrough in the solution of the direct kinematics of platform manipulators of the general type was reported by Raghavan (1993), who resorted to polynomial continuation, a technique already mentioned in Section 8.2, for computing up to 40 poses of M for given leg lengths of a parallel manipulator with legs of the type depicted in Fig. 8.8, but with attachment points at both $\mathcal{M}$ and $\mathcal{B}$ with an arbitrary layout. What Raghavan did not derive is the characteristic 40th-degree polynomial of the general platform manipulator. Independently, Wampler (1996) and Husty (1996) devised procedures to derive this polynomial, although Wampler did not pursue the monovariate polynomial approach and preferred to cast the problem in a form suitable for its solution by means of polynomial continuation. Husty did derive the 40th-degree polynomial for several examples. In the process, he showed that this polynomial is the underlying characteristic polynomial for all manipulators of the platform type, which simplifies to a lower-degree polynomial for simpler architectures. As a matter of fact, Lee and Roth (1993) solved the direct kinematics of platform manipulators for which the attachment points at the base and the moving platforms

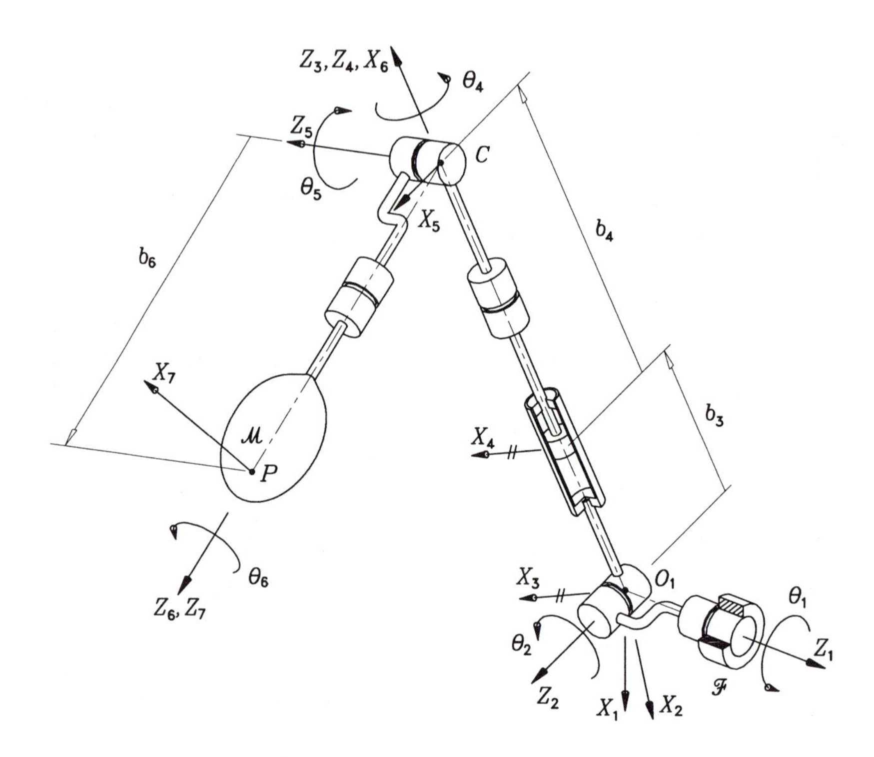

FIGURE 8.8. A layout of a leg of the manipulator of Fig. 8.7.

are located at the vertices of planar, similar hexagons. These researchers showed that the problem here reduces to a cascade of quadratic and linear equations. In the particular case in which both polygons are regular, however, the manipulator degenerates into a movable structure, upon fixing the leg lengths, and hence, the solution set becomes a continuum. Lazard and Merlet (1994), in turn, showed that the platform manipulator originally proposed by Stewart (1965), and known as the *Stewart platform*, has a 12th-degree characteristic polynomial. Interestingly, these mechanical systems were first introduced by Gough (1956–1957) for testing tires; Stewart (1965) suggested their use as flight simulators, an application that is now well established.

Below we analyze the inverse kinematics of one leg of the manipulator of Fig. 8.7a, as depicted in Fig. 8.8. The Denavit-Hartenberg parameters of the leg shown in this figure are given in Table 8.5. It is apparent that the leg under study is a decoupled manipulator. Its inverse kinematics can be derived by properly modifying the scheme presented in Section 4.4, for we now have a prismatic joint, which is, in fact, the only active joint of this manipulator. Moreover, by virtue of the underlying design, the active joint variable, b_3, can take on only positive values.

TABLE 8.5. DH Parameters of the Leg of Fig. 8.8

i	a_i	b_i	α_i
1	0	0	90°
2	0	0	90°
3	0	b_3	0°
4	0	b_4 (const)	90°
5	0	0	90°
6	0	b_6 (const)	0°

In view of the DH parameters of this manipulator, eq.(4.16) reduces to

$$\mathbf{Q}_1\mathbf{Q}_2(\mathbf{a}_3 + \mathbf{a}_4) = \mathbf{c} \tag{8.44}$$

where $\mathbf{c}$ denotes the position vector of the center C of the spherical wrist. Upon equating the squares of the Euclidean norms of both sides of the foregoing equation, we obtain

$$\|\mathbf{a}_3 + \mathbf{a}_4\|^2 = \|\mathbf{c}\|^2 \tag{8.45}$$

where by virtue of the DH parameters of Table 8.5,

$$\|\mathbf{a}_3 + \mathbf{a}_4\|^2 = (b_3 + b_4)^2$$

Now, since both b_3 and b_4 are positive by construction, eq.(8.45) readily leads to the desired inverse kinematics solution, namely,

$$b_3 = \|\mathbf{c}\| - b_4 > 0 \tag{8.46}$$

a result that could have been derived by inspection from Fig. 8.8.

Note that the remaining five joint variables of the leg under study are not needed for purposes of inverse kinematics, and hence, their calculation could be skipped. However, in studying the differential kinematics of these manipulators, these variables will be needed, and so it is convenient to solve for them now. This is straightforward, as shown below: Upon expansion of eq.(8.44), we derive three scalar equations in two unknowns, θ_1 and θ_2, namely,

$$(b_3 + b_4)s_2 = x_C c_1 + y_C s_1 \tag{8.47a}$$

$$-(b_3 + b_4)c_2 = z_C \tag{8.47b}$$

$$0 = x_C s_1 - y_C c_1 \tag{8.47c}$$

in which c_i and s_i stand for $\cos\theta_i$ and $\sin\theta_i$, respectively, while b_3 occurring in the above equations is available in eq.(8.46). From eq.(8.47c), θ_1 is derived as

$$\theta_1 = \tan^{-1}\left(\frac{y_C}{x_C}\right) \tag{8.48a}$$

which yields a unique value of θ_1 rather than the two lying π radians apart, for the two coordinates x_C and y_C determine the quadrant in which θ_1 lies. Once θ_1 is known, θ_2 is derived uniquely from the remaining two equations through its cosine and sine functions, i.e.,

$$c_2 = -\frac{z_C}{b_3 + b_4}, \quad s_2 = \frac{x_C c_1 + y_C s_1}{b_3 + b_4} \tag{8.48b}$$

With the first three joint variables of this leg known, the remaining ones, i.e., those of the "wrist," are calculated as described in Subsection 4.4.2. Therefore, the inverse kinematics of each leg admits two solutions, one for the first three variables and two for the last three. Moreover, since the only actuated joint is one of the first three, which of the two wrist solutions is chosen does not affect the value of b_3, and hence, each manipulator leg admits only one inverse kinematics solution.

While the inverse kinematics of this leg is quite straightforward, its direct kinematics is not. Below we give an outline of the solution procedure for the manipulator under study that follows the procedure proposed by Nanua, Waldron, and Murthy (1990).

In Fig. 8.7a, consider the triangles $A_iS_iB_i$, for $i = 1, 2, 3$, where the subscript i stands for the ith pair of legs. When the lengths of the six legs are fixed and plate $\mathcal{M}$ is removed, triangle $A_iS_iB_i$ can only rotate about the axis A_iB_i. Therefore, we can replace the pair of legs of lengths q_{ia} and q_{ib} by a single leg of length l_i, connected to the base plate $\mathcal{B}$ by a revolute joint with its axis along A_iB_i. The resulting simplified structure, as shown in Fig. 8.9, is kinematically equivalent to the original structure in Fig. 8.7a.

Now we introduce the coordinate frame $\mathcal{F}_i$, with origin at the attachment point O_i of the ith leg with the base plate $\mathcal{B}$, according to the convention below:

For $i = 1, 2, 3$,

O_i is set at the center of the revolute joint;

X_i is directed from A_i to B_i;

Y_i is chosen such that Z_i is perpendicular to the plane of the hexagonal base and points upwards.

Next, we locate the three vertices S_1, S_2, and S_3 of the triangular plate with position vectors stemming from the center O of the hexagon. Furthermore, we need to determine l_i and O_i. Referring to Figs. 8.9 and 8.10, and letting $\mathbf{a}_i$ and $\mathbf{b}_i$ denote the position vectors of points A_i and B_i, respectively, we have

$$d_i = \|\mathbf{b}_i - \mathbf{a}_i\|$$
$$r_i = \frac{d_i^2 + q_{ia}^2 - q_{ib}^2}{2d_i}$$

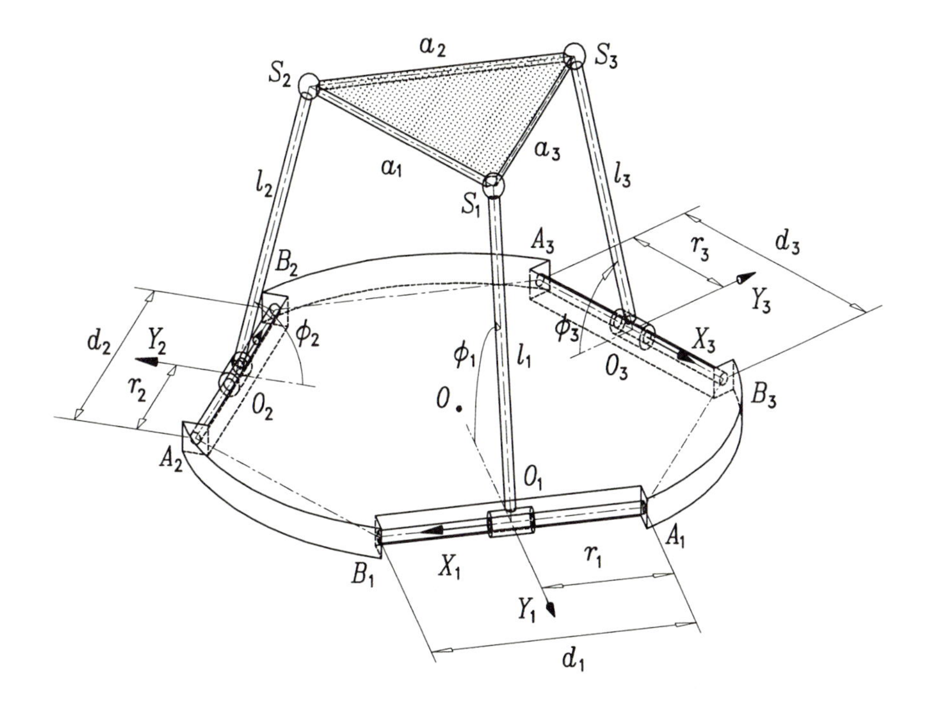

FIGURE 8.9. Equivalent simplified mechanism.

$$l_i = \sqrt{q_{ia}^2 - r_i^2}$$
$$\mathbf{u}_i = \frac{\mathbf{b}_i - \mathbf{a}_i}{d_i}$$

for $i = 1, 2, 3$, and hence, $\mathbf{u}_i$ is the unit vector directed from A_i to B_i. Moreover, the position of the origin O_i is given by vector $\mathbf{o}_i$, as indicated below:

$$\mathbf{o}_i = \mathbf{a}_i + r_i \mathbf{u}_i, \qquad \text{for } i = 1, 2, 3. \tag{8.49}$$

Furthermore, let $\mathbf{s}_i$ be the position vector of S_i in frame $\mathcal{F}_i\,(O_i,\, X_i,\, Y_i,\, Z_i)$. Then

$$\mathbf{s}_i = \begin{bmatrix} 0 \\ -l_i \cos\phi_i \\ l_i \sin\phi_i \end{bmatrix}, \qquad \text{for } i = 1, 2, 3 \tag{8.50}$$

Now a frame $\mathcal{F}_0\,(O,\, X,\, Y,\, Z)$ is defined with origin at O and axes X and Y in the plane of the base hexagon, and related to X_i and Y_i as depicted in Fig. 8.11. When expressed in frame $\mathcal{F}_0$, $\mathbf{s}_i$ takes on the form

$$[\,\mathbf{s}_i\,]_0 = [\,\mathbf{o}_i\,]_0 + [\,\mathbf{R}_i\,]_0 \mathbf{s}_i, \qquad \text{for } i = 1, 2, 3 \tag{8.51}$$

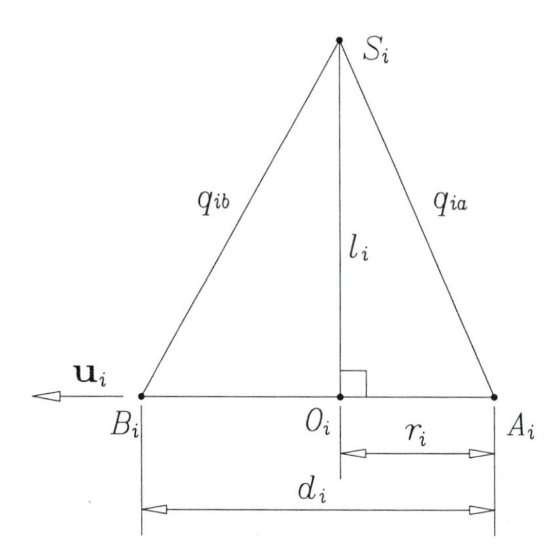

FIGURE 8.10. Replacing each pair of legs with a single leg.

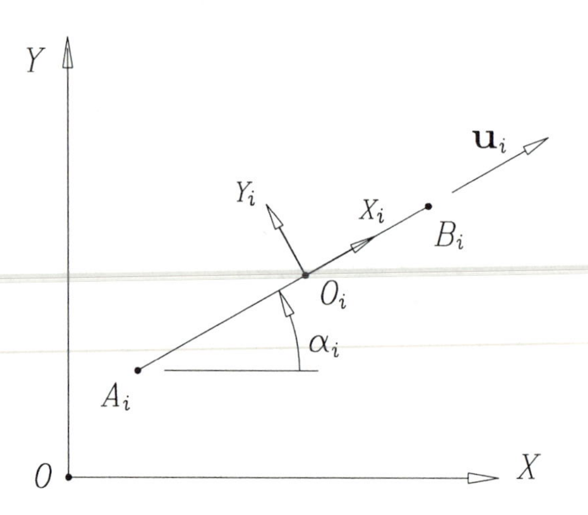

FIGURE 8.11. Relation between frames $\mathcal{F}_0$ and $\mathcal{F}_i$.

where $[\,\mathbf{R}_i\,]_0$ is the matrix that rotates frame $\mathcal{F}_0$ to frame $\mathcal{F}_i$, expressed in $\mathcal{F}_0$, and is given as

$$[\,\mathbf{R}_i\,]_0 = \begin{bmatrix} \cos\alpha_i & -\sin\alpha_i & 0 \\ \sin\alpha_i & \cos\alpha_i & 0 \\ 0 & 0 & 1 \end{bmatrix}, \qquad \text{for } i = 1, 2, 3 \tag{8.52}$$

Referring to Fig. 8.11,

$$\cos\alpha_i = \mathbf{u}_i \cdot \mathbf{i} = u_{ix} \tag{8.53}$$

$$\sin\alpha_i = \mathbf{u}_i \cdot \mathbf{j} = u_{iy} \tag{8.54}$$

After substitution of eqs.(8.52)–(8.54) into eq.(8.51), we obtain

$$[\,\mathbf{s}_i\,]_0 = [\,\mathbf{o}_i\,]_0 + l_i \begin{bmatrix} u_{iy}\cos\phi_i \\ -u_{ix}\cos\phi_i \\ \sin\phi_i \end{bmatrix}, \qquad \text{for } i = 1,2,3 \tag{8.55}$$

where $\mathbf{o}_i$ is given by eq.(8.49).

Since the distances between the three vertices of the triangular plate are fixed, the position vectors $\mathbf{s}_1, \mathbf{s}_2$, and $\mathbf{s}_3$ must satisfy the constraints below:

$$\|\mathbf{s}_2 - \mathbf{s}_1\|^2 = a_1^2 \tag{8.56}$$

$$\|\mathbf{s}_3 - \mathbf{s}_2\|^2 = a_2^2 \tag{8.57}$$

$$\|\mathbf{s}_1 - \mathbf{s}_3\|^2 = a_3^2 \tag{8.58}$$

After expansion, equations (8.56)–(8.58) take on the forms:

$$D_1 c\phi_1 + D_2 c\phi_2 + D_3 c\phi_1 c\phi_2 + D_4 s\phi_1 s\phi_2 + D_5 = 0 \tag{8.59a}$$

$$E_1 c\phi_2 + E_2 c\phi_3 + E_3 c\phi_2 c\phi_3 + E_4 s\phi_2 s\phi_3 + E_5 = 0 \tag{8.59b}$$

$$F_1 c\phi_1 + F_2 c\phi_3 + F_3 c\phi_1 c\phi_3 + F_4 s\phi_1 s\phi_3 + F_5 = 0 \tag{8.59c}$$

where $c(\cdot)$ and $s(\cdot)$ stand for $\cos(\cdot)$ and $\sin(\cdot)$, respectively, while $\{D_i, E_i, F_i\}_1^5$ are functions of the data only and have the forms shown below:

$$D_1 = 2l_2(\mathbf{o}_2 - \mathbf{o}_1)^T \mathbf{E}\mathbf{u}_2$$
$$D_2 = -2l_1(\mathbf{o}_2 - \mathbf{o}_1)^T \mathbf{E}\mathbf{u}_1$$
$$D_3 = -2l_1 l_2 \mathbf{u}_2^T \mathbf{u}_1$$
$$D_4 = -2l_1 l_2$$
$$D_5 = \|\mathbf{o}_2\|^2 + \|\mathbf{o}_1\|^2 - 2\mathbf{o}_1^T \mathbf{o}_2 + l_1^2 + l_2^2 - a_1^2$$

$$E_1 = 2l_3(\mathbf{o}_3 - \mathbf{o}_2)^T \mathbf{E}\mathbf{u}_3$$
$$E_2 = -2l_2(\mathbf{o}_3 - \mathbf{o}_2)^T \mathbf{E}\mathbf{u}_2$$
$$E_3 = -2l_2 l_3 \mathbf{u}_3^T \mathbf{u}_2$$
$$E_4 = -2l_2 l_3$$
$$E_5 = \|\mathbf{o}_3\|^2 + \|\mathbf{o}_2\|^2 - 2\mathbf{o}_3^T \mathbf{o}_2 + l_2^2 + l_3^2 - a_2^2$$

$$F_1 = 2l_1(\mathbf{o}_1 - \mathbf{o}_3)^T \mathbf{E}\mathbf{u}_1$$
$$F_2 = -2l_3(\mathbf{o}_1 - \mathbf{o}_3)^T \mathbf{E}\mathbf{u}_3$$

$$F_3 = -2l_1 l_3 \mathbf{u}_3^T \mathbf{u}_1$$
$$F_4 = -2l_1 l_3$$
$$F_5 = \|\mathbf{o}_3\|^2 + \|\mathbf{o}_1\|^2 - 2\mathbf{o}_3^T \mathbf{o}_1 + l_1^2 + l_3^2 - a_3^2$$

In the above relations the 2×2 matrix $\mathbf{E}$ is defined as in eq.(4.102), and the frame in which the vectors are expressed is immaterial, as long as all vectors appearing in the same scalar product are expressed in the same frame. Since expressions for these vectors in $\mathcal{F}_0$ have already been derived, it is just simpler to perform those computations in this frame.

Our next step is to reduce the foregoing system of three equations in three unknowns to two equations in two unknowns, and hence, obtain two contours in the plane of two of the three unknowns, the desired solutions being determined as the intersections of the two contours. Since eq.(8.59a) is already free of ϕ_3, all we have to do is eliminate ϕ_3 from equations (8.59b) and (8.59c). To do this, we resort to the usual trigonometric identities relating $c\phi_3$ and $s\phi_3$ with $\tan(\phi_3/2)$, in eqs.(8.59b) and (8.59c). After we have cleared the denominators by multiplying the two foregoing equations by $(1 + \tau_3^2)$, the equations thus resulting take on the forms

$$k_1\tau_3^2 + k_2\tau_3 + k_3 = 0 \tag{8.60a}$$
$$m_1\tau_3^2 + m_2\tau_3 + m_3 = 0 \tag{8.60b}$$

where k_1, k_2, and k_3 are linear combinations of $s\phi_2$, $c\phi_2$, and 1. Likewise, m_1, m_2, and m_3 are linear combinations of $s\phi_1$, $c\phi_1$, and 1, namely,

$$k_1 = E_1 c\phi_2 - E_2 - E_3 c\phi_2 + E_5$$
$$k_2 = 2E_4 s\phi_2$$
$$k_3 = E_1 c\phi_2 + E_2 + E_3 c\phi_2 + E_5$$
$$m_1 = F_1 c\phi_1 - F_2 - F_3 c\phi_1 + F_5$$
$$m_2 = 2F_4 s\phi_1$$
$$m_3 = F_1 c\phi_1 + F_2 + F_3 c\phi_1 + F_5$$

Next, we eliminate τ_3 from the above equations dialytically, as we did in Subsection 4.5.3 to find the workspace of a three-axis serial manipulator. We proceed now by multiplying each of the above equations by τ_3 to obtain two more equations, namely,

$$k_1\tau_3^3 + k_2\tau_3^2 + k_3\tau_3 = 0 \tag{8.60c}$$
$$m_1\tau_3^3 + m_2\tau_3^2 + m_3\tau_3 = 0 \tag{8.60d}$$

Further, we write eqs.(8.60a)–(8.60d) in homogeneous form, i.e., as

$$\mathbf{\Phi}\boldsymbol{\tau}_3 = \mathbf{0} \tag{8.61a}$$

with the 4×4 matrix $\mathbf{\Phi}$ and the 4-dimensional vector $\boldsymbol{\tau}_3$ defined as

$$\mathbf{\Phi} \equiv \begin{bmatrix} k_1 & k_2 & k_3 & 0 \\ m_1 & m_2 & m_3 & 0 \\ 0 & k_1 & k_2 & k_3 \\ 0 & m_1 & m_2 & m_3 \end{bmatrix}, \quad \boldsymbol{\tau}_3 \equiv \begin{bmatrix} \tau_3^3 \\ \tau_3^2 \\ \tau_3 \\ 1 \end{bmatrix} \tag{8.61b}$$

Equation (8.61a) constitutes a linear homogeneous system. Moreover, in view of the form of vector $\boldsymbol{\tau}_3$, we are interested only in nontrivial solutions, which exist only if $\det(\mathbf{\Phi})$ vanishes. We thus have the condition

$$\det(\mathbf{\Phi}) = 0 \tag{8.61c}$$

Equations (8.59a) and (8.61c) form a system of two equations in two unknowns, ϕ_1 and ϕ_2. These two equations can be further reduced to a single 16th-degree polynomial equation (Nanua, Waldron and Murthy, 1990), as discussed later on.

In the spirit of the contour method introduced earlier, we plot these two equations as two contours in the ϕ_1-ϕ_2 plane and determine the desired solutions at points where the two contours intersect. Once a pair of (ϕ_1, ϕ_2) values is found, ϕ_3 can be uniquely determined from eqs.(8.59b & c). Indeed, these equations can be arranged in the form:

$$\begin{bmatrix} E_4 s\phi_2 & E_2 + E_3 c\phi_2 \\ F_4 s\phi_1 & F_2 + F_3 c\phi_1 \end{bmatrix} \begin{bmatrix} s\phi_3 \\ c\phi_3 \end{bmatrix} = \begin{bmatrix} -E_1 c\phi_2 - E_5 \\ -F_1 c\phi_1 - F_5 \end{bmatrix}$$

From the above equation, both $c\phi_3$ and $s\phi_3$ can be found *uniquely*; with the foregoing unique values, ϕ_3 is determined uniquely as well.

Knowing the angles ϕ_1, ϕ_2, and ϕ_3 allows us to determine the position vectors of the three vertices of the mobile plate, $\mathbf{s}_1$, $\mathbf{s}_2$, and $\mathbf{s}_3$, whose expressions are given by eq.(8.55). Since three points define a plane, the pose of the end-effector is uniquely determined by the positions of its three vertices. We illustrate the foregoing procedure with a numerical example below:

Example 8.3.1 (A Contour-Intersection Approach) *Nanua, Waldron, and Murthy (1990) studied the direct kinematics of a manipulator of the kind under analysis. This is a platform manipulator whose base plate has six vertices with coordinates expressed with respect to the fixed reference frame $\mathcal{F}_0$ as given below, with all data given in meters:*

$$\begin{aligned} A_1 &= (-2.9, -0.9), & B_1 &= (-1.2, \ \ 3.0) \\ A_2 &= (\ \ 2.5, \ \ 4.1), & B_2 &= (\ \ 3.2, \ \ 1.0) \\ A_3 &= (\ \ 1.3, -2.3), & B_3 &= (-1.2, -3.7) \end{aligned}$$

The dimensions of the movable triangular plate are, in turn,

$$a_1 = 2.0, \qquad a_2 = 2.0, \qquad a_3 = 3.0$$

Determine all possible poses of the moving plate for the dimensions of the six legs given below:

$$q_{1a} = 5.0, \qquad q_{1b} = 4.5$$
$$q_{2a} = 5.5, \qquad q_{2b} = 5.0$$
$$q_{3a} = 5.7, \qquad q_{3b} = 5.5$$

Solution: After substitution of the given numerical values, eqs.(8.59a) and (8.61c) become, with c_i and s_i standing for $\cos\phi_i$ and $\sin\phi_i$, respectively,

$$61.848 - 36.9561c_1 - 47.2376c_2 + 33.603c_1c_2 - 41.6822s_1s_2 = 0$$

$$\begin{aligned}-28.5721 + 48.6506c_1 - 20.7097c_1^2 + 68.7942c_2 - 100.811c_1c_2\\ +35.9634c_1^2c_2 - 41.4096c_2^2 + 50.8539c_1c_2^2 - 15.613c_1^2c_2^2 - 52.9789s_1^2\\ +67.6522c_2s_1^2 - 13.2765c_2^2s_1^2 + 74.1623s_1s_2 - 25.6617c_1s_1s_2\\ -67.953c_2s_1s_2 + 33.9241c_1c_2s_1s_2 - 13.202s_2^2\\ -3.75189c_1s_2^2 + 6.13542c_1^2s_2^2 = 0\end{aligned}$$

The foregoing equations determine contours C_1 and C_2 in the ϕ_1-ϕ_2 plane, which are plotted in Figs. 8.12. Four real solutions are found by superimposing C_1 and C_2, as shown in the above figure. The numerical values of the solutions, listed in Table 8.6, agree with the published results. Solutions 1 and 2 represent two poses of the triangular plate over the base, while solutions 3 and 4 are just the reflections of solutions 1 and 2 with respect to the plane of the base plate. Hence, the geometric symmetry gives rise to an algebraic symmetry of the solutions.

Example 8.3.2 (The Monovariate Polynomial Approach) *Reduce the two equations found in Example 8.3.1, eqs.(8.59a) and (8.61c), to a single monovariate polynomial equation.*

Solution: We first substitute the trigonometric identities relating $c\phi_i$ and $s\phi_i$ with $\tau_i \equiv \tan(\phi_i/2)$, for $i = 1,\ 2$, into eqs.(8.59a) and (8.61c). Upon

TABLE 8.6. Solutions for Nanua, Waldron, and Murthy's Example

No.	ϕ_1 (rad)	ϕ_2 (rad)	ϕ_3 (rad)
1	0.8335	0.5399	0.8528
2	1.5344	0.5107	0.2712
3	-0.8335	-0.5399	-0.8528
4	-1.5344	-0.5107	-0.2712

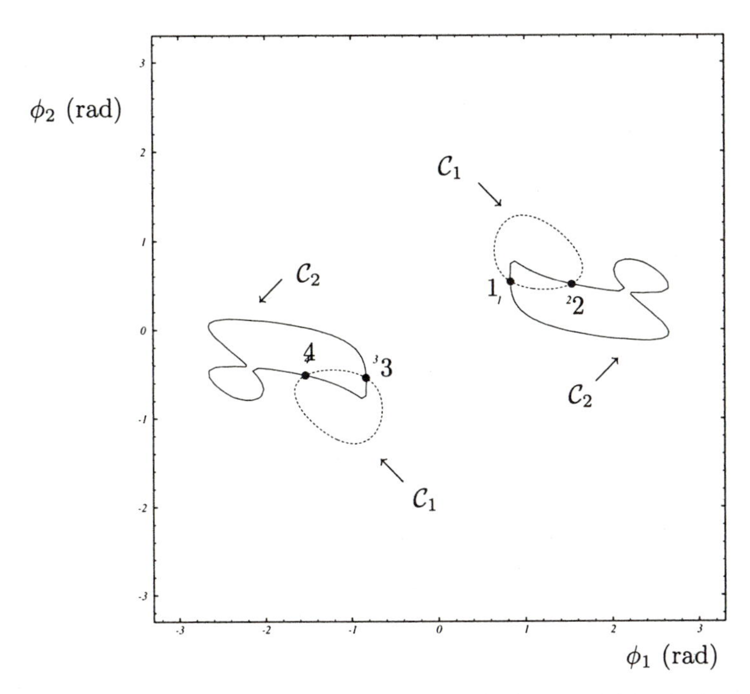

FIGURE 8.12. Contours $\mathcal{C}_1$ and $\mathcal{C}_2$ for Nanua, Waldron, and Murthy's example.

clearing the denominators by multiplying those equations by $(1+\tau_1^2)(1+\tau_2^2)$, we obtain two polynomial equations in τ_1, namely,

$$G_1\tau_1^4 + G_2\tau_1^3 + G_3\tau_1^2 + G_4\tau_1 + G_5 = 0 \tag{8.62}$$

$$H_1\tau_1^2 + H_2\tau_1 + H_3 = 0 \tag{8.63}$$

where

$$G_1 = K_1\tau_2^4 + K_2\tau_2^2 + K_3$$
$$G_2 = K_4\tau_2^3 + K_5\tau_2$$
$$G_3 = K_6\tau_2^4 + K_7\tau_2^2 + K_8$$
$$G_4 = K_9\tau_2^3 + K_{10}\tau_2$$
$$G_5 = k_{11}\tau_2^4 + K_{12}\tau_2^2 + K_{13}$$

and

$$H_1 = L_1\tau_2^2 + L_2$$
$$H_2 = L_3\tau_2$$
$$H_3 = L_4\tau_2^2 + L_5$$

In the above relations, $\{K_i\}_1^{13}$ and $\{L_i\}_1^5$ are all functions of the data. We now eliminate τ_1 from eqs.(8.62) and (8.63), following Bezout's method, as given in (Salmon, 1964). To do this, we multiply eq.(8.62) by H_1 and eq.(8.63) by $G_1\tau_1^2$ and subtract the two equations thus resulting, which leads to a cubic equation in τ_1, namely,

$$(G_2H_1 - G_1H_2)\tau_1^3 + (G_3H_1 - G_1H_3)\tau_1^2 + G_4H_1\tau_1 + G_5H_1 = 0 \quad (8.64a)$$

Likewise, if eq.(8.62) is multiplied by $H_1\tau_1 + H_2$ and eq.(8.63) by $G_1\tau_1^3 + G_2\tau_1^2$ and the equations thus resulting are subtracted from each other, one more cubic equation in τ_1 is obtained, namely,

$$\begin{aligned}(G_1H_3 - G_3H_1)\tau_1^3 &+ (G_4H_1 + G_3H_2 - G_2H_3)\tau_1^2 \\ &+(G_5H_1 + G_4H_2)\tau_1 + G_5H_2 = 0 \end{aligned} \quad (8.64b)$$

Moreover, if we multiply eq.(8.63) by τ_1, a third cubic equation in τ_1 can be derived, i.e.,

$$H_1\tau_1^3 + H_2\tau_1^2 + H_3\tau_1 = 0 \quad (8.64c)$$

Now, eqs.(8.63) and (8.64a–c) constitute a homogeneous linear system of four equations in the first four powers of τ_1, which can be cast in the form

$$\mathbf{H}\boldsymbol{\tau}_1 = \mathbf{0} \quad (8.65)$$

where $\boldsymbol{\tau}_1 \equiv [\tau_1^3 \quad \tau_1^2 \quad \tau_1 \quad 1]^T$ and

$$\mathbf{H} \equiv \begin{bmatrix} G_2H_1 - G_1H_2 & G_3H_1 - G_1H_3 & G_4H_1 & G_5H_1 \\ G_3H_1 - G_1H_3 & G_3H_2 - G_2H_3 + G_4H_1 & G_4H_2 + G_5H_1 & G_5H_2 \\ H_1 & H_2 & H_3 & 0 \\ 0 & H_1 & H_2 & H_3 \end{bmatrix}$$

In order for eq.(8.65) to admit a nontrivial solution, the determinant of its coefficient matrix must vanish, i.e.,

$$\det(\mathbf{H}) = 0 \quad (8.66)$$

Now, we expand $\det(\mathbf{H})$ in the form

$$\det(\mathbf{H}) = H_1\Delta_1 + H_2\Delta_2 + H_3\Delta_3$$

where Δ_i, for $i = 1, 2, 3$, is the cofactor of entry H_i in the last row of matrix $\mathbf{H}$ above, i.e.,

$$\Delta_1 = -\det\left(\begin{bmatrix} G_2H_1 - G_1H_2 & G_4H_1 & G_5H_1 \\ G_3H_1 - G_1H_3 & G_4H_2 + G_5H_1 & G_5H_2 \\ H_1 & H_3 & 0 \end{bmatrix}\right)$$

$$\Delta_2 = \det\left(\begin{bmatrix} G_2H_1 - G_1H_2 & G_3H_1 - G_1H_3 & G_5H_1 \\ G_3H_1 - G_1H_3 & G_3H_2 - G_2H_3 + G_4H_1 & G_5H_2 \\ H_1 & H_2 & 0 \end{bmatrix}\right)$$

$$\Delta_3 = -\det\left(\begin{bmatrix} G_2H_1 - G_1H_2 & G_3H_1 - G_1H_3 & G_4H_1 \\ G_3H_1 - G_1H_3 & G_3H_2 - G_2H_3 + G_4H_1 & G_4H_2 + G_5H_1 \\ H_1 & H_2 & H_3 \end{bmatrix}\right)$$

Given the definitions of $\{G_k\}_1^5$ and $\{H_k\}_1^3$, it is apparent that G_1, G_3, and G_5 are quartic, while G_2 and G_4 are cubic polynomials in τ_2. Likewise, H_1 and H_3 are quadratic, while H_2 is linear in τ_2 as well. As a result, Δ_1 and Δ_3 are 14th-degree polynomials, and Δ_2 is a 13th-degree polynomial in τ_2. Hence, $\det(\mathbf{H})$ is a 16th-degree polynomial in τ_2. This equation, in general, admits up to 16 different solutions. Moreover, the roots of the polynomial are either in the form of complex conjugate pairs or real pairs. In the latter case, each pair represents two symmetrical positions of the mobile platform with respect to the base, i.e., for each solution found above the base, another, mirror-imaged, solution exists below it. This symmetry exists, in general, as long as the six base attachment points are coplanar.

Other parallel manipulators are the planar and spherical counterparts of that studied above, and sketched in Figs. 8.13 and 8.14. The direct kinematics of the manipulator of Fig. 8.13 was found to admit up to six real solutions (Gosselin, Sefrioui and Richard, 1992), while the spherical manipulator of Fig. 8.14 has been found to admit up to eight direct kinematic solutions (Gosselin, Sefrioui and Richard, 1994a, b). A comprehensive account of the simulation and design of three-dof spherical parallel manipulators, which includes workspace analysis as well, is included in (Gosselin, Perreault, and Vaillancourt, 1995).

8.3.1 Velocity and Acceleration Analyses of Parallel Manipulators

Now we proceed to the velocity analysis of the manipulator of Fig. 8.7a. The inverse velocity analysis of this manipulator consists in determining the six rates of the active joints, $\{\dot{b}_k\}_1^6$, given the twist of the moving platform, $\mathbf{t}$. The velocity analysis of a typical leg leads to a relation of the form of eq.(4.54), namely,

$$\mathbf{J}_J\dot{\boldsymbol{\theta}}_J = \mathbf{t}_J, \quad J = I, II, \ldots, VI \tag{8.67a}$$

where $\mathbf{J}_J$ is the Jacobian of the Jth leg, $\dot{\boldsymbol{\theta}}_J$ is the 6-dimensional joint-rate vector of the same leg, and $\mathbf{t}_J$ is the twist of the moving platform $\mathcal{M}$, with its operation point defined as the point C_J of concurrency of the three revolutes composing the spherical joint of attachment of the leg to the moving platform $\mathcal{M}$, and shown in Fig. 8.8 as C, subscript J indicating that point C of that figure is different for different legs. We thus have

$$\mathbf{J}_J \equiv \begin{bmatrix} \mathbf{e}_1 & \mathbf{e}_2 & \mathbf{0} & \mathbf{e}_4 & \mathbf{e}_5 & \mathbf{e}_6 \\ \mathbf{e}_1 \times (b_3 + b_4)\mathbf{e}_3 & \mathbf{e}_2 \times (b_3 + b_4)\mathbf{e}_3 & \mathbf{e}_3 & \mathbf{0} & \mathbf{0} & \mathbf{0} \end{bmatrix}_J \tag{8.67b}$$

$$\mathbf{t}_J = \begin{bmatrix} \boldsymbol{\omega} \\ \dot{\mathbf{c}}_J \end{bmatrix} \tag{8.67c}$$

where the leg geometry has been taken into account.

Furthermore, from Fig. 8.8, it is apparent that

$$\dot{\mathbf{c}}_J = \dot{\mathbf{p}} - \boldsymbol{\omega} \times \mathbf{r}_J \tag{8.68}$$

with $\mathbf{r}_J$ defined as the vector directed from C_J to the operation point P of the moving platform.

Upon multiplication of both sides of the velocity relation of this leg, eq.(8.67a), by $\mathbf{k}_J^T$ from the left, with $\mathbf{k}_J$ suitably defined, we obtain a relation free of all unactuated joint rates. Indeed, a suitable definition of $\mathbf{k}_J$ is shown below:

$$\mathbf{k}_J \equiv [\,\mathbf{0}^T \quad \mathbf{e}_3^T\,]_J^T$$

and hence, on the one hand,

$$\mathbf{k}_J^T \mathbf{J}_J \dot{\boldsymbol{\theta}}_J = (\dot{b}_3)_J$$

where the subscript J reminds us that $\dot{b}_3$ is different for each leg. In order to ease the notation, and since we have a single variable b_3 per leg, we define henceforth

$$b_J \equiv (b_3)_J \tag{8.69a}$$

and hence, the above relation between $\mathbf{t}_J$ and the actuated joint rate of the Jth leg takes the form

$$\mathbf{k}_J^T \mathbf{J}_J \dot{\boldsymbol{\theta}}_J = \dot{b}_J \tag{8.69b}$$

On the other hand,

$$\mathbf{k}_J^T \mathbf{t}_J = (\mathbf{e}_3^T)_J \dot{\mathbf{c}}_J$$

Likewise, we define

$$(\mathbf{e}_3)_J \equiv \mathbf{e}_J \tag{8.70a}$$

the foregoing relation thus yielding

$$\mathbf{k}_J^T \mathbf{t}_J \equiv \mathbf{e}_J^T \dot{\mathbf{c}}_J \tag{8.70b}$$

Note that vectors $\mathbf{e}_J$ and $\mathbf{r}_J$ define uniquely the line along the two attachment points of the Jth leg. Henceforth, this line will be termed the axis of the Jth leg.

Upon equating the right-hand sides of eqs.(8.69b) and (8.70b), the desired expression for the actuated joint rate is derived, namely,

$$\dot{b}_J = \mathbf{e}_J^T \dot{\mathbf{c}}_J, \quad J = I, II, \ldots, VI \tag{8.71a}$$

That is, the Jth joint rate is nothing but the projection onto the Jth leg axis of the velocity of point C_J. Furthermore, upon substituting eq.(8.68) in eq.(8.71a) above, we obtain the relations between the actuated joint rates and the twist of the moving platform, namely,

$$\dot{b}_J = [\,(\mathbf{e}_J \times \mathbf{r}_J)^T \quad \mathbf{e}_J^T\,] \begin{bmatrix} \boldsymbol{\omega} \\ \dot{\mathbf{p}} \end{bmatrix}, \quad J = I, II, \ldots, VI \tag{8.71b}$$

for all six actuated joint rates. Upon assembling all six leg-equations of eq.(8.71b), we obtain the desired relation between the vector of actuated joint rates and the twist of the moving platform, namely,

$$\dot{\mathbf{b}} = \mathbf{K}\mathbf{t} \tag{8.72a}$$

with the 6-dimensional vectors $\mathbf{b}$ and $\mathbf{t}$ defined as the vector of joint variables and the twist of the platform at the operation point, respectively. Moreover, the 6×6 matrix $\mathbf{K}$ is the Jacobian of the manipulator at hand. These quantities are displayed below:

$$\mathbf{b} \equiv \begin{bmatrix} b_I \\ b_{II} \\ \vdots \\ b_{VI} \end{bmatrix}, \quad \mathbf{t} \equiv \begin{bmatrix} \boldsymbol{\omega} \\ \dot{\mathbf{p}} \end{bmatrix}, \quad \mathbf{K} \equiv \begin{bmatrix} (\mathbf{e}_I \times \mathbf{r}_I)^T & \mathbf{e}_I^T \\ (\mathbf{e}_{II} \times \mathbf{r}_{II})^T & \mathbf{e}_{II}^T \\ \vdots & \\ (\mathbf{e}_{VI} \times \mathbf{r}_{VI})^T & \mathbf{e}_{VI}^T \end{bmatrix} \tag{8.72b}$$

From the above display, it is apparent that each row of $\mathbf{K}$ is the transpose of the Plücker array of the corresponding leg axis, although in axis coordinates, as opposed to the Jacobian matrix $\mathbf{J}$ of serial manipulators, whose columns are the Plücker coordinates of the corresponding joint axis in ray coordinates. Moreover, in these coordinates, the moment of the leg axis is taken with respect to the operation point P of M. One more difference between the velocity analysis of serial and parallel manipulators is the role played by the actuator joint rates in the underlying forward and direct kinematics. In the case of parallel manipulators, this role is changed, for now we have that the actuator joint rates are given by explicit formulas in terms of the twist of the moving platform, along with the manipulator architecture and configuration. Finding the platform twist requires inverting matrix $\mathbf{K}$. Moreover, the significance of singularities also changes: When $\mathbf{K}$ becomes singular, some instantaneous motions of the platform are possible even if all actuated joints are kept locked. That is, a singularity of $\mathbf{K}$ is to be interpreted now as the inability of the manipulator to withstand a certain static wrench. An extensive analysis of the singularities of parallel manipulators using line geometry in a form that is known as *Grassmann geometry* was reported by Merlet (1989).

Now, the acceleration analysis of the same leg is straightforward. Indeed, upon differentiation of both sides of eq.(8.72a) with respect to time, one obtains

$$\ddot{\mathbf{b}} = \mathbf{K}\dot{\mathbf{t}} + \dot{\mathbf{K}}\mathbf{t} \tag{8.73a}$$

where $\dot{\mathbf{K}}$ takes the form

$$\dot{\mathbf{K}} = \begin{bmatrix} \dot{\mathbf{u}}_I^T & \dot{\mathbf{e}}_I^T \\ \dot{\mathbf{u}}_{II}^T & \dot{\mathbf{e}}_I^T \\ \vdots & \\ \dot{\mathbf{u}}_{VI}^T & \dot{\mathbf{e}}_{VI}^T \end{bmatrix} \tag{8.73b}$$

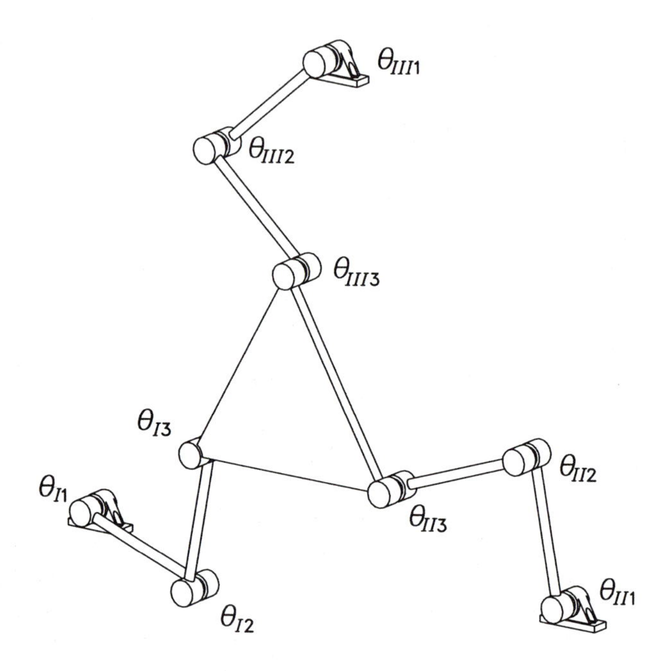

FIGURE 8.13. A planar parallel manipulator.

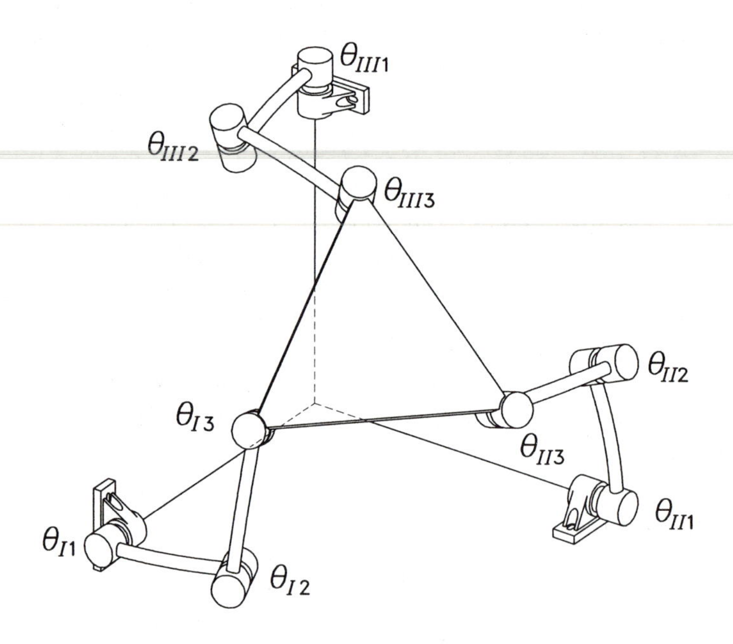

FIGURE 8.14. A spherical parallel manipulator.

and $\mathbf{u}_J$ is defined as

$$\mathbf{u}_J \equiv \mathbf{e}_J \times \mathbf{r}_J \tag{8.73c}$$

Therefore,

$$\dot{\mathbf{u}}_J = \dot{\mathbf{e}}_J \times \mathbf{r}_J + \mathbf{e}_J \times \dot{\mathbf{r}}_J \tag{8.73d}$$

Now, since vectors $\mathbf{r}_J$ are fixed to the moving platform, their time-derivatives are simply given by

$$\dot{\mathbf{r}}_J = \boldsymbol{\omega} \times \mathbf{r}_J \tag{8.73e}$$

On the other hand, vector $\mathbf{e}_J$ is directed along the leg axis, and so, its time-derivative is given by

$$\dot{\mathbf{e}}_J = \boldsymbol{\omega}_J \times \mathbf{e}_J$$

with $\boldsymbol{\omega}_J$ defined as the angular velocity of the third link of the leg, i.e.,

$$\boldsymbol{\omega}_J = (\dot{\theta}_1 \mathbf{e}_1 + \dot{\theta}_2 \mathbf{e}_2)_J$$

the subscript J of the above parentheses reminding us that this angular velocity differs from leg to leg. Clearly, we need expressions for the rates of the first two joints of each leg. Below we derive the corresponding expressions. In order to simplify the notation, we start by defining

$$\mathbf{f}_J \equiv (\mathbf{e}_1)_J, \quad \mathbf{g}_J \equiv (\mathbf{e}_2)_J \tag{8.73f}$$

Now we write the second vector equation of eq.(8.67a) using the foregoing definitions, which yields

$$(\dot{\theta}_1)_J \mathbf{f}_J \times (b_J + b_4)\mathbf{e}_J + (\dot{\theta}_2)_J \mathbf{g}_J \times (b_J + b_4)\mathbf{e}_J + \dot{b}_J \mathbf{e}_J = \dot{\mathbf{c}}_J$$

where b_4 is the same for all legs, since all have identical architectures. Now we can eliminate $(\dot{\theta}_2)_J$ from the foregoing equation by dot-multiplying both of its sides by $\mathbf{g}_J$, thereby producing

$$(\dot{\theta}_1)_J \mathbf{g}_J \times \mathbf{f}_J \cdot (b_J + b_4)\mathbf{e}_J + +\mathbf{g}_J^T(\mathbf{e}_J \mathbf{e}_J^T)\dot{\mathbf{c}}_J = \mathbf{g}^T \dot{\mathbf{c}}_J$$

where an obvious exchange of the cross and the dot in the above equation has taken place, and expression (8.71a) for $\dot{b}_J$ has been recalled. Now it is a simple matter to solve for $(\dot{\theta}_1)_J$ from the above equation, which yields

$$(\dot{\theta}_1)_J = -\frac{\mathbf{g}_J^T(\mathbf{1} - \mathbf{e}_J \mathbf{e}_J^T)\dot{\mathbf{c}}_J}{\Delta_J}$$

with Δ_J defined as

$$\Delta_J \equiv (b_J + b_4)\mathbf{e}_J \times \mathbf{f}_J \cdot \mathbf{g}_J \tag{8.74}$$

Moreover, we can obtain the above expression for $(\dot{\theta}_1)_J$ in terms of the platform twist by recalling eq.(8.68), which is reproduced below in a more suitable form for quick reference:

$$\dot{\mathbf{c}}_J = \mathbf{C}_J \mathbf{t}$$

where $\mathbf{t}$ is the twist of the platform, the 3×6 matrix $\mathbf{C}_J$ being defined as

$$\mathbf{C}_J \equiv [\mathbf{R}_J \quad \mathbf{1}]$$

in which $\mathbf{R}_J$ is the cross-product matrix of $\mathbf{r}_J$ and $\mathbf{1}$ is the 3×3 identity matrix. Therefore, the expression sought for $(\dot{\theta}_1)_J$ takes the form

$$(\dot{\theta}_1)_J = -\frac{1}{\Delta_J}\mathbf{g}_J^T(\mathbf{1} - \mathbf{e}_J\mathbf{e}_J^T)\mathbf{R}_J\mathbf{t}, \quad J = I, II, \ldots, VI \tag{8.75a}$$

A similar procedure can be followed to find $(\dot{\theta}_2)_J$, the final result being

$$(\dot{\theta}_2)_J = \frac{1}{\Delta_J}\mathbf{f}_J^T(\mathbf{1} - \mathbf{e}_J\mathbf{e}_J^T)\mathbf{R}_J\mathbf{t}, \quad J = I, II, \ldots, VI \tag{8.75b}$$

thereby completing the calculations required to obtain the rates of all actuated joints. Note that the unit vectors involved in those calculations, $\mathbf{e}_J$, $\mathbf{f}_J$, and $\mathbf{g}_J$, are computed from the leg inverse kinematics, as discussed above.

The velocity analysis of the planar and spherical parallel manipulators of Figs. 8.13 and 8.14 are outlined below: Using the results of Subsection 4.8.2, the velocity relations of the Jth leg of the planar manipulator take the form

$$\mathbf{J}_J\dot{\boldsymbol{\theta}}_J = \mathbf{t}, \quad J = I,\ II,\ III \tag{8.76}$$

where $\mathbf{J}_J$ is the Jacobian matrix of this leg, as given by eq.(4.107), while $\dot{\boldsymbol{\theta}}_J$ is the 3-dimensional vector of joint rates of this leg, i.e.,

$$\mathbf{J}_J \equiv \begin{bmatrix} 1 & 1 & 1 \\ \mathbf{E}\mathbf{r}_{J1} & \mathbf{E}\mathbf{r}_{J2} & \mathbf{E}\mathbf{r}_{J3} \end{bmatrix}, \quad \dot{\boldsymbol{\theta}}_J \equiv \begin{bmatrix} \dot{\theta}_{J1} \\ \dot{\theta}_{J2} \\ \dot{\theta}_{J3} \end{bmatrix}, \quad J = I,\ II,\ III$$

For purposes of kinematic velocity control, however, we are interested only in the first joint rate of each leg; i.e., all we need to determine in order to produce a desired twist of the end-effector is not all of the foregoing nine joint rates, but only $\dot{\theta}_{I1}$, $\dot{\theta}_{II1}$, and $\dot{\theta}_{III1}$. Thus, we want to eliminate from eq.(8.76) the unactuated joint rates $\dot{\theta}_{J2}$ and $\dot{\theta}_{J3}$, which can be readily done if we multiply both sides of the said equation by a 3-dimensional vector $\mathbf{n}_J$ perpendicular to the second and the third columns of $\mathbf{J}_J$. This vector can be most easily determined as the cross product of those two columns, namely, as

$$\mathbf{n} \equiv \mathbf{j}_{J2} \times \mathbf{j}_{J3} = \begin{bmatrix} -\mathbf{r}_{J2}^T\mathbf{E}\mathbf{r}_{J3} \\ \mathbf{r}_{J2} - \mathbf{r}_{J3} \end{bmatrix}$$

Upon multiplication of both sides of eq.(8.76) by $\mathbf{n}_J^T$, we obtain

$$\left[-\mathbf{r}_{J2}^T\mathbf{E}\mathbf{r}_{J3} + (\mathbf{r}_{J2} - \mathbf{r}_{J3})^T\mathbf{E}\mathbf{r}_{J1}\right]\dot{\theta}_{J1} = -(\mathbf{r}_{J2}^T\mathbf{E}\mathbf{r}_{J3})\omega + (\mathbf{r}_{J2} - \mathbf{r}_{J3})^T\dot{\mathbf{c}} \tag{8.77}$$

and hence, we can solve directly for $\dot{\theta}_{J1}$ from the foregoing equation, thereby deriving

$$\dot{\theta}_{J1} = \frac{-(\mathbf{r}_{J2}^T \mathbf{E} \mathbf{r}_{J3})\omega + (\mathbf{r}_{J2} - \mathbf{r}_{J3})^T \dot{\mathbf{c}}}{-\mathbf{r}_{J2}^T \mathbf{E} \mathbf{r}_{J3} + (\mathbf{r}_{J2} - \mathbf{r}_{J3})^T \mathbf{E} \mathbf{r}_{J1}} \tag{8.78a}$$

Note that eq.(8.77) can be written in the form

$$j_J \dot{\theta}_{J1} = \mathbf{k}_J^T \mathbf{t}, \quad J = I,\, II,\, III \tag{8.78b}$$

with j_J and $\mathbf{k}_J$ defined, for $J = I,\, II,\, III$, as

$$j_J \equiv (\mathbf{r}_{J2} - \mathbf{r}_{J3})^T \mathbf{E} \mathbf{r}_{J1} - \mathbf{r}_{J2}^T \mathbf{E} \mathbf{r}_{J3}, \quad \mathbf{k}_J \equiv [\, \mathbf{r}_{J2}^T \mathbf{E} \mathbf{r}_{J3} \quad (\mathbf{r}_{J2} - \mathbf{r}_{J3})^T \,]^T \tag{8.78c}$$

If we now define further

$$\dot{\boldsymbol{\theta}} \equiv [\, \dot{\theta}_{J1} \quad \dot{\theta}_{J2} \quad \dot{\theta}_{J3} \,]^T$$

and assemble all three foregoing joint-rate-twist relations, we obtain

$$\mathbf{J} \dot{\boldsymbol{\theta}} = \mathbf{K} \mathbf{t} \tag{8.79}$$

where $\mathbf{J}$ and $\mathbf{K}$ are the two manipulator Jacobians defined as

$$\mathbf{J} \equiv \mathrm{diag}(j_I,\, j_{II},\, j_{III}), \quad \mathbf{K} \equiv \begin{bmatrix} \mathbf{k}_I^T \\ \mathbf{k}_{II}^T \\ \mathbf{k}_{III}^T \end{bmatrix} \tag{8.80}$$

Expressions for the joint accelerations can be readily derived by differentiation of the foregoing expressions with respect to time.

The velocity analysis of the spherical parallel manipulator of Fig. 8.14 can be accomplished similarly. Thus, the velocity relations of the Jth leg take on the form

$$\mathbf{J}_J \dot{\boldsymbol{\theta}}_J = \boldsymbol{\omega}, \quad J = I,\, II,\, III \tag{8.81}$$

where the Jacobian of the Jth leg, $\mathbf{J}_J$, is defined as

$$\mathbf{J}_J \equiv [\, \mathbf{e}_{J1} \quad \mathbf{e}_{J2} \quad \mathbf{e}_{J3} \,]$$

while the joint-rate vector of the Jth leg, $\dot{\boldsymbol{\theta}}_J$, is defined exactly as in the planar case analyzed above. Again, for kinematic velocity control purposes, we are interested only in the actuated joint rates, namely, $\dot{\theta}_{I1}$, $\dot{\theta}_{II1}$, and $\dot{\theta}_{III1}$. As in the planar case, we can eliminate $\dot{\theta}_{J2}$ and $\dot{\theta}_{J3}$ upon multiplication of both sides of eq.(8.81) by a vector $\mathbf{n}_J$ perpendicular to the second and the third columns of $\mathbf{J}_J$. An obvious definition of this vector is, then,

$$\mathbf{n}_J \equiv \mathbf{e}_{J2} \times \mathbf{e}_{J3}$$

The desired joint-rate relation is thus readily derived as

$$j_J \dot{\theta}_{J1} = \mathbf{k}_J^T \boldsymbol{\omega}, \quad J = I,\, II,\, III \tag{8.82}$$

where j_J and $\mathbf{k}_J$ are now defined as

$$j_J \equiv \mathbf{n}\mathbf{e}_{J1} \times \mathbf{e}_{J2} \cdot \mathbf{e}_{J3} \tag{8.83a}$$

$$\mathbf{k}_J \equiv \mathbf{e}_{J2} \times \mathbf{e}_{J3} \tag{8.83b}$$

The accelerations of the actuated joints can be derived, again, by differentiation of the foregoing expressions.

We can then say that in general, parallel manipulators, as opposed to serial ones, have two Jacobian matrices.

8.4 Multifingered Hands

Shown in Fig. 8.15 is a three-fingered hand with fingers $\mathcal{A}$, $\mathcal{B}$, and $\mathcal{C}$, each finger supplied with three revolute joints. Each finger of this hand is supplied with two revolutes of parallel axes that are normal to the axis of the third one. Thus, each finger comprises three links, the one closest to the palm $\mathcal{P}$ being of virtually zero length. Of the other two, that in contact with the object is called the *distal phalanx*; its neighboring link is called the *proximal phalanx*. Moreover, each finger is coupled to the hand palm $\mathcal{P}$ via a revolute, while the contact between the finger tip and the hand-held object $\mathcal{O}$ is assumed to take place at one point, i.e., the fingers are assumed to be *hard*, as opposed to *soft*; for the latter, contact takes place over a finite area. Thus, while hard fingers can exert only force and no moment on the manipulated object, soft fingers can exert both force and moment. For the sake of conciseness, we will deal only with hard fingers here. Let the contact points of fingers $\mathcal{A}$, $\mathcal{B}$, and $\mathcal{C}$ with $\mathcal{O}$ be denoted by A_O, B_O, and C_O, respectively. The purpose of the hand is to manipulate $\mathcal{O}$ with respect to $\mathcal{P}$. The motion of $\mathcal{O}$, moreover, can be specified through its pose, given in turn by the position vector $\mathbf{o}$ of one of its points, O, and its orientation matrix $\mathbf{Q}$ with respect to a frame fixed to $\mathcal{P}$. Let $\mathbf{a}$, $\mathbf{b}$, and $\mathbf{c}$ denote the vectors directed from O to A_O, B_O, and C_O, respectively, when $\mathcal{O}$ is in its reference pose, at which $\mathbf{Q} = \mathbf{1}$. Hence, the position vectors $\mathbf{a}_O$, $\mathbf{b}_O$, and $\mathbf{c}_O$ of A_O, B_O, and C_O are given by

$$\mathbf{a}_O = \mathbf{o} + \mathbf{Q}\mathbf{a} \tag{8.84a}$$

$$\mathbf{b}_O = \mathbf{o} + \mathbf{Q}\mathbf{b} \tag{8.84b}$$

$$\mathbf{c}_O = \mathbf{o} + \mathbf{Q}\mathbf{c} \tag{8.84c}$$

Thus, the location of the three contact points is fully determined if the pose of $\mathcal{P}$ and the locations of A_O, B_O, and C_O in $\mathcal{O}$ are given. Once the position vectors of the three contact points are known, determining the joint variables to take $\mathcal{O}$ to the desired pose reduces to solving a 3-dimensional positioning problem for each finger, with three joints—a problem already

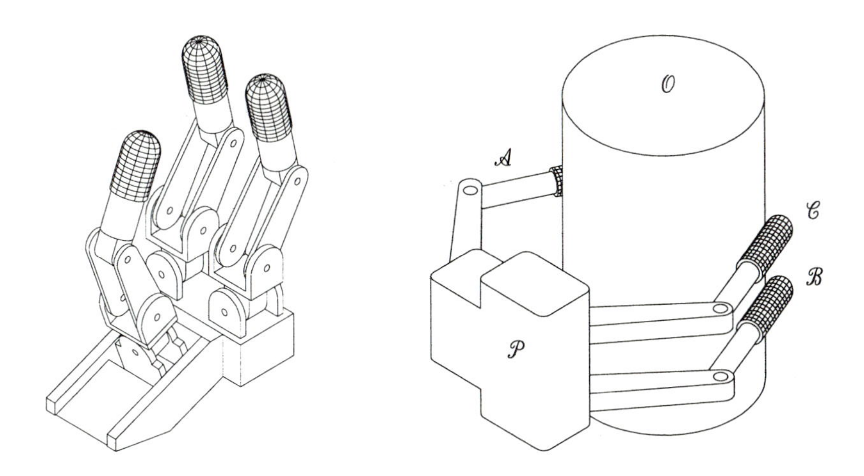

FIGURE 8.15. A three-fingered hand.

discussed in Subsection 4.4.1. The joint rates and accelerations are then determined as in Sections 4.4 and 4.6.

While the mechanics of grasping is quite elaborate, due to the deformation of both fingers and object, some assumptions will be introduced here to produce a simple model. One such assumption is rigidity; a second is *smoothness*, under which each finger is capable of exerting only normal force on the object. Moreover, this force is *unidirectional*, for the finger cannot exert a *pull* on the object. The smoothness and rigidity assumption bring about limitations, for they require a rather large number of fingers to exert an arbitrary wrench on the grasped object, as shown below.

We assume that we have a rigid object $\mathcal{O}$ bounded by a surface $\mathcal{S}$ that is smooth *almost everywhere*, i.e., it has a well-defined normal $\mathbf{n}$ everywhere except at either isolated points or isolated curves on $\mathcal{S}$. Below we show that in order to exert an arbitrary wrench $\mathbf{w}$ onto $\mathcal{O}$, a hand with rigid and smooth fingers should have at least seven fingers. Assume that the n contact points on $\mathcal{S}$ are $\{ P_i \}_1^n$ and that we want to find n pressure values $\{ \lambda_i \}_1^n$ at the contact points that will produce the desired wrench $\mathbf{w}$ onto $\mathcal{O}$.

Moreover, let the unit normal at P_i be denoted by $\mathbf{n}_i$ and the vector directed from O to P_i be denoted by $\mathbf{p}_i$, as shown in Fig. 8.16

The wrench $\mathbf{w}_i$ exerted by each finger onto $\mathcal{O}$ at P_i is clearly

$$\mathbf{w}_i = \lambda_i \begin{bmatrix} \mathbf{p}_i \times (-\mathbf{n}_i) \\ -\mathbf{n}_i \end{bmatrix}, \quad \lambda_i \geq 0$$

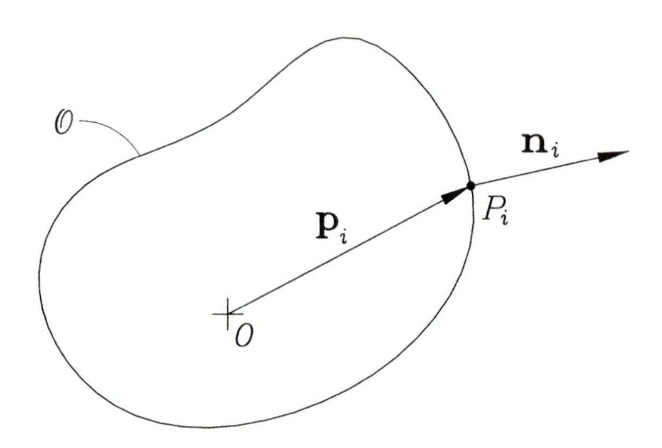

FIGURE 8.16. Geometry of grasped object $\mathcal{O}$.

Upon equating the resultant wrench with the desired wrench, we obtain

$$\sum_{1}^{n} \begin{bmatrix} -\mathbf{p}_i \times \mathbf{n}_i \\ -\mathbf{n}_i \end{bmatrix} \lambda_i = \mathbf{w}$$

or in compact form, as

$$\mathbf{G}\boldsymbol{\lambda} = -\mathbf{w} \tag{8.85a}$$

where $\mathbf{G}$ is the $6 \times n$ *grasping matrix* and $\boldsymbol{\lambda}$ is the n-dimensional vector of pressure values, i.e.,

$$\mathbf{G} \equiv \begin{bmatrix} \mathbf{p}_1 \times \mathbf{n}_1 & \cdots & \mathbf{p}_n \times \mathbf{n}_n \\ \mathbf{n}_1 & \cdots & \mathbf{n}_n \end{bmatrix}, \qquad \boldsymbol{\lambda} \equiv \begin{bmatrix} \lambda_1 \\ \vdots \\ \lambda_n \end{bmatrix} \tag{8.85b}$$

Note that the ith column of the grasping matrix is nothing but the array of Plücker coordinates of the line of action of the force exerted by the ith finger on the object, in ray coordinates—see Subsection 3.2.2.

Thus, for $n = 6$, a unique pressure vector is obtained as long as $\mathbf{G}$ is nonsingular. However, negative values of $\{\lambda_i\}_1^n$ are not allowed, and since nothing prevents these values from becoming negative, six fingers of the type considered here are not enough. We must thus have at least seven such fingers in order to be able to apply an arbitrary wrench onto the body. For $n = 7$ and a full-rank 6×7 grasping matrix, we can generate nonnegative values of $\{\lambda_i\}_1^n$ as shown below: Let $\mathbf{u}$ be a unit vector spanning the nullspace of $\mathbf{G}$. Then an arbitrary $\boldsymbol{\lambda}$ can be expressed as

$$\boldsymbol{\lambda} = \boldsymbol{\lambda}_0 + \mu\mathbf{u}$$

where $\boldsymbol{\lambda}_0$ is a *particular solution* of eq.(8.85a) and μ is a scalar, as yet to be determined. For example, if $\boldsymbol{\lambda}_0$ is chosen as the *minimum-norm solution*

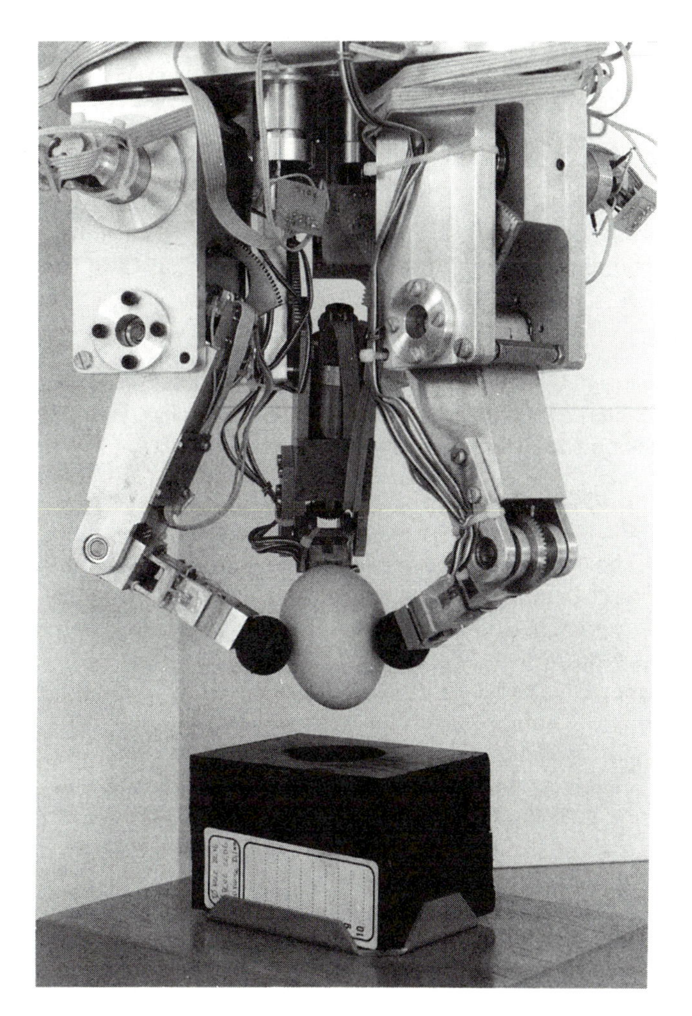

FIGURE 8.17. A prototype of the KU Leuven three-fingered hand (Courtesy of Prof. H. van Brussel).

of eq.(8.85a), then we have, explicitly,

$$\boldsymbol{\lambda}_0 = -\mathbf{G}^\dagger \mathbf{w}$$

where $\mathbf{G}^\dagger$ is the *generalized inverse* of $\mathbf{G}$, defined as

$$\mathbf{G}^\dagger \equiv \mathbf{G}^T(\mathbf{G}\mathbf{G}^T)^{-1}$$

Note that for n hard fingers, the 6×6 product $\mathbf{G}\mathbf{G}^T$ has the general form

$$\mathbf{G}\mathbf{G}^T = \begin{bmatrix} \sum_1^n (\mathbf{p}_i \times \mathbf{n}_i)(\mathbf{p}_i \times \mathbf{n}_i)^T & \sum_1^n (\mathbf{p}_i \times \mathbf{n}_i)\mathbf{n}_i^T \\ \sum_1^n \mathbf{n}_i(\mathbf{p}_i \times \mathbf{n}_i)^T & \mathbf{n}_i\mathbf{n}_i^T \end{bmatrix}$$

Although a symbolic expression for the inverse of $\mathbf{H}$ is not possible in the general case, we can always express this inverse in block form, namely,

$$(\mathbf{G}\mathbf{G}^T)^{-1} = \begin{bmatrix} \mathbf{H}_{11} & \mathbf{H}_{12} \\ \mathbf{H}_{12}^T & \mathbf{H}_{22} \end{bmatrix}$$

where consistently, $\mathbf{H}_{11}$ has units of meter^{-2}, $\mathbf{H}_{12}$ has units of meter^{-1}, and $\mathbf{H}_{22}$ is dimensionless. Moreover, we can partition $\mathbf{G}$ in the form

$$\mathbf{G} \equiv \begin{bmatrix} \mathbf{A} \\ \mathbf{B} \end{bmatrix}$$

in which $\mathbf{A}$ has units of meter, while $\mathbf{B}$ is dimensionless. Hence, the product $\mathbf{G}^T\mathbf{H}$ takes on the form

$$\mathbf{G}^T\mathbf{H} = [\,\mathbf{A}^T\mathbf{H}_{11} + \mathbf{B}^T\mathbf{H}_{12}^T \quad \mathbf{A}^T\mathbf{H}_{12} + \mathbf{B}^T\mathbf{H}_{22}\,]$$

and hence, the left-hand block of the foregoing product has units of meter^{-1}, while the right-hand block is dimensionless. Upon multiplying the desired wrench $\mathbf{w}$ from the left by this product, the result, $\boldsymbol{\lambda}_0$, has consistently units of Newton.

Furthermore, let g_i be the ith component of $\mathbf{G}^{\dagger}\mathbf{w}$ and let g_m be defined as

$$g_m \equiv \min_i\{g_i\}$$

Now, if $g_m \geq 0$, then $\mu = 0$. Otherwise, we make

$$g_m + \mu u_m = 0$$

with u_m denoting the mth component of $\mathbf{u}$, and hence,

$$\mu = -\frac{g_m}{u_m}$$

thereby guaranteeing that the components of vector $\boldsymbol{\lambda}$ are nonnegative.

In the presence of friction, however, fewer than six fingers suffice to grasp an object. Moreover, in the presence of friction, the force transmitted by a finger has, in addition to its normal component, a tangential component that, hence, gives rise to a contact force making a nonzero angle with the normal $\mathbf{n}_i$ to the object surface at the ith contact point. Therefore, by virtue of the linear relation between the normal and the tangential components of the transmitted force, given by the coefficient of friction μ, this force is constrained to lie within the *friction cone*. This cone has its apex at the contact point P_i, its elements making an angle α with the normal, that is given by $\alpha = \arctan(\mu)$. Moreover, by virtue of the fundamental assumption of Coulomb friction analysis, μ lies between 0 and 1, and hence, α is constrained to lie between 0° and 45°.

Now, as long as the normal force exerts a push on the object, the coupling between the distal phalanx and the object can be modeled with a spherical joint, the underlying kinematic analysis thus becoming that of a parallel manipulator, as studied in Subsection 8.3.

Shown in Fig. 8.17 is an example of a three-fingered hand. This hand was developed at the Katholieke Universiteit Leuven (Van Brussel et al., 1989).

The literature on multifingered hands and the problem of grasping is far richer than we can afford to describe here. Extensive studies on these subjects have been recently reported by Raynaerts (1995) and Teichmann (1995.)

8.5 Walking Machines

Besides the walking machines introduced in Chapter 1, namely, the OSU Adaptive Suspension Vehicle and the TUM Hexapod, other legged machines or leg designs are emerging with special features. For example, CARL, shown in Fig. 8.18, is a compliant articulated robot leg that has been designed at McGill University's Centre for Intelligent Machines (CIM) by Prof. Buehler and his team (Menitto and Buehler, 1996). This leg contains an actuation package with a high load-carrying capacity (ATLAS) and an antagonistic pair of concentric linear-to-angular displacement devices. The leg has four degrees of freedom, of which two are actuated by ATLAS and one by a harmonic drive motor, while one is unactuated. This leg design is intended to provide locomotion to a quadruped that is currently under design at CIM.

As nature shows in mammals, four legs are necessary to guarantee the static equilibrium of the body while one leg is in the swing phase. Static equilibrium is achieved as long as the horizontal projection of the mass center of the overall body-legs system lies within the triangle defined by the contact points of the three legs that are in the stance phase. More than four legs would allow for greater mobility. For purposes of symmetry, some walking machines are designed as hexapods, so as to allow for an equal number of legs in the swing and the stance phases.

We undertake the kinematic analysis of walking machines using the hexapod displayed in Fig. 8.19.

Furthermore, contact with the ground is assumed to take place such that the ground can exert only a "pushing" force on each leg but no moment. Thus, while we can model the contact between leg and ground as a spherical joint, care must be taken so that no pulls of the ground on the leg are required for a given gait.

Additionally, we shall assume that the leg is actuated by three revolutes, namely, those with variables θ_4, θ_5, and θ_6 in Fig. 8.20, where $\mathcal{G}$ denotes

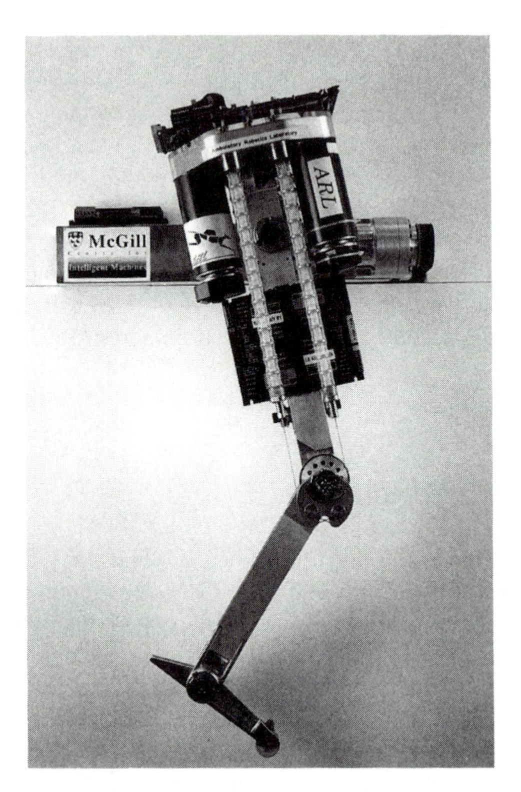

FIGURE 8.18. The compliant articulated robot leg (Courtesy of Prof. M. Buehler).

the ground and $\mathcal{B}$ the body of the machine. A photograph of one of the six identical legs of the walking machine developed at the Technical University of Munich, introduced in Fig. 1.9, is included in Fig. 8.21. The Denavit-Hartenberg parameters of this leg, proceeding from the ground upwards, are displayed in Table 8.7. Note that the architecture of this leg is simply that of a three-revolute manipulator carrying a spherical joint at its end-effector, similar to that of the decoupled manipulators studied in Section 4.4. The spherical joint accounts for the coupling of the leg with the ground. We are thus assuming that when a leg is in contact with the ground, the contact point of the leg is immobile. At the same time, the motion of the body $\mathcal{B}$ is prescribed through the motion of a point on the axis of the revolute coupled to the body. Such a point is indicated by P_J for the Jth leg. Moreover, the point of the Jth leg in contact with the ground will be denoted by O_J. Thus, when prescribing the motion of the body through that of each of the six points P_I, P_{II}, ..., P_{VI}, the rigid-body compatibility conditions of eqs.(7.14), (7.15), and (7.25) must be observed. The pose of the body $\mathcal{B}$ is thus specified by the position of a point C of the body and the orientation matrix $\mathbf{Q}$ of the body with respect to a frame fixed to the ground, the position vector of C in that frame being denoted by $\mathbf{c}$. The specification

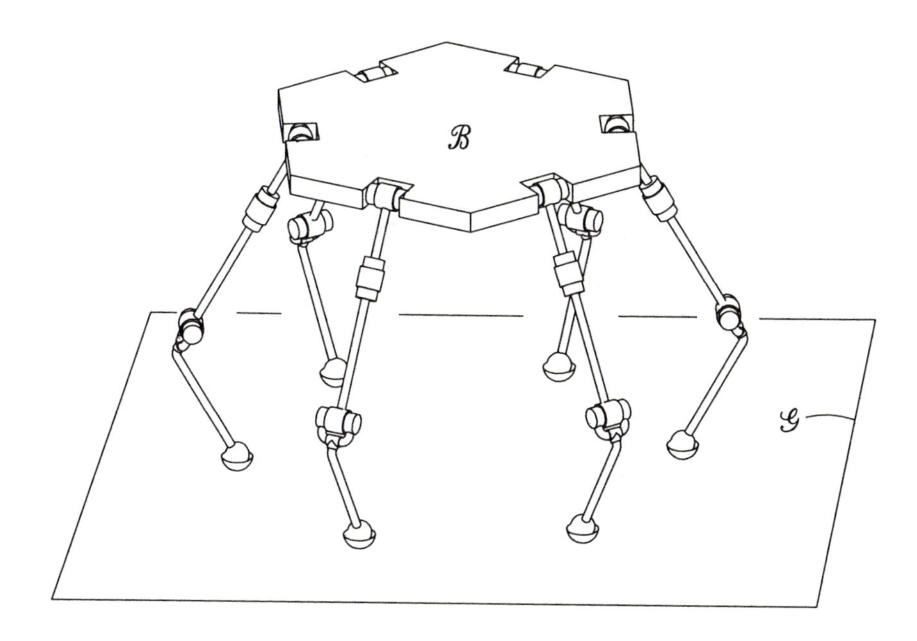

FIGURE 8.19. A general hexapod.

of points P_I to P_{VI} thus follows from the knowledge of $\mathbf{c}$ and $\mathbf{Q}$, thereby guaranteeing compliance with the above-mentioned constraints.

Furthermore, a maneuver of $\mathcal{B}$, given by a prescribed pose, can be achieved by suitable values of the variables associated with the actuated joints, which thus leads to a problem of inverse kinematics associated, again, with a parallel manipulator.

The mechanical system that results from the kinematic coupling of the machine legs with the ground is thus equivalent to a parallel manipulator. The essential difference between a walking machine and a parallel manipulator is that the former usually involves more actuators than degrees of freedom. This feature is known as *redundant actuation* and will not be pursued here.

TABLE 8.7. DH Parameters of the Leg of the TU-Munich Walking Machine

i	a_i (mm)	b_i (mm)	α_i
1	17	0	90°
2	123	0	180°
3	116	0	0°
4	0	0	90°
5	0	0	90°
6	0	0	0°

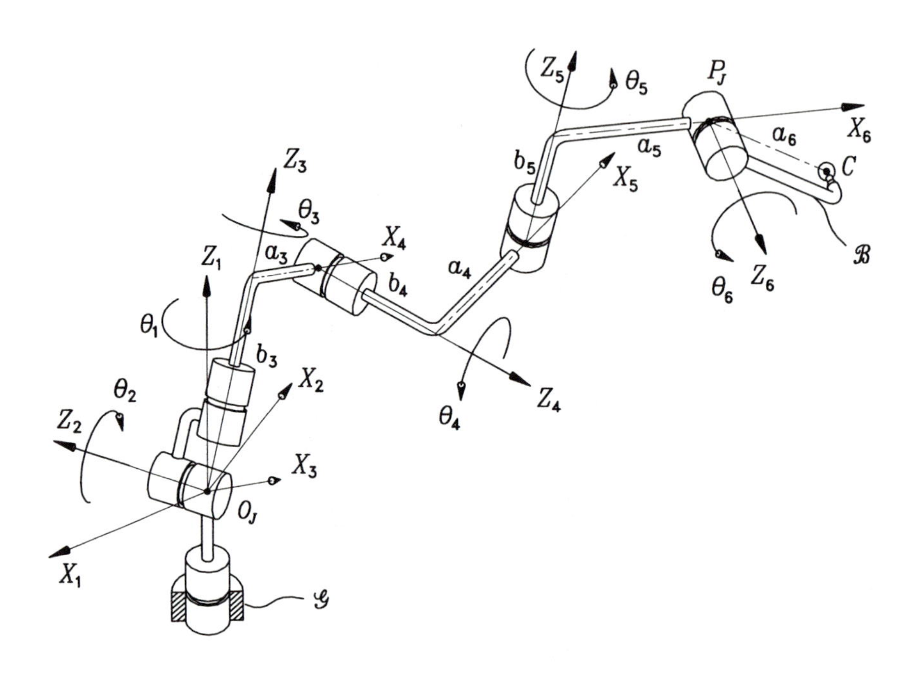

FIGURE 8.20. One of the legs of a walking machine with three actuated revolutes.

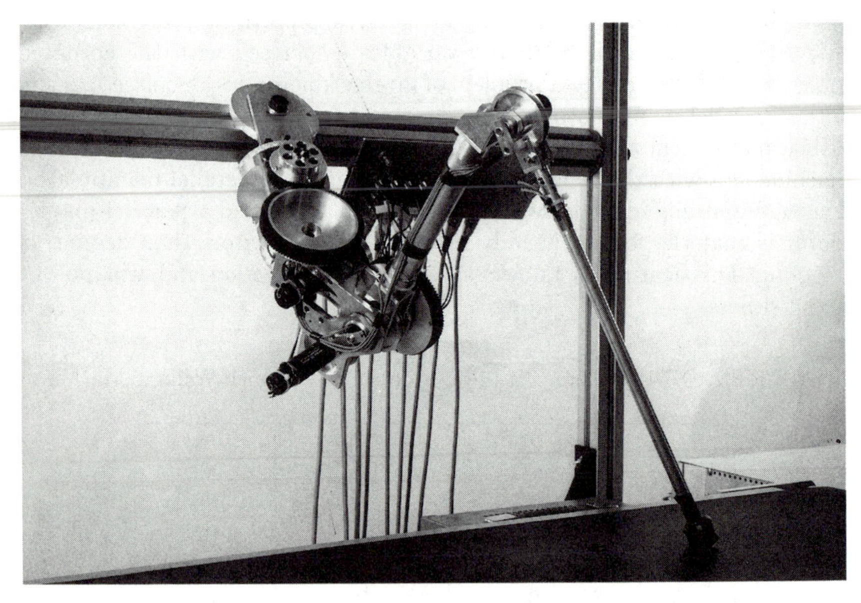

FIGURE 8.21. One of the six identical legs of the TU Munich Hexapod (Courtesy of Prof. F. Pfeiffer. Reproduced with permission of TSI Enterprises, Inc.)

8.6 Rolling Robots

While rolling robots are currently under development for autonomous operation on rough terrain, we focus here on the simplest ones, i.e., robots meant for tasks on horizontal surfaces, and so, their platforms undergo planar motion, which greatly simplifies their kinematics.

Rolling robots are basically of two kinds, depending on whether they are supplied with conventional or with omnidirectional wheels. Robots with conventional wheels are capable only of 2-dof motions, and hence, are kinematically equivalent to conventional terrestrial vehicles. However, robots with omnidirectional wheels (ODWs) are capable of 3-dof motions, which increases substantially their maneuverability. Below we outline the kinematics of the two kinds of robots.

8.6.1 Robots with Conventional Wheels

We begin with robots rolling on conventional wheels. Since these have two degrees of freedom, they need only two actuators, the various designs available varying essentially in where these actuators are located. The basic architecture of this kind of robot is displayed in Fig. 8.22a, in which we distinguish a chassis, or robot body, depicted as a triangular plate in that figure: two coaxial wheels that are coupled to the chassis by means of revolutes of axes passing through points O_1 and O_2; and a third wheel mounted on a bracket.

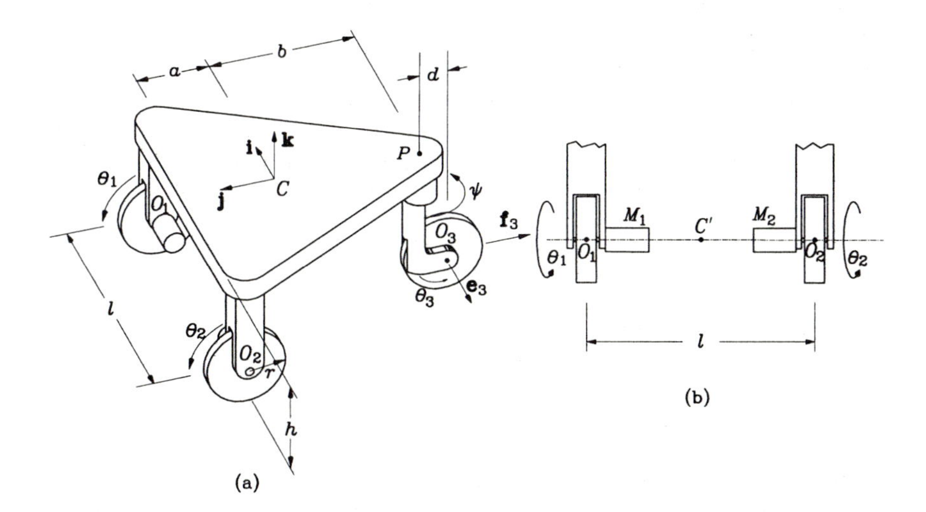

FIGURE 8.22. A 2-dof rolling robot: (a) its general layout; and (b) a detail of its actuated wheels.

Now, the two actuators can be placed in two essentially different arrays. In the first array, not shown in the figure, one actuator is used for propulsion and the other for steering, the former being used to provide locomotion power to the common two-wheel axle via a differential gear train. This train is required to allow for different angular velocities of the two coaxial wheels. Moreover, the orientation of the mid-plane of the steering wheel, defined by angle ψ, is controlled with the second actuator. This design has some drawbacks, namely, (*i*) the two motors serving two essentially different tasks call for essentially different operational characteristics, to the point that both may not be available from the same manufacturer; (*ii*) the power motor calls for velocity control, the steering motor for position control, thereby giving rise to two independent control systems that may end up by operating in an uncoordinated fashion; and finally, (*iii*) the use of a differential gear train increases costs, weight, and brings about the inherent backlash of gears.

In the second actuation array, shown in Fig. 8.22, the two coaxial wheels are powered independently, thereby doing away with the differential train and its undesirable side effects, the third wheel becoming a caster wheel. Moreover, the orientation of the latter is determined by friction and inertia forces, thereby making unnecessary the steering control system of the first array. Below we analyze the kinematics of a robot with this form of actuation.

Let point C of the platform be the *operation point*, its position vector in a frame fixed to the ground being denoted by $\mathbf{c}$. Additionally, let ω be the scalar angular velocity of the platform about a vertical axis. By virtue of the 2-dof motion of this robot, we can control either the velocity $\dot{\mathbf{c}}$ of C or a combination of ω and a scalar function of $\dot{\mathbf{c}}$ by properly specifying the two joint rates $\dot{\theta}_1$ and $\dot{\theta}_2$. However, we cannot control the two components of $\dot{\mathbf{c}}$ and ω simultaneously.

In order to proceed with the kinematic analysis of the system at hand, we define an orthonormal triad of vectors whose orientation is fixed with respect to the chassis. Let this triad be denoted by $\{\mathbf{i}, \mathbf{j}, \mathbf{k}\}$, with $\mathbf{k}$ pointing in the upward vertical direction. Thus, the velocities $\dot{\mathbf{o}}_i$ of points O_i, for $i = 1, 2$, are given by

$$\dot{\mathbf{o}}_i = r\dot{\theta}_i \mathbf{j}, \quad i = 1, 2 \tag{8.86a}$$

and moreover, the angular velocity ω of line O_1O_2 in planar motion, which is the same as that of the platform, can be readily expressed as

$$\omega = \frac{r}{l}(\dot{\theta}_1 - \dot{\theta}_2) \tag{8.86b}$$

its positive direction being that of $\mathbf{k}$.

Furthermore, the velocity of C can now be written in 2-dimensional form as

$$\dot{\mathbf{c}} = \dot{\mathbf{o}}_1 + \omega \mathbf{E}(\mathbf{c}' - \mathbf{o}_1) \tag{8.86c}$$

with $\mathbf{c}'$ denoting the position vector of point C', the orthogonal projection of C onto the horizontal plane of O_1 and O_2, while $\mathbf{E}$ is as defined in eq.(4.102). Thus, all vectors of eq.(8.86c) are 2-dimensional. Upon substitution of eqs.(8.86a & b) into eq.(8.86c), we obtain an expression for $\dot{\mathbf{c}}$ in terms of the joint rates, namely,

$$\dot{\mathbf{c}} = a\frac{r}{l}(\dot{\theta}_1 - \dot{\theta}_2)\mathbf{i} + \frac{r}{2}(\dot{\theta}_1 + \dot{\theta}_2)\mathbf{j} \tag{8.86d}$$

Equations (8.86b & d) express now the differential direct kinematics relations of the robot under study. In compact form, these relations are expressed as

$$\mathbf{t} = \mathbf{L}\dot{\boldsymbol{\theta}}_a \tag{8.86e}$$

with the 3×2 matrix $\mathbf{L}$ defined as

$$\mathbf{L} \equiv \begin{bmatrix} r/l & -r/l \\ (ar/l)\mathbf{i} + (r/2)\mathbf{j} & -(ar/l)\mathbf{i} + (r/2)\mathbf{j} \end{bmatrix} \tag{8.86f}$$

Moreover, the *planar twist* $\mathbf{t}$ of the platform and the 2-dimensional vector $\dot{\boldsymbol{\theta}}_a$ of actuated joint rates are defined as

$$\mathbf{t} \equiv \begin{bmatrix} \omega \\ \dot{\mathbf{c}} \end{bmatrix}, \quad \dot{\boldsymbol{\theta}}_a \equiv \begin{bmatrix} \dot{\theta}_1 \\ \dot{\theta}_2 \end{bmatrix} \tag{8.86g}$$

Computing the joint rates from the foregoing equations, i.e., solving the associated inverse kinematics problem, is now a trivial task. The inverse kinematics relations are computed below by noticing that eq.(8.86b) provides an equation for the joint-rate difference. Thus, all we need now is a second equation for the joint-rate sum. By inspection of eq.(8.86d), it is apparent that we can derive this relation by dot-multiplying both sides of this equation by $\mathbf{j}$, thereby obtaining

$$\dot{\mathbf{c}} \cdot \mathbf{j} = \frac{r}{2}(\dot{\theta}_1 + \dot{\theta}_2) \tag{8.87}$$

The two equations (8.86b) and (8.87) can now be cast into the usual form

$$\mathbf{J}\dot{\boldsymbol{\theta}}_a = \mathbf{K}\mathbf{t} \tag{8.88a}$$

where the two robot Jacobians $\mathbf{J}$ and $\mathbf{K}$ are given below:

$$\mathbf{J} \equiv \begin{bmatrix} 1 & -1 \\ 1 & 1 \end{bmatrix}, \quad \mathbf{K} \equiv \begin{bmatrix} (l/r) & \mathbf{0}^T \\ 0 & (2/r)\mathbf{j}^T \end{bmatrix} \tag{8.88b}$$

Note that $\mathbf{J}$ is a 2×2 matrix, but $\mathbf{K}$ is a 2×3 matrix.

The inverse kinematics relations are readily derived from eq.(8.88a), namely,

$$\dot{\theta}_1 = \frac{1}{2}\left(\frac{l}{r}\omega + \frac{2}{r}\dot{y}\right)$$

$$\dot{\theta}_2 = -\frac{1}{2}\left(\frac{l}{r}\omega - \frac{2}{r}\dot{y}\right)$$

where $\dot{y} \equiv \dot{\mathbf{c}} \cdot \mathbf{j}$.

Now, in order to complete the kinematic analysis of the robot at hand, we calculate the rates of the unactuated joints, $\dot{\theta}_3$ and $\dot{\psi}$. To this end, let $\boldsymbol{\omega}_i$, for $i = 1, \ldots, 3$, and $\dot{\mathbf{o}}_3$ denote the 3-dimensional angular velocity vector of the ith wheel and the 3-dimensional velocity vector of the center of the caster wheel. Likewise, ω_4 denotes the scalar angular velocity of the bracket.

We thus have, for the angular velocity vectors of the two actuated wheels,

$$\boldsymbol{\omega}_1 = -\dot{\theta}_1\mathbf{i} + \omega\mathbf{k} = -\dot{\theta}_1\mathbf{i} + \frac{r}{l}(\dot{\theta}_1 - \dot{\theta}_2)\mathbf{k} = \begin{bmatrix} -\mathbf{i} + (r/l)\mathbf{k} & -(r/l)\mathbf{k} \end{bmatrix} \begin{bmatrix} \dot{\theta}_1 \\ \dot{\theta}_2 \end{bmatrix}$$

$$\boldsymbol{\omega}_2 = -\dot{\theta}_2\mathbf{i} + \omega\mathbf{k} = -\dot{\theta}_2\mathbf{i} + \frac{r}{l}(\dot{\theta}_1 - \dot{\theta}_2)\mathbf{k} = \begin{bmatrix} (r/l)\mathbf{k} & -\mathbf{i} - (r/l)\mathbf{k} \end{bmatrix} \begin{bmatrix} \dot{\theta}_1 \\ \dot{\theta}_2 \end{bmatrix}$$

In the ensuing derivations, we will need the velocities of the centers of the two actuated wheels, which are given by

$$\dot{\mathbf{o}}_1 = \boldsymbol{\omega}_1 \times r\mathbf{k} = r\dot{\theta}_1\mathbf{j}$$
$$\dot{\mathbf{o}}_2 = \boldsymbol{\omega}_2 \times r\mathbf{k} = r\dot{\theta}_2\mathbf{j}$$

Moreover, the angular velocity of the caster wheel can be written most easily in the frame fixed to the bracket, $\{\mathbf{e}_3, \mathbf{f}_3, \mathbf{k}\}$, namely,

$$\boldsymbol{\omega}_3 = \dot{\theta}_3\mathbf{e}_3 + (\omega + \dot{\psi})\mathbf{k} \tag{8.89}$$

with ψ denoting the angle between vectors $\mathbf{j}$ and $\mathbf{e}_3$ of Fig. 10.6a, measured in the positive direction of $\mathbf{k}$, as indicated in the layout of Fig. 8.23.

Note that vector $\mathbf{e}_3$ is parallel to the axis of rolling of the caster wheel, while $\mathbf{f}_3$ is a horizontal vector perpendicular to $\mathbf{e}_3$. These two sets of unit vectors are related by

$$\mathbf{e}_3 = -\sin\psi\mathbf{i} + \cos\psi\mathbf{j} \tag{8.90a}$$
$$\mathbf{f}_3 = -\cos\psi\mathbf{i} - \sin\psi\mathbf{j} \tag{8.90b}$$

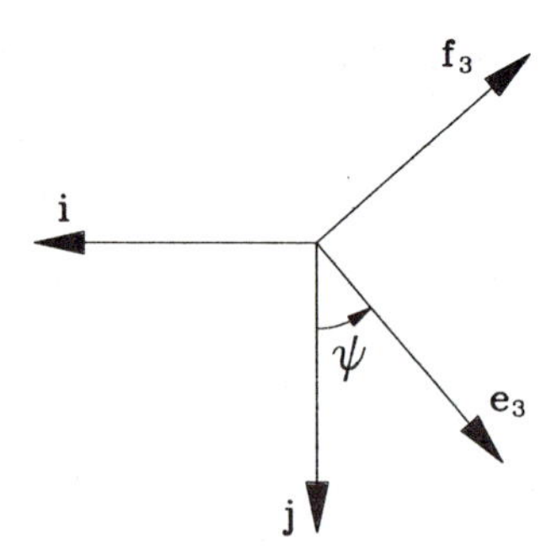

FIGURE 8.23. Layout of the unit vectors fixed to the platform and to the bracket.

their inverse relations being

$$\mathbf{i} = -\sin\psi\mathbf{e}_3 - \cos\psi\mathbf{f}_3 \tag{8.90c}$$

$$\mathbf{j} = \cos\psi\mathbf{e}_3 - \sin\psi\mathbf{f}_3 \tag{8.90d}$$

Furthermore, the velocity of the center of the caster wheel is derived as

$$\dot{\mathbf{o}}_3 = \boldsymbol{\omega}_3 \times r\mathbf{k} = -r\dot{\theta}_3\mathbf{f}_3$$

while the scalar angular velocity of the bracket, ω_4, is given by

$$\omega_4 = \omega + \dot{\psi} = \frac{r}{l}(\dot{\theta}_1 - \dot{\theta}_2) + \dot{\psi} \tag{8.91}$$

In the sequel, we shall need $\dot{\mathbf{c}}$ in bracket coordinates. Such an expression is obtained from eqs.(8.86d) and (8.90c & d), namely,

$$\begin{aligned}\dot{\mathbf{c}} = &[-a\frac{r}{l}(\dot{\theta}_1 - \dot{\theta}_2)\sin\psi + \frac{r}{2}(\dot{\theta}_1 + \dot{\theta}_2)\cos\psi]\mathbf{e}_3 \\ &-[a\frac{r}{l}(\dot{\theta}_1 - \dot{\theta}_2)\cos\psi + \frac{r}{2}(\dot{\theta}_1 + \dot{\theta}_2)\sin\psi]\mathbf{f}_3\end{aligned} \tag{8.92}$$

Expressions for the dependent rates in terms of the independent ones, $\dot{\theta}_1$ and $\dot{\theta}_2$, are readily derived. To this end, we express the velocity of P in two independent forms, one in terms of the velocity of O_3 and the other in terms of the velocity of C, i.e.,

$$\dot{\mathbf{p}} = \dot{\mathbf{o}}_3 + \omega_4\mathbf{k} \times (\mathbf{p} - \mathbf{o}_3) \tag{8.93a}$$

$$\dot{\mathbf{p}} = \dot{\mathbf{c}} + \omega\mathbf{k} \times (-b\mathbf{j}) \tag{8.93b}$$

Upon equating the right-hand sides of the above equations, we obtain a 3-dimensional vector equation relating dependent with independent rates, namely,

$$-r\dot{\theta}_3\mathbf{f}_3 + (\omega + \dot{\psi})\mathbf{k} \times (\mathbf{p} - \mathbf{o}_3) = \dot{\mathbf{c}} + b\omega\mathbf{i}$$

where we have recalled the expressions derived above for $\dot{\mathbf{o}}_3$ and ω_4. Further, we rewrite the foregoing equation with the unknown rates, $\dot{\theta}_3$ and $\dot{\psi}$, on the left-hand side, i.e.,

$$-r\dot{\theta}_3\mathbf{f}_3 + \dot{\psi}\mathbf{k} \times (\mathbf{p} - \mathbf{o}_3) = \dot{\mathbf{c}} + b\omega\mathbf{i} - \omega\mathbf{k} \times (\mathbf{p} - \mathbf{o}_3) \tag{8.94}$$

Moreover, we note that, from Fig. 8.22,

$$\mathbf{p} - \mathbf{o}_3 = -d\mathbf{f}_3 + (h - r)\mathbf{k}$$

and hence,

$$\mathbf{k} \times (\mathbf{p} - \mathbf{o}_3) = d\mathbf{e}_3$$

equation (8.94) thus becoming

$$-r\dot{\theta}_3\mathbf{f}_3 + \dot{\psi}d\mathbf{e}_3 = \dot{\mathbf{c}} + \omega(b\mathbf{i} - d\mathbf{e}_3) \tag{8.95}$$

Now it is a simple matter to solve for $\dot\theta_3$ and $\dot\psi$ from eq.(8.95). Indeed, we solve for $\dot\theta_3$ by dot-multiplying both sides of the above equation by $\mathbf{f}_3$. Likewise, we solve for $\dot\psi$ by dot-multiplying both sides of the same equation by $\mathbf{e}_3$, thus obtaining

$$\begin{aligned} -r\dot\theta_3 &= \dot{\mathbf{c}}\cdot\mathbf{f}_3 + \omega b\mathbf{i}\cdot\mathbf{f}_3 \\ d\dot\psi &= \dot{\mathbf{c}}\cdot\mathbf{e}_3 + \omega(b\mathbf{i}\cdot\mathbf{e}_3 - d) \end{aligned}$$

Now, by recalling the expressions derived above for ω and $\dot{\mathbf{c}}$, we obtain

$$\begin{aligned} \dot{\mathbf{c}}\cdot\mathbf{f}_3 &= -a\frac{r}{l}(\dot\theta_1-\dot\theta_2)\cos\psi - \frac{r}{2}(\dot\theta_1+\dot\theta_2)\sin\psi \\ \dot{\mathbf{c}}\cdot\mathbf{e}_3 &= -a\frac{r}{l}(\dot\theta_1-\dot\theta_2)\sin\psi + \frac{r}{2}(\dot\theta_1+\dot\theta_2)\cos\psi \\ \mathbf{i}\cdot\mathbf{f}_3 &= -\cos\psi, \quad \mathbf{i}\cdot\mathbf{e}_3 = -\sin\psi \end{aligned}$$

Therefore,

$$\dot\theta_3 = \alpha\cos\psi(\dot\theta_1-\dot\theta_2) + \frac{1}{2}(\sin\psi)(\dot\theta_1+\dot\theta_2) \tag{8.96a}$$

$$\dot\psi = \rho\left[-(\alpha\sin\psi+\delta)(\dot\theta_1-\dot\theta_2) + \frac{1}{2}(\cos\psi)(\dot\theta_1+\dot\theta_2)\right] \tag{8.96b}$$

with the definitions given below:

$$\alpha \equiv \frac{a+b}{l}, \quad \delta \equiv \frac{d}{l}, \quad \rho \equiv \frac{r}{d} \tag{8.97}$$

Hence, if we let $\dot{\boldsymbol{\theta}}_u$ be the vector of *unactuated joint rates*, $\dot{\boldsymbol{\theta}}_u \equiv [\dot\theta_3 \quad \dot\psi]^T$, then we have

$$\dot{\boldsymbol{\theta}}_u = \boldsymbol{\Theta}\dot{\boldsymbol{\theta}}_a \tag{8.98a}$$

with $\boldsymbol{\Theta}$ defined as

$$\boldsymbol{\Theta} \equiv \begin{bmatrix} \alpha\cos\psi + (\sin\psi)/2 & -\alpha\cos\psi + (\sin\psi)/2 \\ \rho[-\alpha\sin\psi + (\cos\psi)/2 - \delta] & \rho[\alpha\sin\psi + (\cos\psi)/2 + \delta] \end{bmatrix} \tag{8.98b}$$

thereby completing the intended kinematic analysis.

8.6.2 Robots with Omnidirectional Wheels

In general, omnidirectional wheels (ODWs) allow for two independent translational motions on the supporting floor and one independent rotational motion about a vertical axis. Based on the shapes of the wheels, moreover, ODWs can be classified into spherical wheels and *Mekanum* wheels, the latter also being known as *ilonators*. We focus here on ODWs of the Mekanum type and assume that the robot of interest is equipped with n of these.

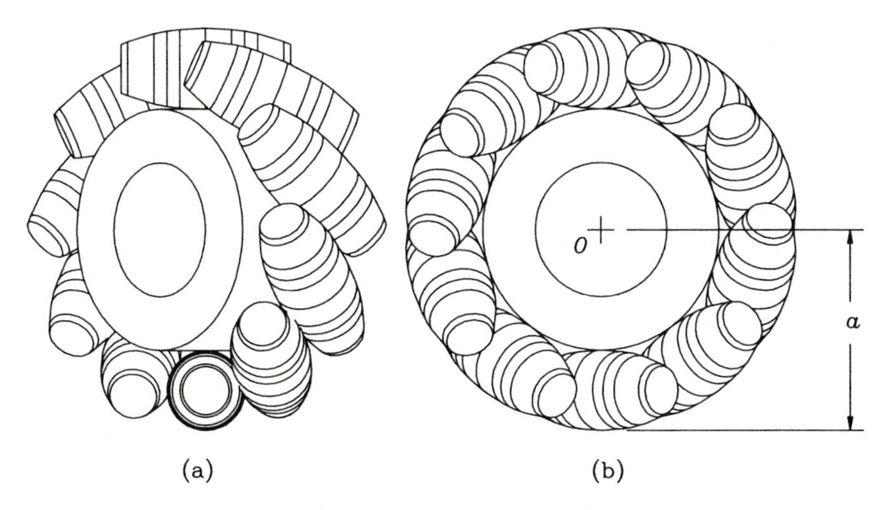

FIGURE 8.24. (a) The Mekanum Wheel; (b) its side view.

The Mekanum wheel bears a set of rollers mounted along the periphery of the wheel hub at a given angle, as shown in Fig. 8.24a. Furthermore, the rollers are shaped so that the wheel appears as circular on its side view, as shown in Fig. 8.24b, in order to ensure a smooth motion. Pairwise orthogonal unit vectors $\mathbf{e}_i$, $\mathbf{f}_i$ and $\mathbf{g}_i$, $\mathbf{h}_i$ are defined on the mid-planes of the wheel hub and on those of the roller in contact with the floor, respectively. Note that the roller in contact with the floor is termed *active* in the discussion below. Now we take to finding the kinematic relation between the wheel joint rates $\{\dot{\theta}_i\}$ and the Cartesian velocity variables of the robot, namely, the scalar angular velocity ω and the 2-dimensional velocity vector $\dot{\mathbf{c}}$ of the mass center of the platform. To this end, we express the velocity $\dot{\mathbf{o}}_i$ of the centroid O_i of the ith wheel in two different forms: first we look at this velocity from the active roller up to the centroid O_i; then from the mass center C of the platform to O_i.

If we relate the velocity of O_i with that of the contact point of the active roller with the ground, with the aid of Fig. 8.25 we can then write

$$\dot{\mathbf{o}}_i = \dot{\mathbf{p}}_i + \mathbf{v}_i \tag{8.99}$$

with $\mathbf{v}_i$ defined as the relative velocity of O_i with respect to P_i. Now let $\boldsymbol{\omega}_h$ and $\boldsymbol{\omega}_r$ denote the angular velocity vectors of the hub and the roller, respectively, i.e.,

$$\boldsymbol{\omega}_h = \omega\mathbf{k} + \dot{\theta}_i\mathbf{e}_i, \quad \boldsymbol{\omega}_r = \boldsymbol{\omega}_h + \dot{\phi}_i\mathbf{g}_i$$

We thus have

$$\dot{\mathbf{p}}_i = \boldsymbol{\omega}_r \times \overrightarrow{Q_iP_i} = (\omega\mathbf{k} + \dot{\theta}_i\mathbf{e}_i + \dot{\phi}_i\mathbf{g}_i) \times b\mathbf{k}$$

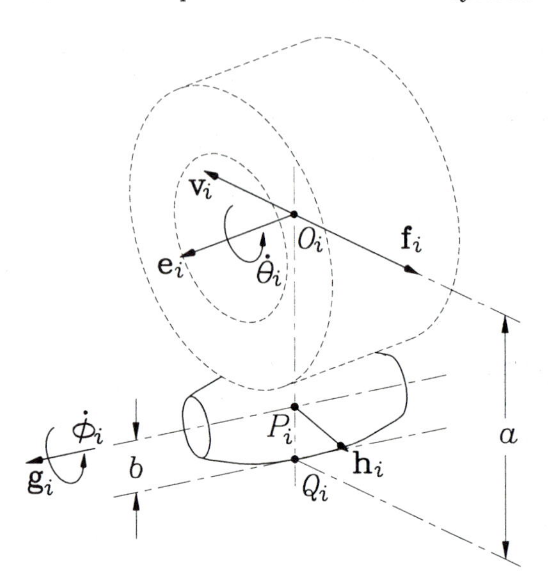

FIGURE 8.25. The active roller of the ith wheel.

where a and b are the height of the axis of the wheel hub, as shown in Fig. 8.24b, and the radius of the rollers, respectively. In addition, $\dot{\theta}_i$ denotes the rate of the wheel hub, while $\dot{\phi}_i$ denotes that of the active roller, which are positive in the directions of vectors $\mathbf{e}_i$ and $\mathbf{g}_i$, respectively. Hence,

$$\dot{\mathbf{p}}_i = -b(\dot{\theta}_i \mathbf{f}_i + \dot{\phi}_i \mathbf{h}_i) \tag{8.100}$$

Moreover,

$$\mathbf{v}_i = \boldsymbol{\omega}_h \times \overrightarrow{P_i O_i} = (\omega \mathbf{k} + \dot{\theta}_i \mathbf{e}_i) \times (a - b)\mathbf{k}$$

so that

$$\mathbf{v}_i = -\dot{\theta}_i (a - b)\mathbf{f}_i \tag{8.101}$$

thereby obtaining the desired expression for $\dot{\mathbf{o}}_i$, namely,

$$\dot{\mathbf{o}}_i = -a\dot{\theta}_i \mathbf{f}_i - b\dot{\phi}_i \mathbf{h}_i \tag{8.102}$$

A general layout of the ith ODW with roller axes at an angle α_i with respect to the normal $\mathbf{e}_i$ to the midplane of the corresponding hub is shown in Fig. 8.26. The subscript i is associated with both the ith wheel and its active roller. Moreover, the velocity of the ith wheel, $\dot{\mathbf{o}}_i$, can be expressed in terms of the Cartesian velocity variables, $\dot{\mathbf{c}}$ and ω, as

$$\dot{\mathbf{o}}_i = \dot{\mathbf{c}} + \omega \mathbf{E} \mathbf{d}_i \tag{8.103}$$

where we have used a 2-dimensional vector representation, with $\mathbf{d}_i$ defined as the vector directed from point C to the centroid O_i of the hub and $\mathbf{E}$ defined as in eq.(4.102). Furthermore, since all rollers are unactuated and

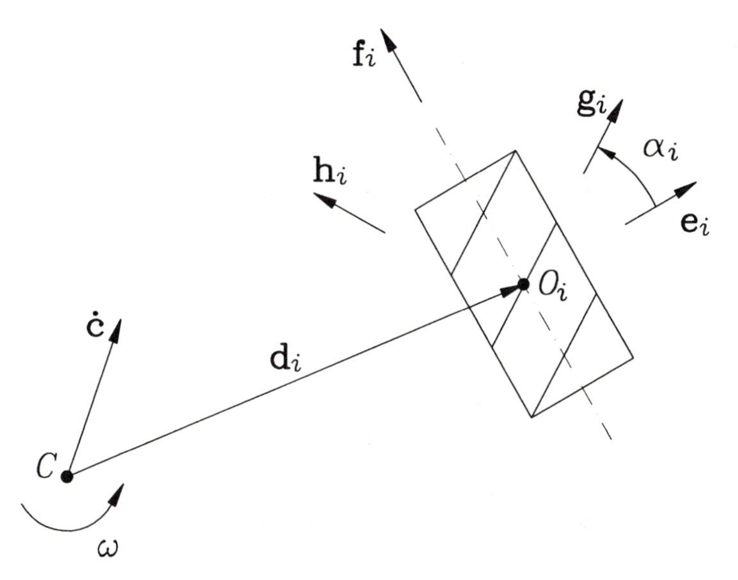

FIGURE 8.26. The layout of the ith wheel with respect to the robot platform.

they rotate idly, the value of $\dot{\phi}_i$ is immaterial to our study. Hence, we take to eliminating this variable from the foregoing equations, which is done by dot-multiplying both sides of eq.(8.102) with $\mathbf{g}_i$, normal to $\mathbf{h}_i$, thereby deriving

$$\mathbf{g}_i^T \dot{\mathbf{o}}_i = -a\dot{\theta}_i \mathbf{g}_i^T \mathbf{f}_i$$

But

$$\mathbf{g}_i^T \mathbf{f}_i = \sin \alpha_i$$

Therefore,

$$\mathbf{g}_i^T \dot{\mathbf{o}}_i = -a(\sin \alpha_i)\dot{\theta}_i \tag{8.104}$$

The same multiplication performed on eq.(8.103) yields

$$\mathbf{g}_i^T \dot{\mathbf{o}}_i = (\mathbf{g}_i^T \mathbf{E} \mathbf{d}_i)\omega + \mathbf{g}_i^T \dot{\mathbf{c}} \tag{8.105}$$

Upon equating the right-hand sides of eqs.(8.104) and (8.105), we derive the desired relation, namely,

$$-a(\sin \alpha_i)\dot{\theta}_i = \mathbf{k}_i^T \mathbf{t}, \quad i = 1, \ldots, n \tag{8.106}$$

where the 3-dimensional vector $\mathbf{k}_i$ is defined as

$$\mathbf{k}_i = \begin{bmatrix} \mathbf{g}_i^T \mathbf{E} \mathbf{d}_i \\ \mathbf{g}_i \end{bmatrix}$$

and the twist vector $\mathbf{t}$ is as defined in eq.(8.86g). We now define the vector of *wheel rates* $\dot{\boldsymbol{\theta}}$ in the form

$$\dot{\boldsymbol{\theta}} \equiv [\,\dot{\theta}_1 \quad \dot{\theta}_2 \quad \cdots \quad \dot{\theta}_n\,] \tag{8.107}$$

If the n equations of eq.(8.106) are now assembled, we obtain

$$\mathbf{J}\dot{\boldsymbol{\theta}} = \mathbf{K}\mathbf{t} \tag{8.108}$$

where if we assume that all angles α_i are identical and labeled α, the $n \times n$ Jacobian $\mathbf{J}$ and the $n \times 3$ Jacobian $\mathbf{K}$ take the forms

$$\mathbf{J} \equiv -a \sin\alpha \mathbf{1} \tag{8.109a}$$

$$\mathbf{K} \equiv \begin{bmatrix} \mathbf{g}_1^T \mathbf{E}\mathbf{d}_1 & \mathbf{g}_1^T \\ \vdots & \vdots \\ \mathbf{g}_n^T \mathbf{E}\mathbf{d}_n & \mathbf{g}_n^T \end{bmatrix} \tag{8.109b}$$

with $\mathbf{1}$ denoting the $n \times n$ identity matrix.

Given eqs.(8.109a) and (8.109b), the differential inverse kinematics can be resolved as

$$\dot{\boldsymbol{\theta}} = -\frac{1}{a \sin\alpha}\mathbf{K}\mathbf{t} \tag{8.110}$$

whence it is apparent that $\sin\alpha$ must be different from zero, i.e., the axes of the rollers must not be parallel to the axis of the hub. If these axes are parallel, then the ODWs reduce to conventional wheels.

On the other hand, the twist can be obtained from eq.(8.108), for $n = 3$, as

$$\mathbf{t} = \mathbf{K}^{-1}\mathbf{J}\dot{\boldsymbol{\theta}} \tag{8.111}$$

where $\mathbf{K}^{-1}$ can be found in closed form as

$$\mathbf{K}^{-1} = \frac{1}{\Delta}\begin{bmatrix} \mathbf{g}_3^T \mathbf{E}\mathbf{g}_2 & \mathbf{g}_1^T \mathbf{E}\mathbf{g}_3 & \mathbf{g}_2^T \mathbf{E}\mathbf{g}_1 \\ \mathbf{E}(r_2\mathbf{g}_3 - r_3\mathbf{g}_2) & \mathbf{E}(r_3\mathbf{g}_1 - r_1\mathbf{g}_3) & \mathbf{E}(r_1\mathbf{g}_2 - r_2\mathbf{g}_1) \end{bmatrix} \tag{8.112a}$$

with Δ and $\{ r_i \}_1^3$ defined as

$$\Delta \equiv \det(\mathbf{K}) = r_1\mathbf{g}_3^T\mathbf{E}\mathbf{g}_2 + r_2\mathbf{g}_1^T\mathbf{E}\mathbf{g}_2 + r_3\mathbf{g}_2^T\mathbf{E}\mathbf{g}_1 \tag{8.112b}$$

$$r_i \equiv \mathbf{g}_i^T\mathbf{E}\mathbf{d}_i, \quad i = 1, 2, 3 \tag{8.112c}$$

which thus completes the kinematic analysis of the system at hand.

9

Trajectory Planning: Continuous-Path Operations

9.1 Introduction

As a follow-up to Chapter 5, where we studied trajectory planning for pick-and-place operations (PPO), we study in this chapter continuous-path operations. In PPO, the pose, twist, and twist-rate of the EE are specified only at the two ends of the trajectory, the purpose of trajectory planning then being to blend the two end poses with a smooth motion. When this blending is done in the joint-variable space, the problem is straightforward, as demonstrated in Chapter 5. There are instances in which the blending must be made in the Cartesian-variable space, in which advanced notions of interpolation in what is known as the *image space* of spatial displacements, as introduced by Ravani and Roth (1984), are needed. The image space of spatial displacements is a projective space with three *dual* dimensions, which means that a point of this space is specified by four coordinates—similar to the homogeneous coordinates introduced in Section 2.5—of the form $x_i + \epsilon\xi_i$, for $i = 1, 2, 3, 4$, where ϵ is the *dual unity*, which has the property that $\epsilon^2 = 0$. The foregoing coordinates are thus dual numbers, their purpose being to represent both rotation and translation in one single quantity. In following Ravani and Roth's work, Ge and Kang (1995) proposed an interpolation scheme that produces curves in the image space with second-order geometric continuity, which are referred to as G^2 curves.

These interpolation techniques lie beyond the scope of the book and will be left aside.

The purpose of this chapter is to develop motion interpolation techniques in Cartesian space that produce smooth motions in both Cartesian and joint spaces. Motion interpolation in joint space was discussed in Chapter 5, the present chapter being devoted to motion interpolation in Cartesian space. To this end, we resort to basic notions of differential geometry.

9.2 Curve Geometry

Continuous-path robotics applications appear in operations such as arc-welding, flame-cutting, deburring, and routing. In these operations, a tool is rigidly attached to the end-effector of a robotic manipulator, the tool being meant to trace a continuous and smooth trajectory in a 6-dimensional configuration space. Three dimensions of this space describe the spatial path followed by the *operation point* of the EE, while the remaining three describe the orientation of the EE. Some applications require that this task take place along a warped curve, such as those encountered at the intersections of warped surfaces, while the path is to be traversed as a prescribed function of time. This function, moreover, is task-dependent; e.g., in arc-welding, the electrode must traverse the path at a constant speed, if no compensation for gravity is taken into account. If gravity compensation is warranted, then the speed varies with the orientation of the path with respect to the vertical. Below we will define this orientation as that of the *Frenet-Serret frame* associated with every point of the path where the path is smooth.

Moreover, for functional reasons, the orientation of the EE is given as a rotation matrix that is, in turn, a prescribed smooth function of time. In arc-welding, for example, the orientation of the electrode with respect to the curve must be constant. The trajectory planning of the configuration subspace associated with the warped path is more or less straightforward, but the planning of the trajectory associated with the orientation subspace is less so.

While most methods of trajectory planning at the Cartesian-coordinate level focus on the path followed by the operation point, the underlying inverse kinematics of a six-axis robotic manipulator requires the specification of the orientation of the EE as well. In the presence of simple manipulators with a spherical wrist, as those studied in Subsection 4.4.2, the positioning and the orientation tasks are readily separable, and hence, the planning of the two tasks can be done one at a time. In other instances, e.g., in most arc-welding robots, such a separation is not possible, and both tasks must be planned concurrently, which is the focus of our discussion below. Here, we follow the technique presented in (Angeles, Rojas, and López-Cajún, 1988).

Crucial to our discussion is the concept of *path orientation*. Let Γ be a warped curve in 3-dimensional space that is smooth in a certain interval of interest for our discussion. Under these conditions, we can associate with every point of this interval an orthonormal triad of vectors, i.e., a set of unit vectors that are mutually orthogonal, namely, the *tangent*, the *normal*, and the *binormal* vectors of Γ. Therefore, when this set of vectors is properly arranged in a 3×3 array, a rotation is obtained. This matrix thus represents the *orientation* of Γ. In order to characterize these vectors, let s be the arc length measured along Γ from a certain reference point on this curve. Below we review the basic differential-geometric concepts pertaining to our discussion.

The tangent, normal, and binormal unit vectors, $\mathbf{e}_t$, $\mathbf{e}_n$, and $\mathbf{e}_b$, respectively, associated with every point of Γ where this curve is smooth, are generically termed here the *Frenet-Serret vectors*. These vectors are defined as

$$\mathbf{e}_t \equiv \mathbf{r}' \tag{9.1a}$$

$$\mathbf{e}_b \equiv \frac{\mathbf{r}' \times \mathbf{r}''}{\|\mathbf{r}' \times \mathbf{r}''\|} \tag{9.1b}$$

$$\mathbf{e}_n \equiv \mathbf{e}_b \times \mathbf{e}_t \tag{9.1c}$$

where $\mathbf{r}'$ stands for $d\mathbf{r}/ds$ and $\mathbf{r}''$ for $d^2\mathbf{r}/ds^2$. Now the Frenet-Serret relations among the three aforementioned unit vectors and the curvature κ and torsion τ of Γ are recalled (Brand, 1965):

$$\frac{d\mathbf{e}_t}{ds} = \kappa \mathbf{e}_n \tag{9.2a}$$

$$\frac{d\mathbf{e}_n}{ds} = -\kappa \mathbf{e}_t + \tau \mathbf{e}_b \tag{9.2b}$$

$$\frac{d\mathbf{e}_b}{ds} = -\tau \mathbf{e}_n \tag{9.2c}$$

Moreover, the curvature and torsion can be calculated with the aid of the formulas

$$\kappa = \|\mathbf{r}' \times \mathbf{r}''\| \tag{9.3a}$$

$$\tau = \frac{\mathbf{r}' \times \mathbf{r}'' \cdot \mathbf{r}'''}{\kappa^2} \tag{9.3b}$$

where $\mathbf{r}'''$ stands for $d^3\mathbf{r}/ds^3$. Furthermore, differentiation of κ and τ, as given above, with respect to s yields

$$\kappa'(s) = (\mathbf{r}' \times \mathbf{r}'') \cdot \frac{(\mathbf{r}' \times \mathbf{r}''')}{\kappa} \tag{9.4a}$$

$$\tau'(s) = \frac{\mathbf{r}' \times \mathbf{r}'' \cdot \mathbf{r}^{(iv)} - 2\tau(\mathbf{r}' \times \mathbf{r}'') \cdot (\mathbf{r}' \times \mathbf{r}''')}{\kappa^2} \tag{9.4b}$$

where $\mathbf{r}^{(iv)}$ stands for $d^4\mathbf{r}/ds^4$. The geometric interpretation of the curvature is the rate of change of orientation of the tangent vector with respect to the arc length; that of the torsion is the rate at which the curve quits the plane of the tangent and normal vectors. Thus, at segments where the curve is straight, the curvature vanishes, whereas at segments where the curve is planar, the torsion vanishes. Now, from the Frenet-Serret formulas and the *chain rule*, we can derive the time-rate of change of the Frenet-Serret vectors, namely,

$$\dot{\mathbf{e}}_t \equiv \frac{d\mathbf{e}_t}{ds}\dot{s} = \dot{s}\kappa\mathbf{e}_n \tag{9.5a}$$

$$\dot{\mathbf{e}}_n \equiv \frac{d\mathbf{e}_n}{ds}\dot{s} = -\dot{s}\kappa\mathbf{e}_t + \dot{s}\tau\mathbf{e}_b \tag{9.5b}$$

$$\dot{\mathbf{e}}_b \equiv \frac{d\mathbf{e}_b}{ds}\dot{s} = -\dot{s}\tau\mathbf{e}_n \tag{9.5c}$$

Furthermore, let $\boldsymbol{\omega}$ be the angular velocity of the Frenet-Serret frame. Then clearly,

$$\dot{\mathbf{e}}_t \equiv \boldsymbol{\omega} \times \mathbf{e}_t \tag{9.6a}$$

$$\dot{\mathbf{e}}_n \equiv \boldsymbol{\omega} \times \mathbf{e}_n \tag{9.6b}$$

$$\dot{\mathbf{e}}_b \equiv \boldsymbol{\omega} \times \mathbf{e}_b \tag{9.6c}$$

Upon equating pairwise the right-hand sides of eqs.(9.5a–c) and eqs.(9.6a–c), we obtain three vector equations determining $\boldsymbol{\omega}$, namely,

$$-\mathbf{E}_t\boldsymbol{\omega} = \dot{s}\kappa\mathbf{e}_n \tag{9.7a}$$

$$-\mathbf{E}_n\boldsymbol{\omega} = -\dot{s}\kappa\mathbf{e}_t + \dot{s}\tau\mathbf{e}_b \tag{9.7b}$$

$$-\mathbf{E}_b\boldsymbol{\omega} = -\dot{s}\tau\mathbf{e}_n \tag{9.7c}$$

where we have introduced the cross-product matrices $\mathbf{E}_t$, $\mathbf{E}_n$, and $\mathbf{E}_b$ of vectors $\mathbf{e}_t$, $\mathbf{e}_n$, and $\mathbf{e}_b$, respectively, thereby obtaining a system of nine scalar equations in three unknowns, namely, the three scalar components of $\boldsymbol{\omega}$, i.e.,

$$\mathbf{A}\boldsymbol{\omega} = \mathbf{b} \tag{9.8a}$$

with $\mathbf{A}$ defined as the 9×3 matrix and $\mathbf{b}$ as the 9-dimensional vector displayed below:

$$\mathbf{A} \equiv -\begin{bmatrix} \mathbf{E}_t \\ \mathbf{E}_n \\ \mathbf{E}_b \end{bmatrix}, \quad \mathbf{b} \equiv \begin{bmatrix} \dot{s}\kappa\mathbf{e}_n \\ \dot{s}(-\kappa\mathbf{e}_t + \tau\mathbf{e}_b) \\ -\dot{s}\tau\mathbf{e}_n \end{bmatrix} \tag{9.8b}$$

Although the foregoing system is overdetermined, it is consistent, and hence it comprises exactly three linearly independent equations, the remaining six being dependent on the former. One way to reduce system (9.8a) to only three equations consists in multiplying both sides of this equation by $\mathbf{A}^T$. Now, the product $\mathbf{A}^T\mathbf{A}$ greatly simplifies because matrix $\mathbf{A}$ turns out to be isotropic, as discussed in Section 4.9, i.e., its three columns are mutually orthogonal and all have the same magnitude. This fact can become apparent if we realize that the three 3×3 blocks of $\mathbf{A}$ are cross-product matrices of three orthonormal vectors. Thus,

$$\mathbf{A}^T\mathbf{A} = \mathbf{E}_t^T\mathbf{E}_t + \mathbf{E}_n^T\mathbf{E}_n + \mathbf{E}_b^T\mathbf{E}_b$$

If we now recall Theorem 2.3.4, the foregoing products take on quite simple forms, namely,

$$\begin{aligned}
\mathbf{E}_t^T\mathbf{E}_t &= -\mathbf{E}_t^2 = -(-\mathbf{1} + \mathbf{e}_t\mathbf{e}_t^T) \\
\mathbf{E}_n^T\mathbf{E}_n &= -\mathbf{E}_n^2 = -(-\mathbf{1} + \mathbf{e}_n\mathbf{e}_n^T) \\
\mathbf{E}_b^T\mathbf{E}_b &= -\mathbf{E}_b^2 = -(-\mathbf{1} + \mathbf{e}_b\mathbf{e}_b^T)
\end{aligned}$$

Moreover, for any 3-dimensional vector $\mathbf{v}$, we have

$$(\mathbf{e}_t\mathbf{e}_t^T + \mathbf{e}_n\mathbf{e}_n^T + \mathbf{e}_b\mathbf{e}_b^T)\mathbf{v} \equiv \mathbf{v}$$

and hence, the above sum in parentheses reduces to the identity matrix, i.e.,

$$\mathbf{e}_t\mathbf{e}_t^T + \mathbf{e}_n\mathbf{e}_n^T + \mathbf{e}_b\mathbf{e}_b^T \equiv \mathbf{1}$$

the product $\mathbf{A}^T\mathbf{A}$ thus reducing to

$$\mathbf{A}^T\mathbf{A} = (2)\mathbf{1}$$

Therefore, $\boldsymbol{\omega}$ takes on the form

$$\boldsymbol{\omega} = \frac{1}{2}[\mathbf{E}_t \quad \mathbf{E}_n \quad \mathbf{E}_b]\begin{bmatrix} \dot{s}\kappa\mathbf{e}_n \\ \dot{s}(-\kappa\mathbf{e}_t + \tau\mathbf{e}_b) \\ -\dot{s}\tau\mathbf{e}_n \end{bmatrix}$$

or upon expansion,

$$\boldsymbol{\omega} = \frac{\dot{s}}{2}[\kappa\mathbf{e}_t \times \mathbf{e}_n + \mathbf{e}_n(\tau\mathbf{e}_b - \kappa\mathbf{e}_t) - \tau\mathbf{e}_b \times \mathbf{e}_n] \tag{9.10}$$

However, since the Frenet-Serret triad is orthonormal, we have

$$\mathbf{e}_t \times \mathbf{e}_n = \mathbf{e}_b, \quad \mathbf{e}_n \times \mathbf{e}_b = \mathbf{e}_t, \quad \mathbf{e}_b \times \mathbf{e}_t = \mathbf{e}_n \tag{9.11}$$

Upon substitution of expressions(9.11) into the expression for $\boldsymbol{\omega}$ given in eq.(9.10), we obtain

$$\boldsymbol{\omega} = \dot{s}\boldsymbol{\delta} \tag{9.12}$$

where $\boldsymbol{\delta}$ is the *Darboux* vector, defined as

$$\boldsymbol{\delta} = \tau \mathbf{e}_t + \kappa \mathbf{e}_b \tag{9.13}$$

Expressions for the curvature and torsion in terms of the time-derivatives of the position vector are readily derived using the chain rule, which leads to

$$\kappa = \frac{\|\dot{\mathbf{r}} \times \ddot{\mathbf{r}}\|}{\|\dot{\mathbf{r}}\|^3} \tag{9.14a}$$

$$\tau = \frac{\dot{\mathbf{r}} \times \ddot{\mathbf{r}} \cdot \dddot{\mathbf{r}}}{\|\dot{\mathbf{r}} \times \ddot{\mathbf{r}}\|^2} \tag{9.14b}$$

Upon differentiation of both sides of eq.(9.12), the angular acceleration $\dot{\boldsymbol{\omega}}$ is derived as

$$\dot{\boldsymbol{\omega}} = \ddot{s}\boldsymbol{\delta} + \dot{s}\dot{\boldsymbol{\delta}} \tag{9.15}$$

where the time-derivative of the Darboux vector is given, in turn, as

$$\dot{\boldsymbol{\delta}} = \dot{\tau}\mathbf{e}_t + \dot{\kappa}\mathbf{e}_b \tag{9.16}$$

in which eqs.(9.5a–c) have contributed to the simplification of the above expression. The time-derivatives of the curvature and torsion are readily derived by application of the chain rule, thereby obtaining

$$\dot{\kappa} \equiv \dot{s}\kappa'(s) = \frac{\dot{s}}{\kappa}(\mathbf{r}' \times \mathbf{r}''') \cdot (\mathbf{r}' \times \mathbf{r}'') \tag{9.17a}$$

$$\dot{\tau} \equiv \dot{s}\tau'(s) = \frac{\dot{s}}{\kappa^2}[\mathbf{r}' \times \mathbf{r}'' \cdot \mathbf{r}^{(iv)} - 2\tau(\mathbf{r}' \times \mathbf{r}''') \cdot (\mathbf{r}' \times \mathbf{r}'')] \tag{9.17b}$$

The time-derivative of the Darboux vector thus reduces to

$$\dot{\boldsymbol{\delta}} = \dot{s}(A\mathbf{e}_t + B\mathbf{e}_b) \tag{9.18a}$$

where scalars A and B are computed as

$$A \equiv \frac{\mathbf{r}' \times \mathbf{r}'' \cdot \mathbf{r}^{(iv)} - 2\tau(\mathbf{r}' \times \mathbf{r}''') \cdot (\mathbf{r}' \times \mathbf{r}'')}{\kappa^2} \tag{9.18b}$$

$$B \equiv \frac{(\mathbf{r}' \times \mathbf{r}''') \cdot (\mathbf{r}' \times \mathbf{r}'')}{\kappa} \tag{9.18c}$$

and hence, the angular acceleration reduces to

$$\dot{\boldsymbol{\omega}} = \ddot{s}\boldsymbol{\delta} + \dot{s}^2(A\mathbf{e}_t + B\mathbf{e}_b) \tag{9.19}$$

From the relations derived above, it is apparent that the angular velocity is a bilinear function of the Darboux vector and $\dot{s}$, while the angular acceleration is linear in $\ddot{s}$ and quadratic in $\dot{s}$. The computational costs involved in the calculation of the angular velocity and its time-derivative amount to 31 multiplications and 13 additions for the former, and 28 multiplications with 14 additions for the latter (Angeles, Rojas, and López-Cajún, 1988). Notice that the angular velocity requires, additionally, one square root.

In the above discussion, it is assumed that explicit formulas for the two time-derivatives of the arc length s are available. This is often not the case, as we show with the examples below, whereby an intermediate parameter, which is easier to handle, is introduced. What we will need are, in fact, alternative expressions for the quantities involved, in terms of kinematic variables; i.e., we need time-derivatives of the position vector $\mathbf{r}$ rather than derivatives of this vector with respect to the arc length s. Below we derive these expressions.

First, note that $\mathbf{e}_t$ can be obtained by simply normalizing the velocity vector $\dot{\mathbf{r}}$, namely, as

$$\mathbf{e}_t = \frac{\dot{\mathbf{r}}}{\|\dot{\mathbf{r}}\|} \tag{9.20}$$

where it is not difficult to realize that

$$\dot{s} = \|\dot{\mathbf{r}}\| \tag{9.21}$$

Moreover, the binormal vector $\mathbf{e}_b$ can be derived by application of the chain rule to vector $\mathbf{r}'$, namely,

$$\mathbf{r}'' = \frac{d\mathbf{r}'}{ds} \equiv \frac{d\mathbf{r}'/dt}{ds/dt} \equiv \frac{1}{\dot{s}}\frac{d}{dt}(\mathbf{r}') \tag{9.22a}$$

But

$$\mathbf{r}'(s) \equiv \frac{d\mathbf{r}}{ds} \equiv \frac{\dot{\mathbf{r}}}{\dot{s}} \tag{9.22b}$$

and hence,

$$\mathbf{r}'' = \frac{1}{\dot{s}}\left[\frac{d}{dt}\left(\frac{\dot{\mathbf{r}}}{\dot{s}}\right)\right] = \frac{\dot{s}\ddot{\mathbf{r}} - \ddot{s}\dot{\mathbf{r}}}{\dot{s}^3} \tag{9.22c}$$

Now, upon substitution of expressions (9.22b & c) into eq.(9.1b), an alternative expression for $\mathbf{e}_b$ is derived, in terms of time-derivatives of the position vector, namely,

$$\mathbf{e}_b = \frac{\dot{\mathbf{r}} \times \ddot{\mathbf{r}}}{\|\dot{\mathbf{r}} \times \ddot{\mathbf{r}}\|} \tag{9.23}$$

Finally, $\mathbf{e}_n$ can be readily computed as the cross product of the first two vectors of the Frenet-Serret triad, namely,

$$\mathbf{e}_n \equiv \mathbf{e}_b \times \mathbf{e}_t = \frac{(\dot{\mathbf{r}} \times \ddot{\mathbf{r}}) \times \dot{\mathbf{r}}}{\|\dot{\mathbf{r}} \times \ddot{\mathbf{r}}\| \|\dot{\mathbf{r}}\|} \tag{9.24}$$

The time-derivatives of the Frenet-Serret vectors can be computed by direct differentiation of the expressions given above, namely, eqs.(9.20), (9.23), and (9.24).

9.3 Parametric Path Representation

Only seldom is an explicit representation of the position vector $\mathbf{r}$ of a geometric curve possible in terms of the arc length. In most practical cases, alternative representations should be used. The representation of the position vector in terms of a parameter σ, whatever its geometric interpretation may be, whether length or angle, will henceforth be termed a *parametric representation* of the curve at hand. The choice of σ is problem-dependent, as we illustrate with examples.

Below we derive expressions for (a) the Frenet-Serret triad; (b) the curvature and torsion; and (c) the derivatives of the latter with respect to the arc length. All these expressions, moreover, will be given in terms of derivatives with respect to the working parameter σ. The key relation that we will use is based on the chain rule, already recalled several times earlier. Thus, for any vector $\mathbf{v}(\sigma)$,

$$\frac{d\mathbf{v}}{ds} = \frac{d\mathbf{v}}{d\sigma}\frac{d\sigma}{ds}$$

However, the foregoing relation is not very useful because we do not have an explicit representation of parameter σ in terms of the arc length. Nevertheless, we will assume that these two variables, s and σ, obey a *monotonic* relation. What this means is that

$$\frac{d\sigma}{ds} > 0 \tag{9.25}$$

which is normally the case. Under this assumption, moreover, we can write the derivative of $\mathbf{v}$ as

$$\frac{d\mathbf{v}}{ds} = \frac{d\mathbf{v}/d\sigma}{ds/d\sigma}$$

where, apparently,

$$\frac{ds}{d\sigma} = \|\frac{d\mathbf{r}}{d\sigma}\| = \|\mathbf{r}'(\sigma)\|$$

Therefore, the derivative sought takes the form

$$\frac{d\mathbf{v}}{ds} = \frac{\mathbf{v}'(\sigma)}{\|\mathbf{r}'(\sigma)\|} \tag{9.26a}$$

It goes without saying that the same relation holds for scalars, i.e.,

$$\frac{dv}{ds} = \frac{v'(\sigma)}{\|\mathbf{r}'(\sigma)\|} \tag{9.26b}$$

Expressions for the Frenet-Serret triad now follow immediately, i.e.,

$$\mathbf{e}_t = \frac{\mathbf{r}'(\sigma)}{\|\mathbf{r}'(\sigma)\|} \tag{9.27a}$$

$$\mathbf{e}_b = \frac{\mathbf{r}'(\sigma) \times \mathbf{r}''(\sigma)}{\|\mathbf{r}'(\sigma) \times \mathbf{r}''(\sigma)\|} \tag{9.27b}$$

$$\mathbf{e}_n = \mathbf{e}_b \times \mathbf{e}_t = \frac{[\mathbf{r}'(\sigma) \times \mathbf{r}''(\sigma)] \times \mathbf{r}'(\sigma)}{\|\mathbf{r}'(\sigma) \times \mathbf{r}''(\sigma)\| \|\mathbf{r}'(\sigma)\|} \tag{9.27c}$$

Now, paraphrasing relations (9.14a & b), we have

$$\kappa = \frac{\|\mathbf{r}'(\sigma) \times \mathbf{r}''(\sigma)\|}{\mathbf{r}'(\sigma)\|^3} \tag{9.28a}$$

$$\tau = \frac{\mathbf{r}'(\sigma) \times \mathbf{r}''(\sigma) \cdot \mathbf{r}'''}{\mathbf{r}'(\sigma) \times \mathbf{r}''(\sigma)\|^2} \tag{9.28b}$$

the partial derivatives of the curvature and torsion with respect to the arc length being computed in terms of the corresponding partial derivatives with respect to the parameter σ, which is done with the aid of the chain rule, i.e.,

$$\kappa'(s) = \frac{\kappa'(\sigma)}{\|\mathbf{r}'(s)\|}, \quad \tau'(s) = \frac{\tau'(\sigma)}{\|\mathbf{r}'(\sigma)\|} \tag{9.29}$$

Expressions for $\kappa'(\sigma)$ and $\tau'(\sigma)$, in turn, are derived by a straightforward differentiation of the expressions for κ and τ in terms of σ, as given in eqs.(9.28a & b). To this end, we first recall a useful expression for the derivative of a rational expression $q(x)$ whose numerator and denominator are denoted by $N(x)$ and $D(x)$, respectively. This expression is

$$q'(x) = \frac{1}{D(x)}[N'(x) - q(x)D'(x)] \tag{9.30a}$$

Note that nothing prevents the numerator of the foregoing rational expression from being a vector, and hence, a similar formula can be applied to vector ratios as well. Let the denominator of a vector rational function $\mathbf{q}(x)$ be $\mathbf{n}(x)$. Under these conditions, then, we have

$$\mathbf{q}'(x) = \frac{1}{D(x)}[\mathbf{n}'(x) - \mathbf{q}(x)D'(x)] \tag{9.30b}$$

As a matter of fact, the above relation can be extended to matrix numerators. Not only is this possible, but the argument can likewise be a vector or a matrix variable, and similar formulas would apply correspondingly.

We thus have, for the derivative of the curvature,

$$\kappa'(\sigma) = \frac{1}{\|\mathbf{r}'(\sigma)\|^3}\left[\frac{d}{d\sigma}\|\mathbf{r}'(\sigma) \times \mathbf{r}''(\sigma)\| - \kappa\frac{d}{d\sigma}\|\mathbf{r}'(\sigma)\|^3\right] \tag{9.31}$$

Now we find the first term inside the brackets of the foregoing expression from the relation

$$\frac{d}{d\sigma}\|\mathbf{r}'(\sigma) \times \mathbf{r}''(\sigma)\|^2 = 2\|\mathbf{r}' \times \mathbf{r}''\|\frac{d}{d\sigma}\|\mathbf{r}' \times \mathbf{r}''\|$$

which yields

$$\frac{d}{d\sigma}\|\mathbf{r}' \times \mathbf{r}''\| = \frac{1}{2\|\mathbf{r}' \times \mathbf{r}''\|}\frac{d}{d\sigma}\|\mathbf{r}'(\sigma) \times \mathbf{r}''(\sigma)\|^2$$

But

$$\begin{aligned}\frac{d}{d\sigma}\|\mathbf{r}'(\sigma) \times \mathbf{r}''(\sigma)\|^2 &= \frac{d}{d\sigma}\{[\mathbf{r}'(\sigma) \times \mathbf{r}''(\sigma)] \cdot [\mathbf{r}'(\sigma) \times \mathbf{r}''(\sigma)]\} \\ &= 2[\mathbf{r}'(\sigma) \times \mathbf{r}''(\sigma)] \cdot \frac{d}{d\sigma}[\mathbf{r}'(\sigma) \times \mathbf{r}''(\sigma)] \qquad (9.32)\end{aligned}$$

the derivative of the above term in brackets reducing to

$$\frac{d}{d\sigma}[\mathbf{r}'(\sigma) \times \mathbf{r}''(\sigma)] = \mathbf{r}'(\sigma) \times \mathbf{r}'''(\sigma)$$

and hence,

$$\frac{d}{d\sigma}\|\mathbf{r}' \times \mathbf{r}''\| = \frac{[\mathbf{r}'(\sigma) \times \mathbf{r}''(\sigma)] \cdot [\mathbf{r}'(\sigma) \times \mathbf{r}'''(\sigma)]}{\|\mathbf{r}' \times \mathbf{r}''\|} \qquad (9.33a)$$

Furthermore,

$$\frac{d}{d\sigma}\|\mathbf{r}'(\sigma)\|^3 = 3\|\mathbf{r}'(\sigma)\|^2\frac{d}{d\sigma}\|\mathbf{r}'(\sigma)\|$$

the last derivative again being found from an intermediate relation, namely,

$$\frac{d}{d\sigma}\|\mathbf{r}'(\sigma)\|^2 = 2\|\mathbf{r}'(\sigma)\|\frac{d}{d\sigma}\|\mathbf{r}'(\sigma)\|$$

whence,

$$\frac{d}{d\sigma}\|\mathbf{r}'(\sigma)\| = \frac{1}{2\|\mathbf{r}'(\sigma)\|}\frac{d}{d\sigma}\|\mathbf{r}'(\sigma)\|^2$$

with

$$\frac{d}{d\sigma}\|\mathbf{r}'(\sigma)\|^2 = \frac{d}{d\sigma}[\mathbf{r}'(\sigma) \cdot \mathbf{r}'(\sigma)] = 2\mathbf{r}'(\sigma) \cdot \mathbf{r}''(\sigma)$$

and so,

$$\frac{d}{d\sigma}\|\mathbf{r}'(\sigma)\| = \frac{\mathbf{r}'(\sigma) \cdot \mathbf{r}''(\sigma)}{\|\mathbf{r}'(\sigma)\|}\frac{d}{d\sigma}\|\mathbf{r}'(\sigma)\|^2$$

Therefore,

$$\frac{d}{d\sigma}\|\mathbf{r}'(\sigma)\|^3 = 3\|\mathbf{r}'(\sigma)\|\mathbf{r}'(\sigma) \cdot \mathbf{r}''(\sigma) \qquad (9.33b)$$

Substitution of eqs.(9.33a & b) into eq.(9.31) yields the desired expression, namely,

$$\kappa'(\sigma) = \frac{[\mathbf{r}'(\sigma) \times \mathbf{r}''(\sigma)] \cdot [\mathbf{r}'(\sigma) \times \mathbf{r}'''(\sigma)]}{\|\mathbf{r}'(\sigma)\|^3\|\mathbf{r}' \times \mathbf{r}''\|} - 3\kappa\frac{\mathbf{r}'(\sigma) \cdot \mathbf{r}''(\sigma)}{\|\mathbf{r}'(\sigma)\|^2} \qquad (9.34)$$

Likewise,

$$\tau'(\sigma) = \frac{N}{D} \tag{9.35a}$$

with N and D defined as

$$N \equiv \frac{d}{d\sigma}[\mathbf{r}'(\sigma) \times \mathbf{r}''(\sigma) \cdot \mathbf{r}'''(\sigma)] - \tau \frac{d}{d\sigma}\|\mathbf{r}'(\sigma) \times \mathbf{r}''(\sigma)\|^2 \tag{9.35b}$$

$$D \equiv \|\mathbf{r}'(\sigma) \times \mathbf{r}''(\sigma)\|^2 \tag{9.35c}$$

The first term of the numerator N of the foregoing expression can be readily calculated as

$$\frac{d}{d\sigma}[\mathbf{r}'(\sigma) \times \mathbf{r}''(\sigma) \cdot \mathbf{r}'''(\sigma)] = \mathbf{r}'(\sigma) \times \mathbf{r}''(\sigma) \cdot \mathbf{r}^{(iv)}(\sigma) \tag{9.35d}$$

while the derivative appearing in the second term of the same numerator was obtained previously, as displayed in eq.(9.32). Upon substitution of the expressions appearing in eqs.(9.32) and (9.35d) into eq.(9.35a), we obtain the desired expression, namely,

$$\tau'(\sigma) = \frac{\mathbf{r}'(\sigma) \times \mathbf{r}''(\sigma) \cdot [\mathbf{r}^{(iv)}(\sigma) - 2\tau\mathbf{r}'(\sigma) \times \mathbf{r}'''(\sigma)]}{\|\mathbf{r}'(\sigma) \times \mathbf{r}''(\sigma)\|^2} \tag{9.35e}$$

thereby completing the desired relations.

Example 9.3.1 (Planning of a gluing operation) *A robot used for a gluing operation is required to guide the glue nozzle fixed to its end-effector through a helicoidal path so that the tip of the nozzle traverses the helix at a constant speed v_0 and the end-effector maintains a fixed orientation with respect to the curve, i.e., with respect to the Frenet-Serret triad of the helix. Determine the orientation matrix $\mathbf{Q}$ of the end-effector with respect to a frame $\{x, y, z\}$ fixed to the robot base, as well as the angular velocity and angular acceleration of the end-effector. The operation is to be performed with a Fanuc S-300 robot, whose Denavit-Hartenberg (DH) parameters are given in Table 9.1, while the axis of the helix is chosen to be parallel to the first axis of the robot with origin at $(2, -2, 1.2)$, the foregoing coordinates being given in meters. Find the joint trajectories of the robot as well as the associated joint rates and joint accelerations from Cartesian position, velocity, and acceleration data. Verify that the joint-rate and joint-acceleration profiles are compatible with those of the joint variables. Assume that the radius of the helix is $a = 1.6$ m and that its pitch is $b = 2.5$ m/turn. Finally, the gluing seam spans through one quarter of a helix turn.*

Solution: We will use a Cartesian frame fixed to the base of the robot such that its z axis coincides with the axis of the first revolute. The helix can then be given in the parametric representation shown below:

$$x = 2 + a\cos\varphi$$

TABLE 9.1. DH Parameters of a Fanuc S-300 Robot

Link	a_i (m)	b_i (m)	α_i (deg)
1	0.0	0.9	90
2	0.9	0.0	0
3	0.95	0.0	90
4	0.0	1.3	-90
5	0.0	0.0	90
6	0.0	0.44	-90

$$y = -2 + a\sin\varphi$$
$$z = 12 + \frac{b\varphi}{2\pi}$$

where the parameter φ is the angle made by the projection, onto the X-Y plane, of the position vector of a point P of the helix with the x axis. In the process, we will need first and second time-derivatives of the foregoing Cartesian coordinates. These are given below for quick reference:

$$\dot{x} = -a\dot{\varphi}\sin\varphi$$
$$\dot{y} = a\dot{\varphi}\cos\varphi$$
$$\dot{z} = \frac{b}{2\pi}\dot{\varphi}$$

and

$$\ddot{x} = -a\dot{\varphi}^2\cos\varphi - a\ddot{\varphi}\sin\varphi$$
$$\ddot{y} = -a\dot{\varphi}^2\sin\varphi + a\ddot{\varphi}\cos\varphi$$
$$\ddot{z} = \frac{b}{2\pi}\ddot{\varphi}$$

We now impose the constant-speed condition, which leads to

$$\dot{x}^2 + \dot{y}^2 + \dot{z}^2 \equiv a^2\dot{\varphi}^2 + \frac{b^2}{4\pi^2}\dot{\varphi}^2 = v_0^2$$

and hence,

$$\dot{\varphi} = c$$

where the constant c is defined as

$$c \equiv v_0\sqrt{\frac{4\pi^2}{4\pi^2a^2 + b^2}}$$

Thus, $\dot{\varphi}$ is constant, and hence,

$$\varphi = ct$$

Moreover, in terms of constant c, the Cartesian coordinates of a point of the helix take on the forms

$$x = 2 + a\cos ct$$
$$y = -2 + a\sin ct$$
$$z = 12 + \frac{bc}{2\pi}t$$

the first time-derivatives of these coordinates becoming

$$\dot{x} = -ac\sin ct$$
$$\dot{y} = ac\cos ct$$
$$\dot{z} = \frac{bc}{2\pi}$$

and the corresponding second time-derivatives

$$\ddot{x} = -ac^2\cos ct$$
$$\ddot{y} = -ac^2\sin ct$$
$$\ddot{z} = 0$$

Now the Frenet-Serret triad is readily calculated as

$$\mathbf{e}_t \equiv \frac{d\mathbf{r}}{ds} \equiv \frac{\dot{\mathbf{r}}}{\dot{s}} = \frac{c}{v_0}\begin{bmatrix} -a\sin ct \\ a\cos ct \\ b/2\pi \end{bmatrix}$$

Furthermore,

$$\frac{d\mathbf{e}_t}{ds} \equiv \frac{\dot{\mathbf{e}}_t}{\dot{s}} = \frac{ac^2}{v_0^2}\begin{bmatrix} -\cos ct \\ -\sin ct \\ 0 \end{bmatrix} \equiv \kappa\mathbf{e}_n$$

from which it is apparent that

$$\kappa = a\frac{c^2}{v_0^2} \equiv \frac{4\pi^2 a}{4\pi^2 a^2 + b^2}, \quad \mathbf{e}_n = -\begin{bmatrix} \cos ct \\ \sin ct \\ 0 \end{bmatrix}$$

Thus, the binormal vector $\mathbf{e}_b$ is calculated simply as the cross product of the first two vectors of the Frenet-Serret triad, namely,

$$\mathbf{e}_b \equiv \mathbf{e}_t \times \mathbf{e}_n = -\frac{c}{v_0}\begin{bmatrix} -(b/2\pi)\sin ct \\ (b/2\pi)\cos ct \\ -a \end{bmatrix}$$

and hence, the orientation matrix $\mathbf{Q}$ of the gluing nozzle, or of the end-effector for that matter, is given by

$$\mathbf{Q} \equiv [\,\mathbf{e}_t \quad \mathbf{e}_n \quad \mathbf{e}_b\,]$$

Hence,

$$\mathbf{Q} = \frac{c}{v_0}\begin{bmatrix} -a\sin ct & -(v_0/c)\cos ct & (b/2\pi)\sin ct \\ a\cos ct & -(v_0/c)\sin ct & -(b/2\pi)\cos ct \\ (b/2\pi) & 0 & a \end{bmatrix}$$

Now, the angular velocity is determined from eq.(9.12), which requires the calculation of the Darboux vector, as given in eq.(9.13). Upon calculation of the Darboux vector and substitution of the expression thus resulting into eq.(9.12), we obtain

$$\boldsymbol{\omega} = \frac{c^3}{v_0^2}\begin{bmatrix} 0 \\ 0 \\ (4\pi^2a^2 + b^2)/4\pi^2 \end{bmatrix} = c\begin{bmatrix} 0 \\ 0 \\ 1 \end{bmatrix}$$

which is thus constant, and hence,

$$\dot{\boldsymbol{\omega}} = \mathbf{0}$$

Now, the coordinates of the center of the wrist, C, are determined with the aid of relation (4.18c), where the operation point is a point on the helix, i.e., $\mathbf{p} = x\mathbf{i} + y\mathbf{j} + z\mathbf{k}$, parameters b_6, λ_6, and μ_6 being obtained from Table 9.1, namely,

$$b_6 = 0.440 \text{ m}, \quad \lambda_6 = \cos\alpha_6 = 0, \quad \mu_6 = \sin\alpha_6 = -1$$

Furthermore, the numerical value of c is obtained from the helix geometry, namely,

$$c = 0.8\sqrt{\frac{4\pi^2}{4\pi^2 \times 1.6^2 + 2.5^2}} = 0.48522 \text{ s}^{-1}$$

Upon substitution in eq.(4.18c) of the entries found above for $\mathbf{Q}$, along with the numerical values, we obtain the Cartesian coordinates of the center C of the spherical wrist of the robot as

$$\begin{bmatrix} x_C \\ y_C \\ z_C \end{bmatrix} = \begin{bmatrix} 2 + 1.16\cos(0.48522t) \\ -2 + 1.16\sin(0.48522t) \\ 12 + 0.19306t \end{bmatrix}$$

in meters. Apparently, point C describes a helicoidal path as well, although of a smaller radius, that is coaxial with the given helix.

Now the time-histories of the joint angles are computed from inverse kinematics. Note that the robot at hand being of the decoupled type, it allows for a simple inverse kinematics solution. The details of the solution were discussed extensively in Section 4.4 and are left as an exercise to the reader.

Of the four inverse kinematics solutions of the arm, three were found to lead to link interferences with the aid of RVS, the package for robot visualization developed at McGill University. Hence, only one such solution

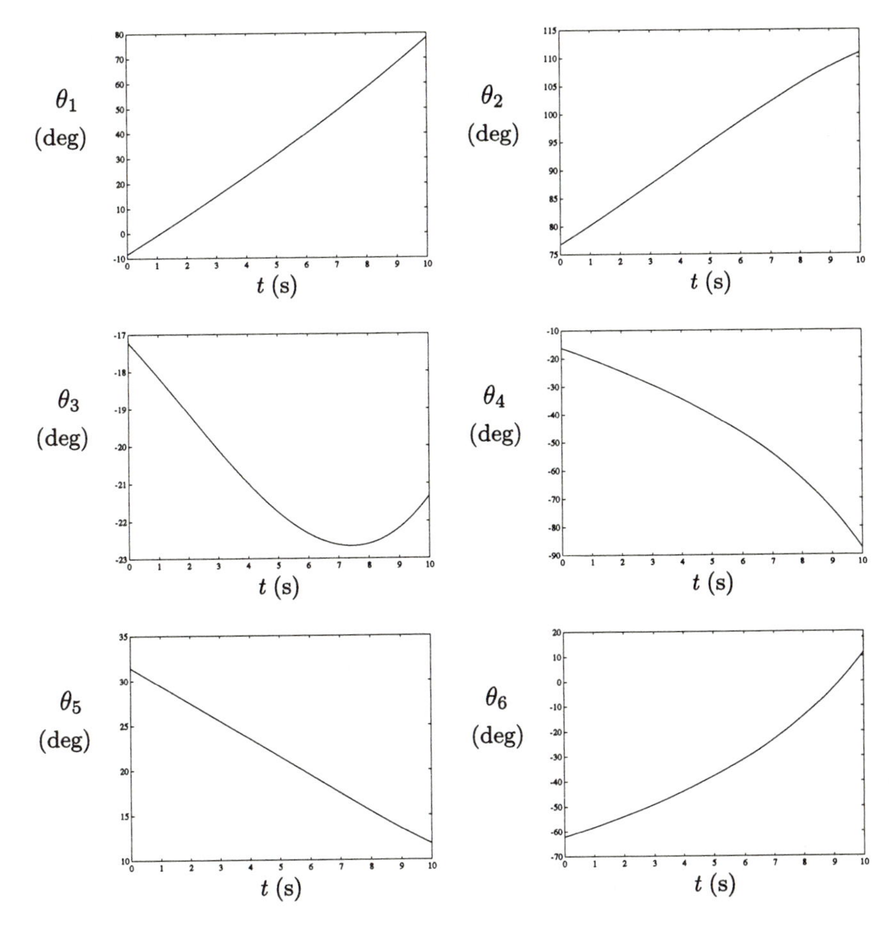

FIGURE 9.1. Joint trajectories for a Fanuc S-300.

is physically possible. This solution, along with one of the two wrist solutions, is plotted in Fig. 9.1, with Figs. 9.2 and 9.3 showing, respectively, the corresponding joint rates and joint accelerations.

Note that the maxima and minima of the joint-variables occur at instants where the corresponding joint rates vanish. Likewise, the maxima and minima of joint rates occur at instants where the associated joint accelerations vanish, thereby verifying that the computed results are compatible. A more detailed verification can be done by numerical differentiation of the joint-variable time-histories.

Example 9.3.2 (Planning of an arc-welding operation) *A spherical reservoir of radius R is to be arc-welded to a cylindrical pipe of radius r, with the axis of the cylinder located a distance d from the center of the sphere, all elements of the cylinder piercing the sphere, i.e., $d+r \leq R$, as shown in*

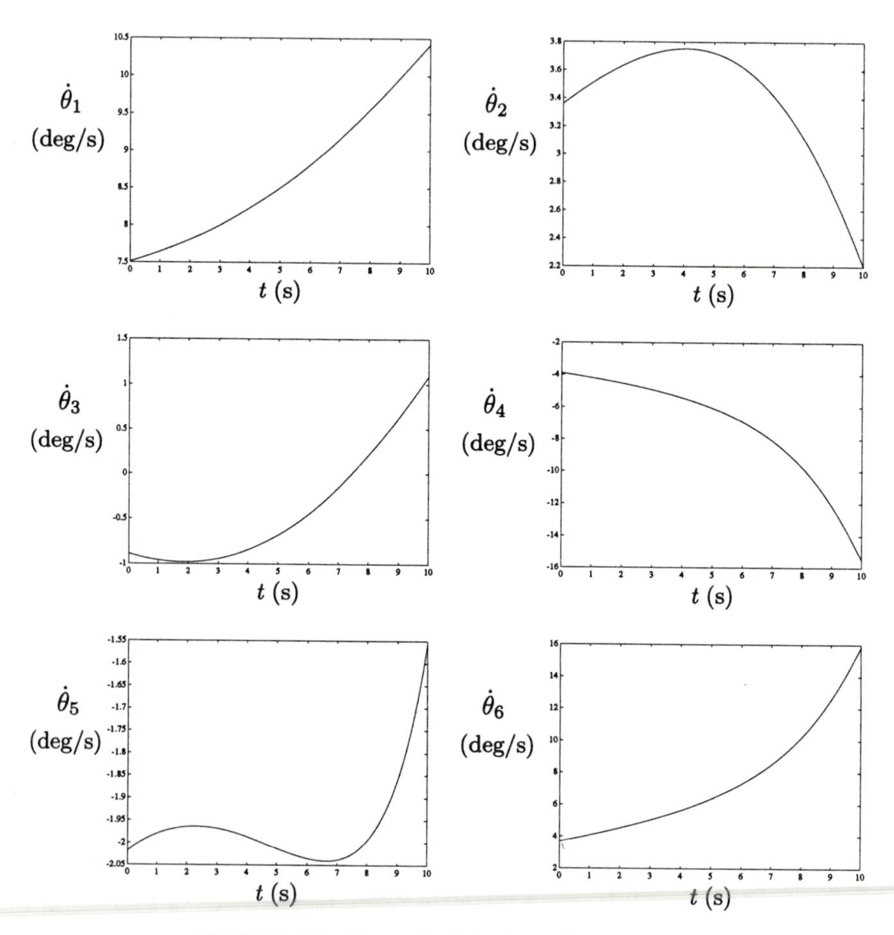

FIGURE 9.2. Joint velocities for a Fanuc S-300.

Fig. 9.4. Note that two intersection curves are geometrically possible, but the welding will take place only along the upper curve. Moreover, the welding electrode is to traverse the intersection curve, while the tool carrying the electrode is to keep a constant orientation with respect to that curve. In the coordinate frame shown in Fig. 9.4, find an expression for the rotation matrix defining the orientation of the end-effector, to which the electrode is rigidly attached.

Solution: Note that the X axis of the coordinate frame indicated in Fig. 9.4 intersects the $\mathcal{A}$ axis of the cylinder, this axis being parallel to the Z axis. Moreover, we define φ as the angle shown in Fig 9.4b. Now, the x and y coordinates of an arbitrary point of the intersection curve are given by

$$x = d + r\cos\varphi \tag{9.36a}$$

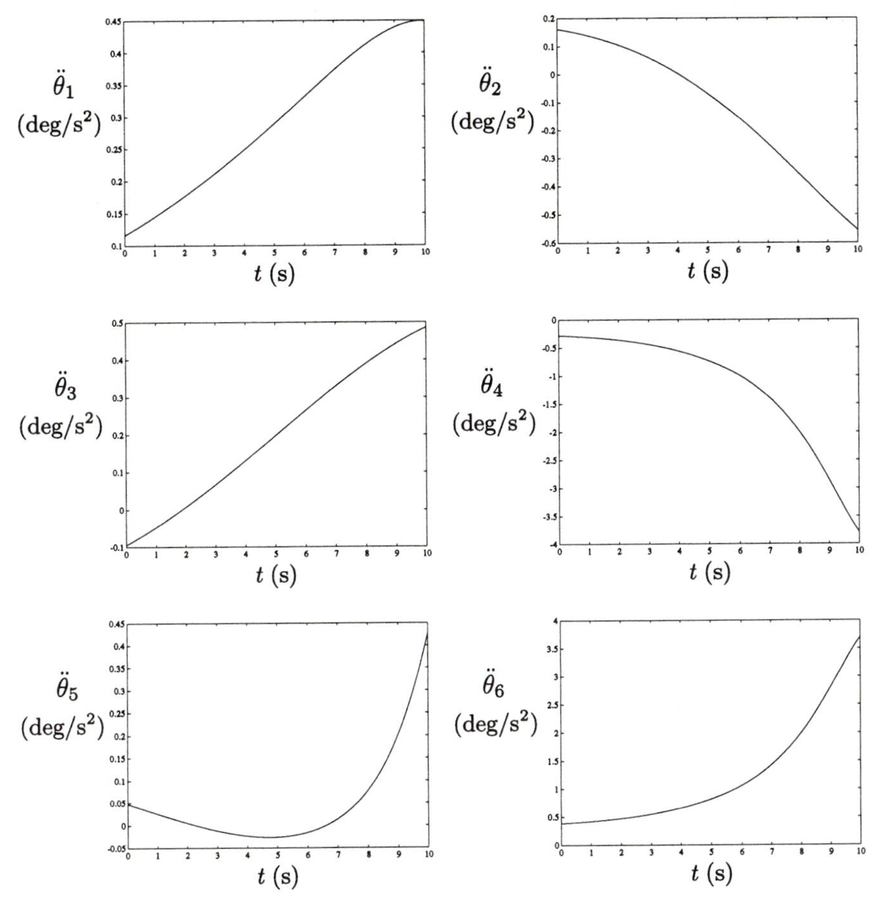

FIGURE 9.3. Joint accelerations for a Fanuc S-300.

$$y = r \sin \varphi \tag{9.36b}$$

Further, in order to find the remaining z coordinate, we use the equation of the sphere, $\mathcal{S}$, namely,

$$\mathcal{S}: \quad x^2 + y^2 + z^2 = R^2$$

If we substitute the x and y coordinates of the intersection curve in the above equation and then solve for the z coordinate in terms of φ, we obtain

$$z = \pm\sqrt{R^2 - r^2 - d^2 - 2dr \cos \varphi} \tag{9.36c}$$

In the above relation, the plus and minus signs correspond to the upper and lower portions of the intersection curve, respectively. Since we are interested in only the upper intersection, we will take only the positive

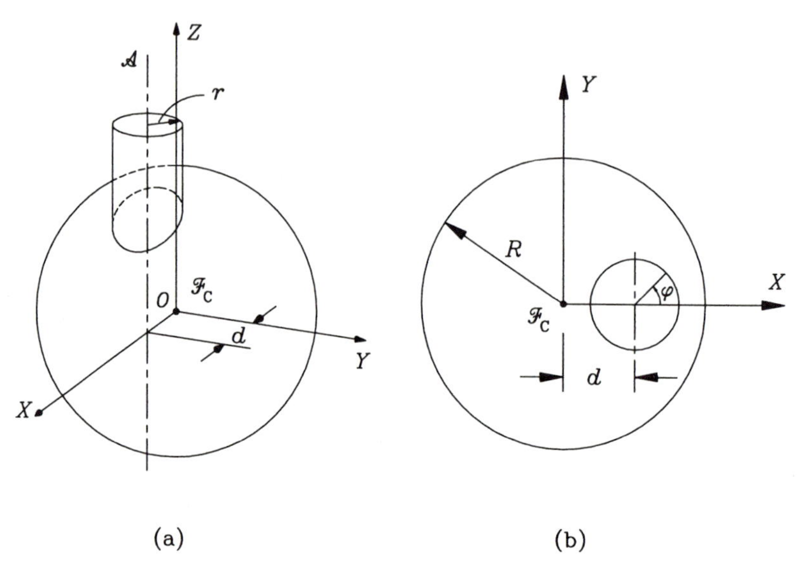

FIGURE 9.4. Intersection curve between a spherical reservoir and a cylindrical pipe.

sign in the above relation. Furthermore, we define

$$d \equiv \lambda r, \quad R \equiv \mu r$$

where λ and μ are nondimensional constants. Moreover, let

$$\rho^2 \equiv \mu^2 - \lambda^2 - 1$$
$$\hat{\varphi} \equiv \frac{1}{\sqrt{\rho^2 - 2\lambda\cos\varphi}}$$

Then the position vector $\mathbf{r}$ of any point on the intersection curve can be expressed in the form

$$\mathbf{r} = r \begin{bmatrix} \lambda + \cos\varphi \\ \sin\varphi \\ 1/\hat{\varphi} \end{bmatrix} \tag{9.37}$$

Now, upon differentiation of $\mathbf{r}$ with respect to φ, we obtain

$$\mathbf{r}'(\varphi) = r \begin{bmatrix} -\sin\varphi \\ \cos\varphi \\ \lambda\hat{\varphi}\sin\varphi \end{bmatrix} \tag{9.38a}$$

$$\mathbf{r}''(\varphi) = r \begin{bmatrix} -\cos\varphi \\ -\sin\varphi \\ \lambda\hat{\varphi}\cos\varphi - (\lambda^2\sin^2\varphi)\hat{\varphi}^3 \end{bmatrix} \tag{9.38b}$$

where we have used the relation

$$\hat{\varphi}'(\varphi) = -(\lambda\sin\varphi)\hat{\varphi}^3$$

In addition, using eqs.(9.38a) and (9.38b), we derive the items needed to compute the Frenet-Serret triad, from which we will derive the required orientation matrix, i.e.,

$$\mathbf{r}'(\varphi) \times \mathbf{r}''(\varphi) = r^2 \begin{bmatrix} \lambda\hat{\varphi} - \lambda^2\hat{\varphi}^3 \cos\varphi \sin^2\varphi \\ -\lambda^2\hat{\varphi}^3 \sin^3\varphi \\ 1 \end{bmatrix} \tag{9.39a}$$

$$\|\mathbf{r}'(\varphi)\| = rG(\varphi) \tag{9.39b}$$

$$\|\mathbf{r}'(\varphi) \times \mathbf{r}''(\varphi)\| = r^2\hat{\varphi}^3\sqrt{D(\varphi)} \tag{9.39c}$$

with functions $D(\varphi)$ and $G(\varphi)$ defined as

$$D \equiv \rho^4\lambda^2 + \lambda^4 + \rho^6 - 6\rho^2\lambda(\lambda^2+\rho^2)\cos\varphi + 6\lambda^2(\lambda^2+2\rho^2)\cos^2\varphi \\ + 2\lambda^3(\rho^2-4)\cos^3\varphi - 4\lambda^4\cos^4\varphi \tag{9.39d}$$

$$G \equiv \sqrt{1+\lambda^2\hat{\varphi}^2\sin^2\varphi} \tag{9.39e}$$

Now $\mathbf{e}_t$, $\mathbf{e}_b$, and $\mathbf{e}_n$ are obtained as

$$\mathbf{e}_t \equiv \frac{\mathbf{r}'(\varphi)}{\|\mathbf{r}'(\varphi)\|} = \frac{1}{G}\begin{bmatrix} -\sin\varphi \\ \cos\varphi \\ \lambda\hat{\varphi}\sin\varphi \end{bmatrix} \tag{9.40a}$$

$$\mathbf{e}_b \equiv \frac{\mathbf{r}'(\varphi)\times\mathbf{r}''(\varphi)}{\|\mathbf{r}'(\varphi)\times\mathbf{r}''(\varphi)\|} = \frac{1}{\hat{\varphi}^3\sqrt{D}}\begin{bmatrix} \lambda\hat{\varphi} - \lambda^2\hat{\varphi}^3\cos\varphi\sin^2\varphi \\ -\lambda^2\hat{\varphi}^3\sin^3\varphi \\ 1 \end{bmatrix} \tag{9.40b}$$

$$\mathbf{e}_n \equiv \frac{1}{\hat{\varphi}^3\sqrt{D}\,G}\begin{bmatrix} -\lambda^3\hat{\varphi}^4\sin^4\varphi - \cos\varphi \\ \lambda\hat{\varphi}^4\cos\varphi\sin^3\varphi - \lambda^2\hat{\varphi}^2\sin\varphi - \sin\varphi \\ \lambda\hat{\varphi}\cos\varphi - \lambda^2\hat{\varphi}^3\sin^2\varphi \end{bmatrix} \tag{9.40c}$$

where $\mathbf{e}_n$ has been calculated as $\mathbf{e}_n = \mathbf{e}_b \times \mathbf{e}_t$.

The orthogonal matrix defining the orientation of the end-effector can now be readily computed as

$$\mathbf{Q} \equiv [\,\mathbf{e}_t \quad \mathbf{e}_n \quad \mathbf{e}_b\,]$$

for we have all the necessary expressions. Note, however, that these expressions allow us to find $\mathbf{Q}$ for any value of φ, but we do not have, as yet, an expression of the form $\varphi(t)$ that would allow us to obtain $\mathbf{Q}(t)$. Such an expression is derived in Example 9.5.1.

Example 9.3.3 (Calculation of torsion, curvature, and Darboux vector) *We refer here to the intersection curve of Example 9.3.2, for which we want to find expressions for its curvature, torsion, and Darboux vector.*

Solution: We can use directly the expressions derived above, eqs.(9.28a & b), to obtain the curvature and torsion in terms of derivatives with respect to parameter φ. With these expressions and those for the Frenet-Serret triad, the Darboux vector would follow. However, we can take shortcuts, for we already have expressions for the Frenet-Serret triad, if we express the curvature and torsion in terms of this triad and its derivatives with respect to φ, as we explain below. Indeed, from the Frenet-Serret relations, eqs.(9.2b), we can express the curvature and torsion in the forms

$$\kappa = \mathbf{e}_t'(s) \cdot \mathbf{e}_n \tag{9.41a}$$

$$\tau = -\mathbf{e}_b'(s) \cdot \mathbf{e}_n \tag{9.41b}$$

and hence, all we need now are the derivatives of the tangent and normal vectors with respect to s. These are readily derived using relation (9.26a), i.e.,

$$\mathbf{e}_t'(s) = \frac{\mathbf{e}_t'(\varphi)}{\|\mathbf{r}'(\varphi)\|} \tag{9.42a}$$

$$\mathbf{e}_b'(s) = \frac{\mathbf{e}_b'(\varphi)}{\|\mathbf{r}'(\varphi)\|} \tag{9.42b}$$

Now, in order to differentiate the Frenet-Serret triad with respect to φ, we first note, from eqs.(9.40a–c), that these three expressions are vector rational functions, and hence, their derivatives with respect to φ are derived by applying eq.(9.30b), thereby obtaining

$$\mathbf{e}_t'(\varphi) = \frac{1}{G}[\mathbf{n}_t'(\varphi) - \mathbf{e}_t G'(\varphi)] \tag{9.43}$$

$$\mathbf{e}_b'(\varphi) = \frac{1}{\hat{\varphi}^3\sqrt{D}}\left\{\mathbf{n}_b'(\varphi) - \mathbf{e}_b\left[3\hat{\varphi}^2\hat{\varphi}'(\varphi) - \frac{D'(\varphi)}{2\sqrt{D}}\right]\right\} \tag{9.44}$$

where $\mathbf{n}_t$ and $\mathbf{n}_b$ are the numerators of the vector rational expressions of $\mathbf{e}_t$ and $\mathbf{e}_b$, respectively, given in eq.(9.40a). Below we calculate the foregoing derivatives with respect to φ:

$$\mathbf{n}_t'(\varphi) = \begin{bmatrix} -\cos\varphi \\ -\sin\varphi \\ \lambda\hat{\varphi}(\cos\varphi - \lambda\hat{\varphi}^2\sin^2\varphi) \end{bmatrix}$$

$$\mathbf{n}_b'(\varphi) = \lambda\begin{bmatrix} \hat{\varphi}' - \lambda\hat{\varphi}^2\sin\varphi[3\hat{\varphi}'\cos\varphi\sin\varphi - \hat{\varphi}(3\cos^2\varphi - 1)] \\ -3\lambda\hat{\varphi}^2\sin^2\varphi[\hat{\varphi}'\sin\varphi - \hat{\varphi}\cos\varphi] \\ 0 \end{bmatrix}$$

$$\hat{\varphi}' \equiv \hat{\varphi}'(\varphi)$$

$$\begin{aligned} D'(\varphi) = {} & 6\rho^2\lambda(\lambda^2 + \rho^4)\sin\varphi - 12\lambda^2(\lambda^2 + 2\rho^2)\cos\varphi\sin\varphi \\ & - 6\lambda^3(\rho^2 - 4)\cos^2\varphi\sin\varphi - 16\lambda^4\cos^3\varphi\sin\varphi \end{aligned}$$

$$G'(\varphi) = \frac{\lambda^2\hat{\varphi}\sin\varphi}{2G}(2\cos\varphi - \lambda\hat{\varphi}^2\sin^2\varphi)$$

and $\|\mathbf{r}'(\varphi)\|$ was already calculated in Example 9.3.2.

If we now substitute all the foregoing expressions into eqs.(9.42a & b), we obtain, after intensive simplification,

$$\kappa = \frac{\sqrt{D}\hat{\varphi}^3}{G^3 r} \tag{9.45a}$$

$$\tau = -3\frac{\lambda^2 \hat{\varphi} E \sin\varphi}{rDG^2} \tag{9.45b}$$

with function $E(\varphi)$ defined, in turn, as

$$E(\varphi) \equiv \frac{1}{\hat{\varphi}^4}[-\lambda^3\hat{\varphi}^4 \sin^4\varphi + \lambda^2\hat{\varphi}^2 \sin^2\varphi(\lambda\cos\varphi - 1) + \cos\varphi] \tag{9.46}$$

With the foregoing expressions for $\mathbf{e}_t$, $\mathbf{e}_b$, τ, and κ, computing the Darboux vector of the intersection curve reduces to a routine substitution of the foregoing expressions into eq.(9.13).

9.4 Parametric Splines in Trajectory Planning

Sometimes the path to be followed by the tip of the end-effector is given only as a discrete set of sampled points $\{P_i\}_1^N$. This is the case, for example, if the path is the intersection of two warped surfaces, as in the arc-welding of two plates of the hull of a vessel or the spot-welding of two sheets of the fuselage of an airplane. In these instances, the coordinates of the sampled points are either calculated numerically via nonlinear-equation solving or estimated using a vision system. In either case, it is clear that only point coordinates are available, while trajectory planning calls for information on derivatives of the position vector of points along the path with respect to the arc length. These derivatives can be estimated via a suitable interpolation of the given coordinates. Various interpolation schemes are available (Foley and Van Dam, 1982; Hoschek and Lasser, 1992), the most widely accepted ones being based on spline functions, which were introduced in Section 5.6. The splines introduced therein are applicable whenever a *function*, not a geometric curve, is to be interpolated. However, in trajectory planning, geometric curves in three-dimensional space come into play, and hence, those splines, termed *nonparametric*, are no longer applicable. What we need here are *parametric* splines, as described below.

Although parametric splines, in turn, can be of various types (Dierckx, 1993), we will focus here on cubic parametric splines because of their simplicity.

Let $P_i(x_i, y_i, z_i)$, for $i = 1, \ldots, N$, be the set of sampled points on the path to be traced by the tip of the end-effector, $\{\mathbf{p}_i\}_1^N$ being the set of corresponding position vectors. Our purpose in this section is to produce a smooth curve Γ that passes through $\{P_i\}_1^N$ and that has a continuous

Frenet-Serret triad. To this end, we will resort to the expressions derived in Section 9.3, in terms of a parameter σ, which we will define presently.

We first introduce a few definitions: Let the kth derivative of the position vector $\mathbf{p}$ of an arbitrary point P of Γ with respect to σ, evaluated at P_i, be denoted by $\mathbf{p}_i^{(k)}$, its components being denoted correspondingly by $x_i^{(k)}$, $y_i^{(k)}$, and $z_i^{(k)}$. Next, the coordinates of P are expressed as piecewise cubic polynomials of σ, namely,

$$x(\sigma) = A_{xi}(\sigma - \sigma_i)^3 + B_{xi}(\sigma - \sigma_i)^2 + C_{xi}(\sigma - \sigma_i) + D_{xi} \quad (9.47a)$$

$$y(\sigma) = A_{yi}(\sigma - \sigma_i)^3 + B_{yi}(\sigma - \sigma_i)^2 + C_{yi}(\sigma - \sigma_i) + D_{yi} \quad (9.47b)$$

$$z(\sigma) = A_{zi}(\sigma - \sigma_i)^3 + B_{zi}(\sigma - \sigma_i)^2 + C_{zi}(\sigma - \sigma_i) + D_{zi} \quad (9.47c)$$

for a real parameter σ, such that $\sigma_i \leq \sigma \leq \sigma_{i+1}$ and $i = 1, \ldots, N-1$, with σ_i defined as

$$\sigma_1 = 0, \quad \sigma_{i+1} \equiv \sigma_i + \Delta\sigma_i, \quad \Delta\sigma_i \equiv \sqrt{\Delta x_i^2 + \Delta y_i^2 + \Delta z_i^2} \quad (9.47d)$$

$$\Delta x_i \equiv x_{i+1} - x_i, \qquad \Delta y_i \equiv y_{i+1} - y_i, \qquad \Delta z_i \equiv z_{i+1} - z_i \quad (9.47e)$$

and hence, $\Delta\sigma_i$ represents the length of the chord subtended by the arc of path between P_i and P_{i+1}. Likewise, σ denotes a path length measured along the spatial polygonal joining the N points $\{P_i\}_1^N$. Thus, the closer the aforementioned points, the closer the approximation of $\Delta\sigma_i$ to the arc length between these two points, and hence, the better the approximations of the curve properties.

The foregoing spline coefficients A_{xi}, A_{yi}, $\ldots$, D_{zi}, for $i = 1, \ldots, N-1$, are determined as explained below. Let us define the N-dimensional vectors

$$\mathbf{x} \equiv [x_1, \ldots, x_N]^T, \qquad \mathbf{x}'' \equiv [x_1'', \ldots, x_N'']^T \quad (9.48a)$$

$$\mathbf{y} \equiv [y_1, \ldots, y_N]^T, \qquad \mathbf{y}'' \equiv [y_1'', \ldots, y_N'']^T \quad (9.48b)$$

$$\mathbf{z} \equiv [z_1, \ldots, z_N]^T, \qquad \mathbf{z}'' \equiv [z_1'', \ldots, z_N'']^T \quad (9.48c)$$

The relationships between $\mathbf{x}$, $\mathbf{y}$, and $\mathbf{z}$ and their counterparts $\mathbf{x}''$, $\mathbf{y}''$, and $\mathbf{z}''$ are the same as those found for nonparametric splines in eq.(5.58a), namely,

$$\mathbf{A}\mathbf{x}'' = 6\mathbf{C}\mathbf{x} \quad (9.49a)$$

$$\mathbf{A}\mathbf{y}'' = 6\mathbf{C}\mathbf{y} \quad (9.49b)$$

$$\mathbf{A}\mathbf{z}'' = 6\mathbf{C}\mathbf{z} \quad (9.49c)$$

which are expressions similar to those of eq.(5.58a), except that the $\mathbf{A}$ and $\mathbf{C}$ matrices appearing in eq.(9.49b) are now themselves functions of the coordinates of the supporting points (SP) of the spline. In fact, the $(N-2) \times N$ matrices $\mathbf{A}$ and $\mathbf{C}$ are now defined exactly as in eqs.(5.58b &

c), repeated below for quick reference:

$$\mathbf{A} = \begin{bmatrix} \alpha_1 & 2\alpha_{1,2} & \alpha_2 & 0 & \cdots & 0 & 0 \\ 0 & \alpha_2 & 2\alpha_{2,3} & \alpha_3 & \cdots & 0 & 0 \\ \vdots & \vdots & \ddots & \ddots & \ddots & \vdots & \vdots \\ 0 & 0 & \cdots & \alpha_{N'''} & 2\alpha_{N''',N''} & \alpha_{N''} & 0 \\ 0 & 0 & 0 & \cdots & \alpha_{N''} & 2\alpha_{N'',N'} & \alpha_{N'} \end{bmatrix} \tag{9.49d}$$

and

$$\mathbf{C} = \begin{bmatrix} \beta_1 & -\beta_{1,2} & \beta_2 & 0 & \cdots & 0 & 0 \\ 0 & \beta_2 & -\beta_{2,3} & \beta_3 & \cdots & 0 & 0 \\ \vdots & \vdots & \ddots & \ddots & \ddots & \vdots & \vdots \\ 0 & 0 & \cdots & \beta_{N'''} & -\beta_{N''',N''} & \beta_{N''} & 0 \\ 0 & 0 & 0 & \cdots & \beta_{N''} & -\beta_{N'',N'} & \beta_{N'} \end{bmatrix} \tag{9.49e}$$

where α_k and β_k are now defined correspondingly, i.e., for $i, j, k = 1, \ldots, N'$,

$$\alpha_k = \Delta\sigma_k, \quad \alpha_{i,j} = \alpha_i + \alpha_j, \quad \beta_k = 1/\alpha_k, \quad \beta_{i,j} = \beta_i + \beta_j \tag{9.50}$$

while N', N'', and N''' are defined as in eq.(5.58f), i.e., as

$$N' \equiv N - 1, \quad N'' \equiv N - 2, \qquad N''' \equiv N - 3 \tag{9.51}$$

Note that the spline $\mathbf{p}(\sigma)$ is fully determined once its coefficients are known. These are computed exactly as their counterparts for nonparametric splines, namely, as in eqs.(5.55a–e). Obviously, different from the aforementioned formulas, the coefficients of the parametric spline pertain to three coordinates, and hence, three sets of such coefficients need be computed in this case. In order to simplify matters, we introduce the vectors below:

$$\mathbf{a}_k \equiv \begin{bmatrix} A_{xk} \\ A_{yk} \\ A_{zk} \end{bmatrix}, \quad \mathbf{b}_k \equiv \begin{bmatrix} B_{xk} \\ B_{yk} \\ B_{zk} \end{bmatrix}, \quad \mathbf{c}_k \equiv \begin{bmatrix} C_{xk} \\ C_{yk} \\ C_{zk} \end{bmatrix}, \quad \mathbf{d}_k \equiv \begin{bmatrix} D_{xk} \\ D_{yk} \\ D_{zk} \end{bmatrix} \tag{9.52}$$

and thus, the position vector of an arbitrary point P on the parametric spline takes on the form

$$\mathbf{p}(\sigma) = \mathbf{a}_k(\sigma - \sigma_k)^3 + \mathbf{b}_k(\sigma - \sigma_k)^2 + \mathbf{c}_k(\sigma - \sigma_k) + \mathbf{d}_k, \quad k = 1, \ldots, N - 1 \tag{9.53a}$$

in the interval $\sigma_k \leq \sigma \leq \sigma_{k+1}$. The counterpart set of eqs.(5.55a–e) is then

$$\mathbf{a}_k = \frac{1}{6\,\Delta\sigma_k}(\mathbf{p}''_{k+1} - \mathbf{p}''_k) \tag{9.53b}$$

$$\mathbf{b}_k = \frac{1}{2}\mathbf{p}''_k \tag{9.53c}$$

$$\mathbf{c}_k = \frac{\Delta\mathbf{p}_k}{\Delta\sigma_k} - \frac{1}{6}\Delta\sigma_k\,(\mathbf{p}''_{k+1} + 2\mathbf{p}''_k) \tag{9.53d}$$

$$\mathbf{d}_k = \mathbf{p}_k \tag{9.53e}$$

$$\Delta\mathbf{p}_k \equiv \mathbf{p}_{k+1} - \mathbf{p}_k \tag{9.53f}$$

where vectors $\mathbf{p}_k$ and $\mathbf{p''}_k$ are defined as

$$\mathbf{p}_k \equiv \begin{bmatrix} x_k \\ y_k \\ z_k \end{bmatrix}, \qquad \mathbf{p}''_k \equiv \begin{bmatrix} x''_k \\ y''_k \\ z''_k \end{bmatrix} \tag{9.54}$$

Note that since $\mathbf{p}$ is piecewise cubic in σ, $\mathbf{p}'$ is piecewise quadratic, whereas $\mathbf{p}''$ is piecewise linear in the same argument, $\mathbf{p}'''$ being piecewise constant; higher-order derivatives vanish. Properly speaking, however, the piecewise constancy of $\mathbf{p}'''$ causes the fourth-order derivative to be discontinuous at the SP, and consequently, all higher-order derivatives are equally discontinuous at those points. In practice, these discontinuities are smoothed out by the inertia of the links and the motors, if the SP are chosen close enough. Obviously, higher-order continuity can be achieved if higher-order splines, e.g., quintic splines, are used instead. For the sake of conciseness, these splines are not discussed here, the interested reader being directed to the specialized literature (Dierckx, 1993).

Further, the $N \times 3$ matrices $\mathbf{P}$ and $\mathbf{P}''$ are defined as

$$\mathbf{P} \equiv \begin{bmatrix} \mathbf{p}_1^T \\ \mathbf{p}_2^T \\ \vdots \\ \mathbf{p}_N^T \end{bmatrix}, \quad \mathbf{P}'' \equiv \begin{bmatrix} (\mathbf{p}''_1)^T \\ (\mathbf{p}''_2)^T \\ \vdots \\ (\mathbf{p}''_N)^T \end{bmatrix} \tag{9.55}$$

which allows us to rewrite eqs.(9.49b) in matrix form as

$$\mathbf{A}\mathbf{P}'' = 6\mathbf{C}\mathbf{P} \tag{9.56}$$

It is now apparent that the spline coefficients $\mathbf{a}_k, \ldots, \mathbf{d}_k$ can be calculated once vectors $\mathbf{p}''_k$ are available. These vectors can be computed via matrix $\mathbf{P}''$ as the solution to eq.(9.56). However, finding this solution requires *inverting* the $(N-2) \times N$ matrix $\mathbf{A}$, which is rectangular and hence cannot be inverted, properly speaking. We thus have an underdetermined system of linear equations, and further conditions are needed in order to render it determined. Such conditions are those defining the type of spline at hand. For example, closed paths call naturally for *periodic splines*, while open paths call for other types such as *natural splines*. The conditions imposed on periodic parametric splines are listed below:

$$\mathbf{p}_N = \mathbf{p}_1, \quad \mathbf{p}'_N = \mathbf{p}'_1, \quad \mathbf{p}''_N = \mathbf{p}''_1 \tag{9.57a}$$

On the other hand, natural parametric splines are obtained under the conditions

$$\mathbf{p}''_1 = \mathbf{p}''_N = \mathbf{0} \tag{9.57b}$$

Thus, if a periodic parametric spline is required, then vectors $\mathbf{p}_N$ and $\mathbf{p}''_N$ can be deleted from matrices $\mathbf{P}$ and $\mathbf{P}''$, respectively, these then becoming

$(N-1) \times 3$ matrices, namely,

$$\mathbf{P} \equiv \begin{bmatrix} \mathbf{p}_1^T \\ \mathbf{p}_2^T \\ \vdots \\ \mathbf{p}_{N-1}^T \end{bmatrix}, \quad \mathbf{P}'' \equiv \begin{bmatrix} (\mathbf{p}_1'')^T \\ (\mathbf{p}_2'')^T \\ \vdots \\ (\mathbf{p}_{N-1}''{}^T) \end{bmatrix} \tag{9.58}$$

Moreover, the first-derivative condition of eq.(9.57a) is added to the $N-2$ continuity conditions of eq.(5.56), thereby obtaining $N-1$ equations of this form. Consequently, $\mathbf{A}$ becomes an $(N-1) \times (N-1)$ matrix. Correspondingly, $\mathbf{C}$ also becomes an $(N-1) \times (N-1)$ matrix, i.e.,

$$\mathbf{A} \equiv \begin{bmatrix} 2\alpha_{1,N'} & \alpha_1 & 0 & 0 & \cdots & \alpha_{N'} \\ \alpha_1 & 2\alpha_{1,2} & \alpha_2 & 0 & \cdots & 0 \\ 0 & \alpha_2 & 2\alpha_{2,3} & \alpha_3 & \cdots & 0 \\ \vdots & \vdots & \ddots & \ddots & \ddots & \vdots \\ 0 & 0 & \cdots & \alpha_{N'''} & 2\alpha_{N''',N''} & \alpha_{N''} \\ \alpha_{N'} & 0 & 0 & \cdots & \alpha_{N''} & 2\alpha_{N'',N'} \end{bmatrix} \tag{9.59a}$$

and

$$\mathbf{C} \equiv \begin{bmatrix} -\beta_{1,N'} & \beta_1 & 0 & 0 & \cdots & \beta_{N'} \\ \beta_1 & -\beta_{1,2} & \beta_2 & 0 & \cdots & 0 \\ 0 & \beta_2 & -\beta_{2,3} & \beta_3 & \cdots & 0 \\ \vdots & \vdots & \ddots & \ddots & \ddots & \vdots \\ 0 & 0 & \cdots & \beta_{N'''} & -\beta_{N''',N''} & \beta_{N''} \\ \beta_{N'} & 0 & 0 & \cdots & \beta_{N''} & -\beta_{N'',N'} \end{bmatrix} \tag{9.59b}$$

Since $\mathbf{A}$ is nonsingular, eq.(9.56) can be solved for $\mathbf{P}''$, namely,

$$\mathbf{P}'' = 6\mathbf{A}^{-1}\mathbf{C}\mathbf{P} \tag{9.60}$$

thereby computing all vectors $\{\mathbf{p}_k''\}_1^{N-1}$, from which $\mathbf{p}_N''$ can be readily obtained. Hence, the spline coefficients follow.

Likewise, if natural parametric splines are used, then $\mathbf{P}''$ becomes an $(N-2) \times 3$ matrix, while $\mathbf{A}$, consequently, becomes an $(N-2) \times (N-2)$ matrix, as given in eq.(5.59).

Example 9.4.1 (Spline-approximation of a warped path) *For the numerical values* $R = 0.6$ m, $r = 0.15$ m, *and* $d = 0.3$ m, *determine the periodic parametric cubic spline approximating the intersection of the sphere and the cylinder of Fig. 9.4, with 12 equally spaced supporting points along the cylindrical coordinate* φ, *i.e., with supporting points distributed along the intersection curve at intervals* $\Delta\varphi = 30°$. *Using the spline, find values of the tangent, normal, and binormal vectors of the curve, as well as the rotation matrix* $\mathbf{Q}$. *In order to quantify the error in this approximation,*

TABLE 9.2. The Cartesian Coordinates of the Supporting Points

φ	0°	30°	60°	90°	120°	150°
x	0.45	0.429904	0.375	0.3	0.225	0.170096
y	0	0.075	0.129904	0.15	0.129904	0.075
z	0.396863	0.411774	0.45	0.497494	0.540833	0.570475
φ	180°	210°	240°	270°	300°	330°
x	0.15	0.170096	0.225	0.3	0.375	0.429904
y	0	-0.075	-0.129904	-0.15	-0.129904	-0.075
z	0.580948	0.570475	0.540833	0.497494	0.45	0.411774

compare (i) the components of the two position vectors, the exact and the spline-generated ones, while normalizing their differences using the radius of the cylinder r; and (ii) the Euler-Rodrigues parameters of the exact and the spline-approximated rotation matrices. Plot these errors vs. φ.

Solution: We use eq.(9.37) to find the Cartesian coordinates of the supporting points. The numerical results are given in terms of the components of $\mathbf{r} \equiv [x,\ y,\ z]^T$ in Table 9.2. Note that this table does not include the Cartesian-coordinate values at 360° because these are identical with those at 0°.

The four Euler-Rodrigues parameters $\{r_i\}_{i=0}^{3}$ of the rotation matrix are most suitably calculated in terms of the linear invariants, i.e., as appearing in eq.(2.77). If we let $\widetilde{\mathbf{p}}$ and $\widetilde{\mathbf{r}}$ denote the estimates of $\mathbf{p}$ and $\mathbf{r}$, respectively, then the orientation error is evaluated via the the four differences $\Delta r_i = r_i - \widetilde{r}_i$, for $i = 0, \ldots, 3$. The positioning error is computed, in turn, as the normalized difference $\boldsymbol{\epsilon} = (\mathbf{p} - \widetilde{\mathbf{p}})/r$ to yield a dimensionless number, its components being denoted by ϵ_x, ϵ_y, and ϵ_z. The aforementioned errors are plotted vs. φ in Figs. 9.5 and 9.6.

Note that the orientation errors are, roughly, one order of magnitude greater than the positioning errors.

9.5 Continuous-Path Tracking

When a continuous trajectory is to be tracked with a robot, the joint angles have to be calculated along a continuous set of poses of the end-effector. In practice, the continuous trajectory is sampled at a discrete set of close-enough poses $\{\,\mathbf{s}_k\,\}_1^N$ along the continuous trajectory. Then in principle, an IKP must be solved at each sampled pose. If the manipulator is of the decoupled type, these calculations are feasible in a fraction of a millisecond, for the solution reduces, in the majority of the cases, to a cascading of quadratic equations. In the worst case, the inverse kinematics of a decoupled

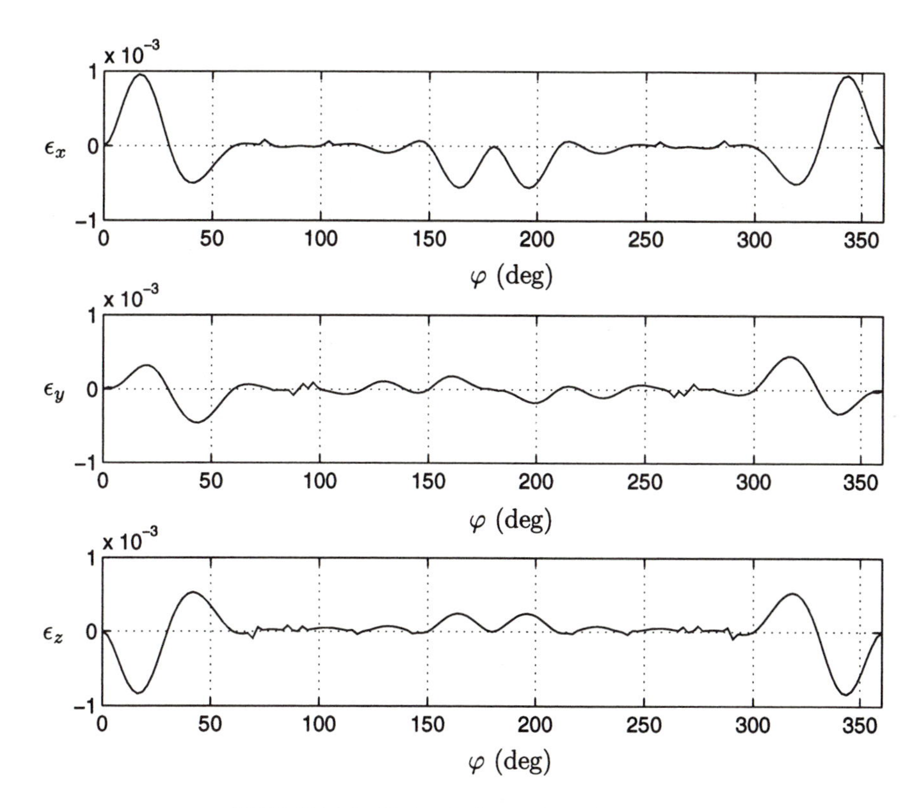

FIGURE 9.5. Plots of the positioning errors.

manipulator requires finding all the roots of a quartic equation at each sampled pose, but this is still feasible in the same time frame, for the four roots of interest can be calculated from formulas. However, if the manipulator has an architecture not lending itself to a simple solution and requires solving polynomials of a degree higher than four, then finding all solutions at each sample pose may require a few milliseconds, which may be too slow in fast operations. Hence, an alternative approach is needed.

The alternative is to solve the IKP *iteratively*. That is, if we have the value of the vector of joint variables $\boldsymbol{\theta}(t_k)$ and want to find its value at t_{k+1}, then we use Algorithm 9.5.1.

Various procedures are available to find the correction $\Delta\boldsymbol{\theta}$ above. The one we have found very convenient is based on the Newton-Raphson method (Dahlquist and Björck, 1974). In the realm of Newton methods—there are several of these, the Newton-Raphson method being only one of this class—the closure equations (4.9a & b) are written in the form

$$\mathbf{f}(\boldsymbol{\theta}) = \mathbf{s}_d \tag{9.61}$$

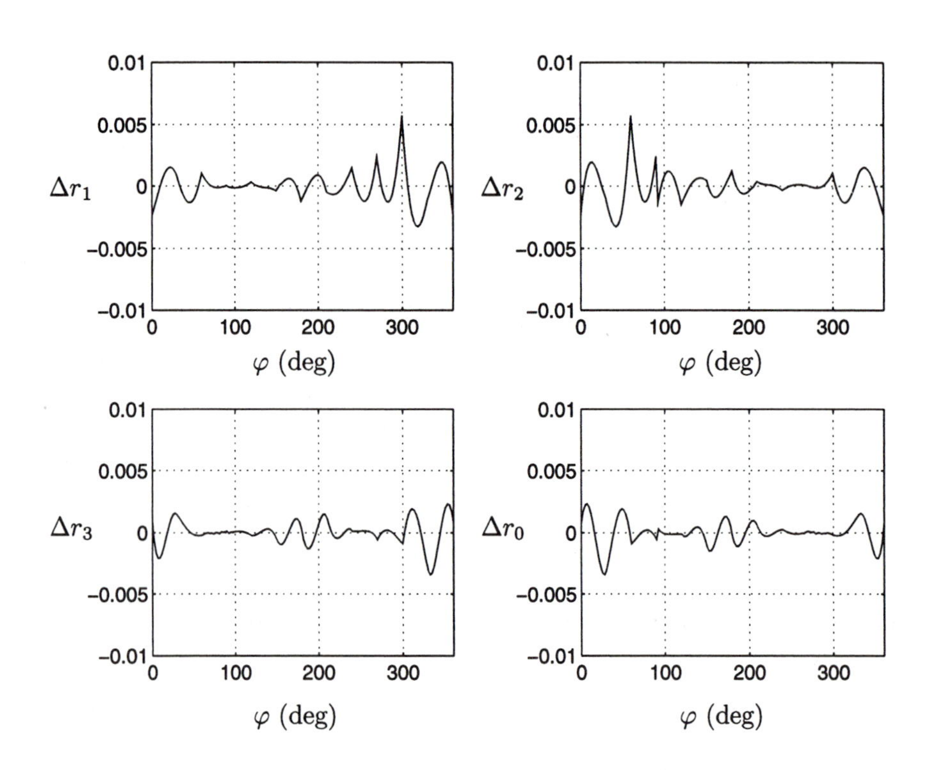

FIGURE 9.6. Plots of the orientation errors.

Algorithm 9.5.1

```
    θ ← θ(t_k)
1   find correction Δθ
    if ‖Δθ‖ ≤ ε, then stop;
    else
        θ ← θ + Δθ
    go to 1
```

where $\mathbf{s}_d$ is the 7-dimensional *prescribed pose array*. We recall here the definition of the pose array introduced in Section 3.2 to represent $\mathbf{s}_d$, namely,

$$\mathbf{s}_d \equiv \begin{bmatrix} \mathbf{q} \\ q_0 \\ \mathbf{p} \end{bmatrix}_d \tag{9.62}$$

with $\mathbf{q}$ and q_0 defined, in turn, as a 3-dimensional vector invariant and its corresponding scalar, respectively, of the rotation $\mathbf{Q}$. Moreover, $\mathbf{p}$ is the position vector of the operation point. Therefore, the 7-dimensional vector $\mathbf{f}$ is defined, correspondingly, as

$$\mathbf{f}(\boldsymbol{\theta}) \equiv \begin{bmatrix} \mathbf{f}_v(\boldsymbol{\theta}) \\ f_0(\boldsymbol{\theta}) \\ \mathbf{f}_p(\boldsymbol{\theta}) \end{bmatrix} \tag{9.63}$$

where $\mathbf{f}_v(\boldsymbol{\theta})$ denotes the counterpart of $\mathbf{q}$ above, as pertaining to the product $\mathbf{Q}_1 \cdots \mathbf{Q}_6$ of eq.(4.9a); $f_0(\boldsymbol{\theta})$ is the counterpart of q_0, as pertaining to the same product; and $\mathbf{f}_p(\boldsymbol{\theta})$ is the sum $\mathbf{a}_1 + \ldots + \mathbf{a}_6$. In principle, any of the three types of rotation invariants introduced in Section 3.2 can be used in the above formulation.

Now, eq.(9.61) represents a nonlinear system of seven equations in six unknowns. The system is thus overdetermined, but since the four rotational equations are consistent, this system should admit an exact solution, even if this solution is complex. For example, if $\mathbf{p}$ is specified in $\mathbf{s}_d$ above as lying outside of the manipulator workspace, then no real solution is possible, and the solution reported by any iterative procedure capable of handling complex solutions will be complex.

Upon application of a Newton-type method to find a solution of eq.(9.61), we assume that we have an initial guess $\boldsymbol{\theta}^0$, and based on this value, we generate a sequence $\boldsymbol{\theta}^1, \ldots, \boldsymbol{\theta}^i, \boldsymbol{\theta}^{i+1}, \ldots$, until either a convergence or an abortion criterion is met. This sequence is generated in the form

$$\boldsymbol{\theta}^{i+1} = \boldsymbol{\theta}^i + \Delta\boldsymbol{\theta}^i \tag{9.64}$$

with $\Delta\boldsymbol{\theta}^i$ calculated from

$$\boldsymbol{\Phi}(\boldsymbol{\theta}^i)\Delta\boldsymbol{\theta}^i = -\mathbf{f}(\boldsymbol{\theta}^i) \tag{9.65}$$

and $\boldsymbol{\Phi}$ defined as the *Jacobian matrix* of $\mathbf{f}(\boldsymbol{\theta})$ with respect to $\boldsymbol{\theta}$. Note that by virtue of its definition, $\boldsymbol{\Phi}$ is a 7×6 matrix. A common misconception in the robotics literature is to confuse this Jacobian matrix with the Jacobian defined by Whitney (1972) and introduced in eq.(4.55a), which maps joint rates into the EE twist. The difference between the two Jacobians is essential and is made clear in the discussion below. First and foremost, $\boldsymbol{\Phi}$ is an actual Jacobian matrix, while Whitney's Jacobian, properly speaking, is not. In fact, $\boldsymbol{\Phi}$ is defined as

$$\boldsymbol{\Phi} \equiv \frac{\partial \mathbf{f}}{\partial \boldsymbol{\theta}} \tag{9.66}$$

In order to find $\mathbf{\Phi}$ in eq.(9.65), we note that by application of the chain rule,

$$\dot{\mathbf{f}} = \frac{\partial \mathbf{f}}{\partial \boldsymbol{\theta}}\dot{\boldsymbol{\theta}} \equiv \mathbf{\Phi}\dot{\boldsymbol{\theta}} \tag{9.67}$$

However, from the definition of $\mathbf{f}$, we have that $\dot{\mathbf{f}}$ is the time-derivative of the pose array of the EE, i.e., $\dot{\mathbf{s}}$. Moreover, by virtue of eq.(3.79), this time-derivative can be expressed as a linear transformation of the twist $\mathbf{t}$ of the EE, i.e.,

$$\dot{\mathbf{f}} = \mathbf{T}\mathbf{t} \tag{9.68a}$$

with $\mathbf{T}$ defined in Section 3.2 as

$$\mathbf{T} \equiv \begin{bmatrix} \mathbf{F} & \mathbf{O}_{43} \\ \mathbf{O} & \mathbf{1} \end{bmatrix} \tag{9.68b}$$

where $\mathbf{O}$ and $\mathbf{O}_{43}$ denote the 3×3 and the 4×3 zero matrices, $\mathbf{1}$ being the 3×3 identity matrix. Moreover, matrix $\mathbf{F}$ takes on various forms, depending on the type of rotation representation adopted, as discussed in Section 3.2.

Moreover, we write the left-hand side of eq.(9.68a) as shown in eq.(9.67) and the twist $\mathbf{t}$ of the right-hand side of eq.(9.68a) in terms of $\dot{\boldsymbol{\theta}}$, as expressed in eq.(4.54), thereby obtaining

$$\mathbf{\Phi}\dot{\boldsymbol{\theta}} \equiv \mathbf{T}\mathbf{J}\dot{\boldsymbol{\theta}} \tag{9.69}$$

which is a relation valid for any value of $\dot{\boldsymbol{\theta}}$. As a consequence, then,

$$\mathbf{\Phi} = \mathbf{T}\mathbf{J} \tag{9.70}$$

whence the relation between the two Jacobians is apparent. Note that eq.(9.68a) allows us to write

$$\dot{\mathbf{f}} = \mathbf{T}\mathbf{J}\dot{\boldsymbol{\theta}} \tag{9.71}$$

Furthermore, if we denote by $\mathbf{t}_d$ the prescribed value of the twist of the EE, then we have

$$\dot{\mathbf{s}}_d = \mathbf{T}\mathbf{t}_d \tag{9.72}$$

Upon equating the right-hand sides of eqs.(9.71) and (9.72), we obtain

$$\mathbf{T}\mathbf{J}\dot{\boldsymbol{\theta}} = \mathbf{T}\mathbf{t}_d \tag{9.73}$$

If linear invariants are used to represent the rotation, then $\mathbf{T}$ becomes rank-deficient if and only if the angle of the rotation becomes π (Tandirci, Angeles, and Darcovich, 1994); otherwise, $\mathbf{T}$ is always of full rank, and eq.(9.73) leads to

$$\mathbf{J}\dot{\boldsymbol{\theta}} = \mathbf{t}_d \tag{9.74}$$

which is exactly the same as eq.(4.54). Now we multiply both sides of the foregoing equation by Δt, thereby obtaining

$$\mathbf{J}\Delta\boldsymbol{\theta} = \mathbf{t}_d \Delta t \tag{9.75}$$

All we need now is, apparently, the product in the right-hand side of the above equation, namely,

$$\mathbf{t}_d \Delta t = \begin{bmatrix} \boldsymbol{\omega}\Delta t \\ \dot{\mathbf{p}}\Delta t \end{bmatrix} = \begin{bmatrix} \boldsymbol{\omega}\Delta t \\ \Delta\mathbf{p} \end{bmatrix} \tag{9.76}$$

The product $\boldsymbol{\omega}\Delta t$ is found below: First and foremost, it is common practice in the realm of Newton methods to assume that a good enough approximation to the root sought is available, and hence, $\Delta\boldsymbol{\theta}$ is "small." That is, we assume that $\|\Delta\boldsymbol{\theta}\|$ is small, where $\|\cdot\|$ denotes *any* vector norm. Moreover, we use the end-effector pose at $t = t_k$ as a reference to describe the desired pose at $t = t_{k+1}$, the rotation sought—that takes the EE to its desired attitude—being assumed about an axis parallel to $\mathbf{e}_d$ and through an as-yet unknown angle $\Delta\phi$. Thus, the rotation needed to take the end-effector to the desired pose is assumed to involve a small angle of rotation $\Delta\phi$ about an axis parallel to the unit vector $\mathbf{e}_d$. Here, $\mathbf{e}_d$ is the unit vector parallel to the axis of the rotation that took the end-effector from the fixed frame $\mathcal{F}_1$, i.e., the frame attached to the robot base, to the attitude it has at $t = t_k$.

Henceforth, we will use linear invariants to represent rotations. The linear invariants of this rotation are $\mathbf{q} = \mathbf{e}_d \sin\Delta\phi$ and $q_0 = \cos\Delta\phi$. Moreover, the natural invariants of the rotation matrix at $t = t_k$ are $\mathbf{e}_d$ and $\phi = 0$. We now have, from eqs.(3.78a & c),

$$\boldsymbol{\omega}\Delta t = \widetilde{\mathbf{L}}\dot{\boldsymbol{\lambda}}\Delta t = \widetilde{\mathbf{L}}\Delta\boldsymbol{\lambda} \tag{9.77}$$

where $\widetilde{\mathbf{L}}$ is computed at $t = t_k$, i.e.,

$$\widetilde{\mathbf{L}} = [\,\mathbf{1} \quad \mathbf{0}\,], \quad \Delta\boldsymbol{\lambda} = \begin{bmatrix} \mathbf{e}_d \sin\Delta\phi \\ \cos\Delta\phi \end{bmatrix} \tag{9.78}$$

whence

$$\boldsymbol{\omega}\Delta t = \mathbf{e}_d \sin\Delta\phi \tag{9.79}$$

In summary, then, the correction $\Delta\boldsymbol{\theta}$ is computed from

$$\mathbf{J}\Delta\boldsymbol{\theta} = \Delta\mathbf{t} \tag{9.80}$$

with $\Delta\mathbf{t}$ defined as

$$\Delta\mathbf{t} \equiv \begin{bmatrix} \mathbf{e}_d \sin\Delta\phi \\ \Delta\mathbf{p} \end{bmatrix} \tag{9.81}$$

and $\Delta\mathbf{p}$ defined, in turn, as the difference between the prescribed value $\mathbf{p}_d$ of the position vector of the operation point and its value at the current

Algorithm 9.5.2

1 $\Delta\phi \leftarrow \phi - \phi_d$

$\Delta\mathbf{p} \leftarrow \mathbf{p} - \mathbf{p}_d$

$\Delta\mathbf{t} \leftarrow \begin{bmatrix} (\sin\Delta\phi)\mathbf{e}_d \\ \Delta\mathbf{p} \end{bmatrix}$

$\Delta\boldsymbol{\theta} \leftarrow \mathbf{J}^{-1}\Delta\mathbf{t}$

`if` $\|\Delta\boldsymbol{\theta}\| \leq \epsilon$`, then stop;`

`else`

  $s \leftarrow \mathbf{f}(\boldsymbol{\theta})$

`go to 1`

iteration. Thus, the numerical path-tracking scheme consists essentially of eqs.(9.80) and (9.81), as first proposed by Pieper (1968). We thus have Algorithm 9.5.2.

When implementing the foregoing procedure, we want to save processing time; hence, we aim at fast computations. The computation of the correction $\Delta\boldsymbol{\theta}$ involves only linear-equation solving, which was discussed at length in Chapter 4 and need not be discussed further here. The only item that still needs some discussion is the calculation of the vector norm $\|\Delta\boldsymbol{\theta}\|$. Since any norm can be used here, we can choose the norm that is fastest to compute, namely, the *maximum norm*, also known as the *Chebyshev norm*, represented as $\|\Delta\boldsymbol{\theta}\|_\infty$, and defined as

$$\|\Delta\boldsymbol{\theta}\|_\infty \equiv \max_i\{\,|\theta_i|\,\} \tag{9.82}$$

Note that this norm only requires comparisons and no floating-point operations. The Euclidean norm, on the contrary, requires n multiplications, $n-1$ additions, and one square root, for an n-dimensional vector.

Example 9.5.1 (Path-tracking for arc-welding) *Here we want to weld the sphere and the cylinder of Example 9.3.2 using a robot for arc welding, e.g., a Fanuc Arc Mate, whose Denavit-Hartenberg parameters are listed in Table 4.2. Furthermore, the welding seam to be tracked is placed well within the workspace of the manipulator. A location found quite suitable for this task was obtained with the aid of RVS, our Robot Visualization System. This location requires that the coordinate frame $\mathcal{F}_C$ of Fig. 9.4 have its axes parallel pairwise to those of the robot base, $\mathcal{F}_1$. The latter is defined according to the Denavit-Hartenberg notation, and so Z_1 coincides with the* *axis of the first revolute; it is, moreover, directed upwards. The position found for the origin O_C of $\mathcal{F}_C$, of position vector* $\mathbf{o}$, *is given in* $\mathcal{F}_1$ *as*

$$[\mathbf{o}]_1 \equiv \begin{bmatrix} x \\ y \\ z \end{bmatrix} = \begin{bmatrix} -1.0 \\ -0.1 \\ 0.5 \end{bmatrix} \text{ m}$$

Find the time-histories of all the joint variables that will perform the desired operation with the tip of the electrode traversing the intersection curve at the constant speed of $v_0 = 0.1$ *m/s. Furthermore, plot the variation of the condition number of the Jacobian matrix along the path.*

Solution: The robot at hand was studied in Subsection 8.2.3, where it was found not to be of the decoupled type. In fact, this robot does not admit a closed-form inverse kinematics solution, and hence, the foregoing iterative procedure is to be used.

At the outset, we calculate all inverse kinematics solutions at the pose corresponding to $\varphi = 0$ using the contour intersection approach of Subsection 8.2.1. This pose is defined by the orthogonal matrix $\mathbf{Q}$ and the position vector $\mathbf{p}$ given below:

$$[\mathbf{Q}]_1 \equiv [\,\mathbf{e}_b \quad \mathbf{e}_t \quad \mathbf{e}_n\,] = \begin{bmatrix} 0.6030 & 0 & -0.7977 \\ 0 & 1 & 0 \\ 0.7977 & 0 & 0.6030 \end{bmatrix}, \quad [\mathbf{p}]_1 = \begin{bmatrix} -0.5500 \\ -0.100 \\ 0.8969 \end{bmatrix} \text{ m}$$

with both $\mathbf{Q}$ and $\mathbf{p}$ given in robot-base coordinates. The contours for the above pose, which were obtained using the procedure of Subsection 8.2.1, are shown in Fig. 9.7, the eight solutions obtained being summarized in Table 9.3, which includes the condition number of the Jacobian, $\kappa(\mathbf{J})$, of each solution. Note that the calculation of $\kappa(\mathbf{J})$ required computing the characteristic length of the robot, as explained in Section 4.9. This length, as calculated in that section, turned out to be $L = 0.3573$ m.

Now, we have eight solutions at our disposal, from which we must choose one for path-tracking. In the absence of any criterion to single out one specific solution, we can pick the solution with the lowest condition number. If we do this, we end up with solution 1 in Table 9.3. However, when we attempted to track the given path with this solution, it turned out that this solution encountered a singularity and was hence discarded. Of the seven remaining solutions, solution 5 has the lowest condition number; this solution led to a singularity-free trajectory.

Once the appropriate solution is chosen, the trajectory can be tracked with the aid of Algorithm 9.5.2. Here, we need a discrete set of poses at equal time-intervals. Note that we can produce such a set at equal intervals of angle φ because we have expressions for the pose variables in terms of this angle. In order to obtain this set at equal time-intervals, then, we need angle φ as a function of time, i.e., $\varphi(t)$. In the sequel, we will also need the

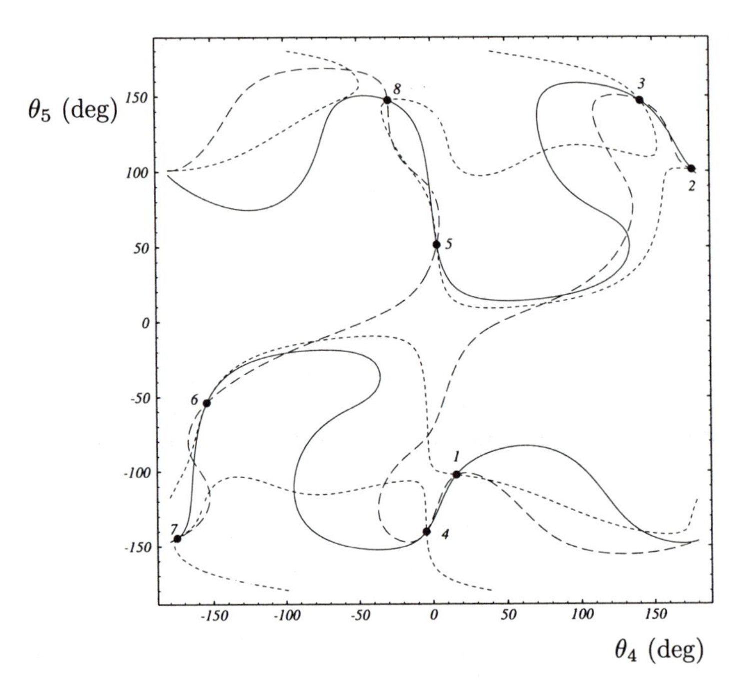

FIGURE 9.7. Contour solutions of the Fanuc Arc Mate robot at the given EE pose.

TABLE 9.3. Inverse Kinematics Solutions of the Fanuc Arc Mate Robot for the Given EE Pose.

i	$\kappa(\mathbf{J})$	θ_1	θ_2	θ_3
1	4.74	19.9039°	124.909°	−176.484°
2	4.85	−3.6664°	124.723°	−173.071°
3	11.12	−154.951°	−67.5689°	−135.549°
4	6.31	−176.328°	−63.4487°	−129.817°
5	4.79	−176.341°	75.1632°	−76.6692°
6	5.20	−153.567°	73.4546°	−72.5407°
7	8.68	−3.6362°	−129.644°	−32.9672°
8	9.94	18.9031°	−131.096°	−26.8084°

i	θ_4	θ_5	θ_6
1	16.1379°	−102.29°	−15.8409°
2	177.019°	101.19°	−177.208°
3	141.716°	146.966°	17.754°
4	−4.5893°	−140.319°	−178.681°
5	3.7343°	51.4104°	−179.877°
6	−153.868°	−53.7328°	−0.5046°
7	−175.011°	−144.428°	178.133°
8	−28.6793°	147.417°	13.0786°

time T needed to complete the task. Now, since the speed of the electrode tip is constant and equal to v_0, the time T is readily obtained by dividing the total length l of the curve by v_0. The length of the curve, in turn, can be computed as $s(2\pi)$, where function $s(\varphi)$ denotes the arc length as a function of angle φ, i.e.,

$$s(\varphi) = \int_0^{\varphi} \|\mathbf{r}'(\varphi)\| d\varphi \tag{9.83}$$

We thus obtain, by numerical quadrature,

$$l \equiv s(2\pi) = 1.0257 \text{ m}$$

Hence, the total time is

$$T \equiv \frac{l}{v_0} = 10.257 \text{ s}$$

Now, in order to obtain $\varphi(t)$, we first calculate $\dot{s}$ as

$$\dot{s} \equiv \frac{ds}{dt} = \frac{ds}{d\varphi}\frac{d\varphi}{dt} = \dot{\varphi}\frac{ds}{d\varphi} \tag{9.84a}$$

Furthermore, we note that $ds/d\varphi = \|\mathbf{r}'(\varphi)\|$, which allows us to write $\dot{s}$ as

$$\dot{s} \equiv \dot{\varphi}\|\mathbf{r}'(\varphi)\|$$

Moreover, $\|\mathbf{r}'(\varphi)\|$ was found in eq.(9.39b) to be

$$\|\mathbf{r}'(\varphi)\| = rG(\varphi)$$

$\dot{s}$ thus becoming

$$\dot{s} = rG\dot{\varphi} \tag{9.84b}$$

Furthermore, we recall the expression derived for $G(\varphi)$ in eq.(9.39e). This expression, along with the constancy condition on $\dot{s}$, i.e., $\dot{s} = v_0$, leads to

$$r\dot{\varphi}\sqrt{1 + (\lambda\hat{\varphi}\sin\varphi)^2} = v_0$$

where r is the radius of the cylinder. Upon solving for $\dot{\varphi}$ from the above equation, we obtain

$$\dot{\varphi} = \frac{v_0}{r}\sqrt{\frac{\rho^2 - 2\lambda\cos\varphi}{\rho^2 - 2\lambda\cos\varphi + \lambda^2\sin^2\varphi}}$$

which is a nonlinear first-order differential equation for $\varphi(t)$. Its initial value can be assigned as $\varphi(0) = 0$, thereby formulating a nonlinear first-order initial-value problem. The numerical solution of the foregoing problem is nowadays routine work, which can be handled with suitable software, e.g.,

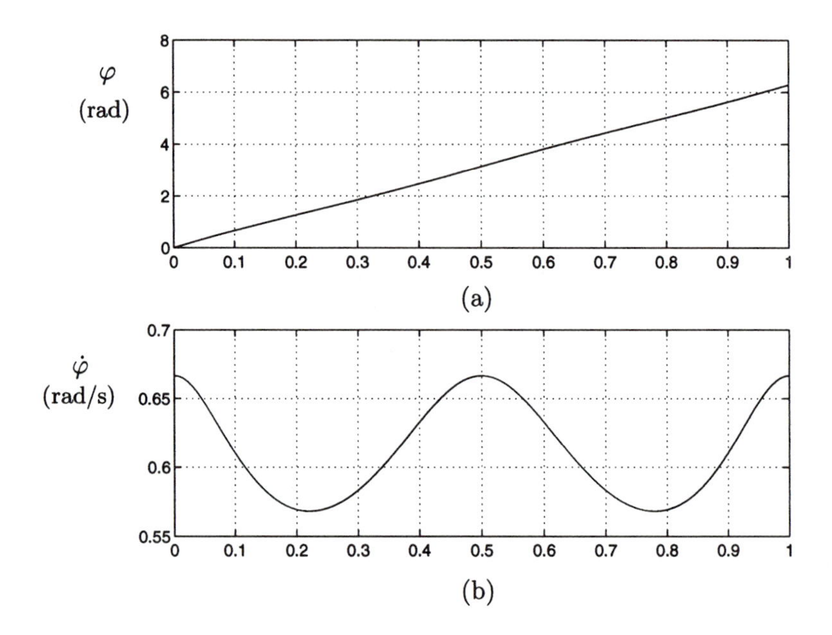

FIGURE 9.8. Plot of φ vs. nondimensional time.

Matlab (Etter, 1993) or IMSL. Upon solving this equation, a data file is produced that contains the time-history of φ. The plot of φ vs. nondimensional time is displayed in Fig. 9.8a. Since the variations of $\varphi(t)$ are relatively small, this plot provides little information on the time-history of interest. A more informative plot, that of $\dot{\varphi}(t)$, is included in Fig. 9.8b for this reason.

With $\varphi(t)$ known as a function of time, we can now specify the pose of the end-effector, i.e., $\mathbf{p}$ and $\mathbf{Q}$, as functions of time.

The whole trajectory was tracked with the robot at hand using the algorithm outlined in this section. With the aid of this algorithm, we produced the plots of Fig. 9.9. Also, the time-history of the condition number of the manipulator Jacobian was computed and plotted in Fig. 9.10. Apparently, the condition number of the Jacobian remains within the same order of magnitude throughout the whole operation, below 10, thereby showing that the manipulator remains far enough from singularities during this task—the condition number becomes very large when a singularity is approached, becoming unbounded at singularities.

A rendering of the welding seam with the Frenet-Serret triad at a sample of points is displayed in Fig. 9.11. It is noteworthy that the torsion of the path is manifested in this figure by virtue of the inclination of the Z axis, which changes from point to point. In a planar curve, this axis would remain at a fixed orientation while traversing the curve.

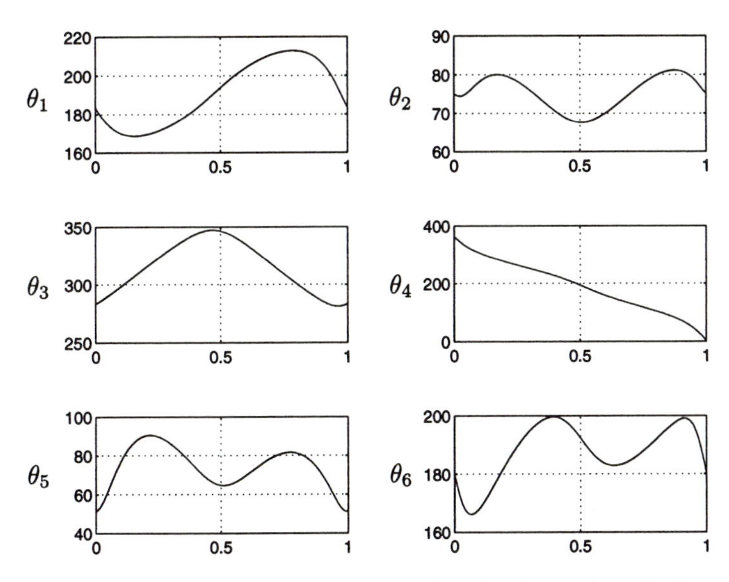

FIGURE 9.9. Time-histories of the joint variables (in degrees) of the Fanuc Arc Mate robot used to track a warped curve for arc-welding vs. nondimensional time.

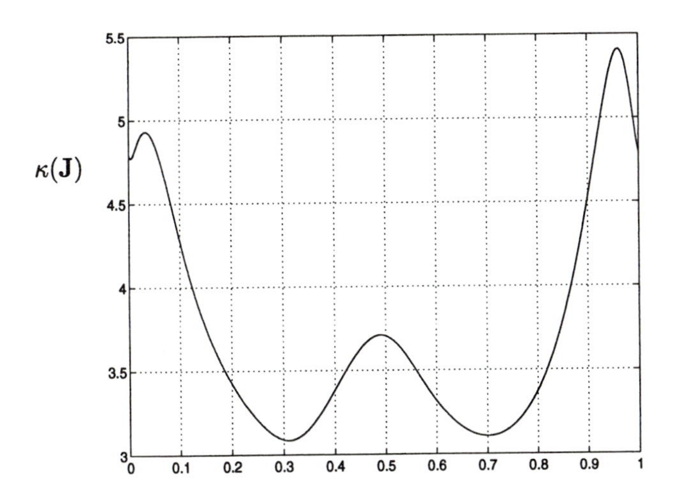

FIGURE 9.10. Time-history of the condition number of the Jacobian matrix during an arc-welding operation vs. nondimensional time.

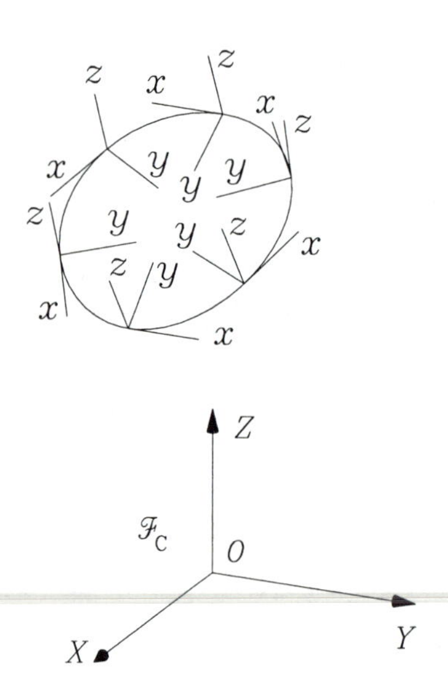

FIGURE 9.11. Welding seam with Frenet-Serret frames.

10
Dynamics of Complex Robotic Mechanical Systems

10.1 Introduction

The subject of this chapter is the dynamics of the class of robotic mechanical systems introduced in Chpater 8 under the generic name of *complex*. However, notice that this class comprises serial manipulators not allowing a decoupling of the orientation from the positioning tasks. For purposes of dynamics, this decoupling is irrelevant and hence, was not a condition in the study of the dynamics of serial manipulators in Chapter 6. Thus, serial manipulators need not be further studied here, the focus being on parallel manipulators and rolling robots. The dynamics of walking machines and multifingered hands involves special features that render these systems more elaborate from the dynamics viewpoint, namely, a time-varying topology. What this means is that these systems include kinematic loops that open when a leg takes off or when a finger releases an object and open chains that close when a leg touches ground or when a finger makes contact with an object. The implication here is that the degree of freedom of these systems is time-varying. The derivation of such a mathematical model is outlined in (Pfeiffer, Eltse, and Weidemann, 1995).

The degree of freedom (dof) of the mechanical systems studied here is thus constant. Now, the two kinds of systems studied here pertain to very different types, for parallel manipulators fall into the realm of *holonomic*, while rolling robots into that of *nonholonomic*, mechanical systems. In order to better understand this essential difference between these two types of systems, we give below a summary of the classification of mechanical systems at large.

10.2 Classification of Robotic Mechanical Systems with Regard to Dynamics

Because robotic mechanical systems are a class of general mechanical systems, a classification of the latter will help us focus on the systems motivating this study. Mechanical systems can be classified according to various criteria, the most common one being based on the type of constraints to which these systems are subjected. In this context we find *holonomic* vs. *nonholonomic* and *scleronomic* vs. *rheonomic* constraints. Holonomic constraints are those that are expressed either as a system of algebraic equations in displacement variables, whether angular or translational, not involving any velocity variables, or as a system of equations in velocity variables that nevertheless can be integrated *as a whole* to produce a system of equations of the first type. Note that it is not necessary that every single scalar equation of velocity constraints be integrable; rather, the whole system must be integrable for the system of velocity constraints to lead to a system of displacement constraints. If the system of velocity constraints is not integrable, the constraints are said to be nonholonomic. Moreover, if a mechanical system is subject only to holonomic constraints, it is said to be holonomic; otherwise, it is nonholonomic. Manipulators composed of revolute and prismatic pairs are examples of holonomic systems, while wheeled robots are usually nonholonomic systems. On the other hand, if a mechanical system is subject to constraints that are not explicit functions of time, these constraints are termed scleronomic, while if the constraints are explicit functions of time, they are termed rheonomic. For our purposes, however, this distinction is irrelevant.

In order to understand better one more classification of mechanical systems, we recall the concepts of generalized coordinate and generalized speed that were introduced in Subsection 6.3.2. The generalized coordinates of a mechanical system are all those displacement variables, whether rotational or translational, that determine uniquely a configuration of the system. Note that the set of generalized coordinates of a system is not unique. Moreover, various sets of generalized coordinates of a mechanical system need not have the same number of elements, but there is a minimum number below which the set of generalized coordinates cannot define the configuration of the system. This minimum number corresponds, in the case of holonomic systems, to the *degree of freedom* of the system. Serial and parallel manipulators coupled only by revolute or prismatic pairs are holonomic, their joint variables, grouped in vector $\boldsymbol{\theta}$, playing the role of generalized coordinates, while their joint rates, grouped in vector $\dot{\boldsymbol{\theta}}$, in turn, play the role of generalized speeds. Note that in the case of parallel manipulators, not all joint variables are independent generalized coordinates. In the case of nonholonomic systems, on the other hand, the number of generalized coordinates needed to fully specify their configuration exceeds their degree of freedom by virtue of the lack of integrability of their kinematic constraints. This concept is best illustrated with the aid of examples, which are included in Section 10.5. Time-derivatives of the generalized coordinates, or linear combinations thereof, are termed the *generalized speeds* of the system. If the kinetic energy of a mechanical system is zero when all its generalized speeds are set equal to zero, the system is said to be *catastatic*. If, on the contrary, the kinetic energy of the system is nonzero even if all the generalized speeds are set equal to zero, the system is said to be *acatastatic*. All the systems that we will study in this chapter are catastatic. A light robot mounted on a heavy noninertial base that undergoes a *controlled* motion is an example of an acatastatic system, for the motion of the base can be assumed to be insensitive to the dynamics of the robot; however, the motion of the base does affect the dynamics of the robot.

Another criterion used in classifying mechanical systems, which pertains specifically to robotic mechanical systems, is based on the type of actuation. In general, a system needs at least as many independent actuators as degrees of freedom. However, instances arise in which the number of actuators is greater than the degree of freedom of the system. In these instances, we speak of *redundantly-actuated systems*. In view of the fundamental character of this book, we will not study redundant actuation here; we will thus assume that the number of independent actuators equals the degree of freedom of the system.

The main results of this chapter are applicable to robotic mechanical systems at large. For brevity, we will frequently refer to the objects of our study simply as *systems*.

10.3 The Structure of the Dynamics Models of Holonomic Systems

We saw in Section 6.6 that the mathematical model of a manipulator of the serial type contains basically three terms, namely, one linear in the joint accelerations, one that is quadratic in the joint rates, and one that arises from the environment, i.e., from actuators, dissipation, and potential fields such as gravity. We show in this section that in fact, the essential structure of this model still holds in the case of more general mechanical systems, if we regard the rates of the actuated joints as the independent generalized speeds of the system.

First, we will assume that the mechanical system at hand is composed of r rigid bodies and its degree of freedom is n. Henceforth, we assume that these bodies are coupled in such a way that they may form kinematic loops; for this reason, such systems contain some joints that are not active, i.e., that are not driven by any actuator. For the sake of simplicity, the vector of joint rates that we will consider contains only those rates that are associated with actuated joints. The kinetic energy of the system at hand then takes the form of eq.(6.16), and all other definitions of Subsection 6.3.1 hold.

Upon differentiation of eq.(6.16) with respect to the vector of actuated joint rates, we obtain

$$\frac{\partial T}{\partial \dot{\boldsymbol{\theta}}} = \frac{1}{2}\left(\frac{\partial \mathbf{t}}{\partial \dot{\boldsymbol{\theta}}}\right)^T \boldsymbol{\mu} + \frac{1}{2}\left(\frac{\partial \boldsymbol{\mu}}{\partial \dot{\boldsymbol{\theta}}}\right)^T \mathbf{t} \tag{10.1}$$

We now recall the twist-rate relation of eq.(6.55), which leads to

$$\frac{\partial \mathbf{t}}{\partial \dot{\boldsymbol{\theta}}} = \mathbf{T} \tag{10.2}$$

Moreover, from eq.(6.14),

$$\frac{\partial \boldsymbol{\mu}}{\partial \dot{\boldsymbol{\theta}}} = \mathbf{M}\frac{\partial \mathbf{t}}{\partial \dot{\boldsymbol{\theta}}} = \mathbf{MT} \tag{10.3}$$

which follows because $\mathbf{M}$ is rate-independent.

Furthermore, upon substitution of eqs.(10.2) and (10.3) into eq.(10.1), we obtain

$$\frac{\partial T}{\partial \dot{\boldsymbol{\theta}}} = \mathbf{T}^T \boldsymbol{\mu} \tag{10.4}$$

Hence,

$$\frac{d}{dt}\left(\frac{\partial T}{\partial \dot{\boldsymbol{\theta}}}\right) = \dot{\mathbf{T}}^T \boldsymbol{\mu} + \mathbf{T}^T \dot{\boldsymbol{\mu}} \tag{10.5}$$

Further, upon substitution of eqs.(6.14) and (6.15) into eq.(10.5), we derive a more useful expression for the foregoing derivative, namely,

$$\frac{d}{dt}\left(\frac{\partial T}{\partial \dot{\boldsymbol{\theta}}}\right) = \dot{\mathbf{T}}^T \mathbf{M}\mathbf{t} + \mathbf{T}^T \mathbf{M}\dot{\mathbf{t}} + \mathbf{T}^T \mathbf{W}\mathbf{M}\mathbf{t} \tag{10.6}$$

or in terms of independent speeds and accelerations,

$$\frac{d}{dt}\left(\frac{\partial T}{\partial \dot{\boldsymbol{\theta}}}\right) = \dot{\mathbf{T}}^T \mathbf{M}\mathbf{T}\dot{\boldsymbol{\theta}} + \mathbf{T}^T \mathbf{M}\dot{\mathbf{T}}\dot{\boldsymbol{\theta}} + \mathbf{T}^T \mathbf{M}\mathbf{T}\ddot{\boldsymbol{\theta}} + \mathbf{T}^T \mathbf{W}\mathbf{M}\mathbf{T}\dot{\boldsymbol{\theta}} \tag{10.7}$$

thereby completing the computation of the first term of the left-hand side of the Euler-Lagrange equations of the system, eq.(6.6). Because the second term is more elusive, it is computed in two steps. First, we find an expression for the time-rate of change of the kinetic energy, as given by eq.(6.16), which leads to

$$\dot{T} = \frac{1}{2}\boldsymbol{\mu}^T\dot{\mathbf{t}} + \frac{1}{2}\mathbf{t}^T\dot{\boldsymbol{\mu}} \tag{10.8}$$

where $\dot{\mathbf{t}}$ is derived upon differentiation of the general relation

$$\mathbf{t} = \mathbf{T}\dot{\boldsymbol{\theta}}$$

i.e.,

$$\dot{\mathbf{t}} = \mathbf{T}\ddot{\boldsymbol{\theta}} + \dot{\mathbf{T}}\dot{\boldsymbol{\theta}} \tag{10.9}$$

while $\dot{\boldsymbol{\mu}}$ is derived upon differentiation of expression (6.14), i.e.,

$$\dot{\boldsymbol{\mu}} = \mathbf{M}\dot{\mathbf{t}} + \mathbf{W}\mathbf{M}\mathbf{t} \equiv \mathbf{M}\mathbf{T}\ddot{\boldsymbol{\theta}} + \mathbf{M}\dot{\mathbf{T}}\dot{\boldsymbol{\theta}} + \mathbf{W}\mathbf{M}\mathbf{T}\dot{\boldsymbol{\theta}} \tag{10.10}$$

Upon substitution of expressions (10.9) and (10.10) into eq.(10.8), we obtain

$$\dot{T} = \mathbf{t}^T\mathbf{M}\mathbf{T}\ddot{\boldsymbol{\theta}} + \mathbf{t}^T\mathbf{M}\dot{\mathbf{T}}\dot{\boldsymbol{\theta}} + \frac{1}{2}\mathbf{t}^T\mathbf{W}\mathbf{M}\mathbf{T}\dot{\boldsymbol{\theta}}$$

However, the third term of the above expression vanishes because as we will show below, the product $\mathbf{W}\mathbf{t}$ vanishes. Indeed, we can write, in general,

$$\mathbf{W}\mathbf{t} = \begin{bmatrix} \mathbf{W}_1 & \mathbf{O} & \cdots & \mathbf{O} \\ \mathbf{O} & \mathbf{W}_2 & \cdots & \mathbf{O} \\ \vdots & \vdots & \ddots & \vdots \\ \mathbf{O} & \mathbf{O} & \cdots & \mathbf{W}_r \end{bmatrix} \begin{bmatrix} \mathbf{t}_1 \\ \mathbf{t}_2 \\ \vdots \\ \mathbf{t}_r \end{bmatrix} = \begin{bmatrix} \mathbf{W}_1\mathbf{t}_1 \\ \mathbf{W}_2\mathbf{t}_2 \\ \vdots \\ \mathbf{W}_r\mathbf{t}_r \end{bmatrix} \tag{10.11}$$

where $\mathbf{O}$ denotes the 6×6 zero matrix. Now, if we recall eq.(3.146), each of the 6-dimensional blocks $\mathbf{W}_i\mathbf{t}_i$ of $\mathbf{W}\mathbf{t}$ vanishes. Therefore, $\dot{T}$ reduces to

$$\dot{T} = \mathbf{t}^T\mathbf{M}\dot{\mathbf{T}}\dot{\boldsymbol{\theta}} + \mathbf{t}^T\mathbf{M}\mathbf{T}\ddot{\boldsymbol{\theta}} \tag{10.12}$$

On the other hand, $\dot{T}$ can be regarded as a function of $\boldsymbol{\theta}$ and $\dot{\boldsymbol{\theta}}$, i.e., as $\dot{T} = \dot{T}(\boldsymbol{\theta}, \dot{\boldsymbol{\theta}})$. Hence, by application of the chain rule,

$$\dot{T} = \left(\frac{\partial T}{\partial \boldsymbol{\theta}}\right)^T \dot{\boldsymbol{\theta}} + \left(\frac{\partial T}{\partial \dot{\boldsymbol{\theta}}}\right)^T \ddot{\boldsymbol{\theta}} \tag{10.13}$$

Upon comparison of eqs.(10.12) and (10.13), we obtain

$$\frac{\partial T}{\partial \boldsymbol{\theta}} = \dot{\mathbf{T}}^T\mathbf{M}\mathbf{t} = \dot{\mathbf{T}}^T\mathbf{M}\mathbf{T}\dot{\boldsymbol{\theta}} \tag{10.14}$$

We now set out to calculate the right-hand side of the Euler-Lagrange equations. First, the partial derivative of the power supplied to the system by driving forces with respect to the vector of independent generalized speeds takes the form

$$\frac{\partial \Pi}{\partial \dot{\boldsymbol{\theta}}} = \frac{\partial}{\partial \dot{\boldsymbol{\theta}}}(\mathbf{t}^T\mathbf{w}^A) = \left(\frac{\partial \mathbf{t}}{\partial \dot{\boldsymbol{\theta}}}\right)^T \mathbf{w}^A \tag{10.15}$$

Upon substitution of eq.(10.2) into eq.(10.14), we obtain

$$\frac{\partial \Pi}{\partial \dot{\boldsymbol{\theta}}} = \mathbf{T}^T \mathbf{w}^A \tag{10.16}$$

Similarly, under the assumption that a dissipation function is available,

$$\frac{\partial \Delta}{\partial \dot{\boldsymbol{\theta}}} = \mathbf{T}^T \mathbf{w}^D \tag{10.17}$$

Now, the $n \times n$ matrix associated with $\ddot{\boldsymbol{\theta}}$ in eq.(10.7) is defined as the *generalized inertia matrix* of the system and is denoted by $\mathbf{I}$, i.e.,

$$\mathbf{I}(\boldsymbol{\theta}) = \mathbf{T}^T \mathbf{M} \mathbf{T} \tag{10.18}$$

Upon substitution of eqs.(10.7), (10.14), (10.16), (10.17), and (10.18) into the Euler-Lagrange equations, eq.(6.6), the desired dynamics model is derived, namely,

$$\mathbf{I}\ddot{\boldsymbol{\theta}} + \mathbf{T}^T \mathbf{M} \dot{\mathbf{T}} \dot{\boldsymbol{\theta}} + \mathbf{T}^T \mathbf{W} \mathbf{M} \mathbf{T} \dot{\boldsymbol{\theta}} = \mathbf{T}^T (\mathbf{w}^A - \mathbf{w}^D) \tag{10.19a}$$

or

$$\mathbf{I}\ddot{\boldsymbol{\theta}} = -\mathbf{T}^T \mathbf{M} \dot{\mathbf{T}} \dot{\boldsymbol{\theta}} - \mathbf{T}^T \mathbf{W} \mathbf{M} \mathbf{T} \dot{\boldsymbol{\theta}} + \mathbf{T}^T (\mathbf{w}^A - \mathbf{w}^D) \tag{10.19b}$$

It is important to note that the nonworking constraint wrench $\mathbf{w}^C$ does not appear in the Euler-Lagrange equations, which is one of the attractive features of these equations for simulation and control purposes. On the other hand, the Newton-Euler equations involve nonworking constraint wrenches, which are needed to calculate the design loads acting on each link.

10.4 Dynamics of Parallel Manipulators

We illustrate the modeling techniques of mechanical systems with kinematic loops via a class of systems known as *parallel manipulators.* While parallel manipulators can take on a large variety of forms, we focus here on those termed *platform manipulators*, with an architecture similar to that of flight simulators. In platform manipulators we can distinguish two special links, namely, the base $\mathcal{B}$ and the moving platform $\mathcal{M}$. Moreover, these two links are coupled via six *legs*, with each leg constituting a six-axis kinematic chain of the serial type, as shown in Fig. 10.1, whereby a wrench $\mathbf{w}^W$, represented by a double-headed arrow, acts on $\mathcal{M}$ and is applied at $C_{\mathcal{M}}$, the mass center of $\mathcal{M}$. This figure shows the axes of the revolutes coupling the legs to the two platforms as forming regular polygons. However, the modeling discussed below is not restricted to this particular geometry. As a matter of fact, these axes need not even be coplanar. On the other

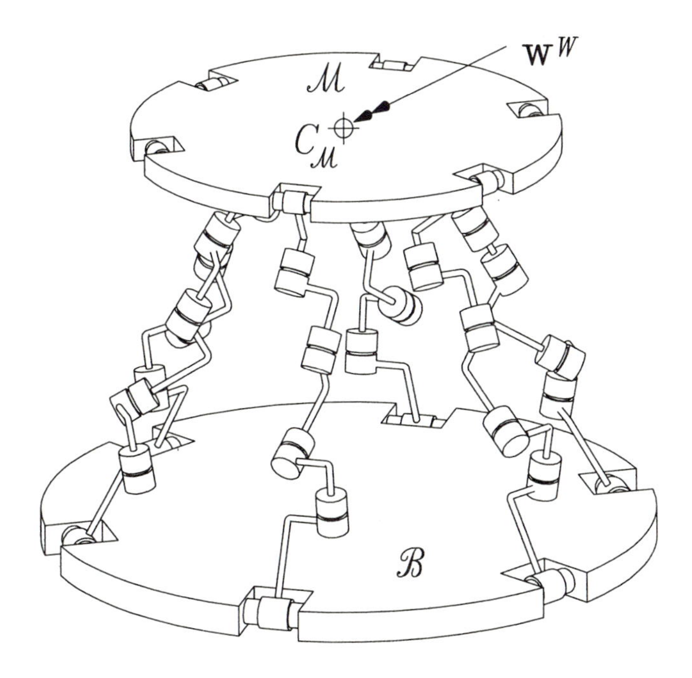

FIGURE 10.1. A platform-type parallel manipulator.

hand, the architecture of Fig. 10.1 is very general, for it includes more specific types of platform manipulators, such as flight simulators. In these, the first three revolute axes stemming from the base platform have intersecting axes, thereby giving rise to a spherical kinematic pair, while the upper two axes intersect at right angles, thus constituting a universal joint. Moreover, the intermediate joint in flight simulators is not a revolute, but rather a prismatic pair, which is the actuated joint of the leg. A leg kinematically equivalent to that of flight simulators can be obtained from that of the manipulator of Fig. 10.1, if the intermediate revolute has an axis perpendicular to the line connecting the centers of the spherical and the universal joints of the corresponding leg, as shown in Fig. 10.2. In flight simulators, the pose of the moving platform is controlled by hydraulic actuators that vary the distance between these two centers. In the revolute-coupled equivalent leg, the length of the same line is controlled by the rotation of the intermediate revolute.

Shown in Fig. 10.3 is the graph of the system depicted in Fig. 10.1. In that graph, the nodes denote rigid links, while the edges denote joints. By application of *Euler's formula* for graphs (Harary, 1972), the number ι of independent loops of a system with many kinematic loops is given by

$$\iota = j - l + 1 \tag{10.20}$$

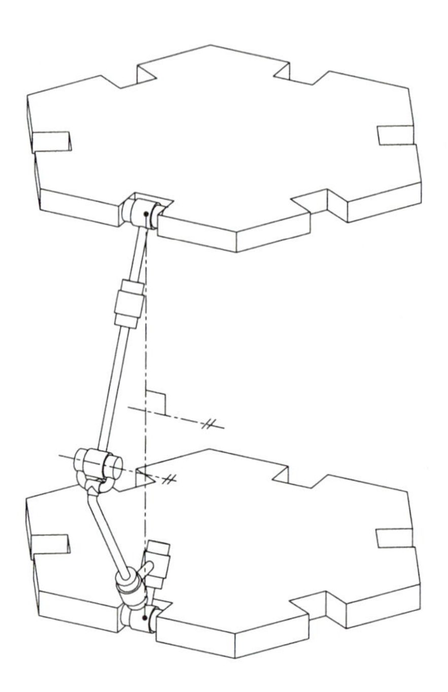

FIGURE 10.2. A leg of a simple platform-type parallel manipulator.

where j is the number of revolute and prismatic joints and l is the number of links.

Thus, if we apply Euler's formula to the system of Fig. 10.1, we conclude that its kinematic chain contains five independent loops. Hence, while the chain apparently contains six distinct loops, only five of these are independent. Moreover, the degree of freedom of the manipulator is six. Indeed, the total number of links of the manipulator is $l = 6 \times 5 + 2 = 32$. Of these, one is fixed, and hence, we have 31 moving links, each with six degrees of freedom prior to coupling. Thus, we have a total of $31 \times 6 = 186$ degrees of freedom at our disposal. Upon coupling, each revolute removes five degrees of freedom, and hence, the 36 kinematic pairs remove 180 degrees of freedom, the manipulator thus being left with 6 degrees of freedom. We derive below the mathematical model governing the motion of the overall system in terms of the independent generalized coordinates associated with the actuated joints of the legs.

We assume, henceforth, that each leg is a six-axis open kinematic chain with either revolute or prismatic pairs, only one of which is actuated, and we thus have as many actuated joints as degrees of freedom. Furthermore, we label the legs with Roman numerals I, II, ..., VI and denote the mass center of the mobile platform $\mathcal{M}$ by $C_{\mathcal{M}}$, with the twist of $\mathcal{M}$ denoted by $\mathbf{t}_{\mathcal{M}}$ and defined at the mass center. That is, if $\mathbf{c}_{\mathcal{M}}$ denotes the position vector of $C_{\mathcal{M}}$ in an inertial frame and $\dot{\mathbf{c}}_{\mathcal{M}}$ its velocity, while $\boldsymbol{\omega}_{\mathcal{M}}$ the angular

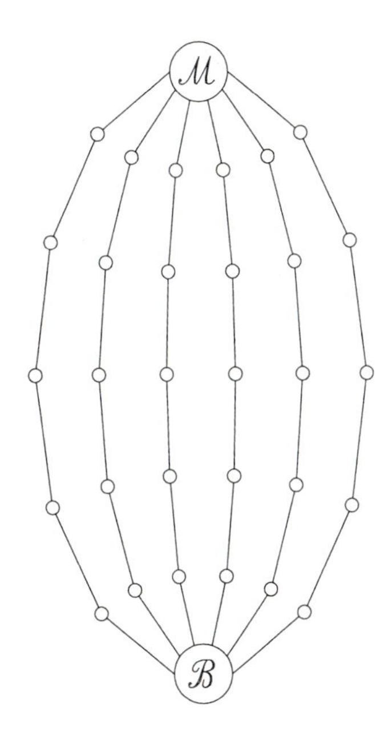

FIGURE 10.3. The graph of the flight simulator.

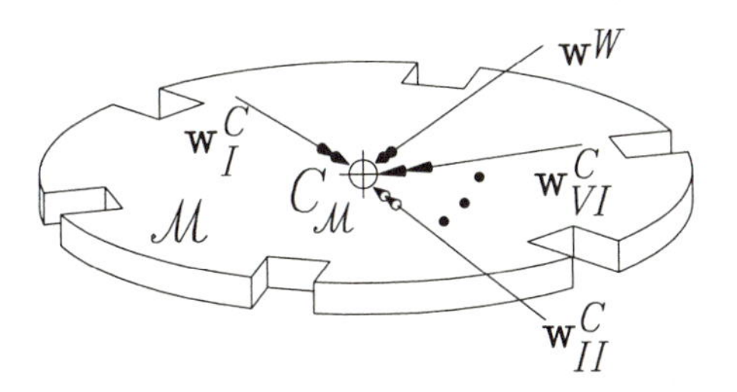

FIGURE 10.4. The free-body diagram of $\mathcal{M}$.

velocity of $\mathcal{M}$, then

$$\mathbf{t}_\mathcal{M} \equiv \begin{bmatrix} \boldsymbol{\omega}_\mathcal{M} \\ \dot{\mathbf{c}}_\mathcal{M} \end{bmatrix} \tag{10.21}$$

Next, the Newton-Euler equations of $\mathcal{M}$ are derived from the free-body diagram shown in Fig. 10.4. In this figure, the legs have been replaced by the *constraint wrenches* $\{\mathbf{w}_J^C\}_I^{VI}$ acting at point $C_\mathcal{M}$, the governing equation taking the form of eq.(10.19a), namely,

$$\mathbf{M}_\mathcal{M}\dot{\mathbf{t}}_\mathcal{M} = -\mathbf{W}_\mathcal{M}\mathbf{M}_\mathcal{M}\mathbf{t}_\mathcal{M} + \mathbf{w}^W + \sum_{J=I}^{VI} \mathbf{w}_J^C \tag{10.22}$$

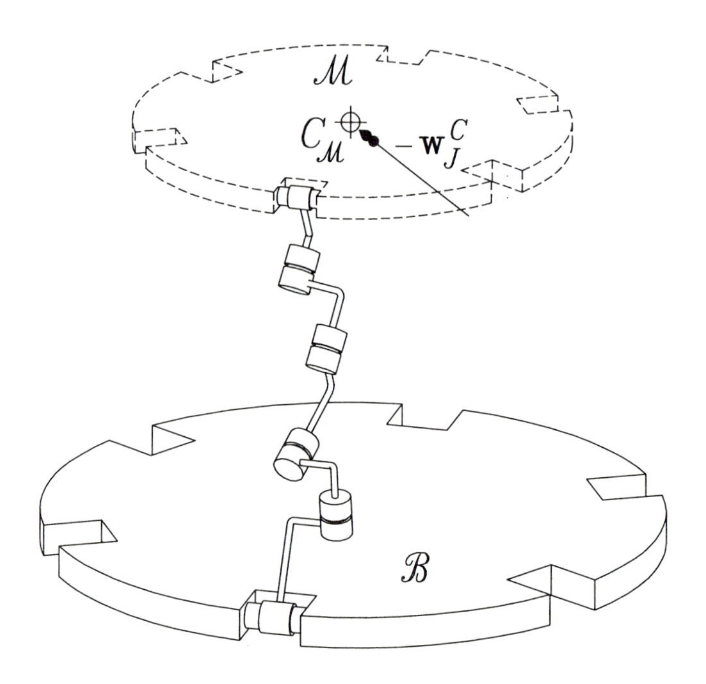

FIGURE 10.5. The serial manipulator of the Jth leg.

with $\mathbf{w}^W$ denoting the external wrench acting on $\mathcal{M}$. Furthermore, let us denote by q_J the variable of the actuated joint of the Jth leg, all variables of the six actuated joints being grouped in the 6-dimensional array $\mathbf{q}$, i.e.,

$$\mathbf{q} \equiv [q_I \quad q_{II} \quad \cdots \quad q_{VI}]^T \tag{10.23}$$

Now, we derive a relation between the twist $\mathbf{t}_{\mathcal{M}}$ and the active joint rates, $\dot{q}_J$, for $J = I, II, \ldots, VI$. To this end, we resort to Fig. 10.5, depicting the Jth leg as a serial-type, six-axis manipulator, whose twist-shape relations are readily expressed as in eq.(4.55a), namely,

$$\mathbf{J}_J \dot{\boldsymbol{\theta}}_J = \mathbf{t}_{\mathcal{M}}, \quad J = I, II, \ldots, VI \tag{10.24}$$

where $\mathbf{J}_J$ is the 6×6 Jacobian matrix of the Jth leg.

In Fig. 10.5, the moving platform $\mathcal{M}$ has been replaced by the constraint wrench transmitted by the moving platform onto the end link of the Jth leg, $-\mathbf{w}_J^C$, whose sign is the opposite of that transmitted by this leg onto $\mathcal{M}$ by virtue of Newton's third law. The dynamics model of the manipulator of Fig. 10.5 then takes the form

$$\mathbf{I}_J \ddot{\boldsymbol{\theta}}_J + \mathbf{C}_J(\boldsymbol{\theta}_J, \dot{\boldsymbol{\theta}}_J)\dot{\boldsymbol{\theta}}_J = \boldsymbol{\tau}_J - \mathbf{J}_J^T \mathbf{w}_J^C, \quad J = I, II, \ldots, VI \tag{10.25}$$

where $\mathbf{I}_J$ is the 6×6 inertia matrix of the manipulator, while $\mathbf{C}_J$ is the matrix coefficient of the inertia terms that are quadratic in the joint rates.

Moreover, $\boldsymbol{\theta}_J$ and $\boldsymbol{\tau}_J$ denote the 6-dimensional vectors of joint variables and joint torques, namely,

$$\boldsymbol{\theta}_J \equiv \begin{bmatrix} \theta_{J1} \\ \theta_{J2} \\ \vdots \\ \theta_{J6} \end{bmatrix}, \quad \boldsymbol{\tau}_J \equiv \begin{bmatrix} 0 \\ \vdots \\ \tau_{Jk} \\ 0 \\ \vdots \\ 0 \end{bmatrix} \tag{10.26}$$

with subscript Jk denoting in turn the only actuated joint of the Jth leg, namely, the kth joint of the leg. If we now introduce $\mathbf{e}_{Jk}$, defined as a unit vector all of whose entries are zero except for the kth entry, which is unity, then we can write

$$\boldsymbol{\tau}_J = f_J \mathbf{e}_{Jk} \tag{10.27}$$

If the actuated joint is prismatic, as is the case in flight simulators, f_J is a force; if this joint is a revolute, then f_J is a torque.

Now, since the dimension of $\mathbf{q}$ coincides with the degree of freedom of the manipulator, it is possible to find, within the framework of the natural orthogonal complement, a 6×6 matrix $\mathbf{L}_J$ mapping the vector of actuated joint rates $\dot{\mathbf{q}}$ into the vector of Jth-leg joint-rates, namely,

$$\dot{\boldsymbol{\theta}}_J = \mathbf{L}_J \dot{\mathbf{q}}, \quad J = I,\ II,\ \ldots,\ VI \tag{10.28}$$

The calculation of $\mathbf{L}_J$ will be illustrated with an example.

Moreover, if the manipulator of Fig. 10.5 is not at a singular configuration, then we can solve for $\mathbf{w}_J^C$ from eq.(10.25), i.e.,

$$\mathbf{w}_J^C = \mathbf{J}_J^{-T} \left[\boldsymbol{\tau}_J - \mathbf{I}_J \ddot{\boldsymbol{\theta}}_J - \mathbf{C}_J \dot{\boldsymbol{\theta}}_J \right] \tag{10.29}$$

in which the superscript $-T$ stands for the transpose of the inverse, or equivalently, the inverse of the transpose, while $\mathbf{I}_J = \mathbf{I}_J(\boldsymbol{\theta}_J)$ and $\mathbf{C}_J = \mathbf{C}_J(\boldsymbol{\theta}_J, \dot{\boldsymbol{\theta}}_J)$. Further, we substitute $\mathbf{w}_J^C$ as given by eq.(10.29) into eq.(10.22), thereby obtaining the Newton-Euler equations of the moving platform free of constraint wrenches. Additionally, the equations thus resulting now contain inertia terms and joint torques pertaining to the Jth leg, namely,

$$\mathbf{M}_{\mathcal{M}} \dot{\mathbf{t}}_{\mathcal{M}} = -\mathbf{W}_{\mathcal{M}} \mathbf{M}_{\mathcal{M}} \mathbf{t}_{\mathcal{M}} + \mathbf{w}^W + \sum_{J=I}^{VI} \mathbf{J}_J^{-T} \left[\boldsymbol{\tau}_J - \mathbf{I}_J \ddot{\boldsymbol{\theta}}_J - \mathbf{C}_J \dot{\boldsymbol{\theta}}_J \right] \tag{10.30}$$

Still within the framework of the natural orthogonal complement, we set up the relation between the twist $\mathbf{t}_{\mathcal{M}}$ and the vector of actuated joint rates $\dot{\mathbf{q}}$ as

$$\mathbf{t}_{\mathcal{M}} = \mathbf{T} \dot{\mathbf{q}} \tag{10.31}$$

which upon differentiation with respect to time, yields

$$\dot{\mathbf{t}}_{\mathcal{M}} = \mathbf{T}\ddot{\mathbf{q}} + \dot{\mathbf{T}}\dot{\mathbf{q}} \tag{10.32}$$

In the next step, we substitute $\mathbf{t}_{\mathcal{M}}$ and its time-derivative as given by eqs.(10.31 & 10.32) into eq.(10.30), thereby obtaining

$$\begin{aligned} &\mathbf{M}_{\mathcal{M}}(\mathbf{T}\ddot{\mathbf{q}} + \dot{\mathbf{T}}\dot{\mathbf{q}}) + \mathbf{W}_{\mathcal{M}}\mathbf{M}_{\mathcal{M}}\mathbf{T}\dot{\mathbf{q}} \\ &\quad + \sum_{J=I}^{VI} \mathbf{J}_J^{-T}\left[\mathbf{I}_J\ddot{\boldsymbol{\theta}}_J + \mathbf{C}_J\dot{\boldsymbol{\theta}}_J\right] = \mathbf{w}^W + \sum_{J=I}^{VI}\mathbf{J}_J^{-T}\boldsymbol{\tau}_J \end{aligned} \tag{10.33}$$

Further, we recall relation (10.28), which upon differentiation with respect to time, yields

$$\ddot{\boldsymbol{\theta}}_J = \mathbf{L}_J\ddot{\mathbf{q}} + \dot{\mathbf{L}}_J\dot{\mathbf{q}} \tag{10.34}$$

Next, relations (10.28 & 10.34) are substituted into eq.(10.33), thereby obtaining the model sought in terms only of actuated joint variables. After simplification, this model takes the form

$$\begin{aligned} &\mathbf{M}_{\mathcal{M}}\mathbf{T}\ddot{\mathbf{q}} + \mathbf{M}_{\mathcal{M}}\dot{\mathbf{T}}\dot{\mathbf{q}} + \mathbf{W}_{\mathcal{M}}\mathbf{M}_{\mathcal{M}}\mathbf{T}\dot{\mathbf{q}} \\ &\quad + \sum_{J=I}^{J=VI} \mathbf{J}_J^{-T}\left[\mathbf{I}_J\mathbf{L}_J\ddot{\mathbf{q}} + \mathbf{I}_J\dot{\mathbf{L}}_J\dot{\mathbf{q}} + \mathbf{C}_J\mathbf{L}_J\dot{\mathbf{q}}\right] = \mathbf{w}^W + \sum_{J=I}^{VI}\mathbf{J}_J^{-T}\boldsymbol{\tau}_J \end{aligned} \tag{10.35}$$

where now $\mathbf{I}_J = \mathbf{I}_J(\mathbf{q})$ and $\mathbf{C}_J = \mathbf{C}_J(\mathbf{q}, \dot{\mathbf{q}})$.

Our final step in this formulation consists of deriving a reduced 6×6 model in terms only of actuated joint variables. Prior to this step, we note that from eqs.(10.24), (10.28), and (10.31),

$$\mathbf{L}_J = \mathbf{J}_J^{-1}\mathbf{T} \tag{10.36}$$

Upon substitution of the above relation into eq.(10.35) and multiplication of both sides of eq.(10.35) by $\mathbf{T}^T$ from the left, we obtain the desired model in the form of eqs.(10.19a), namely,

$$\mathbf{M}(\mathbf{q})\ddot{\mathbf{q}} + \mathbf{N}(\mathbf{q}, \dot{\mathbf{q}})\dot{\mathbf{q}} = \boldsymbol{\tau}^W + \sum_{J=I}^{VI}\mathbf{L}_J\boldsymbol{\tau}_J \tag{10.37}$$

with the 6×6 matrices $\mathbf{M}(\mathbf{q})$, $\mathbf{N}(\mathbf{q}, \dot{\mathbf{q}})$, and vector $\boldsymbol{\tau}^W$ defined as

$$\mathbf{M}(\mathbf{q}) \equiv \mathbf{T}^T\mathbf{M}_{\mathcal{M}}\mathbf{T} + \sum_{J+I}^{VI}\mathbf{L}_J^T\mathbf{I}_J\mathbf{L}_J \tag{10.38a}$$

$$\begin{aligned} \mathbf{N}(\mathbf{q}, \dot{\mathbf{q}}) &\equiv \mathbf{T}^T(\mathbf{M}_{\mathcal{M}}\dot{\mathbf{T}} + \mathbf{W}_{\mathcal{M}}\mathbf{M}_{\mathcal{M}}\mathbf{T}) \\ &\quad + \sum_{J=I}^{VI}\mathbf{L}_J^T\left[\mathbf{I}_J\dot{\mathbf{L}}_J + \mathbf{C}_J\mathbf{L}_J\right] \end{aligned} \tag{10.38b}$$

$$\boldsymbol{\tau}^W \equiv \mathbf{T}^T\mathbf{w}^W \tag{10.38c}$$

Alternatively, the foregoing variables can be expressed in a more compact form that will shed more light on the above model. To do this, we define the 36×36 matrices $\mathbf{I}$ and $\mathbf{C}$ as well as the 6×36 matrix $\mathbf{L}$, the 6×6 matrix $\mathbf{\Lambda}$, and the 6-dimensional vector $\boldsymbol{\phi}$ as

$$\mathbf{I} \equiv \text{diag}(\mathbf{I}_I, \mathbf{I}_{II}, \ldots, \mathbf{I}_{VI}) \tag{10.39a}$$

$$\mathbf{C} \equiv \text{diag}(\mathbf{C}_I, \mathbf{C}_{II}, \ldots, \mathbf{C}_{VI}) \tag{10.39b}$$

$$\mathbf{L} \equiv [\mathbf{L}_I \quad \mathbf{L}_{II} \quad \ldots \quad \mathbf{L}_{VI}] \tag{10.39c}$$

$$\mathbf{\Lambda} \equiv [\mathbf{L}_I \mathbf{e}_{Ik} \quad \mathbf{L}_{II} \mathbf{e}_{IIk} \quad \ldots \quad \mathbf{L}_{VI} \mathbf{e}_{VIk}] \tag{10.39d}$$

$$\boldsymbol{\phi} \equiv [f_I \quad f_{II} \quad \ldots \quad f_{VI}]^T \tag{10.39e}$$

and hence,

$$\mathbf{M}(\mathbf{q}) \equiv \mathbf{T}^T \mathbf{M}_{\mathcal{M}} \mathbf{T} + \mathbf{L}^T \mathbf{I} \mathbf{L} \tag{10.40a}$$

$$\mathbf{N}(\mathbf{q}, \dot{\mathbf{q}}) \equiv \mathbf{L}^T \mathbf{I} \dot{\mathbf{L}} + \mathbf{L}^T \mathbf{C}(\mathbf{q}, \dot{\mathbf{q}}) \mathbf{L} \tag{10.40b}$$

$$\sum_{J=I}^{VI} \mathbf{L}_J \tau_J \equiv \mathbf{\Lambda} \boldsymbol{\phi} \tag{10.40c}$$

whence the mathematical model of eq.(10.37) takes on a more familiar form, namely,

$$\mathbf{M}(\mathbf{q})\ddot{\mathbf{q}} + \mathbf{N}(\mathbf{q}, \dot{\mathbf{q}})\dot{\mathbf{q}} = \boldsymbol{\tau}^W + \mathbf{\Lambda}\boldsymbol{\phi} \tag{10.41}$$

Thus, for inverse dynamics, we want to determine $\boldsymbol{\phi}$ for a motion given by $\mathbf{q}$ and $\dot{\mathbf{q}}$, which can be done from the above equation, namely,

$$\boldsymbol{\phi} = \mathbf{\Lambda}^{-1} \left[\mathbf{M}(\mathbf{q})\ddot{\mathbf{q}} + \mathbf{N}(\mathbf{q}, \dot{\mathbf{q}})\dot{\mathbf{q}} - \boldsymbol{\tau}^W\right] \tag{10.42}$$

Notice, however, that the foregoing solution is not recursive, and since it requires linear-equation solving, it is of order n^3, which thus yields a rather high numerical complexity. It should be possible to produce a recursive algorithm for the computation of $\boldsymbol{\phi}$, but this issue will not be pursued here. Moreover, given the parallel structure of the manipulator, the associated recursive algorithm should be parallelizable with multiple processors.

For purposes of direct dynamics, on the other hand, we want to solve for $\ddot{\mathbf{q}}$ from eq.(10.41). Moreover, for simulation purposes, we need to derive the state-variable equations of the system at hand. This can be readily done if we define $\mathbf{r} \equiv \dot{\mathbf{q}}$, the state-variable model thus taking on the form

$$\dot{\mathbf{q}} = \mathbf{r} \tag{10.43a}$$

$$\dot{\mathbf{r}} = \mathbf{M}^{-1} \left[-\mathbf{N}(\mathbf{q}, \mathbf{r})\mathbf{r} + \boldsymbol{\tau}^W + \mathbf{\Lambda}\mathbf{f}\right] \tag{10.43b}$$

In light of the matrix inversion of the foregoing model, then, the complexity of the forward dynamics computations is also of order n^3.

Example 10.4.1 *Derive matrix $\mathbf{L}_J$ of eq.(10.28) for a manipulator having six identical legs like that of Fig. 10.2.*

Solution: We attach coordinate frames to the links of the serial chain of the Jth leg following the Denavit-Hartenberg notation, while noting that the first three joints intersect at a common point, and hence, we have $\mathbf{r}_1 = \mathbf{r}_2 = \mathbf{r}_3$. According to this notation, we recall, vector $\mathbf{r}_i$ is directed from the origin O_i of the ith frame to the operation point of the manipulator, which in this case, is $C_{\mathcal{M}}$. The Jacobian matrix of the Jth leg then takes the form

$$\mathbf{J}_J = \begin{bmatrix} \mathbf{e}_1 & \mathbf{e}_2 & \mathbf{e}_3 & \mathbf{e}_4 & \mathbf{e}_5 & \mathbf{e}_6 \\ \mathbf{e}_1 \times \mathbf{r}_1 & \mathbf{e}_2 \times \mathbf{r}_1 & \mathbf{e}_3 \times \mathbf{r}_1 & \mathbf{e}_4 \times \mathbf{r}_4 & \mathbf{e}_5 \times \mathbf{r}_5 & \mathbf{e}_6 \times \mathbf{r}_5 \end{bmatrix}_J$$

the subscript J of the array in the right-hand side reminding us that the vectors inside it pertain to the Jth leg. Thus, matrix $\mathbf{J}_J$ maps the joint-rate vector of the Jth leg, $\dot{\boldsymbol{\theta}}_J$, into the twist $\mathbf{t}_{\mathcal{M}}$ of the platform, i.e.,

$$\mathbf{J}_J \dot{\boldsymbol{\theta}}_J = \mathbf{t}_{\mathcal{M}}$$

Clearly, the joint-rate vector of the Jth leg is defined as

$$\dot{\boldsymbol{\theta}}_J \equiv [\dot{\theta}_{J1} \quad \dot{\theta}_{J2} \quad \dot{\theta}_{J3} \quad \dot{\theta}_{J4} \quad \dot{\theta}_{J5} \quad \dot{\theta}_{J6}]^T$$

Now, note that except for $\dot{\theta}_{J4}$, all joint-rates of this leg are passive and thus need not appear in the mathematical model of the whole manipulator. Hence, we should aim at eliminating all joint-rates from the above twist-rate relation, except for the one associated with the active joint. We can achieve this if we realize that

$$\mathbf{r}_{J1} \times \mathbf{e}_{Ji} + \mathbf{e}_{Ji} \times \mathbf{r}_{J1} = \mathbf{0}, \quad i = 1, 2, 3$$

Further, we define a 3×6 matrix $\mathbf{A}_J$ as

$$\mathbf{A}_J \equiv [\mathbf{R}_{J1} \quad \mathbf{1}]$$

with $\mathbf{R}_{J1}$ defined, in turn, as the cross-product matrix of $\mathbf{r}_{J1}$. Now, upon multiplication of $\mathbf{J}_J$ by $\mathbf{A}_J$ from the right, we obtain a 3×6 matrix whose first three columns vanish, namely,

$$\mathbf{A}_J \mathbf{J}_J = [\mathbf{0} \quad \mathbf{0} \quad \mathbf{0} \quad \mathbf{e}_4 \times (\mathbf{r}_4 - \mathbf{r}_1) \quad \mathbf{e}_5 \times (\mathbf{r}_5 - \mathbf{r}_1) \quad \mathbf{e}_6 \times (\mathbf{r}_5 - \mathbf{r}_1)]_J$$

and hence, if we multiply both sides of the above twist-shape equation by $\mathbf{A}_J$ from the right, we will obtain a new twist-shape equation that is free of the first three joint rates. Moreover, this equation is 3-dimensional, i.e.,

$$[\mathbf{e}_4 \times (\mathbf{r}_4 - \mathbf{r}_1)\dot{\theta}_4 + \mathbf{e}_5 \times (\mathbf{r}_5 - \mathbf{r}_1)\dot{\theta}_5 + \mathbf{e}_6 \times (\mathbf{r}_5 - \mathbf{r}_1)\dot{\theta}_6]_J = \boldsymbol{\omega}_{\mathcal{M}} \times \mathbf{r}_{J1} + \dot{\mathbf{c}}_{\mathcal{M}}$$

where the subscript J attached to the brackets enclosing the whole left-hand side again reminds us that all quantities therein are to be understood as pertaining to the Jth leg. For example, $\mathbf{e}_4$ is to be read $\mathbf{e}_{J4}$. Furthermore, only $\dot{\theta}_{J4}$ is associated with an active joint and denoted, henceforth, by q_J, i.e.,

$$q_J \equiv \theta_{J4} \tag{10.44}$$

It is noteworthy that the foregoing method of elimination of passive joint rates is not ad hoc at all. It has been formalized and generalized to all six lower kinematic pairs (Angeles, 1994).

We have now to eliminate both $\dot{\theta}_{J5}$ and $\dot{\theta}_{J6}$ from the foregoing equation. This can be readily accomplished if we dot-multiply both sides of the same equation by vector $\mathbf{u}_J$ defined as the cross product of the vector coefficients of the two passive joint rates, i.e.,

$$\mathbf{u}_J \equiv [\mathbf{e}_5 \times (\mathbf{r}_5 - \mathbf{r}_1)\dot{\theta}_5]_J \times [\mathbf{e}_6 \times (\mathbf{r}_5 - \mathbf{r}_1)\dot{\theta}_6]_J$$

We thus obtain a third twist-shape relation that is scalar and free of passive joint rates, namely,

$$\mathbf{u}_J \cdot [\mathbf{e}_4 \times (\mathbf{r}_4 - \mathbf{r}_1)\dot{\theta}_4]_J = \mathbf{u}_J \cdot (\boldsymbol{\omega}_{\mathcal{M}} \times \mathbf{r}_{J1} + \dot{\mathbf{c}}_{\mathcal{M}})$$

The above equation is clearly of the form

$$\zeta_J \dot{q}_J = \mathbf{y}_J^T \mathbf{t}_{\mathcal{M}}, \quad J = I,\, II,\, \ldots,\, VI$$

with ζ_J and $\mathbf{y}_J$ defined, in turn, as

$$\zeta_J \equiv \mathbf{u}_J \cdot \mathbf{e}_{J4} \times (\mathbf{r}_{J4} - \mathbf{r}_{J1}) \tag{10.45a}$$

$$\mathbf{y}_J \equiv \begin{bmatrix} \mathbf{r}_{J1} \times \mathbf{u}_J \\ \mathbf{u}_J \end{bmatrix} \tag{10.45b}$$

Upon assembling the foregoing six scalar twist-shape relations, we obtain a 6-dimensional twist-shape relation between the active joint rates of the manipulator and the twist of the moving platform, namely,

$$\mathbf{Z}\dot{\mathbf{q}} = \mathbf{Y}\mathbf{t}_{\mathcal{M}}$$

with the obvious definitions for the two 6×6 matrices $\mathbf{Y}$ and $\mathbf{Z}$ given below:

$$\mathbf{Y} \equiv \begin{bmatrix} \mathbf{y}_I^T \\ \mathbf{y}_{II}^T \\ \vdots \\ \mathbf{y}_{VI}^T \end{bmatrix}, \quad \mathbf{Z} \equiv \mathrm{diag}(\zeta_I,\, \zeta_{II},\, \ldots,\, \zeta_{VI})$$

We now can determine matrix $\mathbf{T}$ of the procedure described above, as long as $\mathbf{Y}$ is invertible, in the form

$$\mathbf{T} = \mathbf{Y}^{-1}\mathbf{Z}$$

whence the leg-matrix $\mathbf{L}_J$ of the same procedure is readily determined, namely,

$$\mathbf{L}_J = \mathbf{J}_J^{-1}\mathbf{T}$$

Therefore, all we need now is an expression for the inverse of the leg Jacobian $\mathbf{J}_J$. This Jacobian is clearly full, which might discourage the reader from attempting its closed-form inversion. However, a closer look reveals that this Jacobian is similar to that of decoupled manipulators, studied in Section 4.5, and hence, its closed-form inversion should be reducible to that of a 3×3 matrix. Indeed, if we recall the twist-transfer formula of eqs.(4.63a & b), we can then write $\mathbf{J}_J$ as

$$\mathbf{J}_J \equiv \mathbf{U}_J\mathbf{K}_J$$

where $\mathbf{U}_J$ is a unimodular 6×6 matrix and $\mathbf{K}_J$ is the Jacobian of the same Jth leg, but now defined with its operation point located at the center of the spherical joint. Thus,

$$\mathbf{U}_J \equiv \begin{bmatrix} \mathbf{1} & \mathbf{O} \\ \mathbf{O}_{J1} - \mathbf{C}_{\mathcal{M}} & \mathbf{1} \end{bmatrix}, \quad \mathbf{K}_J \equiv \begin{bmatrix} \mathbf{K}_{11} & \mathbf{K}_{12} \\ \mathbf{O} & \mathbf{K}_{22} \end{bmatrix}_J$$

the superscript J indicating the Jth leg and with the definitions below:

$\mathbf{O}$: the 3×3 zero matrix;

$\mathbf{1}$: the 3×3 identity matrix;

$\mathbf{O}_{J1}$: the cross-product matrix of $\mathbf{o}_{J1}$, the position vector of the center of the spherical joint;

$\mathbf{C}_{\mathcal{M}}$: the cross product matrix of $\mathbf{c}_{\mathcal{M}}$, the position vector of $C_{\mathcal{M}}$.

Furthermore, the 3×3 blocks of $\mathbf{K}_J$ are defined, in turn, as

$$(\mathbf{K}_{11})_J \equiv [\mathbf{e}_1 \quad \mathbf{e}_2 \quad \mathbf{e}_3]_J$$
$$(\mathbf{K}_{12})_J \equiv [\mathbf{e}_4 \quad \mathbf{e}_5 \quad \mathbf{e}_6]_J$$
$$(\mathbf{K}_{22})_J \equiv [\mathbf{e}_4 \times (\mathbf{r}_4 - \mathbf{r}_1) \quad \mathbf{e}_5 \times (\mathbf{r}_5 - \mathbf{r}_1) \quad \mathbf{e}_6 \times (\mathbf{r}_5 - \mathbf{r}_1)]_J$$

Now, if the inverse of a block matrix is recalled, we have

$$\mathbf{K}_J^{-1} = \begin{bmatrix} \mathbf{K}_{11}^{-1} & -\mathbf{K}_{11}^{-1}\mathbf{K}_{12}\mathbf{K}_{22}^{-1} \\ \mathbf{O} & \mathbf{K}_{22}^{-1} \end{bmatrix}_J$$

where the superscript of the blocks has been transferred to the whole matrix, in order to ease the notation. The problem of inverting $\mathbf{K}_J$ has now been reduced to that of inverting two of its 3×3 blocks. These can be inverted explicitly if we recall the concept of *reciprocal bases* (Brand, 1965).

Thus,

$$(\mathbf{K}_{11}^{-1})_J = \frac{1}{\Delta_{11}^J}\begin{bmatrix}(\mathbf{e}_2 \times \mathbf{e}_3)^T\\ (\mathbf{e}_3 \times \mathbf{e}_1)^T\\ (\mathbf{e}_1 \times \mathbf{e}_2)^T\end{bmatrix}_J$$

$$(\mathbf{K}_{22}^{-1})_J = \frac{1}{\Delta_{22}^J}\begin{bmatrix}[(\mathbf{e}_5 \times \mathbf{s}_5) \times (\mathbf{e}_6 \times \mathbf{s}_5)]^T\\ [(\mathbf{e}_6 \times \mathbf{s}_5) \times (\mathbf{e}_4 \times \mathbf{s}_4)]^T\\ [(\mathbf{e}_4 \times \mathbf{s}_4) \times (\mathbf{e}_5 \times \mathbf{s}_5)]^T\end{bmatrix}_J$$

with $\mathbf{s}_{J4}$, $\mathbf{s}_{J5}$, Δ_{11}^J, and Δ_{22}^J defined as

$$\begin{aligned}
\mathbf{s}_{J4} &\equiv \mathbf{r}_{J4} - \mathbf{r}_{J1}\\
\mathbf{s}_{J5} &\equiv \mathbf{r}_{J5} - \mathbf{r}_{J1}\\
\Delta_{11}^J &\equiv \det(\mathbf{K}_{11}^J) = (\mathbf{e}_1 \times \mathbf{e}_2 \cdot \mathbf{e}_3)_J\\
\Delta_{22}^J &\equiv \det(\mathbf{K}_{22}^J) = [(\mathbf{e}_4 \times \mathbf{s}_4) \times (\mathbf{e}_{\times}\mathbf{s}_5) \cdot (\mathbf{e}_6 \times \mathbf{s}_5)]_J
\end{aligned}$$

the subscripted brackets and parentheses still reminding us that all vectors involved pertain to the Jth leg. Moreover, since $\mathbf{U}_J$ is unimodular, its inverse is simply

$$\mathbf{U}_J^{-1} = \begin{bmatrix}\mathbf{1} & \mathbf{O}\\ \mathbf{C}_{\mathcal{M}} - \mathbf{O}_{J1} & \mathbf{1}\end{bmatrix}$$

and hence,

$$\mathbf{J}_J^{-1} = \begin{bmatrix}\mathbf{K}_{11}^{-1} - \mathbf{K}_{11}^{-1}\mathbf{K}_{12}\mathbf{K}_{22}^{-1}(\mathbf{C}_{\mathcal{M}} - \mathbf{O}_{J1}) & -\mathbf{K}_{11}^{-1}\mathbf{K}_{12}\mathbf{K}_{22}^{-1}\\ \mathbf{K}_{22}\mathbf{C}_{\mathcal{M}} & \mathbf{K}_{22}^{-1}\end{bmatrix}_J$$

the matrix sought, $\mathbf{L}_J$, then being calculated as

$$\mathbf{L}_J = \mathbf{J}_J^{-1}\mathbf{Y}^{-1}\mathbf{Z}$$

While we have a closed-form inverse of $\mathbf{J}_J$, we do not have one for $\mathbf{Y}$, which is full and does not bear any particular structure that would allow us its inversion explicitly. Therefore, matrix $\mathbf{L}_J$ should be calculated numerically.

10.5 Dynamics of Rolling Robots

The dynamics of rolling robots, similar to that of other robotic mechanical systems, comprises two main problems, inverse and direct dynamics. We will study both using the same mathematical model. Hence, the main task here is to derive this model. It turns out that while rolling robots usually are nonholonomic mechanical systems, their mathematical models are formally identical to those of holonomic systems, the sole difference being that in the case under discussion, extra variables, besides the independent ones,

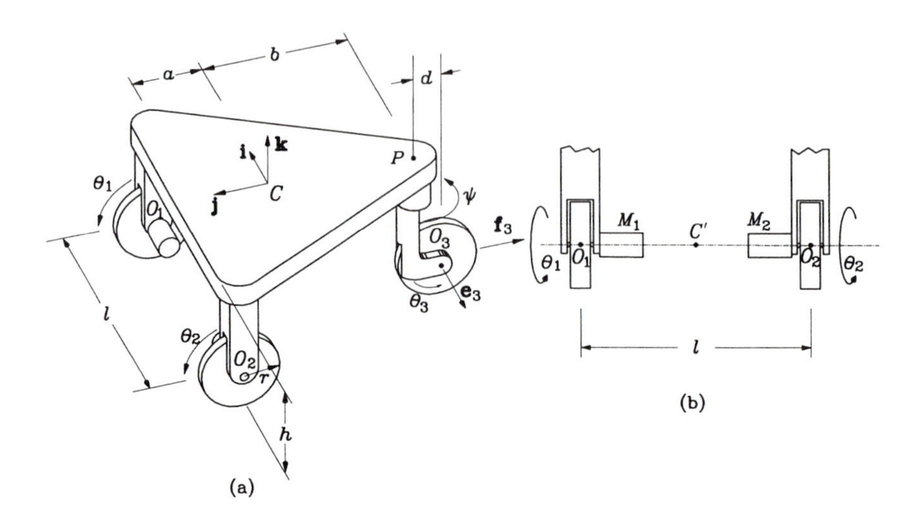

FIGURE 10.6. A 2-dof rolling robot: (a) its general layout; and (b) a detail of its actuated wheels.

are needed to fully describe the configuration of the system. Therefore, relations between these dependent variables and the independent ones will be needed and will be derived in the course of our discussion. Moreover, we will study robots with both conventional and omnidirectional wheels. Of the latter, we will focus on robots with Mekanum wheels.

10.5.1 Robots with Conventional Wheels

We study here the robot in Fig. 8.22, under the assumption that it is driven by motors collocated at the axes of its two coaxial wheels, indicated as M_1 and M_2 in Fig. 8.22b. For quick reference, we repeat this figure here as Fig. 10.6.

Our approach will be one of multibody dynamics, and for this reason, we distinguish five rigid bodies composing the robotic mechanical system at hand. These are the three wheels (two actuated and one caster wheel), the bracket carrying the caster wheel, and the platform. We label these bodies with numbers from 1 to 5, in the foregoing order, while noticing that bodies 4 and 5, the bracket and the platform, undergo planar motion, and hence, deserve special treatment. The 6×6 mass matrices of the first three bodies are labeled $\mathbf{M}_1$ to $\mathbf{M}_3$, with a similar labeling for their corresponding 6-dimensional twists, the counterpart items for bodies 4 and 5 being denoted by $\mathbf{M}'_4$, $\mathbf{M}'_5$, $\mathbf{t}'_4$, and $\mathbf{t}'_5$, the primes denoting 3×3—as opposed to 6×6 in the general case—mass matrices and 3-dimensional—as opposed to 6-dimensional in the general case—twist arrays.

We undertake to formulate the mathematical model of the mechanical system under study, which is of the general form of eq.(10.19a) derived for holonomic systems. The nonholonomy of the system brings about special features that will be highlighted in the derivations below.

As a first step in our formulation, we distinguish between *actuated* and *unactuated joint variables*, grouped into vectors $\boldsymbol{\theta}_a$ and $\boldsymbol{\theta}_u$, respectively, their time-derivatives being the *actuated* and *unactuated joint rates*, $\dot{\boldsymbol{\theta}}_a$ and $\dot{\boldsymbol{\theta}}_u$, respectively. From the kinematic analysis of this system in Subsection 8.6.1, it is apparent that the foregoing vectors are all 2-dimensional, namely,

$$\boldsymbol{\theta}_a \equiv \begin{bmatrix} \theta_1 \\ \theta_2 \end{bmatrix}, \quad \boldsymbol{\theta}_u \equiv \begin{bmatrix} \theta_3 \\ \psi \end{bmatrix} \tag{10.46}$$

Further, we set to deriving expressions for the twists of the five moving bodies in terms of the actuated joint rates, i.e., we write those twists as linear transformations of $\dot{\boldsymbol{\theta}}_a$, i.e.,

$$\mathbf{t}_i = \mathbf{T}_i \dot{\boldsymbol{\theta}}_a, \quad i = 1, 2, 3 \tag{10.47a}$$

and

$$\mathbf{t}'_i = \mathbf{T}'_i \dot{\boldsymbol{\theta}}_a, \quad i = 4, 5 \tag{10.47b}$$

where

$$\mathbf{T}_1 = \begin{bmatrix} -\mathbf{i} + \rho\delta\mathbf{k} & -\rho\delta\mathbf{k} \\ r\mathbf{j} & \mathbf{0} \end{bmatrix} \tag{10.48}$$

$$\mathbf{T}_2 = \begin{bmatrix} \rho\delta\mathbf{k} & -(\mathbf{i} + \rho\delta\mathbf{k}) \\ \mathbf{0} & r\mathbf{j} \end{bmatrix} \tag{10.49}$$

$$\mathbf{T}_3 = \begin{bmatrix} \boldsymbol{\Theta}_3 \\ \mathbf{C}_3 \end{bmatrix} \tag{10.50}$$

$$\mathbf{T}'_4 = \begin{bmatrix} \boldsymbol{\theta}_4^T \\ \mathbf{C}_4 \end{bmatrix} \tag{10.51}$$

$$\mathbf{T}'_5 = \begin{bmatrix} \rho\delta & -\rho\delta \\ r(\lambda\mathbf{i} + (1/2)\mathbf{j}) & r(-\lambda\mathbf{i} + (1/2)\mathbf{j}) \end{bmatrix} \equiv \begin{bmatrix} \boldsymbol{\theta}_5^T \\ \mathbf{C}_5 \end{bmatrix} \tag{10.52}$$

with α, δ, ρ, and λ defined as

$$\alpha \equiv \frac{a+b}{l}, \quad \delta \equiv \frac{d}{l}, \quad \lambda \equiv \frac{a}{l}, \quad \rho \equiv \frac{r}{d} \tag{10.53}$$

In the derivations below, we resort to the notation introduced in Subsection 8.6.1. Expressions for $\boldsymbol{\Theta}_3$ and $\mathbf{C}_3$ are derived below. First, we note that, from eq.(8.89), we can write

$$\boldsymbol{\omega}_3 = (\theta_{11}\dot{\theta}_1 + \theta_{12}\dot{\theta}_2)\mathbf{e}_3 + [\rho\delta(\dot{\theta}_1 - \dot{\theta}_2) + \theta_{21}\dot{\theta}_1 + \theta_{22}\dot{\theta}_2]\mathbf{k} \tag{10.54}$$

or

$$\boldsymbol{\omega}_3 = \boldsymbol{\Theta}_3 \dot{\boldsymbol{\theta}}_a \tag{10.55}$$

with $\boldsymbol{\Theta}_3$ defined as

$$\boldsymbol{\Theta}_3 = [\,\theta_{11}\mathbf{e}_3 + (\theta_{21} + \rho\delta)\mathbf{k} \quad \theta_{12}\mathbf{e}_3 + (\theta_{22} - \rho\delta)\mathbf{k}\,]$$

and $\theta_{i,j}$ denoting, of course, the $(i,\, j)$ entry of $\boldsymbol{\Theta}$, as derived in Subsection 8.6.1. In more compact form,

$$\boldsymbol{\Theta}_3 = [\,\theta_{11}\mathbf{e}_3 + \overline{\theta}_{21}\mathbf{k} \quad \theta_{12}\mathbf{e}_3 + \overline{\theta}_{22}\mathbf{k}\,] \tag{10.56a}$$

with $\overline{\theta}_{21}$ and $\overline{\theta}_{22}$ defined as

$$\overline{\theta}_{21} \equiv \theta_{21} + \rho\delta, \quad \overline{\theta}_{22} \equiv \theta_{22} - \rho\delta \tag{10.56b}$$

Moreover,

$$\dot{\mathbf{c}}_3 = -r\dot{\theta}_3\mathbf{f}_3 = -r(\theta_{11}\dot{\theta}_1 + \theta_{12}\dot{\theta}_2)\mathbf{f}_3$$

and hence,

$$\mathbf{C}_3 = r\,[\,-\theta_{11}\mathbf{f}_3 \quad -\theta_{12}\mathbf{f}_3\,] \tag{10.57}$$

Further, it is apparent from Fig. 10.6 that the scalar angular velocity of the bracket, ω_4, is given by

$$\omega_4 = \omega + \dot{\psi}$$

and hence,

$$\omega_4 = \rho\delta(\dot{\theta}_1 - \dot{\theta}_2) + \theta_{21}\dot{\theta}_1 + \theta_{22}\dot{\theta}_2$$

Therefore, we can write

$$\omega_4 = \boldsymbol{\theta}_4^T\dot{\boldsymbol{\theta}}_a \tag{10.58a}$$

where $\boldsymbol{\theta}_4$ is defined as

$$\boldsymbol{\theta}_4 \equiv [\,\theta_{21} + \rho\delta \quad \theta_{22} - \rho\delta\,]^T \tag{10.58b}$$

Now, since we are given the inertial properties of the bracket in bracket coordinates, it makes sense to express $\dot{\mathbf{c}}_4$ in those coordinates. Such an expression can be obtained below:

$$\dot{\mathbf{c}}_4 = \dot{\mathbf{o}}_3 + \boldsymbol{\omega}_4 \times \frac{1}{2}[-d\mathbf{f}_3 + (h - r)\mathbf{k}] = -\dot{\theta}_3\mathbf{f}_3 + \frac{1}{2}(\omega + \dot{\psi})\mathbf{e}_3$$

Upon expressing $\dot{\theta}_3$ and $\dot{\psi}$ in terms of $\dot{\theta}_1$ and $\dot{\theta}_2$, we obtain

$$\dot{\mathbf{c}}_4 = d\left[\frac{1}{2}\overline{\theta}_{21}\mathbf{e}_3 - \rho\theta_{11}\mathbf{f}_3\right]\dot{\theta}_1 + d\left[\frac{1}{2}\overline{\theta}_{22}\mathbf{e}_3 - \rho\theta_{12}\mathbf{f}_3\right]\dot{\theta}_2 \tag{10.59}$$

whence it is apparent that

$$\mathbf{C}_4 = d\,[\,(1/2)\overline{\theta}_{21}\mathbf{e}_3 - \rho\theta_{11}\mathbf{f}_3 \quad (1/2)\overline{\theta}_{22}\mathbf{e}_3 - \rho\theta_{12}\mathbf{f}_3\,] \tag{10.60}$$

Therefore,

$$\mathbf{T}_4' = \begin{bmatrix} \overline{\theta}_{21} & \overline{\theta}_{22} \\ d[(1/2)\overline{\theta}_{21}\mathbf{e}_3 - \rho\theta_{11}\mathbf{f}_3] & d[(1/2)\overline{\theta}_{22}\mathbf{e}_3 - \rho\theta_{12}\mathbf{f}_3] \end{bmatrix} \tag{10.61}$$

thereby completing all needed twist-shaping matrices.

The 2×2 matrix of generalized inertia, $\mathbf{I}(\boldsymbol{\theta})$, is now obtained. Here we have written this matrix as a function of all variables, independent and dependent, arrayed in the 4-dimensional vector $\boldsymbol{\theta}$, because we cannot obtain an expression for $\boldsymbol{\theta}_u$ in terms of $\boldsymbol{\theta}_a$, given the nonholonomy of the system at hand. Therefore, $\mathbf{I}$ is, in general, a function of θ_1, θ_2, θ_3, and ψ. To be sure, from the above expressions for the twist-shaping matrices $\mathbf{T}_i$ and $\mathbf{T}_i'$, it is apparent that the said inertia matrix is an explicit function of ψ only, its dependence on θ_1 and θ_2 being implicitly given via vectors $\mathbf{e}_3$ and $\mathbf{f}_3$. We derive the expression sought for $\mathbf{I}$ starting from the kinetic energy, namely,

$$T = \sum_1^3 \frac{1}{2}\mathbf{t}_i^T\mathbf{M}_i\mathbf{t}_i + \frac{1}{2}\sum_4^5 (\mathbf{t}_i')^T\mathbf{M}_i'\mathbf{t}_i'$$

or

$$T = \frac{1}{2}\sum_1^3 \dot{\boldsymbol{\theta}}_a^T\mathbf{T}_i^T\mathbf{M}_i\mathbf{T}_i\dot{\boldsymbol{\theta}}_a + \frac{1}{2}\sum_4^5 \dot{\boldsymbol{\theta}}_a^T(\mathbf{T}_i')^T\mathbf{M}_i'\mathbf{T}_i'\dot{\boldsymbol{\theta}}_a \tag{10.62}$$

and hence,

$$\mathbf{I} = \sum_1^3 \mathbf{T}_i^T\mathbf{M}_i\mathbf{T}_i + \sum_4^5 (\mathbf{T}_i')^T\mathbf{M}_i'\mathbf{T}_i' \tag{10.63}$$

In order to expand the foregoing expression, we let $\mathbf{J}_w$ and $\mathbf{J}_c$ be the 3×3 inertia matrices of the two actuated wheels and the caster wheel, respectively, the scalar moments of inertia of the bracket and the platform, which undergo planar motion, being denoted by I_b and I_p. Likewise, we let m_w, m_b, m_c, and m_p denote the masses of the corresponding bodies. Therefore,

$$\mathbf{M}_1 = \begin{bmatrix} \mathbf{J}_w & \mathbf{O} \\ \mathbf{O} & m_w\mathbf{1}_3 \end{bmatrix} = \mathbf{M}_2$$

$$\mathbf{M}_3 = \begin{bmatrix} \mathbf{J}_c & \mathbf{O} \\ \mathbf{O} & m_c\mathbf{1}_3 \end{bmatrix}$$

$$\mathbf{M}_4' = \begin{bmatrix} I_b & \mathbf{0}^T \\ \mathbf{0} & m_b\mathbf{1}_2 \end{bmatrix}$$

$$\mathbf{M}_5' = \begin{bmatrix} I_p & \mathbf{0}^T \\ \mathbf{0} & m_p\mathbf{1}_2 \end{bmatrix}$$

with $\mathbf{O}$ and $\mathbf{1}_3$ denoting the 3×3 zero and identity matrices, while $\mathbf{0}$ and $\mathbf{1}_2$ the 2-dimensional zero vector and the 2×2 identity matrix. Furthermore, under the assumption that the actuated wheels are dynamically balanced, we have

$$\mathbf{J}_w = \begin{bmatrix} I & 0 & 0 \\ 0 & H & 0 \\ 0 & 0 & H \end{bmatrix}$$

Moreover, we assume that the caster wheel can be modeled as a rigid disk of uniform material of the given mass m_c and radius r, and hence, in bracket-fixed coordinates $\{\, \mathbf{e}_3, \mathbf{f}_3, \mathbf{k} \,\}$,

$$\mathbf{J}_c = \frac{1}{4} m_c r^2 \begin{bmatrix} 2 & 0 & 0 \\ 0 & 1 & 0 \\ 0 & 0 & 1 \end{bmatrix}$$

It is now a simple matter to calculate

$$\mathbf{T}_1^T \mathbf{M}_1 \mathbf{T}_1 = \begin{bmatrix} I + (\rho\delta)^2 H + m_w r^2 & -(\rho\delta)^2 H \\ -(\rho\delta)^2 H & (\rho\delta)^2 H \end{bmatrix}$$

$$\mathbf{T}_2^T \mathbf{M}_2 \mathbf{T}_2 = \begin{bmatrix} (\rho\delta)^2 H & -(\rho\delta)^2 H \\ -(\rho\delta)^2 H & I + (\rho\delta)^2 H + m_w r^2 \end{bmatrix}$$

where the *symmetry* between the two foregoing expressions is to be highlighted: that is, the second expression is derived if the diagonal entries of the first expression are exchanged, which is physically plausible, because such an exchange is equivalent to a relabeling of the two wheels. The calculation of the remaining products is less straightforward but can be readily obtained. From the expressions for $\mathbf{T}_3$ and $\mathbf{M}_3$, we have

$$\mathbf{T}_3^T \mathbf{M}_3 \mathbf{T}_3 = [\, \mathbf{\Theta}_3^T \;\; \mathbf{C}_3^T \,] \begin{bmatrix} \mathbf{J}_c & \mathbf{O} \\ \mathbf{O} & m_c \mathbf{1}_3 \end{bmatrix} \begin{bmatrix} \mathbf{\Theta}_3 \\ \mathbf{C}_3 \end{bmatrix}$$
$$= \mathbf{\Theta}_3^T \mathbf{J}_c \mathbf{\Theta}_3 + m_c \mathbf{C}_3^T \mathbf{C}_3$$

In order to calculate the foregoing products, we write $\mathbf{\Theta}_3$ and $\mathbf{C}_3$ in component form, i.e.,

$$\mathbf{J}_c \mathbf{\Theta}_3 = \frac{1}{4} m_c r^2 \begin{bmatrix} 2 & 0 & 0 \\ 0 & 1 & 0 \\ 0 & 0 & 1 \end{bmatrix} \begin{bmatrix} \theta_{11} & \theta_{12} \\ 0 & 0 \\ \overline{\theta}_{21} & \overline{\theta}_{22} \end{bmatrix}$$
$$= \frac{1}{4} m_c r^2 \begin{bmatrix} 2\theta_{11} & 2\theta_{12} \\ 0 & 0 \\ \overline{\theta}_{21} & \overline{\theta}_{22} \end{bmatrix}$$

and hence,

$$\mathbf{\Theta}_3^T \mathbf{J}_c \mathbf{\Theta}_3 = \frac{1}{4} m_c r^2 \begin{bmatrix} 2\theta_{11}^2 + \overline{\theta}_{21}^2 & 2\theta_{11}\theta_{12} + \overline{\theta}_{21}\overline{\theta}_{22} \\ 2\theta_{11}\theta_{12} + \overline{\theta}_{21}\overline{\theta}_{22} & 2\theta_{12}^2 + \overline{\theta}_{22}^2 \end{bmatrix}$$

Likewise,

$$m_3 \mathbf{C}_3^T \mathbf{C}_3 = m_c r^2 \begin{bmatrix} \theta_{11}^2 & \theta_{11}\theta_{12} \\ \theta_{11}\theta_{12} & \theta_{12}^2 \end{bmatrix}$$

Further,

$$(\mathbf{T}_4')^T \mathbf{M}_4' \mathbf{T}_4' = [\, \boldsymbol{\theta}_4 \quad \mathbf{C}_4^T \,] \begin{bmatrix} I_b & \mathbf{0}^T \\ \mathbf{0} & m_b \mathbf{1}_2 \end{bmatrix} \begin{bmatrix} \boldsymbol{\theta}_4^T \\ \mathbf{C}_4 \end{bmatrix}$$
$$= I_b \boldsymbol{\theta}_4 \boldsymbol{\theta}_4^T + m_b \mathbf{C}_4^T \mathbf{C}_4$$

Upon expansion, we have

$$(\mathbf{T}_4')^T \mathbf{M}_4' \mathbf{T}_4' = I_b \begin{bmatrix} \overline{\theta}_{21}^2 & \overline{\theta}_{21}\overline{\theta}_{22} \\ \overline{\theta}_{21}\overline{\theta}_{22} & \overline{\theta}_{22}^2 \end{bmatrix} + \frac{1}{4} m_b d^2 \begin{bmatrix} \overline{\theta}_{21}^2 + 4\rho^2\theta_{11}^2 & \overline{\theta}_{21}\overline{\theta}_{22} + 4\rho^2\theta_{11}\theta_{12} \\ \overline{\theta}_{21}\overline{\theta}_{22} + 4\rho^2\theta_{11}\theta_{12} & \overline{\theta}_{22}^2 + 4\rho^2\theta_{12}^2 \end{bmatrix}$$

Finally,

$$(\mathbf{T}_5')^T \mathbf{M}_5' \mathbf{T}_5' = [\, \boldsymbol{\theta}_5 \quad \mathbf{C}_5^T \,] \begin{bmatrix} I_p & \mathbf{0}^T \\ \mathbf{0} & m_p \mathbf{1}_2 \end{bmatrix} \begin{bmatrix} \boldsymbol{\theta}_5^T \\ \mathbf{C}_5 \end{bmatrix}$$
$$= I_p \boldsymbol{\theta}_5 \boldsymbol{\theta}_5^T + m_p \mathbf{C}_5^T \mathbf{C}_5$$

which can be readily expanded as

$$(\mathbf{T}_5')^T \mathbf{M}_5' \mathbf{T}_5' = I_p (\rho\delta)^2 \begin{bmatrix} 1 & -1 \\ -1 & 1 \end{bmatrix} + m_p r^2 \begin{bmatrix} (1/4) + \lambda^2 & (1/4) - \lambda^2 \\ (1/4) - \lambda^2 & (1/4) + \lambda^2 \end{bmatrix}$$

We can thus express the generalized inertia matrix as

$$\mathbf{I} = \mathbf{I}_w + \mathbf{I}_c + \mathbf{I}_b + \mathbf{I}_p$$

where $\mathbf{I}_w$, $\mathbf{I}_c$, $\mathbf{I}_b$, and $\mathbf{I}_p$ denote the contributions of the actuated wheels, the caster wheel, the bracket, and the platform, respectively, i.e.,

$$\mathbf{I}_w = \sum_1^2 \mathbf{T}_i^T \mathbf{M}_i \mathbf{T}_i = \begin{bmatrix} I + 2(\rho\delta)^2 H + m_w r^2 & -2(\rho\delta)^2 H \\ -2(\rho\delta)^2 H & I + 2(\rho\delta)^2 H + m_w r^2 \end{bmatrix}$$

$$\mathbf{I}_c = \frac{m_c r^2}{4} \begin{bmatrix} 6\theta_{11}^2 + \overline{\theta}_{21}^2 & 6\theta_{11}\theta_{12} + \overline{\theta}_{21}\overline{\theta}_{22} \\ 6\theta_{11}\theta_{12} + \overline{\theta}_{21}\overline{\theta}_{22} & 6\theta_{12}^2 + \overline{\theta}_{22}^2 \end{bmatrix}$$

$$\mathbf{I}_b = I_b \begin{bmatrix} \overline{\theta}_{21}^2 & \overline{\theta}_{21}\overline{\theta}_{22} \\ \overline{\theta}_{21}\overline{\theta}_{22} & \overline{\theta}_{22}^2 \end{bmatrix} + \frac{1}{4} m_b d^2 \begin{bmatrix} \overline{\theta}_{21}^2 + 4\rho^2\theta_{11}^2 & \overline{\theta}_{21}\overline{\theta}_{22} + 4\rho^2\theta_{11}\theta_{12} \\ \overline{\theta}_{21}\overline{\theta}_{22} + 4\rho^2\theta_{11}\theta_{12} & \overline{\theta}_{22}^2 + 4\rho^2\theta_{12}^2 \end{bmatrix}$$

$$\mathbf{I}_p = I_p (\rho\delta)^2 \begin{bmatrix} 1 & -1 \\ -1 & 1 \end{bmatrix} + m_p r^2 \begin{bmatrix} (1/4) + \lambda^2 & (1/4) - \lambda^2 \\ (1/4) - \lambda^2 & (1/4) + \lambda^2 \end{bmatrix}$$

It is now apparent that the contributions of the actuated wheels and the platform are constant, while those of the caster wheel and the bracket are configuration-dependent. Therefore, only the latter contribute to the Coriolis and centrifugal generalized forces. We thus have

$$\mathbf{T}^T\mathbf{M}\dot{\mathbf{T}} = \mathbf{T}_3^T\mathbf{M}_3\dot{\mathbf{T}}_3 + (\mathbf{T}_4')^T\mathbf{M}_4'\dot{\mathbf{T}}_4'$$

From the expression for $\mathbf{T}_3^T\mathbf{M}_3\mathbf{T}_3$, we obtain

$$\mathbf{T}_3^T\mathbf{M}_3\dot{\mathbf{T}}_3 = \mathbf{\Theta}_3^T\mathbf{J}_c\dot{\mathbf{\Theta}}_3 + m_3\mathbf{C}_3^T\dot{\mathbf{C}}_3$$

the time-derivatives being displayed below:

$$\begin{aligned}
\dot{\mathbf{\Theta}}_3 &= [\,\dot{\theta}_{11}\mathbf{e}_3 + \theta_{11}\dot{\psi}\mathbf{f}_3 + \dot{\theta}_{21}\mathbf{k} \quad \dot{\theta}_{12}\mathbf{e}_3 + \theta_{12}\dot{\psi}\mathbf{f}_3 + \dot{\theta}_{22}\mathbf{k}\,]\\
\dot{\mathbf{C}}_3 &= r\,[\,-\dot{\theta}_{11}\mathbf{f}_3 + \theta_{11}\dot{\psi}\mathbf{e}_3 \quad -\dot{\theta}_{12}\mathbf{f}_3 + \theta_{12}\dot{\psi}\mathbf{e}_3\,]
\end{aligned}$$

with the time-derivatives of the entries of $\mathbf{\Theta}$ given as

$$\dot{\mathbf{\Theta}} = \dot{\psi}\begin{bmatrix} -\alpha\sin\psi + (\cos\psi)/2 & \alpha\sin\psi + (\cos\psi)/2 \\ \rho[-\alpha\cos\psi - (\sin\psi)/2] & \rho[\alpha\cos\psi - (\sin\psi)/2] \end{bmatrix} \tag{10.64}$$

Upon expansion, the products appearing in the expression for $\mathbf{T}_3^T\mathbf{M}_3\dot{\mathbf{T}}_3$ become

$$\begin{aligned}
\mathbf{\Theta}_3^T\mathbf{J}_c\dot{\mathbf{\Theta}}_3 &= \frac{m_c r^2}{4}\begin{bmatrix} 2\theta_{11}\dot{\theta}_{11} + \overline{\theta}_{21}\dot{\theta}_{21} & 2\theta_{11}\dot{\theta}_{12} + \overline{\theta}_{21}\dot{\theta}_{22} \\ 2\theta_{12}\dot{\theta}_{11} + \overline{\theta}_{22}\dot{\theta}_{21} & 2\theta_{12}\dot{\theta}_{12} + \overline{\theta}_{22}\dot{\theta}_{22} \end{bmatrix}\\
m_3\mathbf{C}_3^T\dot{\mathbf{C}}_3 &= m_c r^2\begin{bmatrix} \theta_{11}\dot{\theta}_{11} & \theta_{11}\dot{\theta}_{12} \\ \theta_{12}\dot{\theta}_{11} & \theta_{12}\dot{\theta}_{12} \end{bmatrix}
\end{aligned}$$

Therefore,

$$\mathbf{T}_3^T\mathbf{M}_3\dot{\mathbf{T}}_3 = \frac{m_c r^2}{4}\begin{bmatrix} 6\theta_{11}\dot{\theta}_{11} + \overline{\theta}_{21}\dot{\theta}_{21} & 6\theta_{11}\dot{\theta}_{12} + \overline{\theta}_{21}\dot{\theta}_{22} \\ 6\theta_{12}\dot{\theta}_{11} + \overline{\theta}_{22}\dot{\theta}_{21} & 6\theta_{12}\dot{\theta}_{12} + \overline{\theta}_{22}\dot{\theta}_{22} \end{bmatrix}$$

Likewise,

$$(\mathbf{T}_4')^T\mathbf{M}_4'\dot{\mathbf{T}}_4' = I_b\boldsymbol{\theta}_4\dot{\boldsymbol{\theta}}_4^T + m_b\mathbf{C}_4^T\dot{\mathbf{C}}_4$$

the above time-derivatives being

$$\begin{aligned}
\dot{\boldsymbol{\theta}}_4^T &= [\,\dot{\theta}_{21} \quad \dot{\theta}_{22}\,]\\
\dot{\mathbf{C}}_4 &= d\,[\,c_{11}\mathbf{e}_3 + c_{12}\mathbf{f}_3 \quad c_{21}\mathbf{e}_3 + c_{22}\mathbf{f}_3\,]
\end{aligned}$$

with coefficients $c_{i,j}$ given below:

$$\begin{aligned}
c_{11} &= \frac{1}{2}\dot{\theta}_{21} + \rho\theta_{11}\dot{\psi}\\
c_{12} &= \frac{1}{2}\overline{\theta}_{21}\dot{\psi} - \rho\dot{\theta}_{11}\\
c_{21} &= \frac{1}{2}\dot{\theta}_{22} + \rho\theta_{12}\dot{\psi}\\
c_{22} &= \frac{1}{2}\overline{\theta}_{22}\dot{\psi} - \rho\dot{\theta}_{12}
\end{aligned}$$

Hence,

$$I_b \boldsymbol{\theta}_4 \dot{\boldsymbol{\theta}}_4^T = I_b \begin{bmatrix} \theta_{21}\dot{\theta}_{21} & \theta_{21}\dot{\theta}_{22} \\ \theta_{22}\dot{\theta}_{21} & \theta_{22}\dot{\theta}_{22} \end{bmatrix}$$

$$m_b \mathbf{C}_4^T \dot{\mathbf{C}}_4 = \frac{1}{2} m_b d^2 \begin{bmatrix} \overline{\theta}_{21}c_{11} - 2\rho\theta_{11}c_{12} & \overline{\theta}_{21}c_{21} - 2\rho\theta_{11}c_{22} \\ \overline{\theta}_{22}c_{11} - 2\rho\theta_{12}c_{12} & \overline{\theta}_{22}c_{21} - 2\rho\theta_{12}c_{22} \end{bmatrix}$$

Therefore,

$$(\mathbf{T}_4')^T \mathbf{M}_4' \dot{\mathbf{T}}_4' = I_b \begin{bmatrix} \theta_{21}\dot{\theta}_{21} & \theta_{21}\dot{\theta}_{22} \\ \theta_{22}\dot{\theta}_{21} & \theta_{22}\dot{\theta}_{22} \end{bmatrix} + \frac{1}{2} m_b d^2 \begin{bmatrix} \overline{\theta}_{21}c_{11} - 2\rho\theta_{11}c_{12} & \overline{\theta}_{21}c_{21} - 2\rho\theta_{11}c_{22} \\ \overline{\theta}_{22}c_{11} - 2\rho\theta_{12}c_{12} & \overline{\theta}_{22}c_{21} - 2\rho\theta_{12}c_{22} \end{bmatrix}$$

In the final steps, we undertake to calculate $\mathbf{T}^T\mathbf{WMT}$. As we saw earlier, only the caster wheel and the bracket can contribute to this term, for the contributions of the other bodies to the matrix of generalized inertia are constant. However, the bracket undergoes planar motion, and according to Exercise 6.6, its contribution to this term vanishes. Therefore,

$$\mathbf{T}^T\mathbf{WMT} = \mathbf{T}_3^T\mathbf{W}_3\mathbf{M}_3\mathbf{T}_3$$

Upon expansion of the foregoing product, we have

$$\mathbf{T}_3^T\mathbf{W}_3\mathbf{M}_3\mathbf{T}_3 = [\,\boldsymbol{\Theta}_3^T \quad \mathbf{C}_3^T\,] \begin{bmatrix} \boldsymbol{\Omega}_3 & \mathbf{O} \\ \mathbf{O} & \mathbf{O} \end{bmatrix} \begin{bmatrix} \mathbf{I}_c & \mathbf{O} \\ \mathbf{O} & m_c\mathbf{1}_3 \end{bmatrix} \begin{bmatrix} \boldsymbol{\Theta}_3 \\ \mathbf{C}_3 \end{bmatrix} = \boldsymbol{\Theta}_3^T\boldsymbol{\Omega}_3\mathbf{I}_c\boldsymbol{\Theta}_3$$

First, we obtain $\boldsymbol{\Omega}_3$ in bracket coordinates, by recalling eq.(10.54), i.e.,

$$[\,\boldsymbol{\omega}_3\,]_3 = \begin{bmatrix} \omega_e \\ 0 \\ \omega_k \end{bmatrix} \tag{10.65}$$

with ω_e and ω_k denoting the nonzero components of $\boldsymbol{\omega}_3$ in bracket coordinates, i.e.,

$$\omega_e = \theta_{11}\dot{\theta}_1 + \theta_{12}\dot{\theta}_2 \tag{10.66}$$

$$\omega_k = \rho\delta(\dot{\theta}_1 - \dot{\theta}_2) + \theta_{21}\dot{\theta}_1 + \theta_{22}\dot{\theta}_2 \tag{10.67}$$

and hence,

$$\boldsymbol{\Omega}_3 = (\theta_{11}\dot{\theta}_1 + \theta_{12}\dot{\theta}_2)\mathbf{E}_3 + [\rho\delta(\dot{\theta}_1 - \dot{\theta}_2) + \theta_{21}\dot{\theta}_1 + \theta_{22}\dot{\theta}_2]\mathbf{K}$$

with $\mathbf{E}_3$ and $\mathbf{K}$ defined as the cross-product matrices of $\mathbf{e}_3$ and $\mathbf{k}$, respectively, that is,

$$\mathbf{E}_3 = \begin{bmatrix} 0 & 0 & 0 \\ 0 & 0 & -1 \\ 0 & 1 & 0 \end{bmatrix}, \quad \mathbf{K} = \begin{bmatrix} 0 & 1 & 0 \\ 1 & 0 & 0 \\ 0 & 0 & 0 \end{bmatrix} \tag{10.68}$$

Therefore,

$$\boldsymbol{\Omega}_3 = \begin{bmatrix} 0 & -\omega_k & 0 \\ \omega_k & 0 & -\omega_e \\ 0 & \omega_e & 0 \end{bmatrix}$$

After simplification,

$$\boldsymbol{\Omega}_3 \mathbf{I}_c \boldsymbol{\Theta}_3 = \frac{m_c r^2}{4} \begin{bmatrix} 0 & 0 \\ 2\theta_{11}\omega_k - \overline{\theta}_{21}\omega_e & 2\theta_{12}\omega_k - \overline{\theta}_{22}\omega_e \\ 0 & 0 \end{bmatrix}$$

Now it is a simple matter to verify that

$$\boldsymbol{\Theta}_3^T \boldsymbol{\Omega}_3 \mathbf{I}_c \boldsymbol{\Theta}_3 = \mathbf{O}_2$$

with $\mathbf{O}_2$ denoting the 2×2 zero matrix, and hence,

$$\mathbf{T}^T \mathbf{W} \mathbf{T} = \mathbf{O}_2$$

In summary, the Coriolis and centrifugal-force terms of the system at hand take the form

$$\begin{aligned} \mathbf{C}(\boldsymbol{\theta}, \dot{\boldsymbol{\theta}}_a)\dot{\boldsymbol{\theta}}_a = & \frac{m_c r^2}{4} \begin{bmatrix} 6\theta_{11}(\dot{\theta}_{11}\dot{\theta}_1 + \dot{\theta}_{12}\dot{\theta}_2) + \overline{\theta}_{21}(\dot{\theta}_{12}\dot{\theta}_1 + \dot{\theta}_{22}\dot{\theta}_2) \\ 6\theta_{12}(\dot{\theta}_{11}\dot{\theta}_1 + \dot{\theta}_{12}\dot{\theta}_2) + \overline{\theta}_{22}(\dot{\theta}_{12}\dot{\theta}_1 + \dot{\theta}_{22}\dot{\theta}_2) \end{bmatrix} \\ & + I_b(\dot{\theta}_{21}\dot{\theta}_1 + \dot{\theta}_{22}\dot{\theta}_2) \begin{bmatrix} \theta_{21} \\ \theta_{22} \end{bmatrix} + \frac{1}{2} m_b d^2 (c_{11}\dot{\theta}_1 + c_{21}\dot{\theta}_2) \begin{bmatrix} \overline{\theta}_{21} \\ \overline{\theta}_{22} \end{bmatrix} \\ & -2\rho(c_{12}\dot{\theta}_1 + c_{22}\dot{\theta}_2) \begin{bmatrix} \theta_{11} \\ \theta_{12} \end{bmatrix} \end{aligned}$$

If we recall that the c_{ij} coefficients are linear in the joint rates, then the foregoing expression clearly shows the quadratic nature of the Coriolis and centrifugal terms with respect to the joint rates.

The derivation of the forces supplied by the actuators is straightforward:

$$\boldsymbol{\tau} = \begin{bmatrix} \tau_1 \\ \tau_2 \end{bmatrix}$$

The dissipative generalized force is less straightforward, but its calculation is not too lengthy. In fact, if we assume linear dashpots at all joints, then the dissipation function is

$$\Delta = \frac{1}{2} c_1 \dot{\theta}_1^2 + \frac{1}{2} c_2 \dot{\theta}_2^2 + \frac{1}{2} c_3 \dot{\theta}_3^2 + \frac{1}{2} c_4 \dot{\psi}^2 = \frac{1}{2} \dot{\boldsymbol{\theta}}_a^T \mathbf{C}_{12} \dot{\boldsymbol{\theta}}_a + \frac{1}{2} \dot{\boldsymbol{\theta}}_u^T \mathbf{C}_{34} \dot{\boldsymbol{\theta}}_u$$

with $\mathbf{C}_{12}$ and $\mathbf{C}_{34}$ defined as

$$\mathbf{C}_{12} \equiv \begin{bmatrix} c_1 & 0 \\ 0 & c_2 \end{bmatrix}, \quad \mathbf{C}_{34} \equiv \begin{bmatrix} c_3 & 0 \\ 0 & c_4 \end{bmatrix}$$

Now, if we recall the expression for $\dot{\boldsymbol{\theta}}_u$ in terms of $\dot{\boldsymbol{\theta}}_a$, we end up with

$$\Delta = \frac{1}{2}\dot{\boldsymbol{\theta}}_a^T \mathbf{D} \dot{\boldsymbol{\theta}}_a$$

$\mathbf{D}$ being defined, in turn, as the equivalent damping matrix, given by

$$\mathbf{D} = \mathbf{C}_{12} + \boldsymbol{\Theta}^T \mathbf{C}_{34} \boldsymbol{\Theta}$$

the dynamics model under study thus taking the form

$$\mathbf{I}(\boldsymbol{\theta})\ddot{\boldsymbol{\theta}}_a + \mathbf{C}(\boldsymbol{\theta}, \dot{\boldsymbol{\theta}}_a)\dot{\boldsymbol{\theta}}_a = \boldsymbol{\tau} - \mathbf{D}\dot{\boldsymbol{\theta}}_a$$

with $\mathbf{I}$ and $\mathbf{C}(\boldsymbol{\theta}, \dot{\boldsymbol{\theta}}_a)$ given, such as in the case of holonomic systems, as

$$\mathbf{I}(\boldsymbol{\theta}) = \mathbf{T}^T \mathbf{M} \mathbf{T}$$
$$\mathbf{C}(\boldsymbol{\theta}, \dot{\boldsymbol{\theta}}_a) = \mathbf{T}^T \mathbf{M} \dot{\mathbf{T}} + \mathbf{T}^T \mathbf{W} \mathbf{M} \mathbf{T}$$

thereby completing the mathematical model governing the motion of the system at hand. Note here that $\boldsymbol{\theta}$ denotes the 4-dimensional vector of joint variables containing all four angles appearing as components of $\boldsymbol{\theta}_a$ and $\boldsymbol{\theta}_u$. Because of the nonholonomy of the system, an expression for the latter in terms of the former cannot be derived, and thus the whole 4-dimensional vector $\boldsymbol{\theta}$ is left as an argument of both $\mathbf{I}$ and $\mathbf{C}$.

Note that calculating the torque $\boldsymbol{\tau}$ required for a given motion—direct dynamics—of the rolling robot under study is straightforward from the above model. However, given the strong coupling among all variables involved, a recursive algorithm in this case is not apparent. On the other hand, the determination of the motion produced by a given history of joint torques requires (*i*) the calculation of $\mathbf{I}$, which can be achieved symbolically; (*ii*) the inversion of $\mathbf{I}$, which can be done symbolically because this is a 2×2 matrix; (*iii*) the calculation of the Coriolis and centrifugal terms, as well as the dissipative forces; and (*iv*) the integration of the initial-value problem resulting once initial values to $\boldsymbol{\theta}$ and $\dot{\boldsymbol{\theta}}_a$ have been assigned.

10.5.2 Robots with Omnidirectional Wheels

We now consider a 3-dof robot with three actuated wheels of the Mekanum type, as shown in Fig. 8.24, with the configuration of Fig. 10.7, which will be termed, henceforth, the Δ-*array*. This system is illustrated in Fig. 10.8.

Below we will adopt the notation of Subsection 8.6.2, with $\alpha = \pi/2$ and $n = 3$. We now recall that the twist of the platform was represented in planar form as

$$\mathbf{t}' \equiv \begin{bmatrix} \omega \\ \dot{\mathbf{c}} \end{bmatrix} \tag{10.69}$$

where ω is the scalar angular velocity of the platform and $\dot{\mathbf{c}}$ is the 2-dimensional vector of its mass center, which will be assumed to coincide

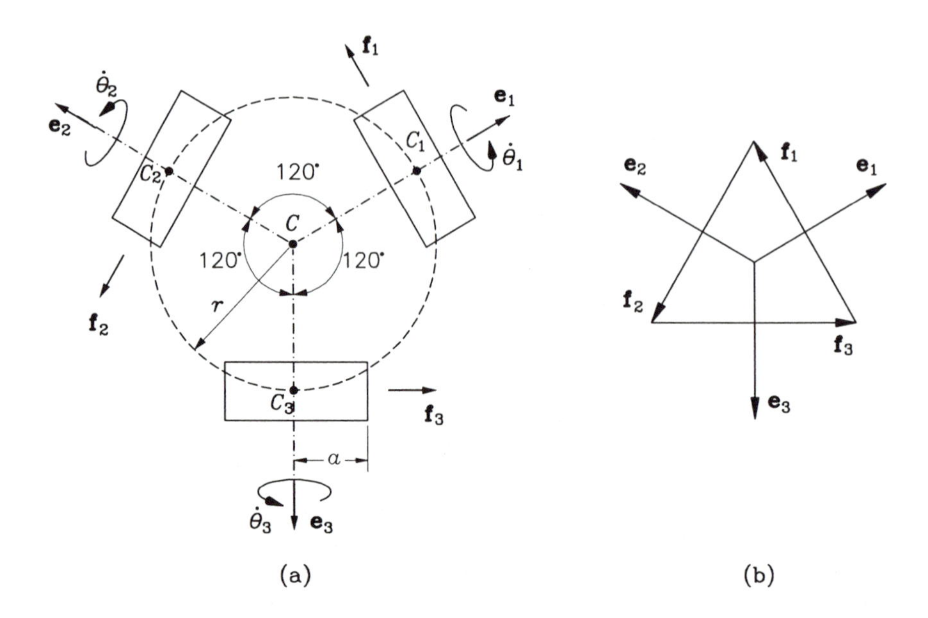

FIGURE 10.7. Rolling robot with ODWs in a Δ-array.

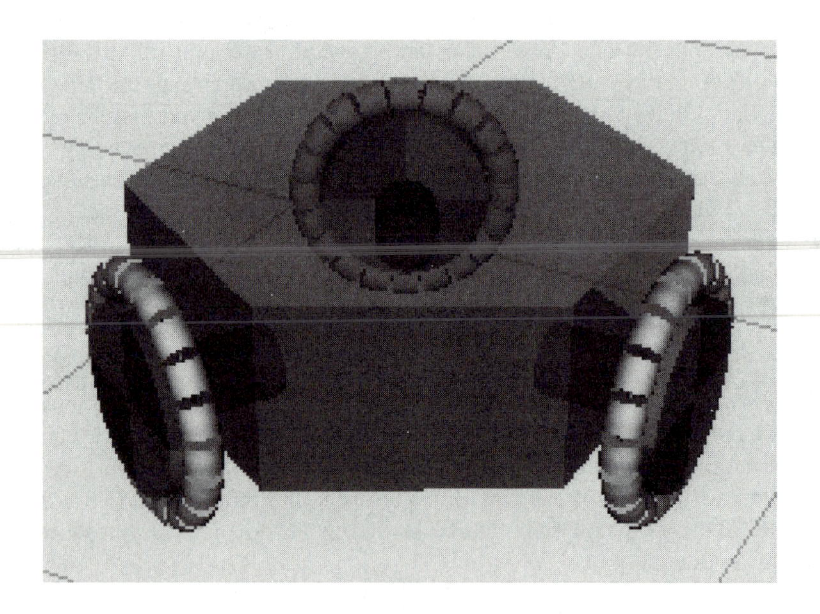

FIGURE 10.8. A view of the three-wheeled robot with Mekanum wheels in a Δ-array.

with the centroid of the set of points $\{ C_i \}_1^3$. Moreover, the three wheels are actuated, and hence, the 3-dimensional vector of actuated joint rates is defined as

$$\dot{\boldsymbol{\theta}}_a \equiv \begin{bmatrix} \dot{\theta}_1 \\ \dot{\theta}_2 \\ \dot{\theta}_3 \end{bmatrix} \tag{10.70}$$

The relation between $\dot{\boldsymbol{\theta}}_a$ and $\mathbf{t}'$ was derived in general in Subsection 8.6.2. As pertaining to the robot of Fig. 10.7, we have

$$\mathbf{J}\dot{\boldsymbol{\theta}}_a = \mathbf{K}\mathbf{t}' \tag{10.71a}$$

with the two 3×3 Jacobians $\mathbf{J}$ and $\mathbf{K}$ defined as

$$\mathbf{J} \equiv -a\mathbf{1}, \quad \mathbf{K} \equiv \begin{bmatrix} r & \mathbf{f}_1^T \\ r & \mathbf{f}_2^T \\ r & \mathbf{f}_3^T \end{bmatrix} \tag{10.71b}$$

where, it is recalled, a is the radius of the wheel hub and r is that of the rollers at the point of contact with the ground. Moreover, vectors $\{ \mathbf{e}_i \}_1^3$ and $\{ \mathbf{f}_i \}_1^3$ were defined in Subsection 8.6.2. Below we derive expressions for ω and $\dot{\mathbf{c}}$ in terms of the joint rates, from eq.(10.71a). First, we expand the three equations of the latter, thus obtaining

$$r\omega + \mathbf{f}_1^T\dot{\mathbf{c}} = -a\dot{\theta}_1 \tag{10.72a}$$
$$r\omega + \mathbf{f}_2^T\dot{\mathbf{c}} = -a\dot{\theta}_2 \tag{10.72b}$$
$$r\omega + \mathbf{f}_3^T\dot{\mathbf{c}} = -a\dot{\theta}_3 \tag{10.72c}$$

Upon adding corresponding sides of the three foregoing equations, we obtain

$$3r\omega + \dot{\mathbf{c}}^T\sum_1^3 \mathbf{f}_i = -a\sum_1^3 \dot{\theta}_i$$

But from Fig. 10.7b, it is apparent that

$$\mathbf{e}_1 + \mathbf{e}_2 + \mathbf{e}_3 = \mathbf{0} \tag{10.73a}$$
$$\mathbf{f}_1 + \mathbf{f}_2 + \mathbf{f}_3 = \mathbf{0} \tag{10.73b}$$

Likewise,

$$\mathbf{e}_1 = \frac{\sqrt{3}}{3}(\mathbf{f}_3 - \mathbf{f}_2), \ \mathbf{e}_2 = \frac{\sqrt{3}}{3}(\mathbf{f}_1 - \mathbf{f}_3), \ \mathbf{e}_3 = \frac{\sqrt{3}}{3}(\mathbf{f}_2 - \mathbf{f}_1) \tag{10.73c}$$
$$\mathbf{f}_1 = \frac{\sqrt{3}}{3}(\mathbf{e}_3 - \mathbf{e}_2), \ \mathbf{f}_2 = \frac{\sqrt{3}}{3}(\mathbf{e}_1 - \mathbf{e}_3), \ \mathbf{f}_3 = \frac{\sqrt{3}}{3}(\mathbf{e}_2 - \mathbf{e}_1) \tag{10.73d}$$

and hence, the above equation for ω and $\dot{\mathbf{c}}$ leads to

$$\omega = -\frac{a}{3r}\sum_1^3 \dot{\theta}_i \tag{10.74}$$

Now we derive an expression for $\dot{\mathbf{c}}$ in terms of the actuated joint rates. We do this by subtracting, sidewise, eqs.(10.72b & c) from eqs.(10.72a & b), respectively, thus obtaining a system of two linear equations in two unknowns, the two components of the 2-dimensional vector $\dot{\mathbf{c}}$, namely,

$$\mathbf{A}\dot{\mathbf{c}} = \mathbf{b}$$

with matrix $\mathbf{A}$ and vector $\mathbf{b}$ defined as

$$\mathbf{A} \equiv \begin{bmatrix} (\mathbf{f}_1 - \mathbf{f}_2)^T \\ (\mathbf{f}_2 - \mathbf{f}_3)^T \end{bmatrix} \equiv \sqrt{3} \begin{bmatrix} \mathbf{e}_3^T \\ \mathbf{e}_1^T \end{bmatrix}, \quad \mathbf{b} \equiv -a \begin{bmatrix} \dot{\theta}_1 - \dot{\theta}_2 \\ \dot{\theta}_2 - \dot{\theta}_3 \end{bmatrix}$$

where we have used relations (10.73c). Since $\mathbf{A}$ is a 2×2 matrix, its inverse can be readily found with the aid of Facts 4.8.3 and 4.8.4, which yields

$$\dot{\mathbf{c}} = \frac{2}{3} a \left[-\mathbf{E}\mathbf{e}_1 \quad \mathbf{E}\mathbf{e}_3 \right] \begin{bmatrix} \dot{\theta}_1 - \dot{\theta}_2 \\ \dot{\theta}_2 - \dot{\theta}_3 \end{bmatrix}$$

Now, from Fig. 10.7,

$$\mathbf{E}\mathbf{e}_1 = \mathbf{f}_1, \quad \mathbf{E}\mathbf{e}_3 = \mathbf{f}_3$$

and hence, $\dot{\mathbf{c}}$ reduces to

$$\dot{\mathbf{c}} = \frac{2}{3} a [(\dot{\theta}_2 - \dot{\theta}_1)\mathbf{f}_1 + (\dot{\theta}_2 - \dot{\theta}_3)\mathbf{f}_3] \equiv \frac{2}{3} a [\dot{\theta}_2(\mathbf{f}_1 + \mathbf{f}_3) - \dot{\theta}_1 \mathbf{f}_1 - \dot{\theta}_2 \mathbf{f}_2]$$

But by virtue of eq.(10.73b),

$$\mathbf{f}_1 + \mathbf{f}_3 = -\mathbf{f}_2$$

the above expression for $\dot{\mathbf{c}}$ thus becoming

$$\dot{\mathbf{c}} = -\frac{2a}{3} \sum_1^3 \dot{\theta}_i \mathbf{f}_i \tag{10.75}$$

Thus, ω is proportional to the mean value of $\{ \dot{\theta}_i \}_1^3$, while $\dot{\mathbf{c}}$ is proportional to the mean value of $\{ \dot{\theta}_i \mathbf{f}_i \}_1^3$. In deriving the Euler-Lagrange equations of the robot at hand, we will resort to the natural orthogonal complement, and therefore, we will require expressions for the twists of all bodies involved in terms of the actuated wheel rates. We start by labeling the wheels as bodies 1, 2, and 3, with the platform being body 4. Moreover, we will neglect the inertia of the rollers, and so no labels need be attached to these. Furthermore, the wheel hubs undergo angular velocities in two orthogonal directions, and hence, a full 6-dimensional twist representation of these will be required. Henceforth, we will regard the angular velocity of the platform and the velocity of its mass center as 3-dimensional vectors. Therefore,

$$\mathbf{t}_4 \equiv \mathbf{T}_4 \dot{\boldsymbol{\theta}}_a, \quad \mathbf{T}_4 \equiv -\lambda \begin{bmatrix} \mathbf{k} & \mathbf{k} & \mathbf{k} \\ 2r\mathbf{f}_1 & 2r\mathbf{f}_2 & 2r\mathbf{f}_3 \end{bmatrix} \tag{10.76}$$

with λ defined, in turn, as the ratio

$$\lambda \equiv \frac{a}{3r} \tag{10.77}$$

Now, the wheel angular velocities are given simply as

$$\boldsymbol{\omega}_i = \dot{\theta}_i \mathbf{e}_i + \omega \mathbf{k} = \dot{\theta}_i \mathbf{e}_i - \lambda (\sum_1^3 \dot{\theta}_i) \mathbf{k} \tag{10.78}$$

or

$$\boldsymbol{\omega}_1 = (\mathbf{e}_1 - \lambda \mathbf{k}) \dot{\theta}_1 - \lambda \dot{\theta}_2 \mathbf{k} - \lambda \dot{\theta}_3 \mathbf{k} \tag{10.79a}$$

$$\boldsymbol{\omega}_2 = -\lambda \dot{\theta}_1 \mathbf{k} + (\mathbf{e}_2 - \lambda \mathbf{k}) \dot{\theta}_2 - \lambda \dot{\theta}_3 \mathbf{k} \tag{10.79b}$$

$$\boldsymbol{\omega}_3 = -\lambda \dot{\theta}_1 \mathbf{k} - \lambda \dot{\theta}_2 \mathbf{k} + (\mathbf{e}_3 - \lambda \mathbf{k}) \dot{\theta}_3 \tag{10.79c}$$

Now, similar expressions are derived for vectors $\dot{\mathbf{c}}_i$. To this end, we resort to the geometry of Fig. 10.7, from which we derive the relations

$$\begin{aligned} \dot{\mathbf{c}}_i &= \dot{\mathbf{c}} + \omega r \mathbf{f}_i \\ &= -2\lambda r \left(\sum_1^3 \dot{\theta}_j \mathbf{f}_j \right) - \lambda r \left(\sum_1^3 \dot{\theta}_j \right) \mathbf{f}_i \end{aligned}$$

and hence,

$$\dot{\mathbf{c}}_1 = -\lambda r [(3\dot{\theta}_1 + \dot{\theta}_2 + \dot{\theta}_3)\mathbf{f}_1 + 2(\dot{\theta}_2 \mathbf{f}_2 + \dot{\theta}_3 \mathbf{f}_3)] \tag{10.80a}$$

$$\dot{\mathbf{c}}_2 = -\lambda r [2\dot{\theta}_1 \mathbf{f}_1 + (\dot{\theta}_1 + 3\dot{\theta}_2 + \dot{\theta}_3)\mathbf{f}_2 + 2\dot{\theta}_3 \mathbf{f}_3] \tag{10.80b}$$

$$\dot{\mathbf{c}}_3 = -\lambda r [2(\dot{\theta}_1 \mathbf{f}_1 + \dot{\theta}_2 \mathbf{f}_2) + (\dot{\theta}_1 + \dot{\theta}_2 + 3\dot{\theta}_3)\mathbf{f}_3] \tag{10.80c}$$

From the foregoing relations, and those for the angular velocities of the wheels, eqs.(10.79a–c), we can now write the twists of the wheels in the form

$$\mathbf{t}_i = \mathbf{T}_i \dot{\boldsymbol{\theta}}_a, \quad i = 1, 2, 3 \tag{10.81}$$

where

$$\mathbf{T}_1 \equiv \begin{bmatrix} \mathbf{e}_1 - \lambda \mathbf{k} & -\lambda \mathbf{k} & -\lambda \mathbf{k} \\ -3\lambda r \mathbf{f}_1 & -\lambda r (\mathbf{f}_1 + 2\mathbf{f}_2) & -\lambda r (\mathbf{f}_1 + 2\mathbf{f}_3) \end{bmatrix}$$

$$\mathbf{T}_2 \equiv \begin{bmatrix} -\lambda \mathbf{k} & \mathbf{e}_2 - \lambda \mathbf{k} & -\lambda \mathbf{k} \\ -\lambda r (2\mathbf{f}_1 + \mathbf{f}_2) & -3\lambda r \mathbf{f}_2 & -\lambda r (\mathbf{f}_2 + 2\mathbf{f}_3) \end{bmatrix}$$

$$\mathbf{T}_3 \equiv \begin{bmatrix} -\lambda \mathbf{k} & -\lambda \mathbf{k} & \mathbf{e}_3 - \lambda \mathbf{k} \\ -\lambda r (2\mathbf{f}_1 + \mathbf{f}_3) & -\lambda r (2\mathbf{f}_2 + \mathbf{f}_3) & -3\lambda r \mathbf{f}_3 \end{bmatrix}$$

On the other hand, similar to what we have in eq.(10.75), an interesting relationship among angular velocities of the the wheels arises here. Indeed, upon adding the corresponding sides of the three equations (10.79a–c), we obtain

$$\sum_1^3 \boldsymbol{\omega}_i = \sum_1^3 \dot{\theta}_i \mathbf{e}_i - 3\lambda\mathbf{k}\sum_1^3 \dot{\theta}_i$$

Further, we dot-multiply the two sides of the foregoing equation by $\mathbf{k}$, which yields, upon rearrangement of terms,

$$3\lambda\sum_1^3 \dot{\theta}_i = -\mathbf{k}\cdot\sum_1^3 \boldsymbol{\omega}_i$$

and by virtue of eq.(10.74),

$$\omega = \mathbf{k}\cdot\overline{\boldsymbol{\omega}}, \quad \overline{\boldsymbol{\omega}} \equiv \frac{1}{3}\sum_1^3 \boldsymbol{\omega}_i \tag{10.82}$$

that is, *the vertical component of the mean wheel angular velocity equals the scalar angular velocity of the platform.*

Now we proceed to establish the mathematical model governing the dynamics of the system under study. To this end, we first label the three wheels as bodies 1, 2, and 3, with the platform being labeled 4. The generalized inertia matrix is then calculated as

$$\mathbf{I} = \sum_1^4 \mathbf{T}_i^T \mathbf{M}_i \mathbf{T}_i \tag{10.83}$$

where

$$\mathbf{M}_i = \begin{bmatrix} \mathbf{I}_w & \mathbf{O} \\ \mathbf{O} & m_w\mathbf{1} \end{bmatrix}, \quad i = 1,2,3, \quad \mathbf{M}_4 = \begin{bmatrix} \mathbf{I}_p & \mathbf{O} \\ \mathbf{O} & m_p\mathbf{1} \end{bmatrix} \tag{10.84}$$

We will also need the angular-velocity dyads, $\mathbf{W}_i$, which are calculated as

$$\mathbf{W}_i = \begin{bmatrix} \mathbf{\Omega}_i & \mathbf{O} \\ \mathbf{O} & \mathbf{O} \end{bmatrix}, \quad i = 1,2,3 \tag{10.85}$$

where $\mathbf{W}_4$ will not be needed, since the platform undergoes planar motion. We have

$$\mathbf{M}_1\mathbf{T}_1 = \begin{bmatrix} \mathbf{I}_w(\mathbf{e}_1 - \lambda\mathbf{k}) & -\lambda\mathbf{I}_w\mathbf{k} & -\lambda\mathbf{I}_w\mathbf{k} \\ -3m_w\lambda r\mathbf{f}_1 & -m_w\lambda r(\mathbf{f}_1 + 2\mathbf{f}_2) & -m_w\lambda r(\mathbf{f}_1 + 2\mathbf{f}_3) \end{bmatrix}$$

Moreover, we assume that in a local coordinate frame $\{\,\mathbf{e}_i,\ \mathbf{f}_i,\ \mathbf{k}\,\}$,

$$\mathbf{I}_w = \begin{bmatrix} I & 0 & 0 \\ 0 & J & 0 \\ 0 & 0 & J \end{bmatrix}$$

and hence,

$$\mathbf{T}_1^T \mathbf{M}_1 \mathbf{T}_1 = \begin{bmatrix} I + \lambda^2 K & \lambda^2 J & \lambda^2 J \\ \lambda^2 J & \lambda^2 L & \lambda^2 M \\ \lambda^2 J & \lambda^2 M & \lambda^2 L \end{bmatrix}$$

where

$$\begin{aligned} K &\equiv J + 9m_w r^2 \\ L &\equiv J + 3m_w r^2 \\ M &\equiv J - 3m_w r^2 \end{aligned}$$

Likewise,

$$\mathbf{T}_2^T \mathbf{M}_2 \mathbf{T}_2 = \begin{bmatrix} \lambda^2 L & \lambda^2 J & \lambda^2 M \\ \lambda^2 J & I + \lambda^2 K & \lambda^2 J \\ \lambda^2 M & \lambda^2 J & \lambda^2 L \end{bmatrix}$$

$$\mathbf{T}_3^T \mathbf{M}_3 \mathbf{T}_3 = \begin{bmatrix} \lambda^2 L & \lambda^2 M & \lambda^2 J \\ \lambda^2 J & \lambda^2 L & \lambda^2 J \\ \lambda^2 J & \lambda^2 J & I + \lambda^2 K \end{bmatrix}$$

Furthermore,

$$\mathbf{M}_4 \mathbf{T}_4 = \lambda \begin{bmatrix} \mathbf{I}_p \mathbf{k} & \mathbf{I}_p \mathbf{k} & \mathbf{I}_p \mathbf{k} \\ 2m_p r \mathbf{f}_1 & 2m_p r \mathbf{f}_2 & 2m_p r \mathbf{f}_3 \end{bmatrix}$$

It is apparent that by virtue of the planar motion undergone by the platform, only its moment of inertia about the vertical passing by its mass center is needed. If we let this moment of inertia be H, then

$$\mathbf{T}_4^T \mathbf{M}_4 \mathbf{T}_4 = \lambda^2 \begin{bmatrix} H + 4m_p r^2 & H - 2m_p r^2 & H - 2m_p r^2 \\ H - 2m_p r^2 & H + 4m_p r^2 & H - 2m_p r^2 \\ H - 2m_p r^2 & H - 2m_p r^2 & H + 4m_p r^2 \end{bmatrix}$$

Upon summing all four products computed above, we obtain

$$\mathbf{I} = \begin{bmatrix} \alpha & \beta & \beta \\ \beta & \alpha & \beta \\ \beta & \beta & \alpha \end{bmatrix}$$

with the definitions below:

$$\begin{aligned} \alpha &\equiv I + \lambda^2 (H + 3J + 14 m_w r^2 + 4 m_p r^2) \\ \beta &\equiv \lambda^2 (H + 3J - \frac{5}{2} m_w r^2 - 2 m_p r^2) \end{aligned}$$

which is a constant matrix. Moreover, note that the geometric and inertial symmetry assumed at the outset is apparent in the form of the foregoing inertia matrix, its inverse being readily obtained in closed form, namely,

$$\mathbf{I}^{-1} = \frac{1}{\Delta}\begin{bmatrix} \alpha+\beta & -\beta & -\beta \\ -\beta & \alpha+\beta & -\beta \\ -\beta & -\beta & \alpha+\beta \end{bmatrix}, \quad \Delta \equiv (\alpha+\beta)\alpha - 2\beta^2$$

Next, we turn to the calculation of the $\mathbf{T}^T\mathbf{M}\dot{\mathbf{T}}$ term. This is readily found to be

$$\mathbf{T}^T\mathbf{M}\dot{\mathbf{T}} = \sum_1^4 \mathbf{T}_i^T\mathbf{M}_i\dot{\mathbf{T}}_i$$

each of the foregoing products being expanded below. We have, first,

$$\dot{\mathbf{T}}_1 = \begin{bmatrix} \omega\mathbf{f}_1 & \mathbf{0} & \mathbf{0} \\ 3\lambda r\omega\mathbf{e}_1 & -\lambda r\omega(\mathbf{e}_3 - \mathbf{e}_2) & \lambda r\omega(\mathbf{e}_3 - \mathbf{e}_2) \end{bmatrix}$$

$$\dot{\mathbf{T}}_2 = \begin{bmatrix} \mathbf{0} & \omega\mathbf{f}_2 & \mathbf{0} \\ \lambda r\omega(\mathbf{e}_1 - \mathbf{e}_3) & 3\lambda r\omega\mathbf{e}_2 & -\lambda r\omega(\mathbf{e}_1 - \mathbf{e}_3) \end{bmatrix}$$

$$\dot{\mathbf{T}}_3 = \begin{bmatrix} \mathbf{0} & \mathbf{0} & \omega\mathbf{f}_3 \\ -\lambda r\omega(\mathbf{e}_2 - \mathbf{e}_1) & \lambda r\omega(\mathbf{e}_2 - \mathbf{e}_1) & 3\lambda r\omega\mathbf{e}_3 \end{bmatrix}$$

$$\dot{\mathbf{T}}_4 = \lambda\begin{bmatrix} \mathbf{0} & \mathbf{0} & \mathbf{0} \\ -2r\omega\mathbf{e}_1 & -2r\omega\mathbf{e}_2 & -2r\omega\mathbf{e}_3 \end{bmatrix}$$

Hence, for the first wheel,

$$\mathbf{M}_1\dot{\mathbf{T}}_1 = \begin{bmatrix} \mathbf{I}_w\omega\mathbf{f}_1 & \mathbf{0} & \mathbf{0} \\ 3\lambda m_w r\omega\mathbf{e}_1 & -\lambda m_w r\omega(\mathbf{e}_3 - \mathbf{e}_2) & \lambda m_w r\omega(\mathbf{e}_3 - \mathbf{e}_2) \end{bmatrix}$$

Therefore,

$$\mathbf{T}_1^T\mathbf{M}_1\dot{\mathbf{T}}_1 = 3\sqrt{3}\lambda^2 m_w r^2\omega\begin{bmatrix} 0 & -1 & 1 \\ 1 & 0 & 0 \\ -1 & 0 & 0 \end{bmatrix}$$

where the skew-symmetric matrix is the cross product matrix of vector $[0,\ 1,\ 1]^T$. By symmetry, the following two products take on a similar form, except for the skew-symmetric matrix, which then becomes, correspondingly, the cross-product matrix of vectors $[1,\ 0,\ 1]^T$ and $[1,\ 1,\ 0]^T$. This means that the first of these three products is affected by the rotation of the second and the third wheels, but not by that of the first one; the second of those products is affected by the rotation of the first and the third wheels, but not by the second; the third product is affected, in turn, by the rotation of the first two wheels, but not by that of the third wheel. Therefore,

$$\mathbf{T}_2^T\mathbf{M}_2\dot{\mathbf{T}}_2 = 3\sqrt{3}\lambda^2 m_w r^2\omega\begin{bmatrix} 0 & -1 & 0 \\ 1 & 0 & -1 \\ 0 & 1 & 0 \end{bmatrix}$$

$$\mathbf{T}_3^T\mathbf{M}_3\dot{\mathbf{T}}_3 = 3\sqrt{3}\lambda^2 m_w r^2\omega\begin{bmatrix} 0 & 0 & 1 \\ 0 & 0 & -1 \\ -1 & 1 & 0 \end{bmatrix}$$

Furthermore,

$$\mathbf{M}_4\dot{\mathbf{T}}_4 = \lambda \begin{bmatrix} \mathbf{0} & \mathbf{0} & \mathbf{0} \\ -2m_p r\omega\mathbf{e}_1 & -2m_p r\omega\mathbf{e}_2 & -2m_p r\omega\mathbf{e}_3 \end{bmatrix}$$

and hence,

$$\mathbf{T}_4^T\mathbf{M}_4\dot{\mathbf{T}}_4 = 2\sqrt{3}\lambda^2 m_p r^2\omega \begin{bmatrix} 0 & -1 & 1 \\ 1 & 0 & -1 \\ -1 & 1 & 0 \end{bmatrix} \tag{10.87a}$$

whose skew-symmetric matrix is readily identified as the cross-product matrix of vector $[\,1,\,1,\,1\,]^T$, thereby indicating an equal participation of all three wheels in this term, a rather plausible result. Upon adding all four products calculated above, we obtain

$$\mathbf{T}^T\mathbf{M}\dot{\mathbf{T}} = 2\sqrt{3}\lambda^2(3m_w + m_p)r^2\omega \begin{bmatrix} 0 & -1 & 1 \\ 1 & 0 & -1 \\ -1 & 1 & 0 \end{bmatrix} \tag{10.88}$$

The equal participation of all three wheels in the foregoing product is apparent. Moreover, notice that the term in parentheses can be regarded as an equivalent mass, which is merely the sum of all four masses involved, the moments of inertia of the wheels playing no role in this term.

We now turn to the calculation of the $\mathbf{T}^T\mathbf{W}\mathbf{M}\mathbf{T}$ term, which can be expressed as a sum, namely,

$$\mathbf{T}^T\mathbf{W}\mathbf{M}\mathbf{T} = \sum_1^3 \mathbf{T}_i^T\mathbf{W}_i\mathbf{M}_i\mathbf{T}_i$$

where we have not considered the contribution of the platform, because this undergoes planar motion. Moreover, matrices $\mathbf{W}_i$, for $i = 1$, 2, and 3, take the obvious forms

$$\mathbf{W}_i \equiv \begin{bmatrix} \boldsymbol{\Omega}_i & \mathbf{O} \\ \mathbf{O} & \mathbf{O} \end{bmatrix}$$

We then have, for the first wheel,

$$\mathbf{W}_1\mathbf{M}_1\mathbf{T}_1 = \begin{bmatrix} \boldsymbol{\omega}_1 \times [\mathbf{I}_w(\mathbf{e}_1 - \lambda\mathbf{k})] & -\boldsymbol{\omega}_1 \times (\lambda\mathbf{I}_w\mathbf{k}) & -\boldsymbol{\omega}_1 \times (\lambda\mathbf{I}_w\mathbf{k}) \\ \mathbf{0} & \mathbf{0} & \mathbf{0} \end{bmatrix}$$

Now, it does not require too much effort to calculate the complete first product, which merely vanishes, i.e.,

$$\mathbf{W}_1\mathbf{M}_1\mathbf{T}_1 = \mathbf{O}_{66}$$

with $\mathbf{O}_{66}$ defined as the 6×6 zero matrix. By symmetry, the remaining two products also vanish, and hence, the sum also does, i.e.,

$$\mathbf{T}^T\mathbf{W}\mathbf{M}\mathbf{T} = \mathbf{O}_{66} \tag{10.89}$$

Now, calculating the dissipative and active generalized forces is straightforward. We will neglect here the dissipation of energy occurring at the bearings of the rollers, and hence, if we assume that the lubricant of the wheel hubs produces linear dissipative torques, then we have

$$\boldsymbol{\delta} = c \begin{bmatrix} \dot{\theta}_1 \\ \dot{\theta}_2 \\ \dot{\theta}_3 \end{bmatrix}, \quad \boldsymbol{\tau} = \begin{bmatrix} \tau_1 \\ \tau_2 \\ \tau_3 \end{bmatrix} \tag{10.90}$$

where c is the common damping coefficient for all three wheel hubs. We now have all the elements needed to set up the mathematical model governing the dynamics of the robot, namely,

$$\mathbf{I}\ddot{\boldsymbol{\theta}}_a + \mathbf{C}(\sigma, \boldsymbol{\theta}_a, \dot{\boldsymbol{\theta}}_a)\dot{\boldsymbol{\theta}}_a = \boldsymbol{\tau} - \boldsymbol{\delta} \tag{10.91}$$

where we have emphasized that the term containing Coriolis and centrifugal forces is a function not only of the vector of the wheel-hub variables $\boldsymbol{\theta}_a$, but also of the orientation of the platform, given by angle σ. However, we do not have a fourth dynamics equation to compute σ, and hence, its value has to be obtained by numerical quadrature from ω. This feature is common to all nonholonomic systems.

A

Kinematics of Rotations: A Summary

The purpose of this appendix is to outline proofs of some results in the realm of kinematics of rotations that were invoked in the preceding chapters. Further details are available in the literature (Angeles, 1988).

We start by noticing two preliminary facts whose proof is straightforward, as the reader is invited to verify.

Lemma A.1 *The* $(d/dt)(\cdot)$ *and the* $\text{vect}(\cdot)$ *operators, for* 3×3 *matrix operands, commute.*

and

Lemma A.2 *The* $(d/dt)(\cdot)$ *and the* $\text{tr}(\cdot)$ *operators, for* $n \times n$ *matrix operands, commute.*

Furthermore, we have

Theorem A.1 *Let* $\mathbf{A}$ *and* $\mathbf{S}$ *both be* 3×3 *matrices, the former arbitrary, the latter skew-symmetric. Then,*

$$\text{vect}(\mathbf{SA}) = \frac{1}{2}[\text{tr}(\mathbf{A})\mathbf{1} - \mathbf{A}]\mathbf{s}$$

where $\mathbf{s} \equiv \text{vect}(\mathbf{S})$.

Proof: An invariant proof of this theorem appears elusive, but a componentwise proof is straightforward. Indeed, let a_{ij} denote the (i, j) entry of

$\mathbf{A}$, and s_i the ith component of $\mathbf{s}$. Then,

$$\mathbf{SA} = \begin{bmatrix} -a_{21}s_3 + a_{31}s_2 & -a_{22}s_3 + a_{32}s_2 & -a_{23}s_3 + a_{33}s_2 \\ a_{11}s_3 - a_{31}s_1 & a_{12}s_3 - a_{32}s_1 & a_{13}s_3 - a_{33}s_1 \\ -a_{11}s_2 + a_{21}s_1 & -a_{12}s_2 + a_{22}s_1 & -a_{13}s_2 + a_{23}s_1 \end{bmatrix}$$

Hence,

$$\mathrm{vect}(\mathbf{SA}) = \frac{1}{2}\begin{bmatrix} (a_{22} + a_{33})s_1 - a_{12}s_2 - a_{13}s_3 \\ (a_{11} + a_{33})s_2 - a_{21}s_1 - a_{23}s_3 \\ (a_{11} + a_{22})s_3 - a_{31}s_1 - a_{32}s_2 \end{bmatrix}$$

On the other hand,

$$\mathrm{tr}(\mathbf{A})\mathbf{1} - \mathbf{A} = \begin{bmatrix} a_{22} + a_{33} & -a_{12} & -a_{13} \\ -a_{21} & a_{11} + a_{33} & -a_{23} \\ -a_{31} & -a_{32} & a_{11} + a_{22} \end{bmatrix}$$

and hence,

$$\frac{1}{2}[\mathrm{tr}(\mathbf{A})\mathbf{1} - \mathbf{A}]\mathbf{s} = \frac{1}{2}\begin{bmatrix} (a_{22} + a_{33})s_1 - a_{12}s_2 - a_{13}s_3 \\ (a_{11} + a_{33})s_2 - a_{21}s_1 - a_{23}s_3 \\ (a_{11} + a_{22})s_3 - a_{31}s_1 - a_{32}s_2 \end{bmatrix}$$

thereby completing the proof. Moreover, we have

Theorem A.2 *Let* $\mathbf{A}$, $\mathbf{S}$, *and* $\mathbf{s}$ *be defined as in Theorem A.1. Then,*

$$\mathrm{tr}(\mathbf{SA}) = -2\mathbf{s} \cdot [\mathrm{vect}(\mathbf{A})]$$

Proof: From the above expression for $\mathbf{SA}$,

$$\begin{aligned} \mathrm{tr}(\mathbf{SA}) &= -a_{21}s_3 + a_{31}s_2 + a_{12}s_3 - a_{32}s_1 - a_{13}s_2 + a_{23}s_1 \\ &= (a_{23} - a_{32})s_1 + (a_{31} - a_{13})s_2 + (a_{12} - a_{21})s_3 \\ &= [\, s_1 \quad s_2 \quad s_3 \,] \begin{bmatrix} a_{23} - a_{32} \\ a_{31} - a_{13} \\ a_{12} - a_{21} \end{bmatrix} = -2\mathbf{s} \cdot [\mathrm{vect}(\mathbf{A})] \end{aligned} \tag{A.1}$$

q.e.d.

Now we turn to the proof of the relations between the time-derivatives of the rotation invariants and the angular-velocity vector. Thus,

Theorem A.3 *Let* $\boldsymbol{\nu}$ *denote the 4-dimensional array of natural rotation invariants, as introduced in Section 2.3.2 and reproduced below for quick reference:*

$$\boldsymbol{\nu} \equiv \begin{bmatrix} \mathbf{e} \\ \phi \end{bmatrix}$$

Then the relationship between $\dot{\boldsymbol{\nu}}$ *and the angular velocity* $\boldsymbol{\omega}$ *is given by*

$$\dot{\boldsymbol{\nu}} = \mathbf{N}\boldsymbol{\omega}$$

with $\mathbf{N}$ *defined as*

$$\mathbf{N} \equiv \begin{bmatrix} [\sin\phi/(2(1-\cos\phi))](\mathbf{1}-\mathbf{e}\mathbf{e}^T) - (1/2)\mathbf{E} \\ \mathbf{e}^T \end{bmatrix}$$

Proof: Let us obtain first an expression for $\dot{\mathbf{e}}$. This is readily done by recalling that $\mathbf{e}$ is the real eigenvector of $\mathbf{Q}$, i.e.,

$$\mathbf{Q}\mathbf{e} = \mathbf{e}$$

Upon differentiation of both sides of the foregoing equation with respect to time, we have

$$\dot{\mathbf{Q}}\mathbf{e} + \mathbf{Q}\dot{\mathbf{e}} = \dot{\mathbf{e}}$$

i.e.,

$$(\mathbf{1} - \mathbf{Q})\dot{\mathbf{e}} = \dot{\mathbf{Q}}\mathbf{e}$$

An expression for $\dot{\mathbf{Q}}$ can be derived from eq.(3.46), which yields

$$\dot{\mathbf{Q}} = \boldsymbol{\Omega}\mathbf{Q} \tag{A.2}$$

Therefore,

$$\dot{\mathbf{Q}}\mathbf{e} = \boldsymbol{\Omega}\mathbf{e} \equiv \boldsymbol{\omega} \times \mathbf{e}$$

and hence, the above equation for $\dot{\mathbf{e}}$ takes the form

$$(\mathbf{1} - \mathbf{Q})\dot{\mathbf{e}} = \boldsymbol{\omega} \times \mathbf{e}$$

from which it is not possible to solve for $\dot{\mathbf{e}}$ because matrix $(\mathbf{1}-\mathbf{Q})$ is singular. Indeed, since both matrices inside the parentheses have an eigenvalue $+1$, their difference has an eigenvalue 0, which renders this difference singular. Thus, one more relation is needed in order to be able to determine $\dot{\mathbf{e}}$. This relation follows from the condition that $\|\mathbf{e}\|^2 = 1$. Upon differentiation of both sides of this condition with respect to time, we obtain

$$\mathbf{e}^T\dot{\mathbf{e}} = 0$$

the last two equations thus yielding a system of four scalar equations to determine $\dot{\mathbf{e}}$. We now assemble these equations into a single one, namely,

$$\mathbf{A}\dot{\mathbf{e}} = \mathbf{b}$$

where $\mathbf{A}$ is a 4×3 matrix, while $\mathbf{b}$ is a 4-dimensional vector, defined as

$$\mathbf{A} \equiv \begin{bmatrix} \mathbf{1} - \mathbf{Q} \\ \mathbf{e}^T \end{bmatrix}, \quad \mathbf{b} \equiv \begin{bmatrix} \boldsymbol{\omega} \times \mathbf{e} \\ 0 \end{bmatrix}$$

The foregoing overdetermined system of four equations in three unknowns now leads to a system of three equations in three unknowns if we multiply its two sides by $\mathbf{A}^T$ from the right, thereby producing

$$\mathbf{A}^T\mathbf{A}\dot{\mathbf{e}} = \mathbf{A}^T\mathbf{b}$$

We can therefore solve for $\dot{\mathbf{e}}$ from the above equation in the form

$$\dot{\mathbf{e}} = (\mathbf{A}^T\mathbf{A})^{-1}\mathbf{A}^T\mathbf{b}$$

where $\mathbf{A}^T\mathbf{A}$ takes the form

$$\mathbf{A}^T\mathbf{A} = (2)\mathbf{1} - (\mathbf{Q} + \mathbf{Q}^T) + \mathbf{e}\mathbf{e}^T$$

But the sum inside the parentheses is readily identified as twice the symmetric component of $\mathbf{Q}$, if we recall the Cartesian decomposition of matrices introduced in eq.(2.56). Therefore,

$$\mathbf{Q} + \mathbf{Q}^T = 2[(\cos\phi)\mathbf{1} + (1 - \cos\phi)\mathbf{e}\mathbf{e}^T]$$

Hence,

$$\mathbf{A}^T\mathbf{A} = 2(1 - \cos\phi)\mathbf{1} - (1 - 2\cos\phi)\mathbf{e}\mathbf{e}^T$$

As the reader can readily verify, the inverse of this matrix is given by

$$(\mathbf{A}^T\mathbf{A})^{-1} = \frac{1}{2(1-\cos\phi)}\mathbf{1} + \frac{1 - 2\cos\phi}{2(1-\cos\phi)}\mathbf{e}\mathbf{e}^T$$

which fails to exist only in the trivial case in which $\mathbf{Q}$ becomes the identity matrix. Upon expansion of the last expression for $\dot{\mathbf{e}}$, we have

$$\dot{\mathbf{e}} = -\frac{1}{2(1-\cos\phi)}(\mathbf{E} - \mathbf{Q}^T\mathbf{E})\boldsymbol{\omega}$$

Now $\mathbf{Q}^T\mathbf{E}$ is obtained by recalling eq.(2.54), thereby obtaining

$$\mathbf{Q}^T\mathbf{E} = (\cos\phi)\mathbf{E} + (\sin\phi)(\mathbf{1} - \mathbf{e}\mathbf{e}^T)$$

the final expression for $\dot{\mathbf{e}}$ thus being

$$\dot{\mathbf{e}} = -\frac{1}{2(1-\cos\phi)}[(1 - \cos\phi)\mathbf{E} - (\sin\phi)(\mathbf{1} - \mathbf{e}\mathbf{e}^T)]\boldsymbol{\omega}$$

Now, an expression for $\dot{\phi}$ is obtained upon equating the trace of the two sides of eq.(A.2), which yields

$$\mathrm{tr}(\dot{\mathbf{Q}}) = \mathrm{tr}(\boldsymbol{\Omega}\mathbf{Q}) \tag{A.3}$$

From Lemma A.2,

$$\mathrm{tr}(\dot{\mathbf{Q}}) = \frac{d}{dt}\mathrm{tr}(\mathbf{Q}) \tag{A.4}$$

and hence,

$$\operatorname{tr}(\dot{\mathbf{Q}}) = -2\dot{\phi}\sin\phi$$

On the other hand, from Theorem A.2,

$$\operatorname{tr}(\mathbf{\Omega Q}) = -2\boldsymbol{\omega} \cdot (\sin\phi)\mathbf{e}$$

Therefore,

$$\dot{\phi} = \boldsymbol{\omega} \cdot \mathbf{e}$$

Upon assembling the expressions for $\dot{\mathbf{e}}$ and $\dot{\phi}$, we obtain the desired relation, with $\mathbf{N}$ given as displayed above, thereby proving the theorem.

Theorem A.4 *Let $\boldsymbol{\lambda}$ denote the 4-dimensional array of linear rotation invariants, as introduced in Section 2.3.3 and reproduced below for quick reference:*

$$\boldsymbol{\lambda} \equiv \begin{bmatrix} (\sin\phi)\mathbf{e} \\ \cos\phi \end{bmatrix} \equiv \begin{bmatrix} \operatorname{vect}(\mathbf{Q}) \\ [\operatorname{tr}(\mathbf{Q}) - 1]/2 \end{bmatrix}$$

Then the relationship between $\dot{\boldsymbol{\lambda}}$ and the angular velocity is given by

$$\dot{\boldsymbol{\lambda}} = \mathbf{L}\boldsymbol{\omega}$$

with $\mathbf{L}$ defined as

$$\mathbf{L} \equiv \begin{bmatrix} (1/2)[\operatorname{tr}(\mathbf{Q})\mathbf{1} - \mathbf{Q}] \\ -(\sin\phi)\mathbf{e}^T \end{bmatrix}$$

Proof: From Lemma A.1, we have

$$\frac{d}{dt}\operatorname{vect}(\mathbf{Q}) = \operatorname{vect}(\dot{\mathbf{Q}}) \tag{A.5}$$

On the other hand, equating the vectors of the two sides of eq.(A.2) yields

$$\operatorname{vect}(\dot{\mathbf{Q}}) = \operatorname{vect}(\mathbf{\Omega Q})$$

and hence,

$$\frac{d}{dt}\operatorname{vect}(\mathbf{Q}) = \operatorname{vect}(\mathbf{\Omega Q})$$

But, if we recall Theorem A.1, the foregoing relation leads to

$$\frac{d}{dt}\operatorname{vect}(\mathbf{Q}) = \frac{1}{2}[\operatorname{tr}(\mathbf{Q})\mathbf{1} - \mathbf{Q}]\boldsymbol{\omega}$$

Likewise, from Lemma A.2, we have

$$\frac{d}{dt}\operatorname{tr}(\mathbf{Q}) = \operatorname{tr}(\dot{\mathbf{Q}})$$

and hence,

$$\frac{d}{dt}\mathrm{tr}(\mathbf{Q}) = \mathrm{tr}(\mathbf{\Omega Q})$$

Now, if we recall Theorem A.2, the foregoing relation leads to

$$\frac{d}{dt}\mathrm{tr}(\mathbf{Q}) = -2\boldsymbol{\omega} \cdot [\mathrm{vect}(\mathbf{Q})] = -2(\sin\phi)\mathbf{e}^T\boldsymbol{\omega}$$

Hence,

$$\frac{d}{dt}(\cos\phi) = -(\sin\phi)\mathbf{e}^T\boldsymbol{\omega}$$

Upon assembling the last two expressions for the time-derivatives of the vector of $\mathbf{Q}$ and $\cos\phi$, we obtain the desired relation.

Theorem A.5 *Let $\boldsymbol{\eta}$ denote the 4-dimensional array of the Euler-Rodrigues parameters of a rotation, as introduced in Section 2.3.6 and reproduced below for quick reference:*

$$\boldsymbol{\eta} \equiv \begin{bmatrix} [\sin(\phi/2)]\mathbf{e} \\ \cos(\phi/2) \end{bmatrix} \equiv \begin{bmatrix} \mathbf{r} \\ r_0 \end{bmatrix}$$

Then, the relationship between $\dot{\boldsymbol{\eta}}$ and the angular velocity takes the form

$$\dot{\boldsymbol{\eta}} = \mathbf{H}\boldsymbol{\omega}$$

with $\mathbf{H}$ defined as

$$\mathbf{H} \equiv \frac{1}{2}\begin{bmatrix} \cos(\phi/2)\mathbf{1} - \sin(\phi/2)\mathbf{E} \\ -\sin(\phi/2)\mathbf{e}^T \end{bmatrix} \equiv \frac{1}{2}\begin{bmatrix} r_0\mathbf{1} - \mathbf{R} \\ -\mathbf{r}^T \end{bmatrix}$$

where $\mathbf{R}$ is the cross-product matrix of $\mathbf{r}$.

Proof: If we differentiate $\mathbf{r}$, we obtain

$$\dot{\mathbf{r}} = \dot{\mathbf{e}}\sin\left(\frac{\phi}{2}\right) + \mathbf{e}\frac{\dot{\phi}}{2}\cos\left(\frac{\phi}{2}\right)$$

and hence, all we need to derive the desired relations is to find expressions for $\dot{\mathbf{e}}$ and $\dot{\phi}$ in terms of the Euler-Rodrigues parameters. Note that from Theorem A.3, we already have those expressions in terms of the natural invariants. Thus, substitution of the time-derivatives of the natural invariants, as given in Theorem A.3, into the above expression for $\dot{\mathbf{r}}$ leads to

$$\dot{\mathbf{r}} = -\frac{1}{2}\sin\left(\frac{\phi}{2}\right)\mathbf{E}\boldsymbol{\omega} + \frac{1}{2}\sin\left(\frac{\phi}{2}\right)\frac{\sin\phi}{1-\cos\phi}\boldsymbol{\omega} + \frac{1}{2}(\mathbf{e}\cdot\boldsymbol{\omega})\left[\cos\left(\frac{\phi}{2}\right) - \sin\left(\frac{\phi}{2}\right)\frac{\sin\phi}{1-\cos\phi}\right]\mathbf{e} \tag{A.6}$$

Now, by recalling the identities giving the trigonometric functions of ϕ in terms of those of $\phi/2$, we obtain

$$\sin\left(\frac{\phi}{2}\right)\frac{\sin\phi}{1-\cos\phi} = \cos\left(\frac{\phi}{2}\right)$$

and hence, the term in brackets of the above expression vanishes, the expression for $\dot{\mathbf{r}}$ thus reducing to

$$\dot{\mathbf{r}} = \frac{1}{2}\left[\cos\left(\frac{\phi}{2}\right)\mathbf{1} - \sin\left(\frac{\phi}{2}\right)\mathbf{E}\right]\boldsymbol{\omega} \equiv \frac{1}{2}\left[r_0\mathbf{1} - \mathbf{R}\right]\boldsymbol{\omega}$$

thereby completing the proof.

B

The Numerical Solution of Linear Algebraic Systems

In this appendix we consider the solution of the linear algebraic system

$$\mathbf{A}\mathbf{x} = \mathbf{b} \tag{B.1}$$

with $\mathbf{A}$ defined as a *full-rank* $m \times n$ matrix, while $\mathbf{x}$ and $\mathbf{b}$ are n- and m-dimensional vectors, respectively. The case $m = n$ is the most frequently encountered; this case is well documented in the literature (Dahlquist and Björck, 1974; Golub and Van Loan, 1989) and need not be further discussed. We will consider only the following two cases:

(a) overdetermined: $m > n$; and

(b) underdetermined: $m < n$.

The overdetermined case does not admit a solution, unless vector $\mathbf{b}$ happens to lie in the range of $\mathbf{A}$. Besides this special case, then, we must reformulate the problem, and rather than seeking a *solution* of eq.(B.1), we will look for an *approximation* of that system of equations. Moreover, we will seek an approximation that will satisfy an optimality condition.

The underdetermined case, on the contrary, admits infinitely-many solutions, the objective then being to seek one that satisfies the system of equations *and* satisfies an additional optimality condition as well.

We study each of these cases in the sections below.

B.1 The Overdetermined Case

The error $\mathbf{e}$ in the approximation of eq.(B.1) is defined as

$$\mathbf{e} \equiv \mathbf{b} - \mathbf{A}\mathbf{x} \tag{B.2}$$

An obvious way of imposing an optimality condition on the *solution* $\mathbf{x}$ is to specify that this solution minimize a *norm* of $\mathbf{e}$. All norms of $\mathbf{e}$ can be expressed as

$$\|\mathbf{e}\|_p \equiv \left(\frac{1}{m} \sum_1^m e_k^p \right)^{1/p} \tag{B.3}$$

with e_k being understood as the kth component of the m-dimensional vector $\mathbf{e}$. When $p = 2$, the foregoing norm is known as the *Euclidean norm*, which we have used most frequently in this book. When $p \to \infty$, the *infinity norm*, also known as the *Chebyshev norm*, is obtained. It turns out that upon seeking the value of $\mathbf{x}$ that minimizes a norm of $\mathbf{e}$, the simplest is the Euclidean norm, for the minimization of its square leads to a linear system of equations whose solution can be obtained *directly*, as opposed to *iteratively*. Indeed, let us set up the minimization problem below:

$$z(\mathbf{x}) \equiv \frac{1}{2}\|\mathbf{e}\|_2^2 \quad \to \quad \min_{\mathbf{x}} \tag{B.4}$$

The *normality condition* of the minimization problem at hand is derived upon setting the *gradient* of z with respect to $\mathbf{x}$ equal to zero, i.e.,

$$\frac{dz}{d\mathbf{x}} = \mathbf{0} \tag{B.5}$$

Now, the chain rule and the results of Subsection 2.3.1 allow us to write

$$\frac{dz}{d\mathbf{x}} \equiv \left(\frac{d\mathbf{e}}{d\mathbf{x}} \right)^T \frac{dz}{d\mathbf{e}} \equiv -\mathbf{A}^T\mathbf{e} \tag{B.6}$$

and hence, we have the first result:

Theorem B.1.1 *The error in the approximation of eq.(B.1), for the full-rank $m \times n$ matrix $\mathbf{A}$, with $m > n$, is of minimum Euclidean norm if it lies in the nullspace of $\mathbf{A}^T$.*

Furthermore, if eq.(B.2) is substituted into eq.(B.6), and the product thus resulting is substituted, in turn, into the normality condition, we obtain

$$\mathbf{A}^T\mathbf{A}\mathbf{x} = \mathbf{A}^T\mathbf{b} \tag{B.7}$$

which is known as the *normal equation* of the minimization problem at hand. By virtue of the assumption on the rank of $\mathbf{A}$, the product $\mathbf{A}^T\mathbf{A}$ is *positive-definite* and hence, invertible. As a consequence, the value $\mathbf{x}_0$ of $\mathbf{x}$ that minimizes the Euclidean norm of the approximation error of the given system is

$$\mathbf{x}_0 = (\mathbf{A}^T\mathbf{A})^{-1}\mathbf{A}^T\mathbf{b} \tag{B.8}$$

the matrix coefficient of $\mathbf{b}$ being known as a *generalized inverse* of $\mathbf{A}$. The error obtained with this value is known as the *least-square error* of the approximation, i.e.,

$$\mathbf{e}_0 \equiv \mathbf{b} - \mathbf{A}\mathbf{x}_0 \tag{B.9}$$

The reader should be able to prove one more result:

Theorem B.1.2 (Projection Theorem) *The least-square error is orthogonal to* $\mathbf{A}\mathbf{x}_0$.

While the formula yielding the foregoing generalized inverse is quite simple to implement, the number of floating-point operations (flops) it takes to evaluate, along with the ever-present roundoff errors in both the data and the results, renders it not only inefficient, but also unreliable if applied as such. Indeed, if we recall the concept of condition number, introduced in Section 4.9 and recalled in Subsection 8.2.2, along with the definition adopted in the latter for the condition number, it becomes apparent that the condition number of $\mathbf{A}^T\mathbf{A}$ is *exactly* the square of the condition number of $\mathbf{A}$. This result can be best understood if we apply the Polar-Decomposition Theorem, introduced in Section 4.9, to rectangular matrices, but we will not elaborate on this issue here.

As a consequence, then, even if $\mathbf{A}$ is only slightly ill-conditioned, the product $\mathbf{A}^T\mathbf{A}$ can be catastrophically ill-conditioned. Below we outline two procedures to calculate efficiently the least-square approximation of the overdetermined system (B.1) that preserve the condition number of $\mathbf{A}$ and do this with a low number of flops.

B.1.1 The Numerical Solution of an Overdetermined System of Linear Equations

In seeking a numerical solution of the system of equations at hand, we would like to end up with a triangular system, similar to the LU-decomposition applied to solve a system of as many equations as unknowns, and hence, we have to perform some transformations either on the rows of $\mathbf{A}$ or on its columns. A safe numerical procedure should thus preserve (a) the Euclidean norm of the columns of $\mathbf{A}$ and (b) the inner product between any two columns of this matrix. Hence, a triangularization procedure like LU-decomposition would not work, because this does not preserve inner products. Obviously, the transformations that do preserve these inner products are orthogonal, either rotations or reflections. Of these, the most popular methods are (a) the Gram-Schmidt orthogonalization procedure and (b) *Householder reflections*.

The Gram-Schmidt procedure consists in regarding the columns of $\mathbf{A}$ as a set of n m-dimensional vectors $\{\mathbf{a}_k\}_1^n$. From this set, a new set $\{\mathbf{e}_k\}_1^n$ is obtained that is *orthonormal*. The procedure is quite simple and works as follows: Define $\mathbf{e}_1$ as

$$\mathbf{e}_1 = \frac{\mathbf{a}_1}{\|\mathbf{a}_1\|} \tag{B.10}$$

Further, we define $\mathbf{e}_2$ as the *normal component* of $\mathbf{a}_2$ onto $\mathbf{e}_2$, as introduced in eq.(2.6b), i.e.,

$$\mathbf{b}_2 \equiv (\mathbf{1} - \mathbf{e}_1\mathbf{e}_1^T)\mathbf{a}_2 \tag{B.11a}$$

$$\mathbf{e}_2 \equiv \frac{\mathbf{b}_2}{\|\mathbf{b}_2\|} \tag{B.11b}$$

In the next step, we define $\mathbf{e}_3$ as the unit vector normal to the plane defined by $\mathbf{e}_1$ and $\mathbf{e}_2$ and in the direction in which the inner product $\mathbf{e}_3^T \cdot \mathbf{a}_3$ is positive, which is possible because all vectors of the set $\{\mathbf{a}_k\}_1^m$ have been assumed to be linearly independent—remember that $\mathbf{A}$ has been assumed to be of full rank. To this end, we subtract from $\mathbf{a}_3$ its projection onto the plane mentioned above, i.e.,

$$\mathbf{b}_3 = (\mathbf{1} - \mathbf{e}_1\mathbf{e}_1^T - \mathbf{e}_2\mathbf{e}_2^T)\mathbf{a}_3 \tag{B.12a}$$

$$\mathbf{e}_3 \equiv \frac{\mathbf{b}_3}{\|\mathbf{b}_3\|} \tag{B.12b}$$

and so on, until we obtain $\mathbf{e}_{n-1}$, the last unit vector of the orthogonal set, $\mathbf{e}_n$, being obtained as

$$\mathbf{b}_n = (\mathbf{1} - \mathbf{e}_1\mathbf{e}_1^T - \mathbf{e}_2\mathbf{e}_2^T - \cdots - \mathbf{e}_{n-1}\mathbf{e}_{n-1}^T)\mathbf{a}_n \tag{B.13a}$$

Finally,

$$\mathbf{e}_n \equiv \frac{\mathbf{b}_n}{\|\mathbf{b}_n\|} \tag{B.13b}$$

In the next stage, we represent all vectors of the set $\{\mathbf{a}_k\}_1^m$ in *orthogonal coordinates*, i.e., in the base $\mathcal{O} = \{\mathbf{e}_k\}_1^n$, which are then arranged in an $m \times n$ array $\mathbf{A}_o$. By virtue of the form in which the set $\{\mathbf{e}_k\}_1^n$ was defined, the last $m - k$ components of vector $\mathbf{a}_k$ vanish. We thus have, in the said orthonormal basis,

$$[\mathbf{a}_k]_{\mathcal{O}} = \begin{bmatrix} \alpha_{1k} \\ \alpha_{2k} \\ \vdots \\ \alpha_{kk} \\ 0 \\ \vdots \\ 0 \end{bmatrix} \tag{B.14a}$$

Further, we represent $\mathbf{b}$ in $\mathcal{O}$, thus obtaining

$$[\mathbf{b}]_{\mathcal{O}} = \begin{bmatrix} \beta_1 \\ \beta_2 \\ \vdots \\ \beta_m \end{bmatrix} \tag{B.14b}$$

Therefore, eq.(B.1), when expressed in $\mathcal{O}$, becomes

$$\begin{bmatrix} \alpha_{11} & \alpha_{12} & \cdots & \alpha_{1n} \\ 0 & \alpha_{22} & \cdots & \alpha_{2n} \\ \vdots & \vdots & \ddots & \vdots \\ 0 & 0 & \cdots & \alpha_{nn} \\ 0 & 0 & \cdots & 0 \\ \vdots & \vdots & \ddots & \vdots \\ 0 & 0 & \cdots & 0 \end{bmatrix} \begin{bmatrix} x_1 \\ x_2 \\ \vdots \\ x_n \end{bmatrix} = \begin{bmatrix} \beta_1 \\ \beta_2 \\ \vdots \\ \beta_n \\ \beta_{n+1} \\ \vdots \\ \beta_m \end{bmatrix} \tag{B.15}$$

whence $\mathbf{x}$ can be computed by back-substitution. It is apparent, then, that the last $m-n$ equations of the foregoing system are incompatible: their left-hand sides are zero, while their right-hand sides are not necessarily so. What the right-hand sides of these equations represent, then, is the approximation error in orthogonal coordinates. Its Euclidean norm is, then,

$$\|\mathbf{e}_0\| \equiv \sqrt{\beta_{n+1}^2 + \ldots + \beta_m^2} \tag{B.16}$$

The second method discussed here is based on the application of a chain of n reflections $\{\mathbf{H}_k\}_1^n$, known as *Householder reflections*, to both sides of eq.(B.1). The purpose of these reflections is, again, to obtain a representation of matrix $\mathbf{A}$ in upper-triangular form (Golub and Van Loan, 1989). The algorithm proceeds as follows: We assume that we have applied reflections $\mathbf{H}_1$, $\mathbf{H}_2$, ..., $\mathbf{H}_{k-1}$, in this order, to $\mathbf{A}$ that have rendered it in *upper-trapezoidal form*, i.e.,

$$\mathbf{A}_{i-1} \equiv \mathbf{H}_{i-1} \ldots \mathbf{H}_2 \mathbf{H}_1 \mathbf{A}$$
$$= \begin{bmatrix} a_{11}^* & a_{12}^* & \cdots & a_{1,i-1}^* & a_{1i}^* & \cdots & a_{1n}^* \\ 0 & a_{22}^* & \cdots & a_{2,i-1}^* & a_{2i}^* & \cdots & a_{2n}^* \\ 0 & 0 & \cdots & a_{3,i-1}^* & a_{3i}^* & \cdots & a_{3n}^* \\ \vdots & \vdots & \ddots & \vdots & \vdots & \ddots & \vdots \\ 0 & 0 & \cdots & a_{i-1,i-1}^* & a_{i-1,i}^* & \cdots & a_{i-1,n}^* \\ 0 & 0 & \cdots & 0 & a_{i,i}^* & \cdots & a_{i,n}^* \\ \vdots & \vdots & \ddots & \vdots & \vdots & \ddots & \vdots \\ 0 & 0 & \cdots & 0 & a_{m,i}^* & \cdots & a_{mn}^* \end{bmatrix} \tag{B.17}$$

The next Householder reflection, $\mathbf{H}_i$, is determined so as to render the last $m-i$ components of the ith column of $\mathbf{H}_i\mathbf{A}_{i-1}$ equal to zero, while leaving its first $i-1$ columns unchanged. We do this by setting

$$\alpha_i = \text{sgn}(a_{ii}^*)\sqrt{(a_{ii}^*)^2 + (a_{i,i+1}^*)^2 + \cdots + (a_{im}^*)^2}$$
$$\mathbf{u}_i = [0 \quad 0 \quad \cdots \quad 0 \quad a_{ii}^* + \alpha_i \quad a_{i,i+1}^* \quad \cdots \quad a_{im}^*]^T$$
$$\mathbf{H}_i = \mathbf{1} - 2\frac{\mathbf{u}_i\mathbf{u}_i^T}{\|\mathbf{u}_i\|^2}$$

where $\text{sgn}(x)$ is defined as $+1$ if $x > 0$, as -1 if $x < 0$, and is left undefined when $x = 0$. As the reader can readily verify,

$$\frac{1}{2}\|\mathbf{u}_i\|^2 = \alpha_i(\mathbf{u}_i)_i = \alpha_i(a_{ii}^* + \alpha_i) \equiv \beta_i$$

and hence, the denominator appearing in the expression for $\mathbf{H}_i$ is calculated with one single addition and a single multiplication. It is noteworthy that $\mathbf{H}_i$, as defined above, is the $n \times n$ counterpart of the 3×3 *pure reflection* defined in eq.(2.5). As a matter of fact, $\mathbf{H}_i$ reflects vectors in m-dimensional space onto a hyperplane of unit normal $\mathbf{u}_i/\|\mathbf{u}_i\|$.

It is important to realize that

(a) α_i is defined with the sign of a_{ii}^* because β_i is a multiple of the ith component of $\mathbf{u}_i$, which is, in turn, the sum of a_{ii}^* and α_i, thereby guaranteeing that the absolute value of this sum will always be greater than the absolute value of each of its terms. If this provision were not made, then the resulting sum could be of a negligibly small absolute value, which would thus render β_i a very small positive number, thereby introducing unnecessarily an inadmissibly large roundoff-error amplification upon dividing the product $\mathbf{u}_i\mathbf{u}_i^T$ by β_i;

(b) an arbitrary vector $\mathbf{v}$ is transformed by $\mathbf{H}_i$ with unusually few flops, namely,

$$\mathbf{H}_i\mathbf{v} = \mathbf{v} - \frac{1}{\beta_i}(\mathbf{v}^T\mathbf{u}_i)\mathbf{u}_i$$

Upon application of the n Householder reflections thus defined, the system at hand becomes

$$\mathbf{HAx} = \mathbf{Hb} \tag{B.18}$$

with $\mathbf{H}$ defined as

$$\mathbf{H} \equiv \mathbf{H}_n \ldots \mathbf{H}_2\mathbf{H}_1 \tag{B.19}$$

Similar to that in equation (B.15), the matrix coefficient of $\mathbf{x}$ in eq.(B.18), i.e., $\mathbf{HA}$, is in upper-triangular form. That is, we have

$$\mathbf{HA} = \begin{bmatrix} \mathbf{U} \\ \mathbf{O}_{m'n} \end{bmatrix}, \quad \mathbf{Hb} = \begin{bmatrix} \mathbf{b}_U \\ \mathbf{b}_L \end{bmatrix} \tag{B.20}$$

with $\mathbf{O}_{m'n}$ denoting the $(m-n) \times n$ zero matrix, $m' \equiv m - n$, and $\mathbf{b}_U$ and $\mathbf{b}_L$ being n- and m'-dimensional vectors. The unknown $\mathbf{x}$ can thus be calculated from eq.(B.18) by back-substitution.

Note that the last m' components of the left-hand side of eq.(B.18) are zero, while the corresponding components of the right-hand side of the same equation are not necessarily so. This apparent contradiction can be resolved by recalling that the overdetermined system at hand in general has no solution. The lower part of $\mathbf{b}$, $\mathbf{b}_L$, is then nothing but an m'-dimensional array containing the nonzero components of the approximation error in the new coordinates. That is, the least-square error, $\mathbf{e}_0$, in these coordinates takes the form

$$\mathbf{e}_0 = \begin{bmatrix} \mathbf{0}_n \\ \mathbf{b}_L \end{bmatrix} \tag{B.21a}$$

Therefore,

$$\|\mathbf{e}_0\| = \|\mathbf{b}_L\| \tag{B.21b}$$

B.2 The Underdetermined Case

In this section we study the solution of system (B.1) under the assumption that $m < n$ and $\text{rank}(\mathbf{A}) = m$. Now, the system under study admits infinitely-many solutions, which allows us to impose one condition on a specific solution that we may want to obtain. The obvious choice is a minimality condition on a norm of $\mathbf{x}$. As in the previous section, the minimization of the square of the Euclidean norm of $\mathbf{x}$ leads to a linear problem, and hence, a direct solution of the problem at hand is possible. We thus have

$$z(\mathbf{x}) \equiv \frac{1}{2}\|\mathbf{x}\|_2^2 \quad \rightarrow \quad \min_{\mathbf{x}} \tag{B.22}$$

subject to the constraint represented by eq.(B.1). Since we now have a *constrained minimization problem*, we proceed to its solution via *Lagrange multipliers*. That is, we introduce a new objective function $\zeta(\mathbf{x})$, defined as

$$\zeta(\mathbf{x}) \equiv z(\mathbf{x}) + \boldsymbol{\lambda}^T(\mathbf{A}\mathbf{x} - \mathbf{b}) \quad \rightarrow \quad \min_{\mathbf{x}} \tag{B.23}$$

subject to no constraints, with $\boldsymbol{\lambda}$ defined as an m-dimensional vector of Lagrange multipliers, as yet to be determined. We thus have now an unconstrained minimization problem with $m+n$ *design variables*, the m components of $\boldsymbol{\lambda}$ and the n components of $\mathbf{x}$, that we group in the $(m+n)$-dimensional vector $\mathbf{y} \equiv [\mathbf{x}^T \quad \boldsymbol{\lambda}^T]^T$. The normality condition of the foregoing problem can now be stated as

$$\frac{d\zeta}{d\mathbf{y}} = \mathbf{0}_{m+n} \tag{B.24a}$$

with $\mathbf{0}_{m+n}$ defined as the $(m+n)$-dimensional zero vector. The above condition can be broken down into the two conditions below:

$$\frac{d\zeta}{d\mathbf{x}} = \mathbf{0}_n$$
$$\frac{d\zeta}{d\boldsymbol{\lambda}} = \mathbf{0}_m$$

with $\mathbf{0}_m$ and $\mathbf{0}_n$ defined, respectively, as the m- and n-dimensional zero vectors. The above equations thus lead to

$$\frac{d\zeta}{d\mathbf{x}} \equiv \mathbf{x} + \mathbf{A}^T\boldsymbol{\lambda} = \mathbf{0}_n \tag{B.25}$$
$$\frac{d\zeta}{d\boldsymbol{\lambda}} \equiv \mathbf{A}\mathbf{x} - \mathbf{b} = \mathbf{0}_m \tag{B.26}$$

Upon elimination of $\boldsymbol{\lambda}$ from the above system of equations, we obtain

$$\mathbf{x} = \mathbf{A}^T(\mathbf{A}\mathbf{A}^T)^{-1}\mathbf{b} \tag{B.27}$$

which is the *minimum-norm solution* of the proposed problem. Again, the formula yielding the foregoing solution is deceptively simple. If we attempt the calculation of the inverse occurring in it, we risk introducing unnecessarily an inadmissibly ill-conditioned matrix, the product $\mathbf{A}\mathbf{A}^T$. Therefore, an alternative approach to the straightforward implementation of the above formula should be attempted, as we do in the subsection below.

B.2.1 The Numerical Solution of an Underdetermined System of Linear Equations

The simplest way of solving this problem is by introducing the $m \times m$ identity matrix $\mathbf{1}$, in a disguised manner, between the two factors of the left-hand side of eq.(B.1). To this end, we assume that we have an orthogonal $m \times m$ matrix $\mathbf{H}$, so that

$$\mathbf{H}^T\mathbf{H} = \mathbf{1} \tag{B.28}$$

equation (B.1) thus becoming

$$\mathbf{A}\mathbf{H}^T\mathbf{H}\mathbf{x} = \mathbf{b} \tag{B.29a}$$

which can be rewritten in the form

$$\mathbf{A}\mathbf{H}^T\mathbf{v} = \mathbf{b} \tag{B.29b}$$

with $\mathbf{v}$ defined, obviously, as

$$\mathbf{v} \equiv \mathbf{H}\mathbf{x} \tag{B.29c}$$

Now, $\mathbf{H}$ is chosen as the product of m Householder reflections that transforms $\mathbf{A}^T$ into upper-triangular form, i.e., so that

$$\mathbf{H}\mathbf{A}^T = \begin{bmatrix} \mathbf{U} \\ \mathbf{O}_{n'm} \end{bmatrix} \tag{B.30}$$

with $\mathbf{O}_{n'm}$ defined as the $(n-m)\times m$ zero matrix and $n' \equiv n-m$. Moreover, $\mathbf{U}$ is upper-triangular. Further, let us partition $\mathbf{v}$ in the form

$$\mathbf{v} \equiv \begin{bmatrix} \mathbf{v}_U \\ \mathbf{v}_L \end{bmatrix} \tag{B.31}$$

Upon substitution of eqs.(B.30) and (B.31) into eq.(B.29b), we obtain

$$[\,\mathbf{U}^T \quad \mathbf{O}_{mn'}\,] \begin{bmatrix} \mathbf{v}_U \\ \mathbf{v}_L \end{bmatrix} = \mathbf{b}$$

where $\mathbf{O}_{mn'}$ is the $m \times (n-m)$ zero matrix. Hence,

$$\mathbf{U}^T \mathbf{v}_U + \mathbf{O}_{mn'} \mathbf{v}_L = \mathbf{b} \tag{B.32}$$

whence it is apparent that $\mathbf{v}_L$ can attain any value. Now, since $\mathbf{v}$ is the image of $\mathbf{x}$ under an orthogonal transformation, the Euclidean norms of these two vectors are identical, and hence,

$$\|\mathbf{x}\|^2 = \|\mathbf{v}_U\|^2 + \|\mathbf{v}_L\|^2 \tag{B.33}$$

Therefore, if we want to minimize the Euclidean norm of $\mathbf{x}$, the obvious choice of $\mathbf{v}_L$ is zero. Furthermore, from eq.(B.32),

$$\mathbf{v}_U = \mathbf{U}^{-T}\mathbf{b} \tag{B.34}$$

and so,

$$\mathbf{x} = \mathbf{H}^T\mathbf{v} = \mathbf{H}^T \begin{bmatrix} \mathbf{U}^{-T}\mathbf{b} \\ \mathbf{0}_{n'} \end{bmatrix} \tag{B.35}$$

with $\mathbf{0}_{n'}$ denoting the n'-dimensional zero vector, thereby completing the numerical solution of the problem at hand.

Exercises

While the following exercises are ordered by chapter, the ordering within each chapter does not necessarily correspond to that of the sections within the chapter.

Some of the exercises below call for algebraic manipulations that are very cumbersome and error-prone if done by hand. It is strongly recommended that these exercises be worked out using software for computer algebra, which is nowadays readily available (Pattee, 1995). On the other hand, some problems require numerical computations that most of the time can be done by longhand calculations; when these become cumbersome, the reader is advised to resort to suitable software, e.g., Matlab and its toolboxes (Etter, 1993).

1 An Overview of Robotic Mechanical Systems

The exercises listed below are meant to familiarize the uninitiated reader with the issues involved in robotics, especially in the area of robotic mechanical systems. A major issue, regrettably very often overlooked, is the terminology. In attempting to work out these exercises, the beginner should be able to better understand the language of robotics and realize that a common terminology is not yet available.

1.1 List a few definitions of *machine*, say about half a dozen, trying to cover the broadest timespan to date. *Hint: Denavit and Hartenberg (1964) list a few bibliographical references.*

1.2 Try to give an answer to the question: *Are intelligent machines possible?* Express your own ideas and explore what scientists think about this controversial issue. Penrose (1994) addresses this issue.

1.3 What is the difference among *machine*, *mechanism*, and *linkage*? In particular, analyze critically the definitions given by authorities, such as those found in the most respected dictionaries, encyclopedias, and archival documents of learned societies, e.g., the complete issue of Vol. 26, No. 5 (1991) of *Mechanism and Machine Theory* on terminology.

1.4 What is artificial intelligence? What is fuzzy logic? Can the techniques of these fields be applied to robotics?

1.5 What is mechatronics? What is the difference between mechatronics and robotics? Comerford (1994) and Soureshi, Meckl, and Durfee (1994) give an account on this technology.

1.6 What do you understand as *dexterity*? The concept of dexterity is normally applied to persons. Can it be applied to animals as well? What about machines?

1.7 Define the term *algorithm*. In this context, make a clear distinction between *recursion* and *iteration*. Note that in the robotics literature, there is often confusion between these two terms in particular. Make sure that you do not make the same mistake! Again, Penrose (1994) has provided an extensive discussion of the nature of algorithms.

1.8 What is the difference among terms like *real-time*, *on-line*, and *run-time*?

1.9 How fast can two floating-point numbers be multiplied using a personal computer? What about using a UNIX workstation? a supercomputer? Write a piece of code to estimate this time on your computer facility.

1.10 Answer the foregoing question as pertaining to floating-point addition.

1.11 What is the smallest floating-point number on your computer? Rather than looking for the answer in manuals, write a procedure to estimate it.

1.12 What is the difference between *conventional programming* and *object-oriented programming*? In terms of programming languages, what is the difference between C and C++? Rumbaugh, Blaha, Premerlani, Eddy, and Lorensen (1991) provide an introduction to object-oriented programming, while Stroustrup (1991) gives an introduction to C++.

2 Mathematical Background

2.1 Prove that the range and the nullspace of any linear transformation $\mathbf{L}$ of vector space $\mathcal{U}$ into vector space $\mathcal{V}$ are vector spaces as well, the former of $\mathcal{V}$, the latter of $\mathcal{U}$.

2.2 Let $\mathbf{L}$ map $\mathcal{U}$ into $\mathcal{V}$ and $\dim\{\mathcal{U}\} = n$, $\dim\{\mathcal{V}\} = m$. Moreover, let $\mathcal{R}$ and $\mathcal{N}$ be the range and the nullspace of $\mathbf{L}$, their dimensions being ρ and ν, respectively. Show that $\rho + \nu = n$.

2.3 Given two arbitrary nonzero vectors $\mathbf{u}$ and $\mathbf{v}$ in $\mathcal{E}^3$, find the matrix $\mathbf{P}$ representing the projection of $\mathcal{E}^3$ onto the subspace spanned by $\mathbf{u}$ and $\mathbf{v}$.

2.4 Verify that $\mathbf{P}$, whose matrix representation in a certain coordinate system is given below, is a projection. Then, describe it geometrically, i.e., identify the plane onto which the projection takes place. Moreover, find the nullspace of $\mathbf{P}$.

$$[\mathbf{P}] = \frac{1}{3}\begin{bmatrix} 2 & 1 & -1 \\ 1 & 2 & 1 \\ -1 & 1 & 2 \end{bmatrix}$$

2.5 If for any 3-dimensional vectors $\mathbf{a}$ and $\mathbf{v}$, matrix $\mathbf{A}$ is defined as

$$\mathbf{A} \equiv \frac{\partial(\mathbf{a} \times \mathbf{v})}{\partial \mathbf{v}}$$

then we have

$$\mathbf{A}^T \equiv \frac{\partial(\mathbf{v} \times \mathbf{a})}{\partial \mathbf{v}}$$

Show that $\mathbf{A}$ is skew-symmetric *without introducing components.*

2.6 Let $\mathbf{u}$ and $\mathbf{v}$ be any 3-dimensional vectors, and define $\mathbf{T}$ as

$$\mathbf{T} \equiv \mathbf{1} + \mathbf{u}\mathbf{v}^T$$

The (unit) eigenvectors of $\mathbf{T}$ are denoted by $\mathbf{w}_1$, $\mathbf{w}_2$, and $\mathbf{w}_3$. Show that, say, $\mathbf{w}_1$ and $\mathbf{w}_2$ are any unit vectors perpendicular to $\mathbf{v}$ and different from each other, whereas $\mathbf{w}_3 = \mathbf{u}/\|\mathbf{u}\|$. Also show that the corresponding eigenvalues, denoted by λ_1, λ_2, and λ_3, associated with $\mathbf{w}_1$, $\mathbf{w}_2$, and $\mathbf{w}_3$, respectively, are given as

$$\lambda_1 = \lambda_2 = 1, \quad \lambda_3 = 1 + \mathbf{u} \cdot \mathbf{v}$$

2.7 Show that if $\mathbf{u}$ and $\mathbf{v}$ are any 3-dimensional vectors, then

$$\det(\mathbf{1} + \mathbf{u}\mathbf{v}^T) = 1 + \mathbf{u} \cdot \mathbf{v}$$

Hint: Use the results of the foregoing exercise.

2.8 For the two unit vectors $\mathbf{e}$ and $\mathbf{f}$ in 3-dimensional space, define the two reflections

$$\mathbf{R}_1 = \mathbf{1} - 2\mathbf{e}\mathbf{e}^T, \quad \mathbf{R}_2 = \mathbf{1} - 2\mathbf{f}\,\mathbf{f}^T$$

Now, show that $\mathbf{Q} = \mathbf{R}_1\mathbf{R}_2$ is a rigid-body rotation, and find its axis and its angle of rotation in terms of unit vectors $\mathbf{e}$ and $\mathbf{f}$. Again, no components are permitted in this exercise.

2.9 State the conditions on the unit vectors $\mathbf{e}$ and $\mathbf{f}$, of two reflections $\mathbf{R}_1$ and $\mathbf{R}_2$, respectively, under which a given rotation $\mathbf{Q}$ can be factored into the reflections $\mathbf{R}_1$ and $\mathbf{R}_2$ given in the foregoing exercise. In other words, not every rotation matrix $\mathbf{Q}$ can be factored into those two reflections, for fixed $\mathbf{e}$ and $\mathbf{f}$, but special cases can. Identify these cases.

2.10 Prove that the eigenvalues of the cross-product matrix of the unit vector $\mathbf{e}$ are 0, j, and $-j$, where $j = \sqrt{-1}$. Find the corresponding eigenvectors.

2.11 Without resorting to components, prove that the eigenvalues of a proper matrix $\mathbf{Q}$ are 1, $e^{j\phi}$, and $e^{-j\phi}$.
Hint: Use the result of the foregoing exercise and the Cayley-Hamilton Theorem.

2.12 Find the axis and the angle of rotation of the proper orthogonal matrix $\mathbf{Q}$ given below in a certain coordinate frame.

$$[\mathbf{Q}] = \frac{1}{3}\begin{bmatrix} -1 & -2 & 2 \\ -2 & -1 & -2 \\ 2 & -2 & -1 \end{bmatrix}$$

2.13 Find $\mathbf{E}$ and ϕ of the exponential representation of $\mathbf{Q}$, for $\mathbf{Q}$ given as in Exercise 2.12.

2.14 Cayley's Theorem, which is not to be confused with the theorem of Cayley-Hamilton, states that every 3×3 proper orthogonal matrix $\mathbf{Q}$ can be *uniquely* factored as

$$\mathbf{Q} = (\mathbf{1} - \mathbf{S})(\mathbf{1} + \mathbf{S})^{-1}$$

where $\mathbf{S}$ is a skew-symmetric matrix. Find a general expression for $\mathbf{S}$ in terms of $\mathbf{Q}$, and state the condition under which this factoring is not possible.

2.15 Find matrix $\mathbf{S}$ of Cayley's factoring for $\mathbf{Q}$ as given in Exercise 2.12.

2.16 If $\mathbf{Q}$ represents a rotation about an axis parallel to the unit vector $\mathbf{e}$ through an angle ϕ, then the *Rodrigues vector* $\boldsymbol{\rho}$ of this rotation can be defined as

$$\boldsymbol{\rho} \equiv \tan\left(\frac{\phi}{2}\right)\mathbf{e}$$

Note that if $\mathbf{r}$ and r_0 denote the Euler-Rodrigues parameters of the rotation under study, then $\boldsymbol{\rho} = \mathbf{r}/r_0$. Show that

$$\boldsymbol{\rho} = -\mathrm{vect}(\mathbf{S})$$

for $\mathbf{S}$ given in Exercise 2.14.

2.17 The vertices of a cube, labeled A, B, $\ldots$, H, are located so that A, B, C, and D, as well as E, F, G, and H, are in clockwise order when viewed from outside. Moreover, AE, BH, CG, and DF are edges of the cube, which is to be manipulated so that a mapping of vertices takes place as indicated below:

$$\begin{array}{llll} A \rightarrow D, & B \rightarrow C, & C \rightarrow G, & D \rightarrow F \\ E \rightarrow A, & F \rightarrow E, & G \rightarrow H, & H \rightarrow B \end{array}$$

Find the angle of rotation and the angles that the axis of rotation makes with edges AB, AD, and AE.

2.18 (Euler angles) A rigid body can attain an arbitrary configuration starting from any reference configuration, 0, by means of the composition of three rotations about coordinate axes, as described below: Attach axes X_0, Y_0, and Z_0 to the body in the reference configuration and rotate the body through an angle ϕ about Z_0, thus carrying the axes into X_1, Y_1, and Z_1 ($=Z_0$), respectively. Next, rotate the body through an angle θ about axis Y_1, thus carrying the axes into X_2, Y_2, and Z_2, respectively. Finally, rotate the body through an angle ψ about Z_2 so that the axes coincide with their desired final orientation, X_3, Y_3, and Z_3. Angle ψ is chosen so that axis Z_3 lies in the plane of Z_0 and X_1, whereas angle θ is chosen so as to carry axis Z_1 ($=Z_0$) into Z_3 ($=Z_2$). Show that the rotation matrix carrying the body from configuration 0 to configuration 3 is:

$$\mathbf{Q} = \begin{bmatrix} c\theta c\phi c\psi - s\phi s\psi & -c\theta c\phi s\psi - s\phi c\psi & s\theta c\phi \\ c\theta s\phi c\psi + c\phi s\psi & -c\theta s\phi s\psi + c\phi c\psi & s\theta s\phi \\ -s\theta c\psi & s\theta s\psi & c\theta \end{bmatrix}$$

where $c(\cdot)$ and $s(\cdot)$ stand for $\cos(\cdot)$ and $\sin(\cdot)$, respectively. Moreover, show that the angle of rotation of $\mathbf{Q}$ given above, α, obeys the relation

$$\cos\left(\frac{\alpha}{2}\right) = \cos\left(\frac{\psi + \phi}{2}\right)\cos\left(\frac{\theta}{2}\right)$$

2.19 Given an arbitrary rigid-body rotation about an axis parallel to the unit vector **e** through an angle ϕ, it is possible to find both **e** and ϕ using the linear invariants of the rotation matrix, as long as the vector invariant does not vanish. The latter happens if and only if $\phi = 0$ or π. Now, if $\phi = 0$, the associated rotation matrix is the identity, and **e** is any 3-dimensional vector; if $\phi = \pi$, then we have

$$\mathbf{Q}(\pi) \equiv \mathbf{Q}_\pi = -\mathbf{1} + 2\mathbf{e}\mathbf{e}^T$$

whence we can solve for $\mathbf{e}\mathbf{e}^T$ as

$$\mathbf{e}\mathbf{e}^T = \frac{1}{2}(\mathbf{Q}_\pi + \mathbf{1})$$

Now, it is apparent that the three eigenvalues of $\mathbf{Q}_\pi$ are real and the associated eigenvectors are mutually orthogonal. Find these.

2.20 Explain why *all* the off-diagonal entries of a symmetric rotation matrix *cannot* be negative.

2.21 The three entries above the diagonal of a 3×3 matrix $\mathbf{Q}$ that is supposed to represent a rotation are given below:

$$q_{12} = \frac{1}{2}, \quad q_{13} = -\frac{2}{3}, \quad q_{23} = \frac{3}{4}$$

Without knowing the other entries, explain why the above entries are unacceptable.

2.22 Let $\mathbf{p}_1$, $\mathbf{p}_2$, and $\mathbf{p}_3$ be the position vectors of three arbitrary points in 3-dimensional space. Now, define a matrix $\mathbf{P}$ as

$$\mathbf{P} \equiv [\mathbf{p}_1 \quad \mathbf{p}_2 \quad \mathbf{p}_3]$$

Show that $\mathbf{P}$ is not frame-invariant. *Hint: Show, for example, that it is always possible to find a coordinate frame in which* $\text{tr}(\mathbf{P})$ *vanishes.*

2.23 For $\mathbf{P}$ defined as in Exercise 2.22, let

$$q \equiv \text{tr}(\mathbf{P}^2) - \text{tr}^2(\mathbf{P})$$

Show that q vanishes if the three given points are collinear for $\mathbf{P}$ represented in any coordinate frame.

2.24 For $\mathbf{P}$ defined, again, as in Exercise 2.22, show that $\mathbf{P}\mathbf{P}^T$ is frame-invariant and becomes singular if and only if the three given points are collinear. Note that this matrix is more singularity-robust than $\mathbf{P}$.

2.25 The diagonal entries of a rotation matrix are known to be -0.5, 0.25, and -0.75. Find the off-diagonal entries.

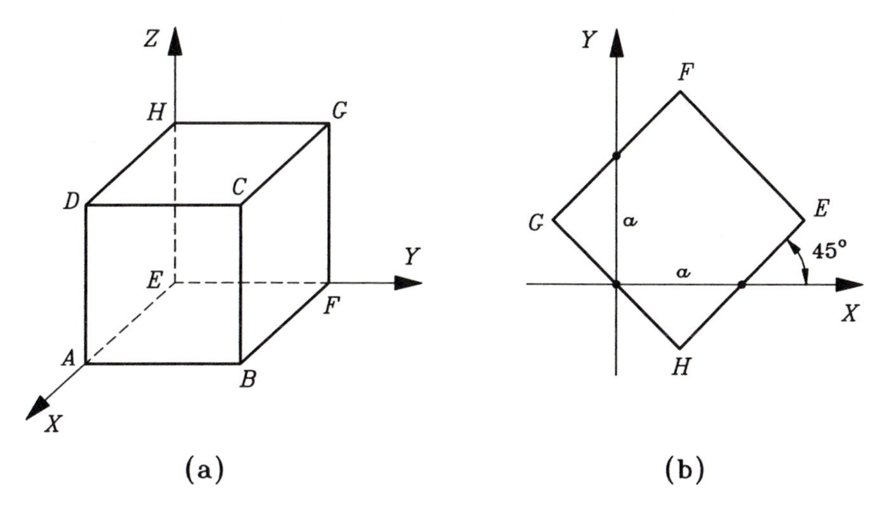

FIGURE 1. A cube in two different configurations.

2.26 As a generalization to the foregoing exercise, discuss how you would go about finding the off-diagonal entries of a rotation matrix whose diagonal entries are known to be a, b, and c. *Hint: This problem can be formulated as finding the intersection of the coupler curve of a four-bar spherical linkage (Chiang, 1988), which is a curve on a sphere, with a certain parallel of the same sphere.*

2.27 Shown in Fig. 1(a) is a cube that is to be displaced in an assembly operation to a configuration in which face $EFGH$ lies in the XY plane, as indicated in Fig. 1(b). Compute the unit vector $\mathbf{e}$ parallel to the axis of the rotation involved and the angle of rotation ϕ, for $0 \leq \phi \leq \pi$.

2.28 The axes X_1, Y_1, Z_1 of a frame $\mathcal{F}_1$ are attached to the base of a robotic manipulator, whereas the axes X_2, Y_2, Z_2 of a second frame $\mathcal{F}_2$ are attached to its end-effector, as shown in Fig. 2. Moreover, the origin P of $\mathcal{F}_2$ has the $\mathcal{F}_1$-coordinates $(1, -1, 1)$. Furthermore, the orientation of the end effector with respect to the base is defined by a rotation $\mathbf{Q}$, whose representation in $\mathcal{F}_1$ is given by

$$[\mathbf{Q}]_1 = \frac{1}{3}\begin{bmatrix} 1 & 1-\sqrt{3} & 1+\sqrt{3} \\ 1+\sqrt{3} & 1 & 1-\sqrt{3} \\ 1-\sqrt{3} & 1+\sqrt{3} & 1 \end{bmatrix}$$

(a) What are the end-effector coordinates of point C of Fig. 2?

(b) The end-effector is approaching the ABC plane shown in Fig. 2. What is the equation of the plane in end-effector coordinates? Verify your result by substituting the answer to (a) into this equation.

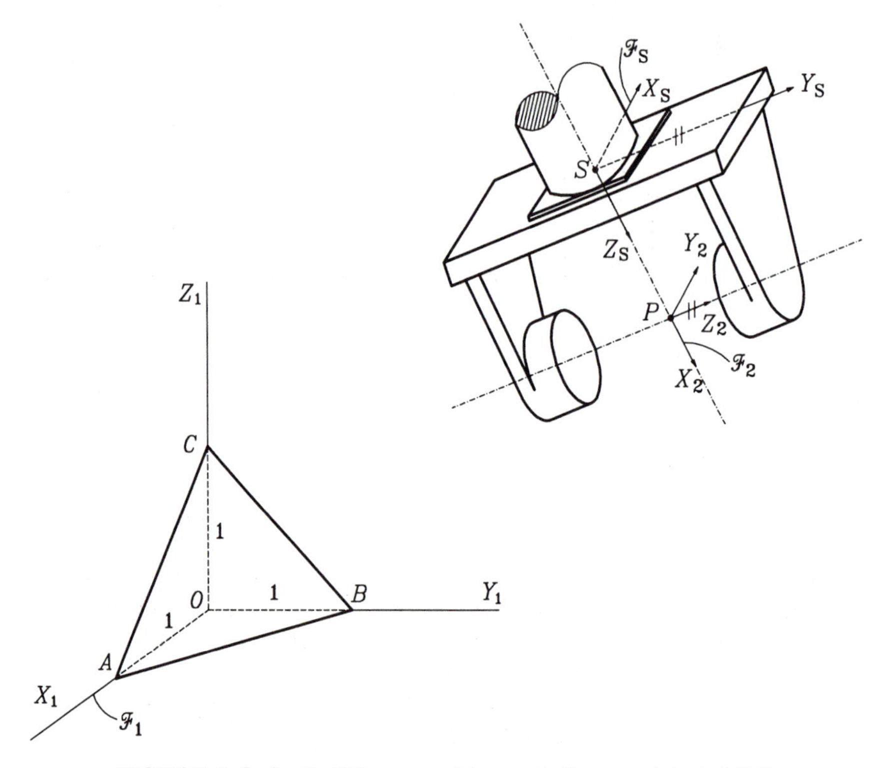

FIGURE 2. Robotic EE approaching a stationary object ABC.

2.29 Shown in Fig. 3 is a cube of unit side, which is composed of two parts. Frames (X_0, Y_0, Z_0) and (X_1, Y_1, Z_1) are attached to each of the two parts, as illustrated in the figure. The second part is going to be picked up by a robotic gripper as the part is transported on a belt conveyor and passes close to the stationary first part. Moreover, the robot is to assemble the cube by placing the second part onto the first one in such a way that vertices A_1, B_1, C_1 are coincident with vertices A_0, B_0, C_0. Determine the axis and the angle of rotation that will carry the second part onto the first one as described above.

2.30 A piece of code meant to produce the entries of rotation matrices is being tested. In one run, the code produced a matrix with diagonal entries -0.866, -0.866, -0.866. Explain how without looking at the other entries, you can decide that the code has a bug.

2.31 Shown in Fig. 4 is a rigid cube of unit side in three configurations. The second and the third configurations are to be regarded as images of the first one. One of the last two configurations is a reflection, and the other is a rotation of the first one. Identify the rotated configuration and find its associated invariants.

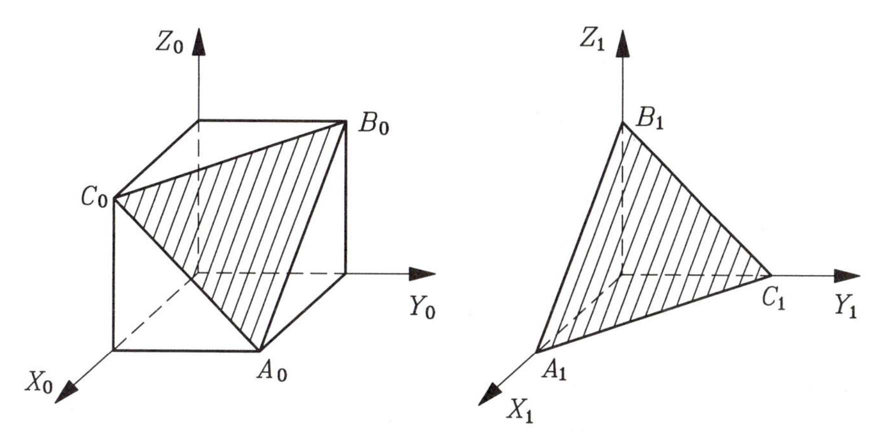

FIGURE 3. Roboticized assembly operation.

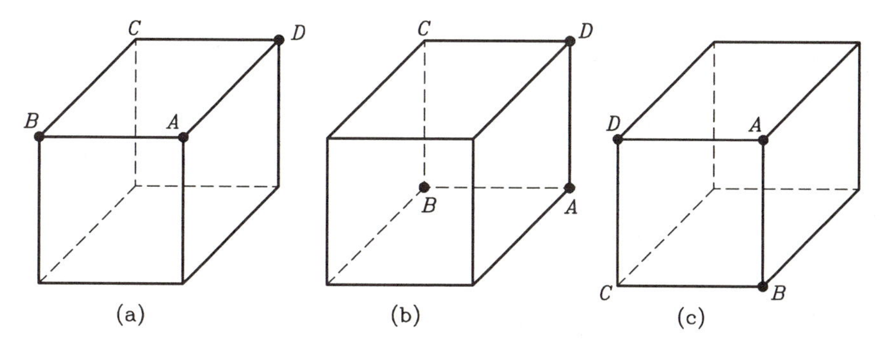

FIGURE 4. Three configurations of a cube.

2.32 Two frames, $\mathcal{G}$ and $\mathcal{C}$, are attached to a robotic gripper and to a camera mounted on the gripper, respectively. Moreover, the camera is rigidly attached to the gripper, and hence, the orientation of $\mathcal{C}$ with respect to $\mathcal{G}$, denoted by $\mathbf{Q}$, remains constant under gripper motions. The orientation of the gripper with respect to a frame $\mathcal{B}$ fixed to the base of the robot was measured in both $\mathcal{G}$ and $\mathcal{C}$. Note that this orientation is measured in $\mathcal{G}$ simply by reading the joint encoders, which report values of the joint variables, as discussed in detail in Chapter 4. The same orientation is measured in $\mathcal{C}$ from estimations of the coordinates of a set of points fixed to $\mathcal{B}$, as seen by the camera.

Two measurements of the above-mentioned orientation, denoted $\mathbf{R}_1$ and $\mathbf{R}_2$, were taken in $\mathcal{G}$ and $\mathcal{C}$, with the numerical values reported below:

$$[\mathbf{R}_1]_{\mathcal{G}} = \begin{bmatrix} 0.667 & 0.333 & 0.667 \\ -0.667 & 0.667 & 0.333 \\ -0.333 & -0.667 & 0.667 \end{bmatrix},$$

$$[\mathbf{R}_1]_{\mathcal{C}} = \begin{bmatrix} 0.500 & 0 & -0.866 \\ 0 & 1.000 & 0 \\ 0.866 & 0 & 0.500 \end{bmatrix},$$

$$[\mathbf{R}_2]_{\mathcal{G}} = \begin{bmatrix} 0.707 & 0.577 & 0.408 \\ 0 & 0.577 & -0.816 \\ -0.707 & 0.577 & 0.408 \end{bmatrix},$$

$$[\mathbf{R}_2]_{\mathcal{C}} = \begin{bmatrix} 1 & 0 & 0 \\ 0 & 0.346 & -0.938 \\ 0 & 0.938 & 0.346 \end{bmatrix}$$

(a) Verify that the foregoing matrices represent rotations.

(b) Verify that the first two matrices represent, in fact, the same rotation $\mathbf{R}_1$, albeit in different coordinate frames.

(c) Repeat item (b) for $\mathbf{R}_2$.

(d) Find $[\mathbf{Q}]_{\mathcal{G}}$. Is your computed $\mathbf{Q}$ orthogonal? If not, what is the error in the computations? Note that you may have encountered here a problem of roundoff error amplification, which can be avoided if a robust computational scheme is used. As a matter of fact, a robust method in this case can be devised by resorting to the *Gram-Schmidt orthogonalization procedure*, as outlined in Appendix B.

2.33 The rotation $\mathbf{Q}$ taking a coordinate frame $\mathcal{B}$, fixed to the base of a robot, into a coordinate frame $\mathcal{G}$, fixed to its gripper and the position vector $\mathbf{g}$ of the origin of $\mathcal{G}$ have the representations in $\mathcal{B}$ given below:

$$[\mathbf{Q}]_{\mathcal{B}} = \frac{1}{3}\begin{bmatrix} 1 & 1-\sqrt{3} & 1+\sqrt{3} \\ 1+\sqrt{3} & 1 & 1-\sqrt{3} \\ 1-\sqrt{3} & 1+\sqrt{3} & 1 \end{bmatrix}, \qquad [\mathbf{g}]_{\mathcal{B}} = \begin{bmatrix} 1-\sqrt{3} \\ \sqrt{3} \\ 1+\sqrt{3} \end{bmatrix}$$

Moreover, let $\mathbf{p}$ be the position vector of any point $\mathcal{P}$ of the 3-dimensional space, its coordinates in $\mathcal{B}$ being (x, y, z). The robot is supported by a cylindrical column C of circular cross section, bounded by planes Π_1 and Π_2. These are given below:

$$C:\ x^2 + y^2 = 4;\ \Pi_1:\ z = 0;\ \Pi_2:\ z = 10$$

Find the foregoing equations in $\mathcal{G}$ coordinates.

2.34 A certain point of the gripper of a robot is to trace an elliptical path of semiaxes a and b, with center at C, the centroid of triangle PQR, as shown in Fig. 5. Moreover, the semiaxis of length a is parallel to edge PQ, while the ellipse lies in the plane of the triangle, and all three vertices are located a unit distance away from the origin.

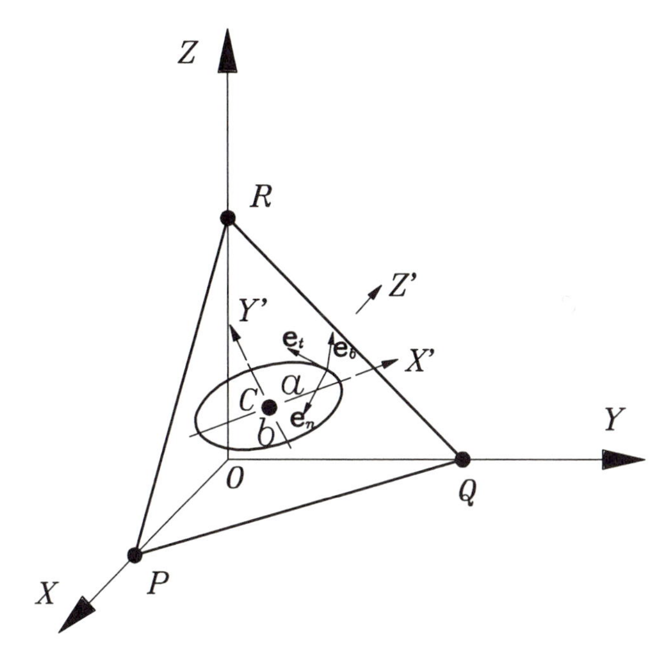

FIGURE 5. An elliptical path on an inclined plane.

(a) For $b = 2a/3$, the gripper is to keep a fixed orientation with respect to the unit tangent, normal, and binormal vectors of the ellipse, denoted by $\mathbf{e}_t$, $\mathbf{e}_n$, and $\mathbf{e}_b$, respectively. Determine the matrix representing the rotation undergone by the gripper from an orientation in which vector $\mathbf{e}_t$ is parallel to the coordinate axis X, while $\mathbf{e}_n$ is parallel to Y and $\mathbf{e}_b$ to Z. Express this matrix in X, Y, Z coordinates, if the equation of the ellipse, in parametric form, is given as

$$x' = a\cos\varphi, \ y' = b\sin\varphi, \ z' = 0$$

the orientation of the gripper thus becoming a function of φ.

(b) Find the value of φ for which the angle of rotation of the gripper, with respect to the coordinate axes X, Y, Z, becomes π.

3 Fundamentals of Rigid-Body Mechanics

3.1 The cube of Fig. 6 is displaced from configuration $AB\ldots H$ into configuration $A'B'\ldots H'$.

(a) Determine the matrix representing the rotation $\mathbf{Q}$ undergone by the cube, in X, Y, Z coordinates.

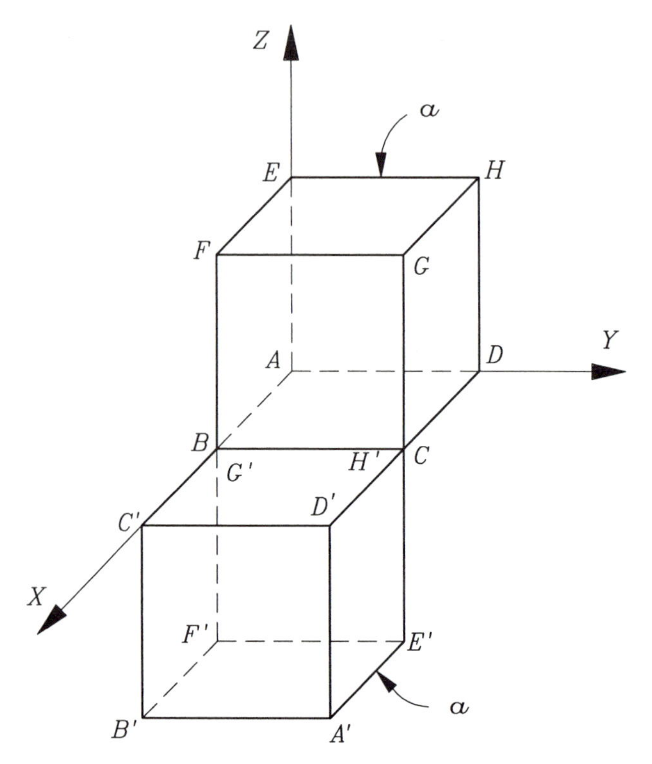

FIGURE 6. Motion of a cube.

(b) Find the Plücker coordinates of line $\mathcal{L}$ of the cube undergoing displacements of minimum magnitude.

(c) Find the intersections of $\mathcal{L}$ with the coordinate planes.

3.2 Two unit forces, $\mathbf{f}_1$ and $\mathbf{f}_2$, are applied to the regular tetrahedron of unit-length edges displayed in Fig. 7 in such a way that $\mathbf{f}_1$ is directed from P_2 to P_3, whereas $\mathbf{f}_2$ is directed from P_4 to P_1. The effect of the foregoing system of forces on the rigid tetrahedron is obtained by application of the resultant of the two forces on a certain point P and a moment $\mathbf{n}$. Find the location of point P lying closest to P_4 that will make the magnitude of $\mathbf{n}$ a minimum.

3.3 The *moment* of a line $\mathcal{L}_1$ about a second line $\mathcal{L}_2$ is a scalar μ defined as

$$\mu = \mathbf{n}_1 \cdot \mathbf{e}_2$$

where $\mathbf{n}_1$ is the moment of $\mathcal{L}_1$ about an arbitrary point P of $\mathcal{L}_2$, while $\mathbf{e}_2$ is a unit vector parallel to line $\mathcal{L}_2$.

Show that the locus of all lines $\mathcal{L}$ intersecting three given lines $\{ \mathcal{L}_k \}_1^3$ is a *quadric* surface, i.e., a surface defined by a function that is

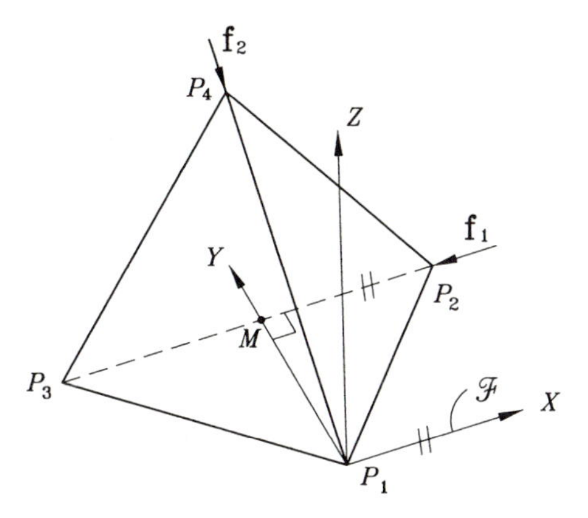

FIGURE 7. A regular tetrahedron.

quadratic in the position vector of a point of the surface. *Hint: Note that the moment of any line of* $\{\mathcal{L}_k\}_1^3$ *with respect to* $\mathcal{L}$ *vanishes.*

Note: The quadric surface above is, in fact, a ruled surface, namely, a one-sheet hyperboloid.

3.4 A robotic gripper is provided with two redundant sensors that are meant to measure a wrench acting on the gripper. The ith sensor, moreover, has its own coordinate frame, labeled $\mathcal{F}_i$, for $i = 1, 2$. Sensor i reported the ith measurement of the wrench $\mathbf{w}_P$, where subscript P indicates that the force is applied at point P, as $[\mathbf{w}_P]_i \equiv [\mathbf{n}^T, \mathbf{f}^T]_i^T$, for $i = 1, 2$. These measurements are given as

$$[\mathbf{n}]_1 = \begin{bmatrix} 0 \\ 0 \\ 5 \end{bmatrix}, \quad [\mathbf{f}]_1 = \begin{bmatrix} 0 \\ 2 \\ 0 \end{bmatrix}$$

$$[\mathbf{n}]_2 = \begin{bmatrix} -5/3 \\ -10/3 \\ 10/3 \end{bmatrix}, \quad [\mathbf{f}]_2 = \begin{bmatrix} -4/3 \\ 4/3 \\ 2/3 \end{bmatrix}$$

(a) Show that the measurements are compatible, based on invariance arguments.

(b) Determine the relative orientation of the two frames, i.e., find the rotation matrix transforming $\mathcal{F}_2$-coordinates into $\mathcal{F}_1$-coordinates.

3.5 In calibrating a robot, the Plücker coordinates of one of its axes are to be determined in a given coordinate frame. To this end, the moment of this axis is measured with respect to two points, A and B, of position vectors $[\mathbf{a}] = [1, 0, 0]^T$ and $[\mathbf{b}] = [0, 1, 1]^T$, respectively.

The said moments, $\mathbf{n}_A$ and $\mathbf{n}_B$, respectively, are measured as

$$[\mathbf{n}_A] = \begin{bmatrix} 0 \\ 2 \\ 0 \end{bmatrix}, \quad [\mathbf{n}_B] = \begin{bmatrix} 0 \\ 1 \\ 1 \end{bmatrix}$$

with all entries given in meters.

(a) Determine the unit vector $\mathbf{e}$ defining the direction of the axis under discussion.

(b) Find the coordinates of the point P^* of the axis that lies closest to the origin

(c) Find the Plücker coordinates of the axis about the origin, i.e., the Plücker coordinates of the axis in which the moment is defined with respect to the origin.

3.6 The gripper $\mathcal{G}$ of a robot is approaching a workpiece $\mathcal{B}$, as indicated in Fig. 8, with planes Π_1 and Π_2 parallel to each other and normal to Π_3. The workpiece is made out of a cube of unit length from which two vertices have been removed, thereby producing the equilateral triangular faces DEF and $D'E'F'$. Moreover, two coordinate frames, $\mathcal{F}$ (X, Y, Z) and $\mathcal{F}'$ (X', Y', Z'), are defined as indicated in the figure, in which Y is, apparently, parallel to line $D'C'$.

It is required to grasp $\mathcal{B}$ with $\mathcal{G}$ in such a way that planes Π_1 and Π_2 coincide with the triangular faces, while carrying the Y' axis to an orientation perpendicular to the diagonal CC' of $\mathcal{B}$. More concretely, in the grasping configuration, frame $\mathcal{F}'$ is carried into $\mathcal{F}''$ (X'', Y'', and Z''), not shown in the figure, in such a way that unit vectors $\mathbf{i}''$, $\mathbf{j}''$, $\mathbf{k}''$, parallel to X'', Y'', Z'', respectively, are oriented so that $\mathbf{i}''$ has all three of its $\mathcal{F}$-components positive, while $\mathbf{j}''$ has its Z-component positive.

(a) Compute the angle of rotation of the motion undergone by $\mathcal{G}$ from a pose in which $\mathcal{F}'$ and $\mathcal{F}$ have identical orientations, termed the *reference pose*, and find the unit vector parallel to the axis of rotation, in frame $\mathcal{F}$.

(b) The position vector of point P of $\mathcal{G}$ is known to be, in the reference pose,

$$[\mathbf{p}]_{\mathcal{F}} = \begin{bmatrix} 2 \\ -1 \\ 0.25 \end{bmatrix}$$

Determine the set of points of $\mathcal{G}$ undergoing a displacement of minimum magnitude, under the condition that P, in the displaced configuration of $\mathcal{G}$, coincides with C'.

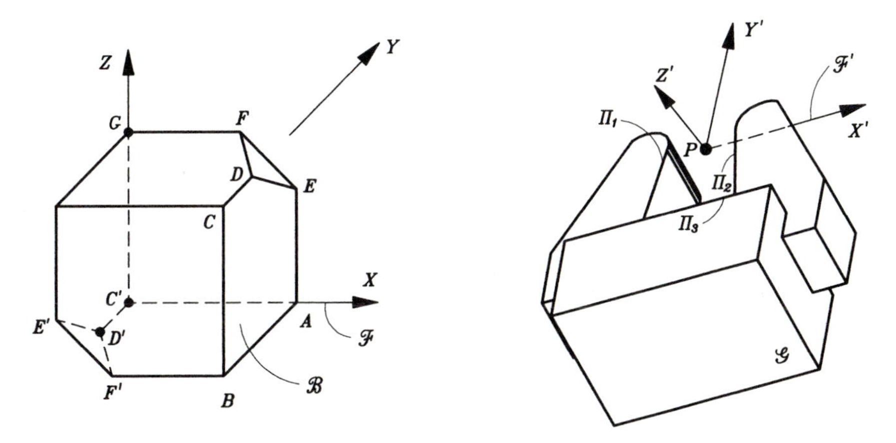

FIGURE 8. A workpiece $\mathcal{B}$ to be grasped by a gripper $\mathcal{G}$.

3.7 A robot-calibration method has been proposed that allows us to determine the location of a joint axis, $\mathcal{L}$, via the Plücker coordinates of the axis in a coordinate frame fixed to the gripper. The Plücker coordinates are given as $\boldsymbol{\pi}_{\mathcal{L}} = [\,\mathbf{e}^T,\, \mathbf{n}^T\,]^T$.

(a) Show that the distance of the axis to the origin of the gripper-fixed coordinate frame, d, can be determined as $d = \|\mathbf{n}\|$.

(b) Show that the point P^* on the axis, which lies closest to the above-mentioned origin, has a position vector $\mathbf{p}^*$ given as

$$\mathbf{p}^* = \mathbf{e} \times \mathbf{n}$$

(c) From measurements on a robot, the Plücker coordinates were estimated, in a gripper-fixed frame $\mathcal{G}$, as

$$[\,\boldsymbol{\pi}_{\mathcal{L}}\,]_{\mathcal{G}} = [-\sqrt{2}/2,\, 0,\, \sqrt{2}/2,\, 0,\, -\sqrt{2},\, 0]^T$$

Find d and $\mathbf{p}^*$ in gripper coordinates

3.8 Prove that for any 3-dimensional vectors $\boldsymbol{\omega}$ and $\mathbf{p}$,

$$\underbrace{\boldsymbol{\omega} \times (\boldsymbol{\omega} \times \cdots (\boldsymbol{\omega} \times (\boldsymbol{\omega}}_{2k \text{ factors}} \times \mathbf{p})) \cdots) = (-1)^k (\|\boldsymbol{\omega}\|^{2k} \mathbf{1} - \|\boldsymbol{\omega}\|^{2(k-1)} \boldsymbol{\omega}\boldsymbol{\omega}^T)\mathbf{p}$$

$$\underbrace{\boldsymbol{\omega} \times (\boldsymbol{\omega} \times \cdots (\boldsymbol{\omega} \times (\boldsymbol{\omega}}_{2k+1 \text{ factors}} \times \mathbf{p})) \cdots) = (-1)^k (\|\boldsymbol{\omega}\|^{2k} \boldsymbol{\omega}) \times \mathbf{p}$$

3.9 A rectangular prism with regular hexagonal bases whose sides are 25 mm long and whose height is 150 mm is to undergo a pick-and-place operation—See Chapter 5 to understand what this means—that requires knowledge of its centroid location and its moment-of-inertia matrix. Find the centroidal principal axes and moments of inertia under the assumption that the prism is made from a homogeneous material.

3.10 The prism of Exercise 3.9 now undergoes a machining process cutting it into two parts, which are separated by a plane that contains one of the edges of the base and makes an angle of 45° with the axis of the prism. Find the centroidal principal axes and moments of inertia of each of the two parts.

3.11 In Exercise 2.22 assume that a mass m is located at every point P_i of position vector $\mathbf{p}_i$. Give a mechanical interpretation of the matrix $m[\mathrm{tr}(\mathbf{P}\mathbf{P}^T)\mathbf{1}-\mathbf{P}\mathbf{P}^T]$, with $\mathbf{P}$ defined in the above-mentioned exercise.

3.12 The centroidal inertia matrix of a rigid body is measured by two observers, who report the two results below:

$$[\mathbf{I}]_{\mathcal{A}} = \begin{bmatrix} 1 & 0 & 0 \\ 0 & 2 & 0 \\ 0 & 0 & 3 \end{bmatrix}, \qquad [\mathbf{I}]_{\mathcal{B}} = \frac{1}{3}\begin{bmatrix} 6 & 2 & 2 \\ 2 & 5 & 0 \\ 2 & 0 & 7 \end{bmatrix}$$

Show that the two measurements are acceptable. (*Hint: Use invariance arguments.*)

3.13 State the conditions under which a point and the mass center of a rigid body share the same principal axes of inertia. In other words, let $\mathbf{I}_P$ and $\mathbf{I}_C$ be the moment-of-inertia matrices of a rigid body about a point P and its mass center, C, respectively. State the conditions under which the two matrices have common eigenvectors. Moreover, under these conditions, what are the relationships between the two sets of principal moments of inertia?

3.14 Show that the smallest principal moment of inertia of a rigid body attains its minimum value at the mass center.

3.15 Show that the time-rate of change of the inertia dyad $\mathbf{M}$ of a rigid body is given by

$$\dot{\mathbf{M}} = \mathbf{W}\mathbf{M} - \mathbf{M}\mathbf{W}$$

Then, recall that the extended momentum $\boldsymbol{\mu}$ is defined as

$$\boldsymbol{\mu} \equiv \mathbf{M}\mathbf{t}$$

where $\mathbf{t}$ is the twist of the body, defined at its mass center. Now, with the above expression for $\dot{\mathbf{M}}$, show that

$$\dot{\boldsymbol{\mu}} = \mathbf{M}\dot{\mathbf{t}} + \mathbf{W}\mathbf{M}\mathbf{t}$$

3.16 A wrench $\mathbf{w}^T = [\,\mathbf{n}^T\ \mathbf{f}^T\,]^T$, with $\mathbf{f}$ acting at point P of the gripper of Fig. 2, is measured by a *six-axis force sensor*, to which a frame $\mathcal{F}_S$ is attached, as indicated in that figure. If points P and S lie a distance of 100 mm apart, find the wrench in $\mathcal{F}_2$, when the readouts of the sensor are

$$[\,\mathbf{n}\,]_S = \begin{bmatrix} 1 \\ 0 \\ 1 \end{bmatrix} \text{ Nm}, \quad [\,\mathbf{f}\,]_S = \begin{bmatrix} 0 \\ 1 \\ 0 \end{bmatrix} \text{ N}$$

4 Kinetostatics of Simple Robotic Manipulators

Exercises 4.22 to 4.27 below pertain to Section 4.9. They are thus to be assigned only if this section was covered either in class or as a reading assignment.

4.1 Shown in Fig. 8.8 is the kinematic chain of one of the six-dof legs of a flight simulator, such as that appearing in Fig. 1.5. The HD parameters of this manipulator are displayed in Table 8.5. In that figure, $\mathcal{M}$ is the moving platform to which an aircraft cockpit is rigidly attached. The six-dof motion of $\mathcal{M}$ is controlled by means of the hydraulic cylinder indicated in the same figure as a prismatic pair. Find all inverse kinematics solutions of this manipulator, relating the pose of $\mathcal{M}$ with all the joint variables.

4.2 Modify the inverse-kinematics solution procedure of Section 4.3 to obtain all the postures of a *PRR* manipulator that give the same EE pose, and show that this problem leads to a quartic polynomial equation.

4.3 Repeat Exercise 4.2 as pertaining to a *PRP* manipulator.

4.4 The manipulator appearing in Fig. 9 is of the orthogonal type, with a decoupled, spherical wrist, and a regional structure consisting of two parallel axes and one axis perpendicular to these two. In that figure, rectangles denote revolutes of axes lying in the X_1-Z_1 plane, while circles with dots indicate revolutes with axes normal to the plane of the figure. Find all inverse kinematics solutions for arbitrary poses of the EE of this manipulator.

4.5 Similar to the manipulator of Fig. 9, that of Fig. 10 is of the orthogonal, decoupled type, except that the latter has a prismatic pair. For an arbitrary pose of its EE, find all inverse kinematics solutions of this manipulator. For a description of the meaning of the rectangles and the circles with dots inside, see Exercise 4.4.

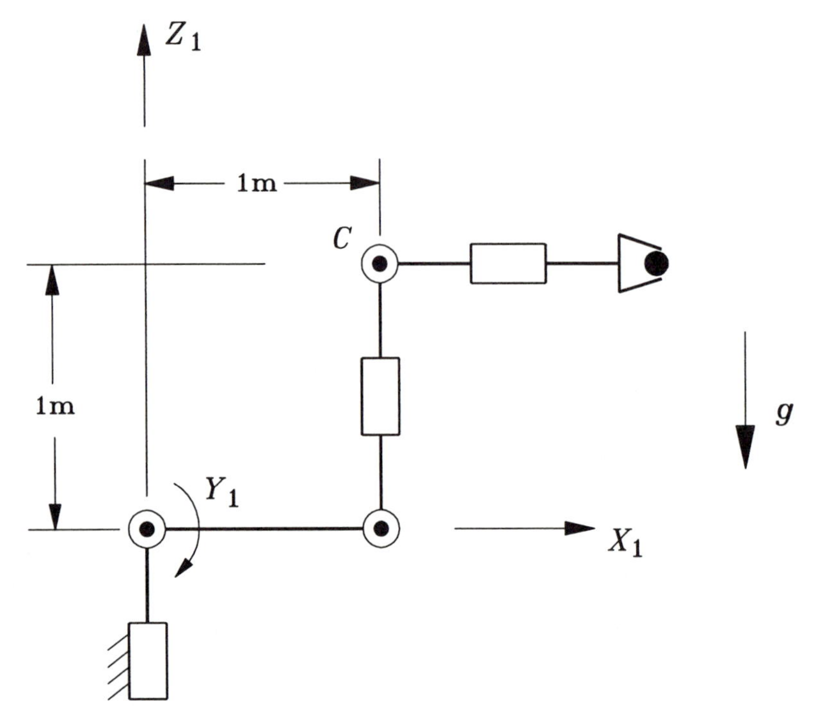

FIGURE 9. A six-revolute robot holding a heavy tool.

4.6 Derive expressions for the angle of rotation and the unit vector parallel to the axis of rotation of matrices $\mathbf{Q}_i$, as introduced in the Denavit-Hartenberg notation of Section 4.2.

4.7 The robotic manipulator of Fig. 9 is instrumented with sensors measuring the torque applied by the motors at the joints. Two readouts are taken of the sensors for the robot in the configuration indicated in the figure. In the first readout, the gripper is empty; in the second, it holds a tool. If the first readout is subtracted from the second, the vector difference $\Delta\boldsymbol{\tau}$ is obtained as

$$\Delta\boldsymbol{\tau} = [0 \quad 2 \quad 1 \quad 0 \quad 1 \quad 0]^T \text{ Nm}$$

With the foregoing information, determine the weight w of the tool and the distance d of its mass center from C, the center of the spherical wrist. For a description of the meaning of the rectangles and the circles with dots inside, see Exercise 4.4.

4.8 A planar three-axis manipulator is shown in Fig. 11, with $a_1 = a_2 = a_3 = 1$ m. When a wrench acts onto the EE of this manipulator, the joint motors exert torques that keep the manipulator under static equilibrium. Readouts of these joint torques are recorded when the

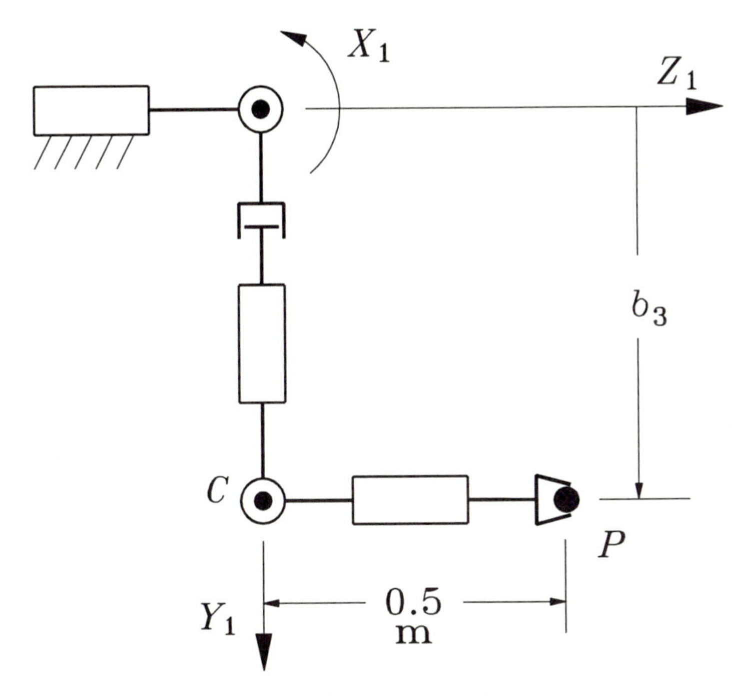

FIGURE 10. ABB-IRB 1000 robotic manipulator.

manipulator is in the posture $\theta_1 = \theta_2 = \theta_3 = 45°$, namely,

$$\tau_1 = -\sqrt{2}\ \text{Nm}, \qquad \tau_2 = -\sqrt{2}\ \text{Nm}, \qquad \tau_3 = 1 - \sqrt{2}\ \text{Nm}$$

Calculate the above-mentioned wrench.

4.9 Shown in Fig. 12 is a computer-generated model of DIESTRO, the robot displayed in Fig. 4.31, with a slightly modified EE. The Denavit-Hartenberg parameters of this robot are given in Table 1. Find the Jacobian matrix of the manipulator at the above configuration.

TABLE 1. DH Parameters of the Modified DIESTRO

i	a_i (mm)	b_i (mm)	α_i	θ_i
1	50	50	90°	90°
2	50	50	−90°	−90°
3	50	50	90°	90°
4	50	50	−90°	−90°
5	50	50	90°	90°
6	0	50	−90°	−90°

4.10 An orthogonal spherical wrist has the architecture shown in Fig. 4.18, with the DH parameters

$$\alpha_4 = 90^\circ, \ \alpha_5 = 90^\circ$$

A frame $\mathcal{F}_7$ is attached to its EE so that Z_7 coincides with Z_6. Find the (Cartesian) orientation that can be attained with two inverse kinematics solutions $\boldsymbol{\theta}_I$ and $\boldsymbol{\theta}_{II}$, defining the two distinct postures, that lie *the farthest* apart. Note that a *distance* between two manipulator postures can be defined as the radical of the quadratic equation yielding the two inverse kinematic solutions of the wrist, whenever the radical is positive. Those postures giving the same EE orientation and lying farthest from each other are thus at the other end of the spectrum from singularities, where the two postures merge into a single one. Hence, the postures lying farthest from each other are singularity-robust.

4.11 For the two postures found in Exercise 4.10, the EE is to move with an angular velocity $\boldsymbol{\omega} = [\omega_1, \ \omega_2, \ \omega_3]^T$ s^{-1}. Show that if $\|\boldsymbol{\omega}\|$ remains

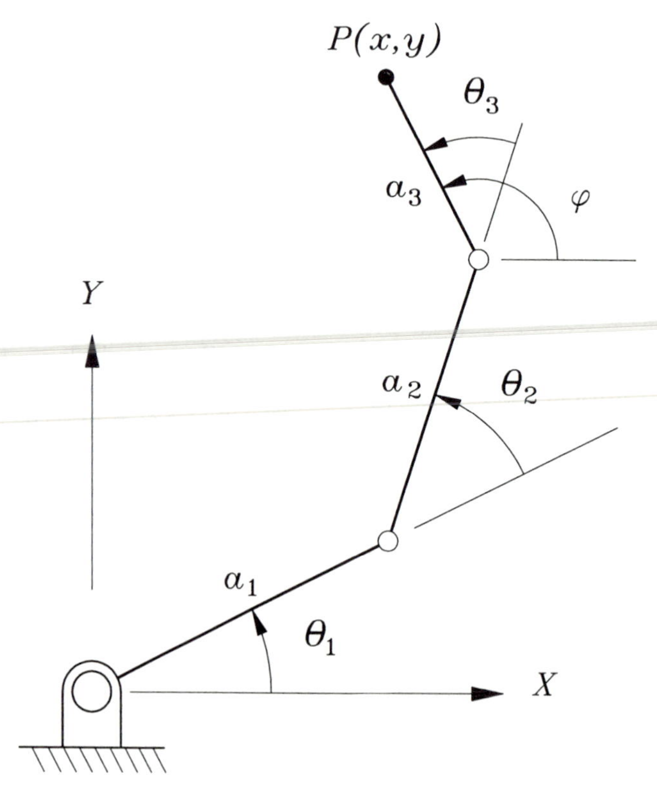

FIGURE 11. A planar three-axis manipulator.

constant, then so does $\|\dot{\boldsymbol{\theta}}\|$, for $\dot{\boldsymbol{\theta}}$ defined as the joint-rate vector of the wrist.

4.12 Point P of the manipulator of Fig. 4.15 is to move with a velocity $\mathbf{v}$ in the posture displayed in that figure. Show that as long as $\|\mathbf{v}\|$ remains constant, so does $\|\dot{\boldsymbol{\theta}}\|$, for $\dot{\boldsymbol{\theta}}$ defined as the joint-rate vector. Moreover, let us assume that in the same posture, point P is to attain a given acceleration $\mathbf{a}$. In general, however, $\|\ddot{\boldsymbol{\theta}}\|$, where $\ddot{\boldsymbol{\theta}}$ denotes the corresponding joint-acceleration vector, does not necessarily remain constant under a constant $\|\mathbf{a}\|$. Under which conditions does $\|\mathbf{a}\|$ remain constant for a constant $\|\ddot{\boldsymbol{\theta}}\|$?

4.13 A load $\mathbf{f}$ is applied to the manipulator of Fig. 4.15 in the posture displayed in that figure. Torque cells at the joints are calibrated to supply torque readouts resulting from this load only, and not from the dead load—its own weight—of the manipulator. Show that under a constant-magnitude load, the magnitude of the joint-torque vector remains constant as well.

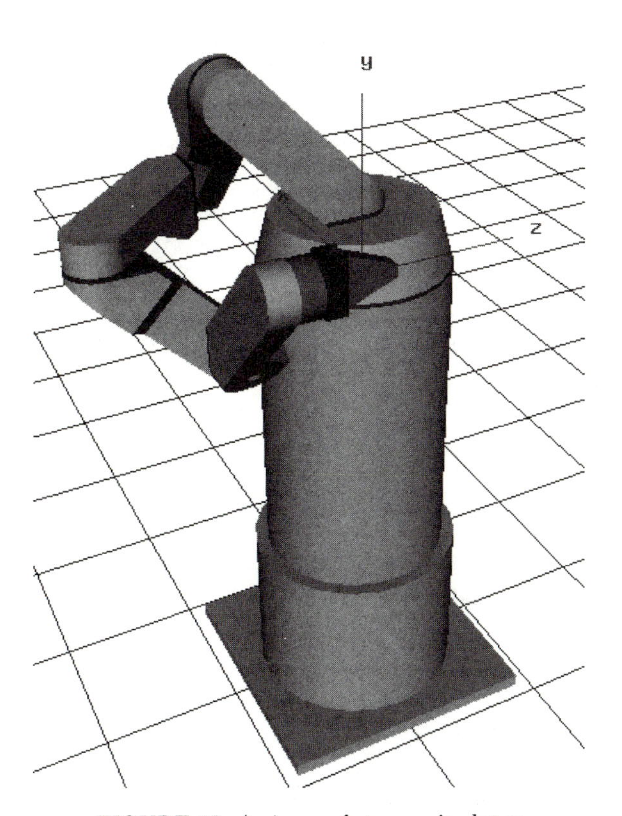

FIGURE 12. A six-revolute manipulator.

4.14 **Dialytic elimination.** The characteristic polynomial of decoupled manipulators for positioning tasks can be derived alternatively via *dialytic elimination*, as introduced in Subsection 4.5.3. It is recalled here that *dialytic elimination* consists in eliminating one unknown from a system of polynomial equations by expressing this system in *linear homogeneous form*, whereby each equation is a linear combination of various successive powers of the unknown to be eliminated, including the zeroth power. This elimination can be achieved as outlined below: Express $\cos\theta_1$ and $\sin\theta_1$ in terms of $\tan(\theta_1/2) \equiv t_1$, thereby obtaining

$$(C\,c_3 + D\,s_3 + E - A)\,t_1^2 + 2\,B\,t_1 + (C\,c_3 + D\,s_3 + E + A) = 0$$
$$(H\,c_3 + I\,s_3 + J - F)\,t_1^2 + 2\,G\,t_1 + (H\,c_3 + I\,s_3 + J + F) = 0$$

which can be further expressed as

$$m\,t_1^2 + 2\,B\,t_1 + n = 0$$
$$p\,t_1^2 + 2\,G\,t_1 + q = 0$$

with obvious definitions for coefficients m, n, p, and q. Next, both sides of the two foregoing equations are multiplied by t_1, thereby producing

$$m\,t_1^3 + 2\,B\,t_1^2 + n\,t_1 = 0$$
$$p\,t_1^3 + 2\,G\,t_1^2 + q\,t_1 = 0$$

Now, the last four equations can be regarded as a system of linear homogeneous equations, namely,

$$\mathbf{M}\mathbf{t}_1 = \mathbf{0}$$

where $\mathbf{0}$ is the 4-dimensional zero vector, while $\mathbf{M}$ is a 4×4 matrix, and $\mathbf{t}_1$ is a 4-dimensional vector. These are defined as

$$\mathbf{M} \equiv \begin{bmatrix} 0 & m & 2B & n \\ 0 & p & 2G & q \\ m & 2B & n & 0 \\ p & 2G & q & 0 \end{bmatrix}, \quad \mathbf{t}_1 \equiv \begin{bmatrix} t_1^3 \\ t_1^2 \\ t_1 \\ 1 \end{bmatrix}$$

Clearly, then, $\mathbf{t}_1 \neq \mathbf{0}$, and hence, $\mathbf{M}$ must be singular. The characteristic polynomial sought, then, can be derived from the condition

$$\det(\mathbf{M}) = 0$$

Show that the last equation is quadratic in $\cos\theta_3$ and $\sin\theta_3$. Hence, the foregoing equation should lead to a quartic equation in $\tan(\theta_3/2)$. Derive the quartic equation involved. *Hint: Do not do this by hand, for it may be too time-consuming and could quickly lead to algebraic mistakes. Use software for symbolic computations instead.*

4.15 Given an arbitrary three-revolute manipulator, as shown in Fig. 4.9, its singular postures are characterized by the existence of a line passing through its operation point about which the moments of its three axes vanish—see Exercise 2.3. Note that this condition can be readily applied to manipulators with a simple architecture, whereby two successive axes intersect at right angles and two others are parallel. However, more complex architectures, like that of the manipulator of Fig. 4.13, are more elusive in this regard. Find the line passing through the operation point and intersecting the three axes of the manipulator of Fig. 4.13 at a singularity. *Hint: A singular posture of this manipulator was found in Example 4.4.2.*

4.16 A robot of the Puma type has the architecture displayed in Fig. 4.3, with the numerical values $a_2 = 0.432$ m, $b_2 = 0.149$ m, $b_4 = 0.432$ m. Find its maximum reach. Then, find the value of the link length a of the manipulator of Fig. 4.15 that gives the same reach as this robot.

4.17 Compute the workspace volume of the manipulator of Fig. 4.3. Here, you can exploit the axial symmetry of the workspace by recalling the *Pappus-Guldinus Theorems*—see any book on multivariable calculus—that yield the volume as $2\pi q$, with q defined as the first moment of the cross-section, which is displayed in Fig. 4.22b, with respect to the axis of symmetry, Z_1. To this end, all you have to do is look for the area of a half-ellipse and the location of its centroid. This information is tabulated in books on elementary mechanics or multivariable calculus (a.k.a. advanced calculus).

4.18 Compute the workspace volume of the manipulator of Fig. 4.15, whose workspace is sketched in Fig. 4.23. Here, you can also use the Pappus-Guldinus Theorem, as suggested in Exercise 4.17. However, the first moment of the cross-section has to be determined numerically, for the area properties of the curve that generates the 3-dimensional workspace are not tabulated. Now, for two manipulators, the Puma-type and the one under discussion, with the same reach, determine which one has the larger workspace. *Note: This exercise is not more difficult than others, but it requires the use of suitable software for the calculation of area properties of planar regions bounded by arbitrary curves. Unless you have access to such a piece of software, do not attempt this exercise.*

4.19 Shown in Fig. 10 is the kinematic chain of the ABB-IRB 1000 robotic manipulator, which contains five revolutes and one prismatic pair. A revolute is represented either as a rectangle or as a circle, depending on whether its axis lies in the plane of the figure or is perpendicular to it. The prismatic pair is represented, in turn, as a dashpot.

(a) Determine the manipulator Jacobian in the X_1, Y_1, Z_1 coordinate frame shown in the figure.

(b) Determine the twist of the end-effector, defined in terms of the velocity of point P, for unit values of all joint-rates, and the posture displayed in the same figure.

(c) Determine the joint accelerations that will produce a vanishing acceleration of the point of intersection, C, of the three wrist axes and a vanishing angular acceleration of the gripper, for the unit joint rates given before.

4.20 The robot in Fig. 10 is now used for a deburring task. When the robot is in the configuration shown in that figure, a static force $\mathbf{f}$ and no moment acts on point P of the deburring tool. This force is sensed by torque sensors placed at the joints of the robot. Assume that the readings of the arm joints are $\tau_1 = 0$, $\tau_2 = 100$ Nm, and $\tau_3 = 50$ N.

(a) Find the force $\mathbf{f}$ acting at P.

(b) Find the readings of the torque sensors placed at the wrist joints.

4.21 A decoupled manipulator is shown in Fig. 8.8 with the DH parameters of Table 8.5 in an arbitrary posture.

(a) Find the Jacobian matrix of this manipulator at a posture with axis X_1 vertical and pointing downwards, while Z_2 and Y_1 make an angle of 180°. Moreover, in this particular posture, Z_3 and Z_4 are vertical and pointing upwards, while Z_7 makes an angle of 180° with Y_1.

(b) At the posture described in item (a), compute the joint-rates that will produce the twist

$$[\boldsymbol{\omega}]_1 = \begin{bmatrix} 1 \\ 1 \\ 1 \end{bmatrix} \omega, \quad [\dot{\mathbf{p}}]_1 = \begin{bmatrix} 1 \\ 1 \\ 1 \end{bmatrix} v$$

(c) A wrench given by a moment $\mathbf{n}$ and a force $\mathbf{f}$ applied at point P acts on the EE of the same manipulator at the posture described in item (a) above. Calculate the joint torques or moments required to balance this wrench, which is given by

$$[\mathbf{n}]_1 = \begin{bmatrix} 1 \\ 1 \\ 1 \end{bmatrix} T, \quad [\mathbf{f}]_1 = \begin{bmatrix} 1 \\ 1 \\ 1 \end{bmatrix} F$$

4.22 Show that the maximum manipulability $\mu = \sqrt{\det(\mathbf{J}\mathbf{J}^T)}$ of an orthogonal spherical wrist is attained when all three of its axes are mutually orthogonal. Find that maximum value.

4.23 Find an expression for the condition number of a three-revolute spherical wrist of twist angles α_1 and α_2 and show that this number depends only on α_1, α_2, and the intermediate joint angle, θ_2. Moreover, find values of these variables that minimize the condition number of the manipulator.

4.24 For the manipulator of Fig. 11, with the dimensions of Exercise 4.8, find the characteristic length, as defined in Section 4.9.

4.25 **Manipulability of decoupled manipulators.** Let μ_a and μ_w represent the manipulability of the arm and the wrist of a decoupled manipulator, i.e.,

$$\mu_a \equiv \sqrt{\det(\mathbf{J}_{21}\mathbf{J}_{21}^T)}, \quad \mu_w \equiv \sqrt{\det(\mathbf{J}_{12}\mathbf{J}_{12}^T)}$$

with $\mathbf{J}_{12}$ and $\mathbf{J}_{21}$ defined in Section 4.5. Show that the manipulability μ of the overall manipulator is the product of the two manipulabilities given above, i.e.,

$$\mu = \mu_a \mu_w$$

4.26 Consider a planar two-revolute manipulator with link lengths a_1 and a_2. Find an expression of the form $\kappa(r, \theta_2)$ for the condition number of its Jacobian, with $r = a_2/a_1$, and establish values of r and θ_2 that minimize κ, which reaches a minimum value of unity.

4.27 Shown in Fig. 4.29 is an orthogonal three-revolute manipulator with an isotropic Jacobian. Find the volume of its workspace. Now consider a second manipulator with a similar orthogonal architecture, but with more common dimensions, i.e., with links of equal length λ. If the two manipulators have the same reach, that is, if

$$\lambda = \frac{1+\sqrt{2}}{2} l$$

find the volume of the workspace of the second manipulator. Finally, determine the KCI—see Section 4.9 for a definition of this term—of the second manipulator. Draw some conclusions with regard to the performance of the two manipulators.

5 Trajectory Planning: Pick-and-Place Operations

5.1 A common joint-rate program for pick-and-place operations is the trapezoidal profile of Fig. 13, whereby we plot $s'(\tau)$ vs. τ, using the notation introduced in Chapter 5, i.e., with $s(\tau)$ and τ defined as dimensionless variables. Here, $s'(\tau)$ starts and ends at 0. From its start to a value τ_1, $s'(\tau)$ grows linearly, until reaching a maximum $s'_{\rm max}$; then, this function remains constant until a value τ_2 is reached, after which the function decreases linearly to zero at the end of the interval.

Clearly, this profile has a discontinuous acceleration and hence, is bound to produce shock and vibration. However, the profile can be smoothed with a spline interpolation as indicated below.

(a) Find the value of $s'_{\rm max}$ in terms of τ_1 and τ_2 so that $s(0) = 0$ and $s(1) = 1$.

(b) Plot $s(\tau)$ with the value of $s'_{\rm max}$ found above and decompose it into a linear part $s_l(\tau)$ and a periodic part $s_p(\tau)$.

(c) Sample $s(\tau)$ with N equally spaced points and find the *periodic* spline that interpolates $s_p(\tau)$, for $\tau_1 = 0.2$ and $\tau_2 = 0.9$. Try various values of N and choose the one that (a) is the smallest possible, 9b) gives a 'good' approximation of the original $s(\tau)$, and (c) yields the best-behaved acceleration program, i.e., an acceleration profile that is smooth and within reasonable bounds. Discuss how you would go about defining a reasonable bound.

5.2 An alternative approach to the solution of the foregoing smoothing problem consists of solving an *inverse* interpolation problem: Plot the acceleration program of the foregoing joint-rate plot, $s''(\tau)$. Now, sample a set of N equally spaced points of $s''(\tau)$ and store them in an N-dimensional array $\mathbf{s}''$. Next, find the ordinates of the supporting points of the interpolating *periodic* spline and store them in an array $\mathbf{s}$ of suitable dimension. Note that $\mathbf{s}''$ does not contain information on the linear part of $s(\tau)$. You will have to modify suitably your array $\mathbf{s}$ so that it will produce the correct abscissa values of the interpolated curve $s(\tau)$, with $s(0) = 0$ and $s(1) = 1$. Moreover, $s(\tau)$

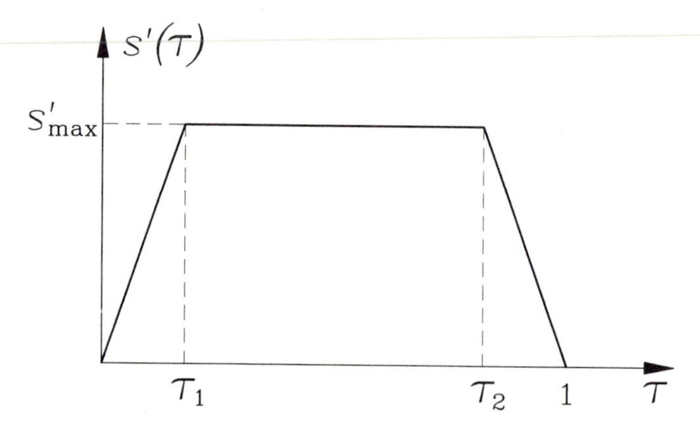

FIGURE 13. A trapezoidal joint-rate profile.

must be monotonic. Try various values of N and choose the smallest one that gives a well-behaved acceleration program, as described in Exercise 5.1.

5.3 One more approach to smoothing the joint-rate profile of Fig. 13 is to use cycloidal motions. To this end, define a segment of a cycloidal-motion function between $\tau = 0$ and $\tau = \tau_1$, so that $s'(\tau_1) = s'_{\max}$, for $s'_{\max}$ as indicated in the same figure. Further, define a similar segment between $\tau = \tau_2$ and $\tau = 1$ so that $s'(\tau_2) = s'_{\max}$ and $s'(1) = 0$. Then, join the two segments with a line of slope $s'_{\max}$. Plot the displacement, velocity, and acceleration of the smoothed motion. Note that the smoothed profile must meet the end conditions $s(0) = 0$ and $s(1) = 1$, and that you may have to introduce a change of variable to shrink the corresponding $s'(\tau)$ segment to meet these conditions.

5.4 A pick-and-place operation involves picking objects from a magazine supplied with an indexing mechanism that presents the objects with a known pose and zero twist, at equal time-intervals T, to a robot, which is to place the objects on a belt conveyor running at a constant speed v_0. Find 5th- and 7th-degree polynomials that can be suitably used to produce the necessary joint-variable time-histories.

5.5 Repeat Exercise 5.4 if now, the objects are to be picked up by the robot from a belt conveyor traveling at a constant velocity $\mathbf{v}_1$ and placed on a second belt conveyor traveling at a constant velocity $\mathbf{v}_2$. Moreover, let $\mathbf{p}_1$ and $\mathbf{p}_2$ designate the position vectors of the points at the pick- and the place poses, respectively. Furthermore, the belts lie in horizontal, parallel planes. Finally, the objects must observe the same attitude with respect to the belt orientation in both the pick- and the place poses.

5.6 Approximate the cycloidal function of Subsection 5.4 using a periodic cubic spline with N subintervals of the same lengths, for various values of N between 5 and 100. Tabulate the approximation error e_N vs. N, with e_N defined as

$$e_N \equiv \max_i \{|e_i|\}_1^{N+1}$$

and

$$e_i \equiv s(\tau_i) - c(\tau_i)$$

in which $s(\tau)$ denotes the spline approximation and $c(\tau)$ the cycloidal function. *Note: the cycloidal function can be decomposed into a linear and a periodic part.*

5.7 From inspection of the plot of the 3-4-5 polynomial and its derivatives displayed in Fig. 5.2, it is apparent that the polynomial can be regarded as the superposition of a linear and a periodic function in the interval $0 \leq \tau \leq 1$. Approximate the underlying periodic function with a periodic cubic spline by subdividing the above-mentioned interval into N equal subintervals, while finding the value of N that will yield a maximum absolute value of less than 10^{-4} in the error in

(a) the function values;

(b) the values of the first derivative; and

(c) the values of the second derivative.

5.8 Repeat Exercise 5.7 for the 4-5-6-7 polynomial of Fig. 5.3.

5.9 A pick-and-place operation is being planned that should observe manufacturer's bounds on the maximum joint rates delivered by the motors of a given robot. To this end, we have the following choices: (a) a 4-5-6-7 polynomial; (b) a *symmetric* trapezoidal speed profile like that of Fig. 13, with $\tau_1 = 0.20$; and (c) a cycloidal motion. Which of these motions produce the minimum time in which the operation can be performed?

5.10 The maximum speed of a cycloidal motion was found to be 2. By noticing that the cycloidal motion is the superposition of a linear and a periodic function, find a cubic-spline motion that will yield a maximum speed of 1.5, with the characteristics of the cycloidal motion at its end points.

5.11 The acceleration of a certain motion $s(\tau)$, for $0 \leq \tau \leq 1$, is given at a sample of instants $\{ \tau_k \}_1^N$ in the form

$$s''(\tau_k) = A\sin(2\pi\tau_k)$$

Find the cubic spline interpolating the given motion so that its second time-derivative will attain those given values, while finding A such that $s(0) = 0$ and $s(1) = 1$. *Hint: A combination of a linear function and a periodic spline can yield this motion. In order to find the function values of the periodic spline, exploit the linear relation between the function values and its second derivatives at the spline supporting points, as discussed in Section 5.6.*

5.12 A robotic joint has been found to require to move, within a time-interval T, with a set of speed values $\{ \dot{\theta}_k \}_1^N$ at equally-spaced instants. Find the natural cubic spline that interpolates the underlying motion so that the angular displacement undergone from beginning to end is a given $\Delta\theta$. *Hint: You will need to establish the linear relation between the spline function values and those of its first derivative.*

6 Dynamics of Serial Robotic Manipulators

6.1 The decoupled robot of Fig. 9 is to undergo a maneuver, at the posture displayed in that figure, that involves the velocity and acceleration specifications given below, in base coordinates:

$$\dot{\mathbf{c}} = \begin{bmatrix} 1 \\ 0 \\ 1 \end{bmatrix} \text{ m/s}, \quad \boldsymbol{\omega} = \begin{bmatrix} 0 \\ 1 \\ 0 \end{bmatrix} \text{ rad/s},$$

$$\ddot{\mathbf{c}} = \begin{bmatrix} 0 \\ 1 \\ 0 \end{bmatrix} \text{ m/s}^2, \quad \dot{\boldsymbol{\omega}} = \begin{bmatrix} 1 \\ 0 \\ 1 \end{bmatrix} \text{ rad/s}^2$$

Compute the joint torques required to drive the robot through the desired maneuver, if the robot is known to have the inertial parameters given below:

$$m_1 = 10.521, \; m_2 = 15.781, \; m_3 = 8.767,$$
$$m_4 = 1.052, \; m_5 = 1.052, \; m_6 = 0.351$$

$$\boldsymbol{\rho}_1 = \begin{bmatrix} 0 \\ -0.054 \\ 0 \end{bmatrix}, \quad \boldsymbol{\rho}_2 = \begin{bmatrix} 0.140 \\ 0 \\ 0 \end{bmatrix}, \quad \boldsymbol{\rho}_3 = \begin{bmatrix} 0 \\ -0.197 \\ 0 \end{bmatrix}$$

$$\boldsymbol{\rho}_4 = \begin{bmatrix} 0 \\ 0 \\ -0.057 \end{bmatrix}, \quad \boldsymbol{\rho}_5 = \begin{bmatrix} 0 \\ -0.007 \\ 0 \end{bmatrix}, \quad \boldsymbol{\rho}_6 = \begin{bmatrix} 0 \\ 0 \\ -0.019 \end{bmatrix}$$

$$\mathbf{I}_1 = \text{diag}\,[\,1.6120 \quad 0.5091 \quad 1.6120\,]$$
$$\mathbf{I}_2 = \text{diag}\,[\,0.4898 \quad 8.0783 \quad 8.2672\,]$$
$$\mathbf{I}_3 = \text{diag}\,[\,3.3768 \quad 0.3009 \quad 3.3768\,]$$
$$\mathbf{I}_4 = \text{diag}\,[\,0.1810 \quad 0.1810 \quad 0.1273\,]$$
$$\mathbf{I}_5 = \text{diag}\,[\,0.0735 \quad 0.0735 \quad 0.1273\,]$$
$$\mathbf{I}_6 = \text{diag}\,[\,0.0071 \quad 0.0071 \quad 0.0141\,]$$

where m_i, $\boldsymbol{\rho}_i$, and $\mathbf{I}_i$ are given in units of Kg, m and Kg m^2, respectively, with the position vectors of the mass centers and the moment-of-inertia matrices given in link-fixed coordinates.

6.2 Derive *homogeneous, linear constraint equations* on the twists of the pairs of coupled bodies appearing in Fig. 14, namely,

(a) two rigid pulleys coupled by an inextensible belt, under no slip;

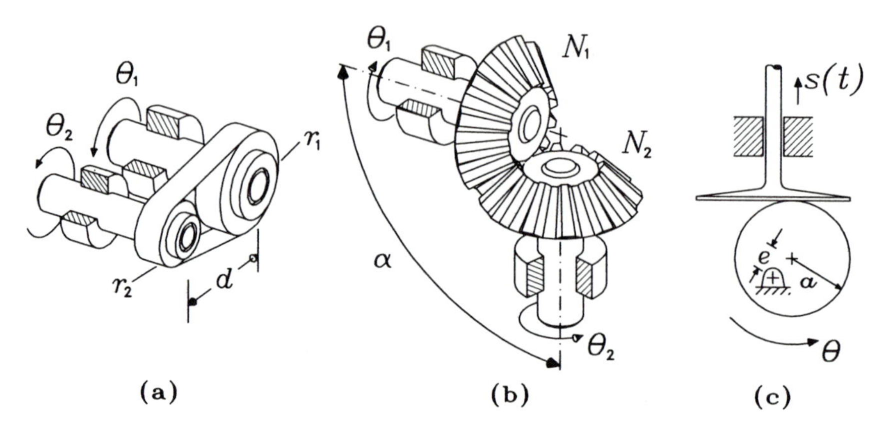

FIGURE 14. Three different pairs of coupled bodies.

(b) the bevel pinion-and-gear train with axes intersecting at an arbitrary angle α;

(c) the cam-and-follower mechanism whose cam disk is an eccentric circular disk.

Notice that the derived constraint equations should have the form:

$$\mathbf{A}\mathbf{t}_1 + \mathbf{B}\mathbf{t}_2 = \mathbf{0}$$

with $\mathbf{t}_1$ and $\mathbf{t}_2$ denoting the twists of bodies 1 and 2, respectively.

6.3 Use the expressions derived in Example 6.6.1 with the aid of the natural orthogonal complement, as pertaining to the planar manipulator of Fig. 6.1, to obtain an expression for the time-derivative of the inertia matrix of this manipulator. Compare the expression thus obtained with that derived in Example 6.3.1, and verify that the difference $\dot{\mathbf{I}} - 2\mathbf{C}$ is skew-symmetric—see Exercise 10.5—where $\mathbf{C}$ is the matrix coefficient of the Coriolis and centrifugal terms.

6.4 A three-revolute spherical wrist with an orthogonal architecture, i.e., with neighboring joint axes at right angles, is shown in Fig. 15. Assume that the moments of inertia of its three links with respect to O, the point of concurrency of the three axes, are given by constant diagonal matrices, in link-fixed coordinates, as

$$\mathbf{I}_4 = \text{diag}(J_1, J_2, J_3)$$
$$\mathbf{I}_5 = \text{diag}(K_1, K_2, K_3)$$
$$\mathbf{I}_6 = \text{diag}(L_1, L_2, L_3)$$

while the potential energy of the wrist is

$$V = -m_6 g a \cos\theta_5$$

Moreover, the motors produce torques τ_4, τ_5, and τ_6, respectively, whereas the power losses can be accounted for via a dissipation function of the form

$$\Delta = \sum_4^6 (\frac{1}{2} b_i \dot{\theta}_i^2 + \tau_i^C |\dot{\theta}_i|)$$

where b_i and τ_i^C, for $i = 4, 5, 6$, are constants.

(a) Derive an expression for the matrix of generalized inertia of the wrist.

(b) Derive an expression for the term of Coriolis and centrifugal forces.

(c) Derive the dynamical model of the wrist.

Hint: The kinetic energy T of a rigid body rotating about a fixed point O with angular velocity $\boldsymbol{\omega}$ can be written as $T = \frac{1}{2}\boldsymbol{\omega}^T \mathbf{I}_O \boldsymbol{\omega}$, where $\mathbf{I}_O$ is the moment-of-inertia matrix of the body with respect to O.

6.5 Shown in Fig. 16 is a two-revolute *pointing manipulator*. The centroïdal inertia matrices of the links are denoted by $\mathbf{I}_1$ and $\mathbf{I}_2$. These are given, in link-fixed coordinates, by:

$$\mathbf{I}_1 = \begin{bmatrix} I_{11} & I_{12} & I_{13} \\ I_{12} & I_{22} & I_{23} \\ I_{13} & I_{23} & I_{33} \end{bmatrix}, \quad \mathbf{I}_2 = \begin{bmatrix} J_{11} & J_{12} & J_{13} \\ J_{12} & J_{22} & J_{23} \\ J_{13} & J_{23} & J_{33} \end{bmatrix}$$

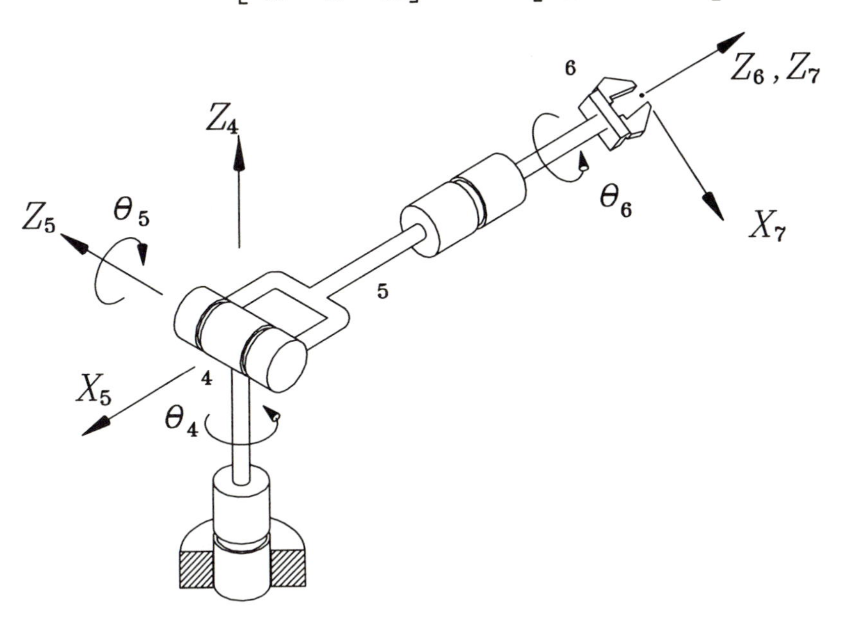

FIGURE 15. A three-revolute spherical wrist.

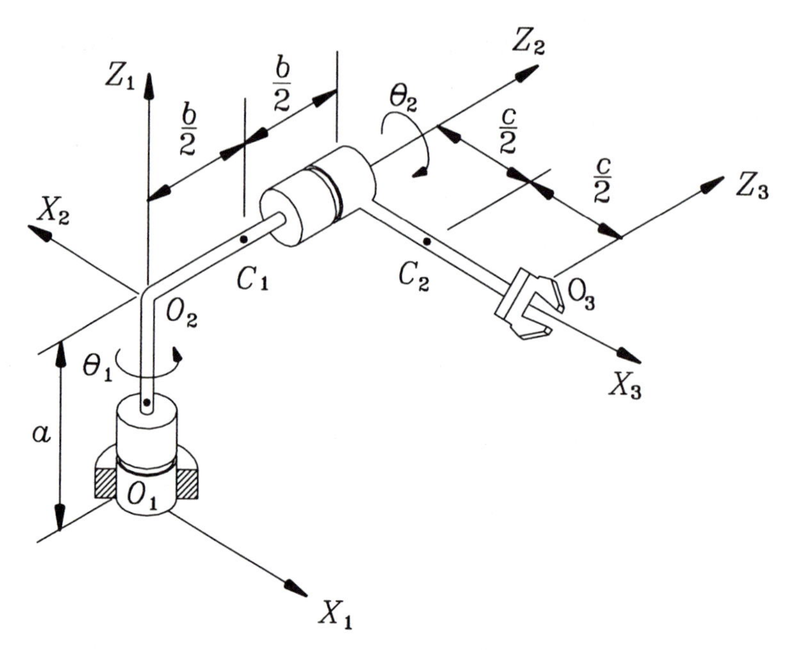

FIGURE 16. A two-revolute pointing manipulator.

Moreover, the mass centers of the links are denoted by C_1 and C_2, respectively, and are shown in the same figure, the masses being denoted by m_1 and m_2.

(a) Determine the kinetic energy of the manipulator as a quadratic function of $\dot{\theta}_1$ and $\dot{\theta}_2$.

(b) Determine the 2×2 matrix of generalized inertia.

(c) Find an expression for the time-rate of change of the matrix of generalized inertia by straightforward differentiation of the expression found in item (b).

(d) Repeat item (c), but now by differentiation of the three factors of $\mathbf{I}$, as given in

$$\mathbf{I} = \mathbf{T}^T \mathbf{M} \mathbf{T}$$

6.6 The twist $\mathbf{t}_i$ of the ith link of an n-dof serial manipulator can be expressed as

$$\mathbf{t}_i = \mathbf{T}_i \dot{\boldsymbol{\theta}}$$

where $\mathbf{T}_i$ is a $6 \times n$ twist-shaping matrix and $\dot{\boldsymbol{\theta}}$ is the n-dimensional vector of actuated joint rates. Moreover, let $\mathbf{M}_i$ and $\mathbf{W}_i$ be the 6×6 matrices defined in Section 6.3. Show that if the link is constrained to undergo planar motion, then the product $\mathbf{T}_i^T \mathbf{W}_i \mathbf{M}_i \mathbf{T}_i$ vanishes.

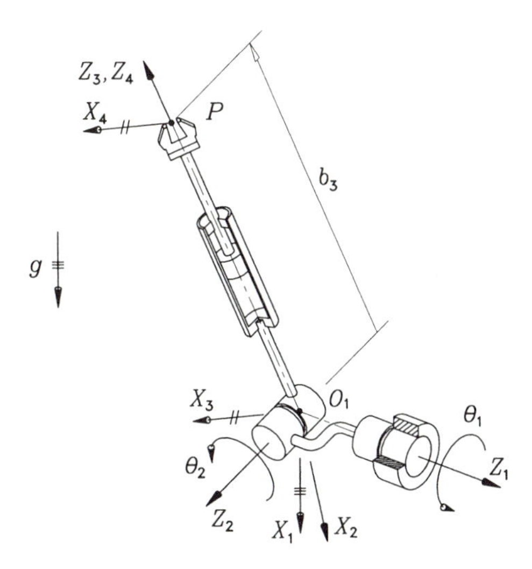

FIGURE 17. An RRP spatial manipulator.

6.7 Devise a recursive algorithm to compute the joint torques required to balance a wrench **w** acting at the EE of a six-revolute manipulator of arbitrary architecture. Then, derive the number of floating-point operations (multiplications and additions) required to compute these torques, and compare your result with the number of floating point operations required to compute the same by matrix-times-vector multiplications, using the transpose Jacobian.

6.8 Establish the computational cost incurred in computing the term of Coriolis and centrifugal forces of an n-revolute serial manipulator, when the Newton-Euler algorithm is used for this purpose.

6.9 Shown in Fig. 17 is an RRP manipulator, whose DH parameters are displayed in Table 2. The masses of its three moving links are denoted by m_1, m_2, and m_3, and the mass center of each of links 1 and 2 coincides with O_1, while the mass center of link 3 is located at P. Moreover, the centroidal moments of inertia of these links are, in link-fixed coordinates,

$$[\mathbf{I}_1]_2 = A\mathbf{1}, \quad [\mathbf{I}_2]_3 = B\mathbf{1}, \quad [\mathbf{I}_3]_4 = C\mathbf{1}$$

where $\mathbf{1}$ denotes the 3×3 identity matrix.

(a) Derive the Euler-Lagrange equations of the manipulator under the assumption that gravity acts in the direction of X_1.

(b) Find the generalized inertia matrix of the manipulator.

TABLE 2. DH Parameters of the RRP Manipulator

i	a_i	b_i	α_i
1	0	0	90°
2	0	0	90°
3	0	b_3	0°

6.10 A link is said to be *inertially isotropic* if its three principal moments of inertia are identical.

(a) Show that *any* direction is a principal axis of inertia of an inertially isotropic link.

(b) Explore the advantages of a manipulator with inertially isotropic links with regard to its real-time control, i.e., find the savings in floating-point operations required to compute the recursive Newton-Euler algorithm of such a manipulator.

6.11 Devise an algorithm similar to Algorithm 6.6.1, but applicable to planar manipulators, and determine the computational costs involved in its implementation.

6.12 Write a piece of code to evaluate numerically the inertia matrix of an n-axis manipulator and test it with the manipulator of Example 6.6.1. For this purpose, assume that $I = ma^2$.

7 Special Topics on Rigid-Body Kinematics

7.1 The regular tetrahedron of Fig. 7, of unit-length edges, moves in such a way that vertex P_1 has a velocity of unit magnitude directed from P_1 to P_4, whereas the velocity of P_2 is parallel to edge P_2P_3. Define a coordinate frame X, Y, Z with origin at P_1, Y axis directed from P_1 to the midpoint M of P_2P_3, and X axis in the plane of P_1, P_2, P_3, as shown in that figure. With the above information,

(a) find the velocity of P_2;

(b) show that the velocity of P_3 cannot be zero;

(c) if the velocity of P_3 lies in the $P_1P_2P_3$ plane, find that velocity;

(d) find the angular velocity of the tetrahedron;

(e) find the set of points of the tetrahedron undergoing a velocity of minimum magnitude.

7.2 The position vectors of three points of a rigid body, $\mathbf{p}_1$, $\mathbf{p}_2$, and $\mathbf{p}_3$, as well as their velocities, $\dot{\mathbf{p}}_1$, $\dot{\mathbf{p}}_2$, and $\dot{\mathbf{p}}_3$, are given below:

$$\mathbf{p}_1 = \begin{bmatrix} 1 \\ 1 \\ 1 \end{bmatrix}, \quad \mathbf{p}_2 = \begin{bmatrix} 1 \\ -1 \\ 1 \end{bmatrix}, \quad \mathbf{p}_3 = \begin{bmatrix} -1 \\ 1 \\ -1 \end{bmatrix}$$

$$\dot{\mathbf{p}}_1 = \begin{bmatrix} 1 \\ 1 \\ 1 \end{bmatrix}, \quad \dot{\mathbf{p}}_2 = \begin{bmatrix} 3 \\ 1 \\ -1 \end{bmatrix}, \quad \dot{\mathbf{p}}_3 = \begin{bmatrix} -1 \\ 1 \\ 3 \end{bmatrix}$$

(a) Is the motion possible?

(b) If the motion is possible, find its angular velocity.

7.3 For matrix $\mathbf{P}$ defined as in eq.(7.4), i.e., as

$$\mathbf{P} \equiv [\,\mathbf{p}_1 - \mathbf{c} \quad \mathbf{p}_2 - \mathbf{c} \quad \mathbf{p}_3 - \mathbf{c}\,]$$

where $\{\,\mathbf{p}_k\,\}_1^3$ are the position vectors of three points of a rigid body, while $\mathbf{c}$ is that of their centroid, prove that $\mathrm{tr}(\mathbf{P}^2) = \mathrm{tr}^2(\mathbf{P})$ if and only if the three given points are collinear.

7.4 With matrix $\mathbf{P}$ defined as in Exercise 7.3 above, prove that $\mathbf{P}$ is orthogonal to $\dot{\mathbf{P}}$, i.e., prove that

$$\mathrm{tr}(\mathbf{P}\dot{\mathbf{P}}^T) = 0$$

7.5 With the notation of Section 7.3, prove that

$$\mathrm{vect}(\mathbf{\Omega}^2\mathbf{P}) = \dot{\mathbf{D}}\boldsymbol{\omega}$$

7.6 Derive the velocity and acceleration compatibility conditions for a body that is known to undergo spherical motion.

7.7 The position vectors of three points of a rigid body, $\mathbf{p}_1$, $\mathbf{p}_2$, and $\mathbf{p}_3$, are given as in Exercise 7.2, and repeated below for quick reference:

$$\mathbf{p}_1 = \begin{bmatrix} 1 \\ 1 \\ 1 \end{bmatrix}, \quad \mathbf{p}_2 = \begin{bmatrix} 1 \\ -1 \\ 1 \end{bmatrix}, \quad \mathbf{p}_3 = \begin{bmatrix} -1 \\ 1 \\ -1 \end{bmatrix}$$

However, the velocities of these points are all zero, while their accelerations are given as

$$\ddot{\mathbf{p}}_1 = \begin{bmatrix} 1 \\ 1 \\ 1 \end{bmatrix}, \quad \ddot{\mathbf{p}}_2 = \begin{bmatrix} 3 \\ 1 \\ -1 \end{bmatrix}, \quad \ddot{\mathbf{p}}_3 = \begin{bmatrix} -1 \\ 1 \\ 3 \end{bmatrix}$$

(a) Show that the motion is compatible.

(b) Find the angular acceleration of the body.

7.8 With the notation of Section 7.2, let

$$\mathbf{R} \equiv \mathbf{P}\mathbf{P}^T$$

(a) Show that the moment of inertia $\mathbf{J}$ of the three given points, which is identical to that of a system of unit masses located at these points, with respect to the given origin, is

$$\mathbf{J} = \mathrm{tr}(\mathbf{R})\mathbf{1} - \mathbf{R}$$

(b) Show that if the three given points move as points of a rigid body undergoing an angular velocity $\boldsymbol{\omega}$ whose cross-product matrix is $\boldsymbol{\Omega}$, then

$$\dot{\mathbf{J}} = \boldsymbol{\Omega}\mathbf{J} - \mathbf{J}\boldsymbol{\Omega}$$

(c) Furthermore, show that if under the conditions of item (b) above, the set of points undergoes an angular acceleration $\dot{\boldsymbol{\omega}}$ of cross-product matrix $\dot{\boldsymbol{\Omega}}$, then

$$\ddot{\mathbf{J}} = \dot{\boldsymbol{\Omega}}\mathbf{J} - \mathbf{J}\dot{\boldsymbol{\Omega}} + \boldsymbol{\Omega}^2\mathbf{J} + \mathbf{J}\boldsymbol{\Omega}^2 - 2\boldsymbol{\Omega}\mathbf{J}\boldsymbol{\Omega}$$

7.9 A wrench of unknown force $\mathbf{f}$ is applied to a rigid body. In order to find this force, its moment with respect to a set of points $\{P_k\}_1^3$, of position vectors $\{\mathbf{p}_k\}_1^3$, is measured and stored in the set $\{\mathbf{n}_k\}_1^3$. Show that $\mathbf{f}$ can be calculated from the relation

$$\mathbf{D}\mathbf{f} = -\mathrm{vect}(\mathbf{M})$$

with $\mathbf{D}$ defined as in Section 7.2, i.e., as

$$\mathbf{D} \equiv \frac{1}{2}[\mathrm{tr}(\mathbf{P})\mathbf{1} - \mathbf{P}]$$

and $\mathbf{M}$ given by

$$\mathbf{M} = [\,\mathbf{n}_1 - \mathbf{n} \quad \mathbf{n}_2 - \mathbf{n} \quad \mathbf{n}_3 - \mathbf{n}\,], \quad \mathbf{n} \equiv \frac{1}{3}\sum_1^3 \mathbf{n}_k$$

Note that $\mathbf{P}$ is defined in Exercise 7.3.

7.10 A wrench is applied to the tetrahedron of Fig. 7. When the force of this wrench acts at point P_k, the resulting moment is $\mathbf{n}_k$, for $k = 1, 2, 3$. For the data displayed below, in frame $\mathcal{F}$ of that figure, find the resultant force $\mathbf{f}$, as well as the line of action of this force that will lead to a moment of minimum magnitude. Determine this moment.

$$\mathbf{n}_1 = -\frac{\sqrt{2}}{4}\begin{bmatrix} 1 \\ 0 \\ 0 \end{bmatrix}, \quad \mathbf{n}_2 = \frac{1}{12}\begin{bmatrix} 3\sqrt{2} \\ -2\sqrt{6} \\ 2\sqrt{3} \end{bmatrix}, \quad \mathbf{n}_3 = \frac{1}{12}\begin{bmatrix} 3\sqrt{2} \\ 2\sqrt{6} \\ -2\sqrt{3} \end{bmatrix}$$

8 Kinematics of Complex Robotic Mechanical Systems

8.1 For the parallel manipulator of Fig. 8.7, find the matrix mapping joint forces into wrenches acting on the moving platform, if actuation is supplied through the prismatic joints.

8.2 We refer to the rolling robot with conventional wheels introduced in Subsection 8.6.1. We would like to study the equivalent concept of manipulability, which here we can call *maneuverability*. This concept refers to the numerical conditioning of the two underlying Jacobian matrices, $\mathbf{J}$ and $\mathbf{K}$, as defined in eqs.(8.88a & b). Clearly, $\mathbf{J}$ is isotropic and hence, optimally conditioned. In attempting to determine the condition number of $\mathbf{K}$, however, we need to order its singular values from smallest to largest.

(a) Show that the two singular values of $\mathbf{K}$ are $\sigma_1 = l/r$ and $\sigma_2 = 2/r$. Obviously, an ordering from smallest to largest is impossible because of the lack of dimensional homogeneity.

(b) In order to cope with the dimensional inhomogeneity of matrix $\mathbf{K}$, we introduce the characteristic length L, which we define below. First, we redefine the Jacobian $\mathbf{K}$ in dimensionless form as

$$\mathbf{K} \leftarrow \begin{bmatrix} (l/r) & \mathbf{0}^T \\ 0 & (2L/r)\mathbf{j}^T \end{bmatrix}$$

Now, L is the value that minimizes the condition number of the dimensionless $\mathbf{K}$. Show that this value is $l/2$ and that it produces a condition number of unity.

8.3 Find an expression for the angular velocity $\dot{\phi}_i$ of the active roller of the ith wheel of the robot with Mekanum wheels introduced in Subsection 8.6.2.

8.4 We refer again to the robot with Mekanum wheels introduced in Subsection 8.6.2. For the case of a three-wheeled robot of this kind, we consider here a design whereby the wheels are equally spaced in a Δ-array. In this array, the centers of the hubs, O_i, lie at the corners of an equilateral triangle of side a; moreover, we assume that $\alpha_i = 90°$, for $i = 1, 2, 3$. Under these conditions, find the characteristic length L of the robot that renders $\mathbf{K}$, as defined in the above-mentioned subsection, dimensionless and of a minimum condition number. Find this minimum as well.

8.5 Find the value of ψ at which the rolling robot of Fig. 8.22 attains a singular configuration. Here, a singularity is understood as a loss of maneuverability in the sense of not being able to drive the unactuated joints by means of the actuated ones. Discuss whether under *reasonable* values of the geometric parameters, this singularity can occur.

8.6 Determine the architecture and the "posture," i.e., the values of the relevant joint variables of the rolling robot of Fig. 8.22 that will render matrix $\mathbf{\Theta}$ isotropic, where $\mathbf{\Theta}$ represents the mapping of actuated joint rates into unactuated ones. Is kinematic isotropy, in this sense, kinematically possible?

8.7 Find a relation among the geometric parameters of the robot of Fig. 8.22 that will allow the steering of the robot along a straight course with the highest possible maneuverability in the sense defined in Exercise 7.5. That is, find a relation among the geometric parameters of this robot that will render $\kappa(\mathbf{\Theta})$ a minimum along a straight course.

8.8 Find the value of ψ under which the robot of Fig. 8.22 performs a maneuver that leaves the midpoint of segment $\overline{O_1O_2}$ stationary. Under this maneuver, state a relationship among the geometric parameters of the robot that minimizes $\kappa(\mathbf{\Theta})$.

8.9 Upon inversion, eq.(8.98a) yields

$$\dot{\boldsymbol{\theta}}_a = \mathbf{U}\dot{\boldsymbol{\theta}}_u$$

(a) Find $\mathbf{U}$.

(b) The above equation can be written as

$$\dot{\theta}_1 = u_{13}\dot{\theta}_3 + u_{1\psi}\dot{\psi} \equiv \mathbf{u}_1^T\dot{\boldsymbol{\theta}}_a$$
$$\dot{\theta}_2 = u_{23}\dot{\theta}_3 + u_{2\psi}\dot{\psi} \equiv \mathbf{u}_2^T\dot{\boldsymbol{\theta}}_a$$

The first of the above equations can be integrated if $\mathbf{u}_1$, which is an implicit function of θ_3 and ψ, is the gradient with respect to $\boldsymbol{\theta}_u \equiv [\theta_3 \quad \psi]^T$ of a scalar function $U_1(\theta_3, \psi)$. Likewise, the second of the above equations can be integrated if a function $U_2(\theta_3, \psi)$ exists, whose gradient with respect to $\boldsymbol{\theta}_u$ is $\mathbf{u}_2$. Further, upon recalling Schwartz's Theorem of multivariable calculus, $\mathbf{u}_i$ is such a gradient if and only if $\nabla\mathbf{u}_i$, i.e., the *Hessian matrix* of U_i with respect to $\boldsymbol{\theta}_u$, is symmetric, for $i = 1, 2$.

Show that the above-mentioned Hessians, for the case at hand, are nonsymmetric, and hence, none of the above differential expressions is integrable. Such expressions are called *nonholonomic.*

Note: To be sure, the above condition is sufficient, but not necessary. It is possible that some individual equations of a system of differential expressions, also called *Pfaffian forms*, are not integrable while the overall system is. An examination of necessary and sufficient condition for integrability falls beyond the scope of this book. Such conditions are best understood with the aid of the Frobenius Theorem (De Luca and Oriolo, 1995).

8.10 For the rolling robot with omnidirectional wheels introduced in Section 8.6.2, with a Δ-array, as described in Exercise 4, show that the equation yielding the angular velocity of the platform in terms of the wheel rates is integrable, but the equations yielding the velocity of the operation point are not.

8.11 **A holonomic rolling robot.** The robot described in Exercise 8.10 can be rendered holonomic at the expense of one degree of freedom. Show that if the three wheel rates are coordinated, either mechanically or electronically so that

$$\dot{\theta}_1 + \dot{\theta}_2 + \dot{\theta}_3 = 0$$

then the platform is constrained to move under pure translation. Under this condition, the robot is holonomic. Find an explicit expression for the position vector $\mathbf{c}$ of the operation point in terms of the wheel angles.

9 Trajectory Planning: Continuous-Path Operations

9.1 A PUMA 560 robot, with the DH parameters of Table 3, is used to perform an arc-welding operation as indicated below: An electrode is rigidly attached to the gripper of the robot. The tip of the electrode, point P, is to trace a helicoidal path at a constant rate of 50 mm/s. Moreover, the center of the wrist is located at a point C, fixed to a Frenet-Serret coordinate frame. In this frame, the coordinates of C

TABLE 3. D-H Parameters of a PUMA 560 Robot

Joint i	α_i (deg)	a_i (m)	b_i (m)
1	90	0	0.660
2	0	0.432	0
3	90	0.020	0.149
4	90	0	0.432
5	90	0	0
6	0	0	0.056

are $(0,\ -50,\ 86.7)$ mm. Moreover, the path to be traced by point P is given as

$$x = a\cos\vartheta, \ y = a\sin\vartheta, \ z = b\vartheta, \ 0 \le \vartheta \le \pi/2$$

with the values $a = 300$ mm, $b = 800/\pi$ mm.

(a) Decide where to locate the robot base with respect to the path so that the latter will lie well within the workspace of the robot. Then, produce plots of θ_i vs. t, for $0 \le t \le T$, where T is the time it takes to traverse the whole trajectory, for $i = 1, 2, \ldots, 6$.

(b) Produce plots of $\dot{\theta}_i$ vs. t in the same time interval for all six joints.

(c) Produce plots of $\ddot{\theta}_i$ vs. t in the same time interval for all six joints.

9.2 A bracket for spot-welding, shown in Fig. 18, is rigidly attached to the end-effector of a robotic manipulator. It is desired that point P of the bracket follow a helicoidal path Γ, while keeping the orientation of the bracket with respect to Γ as indicated below: Let $\mathcal{B} \equiv \{\mathbf{i}_0, \mathbf{j}_0, \mathbf{k}_0\}$ and $\mathcal{F}_7 \equiv \{\mathbf{i}_7, \mathbf{j}_7, \mathbf{k}_7\}$ be triads of unit orthogonal vectors fixed to the base of the robot and to the bracket, respectively. Moreover, let $\mathcal{F} \equiv \{\mathbf{e}_t, \mathbf{e}_n, \mathbf{e}_b\}$ be the Frenet-Serret triad of Γ, given as

$$\begin{aligned}
\mathbf{e}_t &= -0.6\sin\varphi\mathbf{i}_0 + 0.6\cos\varphi\mathbf{j}_0 + 0.8\mathbf{k}_0 \\
\mathbf{e}_n &= -\cos\varphi\mathbf{i}_0 - \sin\varphi\mathbf{j}_0 \\
\mathbf{e}_b &= 0.8\sin\varphi\mathbf{i}_0 - 0.8\cos\varphi\mathbf{j}_0 + 0.6\mathbf{k}_0
\end{aligned}$$

where φ is a given function of time, $\varphi(t)$.

Furthermore, the orientation of the bracket with respect to Γ is to be kept constant and given in terms of the Frenet-Serret triad as

$$\begin{aligned}
\mathbf{i}_7 &= 0.933\mathbf{e}_t + 0.067\mathbf{e}_n - 0.354\mathbf{e}_b \\
\mathbf{j}_7 &= 0.067\mathbf{e}_t + 0.933\mathbf{e}_n + 0.354\mathbf{e}_b \\
\mathbf{k}_7 &= 0.354\mathbf{e}_t - 0.354\mathbf{e}_n + 0.866\mathbf{e}_b
\end{aligned}$$

Additionally, $\mathbf{R}$ and $\mathbf{S}(t)$ denote the rotation matrices defining the orientation of $\mathcal{F}_7$ with respect to $\mathcal{F}$ and of $\mathcal{F}$ with respect to $\mathcal{B}$, respectively.

(a) Find the matrix representation of $\mathbf{S}(t)$ in $\mathcal{B}$.

(b) Find the matrix representation of $\mathbf{R}$ in $\mathcal{F}$.

(c) Let $\mathbf{Q}(t)$ denote the orientation of $\mathcal{F}_7$ with respect to $\mathcal{B}$. Find its matrix representation in $\mathcal{B}$.

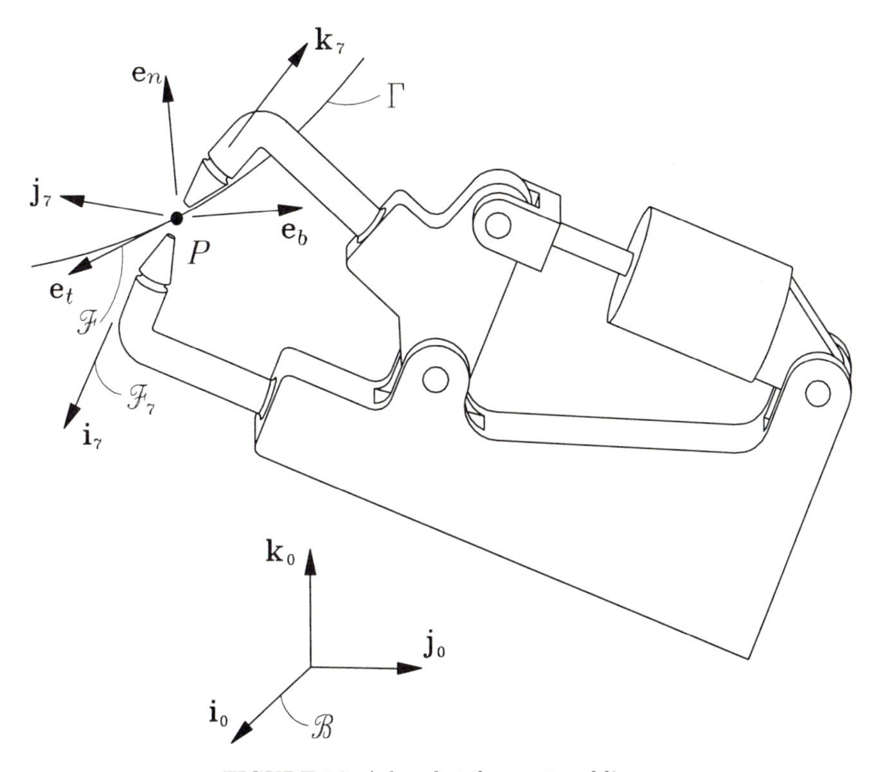

FIGURE 18. A bracket for spot-welding.

(d) Find the Darboux vector $\boldsymbol{\delta}$ of the path, along with its time-derivative, $\dot{\boldsymbol{\delta}}$, in base-fixed coordinates. *Note: You can do this in several ways, as discussed in Section 9.2. Choose the one that will allow you to use previously computed results, thereby simplifying the computations.*

9.3 The parametric equations of a curve are given as

$$x = 2t, \ y = t^2, \ z = t^3/3$$

where t is time. A robotic manipulator is to follow this trajectory so that its gripper keeps a constant orientation with respect to the Frenet-Serret frame of the curve.

(a) Determine the unit vector parallel to the axis of rotation and the angle of rotation of the gripper as functions of time.

(b) Find the angular velocity and angular acceleration of the gripper as functions of time.

9.4 Find the spline approximation of the helix of Example 9.3.1. Then, plot the approximation errors of the Cartesian coordinates of points of the helix, for $N = 5$, 11, and 21 equally spaced supporting points. In order to assess the orientation error, compute the Darboux vectors of the spline, $\boldsymbol{\delta}_s$, and of the helix, $\boldsymbol{\delta}_h$. The approximation error of the orientation is now defined as

$$e_o \equiv \max_{\varphi}\{\|\boldsymbol{\delta}_s(\varphi) - \boldsymbol{\delta}_h(\varphi)\|\}$$

with φ defined as in Example 9.3.1.

9.5 Find the spline approximation of the curvature, torsion, and Darboux vector of the curve introduced in Example 9.3.2. Find expressions for the exact values of these variables and plot the approximation errors, for 5, 10, and 20 equally spaced supporting points vs. φ. In the error definitions given below, subscript e indicates *exact* value, subscript s, *spline* value:

$$\begin{aligned} e_\kappa &\equiv \kappa_s(\varphi) - \kappa_e(\varphi) \\ e_\tau &\equiv \tau_s(\varphi) - \tau_e(\varphi) \\ e_\delta &\equiv \|\boldsymbol{\delta}_s(\varphi) - \boldsymbol{\delta}_e(\varphi)\| \end{aligned}$$

9.6 From the plots of the time-histories of the joint angles calculated in Example 9.5.1, it is apparent that, with the exception of θ_4, which has a linear component, these histories are periodic. Repeat Example 9.5.1, but now using a spline approximation of the welding seam, with $N = 5$, 10, and 20 supporting points. With this spline approximation, calculate the pose, the twist, and the twist-rate at each supporting point. Now, calculate values of $\boldsymbol{\theta}$, $\dot{\boldsymbol{\theta}}$, and $\ddot{\boldsymbol{\theta}}$ at each of these supporting points by means of inverse kinematics. Compare the values thus obtained of $\ddot{\boldsymbol{\theta}}$ with those derived from the linear relation between the function values and the values of its second derivative at the supporting points when using a cubic spline. Tabulate the Euclidean norm of the errors vs. N.

9.7 The decoupled robot of Fig. 9 is to perform an arc-welding operation along a welding seam that requires its wrist center C to travel at a constant speed of 1 m/s along a line joining points A and B, not shown in that figure, while keeping the EE holding the electrode at a constant orientation with respect to the base frame. Moreover, the seam is to be traversed according to the following schedule: With point C located at a point A' on the extension of AB, a distance of 250 mm from A, point C approaches A with a cycloidal motion at the specified speed; upon reaching B, point C decelerates with a cycloidal motion as well, until it reaches a point B' in the other extension of AB, 250 mm from B, with zero speed. The position vectors of points A and B, denoted by $\mathbf{a}$ and $\mathbf{b}$, respectively, are given, in base coordinates, as

$$\mathbf{a} = \begin{bmatrix} 500 \\ -500 \\ 500 \end{bmatrix}, \quad \mathbf{b} = \begin{bmatrix} 1,200 \\ 0 \\ 1,200 \end{bmatrix}$$

in mm. For the above-given data, find the time-histories of all joint variables.

10 Dynamics of Complex Robotic Mechanical Systems

10.1 Find the conditions under which the three-wheeled robot with omnidirectional wheels analyzed in Subsection 10.5.2 has an isotropic inertia matrix. Discuss the advantages of such an inertially-isotropic robot.

10.2 Establish the conditions on the actuated joint rates under which the three-wheeled robot with omnidirectional wheels of Subsection 10.5.2 undergoes pure translation. Under these conditions, the robot has only two degrees of freedom and, hence, a 2×2 inertia matrix. Derive an expression for its inertia matrix. *Hint: The constraint for pure translation can be written as*

$$\mathbf{a}^T \dot{\boldsymbol{\theta}}_a = 0$$

and hence, if the 3×2 *matrix* $\mathbf{L}$ *is an orthogonal complement of* $\mathbf{a}$, *i.e., if* $\mathbf{a}^T\mathbf{L} = \mathbf{0}_2^T$, *where* $\mathbf{0}_2$ *is the 2-dimensional zero vector, then the underlying Euler-Lagrange equations of the constrained system can be derived by multiplying the two sides of the mathematical model found in Subsection 10.5.2 by* $\mathbf{L}^T$:

$$\mathbf{L}^T\mathbf{I}\ddot{\boldsymbol{\theta}}_a + \mathbf{L}^T\mathbf{C}\dot{\boldsymbol{\theta}}_a = \mathbf{L}^T\boldsymbol{\tau} - \mathbf{L}^T\boldsymbol{\delta}$$

Further, upon writing $\dot{\boldsymbol{\theta}}_a$ *as a linear transformation of a 2-dimensional vector* $\mathbf{u}$, *namely, as*

$$\dot{\boldsymbol{\theta}}_a = \mathbf{L}\mathbf{u}$$

we obtain

$$\mathbf{L}^T\mathbf{I}\mathbf{L}\dot{\mathbf{u}} + \mathbf{L}^T\mathbf{C}\mathbf{L}\mathbf{u} = \mathbf{L}^T\boldsymbol{\tau} - \mathbf{L}^T\boldsymbol{\delta}$$

and hence, the generalized inertia matrix under pure translation is $\mathbf{L}^T\mathbf{I}\mathbf{L}$.

10.3 Derive the mathematical model governing the motion of a 2-dof rolling robot with conventional wheels, similar to that of Fig. 8.22, but with two caster wheels instead. The vertical axes of the caster wheels are a distance l apart and a distance $a + b$ from the common axis of the driving wheels. What is the characteristic length of this robot?

10.4 Show that the mathematical model of an arbitrary robotic mechanical system, whether holonomic or nonholonomic, with r rigid bodies and n degrees of freedom, can be cast in the general form

$$\mathbf{I}(\boldsymbol{\theta})\ddot{\boldsymbol{\theta}}_a + \mathbf{C}(\boldsymbol{\theta}, \dot{\boldsymbol{\theta}}_a)\dot{\boldsymbol{\theta}}_a = \boldsymbol{\tau}^A + \boldsymbol{\gamma} + \boldsymbol{\delta}$$

where

$\boldsymbol{\theta}$: the m-dimensional vector of variables associated with all joints, actuated and unactuated;

$\dot{\boldsymbol{\theta}}_a$: the n-dimensional vector of actuated joint variables, $n \leq m$;

$\boldsymbol{\tau}^A$: the n-dimensional vector of actuator torques;

$\boldsymbol{\gamma}$: the n-dimensional vector of gravity torques;

$\boldsymbol{\delta}$: the n-dimensional vector of dissipative torques;

$\mathbf{I}(\boldsymbol{\theta})$: the $n \times n$ matrix of generalized inertia;

$\mathbf{C}(\boldsymbol{\theta}, \dot{\boldsymbol{\theta}}_a)$: the $n \times n$ matrix of Coriolis and centrifugal forces;

with $\mathbf{I}(\boldsymbol{\theta})$ and $\mathbf{C}(\boldsymbol{\theta}, \dot{\boldsymbol{\theta}}_a)$ given by

$$\mathbf{I}(\boldsymbol{\theta}) \equiv \mathbf{T}^T\mathbf{M}\mathbf{T}$$

$$\mathbf{C}(\boldsymbol{\theta}, \dot{\boldsymbol{\theta}}_a) \equiv \mathbf{T}^T\mathbf{M}\dot{\mathbf{T}} + \frac{1}{2}(\mathbf{W}\mathbf{M} - \mathbf{M}\mathbf{W})\mathbf{T}$$

in which

$\mathbf{M}$: the $6r \times 6r$ matrix of system mass;

$\mathbf{T}$: the $n \times 6r$ twist-shaping matrix that maps the n-dimensional vector of actuated joint rates into the $6r$-dimensional vector of system twist $\mathbf{t}$;

$\mathbf{W}$: the $6r \times 6r$ matrix of system angular velocity.

10.5 For the system of Exercise 10.4, show that the matrix difference $\dot{\mathbf{I}}(\boldsymbol{\theta}, \dot{\boldsymbol{\theta}}_a) - 2\mathbf{C}(\boldsymbol{\theta}, \dot{\boldsymbol{\theta}}_a)$ is skew-symmetric. This is a well-known result for serial manipulators (Spong and Vidyasagar, 1989).

10.6 For the rolling robot with conventional wheels of Subsection 10.5.1, find the generalized inertia matrix of the robot under the maneuvers described below:

(a) pure translation;

(b) midpoint of segment O_1O_2 stationary.

In each case, give a physical interpretation of the matrix thus obtained.

10.7 With reference to the same robot of Exercise 10.6, state the conditions on its geometric parameters that yield $\mathbf{I}_w$ and $\mathbf{I}_p$ isotropic, these two 2×2 matrices having been defined in Subsection 10.5.1.

10.8 Find the maneuver(s) under which the Coriolis and centrifugal forces of the robot analyzed in Subsection 10.5.2 vanish. Note that in general, these forces do not vanish, even though the generalized inertia matrix of the robot is constant.

10.9 Find the eigenvalues and eigenvectors of the matrix of generalized inertia of the 3-dof rolling robot with omnidirectional wheels analyzed in Subsection 10.5.2.

10.10 The Euler-Lagrange equations derived for holonomic mechanical systems in Section 10.3, termed the *Euler-Lagrange equations of the second kind*, require that the generalized coordinates describing the system be *independent*. In nonholonomic mechanical systems, a set of kinematic constraints is not integrable, which prevents us from solving for dependent from independent generalized coordinates, the application of the Euler-Lagrange equations as described in that section thus not being possible. However, dependent generalized coordinates can be used if the *Euler-Lagrange equations of the first kind* are recalled. To this end, we let $\mathbf{q}$ be the m-dimensional vector of dependent generalized coordinates that are subject to p differential constraints of the form

$$\mathbf{A}(\mathbf{q})\dot{\mathbf{q}} = \mathbf{b}(\mathbf{q}, t)$$

where $\mathbf{A}$ is a $p \times n$ *matrix of constraints* and $\mathbf{b}$ is a p-dimensional vector depending on the configuration $\mathbf{q}$ and, possibly, on time explicitly. When $\mathbf{b}$ does not contain t explicitly, the constraints are termed *scleronomic*; otherwise, *rheonomic*. Furthermore, let $n \equiv m - p$ be the degree of freedom of the system. The Euler-Lagrange equations of the first kind of the system at hand take on the form

$$\frac{d}{dt}\left(\frac{\partial L}{\partial \dot{\mathbf{q}}}\right) - \frac{\partial L}{\partial \mathbf{q}} = \boldsymbol{\phi} + \mathbf{A}^T\boldsymbol{\lambda}$$

where $\boldsymbol{\lambda}$ is a p-dimensional vector of *Lagrange multipliers* that are chosen so as to satisfy the kinematic constraints. Thus, we regard the m dependent generalized coordinates grouped in vector $\mathbf{q}$ as independent, their constraints giving rise to the constraint forces $\mathbf{A}^T\boldsymbol{\lambda}$.

Use the Euler-Lagrange equations of the first kind to set up the mathematical model of the rolling robot with omnidirectional wheels studied in Subsection 10.5.1.

References

Albala, H., 1982, "Displacement analysis of the N-bar, single- loop, spatial linkage. Part I: Underlying mathematics and useful tables," and "Part II: Basic displacement equations in matrix and algebraic form," *ASME J. Mechanical Design* 104, pp. 504–519, 520–525.

Alizade, R.A., Duffy, J., Hajiyev, E.T., 1983, "Mathematical models for analysis and synthesis of spatial mechanism. Part IV: Seven-link spatial mechanisms," *Mechanism and Machine Theory* 18, no. 5, pp. 323–328.

Allen, P.K., Michelman, P., and Roberts, K.S., 1989, "An integrated system for dextrous manipulation," *Proc. IEEE Int. Conf. on Robotics & Automation* Scottsdale, pp. 612–617.

Altmann, S.L., 1989, "Hamilton, Rodrigues, and the quaternion scandal. What went wrong with mathematical discoveries in the nineteenth century," *Mathematics Magazine* 62, no. 5, pp. 291–308.

Angeles, J., 1982, *Spatial Kinematic Chains. Analysis, Synthesis, Optimization*, Springer-Verlag, Berlin.

Angeles, J., 1985, "On the numerical solution of the inverse kinematic problem," *The Int. J. Robotics Res.* 4, no. 2, pp. 21–37.

Angeles, J., 1988, *Rational Kinematics*, Springer-Verlag, New York.

Angeles, J., 1989, "On the use of invariance in robotics-oriented redundant sensing," *Proc. IEEE Int. Conf. on Robotics & Automation* Scottsdale, pp. 599–603.

Angeles, J., 1994, "On twist and wrench generators and annihilators," in Seabra Pereira, M.F.O. and Ambrósio, J.A.C. (editors), *Computer-Aided Analysis of Rigid and Flexible Mechanical Systems*, Kluwer Academic Publishers, Dordrecht-Boston-London.

Angeles, J., Hommel, G., and Kovács, P., 1993, *Computational Kinematics*, Kluwer Academic Publishers, Dordrecht.

Angeles, J. and Lee, S.S., 1988, "The formulation of dynamical equations of holonomic mechanical systems using a natural orthogonal complement," *ASME J. Applied Mechanics* 55, pp. 243–244.

Angeles, J. and Ma, O., 1988, "Dynamic simulation of n-axis serial robotic manipulators using a natural orthogonal complement," *The Int. J. Robotics Res.* 7, no. 5, pp. 32–47.

Angeles, J., Ma, O., and Rojas, A.A., 1989, "An algorithm for the inverse dynamics of n-axis general manipulators using Kane's formulation of dynamical equations," *Computers and Mathematics with Applications* 17, no. 12, pp. 1545–1561.

Angeles, J., Rojas, A.A., and López-Cajún, C.S., 1988, "Trajectory planning in robotics continuous-path applications," *IEEE J. Robotics & Automation* RA-4, pp. 380–385.

Angeles, J. and Etemadi Zanganeh, K., 1992, "The semigraphical determination of all inverse kinematic solutions of general six-revolute manipulators," *Proc. 9th CISM-IFToMM Symp.on Theory and Practice of Robots and Manipulators*, Sept. 1–4, Udine, pp. 23–32.

Balafoutis, C.A. and Patel, R.V., 1991, *Dynamic Analysis of Robot Manipulators: A Cartesian Tensor Approach*, Kluwer Academic Publishers, Dordrecht.

Borenstein, J., 1993, "Multi-layered control of a four-degree-of-freedom mobile robot with compliant linkage," *Proc. IEEE Int. Conf. on Robotics & Automation* Atlanta, pp. 7–12.

Brand, L., 1965, *Advanced Calculus*, John Wiley & Sons, Inc., New York.

Bricard, R., 1927, *Leçons de Cinématique*, Gauthier-Villars, Paris.

Bryson, Jr., A.E. and Ho, Y-C., 1975, *Applied Optimal Control. Optimization, Estimation and Control*, Hemisphere Publishing Corporation, Washington.

Burdick, J.W., 1995, "A classification of 3R reginal manipulator singularities and geometries," *Mechanism and Machine Theory* 30, no. 1, pp. 71–89.

Ceccarelli, M., 1995, "Screw axis defined by Giulio Mozzi in 1763," *Proc. Ninth World Congress on the Theory of Machines and Mechanisms*, August 29–September 2, vol. 4, pp. 3187–3190.

Ceccarelli, M., 1996, "A formulation for the workspace boundary of general n-revolute manipulators," *Mechanism and Machine Theory* 31, no. 5, pp. 637–646.

Charentus, S. and Renaud, M., 1989, *Calcul du modèle géometrique directe de la plate-forme de Stewart*, LAAS Report # 89260, Laboratoire d'Automatique et d'Analyse de Systèmes, Toulouse.

Chasles, M., 1830, "Notes sur les propriétés générales de deux corps semblables entr'eux et placés d'une manière quelconque dans l'espace, et sur le déplacement fini ou infiniment petit d'un corps solide libre," *Bull. Sci. Math. Ferrusaac* 14, pp. 321–326.

Cheng, H. and Gupta, K.C., 1989, "An historical note on finite rotations," *ASME J. Applied Mechanics* 56, pp. 139–142.

Chiang, C.H., 1988, *Kinematics of spherical Mechanisms*, Cambridge University Press, Cambridge.

Chou, K.C.J. and Kamel, M., 1988, "Quaternions approach to solve the kinematic equation of rotation, $A_a A_x = A_x A_b$, of a sensor-mounted robotic manipulator," *Proc. IEEE Int. Conf. on Robotics & Automation* Philadelphia, pp. 656–662.

Chou, K.C.J. and Kamel, M., 1991, "Finding the position and orientation of a sensor on a robot manipulator using quaternions," *The Int. J. Robotics Res.* 10, no. 3, pp. 240–254.

Clavel, R., 1988, "Delta, a fast robot with parallel geometry," *Proc. 18th Int. Symposium on Industrial Robots*, Lausanne, pp. 91–100.

Comerford, R., 1994, "Mecha...what?," *IEEE Spectrum*, August, pp. 46–49.

Coriolis, G.G., 1835. "Mémoire sur les équations du mouvement relatif des systèmes des corps," *J.Ecole Polytechnique* 15, cahier 24, pp. 142–154.

Craig, J.J., 1989, *Introduction to Robotics: Mechanics and Control*, 2nd Edition, Addison-Wesley Publishing Company, Inc., Reading.

Currie, I.G., 1993, *Fundamental Mechanics of Fluids*, 2nd Edition, McGraw-Hill Book Company, New York.

Dahlquist, G. and Björck, Å., 1974, *Numerical Methods*, Prentice-Hall, Inc., Englewood Cliffs.

De Luca, A. and Oriolo, G., 1995, "Modelling and control of nonholonomic mechanical systems," in Angeles, J. and Kecskeméthy, A. (editors), *Kinematics and Dynamics of Multi-Body Systems*, Springer-Verlag, Vienna-New York, pp. 277–342.

Denavit, J. and Hartenberg, R.S., 1955, "A kinematic notation for lower-pair mechanisms based on matrices," *ASME J. Applied Mechanics* 77, pp. 215–221.

Denavit, J. and Hartenberg, R.S., 1964. *Kinematic Synthesis of Linkages*, McGraw-Hill Book Company, New York.

Dierckx, P., 1993, *Curve and Surface Fitting with Splines*, Clarendon Press, Oxford-New York-Tokyo.

Duffy, J., and Derby, S., 1979, "Displacement analysis of a spatial 7R mechanism—A generalized lobster's arm," *ASME J. Mechanical Design* 101, pp. 224–231.

Duffy, J., 1980, *Analysis of Mechanisms and Robot Manipulators*, Edward Arnold, London.

Duffy, J. and Crane, C., 1980, "A displacement analysis of the general spatial 7-link, 7R mechanism," *Mechanism and Machine Theory* 15, no. 3, pp. 153–169.

Etter, D.M., 1993, *Engineering Problem Solving with Matlab*, Prentice-Hall, Inc., Englewood Cliffs.

Euler, L., 1776, "Nova methodus motum corporum rigidorum determinandi," *Novii Comentarii Academiæ Scientiarum Petropolitanæ*, Vol. 20 (1775) 1776, pp. 208–238 = *Opera Omnia* Vols. 2–9, pp. 99–125.

Everett, J.D., 1875, "On a new method in statics and kinematics," *Messenger of Mathematics* 45, pp. 36–37.

Featherstone, R., 1983, "Position and velocity transformations between robot end-effector coordinates and joint angles," *The Int. J. Robotics Res.* 2, no. 2, pp. 35–45.

Featherstone, R., 1987, *Robot Dynamics Algorithms*, Kluwer Academic Publishers, Dordrecht.

Foley, J.D. and Van Dam, A., 1982, *Fundamentals of Interactive Computer Graphics*, Addison-Wesley Publishing Company, Inc., Reading.

Ge, Q.J. and Kang, D., 1995, "Motion interpolation with G^2 composite Bézier motions," *ASME J. Mechanical Design* 117, pp. 520–525.

Gear, C.W., 1971, *Numerical Initial Value Problems in Ordinary Differential Equations*, Prentice-Hall, Inc., Englewood Cliffs.

Golub, G.H. and Van Loan, C.F., 1989, *Matrix Computations*, The Johns Hopkins University Press, Baltimore.

Gosselin, C.M., Perreault, L., and Vaillancourt, C., 1995, "Simulation and computer-aided kinematic design of three-degree-of-freedom spherical parallel manipulators," *J. Robotic Systems* 12, no. 12, pp. 857–869.

Gosselin, C.M., Sefrioui, J., and Richard, M.J., 1992, "Solutions polynomiales au problème de la cinématique directe des manipulateurs parallèles plans à trois degrés de liberté," *Mechanism and Machine Theory* 27, no. 2, pp. 107–119.

Gosselin, C.M., Sefrioui, J., and Richard, M.J., 1994a, "On the direct kinematics of spherical three-degree-of-freedom parallel manipulators with a coplanar platform," *ASME J. Mechanical Design* 116, pp. 587–593.

Gosselin, C.M., Sefrioui, J., and Richard, M.J., 1994b, "On the direct kinematics of spherical three-degree-of-freedom parallel manipulators of general architecture," *ASME J. Mechanical Design* 116, pp. 594–598.

Gough, V.M., 1956–1957, "Communications", *Proc. Automobile Division of The Institution of Mechanical Engineers*, pp. 166–168. See also "Discussion in London", *Proc. Automobile Division of The Institution of Mechanical Engineers*, pp. 392–394.

Graham, D., 1972, "A behavioural analysis of the temporal organisation of walking movements in the 1st instar and adult stick insect (Carausius morosus)," *J. Comp. Physiol.* 81, pp. 23–25.

Halmos, P., 1974, *Finite-Dimensional Vector Spaces*, Springer-Verlag, New York.

Hamilton, W.R., 1844, "On quaternions: or a new system of imaginaries in algebra," *Phil. Mag.*, 3rd. ser. 25, pp. 489–495.

Harary, F., 1972, *Graph Theory*, Addison-Wesley Publishing Company, Inc., Reading.

Hayward, V., 1994, "Design of hydraulic robot shoulder based on combinatorial mechanism," in Yoshikaswa, T. and Miyazaki, F. (eds.), *Experimental Robotics 3,* Lecture Notes in Control and Information Sciences 200, Springer-Verlag, pp. 297–310. *The Int. J. Robotics Res.* 10, no. 5, pp. 767–790.

Henrici, P., 1964, *Elements of Numerical Analysis*, John Wiley & Sons, Inc., New York.

Hervé, J.M. and Sparacino, F., 1992, "Star, a new concept in robotics," *Proc. 3rd International Workshop on Advances in Robot Kinematics*, September 7–9, Ferrara, pp. 176–183.

Hirose, S., Masui, T., and Kikuchi, H., 1985, "Titan III: A quadruped walking vehicle," in Hanafusa, H. and Inoue, X. (editors), *Robotics Research* 2, pp. 325–331.

Hollerbach, J.M., 1980, "A recursive Lagrangian formulation of manipulator dynamics and a comparitive study of dynamic formulation complexity," *IEEE Trans. Systems, Man, and Cybernetics* SMC-10, no. 11, pp. 730–736.

Hollerbach, J.M. and Sahar, G., 1983, "Wrist-partitioned inverse kinematic accelerations and manipulator dynamics," *The Int. J. Robotics Res.* 2, no. 4, pp. 61–76.

Horaud, R. and Dornaika, F., 1995, "Hand-eye calibration," *The Int. J. Robotics Res.* 14, no. 3, pp. 195–210.

Horn, B.K.P., 1986, *Robot Vision*, The MIT Press, Cambridge (Mass.)

Hoschek, J. and Lasser, D., 1992, *Grundlagen der geometrischen Datenverarbeitung*, B.G. Teubner, Stuttgart.

Hughes, P.C., Sincarsin, W.G., and Carroll, K.A., 1991, "Trussarm—A variable-geometry-truss manipulator," *J. of Intell. Mater. Syst. and Struct.* 2, pp. 148–160.

Husty, M., 1996, "An algorithm for solving the direct kinematics of general Stewart-Gough platforms," *Mechanism and Machine Theory* 31, no. 4, pp. 365–380.

Isidori, A., 1989, *Nonlinear Control Systems*, 2nd Edition, Springer-Verlag, Berlin.

Jacobsen, S.C., Iversen, E.K., Knutti, D.F., and Biggers, K.B., 1984, "The Utah/MIT dextrous hand: Work in progress," *The Int. J. Robotics Res.* 3, no. 4, pp. 21–50.

Jacobsen, S.C., Iversen, E.K., Knutti, D.F., Johnson, R.T., and Biggers, K.B., 1986, "Design of the Utah/MIT dextrous hand," *Proc. IEEE Int. Conf. on Robotics & Automation* San Francisco, pp. 1520–1532.

Kahaner, D., Moler, C., and Nash, S., 1989, *Numerical Methods and Software*, Prentice-Hall, Inc., Englewood Cliffs.

Kahn, M. E. 1969. "The near-minimum time control of open-loop articulated kinematic chains," AI Memo 177, Artificial Intelligence Laboratory, Stanford University, Stanford. Also see Kahn, M.E. and Roth, B., 1971, "The near-minimum-time control of open-loop articulated kinematic chains," *ASME J. Dyn. Systs., Meas., and Control*, 91, pp. 164–172.

Kane, T.R., 1961, "Dynamics of nonholonomic systems," *ASME J. Applied Mechanics* 83, pp. 574–578.

Kane, T.R. and Levinson, D.A., 1983, "The use of Kane's dynamical equations in robotics," *The Int. J. Robotics Res.* 2, no. 3, pp. 3–21.

Khalil, W., Kleinfinger, J.F., and Gautier, M., 1986, "Reducing the computational burden of the dynamical models of robots," *Proc. IEEE Int. Conf. on Robotics & Automation* San Francisco, pp. 525–531.

Killough, S.M. and Pin, F.G., 1992, "Design of an omnidirectional and holonomic wheeled platform prototype," *Proc. IEEE Int. Conf. on Robotics & Automation* Nice, pp. 84–90.

Klein, C.A., Olson, K.W., and Pugh, D.R., 1983, "Use of force and attitude sensors for locomotion of a legged vehicle over irregular terrain," *The Int. J. Robotics Res.* 2, no. 2, pp. 3–17.

Kumar A.V. and Waldron K.J., 1981, "The workspace of a mechanical manipulator," *ASME J. Mechanical Design* 103, pp. 665–672.

Latombe, J.-C., 1991, *Robot Motion Planning*, Kluwer Academic Publishers, Boston-Dordrecht-London.

Lazard, D. and Merlet, J.-P., 1994, "The (true) Stewart platform has 12 configurations," *Proc. IEEE Int. Conf. on Robotics & Automation* San Diego, pp. 2160–2165.

Lee, H.Y. and Liang, C.G., 1988, "Displacement analysis of the general spatial seven-link 7R mechanism," *Mechanism and Machine Theory* 23, no. 3, pp. 219–226.

Lee, H.-Y. and Roth, B., 1993, "A closed-form solution of the forward displacement analysis of a class of in-parallel mechanisms," *Proc. IEEE Int. Conf. on Robotics & Automation* Atlanta, pp. 720–731.

Li, H., Woernle, C., and Hiller, M., 1991, "A complete solution for the inverse kinematic problem of the general 6R robot manipulator," *ASME J. Mechanical Design* 113, pp. 481–486.

Leigh, D.C., 1968, *Nonlinear Continuum Mechanics*, McGraw-Hill Book Co., New York.

Levine, M.D., 1985, *Vision in Man and Machines*, McGraw-Hill Book Co., New York.

Li, H., 1990, *Ein Verfahren zur vollständigen Lösung der Rückwärtstransformation für Industrieroboter mit allgemeiner Geometrie*, doctoral thesis, Universität-Gesamthochschule Duisburg, Duisburg.

Liu, H., Iberall, T., and Bekey, G., 1989, "The multi-dimensional quality of task requirements for dextrous robot hand control," *Proc. IEEE Int. Conf. on Robotics & Automation* Scottsdale, pp. 452–457.

Lozano-Pérez, T., 1981, "Automatic planning of manipulator transfer movements," *IEEE Trans. Systems, Man, and Cybernetics* SMC-11, no. 110, pp. 681–689.

Luh, J.Y.S., Walker, M.W., and Paul, R.P.C., 1980, "On-line computational scheme for mechanical manipulators," *ASME J. Dyn. Systs., Meas., and Control*, 102, pp. 69–76.

Mavroidis, C. and Roth, B., 1992, "Structural parameters which reduce the number of manipulator configurations," *Proc. ASME 22nd Biennial Mechanisms Conference*, September 13–16, Scottsdale, DE-45, pp. 359–366.

Menitto, G. and Buehler, M., 1996, "CARL: A compliant articulated robotic leg for dynamic locomotion," to appear in *Robotics and Autonomous Systems.*

Merlet, J.-P., 1989, "Singular Configurations of Parallel Manipulators and Grassmann Geometry," *The Int. J. Robotics Res.* 8, no. 5, pp. 45–56.

Merlet, J.-P., 1990, *Les Robots Parallèles*, Hermès Publishers, Paris.

Merlet, J.-P., 1991, "An algorithm for the forward kinematics of general parallel manipulators," *Proc. Fifth Int. Conference on Advanced Robotics*, June 19–22, Pisa, pp. 1136–1140.

Morgan, A.P., 1987, *Solving Polynomial Systems Using Continuation for Scientific and Engineering Problems*, Prentice-Hall, Inc., Englewood Cliffs.

Mozzi, G., 1763, *Discorso Matematico Sopra il Rotamento Momentaneo dei Corpi*, Stamperia di Donato Campo, Naples.

Nanua, P., Waldron, K.J., and Murthy, V., 1990, "Direct Kinematic solution of a Stewart platform," *IEEE Trans. Robotics & Automation* 6, pp. 438–444.

Pattee, H.A., 1995, "Selecting computer mathematics," *Mechanical Engineering*, 117, no. 9, pp. 82–84.

Paul, R.P., 1981, *Robot Manipulators. Mathematics, Programming, and Control*, The MIT Press, Cambridge (Mass.)

Paul, R.P. and Stevenson, C.N., 1983, "Kinematics of robot wrists," *The Int. J. Robotics Res.* 2, no. 1, pp. 31–38.

Penrose, R., 1994, *Shadows of the Mind. A Search for the Missing Science of Consciouseness*, Oxford University Press, Oxford.

Pettinato, J.S. and Stephanou, H.E., 1989, "Manipulability and stability of a tentacle based robot manipulator," *Proc. IEEE Int. Conf. on Robotics & Automation* Scottsdale, pp. 458–463.

Pfeiffer, F., Eltze, J., and Weidemann, H.-J., 1995, "The TUM walking machine," *Intelligent Automation and Soft Computing. An International Journal*, 1, no. 3, pp. 307–323.

Phillips J., 1990, *Freedom in Machinery. Vol 2: Screw Theory Exemplified*, Cambridge University Press, Cambridge.

Pieper, D.L., 1968, *The Kinematics of Manipulators Under Computer Control*, Ph.D. thesis, Stanford University, Stanford.

Pierrot, F., Fournier, A., and Dauchez, P., 1991, "Towards a fully-parallel 6 dof robot for high-speed applications," *Proc. IEEE Int. Conf. on Robotics & Automation* Sacramento, pp. 1288–1293.

Primrose, E.J.F., 1986, "On the input-output equation of the general 7R mechanism," *Mechanism and Machine Theory* 21, no. 6, pp. 509–510.

Raghavan, M., 1993, "The Stewart platform of general geometry has 40 configurations," *ASME J. Mechanical Design* 115, pp. 277–282.

Raghavan, M. and Roth, B., 1990, "Kinematic analysis of the 6R manipulator of general geometry," in Miura, H. and Arimoto, S. (editors), *Proc. 5th Int. Symposium on Robotics Research*, MIT Press, Cambridge (Mass.)

Raghavan, M. and Roth, B., 1993, "Inverse kinematics of the general 6R manipulator and related linkages," *ASME J. Mechanical Design* 115, pp. 502–508.

Raibert, M.H., 1986, *Legged Robots That Balance*, MIT Press, Cambridge (Mass.)

Ranjbaran, F., Angeles, J., and Patel, R.V., 1992, "The determination of the Cartesian workspace of general three-axes positioning manipulators", *Proc. Third Int. Workshop on Advances in Robot Kinematics*, Sept.7–9, Ferrara, pp. 318–324.

Ravani, B. and Roth, B., 1984, "Mappings of spatial kinematics," *ASME J. Mechanisms, Transm., and Auto. in Design* 106, pp. 341–347.

Reynaerts, D., 1995, *Control Methods and Actuation Technology for Whole-Hand Dextrous Manipulation*, doctoral dissertation, Fakulteit der Torgepaste Wetenschappen, Katholieke Universiteit Leuven, Leuven.

Roth, B., 1984, "Screws, motors, and wrenches that cannot be bought in a hardware store," in Brady, M. and Paul, R.P. (editors), *Robotics Research. The First International Symposium*, The MIT Press, Cambridge (Mass), pp. 679–693.

Rumbaugh, J., Blaha, M., Premerlani, W., Eddy, F., and Lorensen, W., 1991, *Object-Oriented Modeling and Design*, Prentice-Hall, Inc., Englewood Cliffs.

Rus, D., 1992, "Dextrous rotations of polyhedra," *Proc. IEEE Int. Conf. on Robotics & Automation* Nice, pp. 2758–2763.

Russell, Jr., M., 1983, "Odex I: The first functionoid," *Robotics Age*, 5, pp. 12–18.

Salisbury, J.K. and Craig, J.J., 1982, "Articulated hands: Force and kinematic issues," *The Int. J. Robotics Res.* 1, no. 1, pp. 4–17.

Salmon, G., *Higher Algebra*, 5th Edition (1885), reprinted: 1964, Chelsea Publishing Co., New York.

Samson, C., Le Borgne, M., and Espiau, B., 1991, *Robot Control. The Task Function Approach*, Clarendon Press, Oxford.

Shampine, L.F. and Gear, C.W., 1979, "A user's view of solving stiff ordinary differential equations," *SIAM Review* 21, no. 1, pp. 1–17.

Shiu, Y.C. and Ahmad, S., 1987, "Finding the mounting position of a sensor by solving a homogeneous transform equation of the form AX = XB," *Proc. IEEE Int. Conf. on Robotics & Automation* Raleigh, pp. 1667–1671.

Silver, W.M., 1982, "On the equivalence of Lagrangian and Newton-Euler dynamics for manipulators," *The Int. J. Robotics Res.* 1, no. 2, pp. 118–128.

Song, S.-M. and Waldron, K.J., 1989, *Machines That Walk*, The MIT Press, Cambridge (Mass.)

Soureshi, R., Meckl, P.H., and Durfee, W.M., 1994, "Teaching MEs to use microprocessors," *Mechanical Engineering*, 116, no. 4, pp. 71–76.

Spong, M.W. and Vidyasagar, M.M., 1989, *Robot Dynamics and Control*, John Wiley & Sons, Inc., New York.

Stewart, D., 1965, "A platform with 6 degrees of freedom," *Proc. Institution of Mechanical Engineers* 180, part 1, no. 15, pp. 371–386.

Strang, G., 1988, *Linear Algebra and Its Applications*, Third Edition, Harcourt Brace Jovanovich College Publishers, New York.

Stroustrup, B., 1991, *The C++ Programming Language*, Addison-Wesley Publishing Company, Reading.

Sutherland, I.E. and Ullner, M.K., 1984, "Footprints in the asphalt," *The Int. J. Robotics Res.* 3, no. 2, pp. 29–36.

Synge, J.L., 1960, "Classical Dynamics," in Flügge, S. (editor), *Encyclopedia of Physics*, Vol. III/1, Springer-Verlag, Berlin-Göttingen-Heidelberg, pp. 1–225.

Takano, M., 1985, "A new effective solution for inverse kinematics problem (synthesis) of a robot with any type of configuration," *Journal of the Faculty of Engineering*, The University of Tokyo 38, no. 2, pp. 107–135.

Tandirci, M., Angeles J., and Ranjbaran, F., 1992, "The characteristic point and the characteristic length of robotic manipulators," *ASME 22nd Biennial Mechanisms Conference*, September 13–16, Scottsdale, vol. 45, pp. 203–208.

Tandirci, M., Angeles, J., and Darcovich, J., 1994, "On rotation representations in computational robot kinematics," *J. Intelligent and Robotic Systems* 9, pp. 5–23.

Teichmann, M., 1995, *Grasping and Fixturing: A Geometric Study and Implementation*, Ph.D. thesis, Department of Computer Science, New York University, New York.

Thompson, S.E. and Patel, R.V., 1987, "Formulation of joint trajectories for industrial robots using B-splines," *IEEE Trans. Industrial Electronics* IE-34, pp. 192–199.

Todd, D.J., 1985, *Walking Machines. An Introduction to Legged Robots*, Kogan Page Ltd., London.

Tsai, L.W., and Morgan, A.P., 1985, "Solving the kinematics of the most general six- and five-degree-of-freedom manipulators by continuation methods," *ASME J. Mechanisms, Transm., and Auto. in Design* 107, pp. 189–200.

Uicker, Jr., J.J.1965. *On the Dynamic Analysis of Spatial Linkages Using* 4×4 *Matrices*, Ph.D. thesis, Northwestern University, Evanston.

Van Brussel, H., Santoso, B., and Reynaerts, D., 1989, "Design and control of a multi-fingered robot hand provided with tactile feedback," *Proc. NASA Conf. Space Telerobotics*, Pasadena, Jan.31–Feb.2, Vol. III, pp. 89–101.

Vijaykumar, R., Tsai, M.J., and Waldron, K.J., 1986, "Geometric optimization of serial chain manipulator structures for working volume and dexterity," *The Int. J. Robotics Res.* 5, no. 2, pp. 91–103.

Vinogradov I.B., Kobrinski A.E., Stepanenko, Y.E., and Tives, L.T., 1971, "Details of kinematics of manipulators with the method of volumes," (in Russian), *Mekhanika Mashin*, no. 27–28, pp. 5–16.

von Mises, R., 1924, "Motorrechnung, ein neues Hilfsmittel der Mechanik," *Z. Angewandte Mathematik und Mechanik* 4, no. 2, pp. 155–181.

von Mises, R., 1996, "Motor calculus. A new theoretical device for mechanics," translated from the German by J. E. Baker and K. Wohlhart, Institute for Mechanics, Technical University of Graz, Graz.

Vukobratovic, M. and Stepanenko, J., 1972, "On the stability of anthropomorphic systems," *Math. Biosci.* 15, pp. 1–37.

Wälischmiller, W. and Li, H., 1996, "Development and application of the TELBOT System–A new Tele Robot System," *Proc. 11th CISM–IFToMM Symposium on Theory and Practice of Robots and Manipulators–Ro.Man.-Sy. 11*, Udine (Italy), July 1–4 (in press.)

Walker, M.W. and Orin, D.E., 1982, "Efficient dynamic computer simulation of robotic mechanisms," *ASME J. Dyn. Systs., Meas., and Control*, 104, pp. 205-211.

Wampler, C., 1996, "Forward displacement analysis of general six-in-parallel SPS (Stewart) platform manipulators using soma coordinates," *Mechanism and Machine Theory* 31, no. 3, pp. 331–337.

West, M. and Asada, H., 1995, "Design and control of ball wheel omnidirectional vehicles," *Proc. IEEE Int. Conf. on Robotics & Automation* Nagoya, pp. 1931–1938.

Whitney, D.E., 1972, "The mathematics of coordinated control of prosthetic arms and manipulators," *ASME J. Dyn. Systs., Meas., and Control*, 94, pp. 303–309.

Williams, O., Angeles, J., and Bulca, F., 1993, "Design philosophy of an isotropic six-axis serial manipulator," *Robotics and Computer-Aided Manufacturing* 10, no. 4, pp. 275–286.

Yang, D.C. and Lai, Z.C., 1985, "On the conditioning of robotic manipulators—service angle," *ASME J. Mechanisms, Transm., and Auto. in Design* 107, pp. 262–270.

Yoshikawa, T., 1985, "Manipulability of robotic mechanisms," *The Int. J. Robotics Res.* 4, no. 2, pp. 3–9.

Index

Acatastatic systems, 395
acceleration analysis
  of parallel manipulators, 329
  of rigid bodies, 91
  of serial manipulators, 158
affine transformation, 21
AI (see artificial intelligence)
algorithm definition, 448
angle of rotation, 30
angular acceleration
  computation, 282
  invariant-rate relations, 93
  matrix, 91
  vector, 92
angular velocity
  dyad, 215
  invariant-rate relations, 90, 431–435
  matrix, 83
  vector, 83
Appendix A, 429
Appendix B, 437
arc-welding, 356, 375
  operation, 369
  path tracking, 386
architecture of a manipulator, 109
articulated-body method, 247
artificial intelligence, 4, 448
ASEA-IRB 1000 robot, 469
axial component of a vector, 23
axial vector of a 3×3 matrix, 34

Base frame, 115
basis of a vector space, 23
Bezout's method, 327

C, 448
C++, 448
canonical form of a rotation, 33
Carausius morosus, 15, 497
Cartesian coordinates of a manipulator, 114
Cartesian decomposition, 34
caster wheel, 346
catastatic systems, 395
Cayley-Hamilton theorem, 28
Cayley's theorem, 450
change of basis, 58
characteristic equation, 25, 28, 122
characteristic length, 183
characteristic polynomial, 155, 291

Chasles' theorem (see Mozzi-Chasles' theorem)
Chebyshev norm, 177, 386
Cholesky-decomposition algorithm, 249, 265
closure equations of a manipulator, 116
compatibility conditions
  for acceleration, 283
  for velocity, 279
composite rigid-body method, 247
composition of reflections and rotations, 47
condition number, 177, 303
configuration of a manipulator, 109
continuous path, 192
  operations, 354
  tracking, 380
control vector, 268
coordinate transformation, 48–53
CP (see continuous path)
Coriolis acceleration, 95
Coriolis and centrifugal forces, 240, 247
Couette flow, 272
Coulomb
  dissipation function, 273
  friction, 273
cross-product matrix, 30
curvature, 357
  derivative w. r. t. a parameter, 363
  derivative w. r. t. the arc length, 357
  parametric representation, 363, 365
  time-derivative, 360
cycloidal motion, 201

Darboux vector, 360
  time-derivative, 360
decoupled manipulators, 118
delta-array (Δ-array), 419, 485
Delta robot, 10
Denavit-Hartenberg
  frames, 107
  notation, 107
  parameters, 109
  rotation matrix, 111, 115
  vector joining two frame origins, 112, 115
dexterity, 448
  measures (see kinetostatic performance indices)
dextrous hands (see multifingered hands)
dextrous workspace, 174
dialytic elimination, 157, 468
DH (see Denavit-Hartenberg)
DIESTRO manipulator, 188, 305, 307
  inverse kinematics, 305
  Jacobian, 465
differentiation with respect to vectors, 29
direct kinematics problem of parallel manipulators, 317
displacement equations of a manipulator, 116
dissipation function, 217, 272
dynamic systems, 1
dynamics
  of holonomic systems, 394, 395
  of multibody systems, 215
  of parallel manipulators, 398
  of rigid bodies, 100
  of robotic mechanical systems, 393
  of rolling robots, 409
  of serial manipulators, 213

EE (see end-effector)
end-effector, 107
Euclidean norm, 30
Euler
  angles, 46, 451
  equation (for graphs), 399
  equation (in mechanics), 102
  parameters (see Euler-Rodrigues parameters)
Euler-Lagrange equations, 216
  derived with the NOC, 240, 398
Euler-Rodrigues parameters, 44
Euler's
  theorem, 27

formula for graphs (see Euler equation for graphs)

Fanuc Arc Mate robot
characteristic length, 190
DH parameters, 189
inverse kinematics, 304
KCI, 190
First Law of Thermodynamics, 163
flight simulator, 9, 318
floating-point operation, 161, 247, 448
flop (see floating-point operation)
forward dynamics
of serial manipulators, 246
algorithm (using the NOC), 264
algorithm complexity, 264
Frenet (see Frenet-Serret)
Frenet-Serret
frame, 356
formulas, 357
vectors, 357
friction forces, 217, 240, 271
fuzzy logic, 448

Genealogy of robotic mechanical systems, 3, 4
generalized coordinates, 394
generalized forces, 216, 418, 428
generalized inertia matrix, 219
Cholesky decomposition, 249
factoring, 249
time-rate of change, 219, 222
generalized speeds, 217, 395
gluing operation, 365
grasping matrix, 338
gravity
terms, 236, 270
wrench, 252

Hand-eye calibration, 68
Hexa robot, 10
holonomic systems, 393, 395
homeomorphism, 21
homogeneous coordinates, 54
homotopy, 291

IKP (see inverse kinematics problem)
Index, 505
input, 1, 237
inertia dyad, 103, 215
instant-screw axis, 85
instrument calibration, 66
intelligent machines, 3, 448
intelligent robot, 3
invariance, 63
inverse dynamics
of serial manipulators, 213
recursive, 225
inverse kinematics problem
of a decoupled manipulator, 118
of a general 6R manipulator, 290
of parallel manipulators, 317
inverse vs. forward dynamics, 213
inward recursions, 232, 234
ISA (see instant-screw axis)
isomorphism, 25
isotropic
manipulator, 179
matrix, 176
isotropy, 176
iteration, 381, 448

Jacobian matrix
condition number, 175
evaluation, 147
invertibility, 175
of decoupled manipulators, 145
of serial manipulators, 142
transfer formula, 145
joint
coordinates, 105, 109
parameters, 109
variables, 109

Kane's method, 237
KCI (see kinematic conditioning index)
kernel of a linear transformation, 21
Kinemate, 89
kinematic
chain, 106
conditioning index, 178

constraints, 237
constraints for serial manipulators, 242
pair, 106
kinetostatic performance indices, 174

Lee's manipulator, 307
Lee's procedure, 310
Lee vs. Li, 295
left hand, 10
Li vs. Lee (see Lee vs. Li)
Li's manipulator (see Lee's manipulator)
linear invariants, 34
of a rotation, 35
linear transformations, 20
local structure of a manipulator, 109
lower kinematic pair, 106
LU decomposition, 143

Machine (definition of), 447
main gauche (see left hand)
maneuverability, 483
manipulability, 174
of decoupled manipulators, 471
manipulator
angular velocity matrix, 218
dynamics, 213, 393
kinematics, 105
mass matrix, 218
twist, 218
wrenches, 218
matrix representation, 24
mechanical system, 1
mechatronics, 448
minimum-time trajectory, 241
moment invariants, 64
moment of a line
about a point, 77
about another line, 458
moment of inertia, 102
momentum screw, 103
motor, 89
Mozzi-Chasles' theorem, 73
MSS, 5
multibody system
dynamics, 215
Euler-Lagrange equations, 216
multicubic expression, 117
multifingered hands, 11, 336
multilinear expression, 117
multiquadratic expression, 117
multiquartic expression, 117

Natural orthogonal complement, 236
applied to holonomic systems, 395
applied to parallel manipulators, 398
applied to planar manipulators, 251
applied to rolling robots, 409
Newton
equations, 102
-Euler algorithm, 225
methods, 291, 381
-Raphson method, 69, 381
NOC (see natural orthogonal complement)
nonholonomic systems, 214, 394
noninertial base link, 246
normal component of a vector, 23
nullspace of a linear transformation, 21
numerical conditioning, 303

Object-oriented programming, 448
Odetics series of hexapods, 15
ODW (see omnidirectional wheels)
off-line, 4, 122, 148
omnidirectional wheels, 17
dynamics, 419
kinematics, 350
on-line, 448
operation point, 109
orientation problem, 134
orthogonal complement, 239
orthogonal decomposition of a vector, 23
orthogonal decoupled manipulator, 126

orthogonal projection, 21
orthogonal RRR manipulator
  dynamics, 255
  inverse kinematics, 130, 133
  recursive dynamics, 260
  workspace, 155
OSU ASV, 15
OSU hexapod, 15
outward recursions, 226

Pappus-Guldinus theorems, 469
parallel axes, theorem of, 102
parallel manipulators, 8
  acceleration analysis, 331
  dynamics, 398
  kinematics, 315
  velocity analysis, 329
parametric
  path representation, 362
  representation of curvature, 363
  representation of curvature derivative, 363
  representation of torsion, 363
  representation of torsion derivative, 363
  splines, 375
path-tracking for arc-welding, 386
pick-and-place operations, 191
planar manipulators, 164
  acceleration analysis, 171
  displacement analysis, 165
  dynamics, 251
  static analysis, 173
  velocity analysis, 168
platform manipulators, 8, 315, 399
Plücker coordinates (of a line), 76
  transfer formula, 78, 79
polar-decomposition theorem, 175
polynomial interpolation
  with 3-4-5 polynomial, 195
  with 4-5-6-7 polynomial, 198
pose (of a rigid body), 80
  array, 80
positioning problem, 119
posture of a manipulator, 109
PPO (see pick-and-place operations)
Principle of Virtual Work, 163
prismatic pair, 106, 142, 227
programmable robot, 4
projection, 21
Puma robot, 108, 126
  DH parameters, 109
  inverse kinematics, 126
  workspace, 155
pure reflection, 22

Quaternions, 46

Raghavan and Roth's procedure, 308
range of a linear transformation, 21, 449
Rayleigh dissipation function (see dissipation function)
real time, 213, 448
reciprocal bases, 68, 146, 408
redundant sensing, 66
reflection, 22, 25, 295
  composition with rotations, 47
regional structure of a manipulator, 109
revolute pair, 106
rheonomic systems, 394
robotic hands, 11
  statics, 336
robotic mechanical system, viii, 3
robotic system, 2
Rodrigues (see Euler-Rodrigues)
  vector, 451
rolling robots, 16
  dynamics, 409
  kinematics, 345
rotations, 25
rotation matrix, 30
  exponential representation, 32
RVS, xiii, 193, 368
run-time, 448
Runge-Kutta method, 269

Scleronomic systems, 394
screw
  amplitude, 74, 85, 98
  axis, 74, 85, 97

axis coordinates, 89, 98
motion, 74
pitch, 74, 85, 97
ray coordinates, 89
semigraphical solution of the general IKP, 292
serial manipulators, 6
acceleration, 158
dynamics, 213
kinematics, 105
statics, 162, 173
velocity analysis, 139
workspace, 155
service angle, 174
similarity transformations, 58
simulation, 268
singular-value decomposition, 69, 176
singular values, 176
singularity analysis of decoupled manipulators, 151
SPDM, 5
spherical wrist, 118, 134
workspace, 136
spline(s), 205
natural, 209
nonparametric, 205, 375
parametric, 375
periodic, 209
interpolation of 4-5-6-7 polynomial, 210
square root of a matrix, 44
Star robot, 10
state
variable, 1, 268
-variable equations, 268
-variable model of platform manipulators, 405
vector, 268
static analysis
of rigid bodies, 95
of serial manipulators, 162, 173
Stewart platform (see Stewart-Gough platform)
Stewart-Gough platform, x, 318
direct kinematics, 320
leg kinematics, 318
structure of mechanical systems, 2
structured environment, 3
Sutherland, Sprout & Assoc. Hexapod, 15
system, 1

Telemanipulators, 5
tensors, 19, 215, 237
Titan series of quadrapeds, 15
torsion, 357
derivative w. r. t. the arc length, 357
parametric representation, 363, 365
time-derivative, 360
trace of a square matrix, 34
trajectories with via poses, 203
trajectory planning, 191, 355
trapezoidal velocity profile, 471
truncation error, 269
Trussarm, 10
TU Munich Hand, 12
TU Munich Hexapod, 15
twist
of a rigid body, 88
transfer formula, 91
twist-shape relations, 239
for serial manipulators, 242

Unimodular group (of matrices), 79
unstructured environment, 3
upper kinematic pair, 106

Vector of a 3×3 matrix, 34
vector space, 20
velocity analysis
of parallel manipulators, 329
of rolling robots, 345
of serial manipulators, 139
via poses, 203
virtual work (see Principle of Virtual Work)
viscosity coefficient, 272
viscous forces, 248, 271

Walking machines, 14
kinematics, 341

leg architecture, 342
walking stick, 15
workspace of positioning
manipulators, 155
wrench
acting on a rigid body, 95
axis, 97
pitch, 97
transfer formula, 99

(continued)

Laminar Viscous Flow
V.N. Constantinescu

Thermal Contact Conductance
C.V. Madhusudana

Transport Phenomena with Drops and Bubbles
S.S. Sadhal, P.S. Ayyaswamy, and J.N. Chung

Fundamentals of Robotic Mechanical Systems
J. Angeles